T0180412

# Lecture Notes in Computer Science 13127

More information about this series at https://link.springer.com/bookseries/558

Ivan Lirkov · Svetozar Margenov (Eds.)

# Large-Scale Scientific Computing

13th International Conference, LSSC 2021
Sozopol, Bulgaria, June 7–11, 2021
Revised Selected Papers

 Springer

*Editors*
Ivan Lirkov 🆔
IICT-BAS
Sofia, Bulgaria

Svetozar Margenov 🆔
IICT-BAS
Sofia, Bulgaria

ISSN 0302-9743           ISSN 1611-3349 (electronic)
Lecture Notes in Computer Science
ISBN 978-3-030-97548-7       ISBN 978-3-030-97549-4 (eBook)
https://doi.org/10.1007/978-3-030-97549-4

This Springer imprint is published by the registered company Springer Nature Switzerland AG
The registered company address is: Gewerbestrasse 11, 6330 Cham, Switzerland

# Preface

The 13th International Conference on Large-Scale Scientific Computations (LSSC 2021) was held in Sozopol, Bulgaria, during June 7–11, 2021. The conference was organized by the Institute of Information and Communication Technologies at the Bulgarian Academy of Sciences in cooperation with Sozopol municipality.

The plenary invited speakers and lectures were as follows:

- Y. Efendiev, "Multiscale Modeling: Modeling Subgrid Effects and Temporal Splitting"
- X. Hu, "Graph Laplacians: Algorithms and Applications"
- O. Iliev, T. Prill, and P. Toktaliev, "On Pore Scale Simulation of Reactive Flows in Porous Media"
- M. Ohlberger, "Model Reduction for Large Scale Systems"
- Y. Vassilevski, M. Olshanskii, A. Lozovskiy, and A. Danilov, "Numerical schemes for simulation of incompressible flows in time-dependent domains"

The success of the conference and the present volume are the outcome of the joint efforts of many partners from various institutions and organizations. First, we would like to thank all the members of the Scientific Committee for their valuable contribution forming the scientific face of the conference, as well as for their help in reviewing contributed papers. We especially thank the organizers of the special sessions. We are also grateful to the staff involved in the local organization.

Traditionally, the purpose of the conference is to bring together scientists working with large-scale computational models in natural sciences and environmental and industrial applications, as well as specialists in the field of numerical methods and algorithms for modern high-performance computers. The invited lectures reviewed some of the most advanced achievements in the field of numerical methods and their efficient applications. The conference talks were presented by researchers from academic institutions and practical industry engineers including applied mathematicians, numerical analysts, and computer experts. The general theme for LSSC 2021 was "Large-Scale Scientific Computing" with a particular focus on the organized special sessions.

The special sessions and organizers were as follows:

- Fractional Diffusion Problems: Numerical Methods, Algorithms and Applications— S. Harizanov, R. Lazarov, H. Sun, Z. Zhou, and P. Vabishchevich
- Large-Scale Models: Numerical Methods, Parallel Computations and Applications— K. Georgiev, M. Neytcheva, and Z. Zlatev
- Application of Metaheuristics to Large-Scale Problems—S. Fidanova and G. Luque
- Robust Iterative Solution Methods for Coupled Problems in Poromechanics—J. Kraus and M. Lymbery

- Advanced Discretizations and Solvers for Coupled Systems of Partial Differential Equations—J. Adler, X. Hu, R. Lazarov, and L. Zikatanov
- Optimal Control of ODEs, PDEs and Applications—M. Palladino and T. Scarinci
- Tensor and Matrix Factorization for Big-Data Analysis—B. Alexandrov and H. Djidjev
- Machine Learning and Model Order Reduction for Large Scale Predictive Simulations—O. Iliev and M. Ohlberger
- HPC and Big Data: Algorithms and Applications—A. Karaivanova, T. Gurov, and E. Atanassov

About 150 participants from all over the world attended the conference representing some of the strongest research groups in the field of advanced large-scale scientific computing. This volume contains 62 papers by authors from 19 countries.

The next international conference on LSSC will be organized in June 2023.

December 2021                                                          Ivan Lirkov
                                                                Svetozar Margenov

# Organization

## Scientific Committee

| | |
|---|---|
| James Adler | Tufts University, USA |
| Boian Alexandrov | Los Alamos National Laboratory, USA |
| Emanouil Atanassov | Institute of Information and Communication Technologies, BAS, Bulgaria |
| Pavel Bochev | Sandia National Laboratories, USA |
| Hristo Djidjev | Los Alamos National Laboratory, USA |
| Yalchin Efendiev | Texas A&M University, USA |
| Stefka Fidanova | Institute of Information and Communication Technologies, BAS, Bulgaria |
| Francisco José Gaspar | University of Zaragoza, Spain |
| Ivan Georgiev | Institute of Information and Communication Technologies, BAS, Bulgaria |
| Krassimir Georgiev | Institute of Information and Communication Technologies, BAS, Bulgaria |
| Todor Gurov | Institute of Information and Communication Technologies, BAS, Bulgaria |
| Stanislav Harizanov | Institute of Information and Communication Technologies, BAS, Bulgaria |
| Xiaozhe Hu | Tufts University, USA |
| Oleg Iliev | ITWM, Germany |
| Aneta Karaivanova | Institute of Information and Communication Technologies, BAS, Bulgaria |
| Mikhail Krastanov | Sofia University, Bulgaria |
| Johannes Kraus | University of Duisburg-Essen, Germany |
| Ulrich Langer | Johannes Kepler University Linz, Austria |
| Raytcho Lazarov | Texas A&M University, USA |
| Ivan Lirkov | Institute of Information and Communication Technologies, BAS, Bulgaria |
| Gabriel Luque | University of Málaga, Spain |
| Maria Lymbery | University of Duisburg-Essen, Germany |
| Svetozar Margenov | Institute of Information and Communication Technologies, BAS, Bulgaria |
| Maya Neytcheva | Uppsala University, Sweden |
| Mario Ohlberger | University of Münster, Germany |
| Michele Palladino | GSSI, Italy |
| Carmen Rodrigo | University of Zaragoza, Spain |
| Teresa Scarinci | University of L'Aquila, Italy |

# Contents

**Invited Papers**

Random-Walk Based Approximate $k$-Nearest Neighbors Algorithm
for Diffusion State Distance .......................................... 3
   *Lenore J. Cowen, Xiaozhe Hu, Junyuan Lin, Yue Shen, and Kaiyi Wu*

Model Reduction for Large Scale Systems ............................. 16
   *Tim Keil and Mario Ohlberger*

**Fractional Diffusion Problems: Numerical Methods, Algorithms and
Applications**

Constructions of Second Order Approximations of the Caputo Fractional
Derivative .......................................................... 31
   *Stoyan Apostolov, Yuri Dimitrov, and Venelin Todorov*

Parameter Identification Approach for a Fractional Dynamics Model
of Honeybee Population .............................................. 40
   *Slavi G. Georgiev and Lubin G. Vulkov*

A Newton's Method for Best Uniform Polynomial Approximation ............ 49
   *Irina Georgieva and Clemens Hofreither*

Reduced Sum Implementation of the BURA Method for Spectral
Fractional Diffusion Problems ......................................... 57
   *Stanislav Harizanov, Nikola Kosturski, Ivan Lirkov, Svetozar Margenov,
and Yavor Vutov*

First-Order Reaction-Diffusion System with Space-Fractional Diffusion
in an Unbounded Medium ............................................. 65
   *Dimiter Prodanov*

Performance Study of Hierarchical Semi-separable Compression Solver
for Parabolic Problems with Space-Fractional Diffusion ................... 71
   *Dimitar Slavchev and Svetozar Margenov*

Numerical Solution of Non-stationary Problems with a Rational
Approximation for Fractional Powers of the Operator ..................... 81
   *Petr N. Vabishchevich*

## Large-Scale Models: Numerical Methods, Parallel Computations and Applications

An Exact Schur Complement Method for Time-Harmonic Optimal
Control Problems .................................................    91
  Owe Axelsson, Dalibor Lukáš, and Maya Neytcheva

On the Consistency Order of Runge–Kutta Methods Combined with Active
Richardson Extrapolation ...........................................    101
  Teshome Bayleyegn, István Faragó, and Ágnes Havasi

Study the Recurrence of the Dominant Pollutants in the Formation of AQI
Status over the City of Sofia for the Period 2013–2020 .................    109
  Ivelina Georgieva, Georgi Gadzhev, and Kostadin Ganev

One Solution of Task with Internal Flow in Non-uniform Fluid Using
CABARET Method ..................................................    117
  Valentin A. Gushchin and Vasilii G. Kondakov

Behavior and Scalability of the Regional Climate Model RegCM4 on High
Performance Computing Platforms ....................................    124
  Vladimir Ivanov and Georgi Gadzhev

Quantum Effects on 1/2[111] Edge Dislocation Motion
in Hydrogen-Charged Fe from Ring-Polymer Molecular Dynamics ...........    132
  Ivaylo Katzarov, Nevena Ilieva, and Ludmil Drenchev

Degeneracy of Tetrahedral Partitions Produced by Randomly Generated
Red Refinements ...................................................    140
  Sergey Korotov and Michal Křížek

Effluent Recirculation for Contaminant Removal in Constructed Wetlands
Under Uncertainty: A Stochastic Numerical Approach Based on Monte
Carlo Methodology ................................................    148
  Konstantinos Liolios, Georgios Skodras, Krassimir Georgiev,
  and Ivan Georgiev

Sensitivity Study of Large-Scale Air Pollution Model Based
on Modifications of the Latin Hypercube Sampling Method ...............    156
  Tzvetan Ostromsky, Venelin Todorov, Ivan Dimov, Rayna Georgieva,
  Zahari Zlatev, and Stoyan Poryazov

Sensitivity Operator-Based Approach to the Interpretation
of Heterogeneous Air Quality Monitoring Data .......................    164
  Alexey Penenko, Vladimir Penenko, Elena Tsvetova,
  Alexander Gochakov, Elza Pyanova, and Viktoriia Konopleva

Using the Cauchy Criterion and the Standard Deviation to Evaluate
the Sustainability of Climate Simulations .............................. 172
*Valery Spiridonov and Hristo Chervenkov*

Multidimensional Sensitivity Analysis of an Air Pollution Model Based
on Modifications of the van der Corput Sequence ........................ 180
*Venelin Todorov, Ivan Dimov, Rayna Georgieva, Tzvetan Ostromsky,
Zahari Zlatev, and Stoyan Poryazov*

Running an Atmospheric Chemistry Scheme from a Large Air Pollution
Model by Using Advanced Versions of the Richardson Extrapolation ........ 188
*Zahari Zlatev, Ivan Dimov, István Faragó, Krassimir Georgiev,
and Ágnes Havasi*

**Application of Metaheuristics to Large-Scale Problems**

New Clustering Techniques of Node Embeddings Based on Metaheuristic
Optimization Algorithms .............................................. 201
*Adis Alihodžić, Malek Chahin, and Fikret Čunjalo*

A Comparison of Machine Learning Methods for Forecasting Dow Jones
Stock Index ......................................................... 209
*Adis Alihodžić, Enes Zvorničanin, and Fikret Čunjalo*

Optimal Knockout Tournaments: Definition and Computation ............... 217
*Amelia Bădică, Costin Bădică, Ion Buligiu, Liviu Ion Ciora,
and Doina Logofătu*

Risk Registry Platform for Optimizations in Cases of CBRN and Critical
Infrastructure Attacks ............................................... 226
*Nina Dobrinkova, Evangelos Katsaros, and Ilias Gkotsis*

Influence of the ACO Evaporation Parameter for Unstructured Workforce
Planning Problem .................................................... 234
*Stefka Fidanova and Olympia Roeva*

BINMETA: A New Java Package for Meta-heuristic Searches ............... 242
*Antonio Mucherino*

Synergy Between Convergence and Divergence—Review of Concepts
and Methods ......................................................... 250
*Kalin Penev*

Advanced Stochastic Approaches Based on Optimization of Lattice
Sequences for Large-Scale Finance Problems ............................ 257
  Venelin Todorov, Ivan Dimov, Rayna Georgieva, Stoyan Apostolov,
  and Stoyan Poryazov

Intuitionistic Fuzzy Approach for Outsourcing Provider Selection
in a Refinery ......................................................... 266
  Velichka Traneva and Stoyan Tranev

Quantitative Relationship Between Particulate Matter and Morbidity .......... 275
  Petar Zhivkov and Alexandar Simidchiev

**Advanced Discretizations and Solvers for Coupled Systems of Partial Differential Equations**

Decoupling Methods for Systems of Parabolic Equations .................... 287
  Petr N. Vabishchevich

**Optimal Control of ODEs, PDEs and Applications**

Random Lifting of Set-Valued Maps ...................................... 297
  Rossana Capuani, Antonio Marigonda, and Marta Mogentale

Hölder Regularity in Bang-Bang Type Affine Optimal Control Problems ...... 306
  Alberto Domínguez Corella and Vladimir M. Veliov

Simultaneous Space-Time Finite Element Methods for Parabolic Optimal
Control Problems ...................................................... 314
  Ulrich Langer and Andreas Schafelner

A New Algorithm for the LQR Problem with Partially Unknown Dynamics .... 322
  Agnese Pacifico, Andrea Pesare, and Maurizio Falcone

**Tensor and Matrix Factorization for Big-Data Analysis**

Solving Systems of Polynomial Equations—A Tensor Approach ............. 333
  Mariya Ishteva and Philippe Dreesen

Nonnegative Tensor-Train Low-Rank Approximations
of the Smoluchowski Coagulation Equation ............................. 342
  Gianmarco Manzini, Erik Skau, Duc P. Truong, and Raviteja Vangara

Boolean Hierarchical Tucker Networks on Quantum Annealers .............. 351
  Elijah Pelofske, Georg Hahn, Daniel O'Malley, Hristo N. Djidjev,
  and Boian S. Alexandrov

Topic Analysis of Superconductivity Literature by Semantic Non-negative
Matrix Factorization .............................................. 359
    *Valentin Stanev, Erik Skau, Ichiro Takeuchi, and Boian S. Alexandrov*

## Machine Learning and Model Order Reduction for Large Scale Predictive Simulations

Deep Neural Networks and Adaptive Quadrature for Solving Variational
Problems .......................................................... 369
    *Daria Fokina, Oleg Iliev, and Ivan Oseledets*

A Full Order, Reduced Order and Machine Learning Model Pipeline
for Efficient Prediction of Reactive Flows ............................. 378
    *Pavel Gavrilenko, Bernard Haasdonk, Oleg Iliev, Mario Ohlberger,*
    *Felix Schindler, Pavel Toktaliev, Tizian Wenzel, and Maha Youssef*

A Multiscale Fatigue Model for the Degradation of Fiber-Reinforced
Materials .......................................................... 387
    *N. Magino, J. Köbler, H. Andrä, F. Welschinger, R. Müller,*
    *and M. Schneider*

A Classification Algorithm for Anomaly Detection in Terahertz
Tomography ........................................................ 393
    *Clemens Meiser, Thomas Schuster, and Anne Wald*

Reduced Basis Methods for Efficient Simulation of a Rigid Robot Hand
Interacting with Soft Tissue ......................................... 402
    *Shahnewaz Shuva, Patrick Buchfink, Oliver Röhrle,*
    *and Bernard Haasdonk*

Structured Deep Kernel Networks for Data-Driven Closure Terms
of Turbulent Flows ................................................. 410
    *Tizian Wenzel, Marius Kurz, Andrea Beck, Gabriele Santin,*
    *and Bernard Haasdonk*

## HPC and Big Data: Algorithms and Applications

On the Use of Low-discrepancy Sequences in the Training of Neural
Networks .......................................................... 421
    *E. Atanassov, T. Gurov, D. Georgiev, and S. Ivanovska*

A PGAS-Based Implementation for the Parallel Minimum Spanning Tree
Algorithm ......................................................... 431
    *Vahag Bejanyan and Hrachya Astsatryan*

Comparison of Different Methods for Multiple Imputation by Chain
Equation ........................................................ 439
  *Denitsa Grigorova, Demir Tonchev, and Dean Palejev*

Monte Carlo Method for Estimating Eigenvalues Using Error Balancing ....... 447
  *Silvi-Maria Gurova and Aneta Karaivanova*

Multi-lingual Emotion Classification Using Convolutional Neural
Networks ....................................................... 456
  *Alexander Iliev, Ameya Mote, and Arjun Manoharan*

On Parallel MLMC for Stationary Single Phase Flow Problem ............... 464
  *Oleg Iliev, N. Shegunov, P. Armyanov, A. Semerdzhiev, and I. Christov*

Numerical Parameter Estimates of Beta-Uniform Mixture Models ............ 472
  *Dean Palejev*

Large-Scale Computer Simulation of the Performance of the Generalized
Nets Model of the LPF-algorithm ..................................... 480
  *Tasho D. Tashev, Alexander K. Alexandrov, Dimitar D. Arnaudov,
  and Radostina P. Tasheva*

## Contributed Papers

A New Error Estimate for a Primal-Dual Crank-Nicolson Mixed Finite
Element Using Lowest Degree Raviart-Thomas Spaces for Parabolic
Equations ...................................................... 489
  *Fayssal Benkhaldoun and Abdallah Bradji*

A Finite Volume Scheme for a Wave Equation with Several Time
Independent Delays ............................................... 498
  *Fayssal Benkhaldoun, Abdallah Bradji, and Tarek Ghoudi*

Recovering the Time-Dependent Volatility in Jump-Diffusion Models
from Nonlocal Price Observations .................................... 507
  *Slavi G. Georgiev and Lubin G. Vulkov*

On the Solution of Contact Problems with Tresca Friction
by the Semismooth* Newton Method .................................. 515
  *Helmut Gfrerer, Jiří V. Outrata, and Jan Valdman*

Fitted Finite Volume Method for Unsaturated Flow Parabolic Problems
with Space Degeneration .......................................... 524
  *Miglena N. Koleva and Lubin G. Vulkov*

Minimization of p-Laplacian via the Finite Element Method in MATLAB ..... 533
  *Ctirad Matonoha, Alexej Moskovka, and Jan Valdman*

Quality Optimization of Seismic-Derived Surface Meshes of Geological
Bodies ......................................................... 541
  *P. Popov, V. Iliev, and G. Fitnev*

**Author Index** ....................................................... 553

# Invited Papers

# Random-Walk Based Approximate
# $k$-Nearest Neighbors Algorithm
# for Diffusion State Distance

Lenore J. Cowen[1], Xiaozhe Hu[2(✉)], Junyuan Lin[3], Yue Shen[4], and Kaiyi Wu[2]

[1] Department of Computer Science, Tufts University, Medford, MA 02155, USA
lenore.cowen@tufts.edu
[2] Department of Mathematics, Tufts University, Medford, MA 02155, USA
{xiaozhe.hu,kaiyi.wu}@tufts.edu
[3] Department of Mathematics, Loyola Marymount University, Los Angeles,
CA 90045, USA
junyuan.lin@lmu.edu
[4] Department of Mathematics, Florida State University, Tallahassee, FL 32306, USA
yshen@math.fsu.edu

**Abstract.** Diffusion State Distance (DSD) is a data-dependent metric that compares data points using a data-driven diffusion process and provides a powerful tool for learning the underlying structure of high-dimensional data. While finding the exact nearest neighbors in the DSD metric is computationally expensive, in this paper, we propose a new random-walk based algorithm that empirically finds approximate $k$-nearest neighbors accurately in an efficient manner. Numerical results for real-world protein-protein interaction networks are presented to illustrate the efficiency and robustness of the proposed algorithm. The set of approximate $k$-nearest neighbors performs well when used to predict proteins' functional labels.

**Keywords:** Diffusion State Distance · Random walk · $k$-nearest neighbors

## 1 Introduction

A classical and well-studied problem in bioinformatics involves leveraging neighborhood information in protein-protein interaction (PPI) networks to predict protein functional labels. In a typical setting, a PPI network is provided, where vertices represent proteins and edges are placed between two proteins if there is experimental evidence that they interact in the cell. In addition, some vertices are partially labeled with one or more functional labels representing what is known about their functional role in the cell. These functional labels are derived from some biological ontology (most commonly GO, the Gene Ontology [7]).

The work of Cowen, Hu, and Wu was partially supported by the National Science Foundation under grant DMS-1812503, CCF-1934553, and OAC-2018149.

© Springer Nature Switzerland AG 2022
I. Lirkov and S. Margenov (Eds.): LSSC 2021, LNCS 13127, pp. 3–15, 2022.
https://doi.org/10.1007/978-3-030-97549-4_1

The goal is to use the network structure to impute functional labels for the nodes where they are missing. In 2013, Cao et al. [5] proposed a simple function prediction method based on a novel metric, Diffusion State Distance (DSD). In particular, they proposed having the $k$-nearest neighbors in the DSD vote for their functional label(s) and then assigning the functional label that got the most votes. They showed that the $k$-nearest neighbors in the DSD metric can achieve state-of-the-art results for function prediction. In [4], this approach was generalized in a straightforward way to incorporate confidence weights on the edges of the PPI network. The focus of this paper is on whether this algorithm can be sped up in practice.

Computationally, using the naive algorithm to compute all DSD and then, finding the $k$-nearest neighbors according to DSD in a brute-force way takes $O(n^3)$ time, where $n$ is the number of nodes in the PPI network. In [15], computing DSD of a given undirected but possibly weighted graph is reformulated to solving a series of linear systems of graph Laplacians, which can be approximately solved by existing graph Laplacian solvers in an efficient manner. Furthermore, the authors use Johnson-Lindenstrauss Lemma [11] and random projections [1] to reduce the dimension and, thus, further reduce the computational cost to $\mathcal{O}(n^2 \log n)$.

The next step is to find $k$-nearest neighbors ($k$NN) based on the DSD metric, which is a $k$NN search problem. In general, existing exact and approximate search methods are subject to an unfavorable trade-off: either they need to construct a complex search structure so that the subsequent query retrievals on it are inexpensive, or they build a simpler data structure for which accurate searches remain costly [10,12,13,16]. In this paper, our goal is to develop an accurate approximate $k$NN algorithm, that downstream, allows us to perform functional label prediction. In other words, we develop a method that efficiently computes a set of neighbors that perform well in the $k$NN function prediction task, even if they are not exactly the $k$-nearest neighbors by the DSD metric. To achieve this, we use the special property that DSD is random-walk based, and naturally, adopt a random-walk based approach to approximate $k$-nearest neighbors in the DSD metric. Comparing with the generic $K$-dimensional tree ($K$-$d$ tree) algorithm [2], our random-walk based approach is more efficient for the DSD metric. On the other hand, our random-walk based approximate $k$NN algorithm is capable of finding a set of neighbors that provide good (or occasionally even better) performance for function predictions in practice. Overall, with properly chosen parameters, our random-walk based algorithm's computational complexity is $\mathcal{O}(n \log n)$. Thus, coupling with the algorithm for computing DSD that is developed in [15], we have a function prediction method with competitive function prediction performance comparing with the original method developed in [5], but the running time of the entire procedure is reduced from $\mathcal{O}(n^3)$ to $\mathcal{O}(n \log n)$.

The structure of the paper is organized as follows: Sect. 2 introduces basic matrices related to the graphs and reviews algorithms to compute the exact and the approximate DSD. In Sect. 3, we develop random-walk based approximate $k$NN algorithms and discuss their computational complexity. Numerical experiments are presented in Sect. 4 to show the efficiency and accuracy of our proposed function prediction algorithm for real-life PPI networks. Finally, we give some conclusions in Sect. 5.

## 2   Background

In this section, we review the related background and algorithms of DSD. We follow the notation in [15] for definitions related to graphs and random walks. We also briefly summarize the definitions of DSD and algorithms for computing DSD in this section.

### 2.1   Notation

In general, PPI networks are presented as connected undirected graphs with positive edge weights (where the edge weights most typically represent the confidence in the experimental evidence that indicates the proteins are interacting). We use $G = (V, E, \mathcal{W})$ to denote a connected undirected graph, where $V$ is the vertex set, $E$ is the edge set and $\mathcal{W}$ is the weight set. If two vertices $v_i$ and $v_j$ are incident, we denote an edge $e_{ij} \in E$ with positive weight $w_{ij} > 0$. We denote the number of vertex by $n = |V|$ and the number of edges by $m = |E|$. The degree of a vertex $v_i \in V$ is denoted as $d_i$, which is the number of edges connected to $v_i$ and the weighted degree is defined as $\delta_i = \sum_{v_j \in V, e_{ij} \subset E} w_{ij}$.

Next, we introduce several matrices related to the graphs. The adjacency matrix $A \in \mathbb{R}^{n \times n}$ and the degree matrix $D \in \mathbb{R}^{n \times n}$ are defined as follows,

$$A_{ij} = \begin{cases} w_{ij}, & \text{if } e_{ij} \in E, \\ 0, & \text{otherwise.} \end{cases} \qquad D_{ij} = \begin{cases} \delta_i, & i = j, \\ 0, & i \neq j. \end{cases}$$

Then the graph Laplacian is defined as $L = D - A$ and the normalized graph Laplacian $N$ is defined as $N = D^{-\frac{1}{2}} L D^{-\frac{1}{2}}$. $N$ is positive semi-definite and 0 is an eigenvalue of $N$. The eigenvector $\boldsymbol{d}$ corresponding to the zero eigenvalue is $\boldsymbol{d} := \delta_{\text{total}}^{-1/2} D^{\frac{1}{2}} \mathbf{1}$ where $\delta_{\text{total}} = \sum_{v_i \in V} \delta_i$ is the total weighted degree and $\mathbf{1} \in \mathbb{R}^n$ is the all one vector. Note that $\|\boldsymbol{d}\| = 1$. Since the idea of DSD is based on random walks on the graph, we also introduce the transition matrix $P$ which is defined as $P = D^{-1} A$. Since $P$ is row stochastic and irreducible, the steady state distribution $\boldsymbol{\pi} \in \mathbb{R}^n$ is defined as $\boldsymbol{\pi} = \delta_{\text{total}}^{-1} D \mathbf{1}$, which is the left eigenvector of $P$, i.e. $\boldsymbol{\pi}^T P = \boldsymbol{\pi}^T$. In addition, $\boldsymbol{\pi}$ is normalized as $\boldsymbol{\pi}^T \mathbf{1} = 1$. Finally, we introduce $W = \mathbf{1} \boldsymbol{\pi}^T$ which is the so-called Perron projection.

### 2.2   Diffusion State Distance (DSD)

In this section, we briefly recall the diffusion state distance which was introduced in [5]. We associate each vertex $i$ with a vector $\boldsymbol{h}_i^q \in \mathbb{R}^n$ such that $(\boldsymbol{h}_i^q)_j$ represents the expected number of times for a random walk to start from vertex $i$ and visit vertex $j$ within $q$ steps. Based on the definition of the transition matrix $P$, $\boldsymbol{h}_i^q := (I + P^T + (P^T)^2 + \cdots + (P^T)^q)\boldsymbol{e}_i$, where $\boldsymbol{e}_i$ is the $i$-th column of the identity matrix. Then the $q$-th step DSD between the vertex $i$ and the vertex $j$ is defined as $\mathrm{DSD}^q(i,j) := \|\boldsymbol{h}_i^q - \boldsymbol{h}_j^q\|_p$, where $\|\cdot\|_p$ is the standard $\ell_p$-norm. Then the diffusion state distance is defined by letting $q \to \infty$, i.e., $\mathrm{DSD}(i,j) := \lim_{q\to\infty} \mathrm{DSD}^q(i,j)$. In [5,15], it has been shown that the above limit exists and DSD is well-defined. We only state the convergence theory here.

**Theorem 1 (Convergence for DSD [5,15]).** *Assume that $P$ is row stochastic and irreducible, then $DSD^q(i,j)$ converges as $q \to \infty$, i.e.,*

$$\mathrm{DSD}(i,j) := \lim_{q\to\infty} \mathrm{DSD}^q(i,j) = \|X(\boldsymbol{e}_i - \boldsymbol{e}_j)\|_p,$$

*where $X = (I - P^T + W^T)^{-1} \in \mathbb{R}^{n\times n}$ is called the diffusion state.*

In practice, to compute DSD between all the pairs of vertices, $X$ is precomputed and stored. Furthermore, an alternative formulation, $X = D^{\frac{1}{2}}(N^\dagger + \boldsymbol{dd}^T)D^{-\frac{1}{2}}$, is used in the implementation to improve overall efficiency.

Note that the $i$-th column of the diffusion state $X$ can be viewed as a new coordinate presentation of the vertex $v_i$ in $\mathbb{R}^n$. When $n$ is large, i.e. the coordinates of the new representation are in high dimension, the computational cost is $\mathcal{O}(n^3)$ in general and can be improved to $\mathcal{O}(n^2)$ for sparse graphs. This is still expensive or even infeasible. Therefore, dimension reduction is needed in order to further reduce the computational cost. In [15], based on the well-known Johnson-Lindenstrauss Lemma [1,11], we apply a random matrix $Q \in \mathbb{R}^{s\times n}$ that reduces the dimension of the data from $n$ to $s = \mathcal{O}(\log n)$ in the $l_2$ norm, and with high probability, the projected DSD approximates the exact DSD well. This approach yields the approximate diffusion state $\widetilde{X} := = QD^{\frac{1}{2}}N^\dagger D^{-\frac{1}{2}} \in \mathbb{R}^{s\times n}$. Each column of $\widetilde{X}$ represents the coordinates of each vertex in a vector space of dimension $s = \mathcal{O}(\log n) \ll n$, which naturally reduces the computational cost to $\mathcal{O}(m\log n)$ in general, or $\mathcal{O}(n\log n)$ for sparse graphs [15].

Based on the approximate diffusion state $\widetilde{X}$, the approximate DSD is defined as $\widehat{\mathrm{DSD}}(i,j) = \|\widetilde{X}(\boldsymbol{e}_i - \boldsymbol{e}_j)\|$. The following theorem guarantees the goodness of its approximation property.

**Theorem 2 ([15]).** *Given $\epsilon, \gamma > 0$, let $s \geq \frac{4+2\gamma}{\epsilon^2 - \epsilon^3} \log n$. Then with probability at least $1 - n^{-\gamma}$, for all pair $0 \leq i, j \leq n$,*

$$\sqrt{1-\epsilon}\,\mathrm{DSD}(i,j) \leq \widehat{\mathrm{DSD}}(i,j) \leq \sqrt{1+\epsilon}\,\mathrm{DSD}(i,j).$$

In [8], a spectral dimensional reduction approach was introduce to compute the approximate diffusion state $\widetilde{X}$ as well as $\widehat{\mathrm{DSD}}$. We want to point out that the $k$NN construction algorithms developed in this paper can be applied to that case as well. In fact, our algorithm can be applied to any approximation of the diffusion state $X$.

## 2.3   Construction of $k$ Nearest-Neighbor Set Based on DSD

Once the diffusion state $X$ or its approximation $\widetilde{X}$ has been computed and stored, we want to know, for each vertex, the local $k$NN set in the DSD metric.

The simplest method to find the $k$ closest vertices to each vertex is the brute-force method, which computes the distances between all pairs of vertices and finds the $k$ smallest distances for each vertex. In our case, although this method gives the exact $k$NN neighborhood, the computational cost is $\mathcal{O}(n^3)$ if we use $X$ and $\mathcal{O}(n^2 \log n)$ if we use $\widetilde{X}$. When $n$ is large, the cost of this method is expensive in practice.

There are several methods that find the exact $k$NN more efficiently when the data lie in a low dimensional manifold. For example, the $K\text{-}d$ trees [9], where $K$ is the dimension of the data, has a computational cost of $\mathcal{O}(Kn \log n)$. However, when the data sets are high dimensional, these algorithms are still slow. In our case, if $X$ is used ($K = n$), the cost of the $K\text{-}d$ tree approach is $\mathcal{O}(n^2 \log n)$ and, if $\widetilde{X}$ is used ($K = \mathcal{O}(\log n)$), the cost is $\mathcal{O}(n \log^2 n)$. This is, of course, much better than the brute-force method, but still expensive for our biological application since, as suggested in [9], $K\text{-}d$ trees works the best when $d = \mathcal{O}(1)$.

One natural idea is to restrict the search to a smaller set for each vertex instead of considering all the vertices. This subset should contain most of the target node's $k$-nearest neighbors. Then the $k$-nearest neighbors in this subset can be returned as the approximate $k$NN set. Motivated by the random walk interpretation of DSD, we form the neighborhood subset for each vertex via performing short random walks. We show that the resulting approximate $k$NN sets can be substituted for the exact $k$NN sets for function prediction, without loss of prediction percent accuracy.

## 3   Approximate $k$NN Set Construction

In this section, we develop fast algorithms for constructing approximate $k$NN set. We first present the general algorithm and then discuss how to apply them to the diffusion state $X$ and approximate diffusion state $\widetilde{X}$, respectively.

### 3.1   Random-Walk Based $k$NN Set Construction for $X$

In [5,15], it has been shown that $X = I + P^T + (P^T)^2 + \cdots = \sum_{q=1}^{\infty} (P^T)^q$, which reveals how the diffusion state $X$ is closely related to the random walks on the graph. Based on the fact that the derived infinite series converges, one can approximate $X$ by its partial sums, namely by doing several steps of random walks on the graphs to approximate the diffusion state $X$. Since the goal is to find $k$-nearest neighbors under the DSD metric, if the DSD can be approximated via several steps of random walks, it is natural to assume those $k$-nearest neighbors can also be reached via several steps of random walks with high probability.

To be more specific, from a vertex $v$, if we take one random walk step and get to $u_1$, we put $u_1$ into $v$'s neighborhood set $R(v)$. Then from $u_1$, if we take one more random walk step and get to $u_2$ (which means we take two steps of random walk from $v$ to get to $u_2$), then we add $u_2$ into the set $R(v)$. We continue until $t_1$ steps of a random walk are done and gather all vertices ever reached in $R(v)$. Then, we repeat this process $t_2$ times from vertex $v$ and put all the vertices that are reached by $t_1$ steps of the random walks into the set $R(v)$. Finally, we find $k$-nearest neighbors from $R(v)$ for the vertex $v$. The overall algorithm is presented in Algorithm 1.

---

**Algorithm 1.** Random-walk based $k$NN set for the diffusion state $X$

---

1: **for** all $v \in V$ **do**
2:     $R(v) = \emptyset$.
3:     **for** $j = 1 : t_2$ **do**
4:         **for** $h = 1 : t_1$ **do**
5:             $R(v) = \{u : \text{vertex reached by } j \text{ steps of random walk from } v\} \cup R(v)$.
6:         **end for**
7:     **end for**
8:     Use $X$ to compute the distances between $v$ and vertices in $R(v)$ and find $k$-nearest neighbors of $v$.
9: **end for**

---

The computational complexity of Algorithm 1 is discussed next.

**Theorem 3.** *The overall computational cost of Algorithm 1 is at most* $3t_1t_2n^2 + (k+1)t_1t_2n$.

*Proof.* If we take $t_1$ steps of a random walk and repeat this $t_2$ times to collect $R(v)$ for one vertex $v$, the complexity of performing the random walk is $t_1t_2$ and the size of $R(v)$ is at most $t_1t_2$. Therefore, the computational cost for performing all random walks over the $n$ vertices is bounded by $t_1t_2n$.

Since we are using the diffusion state $X$ here, each vertex is represented by one column of $X$ whose dimension is $n$, computing DSD between one pair of vertex costs $3n$ and computing DSD between $v$ and vertexes in $R(v)$ costs at most $3t_1t_2n$. Therefore, for all $n$ vertices, the overall cost for computing DSD is $3t_1t_2n^2$.

For finding the $k$-nearest neighbors, since $|R(v)| \leq t_1t_2 \ll n$ in practice, we use the naive approach to find $k$-nearest neighbors for each vertex and, the computational cost is at most $kt_1t_2n$.

To summarize, the overall computational cost of Algorithm 1 is at most $3t_1t_2n^2 + (k+1)t_1t_2n$, which completes the proof.

## 3.2   Random-Walk Based $k$NN Set Construction for $\widetilde{X}$

In Algorithm 1, using random walks, we avoid computing too many distances, which reduces the overall computational cost from $\mathcal{O}(n^3)$ to roughly $\mathcal{O}(n^2)$. This is the best we can do for the exact diffusion states, because the diffusion state

$X$ represents each vertex in $\mathbb{R}^n$. Therefore, to further reduce the computational cost, the approximate diffusion state $\widetilde{X}$ should be used since it embeds each vertex in $\mathcal{R}^s$ where $s = \mathcal{O}(\log n) \ll n$. This can be done by replacing $X$ with $\widetilde{X}$ in Algorithm 1. However, note that $\widetilde{X}$ is an approximation of $X$, simply using $\widetilde{X}$ might affect the accuracy of the resulting approximate $k$NN graphs. To mitigate the effect, since random projection is used to compute $\widetilde{X}$, we average over multiple random projections to further improve the quality of the approximate $k$NN graph.

The overall algorithm based on $\widetilde{X}$ is presented in Algorithm 2 and we analyze its computational cost in Theorem 4.

---

**Algorithm 2.** Random-walk based $k$NN set for $\widetilde{X}$

---

1: Given $t_3$ copies of different $\{\widetilde{X}^i\}_{i=1}^{t_3}$.
2: **for** all $v \in V(G)$ **do**
3:    $R(v) = \emptyset$.
4:    **for** $j = 1 : t_2$ **do**
5:       **for** $h = 1 : t_1$ **do**
6:          $R(v) = \{u : \text{vertex reached by } j \text{ steps of random walk from } v\} \cup R(v)$.
7:       **end for**
8:    **end for**
9:    **for** $i = 1 : t_3$ **do**
10:       Compute DSD between $v$ and vertices in $R(v)$ by using $\widetilde{X}^i$.
11:    **end for**
12: **end for**
13: Compute the average DSD between $v$ and vertices in $R(v)$ and find $k$-nearest neighbors of $v$.

---

**Theorem 4.** *The overall computational cost of Algorithm 2 is at most $3t_1t_2t_3sn + (k+1)t_1t_2n + t_3$.*

*Proof.* Following the same argument as in the proof of Theorem 3, performing random walks to find the set $R(v)$ for all vertices costs $t_1t_2n$. For each approximate diffusion state $\widetilde{X}^i$, computing DSD costs at most $3t_1t_2sn$. Since we have $t_3$ copies of $\widetilde{X}$, the total cost for this part is $3t_1t_2t_3sn$. In the final step, computing the average DSD costs $t_3$ and finding the $k$-nearest neighbors cost at most $kt_1t_2n$. Therefore, the overall cost of Algorithm 2 is at most $3t_1t_2t_3sn + (k+1)t_1t_2n + t_3$.

*Remark 1.* In our numerical experiments, we choose $t_1 = \mathcal{O}(1)$, $t_2 = \mathcal{O}(1)$, $t_3 = \mathcal{O}(1)$, $k = \mathcal{O}(1)$, and $s = \mathcal{O}(\log n)$. Therefore, the computational costs for Algortihms 1 and 2 are $\mathcal{O}(n^2)$ and $\mathcal{O}(n \log n)$, respectively. Comparing with $K$-$d$ trees, Algorithm 2 has a slightly better computational complexity. In the numerical experiments, we focus on Algorithm 2 and demonstrate that it efficiently computes approximate $k$NN sets with good approximation qualities, and that preserve function prediction accuracy.

# 4 Numerical Experiments

In this section, we present several numerical experiments to show the efficiency and accuracy of the proposed random-walk based construction of approximate $k$NN graphs.

## 4.1 Information About the Data Sets

The Disease Module Identification DREAM Challenge [6] released a heterogeneous collection of six different human protein-protein association networks with different biological criteria for placing edges between different protein pairs, in order to benchmark different methods for unsupervised network clustering. Here, we use the first two of the DREAM networks, DREAM1 and DREAM2, to additionally benchmark our exact and approximate function prediction methods, where we briefly describe how they were constructed, below. DREAM1 was derived from the STRING database [17], which integrates known and predicted protein-protein interactions across multiple resources, including both direct (physical) and indirect (functional) associations. The edge weights are derived from the STRING association score [17]. DREAM2 was produced from the InWeb database [14], which aggregates evidence that pairs of proteins physically interact in the cell from different databases. Interaction edges supported by multiple sources are given higher confidence. Both DREAM1 and DREAM2 are of low diameter and highly connected (see Table 1); note that we remove some small number of isolated nodes and node-pairs and restrict our study only to the largest connected component of each network (see Table 1) to guarantee the irreducible assumption for convergence. We further remark that the typical "small-world" properties of these networks make all the shortest-path-based distance metrics very uninformative in differentiating meaningful neighborhoods, and we expect that nearby nodes according to the DSD metric will be more relevant for understanding function (see discussion in [5]).

**Table 1.** Largest connected components of the DREAM networks

|           | Vertices | Edges     | Type | Edge weight      |
|-----------|----------|-----------|------|------------------|
| Network 1 | 17,388   | 2,232,398 | PPI  | Confidence score |
| Network 2 | 12,325   | 397,254   | PPI  | Confidence score |

## 4.2 Numerical Results on Approximate $k$NN Set Construction

To test if the constructed approximate $k$NN graph is both efficient and accurate in preserving distance in different networks, we compare it with the standard $K$-$d$ tree approach. As mentioned in Sect. 1, $K$-$d$ trees require $n \gg 2^K$ to achieve the best performance, where $K$ is the dimension of a data point. Therefore, we focus on the approximate diffusion state $\widetilde{X}$ in our tests. Note that, for $\widetilde{X}$, the computational complexity of the $K$-$d$ tree is $\mathcal{O}(n\log^2 n)$ while our proposed Algorithm 2 costs $\mathcal{O}(n\log n)$.

The numerical tests are conducted on a 2.8 GHz Intel Core i7 CPU with 16 GB of RAM. The $K$-$d$ tree approach is implemented based on the built-in MATLAB functions KDTreeSearcher and knnsearch based on the approximate DSD computed by $\widetilde{X}$. For comparison, Algorithm 2 is implemented in MATLAB as well. In our comparison, we fix $\epsilon = 0.5$ (tolerance for generating approximate diffusion state $\widetilde{X}$), $k = 10$ (the number of nearest neighbors for the $k$NN neighborhood – matching the recommended choice of this parameter in [5]), and fix $t_3 = 1$ (number of copies of $\widetilde{X}$ used in Algorithm 2). We first perform a parameter study to find suitable choices of $t_1$ and $t_2$ and compare the accuracy and efficiency of Algorithm 2 with the $K$-$d$ tree approach. The accuracy is measured by $\dfrac{\sum_{i=1}^{n} |\{kNN_{exact}(i)\} \cap \{kNN_{app}(i)\}|}{k \cdot n}$, where $kNN_{exact}(i)$ is the set of k-nearest neighbors of node $i$ using the exact DSD computed by the exact diffusion state $X$. $kNN_{app}(i)$ is the set of approximate k-nearest neighbors of node $i$ found by either the $K$-$d$ tree approach or our method (Algorithm 2) using the approximate DSD computed by the approximate diffusion state $\widetilde{X}$.

**Table 2.** CPU time in seconds and $k$NN accuracy of applying Algorithm 2 on DREAM1.

| CPU time of $K$-$d$ tree: **990.63** | | | | | Accuracy with $K$-$d$ tree: **0.1873** | | | | |
|---|---|---|---|---|---|---|---|---|---|
| $t_2$ \ $t_1$ | 1 | 5 | 10 | 20 | 30 | | | | |
| 40 | 36.81 | 44.07 | 57.48 | 86.01 | 109.76 | | | | |
| 50 | 34.25 | 48.85 | 66.45 | 101.53 | 130.07 | | | | |
| 60 | 35.81 | 50.71 | 70.65 | 110.53 | 146.92 | | | | |
| 70 | 38.84 | 54.02 | 76.30 | 122.21 | 165.19 | | | | |

| $t_2$ \ $t_1$ | 1 | 5 | 10 | 15 | 20 |
|---|---|---|---|---|---|
| 40 | 0.3258 | 0.2687 | 0.2399 | 0.2171 | 0.2064 |
| 50 | 0.3321 | 0.2678 | 0.2401 | 0.2170 | 0.2063 |
| 60 | 0.3367 | 0.2672 | 0.2403 | 0.2163 | 0.2051 |
| 70 | 0.3377 | 0.2664 | 0.2380 | 0.2155 | 0.2043 |

(a) CPU time in seconds          (b) $k$NN accuracy

**Table 3.** CPU time in seconds and $k$NN accuracy of applying Algorithm 2 on DREAM2.

| CPU time of $K$-$d$ tree: **442.07** | | | | | Accuracy of $K$-$d$ tree: **0.4565** | | | | |
|---|---|---|---|---|---|---|---|---|---|
| $t_2$ \ $t_1$ | 1 | 2 | 3 | 4 | 5 | | | | |
| 110 | 15.46 | 19.83 | 23.72 | 20.63 | 30.74 | | | | |
| 120 | 15.68 | 20.42 | 24.39 | 30.21 | 32.22 | | | | |
| 130 | 15.57 | 20.77 | 25.14 | 29.06 | 33.92 | | | | |
| 140 | 15.72 | 21.62 | 26.04 | 30.19 | 35.19 | | | | |

| $t_2$ \ $t_1$ | 1 | 2 | 3 | 4 | 5 |
|---|---|---|---|---|---|
| 110 | 0.4591 | 0.5643 | 0.5440 | 0.5201 | 0.5215 |
| 120 | 0.4607 | 0.5660 | 0.5455 | 0.5318 | 0.5224 |
| 130 | 0.4596 | 0.5671 | 0.5452 | 0.5332 | 0.5221 |
| 140 | 0.4608 | 0.5657 | 0.5446 | 0.5331 | 0.5217 |

(a)CPU time in seconds          (b) $k$NN accuracy

For DREAM1 (Table 2), as $t_2$ (times of repeating the random walk process) increases, the CPU time needed increases, and the accuracy gets better slightly as expected. On the other hand, the accuracy gradually decreases as $t_1$ (steps of random walk) gets larger, which means, in practice, we actually do not need to take long random walks to find $k$-NN, especially for clustered networks such as DREAM 1. We observe similar results for DREAM2, see Table 3. Basically, for high-density networks, relatively shorter random walks should be used and we do not need to repeat the random walks many times. In addition, for both DREAM1 and DREAM2, comparing with the $K$-$d$ tree approach, our proposed Algorithm 2 is consistently faster (about 6× to 29× speedup for DREAM1 and 13× to 29× speedup for DREAM2) while achieves higher $k$NN accuracy. This demonstrates the efficiency and effectiveness of Algorithm 2.

## 4.3   Numerical Results on Function Prediction

Predicting proteins' functions based on proteins with known labels and the network structure is a classical and well-studied problem in computational biology. We can make inferences for the proteins with unknown labels because proteins that are close to each other metrics tend to share similar functions. In this section, we want to show that the approximate $k$NN neighbors produced by Algorithm 2 can provide accurate function prediction results in biological applications.

We carry out the tests on the DREAM1 and DREAM2 networks. The biological function labels were collected from both the Biological Process (BP) and the Molecular Function (MF) hierarchies from the Gene Ontology database [7] via FuncAssociate3.0 [3] on 04/24/2020. On each network, we only consider the GO labels that represent neither too general nor too specific functions by restricting to labels that appear between 100 and 500 times in the largest connected component of that network. The experiments were performed separately for labels in the Biological Process hierarchy and the Molecular Function hierarchy. In our cross-validation experiments, we mark a functional label prediction correct if it matches one of the functional labels that FuncAssociate assigns to that node. The percent accuracy of the function prediction method represents the percent of top predicted GO functional labels that are correct. The number of functional labels to be voted are different across networks and functional hierarchies (BP and MF); the proteins in the largest connected component of DREAM1 have 1007 BP and 165 MF labels that appear between 100 and 500 times. For DREAM2, the numbers are 888 BP and 139 MF labels.

On each network, we carry out five-fold cross-validation on its largest connected component to obtain a percent accuracy score for the top predicted GO functional labels. For every protein $i$ in the test set, we find its $k$-nearest neighbors $\{r_1, r_2, \ldots, r_k\}$ based on Algorithm 2 and use them to perform function prediction by majority voting with weights $\frac{1}{\widetilde{\mathrm{DSD}}(i, r_j)}$, for $j = 1, 2, \ldots, k$, where each protein in the training set votes with this weight for all its functional labels. The label with the most (weighted) votes is then assigned as the top function prediction label for the protein.

The accuracy score of function prediction for the network is the average percentage of correct assignments in the cross-validation test. We compare the function prediction result based on the $k$-nearest neighbors obtained by using Algorithm 2 using approximate diffusion state $\widetilde{X}$ to the one based on the exact $k$NN measured in exact DSD computed by the exact diffusion state $X$.

In our test, we set $k = 10$ (number of the nearest neighbors, as recommended by [5]) and $t_3 = 1$ for Algorithm 2, i.e., only 1 copy of the approximate diffusion state $\widetilde{X}$ (computed by fixing the tolerance $\epsilon = 0.5$) was used. Different combinations of random walk steps $t_1$ and repetitions $t_2$ are tested for both GO Biological Process hierarchy and Molecular Function hierarchy of function labels. The results are averaged over 100 cross-validation trials.

**Table 4.** Accuracy score of function prediction on DREAM1 for GO Biological Process and Molecular Function hierarchy labels. The cells are shaded when the function prediction performance is comparable or even better than the performance obtained by using exact $k$ nearest neighbors measured in exact DSD.

| Accuracy with exact $k$NN: **0.0245** | | | | | Accuracy with exact $k$NN: **0.0236** | | | | |
| $t_2$ \ $t_1$ | 1 | 5 | 10 | 15 | 20 | $t_2$ \ $t_1$ | 1 | 5 | 10 | 15 | 20 |
|---|---|---|---|---|---|---|---|---|---|---|---|
| 1 | 0.0178 | 0.0178 | 0.0225 | 0.0242 | 0.0232 | 1 | 0.0477 | 0.0488 | 0.0315 | 0.029 | 0.0278 |
| 5 | 0.0178 | 0.0238 | 0.0225 | 0.0238 | 0.0246 | 5 | 0.0487 | 0.0296 | 0.0266 | 0.0269 | 0.0308 |
| 10 | 0.0191 | 0.0243 | 0.0244 | 0.0227 | 0.0238 | 10 | 0.0387 | 0.028 | 0.0298 | 0.0275 | 0.0267 |
| 15 | 0.0201 | 0.0238 | 0.0248 | 0.0237 | 0.0242 | 15 | 0.0338 | 0.0277 | 0.0279 | 0.0276 | 0.0270 |
| 20 | 0.0206 | 0.0238 | 0.0255 | 0.0256 | 0.0229 | 20 | 0.0328 | 0.0289 | 0.0296 | 0.0268 | 0.0278 |
| 25 | 0.0228 | 0.0263 | 0.0245 | 0.0255 | 0.0234 | 25 | 0.0325 | 0.0279 | 0.0291 | 0.0263 | 0.0272 |
| 30 | 0.0216 | 0.0247 | 0.0251 | 0.0237 | 0.0244 | 30 | 0.0327 | 0.0274 | 0.0285 | 0.0268 | 0.0284 |

(a) Biological Process        (b) Molecular Function

In Tables 4(a) and 5(a) and (b), the function prediction accuracy increases as $t_2$ increases. In Table 4(b), the trend is the opposite with performance peaks when $t_2 = 1$ or $t_2 = 5$. We observe improved function prediction performance by voting with the approximate $k$-nearest neighbors computed by appropriate choice of $t_1$ and $t_2$ values instead of voting with the exact $k$NN neighbors measured with the exact DSD, for both the Biological Process and Molecular Function labels, in both DREAM1 and DREAM2 networks. We remark that this similar or even improved prediction accuracy is obtained at a much lower computational cost than the exact $k$NN version.

The strong performance of approximate $k$NN on the function prediction task is curious. We hypothesize that in addition to producing a set of neighbors that largely overlaps the kNN neighbors, the approximate $k$NN neighbors produced by averaged local random walks are somehow denoising the signal as well to improve the function prediction accuracy.

**Table 5.** Accuracy score of function prediction on DREAM2 for GO Biological Process and Molecular Function hierarchy labels.The cells are shaded when the function prediction performance is comparable or even better than the performance obtained by using exact $k$-nearest neighbors measured in exact DSD.

| Accuracy with exact $k$NN: **0.1047** | | | | | | Accuracy with exact $k$NN: **0.0991** | | | | |
|---|---|---|---|---|---|---|---|---|---|---|
| $t_2$ \ $t_1$ | 1 | 5 | 10 | 15 | 20 | $t_2$ \ $t_1$ | 1 | 5 | 10 | 15 | 20 |
| 1 | 0.0308 | 0.027 | 0.0873 | 0.1157 | 0.1156 | 1 | 0.0612 | 0.0489 | 0.1112 | 0.1107 | 0.1102 |
| 5 | 0.0303 | 0.2472 | 0.2401 | 0.2330 | 0.2324 | 5 | 0.0563 | 0.1977 | 0.1937 | 0.1889 | 0.1882 |
| 10 | 0.1060 | 0.2821 | 0.2755 | 0.2733 | 0.2670 | 10 | 0.1533 | 0.2214 | 0.2136 | 0.2120 | 0.2077 |
| 15 | 0.1887 | 0.2981 | 0.2885 | 0.2822 | 0.2798 | 15 | 0.1711 | 0.2326 | 0.2264 | 0.2170 | 0.2153 |
| 20 | 0.2055 | 0.3051 | 0.2964 | 0.2893 | 0.2843 | 20 | 0.1791 | 0.2356 | 0.2293 | 0.2215 | 0.2237 |
| 25 | 0.2175 | 0.3059 | 0.2990 | 0.2953 | 0.2941 | 25 | 0.1853 | 0.2407 | 0.232 | 0.2262 | 0.2261 |
| 30 | 0.2207 | 0.3124 | 0.3027 | 0.2992 | 0.2938 | 30 | 0.1852 | 0.2395 | 0.2337 | 0.2315 | 0.2256 |

(a)Biological Process              (b)Molecular Function

## 5   Conclusions

In this paper, we consider the $k$-nearest neighbors problem for the DSD metric. Since DSD is a diffusion-based distance and closely related to the random walks on the graph, we develop approximate $k$NN algorithms that use random walks to find a set of possible neighbors of each vertex and then identify $k$ nearest ones from this set. Our approach provides a good approximation of the $k$NN while reducing the computational cost since the size of the set that is explored is kept small. More precisely, when combined with the approximate DSD computed by the approximate diffusion state $\widetilde{X}$, the computational complexity of our approximate $k$NN algorithm (Algorithm 2) is $\mathcal{O}(n \log n)$. This is not only much better than the naive $k$NN approach which has complexity $\mathcal{O}(n^3)$ for DSD but also slightly better than the $K$-$d$ tree approach which has complexity $\mathcal{O}(n \log^2 n)$ for approximate DSD in theory. In practice, for the DREAM networks, our random-walk based algorithm can achieve about 30 times speed up in time while maintaining or even achieving better $k$NN accuracy. In addition, when applying our algorithm for biological applications such as function predictions on PPI networks, our method provides a competitive prediction performance while significantly reducing the computational cost. The focus in this paper is on the relative performance of exact and approximate measures for function prediction, not the absolute performance. However, we note that in our setting, in an absolute sense, we are doing much better on the function prediction task using DREAM2 than using DREAM1. This is probably because DREAM1 is filled with lots of very low confidence edges, and even the way the confidence weights are incorporated from DREAM1 may be insufficient to denoise the signal, without further thresholding or tuning. DREAM2 only includes edges that pass a confidence filter.

# References

1. Achlioptas, D.: Database-friendly random projections: Johnson-Lindenstrauss with binary coins. J. Comput. Syst. Sci. **66**(4), 671–687 (2003)
2. Bentley, J.L.: Multidimensional binary search trees used for associative searching. Commun. ACM **18**(9), 509–517 (1975)
3. Berriz, G.F., Beaver, J.E., Cenik, C., Tasan, M., Roth, F.P.: Next generation software for functional trend analysis. Bioinformatics **25**(22), 3043–3044 (2009)
4. Cao, M., et al.: New directions for Diffusion-based network prediction of protein function: incorporating pathways with confidence. Bioinformatics **30**(12), i219–i227 (2014)
5. Cao, M., et al.: Going the distance for protein function prediction: a new distance metric for protein interaction networks. PLoS One **8**(10), 1–12 (2013)
6. Choobdar, S., et al.: Assessment of network module identification across complex diseases. Nat. Methods **16**(9), 843–852 (2019)
7. Consortium, T.G.O.: The gene ontology resource: 20 years and still GOing strong. Nucleic Acids Res. **47**(D1), D330–D338 (2018)
8. Cowen, L., Devkota, K., Hu, X., Murphy, J.M., Wu, K.: Diffusion state distances: multitemporal analysis, fast algorithms, and applications to biological networks. SIAM J. Math. Data Sci. **3**(1), 142–170 (2021)
9. Finkel, R., Friedman, J., Bentley, J.: An algorithm for finding best matches in logarithmic expected time. ACM Trans. Math. Softw. **3**, 200–226 (1977)
10. Indyk, P., Motwani, R.: Approximate nearest neighbors: towards removing the curse of dimensionality. In: Proceedings of the Thirtieth Annual ACM Symposium on Theory of Computing, pp. 604–613 (1998)
11. Johnson, W.B., Lindenstrauss, J.: Extensions of Lipschitz mappings into a Hilbert space. Contemp. Math. **26**(189–206), 1 (1984)
12. Kleinberg, J.M.: Two algorithms for nearest-neighbor search in high dimensions. In: Proceedings of the Twenty-ninth Annual ACM Symposium on Theory of Computing, pp. 599–608 (1997)
13. Kushilevitz, E., Ostrovsky, R., Rabani, Y.: Efficient search for approximate nearest neighbor in high dimensional spaces. SIAM J. Comput. **30**(2), 457–474 (2000)
14. Li, T., et al.: A scored human protein-protein interaction network to catalyze genomic interpretation. Nat. Methods **14**(1), 61 (2017)
15. Lin, J., Cowen, L.J., Hescott, B., Hu, X.: Computing the diffusion state distance on graphs via algebraic multigrid and random projections. Numer. Linear Algebra Appl. **25**(3), e2156 (2018)
16. Liu, T., Moore, A.W., Yang, K., Gray, A.G.: An investigation of practical approximate nearest neighbor algorithms. In: Advances in Neural Information Processing Systems, pp. 825–832 (2005)
17. Szklarczyk, D., et al.: STRING v10: protein-protein interaction networks, integrated over the tree of life. Nucleic Acids Res. **43**, D447–D452 (2015)

# Model Reduction for Large Scale Systems

Tim Keil and Mario Ohlberger[✉][iD]

Mathematics Münster, Westfälische Wilhelms-Universität Münster, Einsteinstr. 62,
48149 Münster, Germany
{tim.keil,mario.ohlberger}@wwu.de
https://www.wwu.de/AMM/ohlberger

**Abstract.** Projection based model order reduction has become a
mature technique for simulation of large classes of parameterized systems. However, several challenges remain for problems where the solution manifold of the parameterized system cannot be well approximated
by linear subspaces. While the online efficiency of these model reduction
methods is very convincing for problems with a rapid decay of the Kolmogorov n-width, there are still major drawbacks and limitations. Most
importantly, the construction of the reduced system in the offline phase
is extremely CPU-time and memory consuming for large scale and multi
scale systems. For practical applications, it is thus necessary to derive
model reduction techniques that do not rely on a classical offline/online
splitting but allow for more flexibility in the usage of computational
resources. A promising approach with this respect is model reduction
with adaptive enrichment. In this contribution we investigate Petrov-
Galerkin based model reduction with adaptive basis enrichment within
a Trust Region approach for the solution of multi scale and large scale
PDE constrained parameter optimization.

**Keywords:** PDE constraint optimization · Reduced basis method ·
Trust region method

## 1 Introduction

Model order reduction (MOR) is a very active research field that has seen tremendous development in recent years, both from a theoretical and application point
of view. For an introduction and overview on recent development we refer e.g. to
[4]. A particular promising model reduction approach for parameterized partial
differential equations (pPDEs) is the Reduced Basis (RB) Method that relies on
the approximation of the solution manifold of pPDEs by low dimensional linear
spaces that are spanned from suitably selected particular solutions, called snapshots. For time-dependent problems, the POD-Greedy method [9] defines the

The authors acknowledge funding by the Deutsche Forschungsgemeinschaft for the
project *Localized Reduced Basis Methods for PDE-constrained Parameter Optimization* under contract OH 98/11-1 and by the Deutsche Forschungsgemeinschaft under
Germany's Excellence Strategy EXC 2044 390685587, Mathematics Münster: Dynamics − Geometry − Structure.

© Springer Nature Switzerland AG 2022
I. Lirkov and S. Margenov (Eds.): LSSC 2021, LNCS 13127, pp. 16–28, 2022.
https://doi.org/10.1007/978-3-030-97549-4_2

*Gold-Standard.* As RB methods rely on so called efficient offline/online splitting, they need to be combined with supplementary interpolation methods in case of non-affine parameter dependence or non-linear differential equations. The empirical interpolation method (EIM) [3] and its various generalizations, e.g. [6], are key technologies with this respect. While RB methods are meanwhile very well established and analyzed for scalar coercive problems, there are still major challenges for problems with a slow convergence of the Kolmogorov N-width [13]. Such problems in particular include pPDEs with high dimensional or even infinite dimensional parameter dependence, multiscale problems as well as hyperbolic or advection dominated transport problems. Particular promising approaches for high dimensional parameter dependence and large or multiscale problems are localized model reduction approaches. We refer to [5] for a recent review of such approaches, including the localized reduced basis multiscale method (LRBMS) [15]. Several of these approaches have already been applied in multiscale applications, in particular for battery simulation with resolved electrode geometry and Buttler-Volmer kinetics [7]. Based on efficient localized a posteriori error control and online enrichment, these methods overcome traditional offline/online splitting and are thus particularly well suited for applications in optimization or inverse problems as recently demonstrated in [14,16]. In the context of PDE constrained optimization, a promising Trust Region (TR) – RB approach that updates the reduced model during the trust region iteration has recently been studied in [2,11,17]. In the latter two contributions a new non-conforming dual (NCD) approach has been introduced that improves the convergence of the adaptive TR-RB algorithm in the case of different reduced spaces for the corresponding primal and dual equations of the first order optimality system.

While these contributions were all based on Galerkin projection of the respective equations, we will introduce a new approach based on Petrov-Galerkin (PG) projection in the following. As we will demonstrate in Sect. 2 below, the PG-reduced optimality system is a conforming approximation which allows for more straight forward computation of derivative information and respective a posteriori error estimates. In Sect. 3 we evaluate and compare the Galerkin and Petrov-Galerkin approaches with respect to the error behavior and the resulting TR-RB approaches with adaptive enrichment. Although the convergence of the TR-RB method can be observed for both approaches, the results demonstrate that the PG approach may require further investigation with respect to stabilization.

## 2  Petrov-Galerkin Based Model Reduction for PDE Constrained Optimization

In this contribution we consider the following class of PDE constrained minimization problems:

$$\min_{\mu \in \mathcal{P}} \mathcal{J}(u_\mu, \mu), \qquad \text{with } \mathcal{J}(u, \mu) = \Theta(\mu) + j_\mu(u) + k_\mu(u, u), \qquad \text{(P.a)}$$

subject to $u_\mu \in V$ being the solution of the *state – or primal – equation*

$$a_\mu(u_\mu, v) = l_\mu(v) \qquad \text{for all } v \in V, \qquad \text{(P.b)}$$

where $\Theta \in \mathcal{P} \to \mathbb{R}$ denotes a parameter functional.

Here, $V$ denotes a real-valued Hilbert space with inner product $(\cdot, \cdot)$ and its induced norm $\| \cdot \|$, $\mathcal{P} \subset \mathbb{R}^P$, with $P \in \mathbb{N}$ denotes a compact and convex admissible parameter set and $\mathcal{J} : V \times \mathcal{P} \to \mathbb{R}$ a quadratic continuous functional. In particular, we consider box-constraints of the form

$$\mathcal{P} := \left\{ \mu \in \mathbb{R}^P \mid \mu_a \leq \mu \leq \mu_b \right\} \subset \mathbb{R}^P,$$

for given parameter bounds $\mu_a, \mu_b \in \mathbb{R}^P$, where "$\leq$" has to be understood component-wise.

For each admissible parameter $\mu \in \mathcal{P}$, $a_\mu : V \times V \to \mathbb{R}$ denotes a continuous and coercive bilinear form, $l_\mu, j_\mu : V \to \mathbb{R}$ are continuous linear functionals and $k_\mu : V \times V \to \mathbb{R}$ denotes a continuous symmetric bilinear form. The primal residual of (P.b) is key for the optimization as well as for a posteriori error estimation. We define for given $u \in V$, $\mu \in \mathcal{P}$, the primal residual $r_\mu^{\text{pr}}(u) \in V'$ associated with (P.b) by

$$r_\mu^{\text{pr}}(u)[v] := l_\mu(v) - a_\mu(u, v) \qquad \text{for all } v \in V. \qquad (1)$$

Following the approach of *first-optimize-then-discretize*, we base our discretization and model order reduction approach on the first order necessary optimality system, i.e. (cf. [2] for details and further references)

$$r_{\bar{\mu}}^{\text{pr}}(\bar{u})[v] = 0 \qquad \text{for all } v \in V, \qquad (2a)$$
$$\partial_u \mathcal{J}(\bar{u}, \bar{\mu})[v] - a_\mu(v, \bar{p}) = 0 \qquad \text{for all } v \in V, \qquad (2b)$$
$$(\partial_\mu \mathcal{J}(\bar{u}, \bar{\mu}) + \nabla_\mu r_{\bar{\mu}}^{\text{pr}}(\bar{u})[\bar{p}]) \cdot (\nu - \bar{\mu}) \geq 0 \qquad \text{for all } \nu \in \mathcal{P}. \qquad (2c)$$

From (2b) we deduce the so-called *adjoint – or dual – equation*

$$a_\mu(q, p_\mu) = \partial_u \mathcal{J}(u_\mu, \mu)[q] = j_\mu(q) + 2k_\mu(q, u_\mu) \qquad \text{for all } q \in V, \qquad (3)$$

with solution $p_\mu \in V$ for a fixed $\mu \in \mathcal{P}$ and given the solution $u_\mu \in V$ to the state equation (P.b). For given $u, p \in V$, we introduce the dual residual $r_\mu^{\text{du}}(u, p) \in V'$ associated with (3) as

$$r_\mu^{\text{du}}(u, p)[q] := j_\mu(q) + 2k_\mu(q, u) - a_\mu(q, p) \qquad \text{for all } q \in V. \qquad (4)$$

## 2.1  Petrov-Galerkin Based Discretization and Model Reduction

Assuming $V_h^{\text{pr}}, V_h^{\text{du}} \subset V$ to be a finite-dimensional subspaces, we define a Petrov-Galerkin projection of (P) onto $V_h^{\text{pr}}, V_h^{\text{du}}$ by considering, for each $\mu \in \mathcal{P}$, the solution $u_{h,\mu} \in V_h^{\text{pr}}$ of the *discrete primal equation*

$$a_\mu(u_{h,\mu}, v_h) = l_\mu(v_h) \qquad \text{for all } v_h \in V_h^{\text{du}}, \qquad (5)$$

and for given $\mu, u_{h,\mu}$, the solution $p_{h,\mu} \in V_h^{\mathrm{du}}$ of the *discrete dual equation* as

$$a_\mu(q_h, p_{h,\mu}) = \partial_u \mathcal{J}(u_{h,\mu}, \mu)[q_h] = j_\mu(q_h) + 2k_\mu(q_h, u_{h,\mu}) \quad \forall q_h \in V_h^{\mathrm{pr}}. \quad (6)$$

Note that the test space of one equation corresponds with the ansatz space of the other equation. In order to obtain a quadratic system, we require $\dim V_h^{\mathrm{pr}} = \dim V_h^{\mathrm{du}}$. Note that a Ritz-Galerkin projection is obtained, if the primal and dual discrete spaces coincide, i.e. $V_h^{\mathrm{pr}} = V_h^{\mathrm{du}}$.

Given problem adapted reduced basis (RB) spaces $V_r^{\mathrm{pr}} \subset V_h^{\mathrm{pr}}, V_r^{\mathrm{du}} \subset V_h^{\mathrm{du}}$ of the same low dimension $n := \dim V_r^{\mathrm{pr}} = \dim V_r^{\mathrm{du}}$ we obtain the reduced versions for the optimality system as follows:

-  PG-RB approximation for (2a): For each $\mu \in \mathcal{P}$ the primal variable $u_{r,\mu} \in V_r^{\mathrm{pr}}$ of the *RB approximate primal equation* is defined through

$$a_\mu(u_{r,\mu}, v_r) = l_\mu(v_r) \quad \text{for all } v_r \in V_r^{\mathrm{du}}. \quad (7a)$$

-  PG-RB approximation for (2b): For each $\mu \in \mathcal{P}$, $u_{r,\mu} \in V_r^{\mathrm{pr}}$ the dual/adjoint variable $p_{r,\mu} \in V_r^{\mathrm{du}}$ satisfies the *RB approximate dual equation*

$$a_\mu(q_r, p_{r,\mu}) = \partial_u \mathcal{J}(u_{r,\mu}, \mu)[q_r] = j_\mu(q_r) + 2k_\mu(q_r, u_{r,\mu}) \quad \forall q_r \in V_r^{\mathrm{pr}}. \quad (7b)$$

We define the PG-RB reduced optimization functional by

$$\hat{\mathcal{J}}_r(\mu) := \mathcal{J}(u_{r,\mu}, \mu) \quad (8)$$

with $u_{r,\mu} \in V_r^{\mathrm{pr}}$ being the solution of (7a). We then consider the *RB reduced optimization problem* by finding a locally optimal solution $\bar{\mu}_r$ of

$$\min_{\mu \in \mathcal{P}} \hat{\mathcal{J}}_r(\mu). \quad (\hat{\mathrm{P}}_r)$$

Note that in contrast to the NCD-approach that has been introduced in [2,11], we do not need to correct the reduced functional in our PG-RB approach, as the primal and dual solutions automatically satisfy $r_\mu^{\mathrm{pr}}(u_{r,\mu})[p_{r,\mu}] = 0$. Actually, this is the main motivation for the usage of the Petrov-Galerkin approach in this contribution. In particular this results in the possibility to compute the gradient of the reduced functional with respect to the parameters solely based on the primal and dual solution of the PG-RB approximation, i.e.

$$\left(\nabla_\mu \hat{\mathcal{J}}_r(\mu)\right)_i = \partial_{\mu_i} \mathcal{J}(u_{r,\mu}, \mu) + \partial_{\mu_i} r_\mu^{\mathrm{pr}}(u_{r,\mu})[p_{r,\mu}] \quad (9)$$

for all $1 \le i \le P$ and $\mu \in \mathcal{P}$, where $u_{r,\mu} \in V_r^{\mathrm{pr}}$ and $p_{r,\mu} \in V_r^{\mathrm{du}}$ denote the PG-RB primal and dual reduced solutions of (7a) and (7b), respectively. Note that in [17], the formula in (9) was motivated by replacing the full order functions by their respective reduced counterpart which resulted in an inexact gradient for the non-conforming approach. In the PG setting, (9) instead defines the true gradient of $\hat{\mathcal{J}}_r$ without having to add a correction term as proposed in [2,11]. This also holds for the true Hessian which can be computed by also replacing all full order counterparts in the full order Hessian. In this contribution we however only focus on quasi-Newton methods that do not require a reduced Hessian.

## 2.2  A Posteriori Error Estimation for an Error Aware Algorithm

In order to construct an adaptive trust region algorithm we require a posteriori error estimation for (7a), (7b) and (8). For that, we can utilize standard residual based estimation. For an a posteriori result of the gradient $\nabla_\mu \hat{\mathcal{J}}_r(\mu)$, we also refer to [11].

**Proposition 1 (Upper error bound for the reduced quantities).** *For $\mu \in \mathcal{P}$, let $u_{h,\mu} \in V_h^{\mathrm{pr}}$ and $p_{h,\mu} \in V_h^{\mathrm{du}}$ be solutions of (5) and (6) and let $u_{r,\mu} \in V_r^{\mathrm{pr}}$, $p_{r,\mu} \in V_r^{\mathrm{du}}$ be a solution of (7a), (7b). Then it holds*

*(i)* $\|u_{h,\mu} - u_{r,\mu}\| \le \Delta_{\mathrm{pr}}(\mu) := \alpha_\mu^{-1} \|r_\mu^{\mathrm{pr}}(u_{r,\mu})\|$,

*(ii)* $\|p_{h,\mu} - p_{r,\mu}\| \le \Delta_{\mathrm{du}}(\mu) := \alpha_\mu^{-1}\big(2\gamma_{k_\mu} \Delta_{\mathrm{pr}}(\mu) + \|r_\mu^{\mathrm{du}}(u_{r,\mu}, p_{r,\mu})\|\big)$,

*(iii)* $|\hat{\mathcal{J}}_h(\mu) - \hat{\mathcal{J}}_r(\mu)| \le \Delta_{\hat{\mathcal{J}}_r}(\mu) := \Delta_{\mathrm{pr}}(\mu)\|r_\mu^{\mathrm{du}}(u_{r,\mu}, p_{r,\mu})\| + \Delta_{\mathrm{pr}}(\mu)^2 \gamma_{k_\mu}$

*where $\alpha_\mu$ and $\gamma_{k_\mu}$ define the inf-sup stability constant of $a_\mu$ and the continuity constant of $k_\mu$, respectively.*

*Proof.* For $(i)$ and $(ii)$ we rely on the inf-sup stability of $a_\mu$ on $V_h^{\mathrm{pr}}$ and $V_h^{\mathrm{du}}$ and proceed analogously to [17]. For $(iii)$, we refer to [11] since $r_\mu^{\mathrm{pr}}(u_{r,\mu})[p_{r,\mu}] = 0$.

It is important to mention that the computation of these error estimators include the computation of the (parameter dependent) inf-sup constant of $a_\mu$ which involves an eigenvalue problem on the FOM level. In practice, cheaper techniques such as the successive constraint method [10] can be used. For the conforming approach $V_h^{\mathrm{pr}} = V_h^{\mathrm{du}}$, the inf-sup constant is equivalent to the coercivity constant $\underline{a}_\mu$ of $a_\mu$ which can be cheaply bounded from below with the help of the min-theta approach, c.f. [8].

## 2.3  Trust-Region Optimization Approach and Adaptive Enrichment

Error aware Trust-Region - Reduced Basis methods (TR-RB) with several different advances and features have been extensively studied e.g. in [2,11,17]. They iteratively compute a first-order critical point of problem (P). For each outer iteration $k \ge 0$ of the TR method, we consider a model function $m^{(k)}$ as a cheap local approximation of the quadratic cost functional $\mathcal{J}$ in the so-called trust-region, which has radius $\varrho^{(k)}$ and can be characterized by the a posteriori error estimator. In our approach we choose

$$m^{(k)}(\cdot) := \hat{\mathcal{J}}_r^{(k)}(\mu^{(k)} + \cdot)$$

for $k \ge 0$, where the super-index $(k)$ indicates that we use different RB spaces $V_r^{*,(k)}$ in each iteration. Thus, we can use $\Delta_{\hat{\mathcal{J}}_r}(\mu)$ for characterizing the trust-region. We are therefore interested in solving the following error aware constrained optimization sub-problem

$$\min_{\widetilde{\mu} \in \mathcal{P}} \hat{\mathcal{J}}_r^{(k)}(\widetilde{\mu}) \quad \text{s.t.} \quad \frac{\Delta_{\hat{\mathcal{J}}}(\widetilde{\mu})}{\hat{\mathcal{J}}_r^{(k)}(\widetilde{\mu})} \leq \varrho^{(k)}, \quad \widetilde{\mu} := \mu^{(k)} + s \in \mathcal{P} \tag{10}$$

$$\text{and } r_{\widetilde{\mu}}^{\mathrm{pr}}(u_{\widetilde{\mu}})[v] = 0 \text{ for all } v \in V.$$

In our TR-RB algorithm we build on the algorithm in [11]. In the sequel, we only summarize the main features of the algorithm and refer to the source for more details. We initialize the RB spaces with the starting parameter $u_{\mu^{(0)}}$, i.e. $V_r^{\mathrm{pr},(0)} = \{u_{h,\mu^{(0)}}\}$ and $V_r^{\mathrm{du},(0)} = \{p_{h,\mu^{(0)}}\}$. For every iteration point $\mu_k$, we locally solve (10) with the quasi-Newton projected BFGS algorithm combined with an Armijo-type condition and terminate with a standard reduced FOC termination criteria, modified with a projection on the parameter space $P_{\mathcal{P}}$ to account for constraints on the parameter space. Additionally, we use a second boundary termination criteria for preventing the subproblem from spending too much computational time on the boundary of the trust region. After the next iterate $\mu_{k+1}$ has been computed, the sufficient decrease conditions helps to decide whether to accept the iterate:

$$\hat{\mathcal{J}}_r^{(k+1)}(\mu^{(k+1)}) \leq \hat{\mathcal{J}}_r^{(k)}(\mu_{\mathrm{AGC}}^{(k)}) \qquad \text{for all } k \in \mathbb{N} \tag{11}$$

This condition can be cheaply checked with the help of an sufficient and necessary condition. If $\mu_{k+1}$ is accepted, we use the parameter to enrich the RB spaces in a Lagrangian manner, i.e. $V_r^{\mathrm{pr},k} = V_r^{\mathrm{pr},k-1} \cup \{u_{h,\mu}\}, V_r^{\mathrm{du},k} = V_r^{\mathrm{du},k-1} \cup \{p_{h,\mu}\}$. For more basis constructions, we refer to [2] where also a strategy is proposed to skip an enrichment. After the enrichment, overall convergence of the algorithm can be checked with a FOM-type FOC condition

$$\|\mu^{(k+1)} - P_{\mathcal{P}}(\mu^{(k+1)} - \nabla_{\mu} \hat{\mathcal{J}}_h(\mu^{(k+1)}))\|_2 \leq \tau_{\mathrm{FOC}},$$

where the FOM quantities are available from the enrichment. Moreover, these FOM quantities allow for computing a condition for possibly enlargement of the TR radius if the reduced model is better than expected. For this, we use

$$\varrho^{(k)} := \frac{\hat{\mathcal{J}}_h(\mu^{(k)}) - \hat{\mathcal{J}}_h(\mu^{(k+1)})}{\hat{\mathcal{J}}_r^{(k)}(\mu^{(k)}) - \hat{\mathcal{J}}_r^{(k)}(\mu^{(k+1)})} \geq \eta_{\varrho} \tag{12}$$

For the described algorithm the following convergence result holds.

**Theorem 1 (Convergence of the TR-RB algorithm, c.f. [2]).** *For sufficient assumptions on the Armijo search to solve* (10), *every accumulation point $\bar{\mu}$ of the sequence $\{\mu^{(k)}\}_{k \in \mathbb{N}} \subset \mathcal{P}$ generated by the above described TR-RB algorithm is an approximate first-order critical point for $\hat{\mathcal{J}}_h$, i.e., it holds*

$$\|\bar{\mu} - P_{\mathcal{P}}(\bar{\mu} - \nabla_{\mu} \hat{\mathcal{J}}_h(\bar{\mu}))\|_2 = 0. \tag{13}$$

Note that the above Theorem is not affected from the choice of the reduced primal and dual equations as long as Proposition 1 holds. In fact, the result can also be used for the TR-RB with the proposed PG approximation of these equations.

Let us also mention that it has been shown in [2] that a projected Newton algorithm for the subproblems can enhance the convergence speed and accuracy. Moreover, a reduced Hessian can be used to introduce an a posteriori result for the optimal parameter which can be used as post processing. In the work at hand, we neglect to transfer the ideas from [2] for the PG variant.

## 3    Numerical Experiments

In this section, we analyze the behavior of the proposed PG variant of the TR-RB (BFGS PG TR-RB) algorithm. We aim to compare the computational time and the accuracy of the PG variant to existing approaches from the literature. To this end, we mention the different algorithms that we compare in this contribution:

**Projected BFGS FOM:** As FOM reference optimization method, we consider a standard projected BFGS method, which uses FOM evaluations for all required quantities.

**Non-conforming TR-RB Algorithm From [11] (BFGS NCD TR-RB):** For comparison, we choose the above described TR-RB algorithm but with a non conforming choice of the RB spaces, which means that (7a) and (7b) are Galerkin projected equations where the test space coincides with the respective ansatz space. This requires to use the NCD-corrected reduced objective functional $\hat{\mathcal{J}}_r(\mu) + r_\mu^{\mathrm{pr}}(u_{r,\mu})[p_{r,\mu}]$ and requires additional quantities for computing the Gradient.

For computational details including the choice of all relevant tolerances for both TR-RB algorithms we again refer to [11], where all details can be found. We only differ in the choice of the stopping tolerance $\tau_{\mathrm{FOC}} = 10^{-6}$. Also note that, as pointed out in Sect. 2.2, we use the same error estimators as in [11] which include the coercivity constant instead of the inf-sub constant. The source code for the presented experiments can be found in [12], which is a revised version of the source code for [11] and also contains detailed interactive jupyter-notebooks[1].

### 3.1    Model Problem: Quadratic Objective Functional with Elliptic PDE Constraints

For our numerical evaluation we reconsider Experiment 1 from [2]. We set the objective functional to be a weighted $L^2$-misfit on a domain of interest $D \subseteq \Omega$ with a weighted Tikhonov term, i.e.

$$\mathcal{J}(v, \mu) = \frac{\sigma_d}{2} \int_D (v - u^{\mathrm{d}})^2 + \frac{1}{2} \sum_{i=1}^{M} \sigma_i(\mu_i - \mu_i^{\mathrm{d}})^2 + 1,$$

---

[1] Available at https://github.com/TiKeil/Petrov-Galerkin-TR-RB-for-pde-opt.

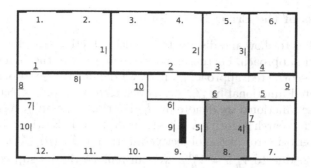

**Fig. 1.** Parameterization based on a blueprint of a building floor, see [11] for details. Numbers indicate potential parameters, where $i.$ is a window or a heater, $\underline{j}$ are doors, and $k|$ are walls. The green region illustrates the domain of interest $D$.

with a desired state $u^{\mathrm{d}}$ and desired parameter $\mu^{\mathrm{d}}$. We added the constant term 1 to verify that $\mathcal{J} > 0$. With respect to the formulation in (P.a), we have $\Theta(\mu) = \frac{\sigma_d}{2} \sum_{i=1}^{M} \sigma_i(\mu_i - \mu_i^{\mathrm{d}})^2 + \frac{\sigma_d}{2} \int_D u^{\mathrm{d}} u^{\mathrm{d}}$, $j_\mu(u) = -\sigma_d \int_D u^{\mathrm{d}} u$, and $k_\mu(u, u) = \frac{\sigma_d}{2} \int_D u^2$. From this general choice we can construct several applications. One instance is to consider the stationary heat equation as equally constraints, i.e. the weak formulation of the parameterized equation

$$-\nabla \cdot (\kappa_\mu \nabla u_\mu) = f_\mu \qquad \text{in } \Omega,$$
$$c_\mu(\kappa_\mu \nabla u_\mu \cdot n) = (u_{\mathrm{out}} - u_\mu) \qquad \text{on } \partial\Omega. \tag{14}$$

with parametric diffusion coefficient $\kappa_\mu$ and source $f_\mu$, outside temperature $u_{\mathrm{out}}$ and Robin function $c_\mu$. From this the bilinear and linear forms $a_\mu$ and $l_\mu$ can easily be deduced and we set the parameter box constraints $\mu_i \in [\mu_i^{\min}, \mu_i^{\max}]$.

As an application of (14), we use the blueprint of a building with windows, heaters, doors and walls, which can be parameterized according to Fig. 1. We picked a certain domain of interest $D$ and we enumerated all windows, walls, doors and heaters separately. We set the computational domain to $\Omega := [0, 2] \times [0, 1] \subset \mathbb{R}^2$ and we model boundary conditions by incorporating all walls and windows that touch the boundary of the blueprint to the Robin function $c_\mu$. All other diffusion components enter the diffusion coefficient $\kappa_\mu$, whereas the heaters are incorporated as a source term on the right hand side $f_\mu$. Moreover, we assume an outside temperature of $u_{\mathrm{out}} = 5$. For our discretization we choose the conforming case for the FOM, i.e. $V_h^{\mathrm{pr}} = V_h^{\mathrm{du}}$, and use a mesh size $h = \sqrt{2}/200$ which resolves all features from the given picture and results in $\dim V_h^{\mathrm{pr}} = 80601$ degrees of freedom.

For the parameterization, we choose 2 doors, 7 heaters and 3 walls which results in a parameter space of 12 dimensions. More details can be viewed in the accompanying source code.

## 3.2    Analysis of the Error Behavior

This section aims to show and discuss the model reduction error for the proposed Petrov–Galerkin approach. Furthermore, we compare it to the Galerkin strategy from the NCD-corrected approach, where we denote the solutions by $u_r^G$, $p_r^G$ and the corrected functional by $\hat{\mathcal{J}}_r^{NCD}$, as well as to the non corrected approach from [17] whose functional we denote by $\hat{J}_r$. For this, we employ a standard goal oriented Greedy search algorithm as also done in [11, Section 4.3.1] with the relative a posteriori error of the objective functional $\Delta_{\hat{\mathcal{J}}_r}(\mu)/\hat{\mathcal{J}}_r(\mu)$. As pointed out in Sect. 2.2, due to $V_h^{pr} = V_h^{du}$, we can replace the inf-sup constant by a lower bound for the coercivity constant of the conforming approach. It is also important to mention that for this experiment, we have simplified our objective functional $\mathcal{J}$ by setting the domain of interest to the whole domain $D \equiv \Omega$. As a result the dual problem is simpler which enhances the stability of the PG approach. The reason for that is further discussed below. Figure 2 shows the difference in the decay and accuracy of the different approaches. It can clearly be seen that the NCD-corrected approach remains the most accurate approach while the objective functional and the gradient of the PG approach shows a better approximation compared to the non corrected version. Clearly the PG approximation of the primal and dual solutions are less accurate and at least the primal error decays sufficiently.

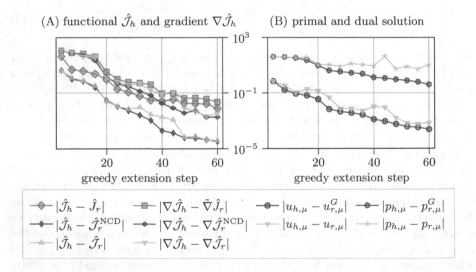

**Fig. 2.** Evolution of the true reduction error in the reduced functional and gradient and its approximations (A) and the primal and dual solutions and its approximations (B), during adaptive greedy basis generation. Depicted is the $L^\infty(\mathcal{P}_{val})$-error for a validation set $\mathcal{P}_{val} \subset \mathcal{P}$ of 100 randomly selected parameters, i.e. $|\hat{\mathcal{J}}_h - \hat{J}_r|$ corresponds to $\max_{\mu \in \mathcal{P}_{val}} |\hat{\mathcal{J}}_h(\mu) - \hat{J}_r(\mu)|$, and so forth.

While producing the shown error study, we experienced instabilities of the PG reduced primal and dual systems. This is due to the fact that for very complex dual problems, the test space of each problem very poorly fits to the respective ansatz space. In fact, the decay of the primal error that can be seen in Fig. 2(B) can not be expected in general and is indeed a consequence of the simplification of the objective functional, where we chose $D \equiv \Omega$. The stability problems can already be seen in the dual error and partly on the primal error for basis size 20. In general it can happen that the reduced systems are highly unstable for specific parameter values. Instead, a sophisticated Greedy Algorithm for deducing appropriate reduced spaces may require to build a larger dual or primal space by adding stabilizing snapshots, i.e. search for supremizers [1]. Since the aim of the work at hand is not to provide an appropriate Greedy based algorithm, we instead decided to reduce the complexity of the functional where stability issues are less present.

## 3.3    TR-RB Algorithm

We now compare the PG variant of the TR-RB with the above mentioned NCD-corrected TR-RB approach. Importantly, we use the original version of the model problem, i.e. we pick the domain of interest to be defined as suggested in Fig. 1. In fact, we accept the possibility of high instabilities in the reduced model. We pick ten random starting parameter, perform both algorithms and compare the averaged result. In Fig. 3 one particular starting parameter is depicted and in Table 1 the averaged results for all ten optimization runs are shown.

It can be seen that the PG variant is a valid approach and converges sufficiently fast with respect to the FOM BFGS method. Surely, it can not be said that this stability issues do not enter the performance of the proposed TR-RB methodology but regardless of the stability of the reduced system, we note that the convergence result in Theorem 1 still holds true. However, the instability of the reduced systems clearly harms the algorithm from iterating as fast as the NCD-corrected approach. One reason for that is that the trust region is much

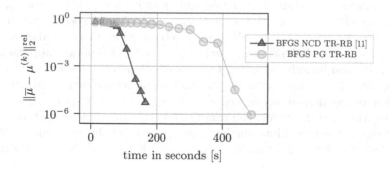

**Fig. 3.** Relative error decay w.r.t. the optimal parameter $\bar{\mu}$ and performance of selected algorithms for a single optimization run with random initial guess $\mu^{(0)}$ for $\tau_{FOC} = 10^{-6}$.

**Table 1.** Performance and accuracy of the algorithms for ten optimization runs with random initial guess $\mu^{(0)}$ and $\tau_{FOC} = 10^{-6}$.

| | Runtime[s] | | Iterations $k$ | Rel. error | FOC cond |
|---|---|---|---|---|---|
| | Avg. (min/max) | Speed-up | Avg. (min/max) | | |
| FOM BFGS | 6955 (4375/15556) | – | 471.44 (349/799) | 3.98e−5 | 3.36e−6 |
| TR NCD-BFGS | 171 (135/215) | 40.72 | 12.56(11/15) | 4.56e−6 | 6.05e−7 |
| TR PG-BFGS | 424 (183/609) | 16.39 | 17.56(11/22) | 4.62e−6 | 7.57e−7 |

larger for the NCD-corrected approach, allowing the method to step faster. We would like to emphasize that the depicted result in Fig. 3 is neither an instance of the worst nor the best performance of the PG approach but rather an intermediate performance. The comparison highly depends on the starting parameter, and the structure of the optimization problem. As discussed above, the suggested PG approach can potentially benefit from more involved enrichment strategies that account for the mentioned stability issues.

Last but not least, it is important to mention that the above experiment showed weaknesses of the chosen projected BFGS approach as FOM method as well as for the TR-RB subproblems which has been extensively studied in [2]. Instead, it is beneficial to choose higher order optimization methods, such as projected Newton type methods, where also the PG variant can be applied to neglect the more involved contributions from the NCD-corrected approach that are entering the Hessian approximation.

## 4    Concluding Remarks

In this contribution we demonstrated, how adaptive enrichment based on rigorous a posteriori error control can be used within a Trust-Region-Reduced-Basis approach to speed up the solution of large scale PDE constrained optimization problems. Within this approach the reduced approximation spaces are tailored towards the solution of the overall optimization problem and thus circumvent the offline construction of a reduced model with good approximation properties with respect to the whole parameter regime. In particular, we compared a new Petrov-Galerkin approach with the NCD-Galerkin approach that has recently been introduced in [2]. Our results demonstrate the benefits of the Petrov-Galerkin approach with respect to the approximation of the objective functional and its derivatives as well as with respect to corresponding a posteriori error estimation. However, the results also show deficiencies with respect to possible stability issues. Thus, further improvements of the enrichment strategy are needed in order to guarantee uniform boundedness of the reduced inf-sup constants.

# References

1. Ballarin, F., Manzoni, A., Quarteroni, A., Rozza, G.: Supremizer stabilization of Pod-Galerkin approximation of parametrized steady incompressible Navier-Stokes equations. Int. J. Numer. Methods Eng. **102**(5), 1136–1161 (2015)
2. Banholzer, S., Keil, T., Mechelli, L., Ohlberger, M., Schindler, F., Volkwein, S.: An adaptive projected newton non-conforming dual approach for trust-region reduced basis approximation of PDE-constrained parameter optimization (2020). arXiv:2012.11653
3. Barrault, M., Maday, Y., Nguyen, N.C., Patera, A.T.: An 'empirical interpolation' method: application to efficient reduced-basis discretization of partial differential equations. C. R. Math. **339**(9), 667–672 (2004)
4. Benner, P., Ohlberger, M., Patera, A., Rozza, G., Urban, K. (eds.): Model Reduction of Parametrized Systems, vol. 17. Springer, Cham (2017). https://doi.org/10.1007/978-3-319-58786-8
5. Buhr, A., Iapichino, L., Ohlberger, M., Rave, S., Schindler, F., Smetana, K.: Localized model reduction for parameterized problems. In: Benner, P., Grivet-Talocia, S., Quarteroni, A., Rozza, G., Schilders, W., Silveira, L.M. (eds.) Snapshot-Based Methods and Algorithms, pp. 245–306. De Gruyter (2021). https://doi.org/10.1515/9783110671490-006
6. Drohmann, M., Haasdonk, B., Ohlberger, M.: Reduced basis approximation for nonlinear parametrized evolution equations based on empirical operator interpolation. SIAM J. Sci. Comput. **34**(2), A937–A969 (2012). https://doi.org/10.1137/10081157X
7. Feinauer, J., et al.: MULTIBAT: unified workflow for fast electrochemical 3D simulations of Lithium-Ion cells combining virtual stochastic microstructures, electrochemical degradation models and model order reduction. J. Comput. Sci. **31**, 172–184 (2019). https://doi.org/10.1016/j.jocs.2018.03.006
8. Haasdonk, B.: Reduced basis methods for parametrized PDEs - a tutorial introduction for stationary and instationary problems. Technical report, University of Stuttgart (2014). http://www.simtech.uni-stuttgart.de/publikationen/prints.php?ID=938, chapter to appear in Benner, P., Cohen, P., Ohlberger, M., Willcox, K.: Model Reduction and Approximation: Theory and Algorithms. SIAM
9. Haasdonk, B., Ohlberger, M.: Reduced basis method for finite volume approximations of parametrized linear evolution equations. Math. Model. Numer. Anal. **42**(2), 277–302 (2008). https://doi.org/10.1051/m2an:2008001
10. Huynh, D., Knezevic, D., Patera, A., Li, H.: Methods and apparatus for constructing and analyzing component-based models of engineering systems (2015). https://www.google.de/patents/US9213788, US Patent 9,213,788
11. Keil, T., Mechelli, L., Ohlberger, M., Schindler, F., Volkwein, S.: A non-conforming dual approach for adaptive trust-region reduced basis approximation of PDE-constrained optimization. ESAIM Math. Model. Numer. Anal. **55**(3), 1239–1269 (2021). https://doi.org/10.1051/m2an/2021019
12. Keil, T., Ohlberger, M.: Software for Model Reduction for Large Scale Systems (2021). https://doi.org/10.5281/zenodo.4627971
13. Ohlberger, M., Rave, S.: Reduced basis methods: success, limitations and future challenges. In: Proceedings of the Conference Algoritmy, pp. 1–12 (2016). http://www.iam.fmph.uniba.sk/amuc/ojs/index.php/algoritmy/article/view/389
14. Ohlberger, M., Schaefer, M., Schindler, F.: Localized model reduction in PDE constrained optimization. Int. Ser. Numer. Math. **169**, 143–163 (2018). https://doi.org/10.1007/978-3-319-90469-6_8

15. Ohlberger, M., Schindler, F.: Error control for the localized reduced basis multiscale method with adaptive on-line enrichment. SIAM J. Sci. Comput. **37**(6), A2865–A2895 (2015). https://doi.org/10.1137/151003660
16. Ohlberger, M., Schindler, F.: Non-conforming localized model reduction with online enrichment: towards optimal complexity in PDE constrained optimization. In: Cancès, C., Omnes, P. (eds.) FVCA 2017. SPMS, vol. 200, pp. 357–365. Springer, Cham (2017). https://doi.org/10.1007/978-3-319-57394-6_38
17. Qian, E., Grepl, M., Veroy, K., Willcox, K.: A certified trust region reduced basis approach to PDE-constrained optimization. SIAM J. Sci. Comput. **39**(5), S434–S460 (2017). https://doi.org/10.1137/16M1081981

# Fractional Diffusion Problems: Numerical Methods, Algorithms and Applications

# Constructions of Second Order Approximations of the Caputo Fractional Derivative

Stoyan Apostolov[1], Yuri Dimitrov[2,3(✉)], and Venelin Todorov[3,4]

[1] Faculty of Mathematics and Informatics, Sofia University, 1164 Sofia, Bulgaria
[2] Department of Mathematics and Physics, University of Forestry,
1756 Sofia, Bulgaria
yuri.dimitrov@ltu.bg
[3] Department of Information Modeling, Institute of Mathematics and Informatics,
Bulgarian Academy of Sciences, 1113 Sofia, Bulgaria
vtodorov@math.bas.bg
[4] Department of Parallel Algorithms, Institute of Information and Communication
Technologies, Bulgarian Academy of Sciences, 1113 Sofia, Bulgaria

**Abstract.** In the present paper we study the properties of the weights of approximations of the second derivative and the Caputo fractional derivative. The approximations of the Caputo derivative are obtained by approximating the second derivative in the expansion formula of the L1 approximation. We show that the properties of their weights are similar to the properties of the weights of the L1 approximation of the Caputo derivative when a suitable choice of the parameters of the approximations is used. The experimental results of applications of the approximations for numerical solution of the two-term ordinary fractional differential equation are given in the paper.

**Keywords:** Fractional derivative · Approximation · Numerical solution

## 1 Introduction

The construction of efficient methods for numerical solution of ordinary and partial fractional differential equations is an active research topic. The Caputo fractional derivative of order $\alpha$, where $0 < \alpha < 1$ is defined as

$$y^{(\alpha)}(x) = \frac{1}{\Gamma(1-\alpha)} \int_{x_0}^{x} \frac{y'(t)}{(x-t)^\alpha} dt.$$

The authors are supported by the Bulgarian National Science Fund under Young Scientists Project KP-06-M32/2 - 17.12.2019. Venelin Todorov is also supported by project No DO1-205/23.11.2018, financed by the Ministry of Education and Science in Bulgaria.

I. Lirkov and S. Margenov (Eds.): LSSC 2021, LNCS 13127, pp. 31–39, 2022.
https://doi.org/10.1007/978-3-030-97549-4_3

Two main approximations of the Caputo derivative are the L1 approximation and the Grünwald difference approximation. Let $n$ be an integer number, representing the number of the mesh points in the discretization of the fractional derivative with an uniform mesh-size $h = (x - x_0)/n$. The L1 approximation has an order $2 - \alpha$ and a second-order expansion formula [3]

$$\frac{1}{\Gamma(2 - \alpha)h^\alpha} \sum_{k=0}^{n} \sigma_k^{(\alpha)} y_{n-k} = y_n^{(\alpha)} + \frac{\zeta(\alpha - 1)}{\Gamma(2 - \alpha)} y_n'' h^{2-\alpha} + O\left(h^2\right), \qquad (1)$$

where $\zeta(x)$ is the Riemann zeta function and $\sigma_0^{(\alpha)} = 1, \sigma_n^{(\alpha)} = (n-1)^{1-\alpha} - n^{1-\alpha}$,

$$\sigma_k^{(\alpha)} = (k - 1)^{1-\alpha} - 2k^{1-\alpha} + (k + 1)^{1-\alpha}, \qquad (k = 1, \cdots, n - 1).$$

The weights of the L1 and Grünwald difference approximations have properties

$$\sigma_0^{(\alpha)} > 0, \quad \sigma_1^{(\alpha)} < \sigma_2^{(\alpha)} < \cdots < \sigma_{n-1}^{(\alpha)} < 0. \qquad (2)$$

The Grünwald difference approximation is a second order shifted approximation of the Caputo derivative and has a generating function $(1 - x)^\alpha$. The L1 approximation has a generating function

$$G(x) = \sum_{k=0}^{\infty} \sigma_k^{(\alpha)} x^k = \frac{(x - 1)^2}{x} Li_{\alpha-1}(x),$$

where $Li_{\alpha-1}(x)$ is the polylogarithm function of order $\alpha - 1$.

Applications of the L1 approximation for numerical solution of fractional differential equations are studied by Jin et al. [7], Lin and Xu [9]. The formulas of the approximations of the Caputo derivative have the form $1/h^\alpha \sum_{k=0}^{n} w_k y_{n-k}$, where $\alpha$ is the order of fractional differentiation. Approximations of order $3 - \alpha$, called L1-2 and L2-1$_\sigma$ formulas are constructed by Gao et al. [6] and Alikhanov [1]. Constructions of approximations of the fractional derivative from the properties of their Fourier transforms and generating functions are studied by Lubich [8], Tian et al. [10], Ding and Li [5]. In [3] we derive second order approximations of the Caputo derivative by substituting the second derivative in the expansion formula (1) with a second order backward difference approximation. In [11] we construct second order shifted approximations of the first derivative with generating functions $\left(1 - e^{A(x-1)}\right)/A$ and $\ln(B + 1 - BX)/B$. The method is used in [4] for construction of approximations of the second derivative which have generating functions $2\left(e^{A(x-1)} - Ax + A - 1\right)/A^2$ and $2\left(B - Bx - \ln(B + 1 - Bx)\right)/B^2$, and second order approximations of the Caputo derivative. The approximations of the second derivative have the property that when the parameters $A$ and $B$ tend to infinity, their weights tend to zero. In Sect. 2 of the present paper we extend the approximations of the second derivative to the class $C^2[x_0, x]$. In Sect. 3 we prove that for every value of the order of fractional differentiation $\alpha$, where $0 < \alpha < 1$, there exist values of the parameters $A$ and $B$ such that the properties of the weights of the second order approximations of the Caputo derivative are similar to the properties (2) of the weights of the L1 approximation. This property of the weights is used in the analysis of the convergence of the numerical solutions of fractional differential equations. In Sect. 4 we consider an application of the second order approximations of the Caputo derivative for numerical solution of the two term ordinary fractional differential equation. The experimental results confirm the convergence order of the approximations of the second derivative and the Caputo derivative studied in the paper.

## 2 Approximations of the Second Derivative

In this section we extend the approximations of the second derivative derived in [4] for all functions in $C^2[x_0, x]$. Let

$$G_1(x) = \frac{2}{A^2} \left( e^{A(x-1)} - Ax + A - 1 \right),$$

$$G_2(x) = \frac{2}{B^2} \left( B - Bx - \ln(B + 1 - Bx) \right).$$

The functions $G_1(x)$ and $H_1(x) = G_1(e^{-x})$ have Maclaurin series expansions

$$G_1(x) = \frac{2}{A^2} \left( e^{-A} + A - 1 \right) - \frac{2}{A} \left( 1 - e^{-A} \right) x + \sum_{k=2}^{\infty} \frac{2A^{k-2}e^{-A}}{k!} x^k, \quad (3)$$

$$H_1(x) = G_1 \left( e^{-x} \right) = x^2 - \frac{3+A}{3} x^3 + O\left( x^4 \right). \quad (4)$$

From (3) and (4) we obtain the second order approximation

$$\mathcal{D}_{1,h}[y] = \frac{1}{h^2} \sum_{k=0}^{n} w_{1,k} y(x - kh) = y''(x) - \frac{3+A}{3} y'''(x) h + O\left( h^2 \right),$$

$$w_{1,0} = \frac{2}{A^2} \left( e^{-A} + A - 1 \right), w_{1,1} = \frac{2}{A} \left( e^{-A} - 1 \right), w_{1,k} = \frac{2A^{k-2}}{k! e^A},$$
$$(2 \leq k \leq n - 2).$$

The functions $G_2(x)$ and $H_2(x) = G_1(e^{-x})$ have Maclaurin series expansions

$$G_2(x) = \frac{2}{B^2} (B - \ln(B + 1)) - \frac{2x}{B+1} + \sum_{k=2}^{\infty} \frac{2B^{k-2}x^k}{k(B+1)^k}, \quad (5)$$

$$H_2(x) = G_2 \left( e^{-x} \right) = x^2 - \frac{3+2B}{3} x^3 + O\left( x^4 \right). \quad (6)$$

From (5) and (6) we obtain

$$\mathcal{D}_{2,h}[y] = \frac{1}{h^2} \sum_{k=0}^{n} w_{2,k} y(x - kh) = y''(x) - \frac{3+2B}{3} y'''(x) h + O\left( h^2 \right),$$

$$w_{2,0} = \frac{2}{B^2}(B - \ln(B+1)), w_{2,1} = -\frac{2}{B+1}, w_{2,k} = \frac{2B^{k-2}}{k(B+1)^k} \quad (2 \leq k \leq n-2).$$

Approximations $\mathcal{D}_{i,h}$ of the Caputo derivative have a second order accuracy when $y(x_0) = y'(x_0) = 0$. Now we extend approximations $\mathcal{D}_{i,h}$ to the class $C^2[x_0, x]$. The function $z(x) = y(x) - y(x_0) - y'(x_0)(x-x_0)$ satisfies the condition $z(x_0) = z'(x_0) = 0$ and $z''(x) = y''(x)$. Approximations $\mathcal{D}_{i,h}$ have a second order accuracy for all functions $y \in C^2[x_0, x]$ when

$$\mathcal{D}_{i,h}[1] = \mathcal{D}_{i,h}[x - x_0] = 0. \tag{7}$$

From (7) we obtain

$$\sum_{k=0}^{n} w_{i,k} = \sum_{i=0}^{n}(n - k - x_0)w_{i,k} = 0,$$

$$\sum_{k=0}^{n} w_{i,k} = \sum_{i=0}^{n} k w_{i,k} = 0. \tag{8}$$

From (8), the weights $w_{i,n-1}$ and $w_{i,n}$ are solutions of the system of equations

$$\left| \begin{array}{l} w_{i,n-1} + w_{i,n} = -\sum_{k=0}^{n-2} w_{i,k}, \\ (n-1)w_{i,n-1} + nw_{i,n} = -\sum_{k=0}^{n-2} k w_{i,k}. \end{array} \right.$$

Approximations $\mathcal{D}_{i,h}$ have a second order accuracy for all functions $y \in C^2[x_0, x]$ when the weights $w_{i,n-1}$ and $w_{i,n}$ have values

$$w_{i,n-1} = \sum_{k=0}^{n-2}(k - n)w_{i,k}, \qquad w_{i,n} = \sum_{k=0}^{n-2}(n - k - 1)w_{i,k}.$$

## 3    Second Order Approximations of the Caputo Derivative

By approximating the second derivative $y_n''$ in expansion formula (1) with $\mathcal{D}_{i,h}[y]$ we obtain second order approximations of the Caputo derivative [4]

$$\mathcal{C}_{i,h}[y] = \frac{1}{\Gamma(2-\alpha)h^\alpha} \sum_{k=0}^{n} \delta_{i,k}^{(\alpha)} y_{n-k} = y_n^{(\alpha)} + O\left(h^2\right),$$

where $\delta_{i,k}^{(\alpha)} = \sigma_{i,k}^{(\alpha)} - \zeta(\alpha - 1)w_{i,k}$ for $k = 0, 1, \cdots, n$. Approximation $\mathcal{C}_{1,h}$ has weights

$$\delta_{1,0}^{(\alpha)} = 1 - \zeta(\alpha-1)\frac{2}{A^2}\left(A - 1 + e^{-A}\right), \delta_{1,1}^{(\alpha)} = 2^{1-\alpha} - 2 + \zeta(\alpha-1)\frac{2}{A}\left(1 - e^{-A}\right),$$

$$\delta_{1,k}^{(\alpha)} = (k-1)^{1-\alpha} - 2k^{1-\alpha} + (k+1)^{1-\alpha} - \frac{2\zeta(\alpha-1)A^{k-2}}{k!e^A} \quad (2 \leq k \leq n-2).$$

Approximation $C_{2,h}$ has weights

$$\delta_{2,0}^{(\alpha)} = 1 - \zeta(\alpha-1)\frac{2}{B^2}\left(B - \ln(B+1)\right), \delta_{2,1}^{(\alpha)} = 2^{1-\alpha} - 2 + \zeta(\alpha-1)\frac{2}{B+1},$$

$$\delta_{2,k}^{(\alpha)} = (k-1)^{1-\alpha} - 2k^{1-\alpha} + (k+1)^{1-\alpha} - \frac{2\zeta(\alpha-1)B^{k-2}}{k(B+1)^k} \qquad (2 \le k \le n-2).$$

The function $-\zeta(\alpha-1)$ is a positive increasing function with values between $-\zeta(-1) = 1/12$ and $-\zeta(0) = 1/2$. The weights $\delta_{i,0}^{(\alpha)}$ are positive and the weights $\delta_{i,1}^{(\alpha)}$ are negative for $i = 1,2$. In Lemma 1 we derive an estimate for the weights of the L1 approximation.

**Lemma 1.**

$$\sigma_k^{(\alpha)} = (k-1)^{1-\alpha} - 2k^{1-\alpha} + (k+1)^{1-\alpha} < -\frac{\alpha(1-\alpha)}{(k+1)^2}.$$

*Proof.* Let $f(x) = x^{1-\alpha}$. From the mean value theorem

$$\sigma_k^{(\alpha)} = f''(\xi) = -\frac{\alpha(1-\alpha)}{\xi^{1+\alpha}},$$

where $k-1 < \xi < k+1$. Therefore

$$\sigma_k^{(\alpha)} < -\frac{\alpha(1-\alpha)}{(k+1)^{1+\alpha}} < -\frac{\alpha(1-\alpha)}{(k+1)^2}.$$

$\square$

In Lemmas 2 and 3 we prove that for every value of the order of fractional differentiation $\alpha \in (0,1)$, there exist values of the parameters $A$ and $B$ such that the weights of approximations $C_{i,h}$ have properties (2).

**Lemma 2.** *Let* $0 < \alpha < 1$ *and* $A > 25e^2/\left(2\pi\alpha^2(1-\alpha)^2\right) + 1$. *Then*

$$\delta_{1,k}^{(\alpha)} < 0, \qquad (2 \le k \le n-2).$$

*Proof.* From Lemma 1

$$\delta_{1,k}^{(\alpha)} < -\frac{\alpha(1-\alpha)}{(k+1)^2} - \frac{2\zeta(\alpha-1)A^{k-2}}{k!e^A} < -\left(\frac{\alpha(1-\alpha)}{(k+1)^2} - \frac{A^{k-2}}{k!e^A}\right),$$

$$\delta_{1,k}^{(\alpha)} < -\frac{1}{e^A(k+1)^2}\left(\alpha(1-\alpha)e^A - \frac{(k+1)^2 A^{k-2}}{k!}\right),$$

$$\delta_{1,k}^{(\alpha)} < -\frac{1}{e^A(k+1)^2}\left(\alpha(1-\alpha)e^A - \frac{5A^{k-2}}{(k-2)!}\right),$$

because $(k+1)^2/(k(k-1)) < 5$ for $k > 1$. The sequence $A^{k-2}/(k-2)!$ has a maximum value when $k-2 = \lfloor A \rfloor$. Then

$$\delta_{1,k}^{(\alpha)} < -\frac{1}{e^A(k+1)^2}\left(\alpha(1-\alpha)e^A - \frac{5A^M}{M!}\right), \qquad (9)$$

where $M = \lfloor A \rfloor$. From the inequality for the factorial function [2]

$$M! > \sqrt{2\pi} M^{M+\frac{1}{2}} e^{-M}$$

we find

$$\frac{A^M}{M!} \leq \frac{A^M}{\sqrt{2\pi} M^{M+\frac{1}{2}} e^{-M}} = \frac{e^M}{\sqrt{2\pi M}} \left(\frac{A}{M}\right)^M.$$

From the inequalities $M \leq A < M+1$ we find

$$\frac{A^M}{M!} \leq \frac{e^M}{\sqrt{2\pi M}} \left(\frac{A}{M}\right)^M < \frac{e^A}{\sqrt{2\pi M}} \left(\frac{M+1}{M}\right)^M < \frac{e^{A+1}}{\sqrt{2\pi (A-1)}}.$$

From (9), the number $\delta_{1,k}^{(\alpha)}$ is negative when

$$\alpha(1-\alpha)e^A > \frac{5e^{A+1}}{\sqrt{2\pi(A-1)}}, \quad \sqrt{2\pi(A-1)} > \frac{5e}{\alpha(1-\alpha)}, \quad A > \frac{25e^2}{2\pi\alpha^2(1-\alpha)^2}+1.$$

□

**Lemma 3.** *Let $0 < \alpha < 1$ and $B > 3/(\alpha(1-\alpha))$. Then*

$$\delta_{2,k}^{(\alpha)} < 0, \qquad (2 \leq k \leq n-2).$$

*Proof.* From Lemma 1

$$\delta_{2,k}^{(\alpha)} < -\frac{\alpha(1-\alpha)}{(k+1)^{1+\alpha}} - \frac{2\zeta(\alpha-1)B^{k-2}}{k(B+1)^k} < -\frac{\alpha(1-\alpha)}{(k+1)^2} + \frac{B^{k-2}}{k(B+1)^k},$$

$$\delta_{2,k}^{(\alpha)} < -\frac{1}{kB^2}\left(\frac{\alpha(1-\alpha)kB^2}{(k+1)^2} - \frac{B^k}{(B+1)^k}\right)$$

$$< -\frac{1}{kB^2}\left(\frac{\alpha(1-\alpha)B^2}{3k} - \left(\frac{B}{B+1}\right)^k\right),$$

because $k/(k+1)^2 > 1/(3k)$.

$$\delta_{2,k}^{(\alpha)} < -\frac{1}{k^2 B}\left(\frac{\alpha(1-\alpha)B}{3} - \frac{\frac{k}{B}}{\left(\left(1+\frac{1}{B}\right)^B\right)^{\frac{k}{B}}}\right).$$

The number $B$ is greater than 1, because $\alpha(1-\alpha) < 1/4$. We use the fact from calculus that the function $(1+1/x)^x$ is increasing for $x > 1$ and converges to $e$. Then

$$\delta_{2,k}^{(\alpha)} < -\frac{1}{k^2 B}\left(\frac{\alpha(1-\alpha)B}{3} - \frac{\frac{k}{B}}{2^{\frac{k}{B}}}\right).$$

The function $f(x) = x/2^x$ has a maximum value at $x = 1/\ln 2$. Hence

$$\frac{\frac{k}{B}}{2^{\frac{k}{B}}} < f(1/\ln 2) = \frac{1}{e\ln 2} < 1.$$

Therefore $\delta_{2,k}^{(\alpha)} < 0$, because $\alpha(1-\alpha)B/3 > 1$.

□

# 4     Numerical Experiments

In this section we consider an application of second order approximations $\mathcal{C}_{i,h}$ of the Caputo derivative for numerical solution of the two-term ordinary fractional differential equation.

$$y^{(\alpha)}(x) + Ky(x) = F(x), \qquad y(0) = y_0. \tag{10}$$

Let $h = 1/N$, where $N$ is a positive integer. By approximating the Caputo derivative at the point $x_n = nh$ with $\mathcal{C}_{i,h}$ we obtain

$$\mathcal{C}_{i,h}[y_n] + Ky_n = F(x_n) + O\left(h^2\right), \qquad n = 2, 3, \cdots, N,$$

Let $\{u_n\}_{n=0}^N$ be the numerical solution of Eq. (10) on the interval $[0, 1]$, where $u_n$ is an approximation of the value of the solution $y_n = y(nh)$. From (11) the sequence $\{u_n\}$ satisfies

$$\frac{1}{\Gamma(2-\alpha)h^\alpha} \sum_{k=0}^n \delta_{i,k}^{(\alpha)} u_{n-k} + Ku_n = F_n, \tag{11}$$

$$u_n = \frac{1}{\delta_{i,0}^{(\alpha)} + K\Gamma(2-\alpha)h^\alpha} \left(\Gamma(2-\alpha)h^\alpha F_n - \sum_{k=1}^n \delta_{i,k}^{(\alpha)} y_{n-k}\right). \tag{NS[i]}$$

Denote by $\bar{u}_1$ the first value of the numerical solution of Eq. (10), which uses the L1 approximation of the Caputo derivative. The number $\bar{u}_1$ is a second order approximation of the value of the solution $y_1 = y(h)$ of Eq. (10) and is computed with the formula [3]

$$\bar{u}_1 = \frac{y_0 + \Gamma(2-\alpha)h^\alpha F(h)}{1 + K\Gamma(2-\alpha)h^\alpha}.$$

The two-term ordinary fractional differential equation,

$$y^{(\alpha)}(x) + Ky(x) = x^\alpha E_{1,2-\alpha}(x) + Ke^x, \qquad y(0) = 1 \tag{12}$$

has a solution $y = e^x$. The experimental results for the error and the order of second order numerical solutions NS[1] and NS[2] of Eq. (12) with initial conditions $u_0 = 1, u_1 = \bar{u}_1$ are given in Tables 1, 2, 3 and 4.

**Table 1.** Maximum error and order of numerical solution NS[1] and $K = 1$.

| $h$ | $\alpha = 0.2, A = 50$ | | $\alpha = 0.5, A = 10$ | | $\alpha = 0.8, A = 1$ | |
|---|---|---|---|---|---|---|
| | *Error* | *Order* | *Error* | *Order* | *Error* | *Order* |
| 0.00125 | $1.5756 \times 10^{-7}$ | 1.8389 | $4.1875 \times 10^{-7}$ | 1.9666 | $1.7914 \times 10^{-7}$ | 1.9896 |
| 0.000625 | $4.1844 \times 10^{-8}$ | 1.9128 | $1.0652 \times 10^{-7}$ | 1.9749 | $4.4969 \times 10^{-8}$ | 1.9940 |
| 0.0003125 | $1.0817 \times 10^{-8}$ | 1.9516 | $2.6973 \times 10^{-8}$ | 1.9815 | $1.1268 \times 10^{-8}$ | 1.9966 |

**Table 2.** Maximum error and order of numerical solution NS[1] and $K = 2$.

| $h$ | $\alpha = 0.2, A = 100$ | | $\alpha = 0.5, A = 20$ | | $\alpha = 0.8, A = 5$ | |
|---|---|---|---|---|---|---|
| | Error | Order | Error | Order | Error | Order |
| 0.00125 | $7.1361 \times 10^{-6}$ | 1.8996 | $5.1827 \times 10^{-7}$ | 1.9220 | $1.0620 \times 10^{-6}$ | 1.9747 |
| 0.000625 | $1.8759 \times 10^{-7}$ | 1.9275 | $1.3502 \times 10^{-7}$ | 1.9405 | $2.6829 \times 10^{-7}$ | 1.9849 |
| 0.0003125 | $4.8497 \times 10^{-7}$ | 1.9516 | $3.4901 \times 10^{-8}$ | 1.9519 | $6.7490 \times 10^{-8}$ | 1.9910 |

**Table 3.** Maximum error and order of numerical solution NS[2] and $K = 1$.

| $h$ | $\alpha = 0.2, B = 30$ | | $\alpha = 0.5, B = 10$ | | $\alpha = 0.8, B = 1$ | |
|---|---|---|---|---|---|---|
| | Error | Order | Error | Order | Error | Order |
| 0.00125 | $1.5346 \times 10^{-7}$ | 1.8083 | $4.7991 \times 10^{-7}$ | 1.9605 | $1.7869 \times 10^{-7}$ | 1.9869 |
| 0.000625 | $4.1229 \times 10^{-8}$ | 1.8961 | $1.2251 \times 10^{-7}$ | 1.9697 | $4.4905 \times 10^{-8}$ | 1.9925 |
| 0.0003125 | $1.0727 \times 10^{-8}$ | 1.9423 | $3.1112 \times 10^{-8}$ | 1.9774 | $1.1259 \times 10^{-8}$ | 1.9957 |

**Table 4.** Maximum error and order of numerical solution NS[2] and $K = 2$.

| $h$ | $\alpha = 0.2, B = 50$ | | $\alpha = 0.5, B = 20$ | | $\alpha = 0.8, B = 5$ | |
|---|---|---|---|---|---|---|
| | Error | Order | Error | Order | Error | Order |
| 0.00125 | $6.0957 \times 10^{-6}$ | 1.8918 | $5.8785 \times 10^{-7}$ | 1.9083 | $1.3907 \times 10^{-6}$ | 1.9619 |
| 0.000625 | $1.6123 \times 10^{-6}$ | 1.9186 | $1.5469 \times 10^{-7}$ | 1.9261 | $3.5343 \times 10^{-7}$ | 1.9764 |
| 0.0003125 | $4.1923 \times 10^{-7}$ | 1.9433 | $4.0217 \times 10^{-8}$ | 1.9435 | $8.9224 \times 10^{-8}$ | 1.9859 |

# References

1. Alikhanov, A.A.: A new difference scheme for the time fractional diffusion equation. J. Comput. Phys. **280**, 424–438 (2015)
2. Batir, N.: Sharp inequalities for factorial n. Proyecciones **27**(1), 97–102 (2008)
3. Dimitrov, Y.: A second order approximation for the Caputo fractional derivative. J. Fract. Calc. Appl. **7**(2), 175–195 (2016)
4. Dimitrov, Y.: Approximations for the second derivative and the Caputo fractional derivative. In: Proceedings of NSFDE&A 2020, Sozopol, Bulgaria (2020)
5. Ding, H., Li, C.: High-order numerical algorithms for Riesz derivatives via constructing new generating functions. J. Sci. Comput. **71**(2), 759–784 (2017)
6. Gao, G.H., Sun, Z.Z., Zhang, H.W.: A new fractional numerical differentiation formula to approximate the Caputo fractional derivative and its applications. J. Comput. Phys. **259**, 33–50 (2014)
7. Jin, B., Lazarov, R., Zhou, Z.: An analysis of the L1 scheme for the subdiffusion equation with nonsmooth data. IMA J. Numer. Anal. **36**(1), 197–221 (2016)
8. Lubich, C.: Discretized fractional calculus. SIAM J. Math. Anal. **17**, 704–719 (1986)
9. Lin, Y., Xu, C.: Finite difference/spectral approximations for the time-fractional diffusion equation. J. Comput. Phys. **225**(2), 1533–1552 (2007)

10. Tian, W.Y., Zhou, H., Deng, W.H.: A class of second order difference approximations for solving space fractional diffusion equations. Math. Comput. **84**, 1703–1727 (2015)
11. Todorov, V., Dimitrov, Y., Dimov, I.: Second order shifted approximations for the first derivative. In: Dimov, I., Fidanova, S. (eds.) HPC 2019. SCI, vol. 902, pp. 428–437. Springer, Cham (2021). https://doi.org/10.1007/978-3-030-55347-0_36

# Parameter Identification Approach for a Fractional Dynamics Model of Honeybee Population

Slavi G. Georgiev$^{(\boxtimes)}$ (ID) and Lubin G. Vulkov

Department of Applied Mathematics and Statistics, FNSE, University of Ruse, Ruse, Bulgaria
{sggeorgiev,lvalkov}@uni-ruse.bg

**Abstract.** In recent years, honeybee losses were reported in many countries such as USA, China, Israel, Turkey, and in Europe, especially, Bulgaria [1].

In order to investigate the colony collapse, many differential equations models were proposed.

Fractional derivatives incorporate the history of the honeybee population dynamics. We study numerically the inverse problem of parameter identification in a model with Caputo differential operator. We use a gradient method of minimizing a quadratic cost functional. Numerical tests with realistic data are performed and discussed.

**Keywords:** Apis Mellifera · Colony losses · Honeybee population dynamics · Parameter estimation · Cost function minimization

## 1  Introduction

The honey bees *Apis mellifera* are the main contributors of pollination and the global honeybee losses lead to disruption of pollination thus causing serious difficulties in *economics, agriculture and ecology* [11].

Some of the main reasons are infections by viruses, Nosema ceranae, Varroa mite [5] and pesticides [7].

On the other hand, fractional-order models have been recognized as a powerful mathematical tool to study anomalous behaviour observed in many physical processes with prominent memory and hereditary properties. For instance, to investigate the causes of colony collapse, the author of [12] proposed a fractional honeybee colony population model. We adopt the model to demonstrate the solution approach to the inverse parameter reconstruction problem.

The rest of the paper is arranged as follows. The next section is devoted to the definition and formulation of the fractional direct and inverse problems. The main novelty of this work, namely the solution to the fractional inverse problem by an adjoint equation optimization method is given in Sect. 3. The computational algorithms for solution to the direct and inverse problems are described in Sect. 4. The simulation results are presented and discussed in Sect. 5.

© Springer Nature Switzerland AG 2022
I. Lirkov and S. Margenov (Eds.): LSSC 2021, LNCS 13127, pp. 40–48, 2022.
https://doi.org/10.1007/978-3-030-97549-4_4

## 2   Fractional Model

Now we will introduce the fractional honeybee population dynamics model, then recall the required facts from the fractional calculus, and finally formulate the inverse modelling problem.

In [12] the following two-order fractional derivatives population model is proposed:

$$D_0^r H = L\frac{H + F}{\omega + H + F} - H\left(\alpha - \sigma\frac{F}{H + F}\right) - nH, \tag{1a}$$

$$D_0^q F = H\left(\alpha - \sigma\frac{F}{H + F}\right) - mF, \tag{1b}$$

$$H(0) = H^0, \ F(0) = F^0, \quad t > 0, \tag{1c}$$

where $D_0^p f$ $(p = r, q)$ denotes the fractional derivative of $f$ of order $0 < p \leqslant 1$. Here, $H$ is the number of hive bees and the ones working outside the hive are the foragers $F$. The total number of bees in the colony is assumed to be $N = H + F$, since bees in compartments other than $H$ and $F$ do not contribute to the work in the hive. The workers are recruited to the forager class from the hive bee class and die with rate $m$. It is assumed that the maximal rate of eclosion is equivalent to the queen's laying rate $L$ and it approaches the maximum as $N$ increases. The constant parameter $\omega$ determines the rate of which $\mathscr{E}(H, F) = \mathscr{E}(N) = LN/(\omega + N)$ approaches $L$ as $N$ gets large.

In the recruitment function $\mathscr{R}(H, F) = \mathscr{R}(N, F) = \alpha - \sigma F/N$ the parameter $\alpha$ represents the maximal rate at which hive bees become foragers when there are no foragers present in the colony, and $\sigma$ is the rate of social inhibition. Following [12], the model (1) do not neglect the death rate $n$ of the hive bees.

### 2.1   Fractional Calculus Background

In the fractional-order extension of the model in [7,8], the first-order derivatives are replaced by Caputo's fractional derivatives of order $0 < p \leqslant 1$. In this subsection we will briefly introduce some results in the fractional calculus, and in particular, for the Caputo's derivative.

Henceforward $\Gamma$ denotes the Gamma function $\Gamma(s) := \int_0^\infty \eta^{s-1}e^{-\eta}d\eta$ for $s > 0$. We remark that $\Gamma(s + 1) = s\Gamma(s)$ for all $s > 0$.

The use of $p < 1$ has just the effect of transforming (1) into a model with memory.

For any function $\nu \in AC[0, T]$, i. e. $\nu$ is absolutely continuous on $[0, T]$, we define the left (forward) Riemann–Liouville integral for $p \in [0, 1]$ [4]:

$$\left(J_{0+}^p \nu\right)(t) := \begin{cases} \nu(t), & p = 0, \\ \dfrac{1}{\Gamma(p)} \displaystyle\int_0^t \frac{\nu(s)}{(t - s)^{1-p}}ds, & 0 < p \leqslant 1, \end{cases} \quad t \in [0, T).$$

Then the left (forward) Caputo derivative as defined above could be expressed in this way:

$$^{\mathcal{C}}D_0^p\nu(t) = \left(J_{0^+}^{1-p}\frac{d\nu}{dt}\right)(t) = \frac{1}{\Gamma(1-p)}\int_0^t \frac{1}{(t-s)^p}\frac{d\nu}{dt}ds.$$

The right (backward) integral is defined as

$$\left(J_{T-}^p\nu\right)(t) := \begin{cases} \nu(t), & p = 0, \\ \frac{1}{\Gamma(p)}\int_t^T \frac{\nu(s)}{(s-t)^{1-p}}ds, 0 < p \leqslant 1, \end{cases} \quad t \in (0,T],$$

and further we define the right (backward) Caputo derivative as

$$^{\mathcal{C}}D_T^p\nu(t) = -\left(J_{T-}^{1-p}\frac{d\nu}{dt}\right)(t) = -\frac{1}{\Gamma(1-p)}\int_t^T \frac{1}{(s-t)^p}\frac{d\nu}{dt}(s)ds.$$

For later use, we recall the following version of integration by parts, see e.g. [4].

**Lemma 1.** *Let $\nu_1(t), \nu_2(t) \in C^1[0,T]$. Then*

$$\int_0^T \left(^{\mathcal{C}}D_0^p\nu_1\right)\nu_2 dt + \nu_1(0)\left(J_{T-}^{1-p}\nu_2\right)(0) = \int_0^T \nu_1\left(^{\mathcal{C}}D_T^p\nu_2\right)dt + \left(J_{T-}^{1-p}\nu_1\right)(T)\nu_2(T).$$

We use the *generalized mean value formula (GMF)* in the following form [4]. Suppose $f(t) \in C[a,b]$ and $D_0^p f(t) \in C[a,b]$ for $0 < p \leqslant 1$, then we have

$$f(t) = f(0) + \frac{1}{\Gamma(p)}D_0^p f(\xi)t^p \text{ with } 0 \leqslant \xi \leqslant t, \ \forall t \in [a,b].$$

It is clear from this formula that if $D_0^p f(t) \geq 0 \ \forall t \in (a,b)$, then the function $f(t)$ is nondecreasing for each $t \in [a,b]$ and if $D_0^p f(t) \leqslant 0 \ \forall t \in (a,b)$ then the function $f(t)$ is nonincreasing $\forall t \in [a,b]$.

**Theorem 1.** *There exists a unique solution $(H,F)^\top \in C^1[0,T]$ to the initial-value problem (1) and it remains positive for $t \in [0,T]$ provided that the initial data is positive.*

*Outline of the Proof.* The existence and uniqueness of the solution to the system (1) could be obtained on the base of the theory in [9] and some results in [10]. We need to show the positivity. On the contrary, let us assume that $H^0 > 0$, $F^0 > 0$ and let $t^1$ be the first time moment in which $H(t^1) = 0$, $F(t^1) > 0$. Then, from (1a) we have $D_0^r H(t^1) = L\frac{F(t^1)}{\omega+F(t^1)} > 0$. But from GMF the function $H(t)$ is non-decreasing for each $t \in [0,t^1]$ which contradicts the assumption. $\square$

## 2.2  Parameter Identification Formulation

The functions $H(t)$, $F(t)$ satisfy the *direct problem* (1) if the coefficients $p^1 = m$, $p^2 = n$, $p^3 = \alpha$, $p^4 = \sigma$, $p^5 = \omega$ are known. In practice, the parameters $m, n, \alpha, \sigma, \omega$ are not known in general and they have to be identified. After their "fair" values are obtained, the model could be used for further robust analysis.

The main question is how to find the coefficients $\boldsymbol{p} \equiv \{m, n, \alpha, \sigma, \omega\}$ for a given honeybee population if we know the population size at certain times:

$$H(t_k; \boldsymbol{p}) = X_k, \ k = 1, \ldots, K_H, \quad F(t_k; \boldsymbol{p}) = Y_k, \ k = 1, \ldots, K_F. \tag{2}$$

The estimation of the parameter $\boldsymbol{p}$ is referred as an *inverse modelling problem*. It means adjusting the parameter values of a mathematical model in such a way to reproduce measured data.

# 3  Solution to the Inverse Problem via the Adjoint Equation Optimization Method

In this section, after formulating the parameter identification problem, we solve it by minimization of a least-square functional using the adjoint equation method.

We solve the point observation problem (1),(2) via minimization of appropriate functionals [6,10]. We are going to minimize the least-square functional

$$J(\boldsymbol{p}) = J(m, n, \alpha, \sigma, \omega) = \sum_{k=1}^{K_H} (H(t_k; \boldsymbol{p}) - X_k)^2 + \sum_{k=1}^{K_F} (F(t_k; \boldsymbol{p}) - Y_k)^2. \tag{3}$$

**Theorem 2.** *The gradient $J'_{\boldsymbol{p}} \equiv (J'_m, J'_n, J'_\alpha, J'_\sigma, J'_\omega)$ of the functional $J(\boldsymbol{p})$ is given by*

$$J'_m(\boldsymbol{p}) = \int_0^T \varphi_F F \, dt, \quad J'_n(\boldsymbol{p}) = \int_0^T \varphi_H H \, dt, \quad J'_\alpha(\boldsymbol{p}) = \int_0^T (\varphi_H - \varphi_F) H \, dt,$$
$$J'_\sigma(\boldsymbol{p}) = \int_0^T (\varphi_F - \varphi_H) \frac{HF}{H+F} \, dt, \quad J'_\omega(\boldsymbol{p}) = L \int_0^T \varphi_H \frac{H+F}{(\omega + H + F)^2} \, dt, \tag{4}$$

*where the functions $\varphi_H = \varphi_H(t)$, $\varphi_F = \varphi_F(t)$ are the unique solutions to the adjoint final-value problem*

$${}^{\mathfrak{C}}\mathrm{D}_T^r \varphi_H = a_{11}(H, F)\varphi_H + a_{12}(H, F)\varphi_F - 2\sum_{k=1}^{K_H} (H(t; p) - X(t))\delta(t - t_k), \ \varphi_H(T) = 0,$$

$${}^{\mathfrak{C}}\mathrm{D}_T^q \varphi_F = a_{21}(H, F)\varphi_H + a_{22}(H, F)\varphi_F - 2\sum_{k=1}^{K_F} (F(t; p) - Y(t))\delta(t - t_k), \ \varphi_F(T) = 0,$$

$$a_{11}(H, F; \boldsymbol{p}) = L\frac{\omega}{(\omega + H + F)^2} + \sigma\frac{F}{H+F} - \sigma\frac{HF}{(H+F)^2} - (\alpha + n), a_{22}(H, F; \boldsymbol{p}) = -\sigma\frac{H^2}{(H+F)^2} - m,$$

$$a_{12}(H, F; \boldsymbol{p}) = \alpha - \sigma\frac{F}{H+F} + \sigma\frac{HF}{(H+F)^2}, a_{21}(H, F; \boldsymbol{p}) = L\frac{\omega}{(\omega + H + F)^2} + \sigma\frac{H^2}{(H+F)^2}$$

$$\tag{5}$$

*and $X(t)$, $Y(t)$ are interpolants of the discrete functions taking values $X_k$ at $t = t_k$, $k = 1, \ldots, K_H$ and $Y_k$ at $t = t_k$, $k = 1, \ldots, K_F$, respectively.*

*Outline of the Proof.* We denote $\delta p = (\delta m, \delta n, \delta \alpha, \delta \sigma, \delta \omega)$, $\delta m = \varepsilon h_1$, $\delta n = \varepsilon h_2$, $\delta \alpha = \varepsilon h_3$, $\delta \sigma = \varepsilon h_4$, $\delta \omega = \varepsilon h_5$ and $\delta H(t; p) = H(t; p + \delta p) - H(t; p)$, $\delta F(t; p) = F(t; p + \delta p) - F(t; p)$.

Then, we write the system (1a)–(1b) at $p := p + \delta p$ for the pair $\{H(t; p + \delta p), F(t; p + \delta p)\}$ with initial $\{H^0, F^0\}$. Next, we perform the differences between the corresponding equations to obtain a system for the pair $\{\delta H(t; p), \delta F(t; p)\}$ with zero initial conditions. After some simple but a bit tedious algebra, we obtain:

$$
{}^{\mathfrak{C}}\mathrm{D}_0^r \delta H = a_{11} \delta H + a_{21} \delta F + \mathscr{O}(\delta H) + \mathscr{O}(\delta F) - H \delta \alpha + \frac{HF}{H + F} \delta \sigma
$$

$$
- L \frac{H + F}{(\omega + H + F)^2} \delta \omega - H \delta n + \mathscr{O}(\delta p), \quad \delta H(0) = 0, \qquad (6a)
$$

$$
{}^{\mathfrak{C}}\mathrm{D}_0^q \delta F = a_{12} \delta H + a_{22} \delta F + \mathscr{O}(\delta H) + \mathscr{O}(\delta F)
$$

$$
- F \delta m + H \delta \alpha - \frac{HF}{H + F} \delta \sigma + \mathscr{O}(\delta p), \quad \delta F(0) = 0. \qquad (6b)
$$

For the increment of the functional $J(p)$ we have:

$$
J(p + \delta p) - J(p) = 2 \sum_{k=1}^{K_H} \int_0^T \delta H(t; p)(H(t; p) - X(t)) \delta(t - t_k) \mathrm{d}t
$$

$$
+ 2 \sum_{k=1}^{K_F} \int_0^T \delta F(t; p)(F(t; p) - Y(t)) \delta(t - t_k) \mathrm{d}t + \mathscr{O}(\varepsilon). \qquad (7)
$$

Following the main idea of the adjoint equation method [10], we multiply Eq. (6a) by a smooth function $\varphi_H(t)$ such that $\varphi_H(T) = 0$, and Eq. (6b) by a function $\varphi_F(t)$ such that $\varphi_F(T) = 0$ (later these functions would be completely reconstructed). We integrate both sides of the results from 0 to $T$ and add them together:

$$
\int_0^T (\varphi_H^{\mathfrak{C}} \mathrm{D}_0^r \delta H + \varphi_F^{\mathfrak{C}} \mathrm{D}_0^q \delta F) \mathrm{d}t = \int_0^T \varphi_H(a_{11} \delta H + a_{21} \delta F) \mathrm{d}t + \int_0^T \varphi_F(a_{12} \delta H + a_{22} \delta F) \mathrm{d}t
$$

$$
- \delta \alpha \int_0^T \varphi_H H \mathrm{d}t + \delta \sigma \int_0^T \varphi_H \frac{HF}{H + F} \mathrm{d}t - L \delta \omega \int_0^T \varphi_H \frac{H + F}{(\omega + H + F)^2} \mathrm{d}t - \delta n \int_0^T \varphi_H H \mathrm{d}t
$$

$$
- \delta m \int_0^T \varphi_F F \mathrm{d}t + \delta \alpha \int_0^T \varphi_F H \mathrm{d}t - \delta \sigma \int_0^T \varphi_F \frac{HF}{H + F} \mathrm{d}t + \mathscr{O}(\varepsilon). \qquad (8)
$$

Integrating by parts the left-hand side using Lemma 1 and using the facts that $\varphi_H(T) = 0$, $\delta H(0) = 0$ and $\varphi_F(T) = 0$, $\delta F(0) = 0$, we get

$$
\int_0^T (\varphi_H^{\mathfrak{C}} \mathrm{D}_0^r \delta H + \varphi_F^{\mathfrak{C}} \mathrm{D}_0^q \delta F) \mathrm{d}t = \int_0^T \delta H^{\mathfrak{C}} \mathrm{D}_T^r \varphi_H \mathrm{d}t + \int_0^T \delta F^{\mathfrak{C}} \mathrm{D}_T^q \varphi_F \mathrm{d}t. \qquad (9)
$$

Then, placing the expressions for ${}^{\mathcal{C}}D_T^r \varphi_H$ and ${}^{\mathcal{C}}D_T^q \varphi_F$ from (5) in (9) and using (7) and (8), after some long manipulations we find

$$J(\boldsymbol{p}+\delta\boldsymbol{p}) - J(\boldsymbol{p}) \equiv J(m+\varepsilon h_1, n+\varepsilon h_2, \alpha+\varepsilon h_3, \sigma+\varepsilon h_4, \omega+\varepsilon h_5) - J(m, n, \alpha, \sigma, \omega)$$

$$= \delta m \int_0^T \varphi_F F \mathrm{d}t + \delta n \int_0^T \varphi_H H \mathrm{d}t + \delta \alpha \int_0^T (\varphi_H - \varphi_F) H \mathrm{d}t$$

$$+ \delta\sigma \int_0^T (\varphi_F - \varphi_H) \frac{HF}{H+F} \mathrm{d}t + L\delta\omega \int_0^T \varphi_H \frac{H+F}{(\omega+H+F)^2} \mathrm{d}t. \quad (10)$$

Now, taking $\varepsilon_2 = \varepsilon_3 = \varepsilon_4 = \varepsilon_5 = 0$, dividing the both sides of (10) by $\varepsilon h_1$ and passing to the limit $\varepsilon \to 0$, we obtain the formula for $J'_m$ in (4). In the same manner one can check the validity of the other formulae in (4).    □

## 4    Numerical Solution to the Direct and Inverse Problems

Let us introduce the following piecewise-uniform mesh:

$$\overline{\omega}_\tau = \{t_0, \ t_i = t_{i-1} + \tau_i J_i, \ t_K = T\} \text{ for } i = 1, \ldots, K-1, \quad (11)$$

for $K \equiv K_N$ and $K \equiv K_F$, and the respective subinterval division $t_i^j = t_{i-1} + j\tau_i$, $j = 1, \ldots, J_i$, where $\forall i = 1, \ldots, K-1$, $t_i$ are the time instances at which observations are taken; $t_i^j$, $j = 1, \ldots, J_i$ and $\tau_i$ are the time nodes and the time step corresponding to $(t_{i-1}, t_i]$.

In case of an integer-order model, it is extensively studied and there are many ways to solve it, see for example [2].

We follow the method described in [12], where the Volterra integral representation of (1a) and (1b) are discretized over the mesh (11) and approximated via the trapezoidal rule to obtain the respective numerical schemes. In an analogous manner, the backward system (5) is solved. In turn, the solution to the inverse problem (1), (2) is sketched with the aid of Algorithm 1.

The numerical approximation of the sought parameter $\boldsymbol{p}$ is $\check{\boldsymbol{p}}$. The user-prescribed tolerance $\varepsilon_{\boldsymbol{p}}$ is chosen according to the practical needs and the value of the descent parameter $\boldsymbol{r}$ (12) is chosen empirically, see the section following.

## 5    Model Simulations

Let us solve the direct problem (1a)–(1c) with data from [12]. We set the number of eggs laid by the queen per day $L = 2\,000$ and the half-saturation constant $\omega = 27\,000$. What is more, the maximal recruitment rate is $\alpha = 0.25$ and the social inhibition coefficient is $\sigma = 0.75$. We assume relatively *low* forager mortality rate $m = 0.154$ and the hive mortality rate is $n = m/18$. We conduct experiments with two types of colonies, where the number of foragers is $F^0 = 0$ or $F^0 = 4\,500$,

---

**Algorithm 1.** Adjoint Equation Optimization Method

Initialize $\boldsymbol{p}_0 \in \mathbb{S}_{\mathrm{adm}}$ and $l = -1$.
**repeat**
    $l := l + 1$
    Solve the direct problem (1) at the current value $\boldsymbol{p}_l$. Then, define

$$H(t_k; \boldsymbol{p}_l) = X_k, \ k = 1, \dots, K_H; \quad F(t_k; \boldsymbol{p}_l) = Y_k, \ k = 1, \dots, K_F$$

    Solve the adjoint problem (5) via the method described in Subsec. 3.2
    Compute $J'(\boldsymbol{p}_l)$ by formulae (4)
    Define the optimization parameter $\boldsymbol{r}_l > 0$ and compute the new parameter value
$\boldsymbol{p}_{l+1}$ by

$$\boldsymbol{p}_{l+1} = \boldsymbol{p}_l - \boldsymbol{r}J'(\boldsymbol{p}_l), \quad \boldsymbol{r} > 0, \ \boldsymbol{r} \in \mathbb{R}_5^+ \tag{12}$$

**until** $\|\triangle\boldsymbol{p}_l\| := \|\boldsymbol{p}_{l+1} - \boldsymbol{p}_l\| < \varepsilon_{\boldsymbol{p}}$
Set $\check{\boldsymbol{p}} := \boldsymbol{p}_{l+1}$

---

while in both cases $H^0 = 4\,500$. The considered time interval is the maximal foraging season from the end of the winter to the end of the summer, which equals $T = 250$ days. Finally, we use high values $r = q = 0.9$. The results could be observed on Fig. 1.

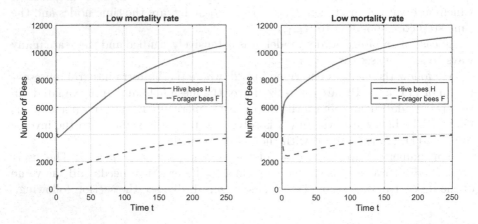

**Fig. 1.** The colony population size for $m = 0.154$, $H^0 = 4\,500$: $F^0 = 0$ (left), $F^0 = 4\,500$ (right)

The colony well survives and it is approaching a disease-free equilibrium state.

Now we are going to solve the inverse problem. We follow the direct problem setting with $H^0 = 4\,500$ and $F^0 = 0$, but now we investigate the beginning of the foraging season as $T = 100$ days. We seek for the unknown parameter $\boldsymbol{p}$, while provided with observations of type (2). Let us set $K_N = K_F = 11$, while the observation times are equidistantly distributed in the interval $[t_0, T]$. The true values of the unknown parameters are $\boldsymbol{p} = (m, n, \alpha, \sigma, \omega)^\top = (0.154, 0.0086, 0.25, 0.75, 27\,000)^\top$. We test Algorithm 1 with initial approximation $\boldsymbol{p}_0 = (0.2, 0.01, 0.3, 0.7, 30\,000)^\top$. The results are given in Table 1.

**Table 1.** Computational test with $\varepsilon_{\boldsymbol{p}} = 2e - 3$

| Parameter | $p_0^i$ | $p^i$ | $\check{p}^i$ | $|p^i - \check{p}^i|$ | $\dfrac{|p^i - \check{p}^i|}{p^i}$ | $r^i$ |
|---|---|---|---|---|---|---|
| $m$ | 0.20 | 0.1540 | 0.1539 | 9.0202e−5 | 5.8573e−4 | 3.47e−13 |
| $n$ | 0.01 | 0.0086 | 0.0086 | 3.1574e−6 | 3.6905e−4 | 3.12e−15 |
| $\alpha$ | 0.30 | 0.2500 | 0.2500 | 3.2130e−5 | 1.2852e−4 | 4.38e−13 |
| $\sigma$ | 0.70 | 0.7500 | 0.7502 | 2.3279e−4 | 3.1038e−4 | 1.59e−12 |
| $\omega$ | 30 000 | 27 000 | 27 008 | 8.2955 | 3.0724e−4 | 2.68e−03 |

The values of $\boldsymbol{p}$ are accurately recovered. The relative errors are of magnitude $1e - 4$, which is acceptable. The gradient method required 25 iterations to converge. The differences (3) between the *observed* values $\{X_k, Y_k\}$ and the *implied* ones $\{H^{\mathrm{obs}}(t_k; \check{\boldsymbol{p}}), F^{\mathrm{obs}}(t_k; \check{\boldsymbol{p}})\}$ are as follows: $J_H(\check{\boldsymbol{p}}) = 3.2064$, $J_F(\check{\boldsymbol{p}}) = 2.7095$. The root mean squared errors are respectively $\mathrm{RMSE}_H(\check{\boldsymbol{p}}) = 0.5399$ and $\mathrm{RMSE}_F(\check{\boldsymbol{p}}) = 0.4963$, which again imply the accuracy of the results.

As every natural phenomenon, honeybee population is associated with an important concern. It is the accuracy of the measurements. The honeybee classes size is quantified by a special hardware, which deviation rate is known a priori. In [2] we have proposed a simple approach to test the algorithm with *perturbed* data. There we have analyzed the integer order counterpart of the model (1). We have found that if the perturbation is small, say less than 5%, then it is possible to cope with the problem without applying further techniques. If this is not the case, we ought to use a kind of a regularization procedure due to the severity of the ill-posedness of the inverse problem. A typical approach is to employ a Tikhonov regularization.

# 6 Conclusion

We have employed a fractional-order model for honeybee population dynamics that balances between simplicity and reality. The proposed adjoint state optimization approach allows one to find the unknown parameters in an accurate and robust way. It is done via solving a linear system of adjoint equations and minimizing a cost functional. The numerical simulations approve the sound foundations of the approach.

A possible continuation of the current work is to employ different fractional operators or more sophisticated models, e.g. [3]. Nevertheless, we believe the current and future algorithms give insight into managing honeybee colonies and help solving contemporary ecological and environmental issues.

**Acknowledgements.** This research is supported by Bulgarian National Science Fund under Project KP-06-PN 46-7 "Design and research of fundamental technologies and methods for precision apiculture".

# References

1. Atanasov, A.Z., Georgiev, I.R.: A multicriteria model for optimal location of honey bee colonies in regions without overpopulation. In: AIP Conference Proceedings, vol. 2333, pp. 090008 (2021)
2. Atanasov, A.Z., Georgiev, S.G.: A numerical parameter estimation approach of the Honeybee population. In: MDIS 2020. CCIS, vol. 1341, pp. 349–362. Springer, Cham (2021). https://doi.org/10.1007/978-3-030-68527-0_22
3. Atanasov, A.Z., Georgiev, S.G., Vulkov, L.G.: Parameter identification of colony collapse disorder in Honeybees as a contagion. In: MDIS 2020. CCIS, vol. 1341, pp. 363–377. Springer, Cham (2021). https://doi.org/10.1007/978-3-030-68527-0_23
4. Baleanu, D., Diethelm, K., Scalas, E., Trujillo, J.J.: Fractional Calculus. Models and Numerical Methods. World Scientific, Singapore (2017)
5. Harbo, J.R.: Effect of brood rearing on honey consumption and the survival of worker Honey Bees. J. Apic. Res. **32**(1), 11–17 (1993)
6. Kabanikhin, S.I.: Inverse and Ill-Posed Problems. De Gruyter, Leipzig (2012)
7. Khoury, D.S., Myerscough, M.R., Barron, A.B.: A quantitative model of Honey Bee colony population dynamics. PLoS One **6**(4), e18491 (2011)
8. Khoury, D.S., Barron, A.B., Meyerscough, M.R.: Modelling food and population dynamics Honey Bee colonies. PLoS One **8**(5), e0059084 (2013)
9. Lin, W.: Global existence theory and chaos control of fractional differential equations. J. Math. Anal. Appl. **332**, 709–726 (2007)
10. Marchuk, G.I., Agoshkov, V.I., Shutyaev, V.P.: Adjoint Equations and Perturbation Algorithms in Nonlinear Problems. CRC Press, Boca Raton (1996)
11. Russel, S., Barron, A.B., Harris, D.: Dynamics modelling of honeybee (Apis Mellifera) colony growth and failure. Ecolog. Model. **265**, 138–169 (2013)
12. Yıldız, T.A.: A fractional dynamical model for Honeybee colony population. Int. J. Biomath. **11**(5), 1850063 (2018)

# A Newton's Method for Best Uniform Polynomial Approximation

Irina Georgieva[1] and Clemens Hofreither[2]([⊠])

[1] Institute of Mathematics and Informatics, Bulgarian Academy of Sciences,
Acad. G. Bonchev, Bl. 8, 1113 Sofia, Bulgaria
`irina@math.bas.bg`
[2] Johann Radon Institute for Computational and Applied Mathematics (RICAM),
Altenberger Street 69, 4040 Linz, Austria
`clemens.hofreither@ricam.oeaw.ac.at`

**Abstract.** We present a novel algorithm, inspired by the recent BRASIL algorithm [10] for rational approximation, for best uniform polynomial approximation based on a formulation of the problem as a nonlinear system of equations and barycentric interpolation. We use results on derivatives of interpolating polynomials with respect to interpolation nodes to compute the Jacobian matrix. The resulting method is fast and stable, can deal with singularities and exhibits superlinear convergence in a neighborhood of the solution.

## 1 Introduction

Motivated by applications in solving fractional diffusion problems [6,7,9], there has recently been renewed interest in the fast and stable computation of best uniform rational approximations. The classical algorithm towards this end is the rational Remez algorithm (see, e.g., [3,4,12]), which is based on the idea of iteratively determining the nodes in which the approximation error equioscillates. Unfortunately, this approach suffers from severe numerical instabilities, which are usually dealt with by using extended precision arithmetic (as in [12]), which in turn significantly slows down the execution of the algorithm. New approaches for stabilizing the Remez algorithm were recently proposed in [5,11] based on so-called barycentric rational representations.

A new algorithm for computing best rational approximations was recently proposed in [10] based on a different idea: observing that the best approximation must interpolate the function to be approximated in a number of nodes, the new approach is to search for these interpolation nodes, rather than the nodes of equioscillation as is done in the Remez algorithm. The BRASIL algorithm proposed in [10] attempts to do so by a simple heuristic, iteratively applied, which attempts to rescale the lengths of the intervals between the interpolation nodes so as to equilibrate the local errors. Rational interpolation is performed using the barycentric formula. This novel approach appears to enjoy excellent numerical stability. However, as a fixed-point iteration, its convergence rate is only linear,

© Springer Nature Switzerland AG 2022
I. Lirkov and S. Margenov (Eds.): LSSC 2021, LNCS 13127, pp. 49–56, 2022.
https://doi.org/10.1007/978-3-030-97549-4_5

whereas the Remez algorithm converges quadratically in a neighborhood of the exact solution. Therefore a natural question is how to construct an algorithm which seeks the interpolation nodes while converging quadratically.

The present work represents the next step towards such an algorithm. We restrict our attention to the somewhat simpler case of best uniform polynomial approximation (note that the Remez algorithm was originally formulated for polynomial approximation as well). We again treat the interpolation nodes as our unknowns and rewrite the best approximation problem as a system of nonlinear equations, for which we then formulate a Newton's method. In order to compute the Jacobian matrix, we derive expressions for the derivative of an interpolating polynomial with respect to an interpolation node.

The content of the paper is laid out as follows: we first formulate the nonlinear system of equations for best uniform polynomial approximation in Sect. 2. We compute derivatives of interpolating polynomials in Sect. 3 and use these results to obtain the Jacobian of the system of equations in Sect. 4. The complete approximation algorithm is formulated in Sect. 5, and numerical experiments are presented in Sect. 6.

## 2   Best Uniform Polynomial Approximation as a System of Nonlinear Equations

We seek to determine a polynomial $p \in \mathcal{P}_n$ which best approximates a given function $f \in C[a, b]$ in the maximum norm. It is a classical result that such a best approximation exists, is unique, and that the best approximation error $f - p$ equioscillates in $n + 2$ distinct nodes $(y_j)_{j=0}^{n+1}$ in $[a, b]$ (see, e.g., [1]). That is, we have

$$f(y_j) - p(y_j) = \lambda(-1)^j, \qquad j = 0, \ldots, n + 1,$$

where $\lambda = \pm \|f - p\|_\infty$. Due to continuity, this implies that there are $n+1$ distinct interpolation nodes $(x_i)_{i=0}^n$ in $(a, b)$ with

$$p(x_i) = f(x_i), \qquad i = 0, \ldots, n,$$

interleaving the equioscillation nodes in the sense

$$a \leq y_0 < x_0 < y_1 < \ldots < y_n < x_n < y_{n+1} \leq b.$$

Let $\mathbf{x} \in \mathcal{X}$ denote a vector of interpolation nodes in the admissible set

$$\mathcal{X} := \{\mathbf{x} \in (a, b)^{n+1} : x_0 < \cdots < x_n\}$$

of nodes in increasing order. We denote by $p[\mathbf{x}] \in \mathcal{P}_n$ the unique polynomial which interpolates $f$ in the nodes $\mathbf{x}$. In each interval $(x_{j-1}, x_j)$, $j = 0, \ldots, n+1$ (letting $x_{-1} = a$ and $x_{n+1} = b$), let

$$y_j := \arg\max_{y \in (x_{j-1}, x_j)} |f(y) - p(y)|, \qquad j = 0, \ldots, n + 1,$$

denote the abscissa where the error $|f - p|$ is largest. Denote

$$\Phi(\mathbf{x}) := (f(y_j) - p[\mathbf{x}](y_j))_{j=0}^{n+1}, \qquad \mathbf{w} := ((-1)^j)_{j=0}^{n+1}.$$

**Theorem 1.** *If there exists $\lambda \in \mathbb{R}$ such that*

$$F(\mathbf{x}, \lambda) := \Phi(\mathbf{x}) - \lambda \mathbf{w} = 0, \qquad F : \mathbb{R}^{n+2} \to \mathbb{R}^{n+2}, \tag{1}$$

*then $p[\mathbf{x}]$ is the best polynomial approximation to $f$ with error $|\lambda| = \|f - p[\mathbf{x}]\|_\infty$.*

*Proof.* By definition, $\|f - p[\mathbf{x}]\|_\infty = \max_{j=0}^{n+1} |f(y_j) - p[\mathbf{x}](y_j)| = |\lambda|$, and thus the error $f - p[\mathbf{x}]$ equioscillates in $(y_j)_{j=0}^{n+1}$. $\qquad\square$

The above result shows that we can view finding the best polynomial approximation as solving the nonlinear Eq. (1).

In the following, we propose a Newton's method to solve this equation. Given initial guesses for the nodes $\mathbf{x}^0 \in \mathcal{X}$ and the signed error $\lambda^0 \in \mathbb{R}$, a Newton step for the solution of (1) is given by

$$\mathbf{d}^0 := (\mathbf{d}_{\mathbf{x}}^0, d_\lambda^0) := -(\nabla F(\mathbf{x}^0, \lambda^0))^{-1} F(\mathbf{x}^0, \lambda^0) \in \mathbb{R}^{n+2}.$$

We must make sure that the interpolation nodes remain within the interval $(a, b)$ and in increasing order, i.e., in the admissible set $\mathcal{X}$. For this purpose, we take a damped step $2^{-k} \mathbf{d}^0$, where $k$ is chosen according to

$$\min\{k \in \mathbb{N}_0 : \mathbf{x}^0 + 2^{-k} \mathbf{d}_{\mathbf{x}}^0 \in \mathcal{X}\}. \tag{2}$$

Since $\mathbf{x}^0 \in \mathcal{X}$ and $\mathcal{X}$ is an open set, such a choice always exists. The updated iterates are then given by

$$\mathbf{x}^1 := \mathbf{x}^0 + 2^{-k} \mathbf{d}_{\mathbf{x}}^0, \qquad \lambda^1 := \lambda^0 + 2^{-k} d_\lambda^0.$$

The main challenge in realizing this Newton's method is the computation of the Jacobian matrix of $F$, which we discuss in the following.

## 3 Derivatives of Polynomial Interpolants

In this section, we compute the derivatives of interpolating polynomials with respect to the interpolation nodes which are required for forming the Jacobian matrix. Let

$$\pi : \mathcal{X} \times \mathbb{R} \to \mathbb{R}, \qquad \pi((\xi_0, \dots, \xi_n), y) := p[\boldsymbol{\xi}](y)$$

denote the unique polynomial of degree at most $n$ in $y$ which interpolates $f : C[a, b] \to \mathbb{R}$ in the nodes $\boldsymbol{\xi}$. Given $\mathbf{x} \in \mathcal{X}$, we can write

$$\pi(\mathbf{x}, y) = \sum_{k=0}^{n} \ell_k(y) f(x_k)$$

with the Lagrange basis polynomials

$$\ell_k(y) = \frac{\omega_k(y)}{\omega_k(x_k)} = \frac{\omega(y)}{(y - x_k)\omega_k(x_k)}, \qquad k = 0, \ldots, n, \qquad (3)$$

where we use the notations

$$\omega(y) = \prod_{k=0}^{n}(y - x_k), \qquad \omega_i(y) = \frac{\omega(y)}{y - x_i}.$$

**Theorem 2.** *The derivative of the polynomial interpolant with respect to the interpolation node $x_i$, $i = 0, \ldots, n$, is given by*

$$\frac{\partial \pi}{\partial \xi_i}(\mathbf{x}, y) = \sum_{k=0}^{n} f(x_k) \frac{\partial}{\partial x_i} \ell_k(y) + \ell_i(y) f'(x_i) = \ell_i(y) \sum_{k=0}^{n} q_{ik} \qquad (4)$$

*with*

$$q_{ik} = \begin{cases} f'(x_k) - f(x_k) \sum_{m \neq k} \frac{1}{x_k - x_m}, & i = k, \\ f(x_k) \frac{\omega_i(x_i)}{\omega_k(x_k)(x_k - x_i)}, & i \neq k. \end{cases}$$

*Proof.* By elementary calculations we have

$$\frac{\partial}{\partial x_i} \ell_i(y) = -\ell_i(y) \sum_{m \neq i} \frac{1}{x_i - x_m},$$

$$\frac{\partial}{\partial x_i} \ell_k(y) = \ell_k(y) \frac{y - x_k}{(y - x_i)(x_k - x_i)} = \ell_i(y) \frac{\omega_i(x_i)}{\omega_k(x_k)(x_k - x_i)}, \qquad i \neq k.$$

The statement directly follows from these identities. □

## 4    Computing the Jacobian

Our aim is to compute the total derivative $\frac{d}{dx_i}\pi(\mathbf{x}, y_j)$, keeping in mind that the local maxima $y_j$ themselves depend on $\mathbf{x}$. In the following we assume sufficient smoothness, in particular, $f \in C^1[a, b]$. We have

$$\frac{d}{dx_i}\pi(\mathbf{x}, y_j) = \frac{\partial \pi}{\partial y}(\mathbf{x}, y_j)\frac{\partial y_j}{\partial x_i} + \frac{\partial \pi}{\partial \xi_i}(\mathbf{x}, y_j),$$

where in the first term the derivative of the interpolating polynomial is taken with respect to the evaluation point $y_j$, whereas in the second term the derivative is taken only with respect to the interpolation node $x_i$, but the evaluation point $y_j$ is considered constant.

Making use of this formula, we obtain

$$\frac{\partial \Phi_j}{\partial x_i}(\mathbf{x}) = \frac{\partial y_j}{\partial x_i}\left(f'(y_j) - p[\mathbf{x}]'(y_j)\right) - \frac{\partial \pi}{\partial \xi_i}(\mathbf{x}, y_j).$$

Since the nodes $y_j$ are local extrema of the error $f - p[\mathbf{x}]$, the term $f'(y_j) - p[\mathbf{x}]'(y_j)$ vanishes whenever $y_j \in (a, b)$. On the other hand, if $y_j \in \{a, b\}$, we have that either $\frac{\partial y_j}{\partial x_i}$ or $(f - p[\mathbf{x}])'(y_j)$ is 0 by a duality argument. Thus,

$$\frac{\partial \Phi_j}{\partial x_i}(\mathbf{x}) = -\frac{\partial \pi}{\partial \xi_i}(\mathbf{x}, y_j),$$

meaning that the dependence of the $y_j$ on $\mathbf{x}$ can be ignored while computing the Jacobian matrix. The term on the right-hand side can be computed using (4). We define the matrix $J(\mathbf{x}) \in \mathbb{R}^{(n+2)\times(n+1)}$ as

$$[J(\mathbf{x})]_{j,i} := -\frac{\partial \pi}{\partial \xi_i}(\mathbf{x}, y_j), \qquad j = 0, \ldots, n+1, \ i = 0, \ldots, n,$$

to be computed by means of formula (4), such that the Jacobian of $F(\mathbf{x}, \lambda)$ is given by

$$\bar{J}(\mathbf{x}, \lambda) := \begin{bmatrix} J(\mathbf{x}) & -\mathbf{w} \end{bmatrix} \in \mathbb{R}^{(n+2)\times(n+2)}. \tag{5}$$

## 5 The Algorithm

The complete method for best uniform polynomial approximation is given in Algorithm 1.

---

**Algorithm 1.** Newton's method for best polynomial approximation

---

**function** BESTPOLY($f \in C[a, b]$, $n \in \mathbb{N}$, $\varepsilon > 0$)

    set initial nodes $\mathbf{x} \in (a, b)^{n+1}$ to Chebyshev nodes of first kind

    **loop**

        set $p \leftarrow$ INTERPOLATE($f, \mathbf{x}$)

        compute abscissae of local maxima

$$y_j = \arg\max_{y \in (x_{j-1}, x_j)} |f(y) - p(y)|, \qquad j = 0, \ldots, n+1$$

        **if** $\frac{\max_j |f(y_j) - p(y_j)|}{\min_j |f(y_j) - p(y_j)|} - 1 < \varepsilon$ **then**

            **return** $p$

        **end if**

        if in first iteration: set $\lambda$ to the mean of the local errors $|f(y_j) - p(y_j)|$

        compute $F(\mathbf{x}, \lambda)$ and Jacobian $\bar{J}(\mathbf{x}, \lambda)$ by (5)

        compute Newton step

$$(\mathbf{d_x}, d_\lambda) \leftarrow -\bar{J}(\mathbf{x}, \lambda)^{-1} F(\mathbf{x}, \lambda)$$

        determine step size $2^{-k}$ by rule (2)

        update

$$\mathbf{x} \leftarrow \mathbf{x} + 2^{-k}\mathbf{d_x}, \quad \lambda \leftarrow \lambda + 2^{-k} d_\lambda$$

    **end loop**

**end function**

---

Some remarks are in order on the implementation of this algorithm:

- The function INTERPOLATE($f, \mathbf{x}$) computes the polynomial interpolant to $f$ in the nodes $\mathbf{x}$ by means of barycentric Lagrange interpolation [2].
- The local maxima $y_j$ may be computed efficiently by means of a golden section search; see [10] for details.
- An initial guess for the error $\lambda$ is obtained in the first iteration of the algorithm by taking the mean of the local error maxima.
- Formula (4) requires the first derivative $f'$, which we assume to be specified along with $f$ itself. If it is not available, finite differences could be used.
- For evaluating $\ell_i(y)$ in (4), we use the so-called barycentric form given by the last expression in (3) for reasons of numerical stability [2,8].
- The algorithm terminates when the equioscillation property of the local maxima is valid up to a user-specified tolerance $\varepsilon$.

## 6  Numerical Examples

We give numerical results for two functions,

$$f_1(x) = \frac{x^{1/4}}{1 + 10x^{1/4}}, \quad x \in [0,1], \qquad f_2(x) = |x|, \quad x \in [-1,1].$$

The function $f_1$ is a challenging example motivated by applications in fractional diffusion; cf. [10]. Using $f_2$ we demonstrate that the algorithm also works in the case of reduced smoothness; we use the sign function $\mathrm{sign}(x)$ in place of $f'(x)$ in this case. We approximate these functions using Algorithm 1 with varying polynomial degree $n$. In all cases, the tolerance for the stopping criterion is $\varepsilon = 10^{-10}$. The results were obtained on a laptop with an AMD Ryzen 5 3500U CPU.

Results for $f = f_1$ are shown in Fig. 1, and for $f = f_2$ in Fig. 2. In both cases, the table on the left shows the degree $n$, the maximum error $\|f - p\|_\infty$, the needed number of iterations, and the computation time (averaged over several runs). The plot on the right shows the convergence history for one particular run, displaying both the residual $\|F(\mathbf{x}, \lambda)\|$ and the deviation from equioscillation $\frac{\max_j |f(y_j) - p(y_j)|}{\min_j |f(y_j) - p(y_j)|} - 1$ (which is used for the stopping criterion) over the iterations. We observe that both of these error measures behave rather similarly and exhibit superlinear convergence during the final iterations. We also remark that the step sizes $2^{-k}$ chosen by (2) are always 1 except for a few initial iterations, thus taking full Newton steps. The convergence rates in the maximum norm are of course rather poor since the functions are not analytic and thus the polynomial approximations converge slowly.

| $n$ | error | iter | time (s) |
|-----|-------------|------|----------|
| 10 | 0.02857802 | 14 | 0.068 |
| 20 | 0.02472576 | 19 | 0.136 |
| 30 | 0.02243189 | 23 | 0.248 |
| 40 | 0.02081294 | 27 | 0.383 |
| 50 | 0.01957241 | 31 | 0.580 |
| 60 | 0.01857363 | 35 | 0.823 |
| 70 | 0.01774225 | 40 | 1.17 |

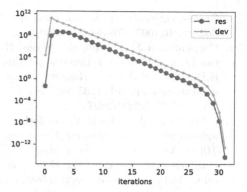

**Fig. 1.** Results for $f = f_1$. *Left:* Maximum error, number of iterations, and CPU time in dependence of degree $n$. *Right:* Convergence history for $n = 50$.

| $n$ | error | iter | time (s) |
|-----|-------------|------|----------|
| 5 | 0.06762090 | 6 | 0.024 |
| 15 | 0.01994878 | 16 | 0.084 |
| 25 | 0.01166106 | 12 | 0.095 |
| 35 | 0.00823581 | 16 | 0.167 |
| 45 | 0.00636543 | 21 | 0.288 |
| 55 | 0.00518721 | 25 | 0.440 |
| 65 | 0.00437698 | 30 | 0.654 |
| 75 | 0.00378564 | 35 | 0.928 |

**Fig. 2.** Results for $f = f_2$. *Left:* Maximum error, number of iterations, and CPU time in dependence of degree $n$. *Right:* Convergence history for $n = 45$.

Since $f_2$ is an even function, the best approximating polynomials for degrees $2n$ and $2n + 1$ are identical. Our algorithm requires the use of the odd degree $2n + 1$ in order to compute this solution.

**Acknowledgments.** This work was supported by the bilateral project KP-06-Austria/8/2019 (WTZ BG 03/2019), funded by Bulgarian National Science Fund and OeAD (Austria). The second author gratefully acknowledges additional support by the Austrian Science Fund (FWF) grant P 33956-NBL.

# References

1. Achieser, N.: Theory of Approximation. Dover Books on Advanced Mathematics. Dover Publications (1992)
2. Berrut, J.P., Trefethen, L.N.: Barycentric Lagrange interpolation. SIAM Rev. **46**(3), 501–517 (2004). https://doi.org/10.1137/s0036144502417715

3. Braess, D.: Nonlinear Approximation Theory. Springer, Heidelberg (1986). https:// doi.org/10.1007/978-3-642-61609-9
4. Carpenter, A.J., Ruttan, A., Varga, R.S.: Extended numerical computations on the 1/9 conjecture in rational approximation theory. In: Graves-Morris, P.R., Saff, E.B., Varga, R.S. (eds.) Rational Approximation and Interpolation. Lecture Notes in Mathematics, vol. 1105, pp. 383–411. Springer, Heidelberg (1984). https://doi. org/10.1007/bfb0072427
5. Filip, S.I., Nakatsukasa, Y., Trefethen, L.N., Beckermann, B.: Rational minimax approximation via adaptive barycentric representations. SIAM J. Sci. Comput. **40**(4), A2427–A2455 (2018). https://doi.org/10.1137/17m1132409
6. Harizanov, S., Lazarov, R., Margenov, S., Marinov, P., Pasciak, J.: Analysis of numerical methods for spectral fractional elliptic equations based on the best uniform rational approximation. J. Comput. Phys. **408**, 109285 (2020). https://doi. org/10.1016/j.jcp.2020.109285
7. Harizanov, S., Lazarov, R., Margenov, S., Marinov, P., Vutov, Y.: Optimal solvers for linear systems with fractional powers of sparse SPD matrices. Numer. Linear Algebra Appl. **25**(5), e2167 (2018). https://doi.org/10.1002/nla.2167
8. Higham, N.J.: The numerical stability of barycentric Lagrange interpolation. IMA J. Numer. Anal. **24**(4), 547–556 (2004). https://doi.org/10.1093/imanum/24.4.547
9. Hofreither, C.: A unified view of some numerical methods for fractional diffusion. Comput. Math. Appl. **80**(2), 332–350 (2020). https://doi.org/10.1016/j.camwa. 2019.07.025
10. Hofreither, C.: An algorithm for best rational approximation based on barycentric rational interpolation. Numer. Algorithms **88**(1), 365–388 (2021). https://doi.org/ 10.1007/s11075-020-01042-0
11. Ioniță, A.C.: Lagrange rational interpolation and its applications to approximation of large-scale dynamical systems. Ph.D. thesis, Rice University, Houston, TY (2013)
12. Varga, R.S., Carpenter, A.J.: Some numerical results on best uniform rational approximation of $x^\alpha$ on $[0, 1]$. Numer. Algorithms **2**(2), 171–185 (1992). https:// doi.org/10.1007/bf02145384

# Reduced Sum Implementation of the BURA Method for Spectral Fractional Diffusion Problems

Stanislav Harizanov, Nikola Kosturski, Ivan Lirkov$^{(\boxtimes)}$ , Svetozar Margenov, and Yavor Vutov

Institute of Information and Communication Technologies,
Bulgarian Academy of Sciences, Sofia, Bulgaria
{sharizanov,kosturski,margenov,yavor}@parallel.bas.bg,
ivan.lirkov@iict.bas.bg

**Abstract.** The numerical solution of spectral fractional diffusion problems in the form $\mathcal{A}^\alpha u = f$ is studied, where $\mathcal{A}$ is a selfadjoint elliptic operator in a bounded domain $\Omega \subset \mathbb{R}^d$, and $\alpha \in (0,1]$. The finite difference approximation of the problem leads to the system $\mathbb{A}^\alpha \mathbf{u} = \mathbf{f}$, where $\mathbb{A}$ is a sparse, symmetric and positive definite (SPD) matrix, and $\mathbb{A}^\alpha$ is defined by its spectral decomposition. In the case of finite element approximation, $\mathbb{A}$ is SPD with respect to the dot product associated with the mass matrix. The BURA method is introduced by the best uniform rational approximation of degree $k$ of $t^\alpha$ in $[0,1]$, denoted by $r_{\alpha,k}$. Then the approximation $\mathbf{u}_k \approx \mathbf{u}$ has the form $\mathbf{u}_k = c_0\mathbf{f} + \sum_{i=1}^k c_i(\mathbb{A} - \tilde{d}_i\mathbb{I})^{-1}\mathbf{f}$, $\tilde{d}_i < 0$, thus requiring the solving of $k$ auxiliary linear systems with sparse SPD matrices. The BURA method has almost optimal computational complexity, assuming that an optimal PCG iterative solution method is applied to the involved auxiliary linear systems. The presented analysis shows that the absolute values of first $\left\{\tilde{d}_i\right\}_{i=1}^{k'}$ can be extremely large. In such a case the condition number of $\mathbb{A} - \tilde{d}_i\mathbb{I}$ is practically equal to one. Obviously, such systems do not need preconditioning. The next question is if we can replace their solution by directly multiplying $\mathbf{f}$ with $-c_i/\tilde{d}_i$. Comparative analysis of numerical results is presented as a proof-of-concept for the proposed RS-BURA method.

## 1 Introduction

We consider the second order elliptic operator $\mathcal{A}$ in the bounded domain $\Omega \subset \mathbb{R}^d$, $d \in \{1,2,3\}$, assuming homogeneous boundary conditions on $\partial\Omega$. Let $\{\lambda_j, \psi_j\}_{j=1}^\infty$ be the eigenvalues and normalized eigenfunctions of $\mathcal{A}$, and let $(\cdot, \cdot)$ stand for the $L^2$ dot product. The spectral fractional diffusion problem $\mathcal{A}^\alpha u = f$ is defined by the equality

$$\mathcal{A}^\alpha u = \sum_{j=1}^\infty \lambda_j^\alpha (u, \psi_j)\psi_j, \quad \text{and therefore} \quad u = \sum_{j=1}^\infty \lambda_j^{-\alpha}(f, \psi_j)\psi_j. \tag{1}$$

© Springer Nature Switzerland AG 2022
I. Lirkov and S. Margenov (Eds.): LSSC 2021, LNCS 13127, pp. 57–64, 2022.
https://doi.org/10.1007/978-3-030-97549-4_6

Now, let a $(2d+1)$-point stencil on a uniform mesh be used to get the finite difference (FDM) approximation of $\mathcal{A}$. Then, the FDM numerical solution of (1) is given by the linear system

$$\mathbb{A}^\alpha \mathbf{u} = \mathbf{f}, \tag{2}$$

where $\mathbf{u}$ and $\mathbf{f}$ are the related mesh-point vectors, $\mathbb{A} \in \mathbb{R}^{N \times N}$ is SPD matrix, and $\mathbb{A}^\alpha$ is defined similarly to (1), using the spectrum $\{\lambda_{j,h}, \Psi_{j,h}\}_{j=1}^N$ of $\mathbb{A}$.

Alternatively, the finite element method (FEM) can be applied for the numerical solution of the fractional diffusion problem, if $\Omega$ is a general domain and some unstructured (say, triangular or tetrahedral) mesh is used. Then, $\mathbb{A} = \mathbb{M}^{-1}\mathbb{K}$, where $\mathbb{K}$ and $\mathbb{M}$ are the stiffness and mass matrices respectively, and $\mathbb{A}$ is SPD with respect to the energy dot product associated with $\mathbb{M}$.

Rigorous error estimates for the linear FEM approximation of (1) are presented in [1]. More recently, the mass lumping case is analyzed in [6]. The general result is that the relative accuracy in $L^2$ behaves essentially as $O(h^{2\alpha})$ for both linear FEM or $(2d+1)$-point stencil FDM discretizations.

A survey on the latest achievements in numerical methods for fractional diffusion problems is presented in [4]. Although there are several different approaches discussed there, all the derived algorithms can be interpreted as rational approximations, see also [7]. In this context, the advantages of the BURA (best uniform rational approximation) method are reported. The BURA method is originally introduced in [5], see [6] and the references therein for some further developments. The method is generalized to the case of Neumann boundary conditions in [3]. The present study is focused on the efficient implementation of the method for large-scale problems, where preconditioned conjugate gradient (PCG) iterative solver of optimal complexity is applied to the arising auxiliary sparse SPD linear systems.

The rest of the paper is organized as follows. The construction and some basic properties of the BURA method are presented in Sect. 2. The numerical stability of computing the BURA parameters is discussed in Sect. 3. The next section is devoted to the question how large can the BURA coefficients be, followed by the introduction of the new RS-BURA method based on reduced sum implementation in Sect. 5. A comparative analysis of the accuracy is provided for the considered test problem ($\alpha = 0.25$, $k = 85$ and reduced sum of 46 terms), varying the parameter $\delta$, corresponding to the spectral condition number of $\mathbb{A}$. Brief concluding remarks are given at the end.

## 2   The BURA Method

Let us consider the min-max problem

$$\min_{r_k(t)\in\mathcal{R}(k,k)} \max_{t\in[0,1]} |t^\alpha - r_k(t)| =: E_{\alpha,k}, \quad \alpha \in (0,1), \tag{3}$$

where $r_k(t) = P_k(t)/Q_k(t)$, $P_k$ and $Q_k$ are polynomials of degree $k$, and $E_{\alpha,k}$ stands for the error of the $k$-BURA element $r_{\alpha,k}$. Following [6] we introduce

$$\mathbb{A}^{-\alpha} \approx \lambda_{1,h}^{-\alpha} r_{\alpha,k}(\lambda_{1,h}\mathbb{A}^{-1}), \quad \text{and respectively} \quad \mathbf{u}_k = \lambda_{1,h}^{-\alpha} r_{\alpha,k}(\lambda_{1,h}\mathbb{A}^{-1})\mathbf{f}, \tag{4}$$

where $\mathbf{u}_k$ is the BURA numerical solution of the fractional diffusion system (2). The following relative error estimate holds true (see [9,10] for more details)

$$\frac{||\mathbf{u} - \mathbf{u}_k||_2}{||\mathbf{f}||_2} \le \lambda_{1,h}^{-\alpha} E_{\alpha,k} \approx \lambda_{1,h}^{-\alpha} 4^{\alpha+1} \sin \alpha\pi \ e^{-2\pi\sqrt{\alpha k}} = O\left(e^{-\sqrt{\alpha k}}\right). \tag{5}$$

Let us denote the roots of $P$ and $Q$ by $\xi_1, \ldots, \xi_k$ and $d_1, \ldots, d_k$ respectively. It is known that they interlace, satisfying the inequalities

$$0 > \xi_1 > d_1 > \xi_2 > d_2 > \ldots > \xi_k > d_k. \tag{6}$$

Using (4) and (6) we write the BURA numerical solution $\mathbf{u}_k$ in the form

$$\mathbf{u}_k = c_0\mathbf{f} + \sum_{i=1}^{k} c_i(\mathbb{A} - \tilde{d}_i\mathbb{I})^{-1}\mathbf{f}, \tag{7}$$

where $\tilde{d}_i = 1/d_i < 0$ and $c_i > 0$. Further $0 > \tilde{d}_k > \ldots > \tilde{d}_1$ and the following property plays an important role in this study

$$\lim_{k\to\infty} \tilde{d}_1 = -\infty. \tag{8}$$

The proof of (8) is beyond the scope of the present article. Numerical data illustrating the behaviour of $\tilde{d}_i$ are provided in Sect. 4.

## 3    Computing the Best Uniform Rational Approximation

Computing the BURA element $r_{\alpha,k}(t)$ is a challenging computational problem on its own. There are more than 30 years modern history of efforts in this direction. For example, in [11] the computed data for $E_{\alpha,k}$ are reported for six values of $\alpha \in (0,1)$ with degrees $k \le 30$ by using computer arithmetic with 200 significant digits. The modified Remez method was used for the derivation of $r_{\alpha,k}(t)$ in [5]. The min-max problem (3) is highly nonlinear and the Remez algorithm is very sensitive to the precision of the computer arithmetic. In particular, this is due to the extreme clustering at zero of the points of alternance of the error function. Thus the practical application of the Remez algorithm is reduced to degrees $k$ up to $10 - 15$ if up to quadruple-precision arithmetic is used.

The discussed difficulties are practically avoided by the recent achievements in the development of highly efficient algorithms that exploit a representation of the rational approximant in barycentric form and apply greedy selection of the support points. The advantages of this approach is demonstrated by the method proposed in [7], which is based on AAA (adaptive Antoulas-Anderson) approximation of $z^{-\alpha}$ for $z \in [\lambda_{1,h}, \lambda_{N,h}]$. Further progress in computational stability has been reported in [8] where the BRASIL algorithm is presented. It is based on the assumption that the best rational approximation must interpolate the function at a certain number of nodes $z_j$, iteratively rescaling the intervals $(z_{j-1}, z_j)$ with the goal of equilibrating the local errors.

**Table 1.** Minimal degree $k$ needed to get the desired accuracy $E_{\alpha,k}$

| $E_{\alpha,k}$ | $\alpha = 0.25$ | $\alpha = 0.50$ | $\alpha = 0.75$ | $\alpha = 0.80$ | $\alpha = 0.90$ |
|---|---|---|---|---|---|
| $10^{-3}$ | 7 | 4 | 3 | 3 | 2 |
| $10^{-4}$ | 12 | 7 | 4 | 4 | 3 |
| $10^{-5}$ | 17 | 9 | 6 | 6 | 5 |
| $10^{-6}$ | 24 | 13 | 9 | 8 | 7 |
| $10^{-7}$ | 31 | 17 | 11 | 11 | 9 |
| $10^{-8}$ | 40 | 21 | 14 | 13 | 11 |
| $10^{-9}$ | 50 | 26 | 18 | 16 | 14 |
| $10^{-10}$ | 61 | 32 | 21 | 20 | 17 |
| $10^{-11}$ | 72 | 38 | 26 | 24 | 20 |
| $10^{-12}$ | 85 | 45 | 30 | 28 | 24 |

In what follows, the approximation $r_{\alpha,k}(t)$ used in BURA is computed utilizing the open access software implementation of BRASIL, developed by Hofreither [2]. In Table 1 we show the minimal degree $k$ needed to get a certain targeted accuracy $E_{\alpha,k} \in (10^{-12}, 10^{-3})$ for $\alpha \in \{0.25, 0.50, 0.75, 0.80, 0.90\}$.

The presented data well illustrate the sharpness of the error estimate (5). The dependence on $\alpha$ is clearly visible.

It is impressive that for $\alpha = 0.25$ we are able to get a relative accuracy of order $O(10^{-12})$ for $k = 85$. Now comes the question how such accuracy can be realized (if possible at all) in real-life applications. Let us note that the published test results for larger $k$ are only for one-dimensional test problems. They are considered representative due to the fact that the error estimates of the BURA method do not depend on the spatial dimension $d \in \{1, 2, 3\}$.

However, as we will see, some new challenges and opportunities appear if $d > 1$ and iterative solution methods have to be used to solve the auxiliary large-scale SPD linear systems that appear in (7). This is due to the fact that some of the coefficients in the related sum are extremely large for large $k$.

## 4    How Large Can the BURA Coefficients Be

The BURA method has almost optimal computational complexity $O((\log N)^2 N)$ [6]. This holds true under the assumption that a proper PCG iterative solver of optimal complexity is used for the systems with matrices $\mathbb{A} - \tilde{d}_i \mathbb{I}$, $\tilde{d}_i < 0$. For example, some algebraic multigrid (AMG) preconditioner can be used as an efficient solver, if the positive diagonal perturbation $-\tilde{d}_i \mathbb{I}$ of $\mathbb{A}$ does not substantially alter its condition number.

The extreme behaviour of the first part of the coefficients $\tilde{d}_i$ is illustrated in Table 2. The case $\alpha = 0.25$ and $k = 85$ is considered in the numerical tests. Based on (8), the expectation is that $-\tilde{d}_1$ can be extremely large. In practice, the displayed coefficients are very large for almost a half of $-\tilde{d}_i$, starting with

**Table 2.** Behaviour of the coefficients $\widetilde{d}_i$ and $c_i$ in (7) for $\alpha = 0.25$ and $k = 85$

| | | |
|---|---|---|
| $\widetilde{d}_1 = -1.7789 \times 10^{45}$ | $c_1 = 1.4698 \times 10^{34}$ | $c_1/\widetilde{d}_1 = -8.2624 \times 10^{-12}$ |
| $\widetilde{d}_2 = -5.0719 \times 10^{42}$ | $c_2 = 1.1593 \times 10^{32}$ | $c_2/\widetilde{d}_2 = -2.2858 \times 10^{-11}$ |
| $\widetilde{d}_3 = -7.1427 \times 10^{40}$ | $c_3 = 3.7598 \times 10^{30}$ | $c_3/\widetilde{d}_3 = -5.2638 \times 10^{-11}$ |
| $\widetilde{d}_4 = -2.1087 \times 10^{39}$ | $c_4 = 2.2875 \times 10^{29}$ | $c_4/\widetilde{d}_4 = -1.0848 \times 10^{-10}$ |
| $\widetilde{d}_5 = -9.7799 \times 10^{37}$ | $c_5 = 2.0285 \times 10^{28}$ | $c_5/\widetilde{d}_5 = -2.0741 \times 10^{-10}$ |
| $\vdots$ | $\vdots$ | $\vdots$ |
| $\widetilde{d}_{35} = -4.7256 \times 10^{17}$ | $c_{35} = 4.3270 \times 10^{12}$ | $c_{35}/\widetilde{d}_{35} = -9.1565 \times 10^{-6}$ |
| $\widetilde{d}_{36} = -1.6388 \times 10^{17}$ | $c_{36} = 1.9278 \times 10^{12}$ | $c_{36}/\widetilde{d}_{36} = -1.1764 \times 10^{-5}$ |
| $\widetilde{d}_{37} = -5.7675 \times 10^{16}$ | $c_{37} = 8.6879 \times 10^{11}$ | $c_{37}/\widetilde{d}_{37} = -1.5064 \times 10^{-5}$ |
| $\widetilde{d}_{38} = -2.0587 \times 10^{16}$ | $c_{38} = 3.9585 \times 10^{11}$ | $c_{38}/\widetilde{d}_{38} = -1.9228 \times 10^{-5}$ |
| $\widetilde{d}_{39} = -7.4487 \times 10^{15}$ | $c_{39} = 1.8227 \times 10^{11}$ | $c_{39}/\widetilde{d}_{39} = -2.4470 \times 10^{-5}$ |

$-\widetilde{d}_1 \approx 10^{45}$. The coefficients $\widetilde{d}_i$ included in Table 2 are representative, taking into account their monotonicity (6). We observe decreasing monotonic behavior for the coefficients $c_i$ and the ratios $c_i/\widetilde{d}_i$. Recall that the former were always positive, while the latter – always negative. The theoretical investigation of this observation is outside the scope of the paper.

The conclusion is, that for such large values of $-\widetilde{d}_i$ the condition number of $\mathbb{A} - \widetilde{d}_i\mathbb{I}$ practically equals 1. Of course, it is a complete nonsense to precondition such matrices. Solving such systems may even not be the right approach. In the next section we will propose an alternative.

## 5    Reduced Sum Method RS-BURA

Let $k' < k$ be chosen such that $|\widetilde{d}_i|$ is very large for $1 \leq i \leq k - k'$. Then, the reduced sum method for implementation of BURA (RS-BURA) for larger $k$ is defined by the following approximation $\mathbf{u} \approx \mathbf{u}_{k,k'}$ of the solution of (2):

$$\mathbf{u}_{k,k'} = \left[ c_0 - \sum_{i=1}^{k-k'} \frac{c_i}{\widetilde{d}_i} \right] \mathbf{f} + \sum_{i=k-k'+1}^{k} c_i (\mathbb{A} - \widetilde{d}_i\mathbb{I})^{-1}\mathbf{f}. \qquad (9)$$

The RS-BURA method can be interpreted as a rational approximation of $\mathbb{A}^{-\alpha}$ of degree $k'$ denoted by $\widetilde{r}_{\alpha,k,k'}(\mathbb{A})$, thus involving solves of only $k'$ auxiliary sparse SPD systems.

Having changed the variable $z := t^{-1}$ we introduce the error indicator

$$\widetilde{E}^\delta_{\alpha,k,k'} = \max_{z \in [1,\delta^{-1}]} |z^{-\alpha} - \widetilde{r}_{\alpha,k,k'}(z)|.$$

Similarly to (5), one can get the relative error estimate

$$\frac{\|\mathbf{u} - \mathbf{u}_{k,k'}\|}{\|\mathbf{f}\|} \leq \widetilde{E}^\delta_{\alpha,k,k'}, \qquad (10)$$

under the assumption $1 \leq \lambda_{1,h} < \lambda_{N,h} \leq 1/\delta$. Let us remember that for the considered second order elliptic (diffusion) operator $\mathcal{A}$, the extreme eigenvalues of $\mathbb{A}$ satisfy the asymptotic bounds $\lambda_{1,h} = O(1)$ and $\lambda_{N,h} = O(h^{-2})$.

Direct computations give rise to

$$\widetilde{r}_{\alpha,k,k'}(z) - r_{\alpha,k}(z) = -\sum_{i=1}^{k-k'} \left[ \frac{c_i}{\widetilde{d}_i} + \frac{c_i}{z - \widetilde{d}_i} \right] = \sum_{i=1}^{k-k'} \left( -\frac{c_i}{\widetilde{d}_i} \right) \frac{z}{z - \widetilde{d}_i} > 0, \quad (11)$$

and since $x/(x-y)$ is monotonically increasing with respect to both arguments, provided $x > 0$ and $y < 0$, we obtain

$$\widetilde{E}^\delta_{\alpha,k,k'} - E_{\alpha,k} \leq \frac{\delta^{-1}}{\delta^{-1} - \widetilde{d}_{k-k'}} \frac{(k-k')c_{k-k'}}{-\widetilde{d}_{k-k'}}.$$

When $k'$ is chosen large enough and $\delta^{-1} \ll -\widetilde{d}_{k-k'}$, the orders of the coefficients $\{\widetilde{d}_i\}_1^{k-k'}$ differ (see Table 1), which allows us to remove the factor $(k-k')$ from the above estimate. Therefore, in such cases the conducted numerical experiments suggest that the following sharper result holds true

$$ord\left( \widetilde{E}^\delta_{\alpha,k,k'} - E_{\alpha,k} \right) \leq ord(\delta^{-1}) + ord\left( c_{k-k'}/(-\widetilde{d}_{k-k'}) \right) - ord(-\widetilde{d}_{k-k'}), \quad (12)$$

providing us with an a priori estimate on the minimal value of $k'$ that will guarantee that the order of $\widetilde{E}^\delta_{\alpha,k,k'} - E_{\alpha,k}$ is smaller than the order of $E_{\alpha,k}$. Again, the theoretical validation of (12) is outside the scope of this paper.

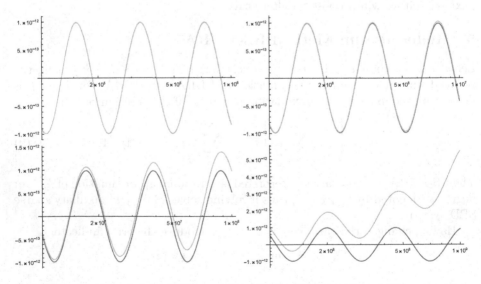

**Fig. 1.** BURA (blue) and RS-BURA (yellow) errors $r_{\alpha,k}(z) - z^{-\alpha}$, respectively $\widetilde{r}_{\alpha,k,k'}(z) - z^{-\alpha}$, $z \in (1, \delta^{-1}]$, $\alpha = 0.25$, $k = 85$, $k' = 46$, $\delta \in \{10^{-6}, 10^{-7}, 10^{-8}, 10^{-9}\}$. (Color figure online)

The next numerical results illustrate the accuracy of RS-BURA. The test problem is defined by the following parameters: $\alpha = 0.25$, $k = 85$, $k' = 46$ and $\delta \in \{10^{-6}, 10^{-7}, 10^{-8}, 10^{-9}\}$. In accordance with Table 1, $E_{0.25,85} \approx 10^{-12}$. In Fig. 1, the behavior of RS-BURA (yellow) and BURA (blue) approximations is compared. The RS-BURA error is always larger than the BURA error, as derived in (11). The largest differences are located at the right part of the interval. As expected, for fixed $(k, k')$, the differences increase with the decrease of $\delta$. However, even in the worst case of $\delta = 10^{-9}$, the accuracy $\widetilde{E}^{\delta}_{0.25,85,46} \approx 5 \times 10^{-12}$. Indeed, according to (12) and Table 1, in this case the order of $\widetilde{E}^{\delta}_{\alpha,k,k'} - E_{\alpha,k}$, thus the order of $\widetilde{E}^{\delta}_{\alpha,k,k'}$ itself, should not exceed $9 + (-5) - 15 = -11$. On the other hand, (12) implies that when $\delta > 10^{-7}$ the order of $\widetilde{E}^{\delta}_{\alpha,k,k'} - E_{\alpha,k}$ is lower than the order of $E_{\alpha,k}$, meaning that the sum reduction does not practically affect the accuracy. This is illustrated on the first row of Fig. 1, where the plots of the two error functions are almost identical. This is a very promising result, taking into account that $\delta$ is controlled by the condition number of $\mathbb{A}$.

From a practical point of view, if $\Omega \subset \mathbb{R}^3$, and the FEM/FDM mesh is uniform or quasi uniform, the size of mesh parameter is limited by $h > 10^{-4}$, corresponding to $N < 10^{12}$. Such a restriction for $N$ holds true even for the nowadays supercomputers. Thus, we obtain that if $h \approx 10^{-4}$, the corresponding $\delta > 10^{-9}$. In general, for real life applications, the FEM mesh is usually unstructured, including local refinement. This leads to some additional increase of the condition number of $\mathbb{A}$. However, the results in Fig. 1 show that RS-BURA has a potential even for such applications.

The final conclusion of these tests is that some additional analysis is needed for better tuning of $k'$ for given $k$, $\alpha$ and a certain targeted accuracy.

# 6   Concluding Remarks

The BURA method was introduced in [5], where the number of the auxiliary sparse SPD system solves was applied as a measure of computational efficiency. This approach is commonly accepted in [6,7], following the unified view of the methods interpreted as rational approximations, where the degree $k$ is used in the comparative analysis.

At the same time, it was noticed that when iterative PCG solvers are used in BURA implementation even for relatively small degrees $k$, the number of iterations is significantly different, depending on the values of $\widetilde{d}_i$ (see [3] and the references therein).

The present paper opens a new discussion about the implementation of BURA methods in the case of larger degrees $k$. It was shown that due to the extremely large values of part of $|\widetilde{d}_i|$, the related auxiliary sparse SPD systems are very difficult (if possible at all) to solve. In this context, the first important question was whether BURA is even applicable to larger $k$ in practice. The proposed RS-BURA method shows one possible way to do this. The presented numerical tests give promising indication for the accuracy of the proposed reduced sum implementation of BURA.

The rigorous analysis of the new method (e.g., the proofs of (8), (11), and (12)) is beyond the scope of this paper.

Also, the results presented in Sect. 5 reopen the topic of computational complexity analysis, including the question of whether the almost optimal estimate $O((\log N)^2 N)$ can be improved.

**Acknowledgements.** We acknowledge the provided access to the e-infrastructure and support of the Centre for Advanced Computing and Data Processing, with the financial support by the Grant No BG05M2OP001-1.001-0003, financed by the Science and Education for Smart Growth Operational Program (2014–2020) and co-financed by the European Union through the European structural and Investment funds.

The presented work is partially supported by the Bulgarian National Science Fund under grant No. DFNI-DN12/1.

# References

1. Bonito, A., Pasciak, J.: Numerical approximation of fractional powers of elliptic operators. Math. Comput. **84**(295), 2083–2110 (2015)
2. Software BRASIL. https://baryrat.readthedocs.io/en/latest/#baryrat.brasil
3. Harizanov, S., Kosturski, N., Margenov, S., Vutov, Y.: Neumann fractional diffusion problems: BURA solution methods and algorithms. Math. Comput. Simul. **189**, 85–98 (2020)
4. Harizanov, S., Lazarov, R., Margenov, S.: A survey on numerical methods for spectral space-fractional diffusion problems. Frac. Calc. Appl. Anal. **23**, 1605–1646 (2020)
5. Harizanov, S., Lazarov, R., Margenov, S., Marinov, P., Vutov, Y.: Optimal solvers for linear systems with fractional powers of sparse SPD matrices. Numer. Linear Algebra Appl. **25**(5), e2167 (2018). https://doi.org/10.1002/nla.2167
6. Harizanov, S., Lazarov, R., Margenov, S., Marinov, P., Pasciak, J.: Analysis of numerical methods for spectral fractional elliptic equations based on the best uniform rational approximation. J. Comput. Phys. **408**, 109285 (2020)
7. Hofreither, C.: A unified view of some numerical methods for fractional diffusion. Comput. Math. Appl. **80**(2), 332–350 (2020)
8. Hofreither, C.: An algorithm for best rational approximation based on Barycentric rational interpolation. Numer. Algorithms **88**(1), 365–388 (2021). https://doi.org/10.1007/s11075-020-01042-0
9. Stahl, H.: Best uniform rational approximation of $x^\alpha$ on [0, 1]. Bull. Amer. Math. Soc. (NS) **28**(1), 116–122 (1993)
10. Stahl, H.: Best uniform rational approximation of $x^\alpha$ on [0, 1]. Acta Math. **190**(2), 241–306 (2003)
11. Varga, R.S., Carpenter, A.J.: Some numerical results on best uniform rational approximation of $x^\alpha$ on [0, 1]. Numer. Algorithms **2**(2), 171–185 (1992)

# First-Order Reaction-Diffusion System with Space-Fractional Diffusion in an Unbounded Medium

Dimiter Prodanov[1,2](✉) (iD)

[1] Neuroscience Research Flanders, IMEC, Kapeldreef 75, 3001 Leuven, Belgium
[2] MMSDP, Institute of Information and Communication Technologies, Bulgarian Academy of Science, Acad. G. Bonchev Street, Block 25A, 1113 Sofia, Bulgaria
dimiter.prodanov@imec.be

**Abstract.** Diffusion in porous media, such as biological tissues, is characterized by deviations from the usual Fick's diffusion laws, which can lead to space-fractional diffusion. The paper considers a simple case of a reaction-diffusion system with two spatial compartments – a proximal one of finite width having a source; and a distal one, which is extended to infinity and where the source is not present but there is a first order decay of the diffusing species. The diffusion term is taken to be proportional to the Riesz fractional Laplacian operator. It is demonstrated that the steady state of the system can be solved in terms of the Hankel transform involving Bessel functions. Methods for numerical evaluation of the resulting integrals are implemented. It is demonstrated that the convergence of the Bessel integrals could be accelerated using standard techniques for sequence acceleration.

**Keywords:** Hankel transform · Riesz Laplacian · Bessel functions

## 1 Introduction

Diffusion in porous media, such as biological tissues, is characterized by deviations from the usual Fick's diffusion laws. In certain cases, such deviations can lead to space-fractional diffusion, notably, when the random displacements follow Lévy flights [2]. The problem arises in the modelling of controlled release of substances and biomedical engineering [6]. The present contribution considers the steady state of a first-order reaction-diffusion system

$$\partial_t c = -D(-\Delta)^\alpha c + s - qc, \quad 0 < \alpha \le 1$$

where the symbol $(-\Delta)^\alpha$ denotes the Riesz fractional Laplacian operator of order $\alpha$ [7], $s$ is a constant spatially-extended source, term $D$ is the diffusion constant, $q$ is an elimination constant, and $c$ is the concentration of the diffusing species. The spatial arrangement is represented in Fig. 1. The schematic demonstrates the semicylindrical section. We will assume cylindrical geometry of infinite

© Springer Nature Switzerland AG 2022
I. Lirkov and S. Margenov (Eds.): LSSC 2021, LNCS 13127, pp. 65–70, 2022.
https://doi.org/10.1007/978-3-030-97549-4_7

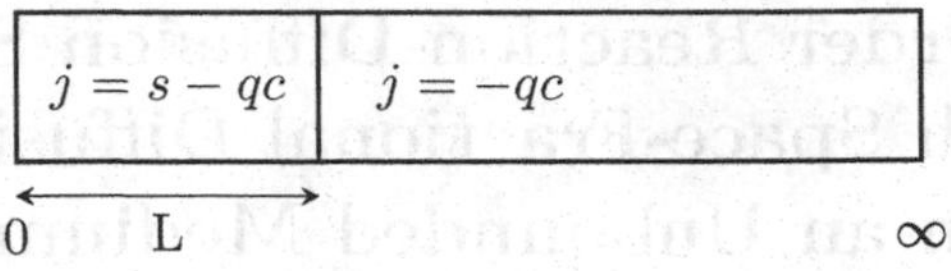

$$\boxed{\begin{array}{c|c} j = s - qc & j = -qc \end{array}}$$

$$\xleftarrow{\qquad} \atop 0 \quad L \qquad\qquad\qquad \infty$$

**Fig. 1.** Schematic of the compartmentalized reaction-diffusion system, $j$ denotes the flux

extent, which effectively renders the problem 2 dimensional. Then $L$ will be interpreted as the radius of the proximal compartment (Fig. 1). The problem was treated initially in [6], however, without presenting a solution in the spatial domain for the cylindrical geometry in the fractional case.

## 2  The Riesz Operator

The Riesz operator can be easily understood in the Fourier domain (see Sect. A for the applicable convention) where

$$-(-\Delta)^\alpha f(x) \mapsto |k|^{2\alpha} \hat{f}(k).$$

That is, we can identify the suggestive algebraic substitution $(-\Delta)^\alpha \mapsto -|k|^{2\alpha}$ [3]. From this perspective, the fractional Laplacian can be considered as the gradient of another operator, that is

$$-|k|^{2\alpha} = i\mathbf{k} \cdot i\mathbf{k}^0 |k|^{2\alpha-1}, \quad \mathbf{k}^0 = \mathbf{k}/k$$

where the dot denotes the scalar product and $\mathbf{k}^0$ is a unit wave vector in the Fourier space. Therefore, the operator can be reparametrized using $\beta = 2\alpha - 1$. Therefore, a Riesz-type of gradient can be defined algebraically as

$$\nabla^\beta \mapsto i\mathbf{k}^0 |k|^{1-\beta} = i\mathbf{k}/|k|^\beta$$

for a suitable function space. This equation can be interpreted physically as a generalized first Fick's law.

## 3  The Reaction-Diffusion System

Consider the compartmentalized system of two spatial compartments – a proximal one of finite width $L$, having a source of intensity $s$; and a distal one, which extends to infinity and where the source is not present but there is only a first-order decay (Fig. 1). The system attains a non-trivial steady state, which will be studied for the cylindrical geometry. From now on we denote the steady state concentration with the same label and assume throughout $0 < \alpha \leq 1$. Let

$$-D(-\Delta)^\alpha c(r) + s - qc(r) = 0 \tag{1}$$

where $r$ is a radial variable. We assume the boundary condition $c(\infty) = 0$. For simplicity of the presentation we assume $D = 1$. For the case $D \neq 1$, the equation can be reparametrized as $s' = s/D$, $q' = q/D$.

For the proximal compartment,

$$-(-\Delta)^\alpha c + s - qc = 0.$$

In the Fourier domain, denoting $|k| = \rho$, the entire system can be transformed as

$$-\rho^{2\alpha}\hat{c} + sL\frac{J_1(\rho L)}{\rho} - q\hat{c} = 0 \implies \hat{c} = sL\frac{J_1(\rho L)}{\rho(\rho^{2\alpha} + q)}.$$

Therefore, the full solution can be recognized as an inverse Hankel transform:

$$c(r) = sL \int_0^\infty \frac{J_0(\rho r)J_1(\rho L)}{\rho^{1+\beta} + q} d\rho, \quad \beta = 2\alpha - 1 \tag{2}$$

## 4  Correspondence with Previous Results

Remarkably, the distal component can be solved also in a different way. For the distal component we assume a fictitious delta source of intensity $s'$ on the boundary of the cylinder, which is also constant in time. That is, the "gluing" boundary conditions read

$$c(L) = c_f(L)$$
$$c(\infty) = c_f(\infty) = 0.$$

If the source is situated at a distance $L$ from the origin we obtain

$$-(-\Delta)^\alpha c - qc + s'\frac{\delta(r - L)}{r} = 0.$$

Therefore,

$$-q\hat{c} - \rho^{2\alpha}\hat{c} = -s'J_0(\rho L) \Rightarrow \hat{c} = \frac{s'J_0(\rho L)}{\rho^{2\alpha} + q}$$

and

$$c_f(r) = s' \int_0^\infty \frac{J_0(\rho r)J_0(\rho L)\rho}{\rho^{1+\beta} + q} d\rho, \quad \beta = 2\alpha - 1 \tag{3}$$

In this case $s'/s = LI_1/I_2$, where $I$ denotes the value of the respective integrals at L.

Remarkably, for $\beta = 1$ and $L = 0$ for the distal component we obtain

$$c_f(r) = s' \int_0^\infty \frac{J_0(\rho r)\rho}{\rho^2 + q} d\rho = s'K_0(qr).$$

where $K$ is the corresponding modified Bessel function. This special case was the solution obtained previously [6]. In this way, the fractional problem leads to a special function, related to the Bessel $K_0$ function.

## 5   Evaluation of the Integrals

The oscillatory integrals above can be computed numerically to a desired order of precision. The section considers the computation of the integral using double-exponential quadrature (DE) integration method [8], compared to the Maxima integration routine QUADPACK [4]. To this end the DE routine was ported to Maxima. Plots of the solution are presented in Fig. 2 for $L = 1$ and $q = 1$ using the DE method for oscillatory integrals.

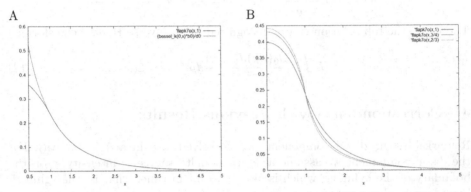

A – The full solution (eq. 2) is compared to the distal one (eq. 3) for $\alpha = 1$. B – Influence of the fractional exponent: the full solution is plotted for $\alpha = 1, 3/4, 2/3$, respectively.

**Fig. 2.** Plots of the solution

**Table 1.** Numerical values of $c(1)$, $\alpha = 1$

| Algorithm/n | | | |
|---|---|---|---|
| DE/OSC | 0.2379457951093593 | | |
| | n = 20 | n = 30 | n = 40 |
| DE/Wynn | 0.2379458091881929 | 0.2379458004645431 | 0.2379457960298526 |
| Q/Wynn | 0.2379458093742743 | 0.2379457998073924 | 0.2379457979814581 |
| DE/Aitken | 0.237945848054538 | 0.2379458651699171 | 0.2379458475394174 |
| Q/Aitken | 0.2379458464074396 | 0.2379458521504544 | 0.2379458175696549 |

Abbreviations: DE – double exponential; Q – Quadpack; n – number of asymptotic Bessel zeroes; OSC – oscillatory integral computed with relative precision $1E - 16$.

The Hankel integrals are oscillatory and converge slowly [5]. A way to improve the convergence is to use convergence acceleration methods for a sequence of integrals between the zeroes of the Bessel $J_0$ function. For fixed integration schema the values can be hard coded in an array. However, for schemata requiring a given precision, the zeroes are computed by approximation, which can lead to higher computational burden. On the other hand, the zeroes can be asymptotically approximated by a closed formula, which could provide an acceptable trade-off between precision and computational complexity. The asymptotic formula of the zeroes of the Bessel $J_\nu(z)$ function reads:

$$r_k(\nu) = \pi \left( k + \frac{\nu}{2} - \frac{1}{4} \right) + O\left( k^{-1} \right), \quad k \to \infty.$$

Convergence acceleration has been implemented by two common algorithms: Wynn's $\varepsilon$ [9] algorithm and Aitken's $\delta^2$ process [1] where the domain of integration is partitioned between the values of $r_k(0)$ for 20, 30 and 40 asymptotic zeroes (see Table 1). From the table it is apparent that both convergence acceleration methods could reach $1E - 8, 1E - 10$ precision compared to the DE oscillatory integral. Presented plots demonstrate smooth behavior of the calculation method.

## 6    Discussion and Conclusion

In summary, the paper demonstrates the solution of a fractional reaction diffusion system in steady state. Presented solution technique could be extended to annular geometries, for example for modeling fractional diffusion on compact domains. The solution is evaluated in terms of numerical quadratures. For the case of Bessel integrals the convergence of the integrals could be successfully accelerated using standard techniques. Presented methods exhibit smooth behavior for practical ranges of the radial variable. It is expected that the solution would be used for data fitting of observed distributions of substances or biological cells as originally intended [6].

## A    Fourier Transform Convention

The Fourier transform will be defined under the engineering convention

$$\hat{f}(k) = \mathcal{F}f(x) := \int_{\mathbb{R}^d} f(x)e^{-ik\cdot x}dx^d$$

with inverse

$$f(x) = \mathcal{F}^{-1}\hat{f}(k) := \frac{1}{(2\pi)^d} \int_{\mathbb{R}^d} \hat{f}(k)e^{ik\cdot x}dx^d.$$

# B    Hankel Transform Convention

The automorphic, positive order, Hankel transform is defined as [5]:

$$\hat{f}(\rho) = \mathcal{H}_\nu f(r) := \int_0^\infty f(z) J_\nu(\rho\, z) z\, dz$$

Remarkably, for 2 spatial dimensions the Laplacian is represented by a monomial factor:

$$\mathcal{H}_0 \Delta f(r) = -\rho^2 \hat{f}(\rho)$$

and in the similar way the Riesz Laplacian is $(-\Delta)^\alpha \mapsto -|\rho|^{2\alpha}$.

# References

1. Aitken, A.: On Bernoulli's numerical solution of algebraic equations. Proc. R. Soc. Edinb. **46**, 289–305 (1926)
2. Gorenflo, R., Mainardi, F.: Random walk models approximating symmetric space-fractional diffusion processes. In: Problems and Methods in Mathematical Physics, pp. 120–145. Birkhäuser, Basel (2001). https://doi.org/10.1007/978-3-0348-8276-7_10
3. Kwaśnicki, M.: Ten equivalent definitions of the fractional laplace operator. Fract. Calc. Appl. Anal. **20**(1), 7–51 (2017). https://doi.org/10.1515/fca-2017-0002
4. Piessens, R., Doncker-Kapenga, E., Überhuber, C.W., Kahaner, D.K.: Quadpack. Springer, Heidelberg (1983). https://doi.org/10.1007/978-3-642-61786-7
5. Poularikas, A.D. (ed.): Hankel transform. In: Electrical Engineering Handbook, vol. 43, 3rd edn. CRC Press, Boca Raton (2010)
6. Prodanov, D., Delbeke, J.: A model of space-fractional-order diffusion in the glial scar. J. Theor. Biol. **403**, 97–109 (2016). https://doi.org/10.1016/j.jtbi.2016.04.031
7. Riesz, M.: Intégrales de Riemann-Liouville et potentiels. Acta Sci. Math. Szeged **9**, 1–42 (1938)
8. Takahasi, H., Mori, M.: Double exponential formulas for numerical integration. Publ. RIMS Kyoto Univ. **9**, 721–741 (1974)
9. Wynn, P.: On the convergence and stability of the epsilon algorithm. SIAM J. Numer. Anal. **3**(1), 91–122 (1966). https://doi.org/10.1137/0703007

# Performance Study of Hierarchical Semi-separable Compression Solver for Parabolic Problems with Space-Fractional Diffusion

Dimitar Slavchev[(✉)] and Svetozar Margenov

Institute of Information and Communication Technologies – Bulgarian Academy of Sciences, Acad. G. Bonchev Street, Block 25A, 1113 Sofia, Bulgaria
{dimitargslavchev,margenov}@parallel.bas.bg

**Abstract.** Equations involving fractional diffusion operators are used to model anomalous processes in which the Brownian motion hypotheses are violated. In this work we utilize the Fractional Laplacian operator, defined through the Riesz potential and homogeneous Dirichlet boundary conditions. We explore a parabolic problem in a model square domain, using a backward Euler scheme for the discretization in time. The resulting series of systems of linear algebraic equations are dense and computationally expensive to solve. When utilizing the traditional Gaussian Elimination, the computational complexity is $O(n^3)$ for the LU factorization and $O(n^2)$ for each time step, where $n$ is the number of unknowns. This can be improved by using Hierarchical Semi-Separable (HSS) compression. With a solver from STRUMPACK, the computational complexity is reduced to $O(n^2 r)$ for the factorization and $O(nr)$ for each time step, where $r$ is the maximum off-diagonal rank of the matrix. The presented numerical experiments show the advantages of the HSS method for the examined problem.

**Keywords:** Anomalous diffusion · Fractional laplacian · Parabolic · HSS compression · STRUMPACK

## 1 Introduction

*Anomalous* (Fractional) diffusion has many applications in modelling physical phenomena – superconductivity, protein diffusion within cells, diffusion through porous media and others. Fractional Diffusion can be modeled with a variety of techniques (see for example [5,6]). In this work we utilize the Riesz potential to discretize a model problem in a square domain $[-1, 1]^2$. We use the method developed by Acosta and Borthagaray in [1] with the MatLab code from [2]. We extend the code by Acosta et al. by adding a calculation of the lumped mass matrix $M_L$ and using it to solve a parabolic in time problem. This requires solving a dense system of linear algebraic equations at each time step. With a

© Springer Nature Switzerland AG 2022
I. Lirkov and S. Margenov (Eds.): LSSC 2021, LNCS 13127, pp. 71–80, 2022.
https://doi.org/10.1007/978-3-030-97549-4_8

uniform time step, the matrix does not change and could be factorized only once, reducing the overall complexity of the problem.

Using a solver based on Gaussian decomposition (LU factorization), the problem is computationally expensive – $O(n^3)$ for the factorization and $O(n^2)$ at each time step, where $n$ is the number of unknowns. This could be reduced significantly by the utilization of Hierarchical methods. The Hierarchically Semi-Separable (HSS) compression in the STRUMPACK package [3] can drop that down to $O(n^2 r)$ for the factorization and $O(nr)$ at each time step, where $r$ is the maximum rank of the off-diagonal blocks of the matrix, as found during the compression. Typically $r \ll n$. This is achieved by first compressing the matrix into an HSS form $H$ (computational complexity $O(n^2 r)$) and then using ULV-like factorization (computational complexity $O(nr^2)$). Solving at each time step has computational complexity $O(nr)$.

## 2    Fractional Diffusion Modeled with the Riesz Potential

The Fractional Laplacian is defined as

$$(-\Delta)^{\alpha} u(x) = C(d,\alpha)\, p.v. \int_{\mathbb{R}^d} \frac{u(x) - u(y)}{|x - y|^{d+2\alpha}} dy, \quad \alpha \in (0,1), \tag{1}$$

where $\alpha$ is the fractional power of the Laplacian operator $\Delta$, $p.v.$ stands for principal value, $d$ is the dimension of the problem (the numerical experiments are carried out with $d = 2$) and $C(d,\alpha)$ is the normalized constant

$$C(d,\alpha) = \frac{2^{2\alpha}\alpha \Gamma\left(\alpha + \frac{d}{2}\right)}{\pi^{d/2}\Gamma(1-\alpha)}.$$

The Fractional Laplacian, defined in Eq. (1) can be regarded as an infinitesimal generator of an $\alpha$-stable Lévi operator and is amongst the simplest pseudo-differential operators. After applying homogeneous Dirichlet boundary conditions, we obtain the parabolic problem

$$\begin{cases} \dfrac{\partial u(x,t)}{\partial t} + \Delta^{\alpha} u(x,t) = f(x,t), & x \in \Omega, \quad t \in [0,T], \\ u(x,t) = 0, & x \in \Omega^c, \quad t \in [0,T], \\ u(x,0) = u^0(x), & x \in \Omega. \end{cases}$$

where $f$ is a function on a bounded domain $\Omega$, $\Omega^c$ is the complement of $\Omega$, $t$ is the time component and $u$ is the unknown.

An admissible triangulation $\mathcal{T}$ of $\Omega$ is considered for the Finite Element discretization. Continuous piecewise linear elements are used over $\mathcal{T}$, forming a discrete space $\mathbb{V}$. The nodal basis $\{\varphi_1, \ldots, \varphi_N\} \in \mathbb{V}$ corresponds to the internal nodes $\{x_1, \ldots, x_N\}$, with $\varphi_i(x_j) = \delta_j^i$. The Cauchy problem, obtained in this way, is

$$M_L \frac{d\mathbf{u}}{dt} + K\mathbf{u} = M_L \mathbf{f}, \quad 0 < t \leq T, \quad \mathbf{u}(0) = \mathbf{u}^0, \tag{2}$$

where the unknown is $\mathbf{u} = (u_j) \in \mathbb{R}^N$, $K = K_{ij} \in \mathbb{R}^{N \times N}$ is the stiffness matrix, $M_L = \text{diag}(m_L^i) \in \mathbb{R}^N$ is the lumped mass matrix ($m_L^i$ is the mass at node $x_i$), $t \in [0, T]$ is the time interval and $\mathbf{f} = f_j \in \mathbb{R}$ is the right hand side of the stationary problem.

The *fractional* stiffness matrix $K$ and right hand side $\mathbf{f}$ can be written as

$$K_{ij} = \frac{C(d,\alpha)}{2} \langle \varphi_i, \varphi_j \rangle_{H^\alpha(\mathbb{R})} \quad \text{and} \quad f_j = \int_\Omega f \varphi_j.$$

A properly chosen ball domain $B$, containing $\Omega$, is introduced for computing $K_{ij}$. For $l, m \in [1, N_{\tilde{T}}]$ ($\tilde{T}$ are the elements on the triangulation of $B$) the elements of $K$ are written in the form

$$K_{ij} = \frac{C(d,\alpha)}{2} \sum_{l=1}^{N_{\tilde{T}}} \left( \sum_{m=1}^{N_{\tilde{T}}} I_{l,m}^{i,j} + 2 J_l^{i,j} \right). \tag{3}$$

The integrals $I_{l,m}^{i,j}$ and $J_l^{i,j}$ denoted as

$$I_{l,m}^{i,j} = \int_{T_l} \int_{T_m} \frac{\left( \varphi_i(x) - \varphi_i(y) \right) \left( \varphi_j(x) - \varphi_j(y) \right)}{|x - y|^{d+2\alpha}} dx dy$$

$$J_l^{i,j} = \int_{T_l} \int_{B^c} \frac{\varphi_i(x) \varphi_j(x)}{|x - y|^{d+2\alpha}} dy dx.$$

Numerical results, including a complete n-dimensional finite element analysis and regularity of solutions in both standard and weighted fractional spaces for the stationary problem $K\mathbf{u} = \mathbf{f}$, can be found in [1]. A MATLAB® code is presented as a supplement work in [2] and we use it to calculate the $K$ and $\mathbf{f}$.

The backwards Euler scheme

$$M_L \frac{\mathbf{u}^{j_t+1} - \mathbf{u}^{j_t}}{\tau_{j_t}} + K\mathbf{u}^{j_t+1} = M_L \mathbf{f}, \quad j_t = 1, \ldots, m \tag{4}$$

is applied for the discretization in time, with $\sum_{j_t=1}^m \tau_{j_t} = T$. We use a uniform step $\tau$ and right hand side $\mathbf{f}$, and will drop the suffix. At each step $j_t$ Eq. (4) is solved as the system of linear equations

$$\tilde{K} \mathbf{u}^{j_t+1} = \tilde{\mathbf{f}}^{j_t}, \quad \text{where} \quad \tilde{K} = \frac{M_L}{\tau} + K \quad \text{and} \quad \tilde{\mathbf{f}}^{j_t} = M_L \left( \mathbf{f} + \frac{\mathbf{u}^{j_t}}{\tau} \right). \tag{5}$$

Both the stiffness matrix and lumped mass matrices are symmetric and positive definite, thus Eq. (5) has a unique solution. Since the matrix $\tilde{K}$ is the same for every time step $j_t$, we can factorize it once and solve (Eq. (5)) with the factorized form for each time step. Using Gaussian Elimination, the computational cost is $O(n^3)$ for the $LU$ factorization, and $O(n^2 m)$ for $m$ time steps.

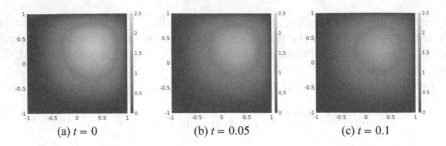

**Fig. 1.** Solution of the backwards Euler scheme at several time points.

For the numerical experiments we use the model 2D problem from [10], by Vabischevich, where the spectral definition of the Fractional Laplacian is used. The problem is defined as

$$u^0 = 100 \left(\frac{x_1+1}{2}\right)^2 \left(1 - \frac{x_1+1}{2}\right) \left(\frac{x_2+1}{2}\right)^2 \left(1 - \frac{x_2+1}{2}\right), \quad x = (x_1, x_2) \in \Omega$$

$$t \in [0, T], \quad T = 0.1$$

$$\tau = T/m, \quad m = 256$$

$$f = \frac{(x_1+1)(x_2+1)}{4}$$

On Fig. 1 we present the solutions obtained for $\alpha = 0.5$ in several points in time. Our results are qualitatively similar to the results in [10].

## 3    Hierarchical Semi-separable Compression Based Solver

Hierarchical compression is introduced by Hackbush in [4], with $\mathcal{H}$-matrices, for matrices arising from Boundary Element Method. The stiffness matrix (3) is produced by similar integral formulations and is also suitable for Hierarchical methods. The gist of the compression is to use the *structure* of the off-diagonal blocks. Here, by matrix *structure*, we mean that the *off-diagonal* blocks have *low rank* and can be represented as a product of smaller matrices. We use the parallel software package STRUctured Matrices PACKage (STRUMPACK) from Rouet et al. [8]. It uses a Hierarchical Semi-Separable compression to reduce the computational complexity of solving systems of linear algebraic equations. The solver works in 3 steps:

1. Hierarchical Semi-Separable compression. The off-diagonal blocks of the matrix $A$ are recursively compressed into HSS form $H$, utilizing randomized sampling. Computational complexity is $O(n^2 r)$, where $r$ is the maximum off-diagonal rank found in the process.
2. ULV-like factorization. Uses the form of the $H$ matrix to apply factorization similar to ULV. Computational complexity is $O(nr^2)$.
3. Solution. Uses the ULV-like factorized form of $A$ and the right hand side to calculate the solution. Computational complexity is $O(nr)$

For the parabolic in time problem (5) we need to apply Steps 1 and 2 on $\tilde{K}$ once and Step 3 for each time step. Computational complexity is $O(n^2 r)$ for the compression and factorization and $O(nrm)$ for the $m$ time steps.

The compression algorithm recursively partitions a matrix $A$ into four blocks

$$A = \begin{bmatrix} A_{1,1} & A_{1,2} \\ A_{2,1} & A_{2,2} \end{bmatrix} = \begin{bmatrix} D_1 & U_1^{\mathrm{big}} B_{1,2} V_2^{\mathrm{big}*} \\ U_2^{\mathrm{big}} B_{2,1} V_1^{\mathrm{big}*} & D_2 \end{bmatrix},$$

where the off-diagonal blocks $A_{1,2}, A_{2,1}$ are *approximated* by a product of the *generator* matrices $U$, $B$, and $V$. These *generators* are calculated by using randomized sampling of the original off-diagonal blocks, and are much smaller than their corresponding blocks for matrices with good *structure*. The second level of recursive compression can be written as

$$A = \begin{bmatrix} \begin{bmatrix} D_1 & U_1^{\mathrm{big}} B_{1,2} V_2^{\mathrm{big}*} \\ U_2^{\mathrm{big}} B_{2,1} V_1^{\mathrm{big}*} & D_2 \end{bmatrix} & U_3^{\mathrm{big}} B_{3,6} V_6^{\mathrm{big}*} \\ U_6^{\mathrm{big}} B_{6,3} V_3^{\mathrm{big}*} & \begin{bmatrix} D_4 & U_4^{\mathrm{big}} B_{4,5} V_5^{\mathrm{big}*} \\ U_5^{\mathrm{big}} B_{5,4} V_4^{\mathrm{big}*} & D_5 \end{bmatrix} \end{bmatrix}$$

This partitioning of the diagonal blocks continues recursively as long as necessary. Note that the *generators* have the recursive property

$$U_3^{\mathrm{big}} = \begin{bmatrix} U_1^{\mathrm{big}} & 0 \\ 0 & U_2^{\mathrm{big}} \end{bmatrix} U_3 \quad \text{and} \quad V_3^{\mathrm{big}} = \begin{bmatrix} V_1^{\mathrm{big}} & 0 \\ 0 & V_2^{\mathrm{big}} \end{bmatrix} V_3,$$

which allows the $^{\mathrm{big}}$ generators to be stored explicitly only at the lowest level of recursion. This property is specific for the HSS and $\mathcal{H}^2$ compressions, and allows more effective storage and computations on the compressed matrix $H$.

Any matrix can be compressed, however the process is effective only when $A$ has *low-rank* off-diagonal blocks. The maximum off-diagonal rank $r$ is calculated during the compression. It is a measure for its effectiveness. Low $r$ signifies a good compression and thus faster overall compression, while high $r$ leads to bad compression and slower execution times. For suitable problems $r \ll n$. It can be a constant (2D Poisson problems), or grow slowly with $n$ (e.g., 3D Helmholtz problems), see [11] for evaluations of $r$ for several problems.

The off-diagonal *low-rank structure* can be destroyed if the rows and columns of the matrix are not properly ordered. Thus, it is important to reorder the matrix $A$ for some problems. This is studied by Rebrova et al. for Kernel Ridge Regression in [7]. In [9] we have compared several reordering schemes for the stationary Fractional Diffusion problem, and, in this paper, we will use the best of them – Recursive Bisection.

The compression algorithm needs two thresholds to be set by the user – the relative $\varepsilon_{\mathrm{rel}}$ and absolute $\varepsilon_{\mathrm{abs}}$. We examine experiments with the default $\varepsilon_{\mathrm{abs}} = 10^8$ and vary $\varepsilon_{\mathrm{rel}} = 10^2, 10^4, 10^6$, and $10^8$.

After the matrix $A$ is compressed into $HSS$ form, it is factorized using ULV-like factorization [3]. The algorithm inside STRUMPACK uses the *generators* unique structure instead of orthogonal transformations (used in the original ULV), hence the name ULV-like factorization. The factorized matrix is then used to solve the original system of linear algebraic equations.

## 4   Numerical Results

We use the supercomputer AVITOHOL at Institute of Information and Communication Technologies, Bulgarian Academy of Sciences for the numerical experiments. AVITOHOL has 150 nodes, each with two Intel Xeon E5-2650v2 8C 2.6 GHz CPUs. We use one node for a total of 16 cores. We use Intel® Compiler Collection and MKL with versions 2017.2.174. The STRUMPACK version is 3.2. We will use 16 threads for the parallel experiments. The code by Acosta [2] is used to generate the stiffness matrix and right hand side data for problems, with $n$ between $\sim$2000 and $\sim$32000. Then the $\tilde{K}$ and $\tilde{f}^{j\iota}$ are assembled and (5) is used as a benchmark.

On Fig. 2 we present the comparison between the execution times of the HSS compression and ULV-like factorization in STRUMPACK and LU decomposition in MKL (Fig. 2a, 2d), the time steps solutions (Fig. 2b, 2e) and the total times (Fig. 2c, 2f). The time of execution of the HSS based algorithm is between $\sim$2.5 to $\sim$5 times faster than MKL® for the largest problem. On Fig. 3a, 3b, 3d and 3e we present the parallel speed up of the solution obtained with the HSS based solver. Using 16 threads, we obtain from $\sim$3 ($n = 2131$) to $\sim$8 ($n = 32302$) times decrease of the run time for all tested settings of $\varepsilon_{rel}$.

The maximum off-diagonal rank $r$, calculated during the HSS compression is presented on Fig. 3c and 3f. The rank is a measure for the effectiveness of the compression and is present in the computational complexity estimates. For our problem, $r$ is significantly smaller than $n$ – between $\sim$20 and $\sim$80 times for the largest relative threshold $\varepsilon_{rel} = 10^2$ and between $\sim$10 and $\sim$30 times for the finest $\varepsilon_{rel} = 10^8$.

The speed-up of the separate parts of the HSS based solver is presented on Fig. 4 for $\varepsilon_{rel} = 10^6$. It is similar for the other relative thresholds. The compression (Fig. 4a) and factorization parts (Fig. 4b) have the same speed-up, that gradually increases from $\sim$3 up to $\sim$8 times for the largest problem. The solutions for each time step (Fig. 4c) have lower parallel speed-up between $\sim$2 and $\sim$6. The time steps, however take much less time (Fig. 2) than the compression and factorization steps for our experiments and are significantly faster than the MKL® solution.

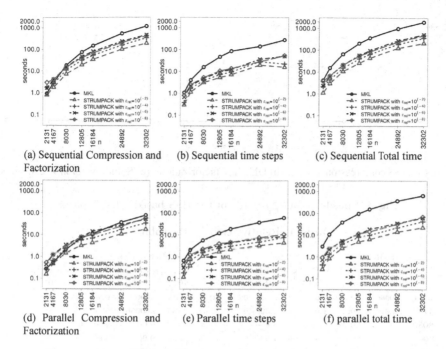

**Fig. 2.** Comparison between solution times obtained with STRUMPACK and MKL with different relative tolerances.

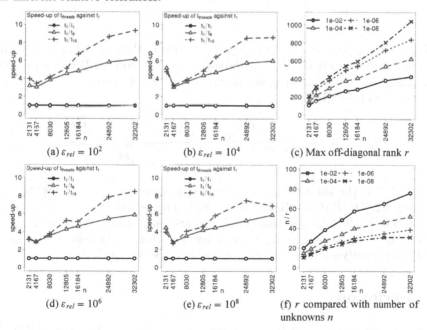

**Fig. 3.** Parallel Speed-up of the STRUMPACK HSS based solver with $m = 256$ time steps (left and center) and the maximum off-diagonal rank $r$ (right)

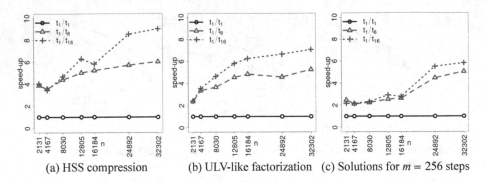

(a) HSS compression    (b) ULV-like factorization    (c) Solutions for $m = 256$ steps

**Fig. 4.** Speed-up for $\varepsilon_{rel} = 10^6$ of the HSS based solver's parts

**Table 1.** Relative error of the solution

| (a) Relative error at $t = 0.05$ | | | | | (b) Relative error at $t = 0.1$ | | | |
|---|---|---|---|---|---|---|---|---|
| n | $R_{relative}$ for STRUMPACK with $r_{tol}$ | | | | n | $R_{relative}$ for STRUMPACK with $r_{tol}$ | | |
|   | $10^{-2}$ | $10^{-4}$ | $10^{-6}$ | $10^{-8}$ |   | $10^{-2}$ | $10^{-4}$ | $10^{-6}$ | $10^{-8}$ |
| 2131 | 0.0066 | 0.00012 | 4.62e−07 | 9.72e−08 | 2131 | 0.01 | 0.00019 | 7.07e−07 | 1.62e−07 |
| 4167 | 0.01 | 0.00023 | 1.03e−06 | 1.36e−07 | 4167 | 0.017 | 0.00036 | 1.55e−06 | 1.98e−07 |
| 8030 | 0.015 | 0.00038 | 2.e−06 | 3.49e−07 | 8030 | 0.026 | 0.0006 | 2.93e−06 | 5.49e−07 |
| 12805 | 0.018 | 0.00048 | 2.49e−06 | 7.56e−07 | 12805 | 0.03 | 0.0008 | 3.6e−06 | 1.16e−06 |
| 16184 | 0.023 | 0.00052 | 2.77e−06 | 7.88e−07 | 16184 | 0.042 | 0.00088 | 4.34e−06 | 1.2e−06 |
| 24892 | 0.035 | 0.00062 | 3.97e−06 | 1.49e−06 | 24892 | 0.065 | 0.00097 | 6.26e−06 | 2.3e−06 |
| 32302 | 0.044 | 0.00054 | 3.97e−06 | 1.76e−06 | 32302 | 0.083 | 0.00083 | 6.41e−06 | 2.58e−06 |

The compressed matrix $H$ is an approximation of the original matrix $A$, thus it is important to study the error of the solution obtained with HSS compression. Using the MKL® LU decomposition solution as a reference, we calculate the relative error of the HSS solution as

$$R_{\text{relative}} = \frac{\left\| x^{\text{Gauss}} - x^{\text{HSS}} \right\|_{l_2}}{\left\| x^{\text{Gauss}} \right\|_{l_2}} = \frac{\sqrt{\sum_{i=1}^{n}(x_i^{\text{Gauss}} - x_i^{\text{HSS}})^2}}{\sqrt{\sum_{i=1}^{n}(x_i^{\text{Gauss}})^2}},$$

where $x^{\text{Gauss}}$ is the solution obtained by the MKL solver and $x^{\text{HSS}}$ is the solution from STRUMPACK's HSS based solver. The relative errors ($R_{\text{relative}}$) are presented in Table 1 for the half (Table 1a) and end (Table 1b). The errors are in the same order of magnitude as the corresponding $\varepsilon_{\text{rel}}$ threshold used in the calculation, signifying that the method has good accuracy.

## 5    Conclusion

We have presented a Hierarchical Semi-Separable Compression solver for parabolic fractional diffusion problems, using the STRUMPACK package. The results are compared with the LU decomposition based solver from Intel®'s MKL library. All tests with the STRUMPACK solver show better times (between $\sim 2.5$ and $\sim 5$ times faster) than the direct LU factorization solver. The maximum off-diagonal rank $r$, calculated during the compression step, is significantly smaller than the number of unknowns and that signifies an effective compression for this problem. This explains the better run times. An avenue for further work is the use of non-uniform time step $\tau_i \neq \tau_j, i \neq j \in [0, m]$.

**Acknowledgments.** The support through the Bulgarian NSF Grant No DFNI-DN12/1 is appreciated. We acknowledge the provided access to the e-infrastructure of the NCHDC - part of the Bulgarian National Roadmap on RIs, with the financial support by the Grant No D01-387/18.12.2020.

The first author is also supported through the Bulgarian Ministry of Education and Science's National Scientific Program "Young Scientists and Postdoctors".

# References

1. Acosta, G., Borthagaray, J.: A fractional Laplace equation: regularity of solutions and finite element approximations. SIAM J. Numer. Anal. **55**(2), 472–495 (2017). https://doi.org/10.1137/15M1033952
2. Acosta, G., Bersetche, F.M., Borthagaray, J.P.: A short FE implementation for a 2D homogeneous Dirichlet problem of a fractional Laplacian. Comput. Math. Appl. **74**(4), 784–816 (2017). https://doi.org/10.1016/j.camwa.2017.05.026
3. Chandrasekaran, S., Gu, M., Pals, T.: A fast $ULV$ decomposition solver for hierarchically semiseparable representations. SIAM J. Matrix Anal. Appl. **28**(3), 603–622 (2006). https://doi.org/10.1137/S0895479803436652
4. Hackbusch, W.: A sparse matrix arithmetic based on $\mathcal{H}$-matrices. Part I: introduction to $\mathcal{H}$-matrices. Computing **62**(2), 89–108 (1999). https://doi.org/10.1007/s006070050015
5. Harizanov, S., Lazarov, R., Margenov, S., Marinov, P., Pasciak, J.: Analysis of numerical methods for spectral fractional elliptic equations based on the best uniform rational approximation. J. Comput. Phys. **408**, 109285 (2020). https://doi.org/10.1016/j.jcp.2020.109285
6. Harizanov, S., Margenov, S., Popivanov, N.: Spectral fractional Laplacian with inhomogeneous Dirichlet data: questions, problems, solutions. In: Georgiev, I., Kostadinov, H., Lilkova, E. (eds.) BGSIAM 2018. SCI, vol. 961, pp. 123–138. Springer, Cham (2021). https://doi.org/10.1007/978-3-030-71616-5_13
7. Rebrova, E., Chávez, G., Liu, Y., Ghysels, P., Li, X.S.: A study of clustering techniques and hierarchical matrix formats for Kernel ridge regression. In: 2018 IEEE International Parallel and Distributed Processing Symposium Workshops (IPDPSW), pp. 883–892, May 2018. https://doi.org/10.1109/IPDPSW.2018.00140
8. Rouet, F.H., Li, X.S., Ghysels, P., Napov, A.: A distributed-memory package for dense hierarchically semi-separable matrix computations using randomization. ACM Trans. Math. Softw. **42**(4), 27:1–27:35 (2016). https://doi.org/10.1145/2930660
9. Slavchev, D., Margenov, S., Georgiev, I.G.: On the application of recursive bisection and nested dissection reorderings for solving fractional diffusion problems using HSS compression. In: AIP Conference Proceedings, vol. 2302, no. 1, p. 120008 (2020). https://doi.org/10.1063/5.0034506

10. Vabishchevich, P.N.: Splitting schemes for non-stationary problems with a rational approximation for fractional powers of the operator. Appl. Numer. Math. **165**, 414–430 (2021). https://doi.org/10.1016/j.apnum.2021.03.006
11. Xia, J.: Efficient structured multifrontal factorization for general large sparse matrices. SIAM J. Sci. Comput. **35**(2), A832–A860 (2013). https://doi.org/10.1137/120867032

# Numerical Solution of Non-stationary Problems with a Rational Approximation for Fractional Powers of the Operator

Petr N. Vabishchevich[1,2]([envelope])[iD]

[1] Nuclear Safety Institute, Russian Academy of Sciences, Moscow, Russia
[2] North-Eastern Federal University, Yakutsk, Russia
vabishchevich@gmail.com

**Abstract.** The numerical solutions of the Cauchy problems for a first and second-order differential-operator equations are discussed. The equation of the problem includes the fractional power of a self-adjoint positive operator. In computational practice, rational approximations of the fractional power operator are widely used. In this work, we construct special time approximations with fractional power operators: the transition to a new level in time provides a set of standard problems for the operator and not for the fractional power operator. Our approach utilizes stable splitting schemes with weights parameters for the additive representation of the rational approximation for the fractional power operator.

**Keywords:** Fractional powers of the operator · Rational approximation · Differential-operator equation · Splitting scheme

## 1 Introduction

Applied mathematical models describing nonlocal processes characterized by fractional derivatives have recently been actively discussed. For example, in reference [6], the corresponding boundary value problems for anomalous diffusion models that include fractional powers of elliptic operators are posed and solved. However, non-stationary problems with a fractional power operator require special attention.

Typically, a finite element or finite volume approximation of an elliptic operator is used in the approximate solution of a standard multidimensional problem. In this case, we have a sparse matrix $A$. The primary efforts to build computational algorithms for solving problems with a fractional power operator are associated with approximations for the full matrix $A^\alpha$, which are more convenient for implementation.

For an approximate solution of spectral space-fractional diffusion problems, various approaches are used [4]. In practice, many computational algorithms can

Supported by the mega-grant of the Russian Federation Government 14.Y26.31.0013 and the research grant 20-01-00207 of Russian Foundation of Basic Research.

I. Lirkov and S. Margenov (Eds.): LSSC 2021, LNCS 13127, pp. 81–88, 2022.
https://doi.org/10.1007/978-3-030-97549-4_9

be interpreted as special variants of the rational approximations of a fractional power operator.

The most interesting approach to build an approximate solution of non-stationary problems for equations with a fractional power operator is unconditionally stable implicit schemes [7]. The problem on a new level in time includes the diffusion operator with fractional power and reaction operator. Here we discuss the possibility of using rational approximations of a fractional power operator.

We consider Cauchy problems for first and second-order evolution equations with a fractional power operator. The new level's problem includes the sum of pairwise permutation operators used with rational approximation. We construct unconditionally stable additive difference schemes when the transition to a new level in time is provided by solving a set of everyday problems.

## 2   Evolutionary Problems with a Fractional Power Operator

Let $H$ be a finite-dimensional Hilbert space. The scalar product for $u, v \in H$ is $(u, v)$, and the norm is $\|u\| = (u, u)^{1/2}$. For a self-adjoint and positive operator $B$, we define the Hilbert space $H_B$ with scalar product and norm $(u, v)_B = (Bu, v)$, $\|u\|_B = (u, v)_B^{1/2}$.

The first example of a problem with a fractional power operator is related to the Cauchy problem for the first-order evolution equation:

$$\frac{du}{dt} + A^\alpha u = 0, \quad 0 < t \leq T, \tag{1}$$

$$u(0) = u^0, \tag{2}$$

when $\alpha \in (0, 1)$. The second example is the Cauchy problem for the second-order evolution equation:

$$\frac{d^2 u}{dt^2} + A^\alpha u = 0, \quad 0 < t \leq T, \tag{3}$$

$$u(0) = u^0, \quad \frac{du}{dt}(0) = \widetilde{u}^0. \tag{4}$$

The linear operator $A : H \mapsto H$ is constant (independent of $t$), self-adjoint and positively definite:

$$\frac{d}{dt} A = A \frac{d}{dt}, \quad A = A^* \geq \delta I, \quad \delta > 0, \tag{5}$$

where $I$ is the identity operator in $H$.

When approximating in time, most attention is paid to the problem of the stability of the approximate solution with respect to the initial data. Let us present the simplest a priori estimates for the solution of the problems under consideration. Taking into account the nonnegativity of $A^\alpha$, for solving the problem (1), (2) and (5) we have

$$\|u(t)\| \leq \|u^0\|, \quad 0 < t \leq T. \tag{6}$$

For the problem (3)–(5), the equality

$$\left\|\frac{du}{dt}(t)\right\|^2 + \|u(t)\|_A^2 = \|v^0\|^2 + \|u^0\|_A^2. \quad 0 < t \le T, \tag{7}$$

is valid.

The time approximation is carried out on a uniform grid with a step $\tau$. Let $y^n = y(t^n)$, $t^n = n\tau$, $n = 0, \ldots, N$, $N\tau = T$. For the problem (1) and (2), we will use a two-level scheme with the weight $\sigma \in (0, 1]$, when

$$\frac{y^{n+1} - y^n}{\tau} + A^\alpha(\sigma y^{n+1} + (1 - \sigma)y^n) = 0, \quad n = 0, \ldots, N - 1, \tag{8}$$

$$y^0 = u^0. \tag{9}$$

The difference scheme (8) and (9) approximates (1) and (2) with sufficient smoothness of the solution $u(t)$ with first order in $\tau$ for $\sigma \ne 0.5$ and with the second—for $\sigma = 0.5$ (Crank-Nicolson scheme). The scheme (5), (8) and (9) is unconditionally stable [7,8] at $\sigma \ge 0.5$. Under these conditions, similarly to (6) we have the a priori estimate

$$\|y^{n+1}\| \le \|u^0\|, \quad n = 0, \ldots, N - 1, \tag{10}$$

for the discrete solution.

For problem (3) and (4), we use a three-level scheme

$$\frac{y^{n+1} - 2y^n + y^{n-1}}{\tau^2} + A(\sigma y^{n+1} + (1 - 2\sigma)y^n + \sigma y^{n-1}) = 0, \tag{11}$$

$$n = 1, \ldots, N - 1,$$

with the given initial conditions

$$y^0 = u^0, \quad y^1 = \overline{u}^1. \tag{12}$$

The second initial condition on the solutions to Eqs. (3) is determined with the second order of approximation, for example, from equation

$$\left(I + \frac{\tau^2}{2}A\right)\overline{u}^1 = u^0 + \tau\widetilde{u}^0.$$

Difference scheme (11) and (12) approximates (3) and (4) with second order in $\tau$ with sufficient smoothness of the solution $u(t)$. Under the usual constraints $\sigma \ge 0.25$, the scheme (11) and (12) is stable [7,8], and equality

$$\left\|\frac{y^{n+1} - y^n}{\tau}\right\|_G^2 + \left\|\frac{y^{n+1} + y^n}{2}\right\|_A^2 = \left\|\frac{\overline{u}^1 - u^0}{\tau}\right\|_G^2 + \left\|\frac{\overline{u}^1 + u^0}{2}\right\|_A^2,$$

$$G = I + \left(\sigma - \frac{1}{4}\right)\tau^2 A, \quad n = 1, \ldots, N - 1, \tag{13}$$

is true.

From (8) and (9) for the solution on a new level in time, we get the equation

$$(I + \sigma\tau A^{\alpha})y^{n+1} = y^n - (1-\sigma)\tau A^{\alpha}y^n.$$

We have two difficulties in solving this problem: (i) computing the right-hand side, which includes the problematic term with $A^{\alpha}y^n$, (ii) solving the equation with the operator $I + \sigma\tau A^{\alpha}$. Simplification of the problem is achieved by introducing a new value

$$y^{n+\sigma} = \sigma y^{n+1} + (1-\sigma)y^n.$$

In this case, Eq. (8) can be conveniently written in the form

$$\frac{y^{n+\sigma} - y^n}{\sigma\tau} + A^{\alpha}y^{n+\sigma} = 0, \quad n = 0, \ldots, N-1. \tag{14}$$

The solution on the new level is provided by finding $y^{n+\sigma}$ from

$$(I + \sigma\tau A^{\alpha})y^{n+\sigma} = y^n,$$

and then from

$$y^{n+1} = \frac{1}{\sigma}(y^{n+\sigma} - y^n) + y^n.$$

Similarly, for the Eq. (11) we use

$$y^{n+\sigma} = \sigma y^{n+1} + (1-2\sigma)y^n + \sigma y^{n-1}$$

and rewrite it as

$$\frac{y^{n+\sigma} - y^n}{\sigma\tau^2} + A^{\alpha}y^{n+\sigma} = 0, \quad n = 1, \ldots, N-1. \tag{15}$$

The solution on a new time level is

$$y^{n+1} = \frac{1}{\sigma}(y^{n+\sigma} - y^n) + 2y^n - y^{n-1}.$$

We have well-developed various computational algorithms to solve stationary problems with a fractional power operator based on various approximations of $A^{-\alpha}$. We want to use them for non-stationary problems without explicit construction of approximation for $(I + \sigma\tau A^{\alpha})^{-1}$ or $(I + \sigma\tau^2 A^{\alpha})^{-1}$ (see (14) and (15)).

## 3 Rational Approximation for the Fractional Power Operator

We use rational approximation

$$A^{-\alpha} \approx R_m(A; \alpha), \quad 0 < \alpha < 1, \tag{16}$$

when for $R_m(A; \alpha)$ we have representation

$$R_m(A; \alpha) = \sum_{i=1}^{m} a_i(\alpha)(b_i(\alpha)I + A)^{-1}. \tag{17}$$

There are many options (see for example [1–3, 10]) for setting coefficients $a_i(\alpha)$, $b_i(\alpha)$, $i = 1, \ldots, m$. For example, note various variants of quadrature formulas for

$$A^{-\alpha} = \frac{\sin(\alpha\pi)}{\pi} \int_0^{\infty} \theta^{-\alpha}(A + \theta I)^{-1}d\theta.$$

The main limitation is associated with the positiveness of the coefficients in the representation (17):

$$a_i(\alpha) > 0, \quad b_i(\alpha) > 0, \quad i = 1, \ldots, m.$$

In this case, for the approximating operator (16), we obtain

$$R_m(A; \alpha) = R_m^*(A; \alpha) > 0, \quad R_m(A; \alpha)A = AR_m^*(A; \alpha).$$

For example, we can apply the approximation (16) directly, if we use equation

$$A^{-\alpha}\frac{du}{dt} + u = 0, \quad 0 < t \leq T,$$

instead of Eq. (1). In this case, the approximate solution $v(t)$ is the solution of the Cauchy problem

$$R_m(A; \alpha)\frac{dv}{dt} + v = 0, \quad 0 < t \leq T,$$

$$v(0) = u^0. \tag{18}$$

The second possibility we consider the Eq. (1) in form of

$$\frac{du}{dt} + A^{-\beta}Au = 0, \quad 0 < t \leq T,$$

where $\beta = 1 - \alpha$, $\beta \in (0, 1)$. For $v(t)$ we will use the equation

$$\frac{dv}{dt} + R_m(A; \beta)Av = 0, \quad 0 < t \leq T. \tag{19}$$

Similarly, an approximate solution to the problem (3) and (4) is determined from

$$\frac{d^2v}{dt^2} + R_m(A; \beta)Av = 0, \quad 0 < t \leq T, \tag{20}$$

$$v(0) = u^0, \quad \frac{dv}{dt}(0) = \tilde{u}^0. \tag{21}$$

Thus, it remains to construct time approximations that are convenient for computational implementation.

## 4   Splitting Scheme for First Order Equations

In the problem (16)–(20) transition to a new level in time is provided by equation

$$\frac{dv}{dt} + Dv = 0, \quad t^n < t \le t^{n+1}, \quad D = R_m(A; \beta)A, \quad \beta = 1 - \alpha. \tag{22}$$

Taking into account (17), we have an additive representation of the operator $D$:

$$D = \sum_{i=1}^{m} D_i, \quad D_i = a_i(\beta)(b_i(\beta)I + A)^{-1}A, \quad i = 1, \ldots, m. \tag{23}$$

We have

$$D_i = D_i^* > 0, \quad D_i D_j = D_j D_i, \quad i, j = 1, \ldots, m, \tag{24}$$

for individual operator terms in (23).

It is convenient to represent the solution of Eq. (22) in the form

$$v^{n+1} = Sv^n, \quad S = \exp(-\tau D).$$

Taking into account the (23) and (24), we obtain the multiplicative representation of the transition operator:

$$S = \prod_{i=1}^{m} S_i, \quad S_i = \exp(-\tau D_i), \quad i = 1, \ldots, m.$$

Instead of Eq. (22), we can use

$$\frac{dv_i}{dt} + D_i v_i = 0, \quad t^n < t \le t^{n+1},$$

$$v_i(t^n) = \begin{cases} v(t^n), & i = 1, \\ v_{i-1}(t^{n+1}), & i = 2, \ldots, m, \end{cases} \tag{25}$$

$$v(t^{n+1}) = v_m(t^{n+1}).$$

Most importantly, the system of Eqs. (25) gives an exact solution to Eq. (22) at times $t^n$, $n = 1, \ldots, N$.

Then we use a two-level splitting scheme [5,9] for an approximate solution of (25):

$$\frac{w^{n+i/m} - w^{n+(i-1)/m}}{\tau} + D_i(\sigma w^{n+i/m} + (1 - \sigma)w^{n+(i-1)/m}) = 0, \tag{26}$$

$$i = 1, \ldots, m, \quad n = 0, \ldots, N - 1,$$

with notation $w^{n+i/m} \approx v_i(t^{n+1})$, $i = 1, \ldots, m$, when setting the initial condition

$$w^0 = u^0. \tag{27}$$

The main result is formulated as follows.

**Theorem 1.** *The additive operator-difference scheme of component-wise splitting (23), (24), (26) and (27) is unconditionally stable for $\sigma \geq 0.5$ and for an approximate solution the a priori estimate*

$$\|w^{n+1}\| \leq \|u^0\|, \quad n = 0, \ldots, N - 1, \tag{28}$$

*is valid.*

To find a solution on a new level in time, equations

$$(b_i(\beta)I + (1 + a_i(\beta)\sigma\tau)A)w^{n+i/m} = \chi^{n+(i-1)/m}, \quad i = 1, \ldots, m,$$

are solved for given right-hand sides

$$\chi^{n+(i-1)/m} = (b_i(\beta)I + (1 - a_i(\beta)(1 - \sigma)\tau)A)w^{n+(i-1)/m}, \quad i = 1, \ldots, m.$$

We have $m$ standard problems with the operators $A + cI$, $c = \text{const} > 0$.

## 5   Factorized Splitting Schemes

Let's consider other variants of splitting schemes [5,9], which, on the one hand, are preferable than (26) and (27), and, on the other hand, are easily generalized to problems for second-order evolution equations (20) and (21).

Let us write the two-level scheme for Eq. (22) in the form

$$B\frac{w^{n+1} - w^n}{\tau} + Dw^n = 0, \quad n = 0, \ldots, N - 1. \tag{29}$$

We define the factorized operator $B$ by the relation

$$B = \prod_{i=1}^{m}(I + \sigma\tau D_i). \tag{30}$$

Taking into account (23) and (24), we have

$$B = B^* > I + \sigma\tau D.$$

**Theorem 2.** *The factorized operator-difference scheme (23), (24), (27), (29) and (30) is unconditionally stable for $\sigma \geq 0.5$ and for an approximate solution a priori estimate*

$$\|w^{n+1}\|_B \leq \|u^0\|_B, \quad n = 0, \ldots, N - 1,$$

*is valid.*

For the problem (20) and (21) we will use the scheme

$$B\frac{w^{n+1} - 2w^n + w^{n-1}}{\tau} + Dw^n = 0, \quad n = 1, \ldots, N - 1, \tag{31}$$

$$w^0 = u^0, \quad w^1 = \overline{u}^1. \tag{32}$$

We define the factorized operator $B$ in the form

$$B = \prod_{i=1}^{m}(I + \sigma\tau^2 D_i). \tag{33}$$

Under conditions (23) and (24) we get

$$B = B^* > I + \sigma\tau^2 D.$$

The following statement is formulated similarly to Theorem 2.

**Theorem 3.** *The factorized tree-level operator-difference scheme (23), (24), (31)–(33) is unconditionally stable for $\sigma \geq 0.25$ and for an approximate solution the equality*

$$\left\|\frac{w^{n+1} - w^n}{\tau}\right\|_G^2 + \left\|\frac{w^{n+1} + w^n}{2}\right\|_D^2 = \left\|\frac{\overline{u}^1 - u^0}{\tau}\right\|_G^2 + \left\|\frac{\overline{u}^1 + u^0}{2}\right\|_D^2,$$

$$G = B - \frac{1}{4}\tau^2 D, \quad n = 1, \ldots, N - 1,$$

*is valid.*

# References

1. Aceto, L., Novati, P.: Rational approximations to fractional powers of self-adjoint positive operators. Numerische Mathematik **143**(1), 1–16 (2019). https://doi.org/10.1007/s00211-019-01048-4
2. Bonito, A., Pasciak, J.: Numerical approximation of fractional powers of elliptic operators. Math. Comput. **84**(295), 2083–2110 (2015)
3. Harizanov, S., Lazarov, R., Margenov, S., Marinov, P.: Numerical solution of fractional diffusion-reaction problems based on BURA. Comput. Math. Appl. **80**(2), 316–331 (2020)
4. Harizanov, S., Lazarov, R., Margenov, S.: A survey on numerical methods for spectral space-fractional diffusion problems. Fract. Calc. Appl. Anal. **23**(6), 1605–1646 (2021)
5. Marchuk, G.I.: Splitting and alternating direction methods. In: Ciarlet, P.G., Lions, J.L. (eds.) Handbook of Numerical Analysis, vol. I, pp. 197–462. North-Holland (1990)
6. Pozrikidis, C.: The Fractional Laplacian. CRC Press, Boca Raton (2018)
7. Samarskii, A.A.: The Theory of Difference Schemes. Dekker, New York (2001)
8. Samarskii, A.A., Matus, P.P., Vabishchevich, P.N.: Difference Schemes with Operator Factors. Kluwer (2002)
9. Vabishchevich, P.N.: Additive Operator-Difference Schemes: Splitting Schemes. de Gruyter, Berlin (2013)
10. Vabishchevich, P.N.: Approximation of a fractional power of an elliptic operator. Linear Algebra Appl. **27**(3), e2287 (2020)

# Large-Scale Models: Numerical Methods, Parallel Computations and Applications

# An Exact Schur Complement Method for Time-Harmonic Optimal Control Problems

Owe Axelsson[1,2,3], Dalibor Lukáš[1,2,3], and Maya Neytcheva[1,2,3](✉)

[1] The Czech Academy of Sciences, Institute of Geonics, Ostrava, Czech Republic
[2] Department of Information Technology, Uppsala University, Uppsala, Sweden
{owe.axelsson,maya.neytcheva}@it.uu.se
[3] VSB - University of Technology, Ostrava, Czech Republic
dalibor.lukas@vsb.cz

**Abstract.** By use of Fourier time series expansions in an angular frequency variable, time-harmonic optimal control problems constrained by a linear differential equation decouples for the different frequencies. Hence, for the analysis of a solution method one can consider the frequency as a parameter. There are three variables to be determined, the state solution, the control variable, and the adjoint variable.

The first order optimality conditions lead to a three-by-three block matrix system where the adjoint optimality variable can be eliminated. For the so arising two-by-two block system, in this paper we study a factorization method involving an exact Schur complement method and illustrate the performance of an inexact version of it.

**Keywords:** PDE-constrained optimal control problems · Distributed control · Preconditioning · Time-harmonic maxwell equations

## 1 Introduction

In an optimal control problem for time-dependent differential equations, see e.g. [1], we seek the solution of the state variable and the control variable that minimizes

$$\mathcal{J}(u,v) = \frac{1}{2} \int_{\Omega \times [0,T]} \|u - u_d\|^2 dx dt + \frac{1}{2}\beta \int_{\Omega \times [0,T]} \|v\|^2 dx dt,$$

subject to a differential equation $\mathcal{L}u = g$, defined in a bounded domain $\Omega \subset \mathbb{R}^m$, $m = 2,3$ and time interval $[0,T]$. Here $u$ and $v$ can be complex. The control

Supported by EU Horizon 2020 research and innovation programme, grant agreement number 847593; the Czech Radioactive Waste Repository Authority (SÚRAO), grant agreement number SO2020-017; VR Grant 2017-03749 *Mathematics and numerics in PDE-constrained optimization problems with state and control constraints*, 2018-2022.

I. Lirkov and S. Margenov (Eds.): LSSC 2021, LNCS 13127, pp. 91–100, 2022.
https://doi.org/10.1007/978-3-030-97549-4_10

variable acts as an additional source function to the differential equation and $\beta > 0$ is a regularization parameter that determines the control cost. Further $u_d$ is a given target solution. Hence, the corresponding Lagrange optimality functional incorporating the adjoint variable $w$ has the form

$$\mathcal{L}(u,v,w) = \mathcal{J}(u,v) + \int_{\Omega \times [0,T]} (\mathcal{L}u(\mathbf{x},t) - g - v)w(\mathbf{x},t)\, d\mathbf{x}\, dt. \tag{1}$$

We assume periodic boundary conditions $u(\mathbf{x},0) = u(\mathbf{x},T)$ and consider time-harmonic solutions $u(\mathbf{x},t) = \tilde{u}(\mathbf{x})e^{i\omega t/T}$ for a given frequency $\omega$, being a multiple of $2\pi$.

Using appropriate finite elements, after space discretization and using the time-harmonic setting, we can formulate the first order necessary optimality conditions. The so arising system is of a saddle point form with a three-by-three block matrix as follows,

$$\begin{bmatrix} A & 0 & B - i\omega A \\ 0 & \beta A & -A \\ B + i\omega A & -A & 0 \end{bmatrix} \begin{bmatrix} \mathbf{u} \\ \mathbf{v} \\ \mathbf{w} \end{bmatrix} = \begin{bmatrix} A u_d \\ 0 \\ g \end{bmatrix}, \tag{2}$$

where $A$ and $B$ are real-valued matrices, $A$ is typically a mass matrix.

When the heat equation is the state constraint, see e.g. [2,3], then $u$ and $v$ are scalar functions and $A$ and $B$ are matrices of size $n$, equal to the number of space degrees of freedom. For the case of eddy current curl-curl (see e.g. [4]) equations, $u$ and $v$ are vector variables with two or three components per space discretization point and matrices $A$ and $B$ are block matrices of size $2n$ or $3n$. We assume that $A$ is symmetric positive definite (spd), $B$ is symmetric positive semi-definite (spsd). After elimination of the adjoint variable $\mathbf{w}$ we obtain the system

$$\begin{bmatrix} A & \beta C^* \\ C & -A \end{bmatrix} \begin{bmatrix} \mathbf{u} \\ \mathbf{v} \end{bmatrix} = \begin{bmatrix} A u_d \\ g \end{bmatrix},$$

with $C = B + i\omega A$, $C^* = B - i\omega A$. We scale the system and introduce $\tilde{\mathbf{v}} = -\sqrt{\beta}\mathbf{v}$ to get

$$\mathcal{A} \begin{bmatrix} \mathbf{u} \\ \tilde{\mathbf{v}} \end{bmatrix} = \begin{bmatrix} A & -\tilde{C}^* \\ \tilde{C} & A \end{bmatrix} \begin{bmatrix} \mathbf{u} \\ \tilde{\mathbf{v}} \end{bmatrix} = \begin{bmatrix} A u_d \\ \tilde{g} \end{bmatrix}, \tag{3}$$

where $\tilde{C} = \sqrt{\beta}C = \tilde{B} + i\tilde{\omega}A$, $\tilde{B} = \sqrt{\beta}B$, $\tilde{\omega} = \sqrt{\beta}\omega$ and $\tilde{g} = \sqrt{\beta}g$.

We focus on the solution of (3) via its Schur complement of $\mathcal{A}$, that is,

$$S = A + \tilde{C}A^{-1}\tilde{C}^* = (1 + \tilde{\omega}^2)A + \tilde{B}A^{-1}\tilde{B}$$

being spd, thus $\mathcal{A}$ is regular. We assume that the problem sizes are large, which motivates the use of preconditioned iterative methods of conjugate gradient type.

As we shall see, the method involves solving two complex valued systems. For this a complex-to-real (C-to-R) method, see e.g. [5], can be used, see next section.

There are two other methods with a similar performance. One is based on the PRESB method (see, e.g. [6]) and one is based on an approximate factorization of the Schur complement in real-valued factors, see [4,7]. The PRESB method involves solving two complex valued systems at each iteration step. These methods are based on preconditioned outer iteration methods of Krylov subspace type which should be flexible, because the inner iterations are not solved exactly, e.g. FGMRES, cf. [8] and GCG, cf. [9].

In practice, the systems in the exact Schur complement method are normally also not solved exactly. Hence, this method needs also an outer iteration method, which however can be a simple defect-correction method. The method has been dealt with in [10] but is here presented with another shorter way of derivation and a more difficult test problem. A numerical comparison of the three methods can be found in [13].

In the next section the method is presented. A method to solve the arising complex valued systems is presented in Sect. 3. In Sect. 4 two related methods are shortly presented and Section 5 shows numerical tests.

## 2   An Exact Schur Complement Method

The system in (3) can be factorized as

$$
\mathcal{A} = \begin{bmatrix} A & 0 \\ \widetilde{B} + i\widetilde{\omega}A & H_1 \end{bmatrix} \begin{bmatrix} A^{-1} & 0 \\ 0 & A^{-1} \end{bmatrix} \begin{bmatrix} A & -(\widetilde{B} - i\widetilde{\omega}A) \\ 0 & H_2 \end{bmatrix},
$$

where $H_1 A^{-1} H_2 = S = (1 + \widetilde{\omega}^2)A + \widetilde{B}A^{-1}\widetilde{B}$, i.e., $H_1 = \sqrt{1 + \widetilde{\omega}^2}A - i\widetilde{B}$, $H_2 = \sqrt{1 + \widetilde{\omega}^2}A + i\widetilde{B} = H_1^*$ are factors in the complex-valued exact factorization of the Schur complement matrix $S$ of $\mathcal{A}$. Then

$$
\mathcal{A}^{-1} = \begin{bmatrix} A^{-1} & A^{-1}(\widetilde{B} - i\widetilde{\omega}A)H_2^{-1} \\ 0 & H_2^{-1} \end{bmatrix} \begin{bmatrix} I & 0 \\ -AH_1^{-1}(\widetilde{B} + i\widetilde{\omega}A)A^{-1} & AH_1^{-1} \end{bmatrix}. \quad (4)
$$

The computation of a solution of the system $\mathcal{A} \begin{bmatrix} \mathbf{u} \\ \widetilde{\mathbf{v}} \end{bmatrix} = \begin{bmatrix} \mathbf{f} \\ \mathbf{g} \end{bmatrix}$ can be done by direct use of (4) but, besides solutions with the matrices $H_1$ and $H_2$, it would also involve two solutions with $A$. We show now that the latter solutions with $A$ can be avoided. Consider first the matrix-vector multiplication arising from the second factor in (4):

$$
\begin{bmatrix} I & 0 \\ -AH_1^{-1}(\widetilde{B} + i\widetilde{\omega}A)A^{-1} & AH_1^{-1} \end{bmatrix} \begin{bmatrix} \mathbf{f} \\ \mathbf{g} \end{bmatrix} = \begin{bmatrix} \mathbf{f} \\ AH_1^{-1}(\mathbf{g} - (\widetilde{B} + i\widetilde{\omega}A)A^{-1}\mathbf{f}) \end{bmatrix}.
$$

Since $H_1 = \sqrt{1 + \widetilde{\omega}^2} A - i\widetilde{B}$ it follows that

$$-AH_1^{-1}(\widetilde{B} + i\widetilde{\omega}A)A^{-1} = -iAH_1^{-1}(-i\widetilde{B} + \sqrt{1 + \widetilde{\omega}^2}A - (\sqrt{1 + \widetilde{\omega}^2} - \widetilde{\omega})A)A^{-1}$$
$$= -i(I - b_{\widetilde{\omega}}AH_1^{-1}),$$

where $b_{\widetilde{\omega}} = \sqrt{1 + \widetilde{\omega}^2} - \widetilde{\omega} = \dfrac{1}{\sqrt{1 + \widetilde{\omega}^2} + \widetilde{\omega}} \leq 1$. Hence,

$$\begin{bmatrix} I & 0 \\ -AH_1^{-1}(\widetilde{B} + i\widetilde{\omega}A)A^{-1} & AH_1^{-1} \end{bmatrix} \begin{bmatrix} \mathbf{f} \\ \mathbf{g} \end{bmatrix} = \begin{bmatrix} \mathbf{f} \\ -i\mathbf{f} + ib_{\widetilde{\omega}}AH_1^{-1}\mathbf{f} + AH_1^{-1}\mathbf{g} \end{bmatrix} = \begin{bmatrix} \mathbf{f} \\ -i(\mathbf{f} - A\mathbf{h}), \end{bmatrix}$$
$$(5)$$

where $\mathbf{h} = H_1^{-1}(b_{\widetilde{\omega}}\mathbf{f} - i\mathbf{g})$. Further, it follows from (4) that $\widetilde{\mathbf{v}} = -iH_2^{-1}(\mathbf{f} - A\mathbf{h})$. To find the component $\mathbf{u}$, note now that

$$A^{-1}(\widetilde{B} - i\widetilde{\omega}A)H_2^{-1} = iA^{-1}(-i\widetilde{B} - \sqrt{1 + \widetilde{\omega}^2}A + b_{\widetilde{\omega}}A)H_2^{-1} = -iA^{-1} + ib_{\widetilde{\omega}}H_2^{-1}.$$

Hence, it follows from (4) and (5) that

$$\mathbf{u} = \begin{bmatrix} A^{-1} & A^{-1}(\widetilde{B} - i\widetilde{\omega}A)H_2^{-1} \end{bmatrix} \begin{bmatrix} A\mathbf{h} + \mathbf{f} - A\mathbf{h} \\ -i(\mathbf{f} - A\mathbf{h}) \end{bmatrix}$$
$$= \mathbf{h} + A^{-1}(\mathbf{f} - A\mathbf{h}) - A^{-1}(\mathbf{f} - A\mathbf{h}) + b_{\widetilde{\omega}}H_2^{-1}(\mathbf{f} - A\mathbf{h}),$$

that is, $\mathbf{u} = \mathbf{h} + b_{\widetilde{\omega}}H_2^{-1}(\mathbf{f} - A\mathbf{h}) = \mathbf{h} + ib_{\widetilde{\omega}}\widetilde{\mathbf{v}}$. The above relations show that, besides a matrix-vector multiplication with $A$ and some complex vector additions, the computation of the vectors $\mathbf{u}$ and $\widetilde{\mathbf{v}}$ requires one solution with $H_1$ to compute the vector $\mathbf{h}$ and one solution with $H_2$ to compute $\widetilde{\mathbf{v}}$. Both systems involve complex matrices, which can be solved by an iterative method using the method in Sect. 3. In Algorithm 1 we summarize the computations, needed to compute the action of $\mathcal{A}^{-1}$ on a vector $\begin{bmatrix} \mathbf{f}^T, & \widetilde{\mathbf{g}}^T \end{bmatrix}$.

---

**Algorithm 1**

---

1: Let $H_1 = \sqrt{1 + \widetilde{\omega}^2}A - i\widetilde{B}$, $b_{\widetilde{\omega}} = \dfrac{1}{\sqrt{1 + \widetilde{\omega}^2} + \widetilde{\omega}}$, $H_2 = H_1^* = \sqrt{1 + \widetilde{\omega}^2}A + i\widetilde{B}$

2: Solve $H_1\mathbf{h} = b_{\widetilde{\omega}}\mathbf{f} - i\widetilde{\mathbf{g}}$

3: Solve $H_2\widetilde{\mathbf{v}} = -i(\mathbf{f} - A\mathbf{h})$

4: Compute $\mathbf{u} = \mathbf{h} + ib_{\widetilde{\omega}}\widetilde{\mathbf{v}}$

---

Clearly, if we solve exactly with $H_1$ and $H_2$, we have a direct solution method for $\mathcal{A}$. In practice they are solved to some limited tolerance. It can then be efficient to imbed the method in a defect-correction framework, applied at least one step.

Namely, after one step of Algorithm 1, where we compute an approximate solution $\begin{bmatrix} \mathbf{u}_0 \\ \widetilde{\mathbf{v}}_0 \end{bmatrix}$ of $\mathcal{A}\begin{bmatrix} \mathbf{u} \\ \widetilde{\mathbf{v}} \end{bmatrix} = \begin{bmatrix} \mathbf{f} \\ \mathbf{g} \end{bmatrix}$, we solve

$$\mathcal{A}\left(\begin{bmatrix} \delta\mathbf{u} \\ \delta\widetilde{\mathbf{v}} \end{bmatrix}\right) = \begin{bmatrix} \mathbf{f} \\ \mathbf{g} \end{bmatrix} - \mathcal{A}\begin{bmatrix} \mathbf{u}_0 \\ \widetilde{\mathbf{v}}_0 \end{bmatrix},$$

using again Algorithm 1. Here $\delta\mathbf{u} = \mathbf{u} - \mathbf{u}_0$ and $\delta\widetilde{\mathbf{v}} = \widetilde{\mathbf{v}} - \widetilde{\mathbf{v}}_0$. Depending on the size of the residual, it can be repeated once more.

## 3  A Complex-to-Real Solution Method

To avoid complex arithmetics in the solution with matrices $H_i$, $i = 1, 2$ in Algorithm 1 we can use a complex-to-real method, that is rewrite, say the system with $H_2$,

$$(\sqrt{1 + \widetilde{\omega}^2}A + i\widetilde{B})(x + iy) = a + ib$$

in real-valued form, where $a, b$ are defined by Step 3 of Algorithm 1. Hence,

$$\begin{bmatrix} \widetilde{A} & -\widetilde{B} \\ \widetilde{B} & \widetilde{A} \end{bmatrix} \begin{bmatrix} x \\ y \end{bmatrix} = \begin{bmatrix} a \\ b \end{bmatrix},$$

where $\widetilde{A} = \sqrt{1 + \widetilde{\omega}^2}A$. This can be solved efficiently by use of the PRESB preconditioning method, see e.g. [5,12], that is preconditioned by

$$\begin{bmatrix} \widetilde{A} & -\widetilde{B} \\ \widetilde{B} & \widetilde{A} + 2\widetilde{B} \end{bmatrix} = \begin{bmatrix} I & -I \\ 0 & I \end{bmatrix} \begin{bmatrix} \widetilde{A} + \widetilde{B} & 0 \\ \widetilde{B} & \widetilde{A} + \widetilde{B} \end{bmatrix} \begin{bmatrix} I & I \\ 0 & I \end{bmatrix}, \qquad (6)$$

Hence, each iteration step involves two solutions with the real-valued matrix $\widetilde{A} + \widetilde{B}$. To analyse the rate of convergence of this method, we include, also for completeness of the paper, the spectral eigenvalue analysis of the preconditioned matrix. This would suffice as the spectrum turns out to be tightly clustered. Accordingly, consider

$$\mathcal{A}\begin{bmatrix} \mathbf{x} \\ \mathbf{y} \end{bmatrix} = \lambda\mathcal{P}\begin{bmatrix} \mathbf{x} \\ \mathbf{y} \end{bmatrix} \quad \text{that is,} \quad \begin{bmatrix} \widetilde{A} & -\widetilde{B} \\ \widetilde{B} & \widetilde{A} \end{bmatrix}\begin{bmatrix} \mathbf{x} \\ \mathbf{y} \end{bmatrix} = \lambda\begin{bmatrix} \widetilde{A} & -\widetilde{B} \\ \widetilde{B} & \widetilde{A} + 2\widetilde{B} \end{bmatrix}\begin{bmatrix} \mathbf{x} \\ \mathbf{y} \end{bmatrix}.$$

A computation shows that

$$(1 - \lambda)\begin{bmatrix} \widetilde{A} & -\widetilde{B} \\ \widetilde{B} & \widetilde{A} + 2\widetilde{B} \end{bmatrix}\begin{bmatrix} \mathbf{x} \\ \mathbf{y} \end{bmatrix} = \begin{bmatrix} 0 & 0 \\ 0 & 2\widetilde{B} \end{bmatrix}\begin{bmatrix} \mathbf{x} \\ \mathbf{y} \end{bmatrix}.$$

Clearly, for $\mathbf{y} = \mathcal{N}(\widetilde{B})$, $\mathcal{N}(\widetilde{B})$ being the null space of $\widetilde{B}$, and any $x$ we see that $\lambda = 1$. For $\lambda \neq 1$ it holds that

$$(1 - \lambda)\mathbf{y}^T(\widetilde{B}\widetilde{A}^{-1}\widetilde{B} + \widetilde{A} + 2\widetilde{B})\mathbf{y} = 2\mathbf{y}^T\widetilde{B}\mathbf{y}.$$

Transforming the latter equality by multiplying by $\widetilde{A}^{-1/2}$ from left and right and denoting $\widehat{B} = \widetilde{A}^{-1/2}\widetilde{B}\widetilde{A}^{-1/2}$ we obtain

$$(1 - \lambda)\widetilde{\mathbf{y}}^T(\widehat{B}^2 + I + 2\widehat{B})\widetilde{\mathbf{y}} = 2\widetilde{\mathbf{y}}^T\widehat{B}\widetilde{\mathbf{y}}, \quad \text{i.e.,} \quad 1 - \lambda = \frac{2\widetilde{\mathbf{y}}^T\widehat{B}\widetilde{\mathbf{y}}}{\widetilde{\mathbf{y}}^T(\widehat{B} + I)^2\widetilde{\mathbf{y}}} \leq \frac{1}{2},$$

that is, $\frac{1}{2} \le \lambda \le 1$, where $\widetilde{\mathbf{y}} = \widetilde{A}^{-1/2}\mathbf{y}$. Hence, all eigenvalues are real and tightly clustered, and Krylov subspace methods, applied to $\mathcal{A}$, preconditioned by $\mathcal{P}$, converge rapidly and with a convergence speed that holds uniformly with respect to all parameters, including the discretization parameter. For small values of $\widetilde{\omega}$, when $\|\widehat{B}\|$ can become small, the eigenvalues cluster at unity. Note that $\widetilde{\omega} = \sqrt{\beta}\omega$, where $\beta$ is normally small.

## 4    Two Related Methods

Assume now that $B$ is not necessarily symmetric. There are two related methods to mention. One is based directly on the use of the PRESB method for Eq. (3), that is the preconditioner equals

$$\mathcal{P} = \begin{bmatrix} A & -\widetilde{C}^* \\ \widetilde{C} & A + \widetilde{C} + \widetilde{C}^* \end{bmatrix} = \begin{bmatrix} A & -\widetilde{C}^* \\ \widetilde{C} & A + \widetilde{B} + \widetilde{B}^T \end{bmatrix},$$

which can be factorized as in (6), so it involves solving inner systems with $A + \widetilde{C}$ and $A + \widetilde{C}^*$. Here

$$A + \widetilde{C} = A + \widetilde{B} + i\widetilde{\omega}A, \quad A + \widetilde{C}^* = A + \widetilde{B}^T - i\widetilde{\omega}A.$$

Thus, the systems are similar to the ones for $H_1$ and $H_2$ in Sect. 2. They are solved in a similar way as in Sect. 3. Hence, this method is based on coupled inner-outer iterations, but where the outer iterations should best also involve a flexible conjugate gradient acceleration method. To sum up, this method uses preconditioned iteration methods twice, while for the exact Schur complement method one can use a simple defect-correction method for the outer iterations.

In the other related method an approximation of $S$ in real-valued factors, $GA^{-1}G^T = (\sqrt{1+\widetilde{\omega}^2}A + \widetilde{B})A^{-1}(\sqrt{1+\widetilde{\omega}^2}A + \widetilde{B}^T) = (1+\widetilde{\omega}^2)A + \widetilde{B}A^{-1}\widetilde{B}^T + \sqrt{1+\widetilde{\omega}^2}(\widetilde{B} + \widetilde{B}^T)$ is used. The corresponding approximate factorization of $\mathcal{A}$ becomes

$$\mathcal{A} \approx \mathcal{B} = \begin{bmatrix} A & 0 \\ C & G \end{bmatrix} \begin{bmatrix} A^{-1} & 0 \\ 0 & A^{-1} \end{bmatrix} \begin{bmatrix} A & -C^* \\ 0 & G^T \end{bmatrix}.$$

Thus, $\begin{bmatrix} \mathbf{u} \\ \mathbf{v} \end{bmatrix} = \mathcal{B}^{-1} \begin{bmatrix} \mathbf{f} \\ \mathbf{g} \end{bmatrix} = \begin{bmatrix} A^{-1} & A^{-1}C^*G^{-T} \\ 0 & G^{-T} \end{bmatrix} \begin{bmatrix} I & 0 \\ -AG^{-1}CA^{-1} & AG^{-1} \end{bmatrix} \begin{bmatrix} \mathbf{f} \\ \mathbf{g} \end{bmatrix}.$

A computation shows that the corresponding algorithm takes the following form.

---

**Algorithm 2**

---

1: Let $G_1 = \sqrt{1+\widetilde{\omega}^2}A + \widetilde{B}$, $G_2 = G_1^T = \sqrt{1+\widetilde{\omega}^2}A + \widetilde{B}^T$, $c_{\widetilde{\omega}}^* = \sqrt{1+\widetilde{\omega}^2} + i\widetilde{\omega}$
2: Solve $G_1\mathbf{h} = c_{\widetilde{\omega}}\mathbf{f} + \mathbf{g}$
3: Solve $G_2\widetilde{\mathbf{v}} = A\mathbf{h} - \mathbf{f}$
4: Compute $\mathbf{u} = \mathbf{h} - c_{\widetilde{\omega}}^*\widetilde{\mathbf{v}}$

---

Thus, in this method we solve two inner systems with real matrices, but the outer iteration may need more iterations. A spectral analysis shows that, denoting $\widehat{B} = A^{-1/2}\widetilde{B}A^{-1/2}$ and

$$\widehat{\alpha} = \frac{\mathbf{y}^T\sqrt{1+\widetilde{\omega}^2}(\widehat{B} + \widehat{B}^T)\mathbf{y}}{\mathbf{y}^T((1+\widetilde{\omega}^2)I + \widehat{B}\widehat{B}^T)\mathbf{y}}, \quad \text{i.e.,} \quad 0 \leq \widehat{\alpha} \leq 1$$

it follows that

$$\frac{1}{\lambda} - 1 \leq \widehat{\alpha}, \quad \text{that is,} \quad \frac{1}{2} \leq \frac{1}{1+\widehat{\alpha}} \leq \lambda \leq 1.$$

As remarked in Sect. 3, for such a tight eigenvalue interval the actual eigenvectors play a minor role. The rate of convergence is hence similar to that of PRESB but we solve two real-valued inner systems instead of two complex-valued, which corresponds to about half the cost.

## 5   Numerical Illustrations

The performance of the preconditioning method from Algorithm 1 in its inexact version is illustrated numerically on the discrete optimal control problem from (1) with a constraint, given by the time-harmonic eddy current problem in $\Omega = [0,1]^3$

$$\mathfrak{L}u \equiv \sigma\frac{\partial u}{\partial t} + \nabla \times (\frac{1}{\mu} \nabla \times u) = \mathfrak{g} \quad \text{in } \Omega \times [0,T] \tag{7}$$

with proper boundary and initial conditions. Here $u$ is the magnetic potential, the problem parameter $\mu$ is the permeability of the media and $\sigma$ is the conductivity. In the numerical experiments it is assumed that $\sigma = 1$.

To numerically solve (1) we follow the 'discretize-then-optimize' framework. The discretization is done using Nédélec finite elements (cf. e.g. [11]) on a regular tetrahedral mesh. We test Algorithm 1 with inexact solution of the systems $H_1$ and $H_2$ and one defect-correction step. The systems with matrices $H_1$ and $H_2$ are solved via their equivalent twice larger real formulations using PRESB-preconditioned Generalized Conjugate Gradient (GCR) method [8] and a (relative) stopping criterion $10^{-8}$. The inner systems in PRESB are solved again via GCR, preconditioned by a V-cycle algebraic multigrid AGMG, see [14]. The (relative) stopping criterion for the inner AGMG solvers is chosen as $10^{-6}$ for the initial step and $10^{-3}$ for the defect-correction step.

All experiments are performed in Matlab R2019a on Lenovo ThinkPad T450s with 12 GB of memory.

**Table 1.** Algorithm 1 with inexactly solved $H_1$ and $H_2$ for various values of $\beta$ and $\omega$

| $\omega$ | $\beta$ | | | | |
|---|---|---|---|---|---|
| | $10^{-2}$ | $10^{-4}$ | $10^{-6}$ | $10^{-8}$ | $10^{-10}$ |
| $nDOFs = 293$ | | | | | |
| $10^{-8}$ | 5(9)/3(12) | 5(9)/3(12) | 5(9)/4(12) | 3(3)/2(4) | 2(3)/1(4) |
| $10^{-4}$ | 5(4)/3(5) | 5(4)/3(5) | 5(4)/4(5) | 3(3)/2(4) | 2(3)/1(4) |
| 1 | 5(3)/3(3) | 5(3)/3(3) | 5(3)/3(3) | 3(3)/2(4) | 2(3)/1(4) |
| $10^4$ | 3(3)/1(4) | 3(3)/1(4) | 3(3)/2(4) | 3(3)/2(4) | 2(3)/1(4) |
| $10^8$ | 2(3)/1(4) | 2(3)/1(4) | 2(3)/1(4) | 2(3)/1(4) | 2(3)/1(4) |
| $nDOFs = 2903$ | | | | | |
| $10^{-8}$ | 4(19)/4(19) | 4(19)/4(19) | 4(19)/4(20) | 4(3)/3(3) | 1(4)/1(4) |
| $10^{-4}$ | 4(7)/4(8) | 4(7)/4(8) | 5(7)/4(8) | 4(3)/3(3) | 1(4)/1(4) |
| 1 | 4(4)/3(4) | 4(4)/3(4) | 4(4)/4(4) | 4(3)/3(3) | 1(4)/1(4) |
| $10^4$ | 4(4)/3(4) | 4(4)/3(4) | 4(4)/3(3) | 4(3)/3(3) | 1(4)/1(4) |
| $10^8$ | 2(4)/1(4) | 2(4)/1(4) | 2(4)/1(4) | 2(4)/2(4) | 1(4)/1(4) |
| $nDOFs = 25602$ | | | | | |
| $10^{-8}$ | 4(44)/4(43) | 4(44)/4(43) | 4(44)/4(44) | 4(4)/4(4) | 1(5)/1(4) |
| $10^{-4}$ | 5(16)/4(16) | 5(16)/4(16) | 5(16)/4(16) | 4(4)/4(4) | 1(5)/1(4) |
| 1 | 4(7)/4(7) | 4(7)/4(7) | 4(7)/4(7) | 4(4)/4(4) | 1(5)/1(4) |
| $10^4$ | 4(4)/3(4) | 4(4)/3(4) | 4(4)/3(4) | 4(4)/4(4) | 1(5)/1(4) |
| $10^8$ | 3(4)/2(4) | 3(4)/2(4) | 3(4)/2(4) | 3(4)/2(4) | 1(5)/1(4) |

The iteration counts of the test runs are shown in Table 1. Each column contains two groups of numbers of the form $X(Y)$, separated by /. Here $X$ denotes the average number of PRESB-preconditioned GCR iterations when solving $H_1$ and $H_2$ and $Y$ denotes the average number of AGMG-preconditioned GCR to solve the blocks, arising in the PRESB preconditioning. The left $X(Y)$ group shows the iteration counts for the initial application of Algorithm 1 and on the right of / are the corresponding results of the defect correction step. We see that the convergence of AGMG somewhat deteriorates for the larger values of the regularization parameter $\beta$ and for small $\omega$.

In Fig. 1 we show the effect of the proposed defect-correction solution scheme. We plot the difference $\mathbf{e} = \mathbf{z}_{ex} - \mathbf{z}_{it}$ (top) and the computed solution $\mathbf{z}_{it}$ (bottom). Here $\mathbf{z}_{ex}$ is obtained by a direct solution of the system in (3). Note that the solution is complex. The top plots show that just one defect-correction step substantially decreases the error by some orders of magnitude, and improves the quality of the numerically computed solution, to be explained by less influence of arising rounding errors.

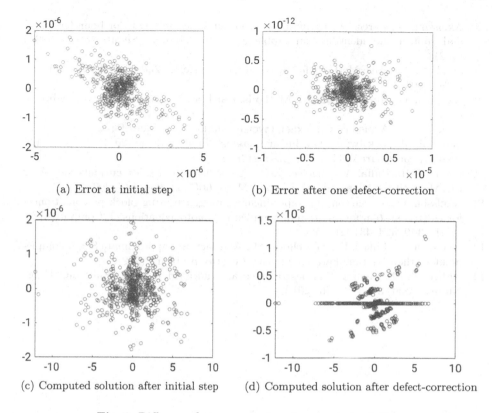

(a) Error at initial step        (b) Error after one defect-correction

(c) Computed solution after initial step      (d) Computed solution after defect-correction

**Fig. 1.** Difference between exact and computed solution

# References

1. Tröltzsch, F.: Optimal Control of Partial Differential Equations: Theory, Methods, and Applications. Graduate Studies in Mathematics, vol. 112. American Mathematical Society, Providence, RI (2010)
2. Axelsson, O., Neytcheva, M., Ström, A.: An efficient preconditioning method for the state box-constrained optimal control problem. J. Num. Math. **26**, 185–207 (2018)
3. Liang, Z.-Z., Axelsson, O., Neytcheva, M.: A robust structured preconditioner for time-harmonic parabolic optimal control problems. Numer. Algorithms **79**(2), 575–596 (2017). https://doi.org/10.1007/s11075-017-0451-5
4. Axelsson, O., Lukáš, D.: Preconditioning methods for Eddy-current optimally controlled time-harmonic electromagnetic problems. J. Numer. Math. **27**, 1–21 (2019)
5. Axelsson, O., Neytcheva, M., Ahmad, B.: A comparison of iterative methods to solve complex valued linear algebraic systems. Numer. Algorithms **66**(4), 811–841 (2013). https://doi.org/10.1007/s11075-013-9764-1
6. Axelsson, O., Neytcheva, M.: Preconditioners for two-by-two block matrices with square blocks, TR 2018-010. Department of Information Uppsala University Technology (2018)

7. Axelsson, O., Béreš, M., Blaheta, R.: Computational methods for boundary optimal control and identification problems. Math. Comput. Simul. **189**, 276–290 (2021)
8. Saad, Y.: Iterative Methods for Sparse Linear Systems, 2nd edn. SIAM, Philadelphia (2003)
9. Axelsson, O.: Iterative Solution Methods. Cambridge University Press, Cambridge (1994)
10. Liang, Z.-Z., Axelsson, O.: Exact inverse solution techniques for a class of complex valued block two-by-two linear systems. Numer. Algor. (2021). https://link.springer.com/journal/11075/online-first?page=2
11. Ammari, H., Buffa, A., Nédélec, J.-C.: A justification of Eddy currents model for the Maxwell equations. SIAM J. Appl. Math. **60**(5), 1805–1823 (2000)
12. Axelsson, O., Karatsson, J.: Superlinear convergence using block preconditioners for the real system formulation of complex Helmholtz equations. J. Comput. Appl. Math. **340**, 424–431 (2018)
13. Axelsson, O., Lukáš, D., Neytcheva, M.: An exact and approximate Schur complement method for time-harmonic optimal control problems (2021)
14. Notay, Y.: An aggregation-based algebraic multigrid method. Electron. Trans. Numer. Anal. **37**, 123–146 (2010)

# On the Consistency Order
# of Runge–Kutta Methods Combined
# with Active Richardson Extrapolation

Teshome Bayleyegn[1] , István Faragó[1,2,3] , and Ágnes Havasi[1,2(✉)]

[1] ELTE Eötvös Loránd University, Pázmány Péter s. 1/C,
Budapest 1117, Hungary
[2] MTA-ELTE Numerical Analysis and Large Networks Research Group,
Pázmány Péter s. 1/C, Budapest 1117, Hungary
`agnes.havasi@ttk.elte.hu`
[3] Institute of Mathematics, Budapest University of Technology and Economics,
Egry J. u. 1., Budapest 1111, Hungary

**Abstract.** Passive and active Richardson extrapolations are robust devices to increase the rate of convergence of time integration methods. While the order of convergence is shown to increase by one under rather natural smoothness conditions if the passive Richardson extrapolation is used, for the active Richardson extrapolation the increase of the order has not been generally proven. It is known that the Lipschitz property of the right-hand side function of the differential equation to be solved yields convergence of order $p$ if the method is consistent in order $p$. In this paper it is shown that the active Richardson extrapolation increases the order of consistency by one when the underlying method is any Runge–Kutta method of order $p = 1, 2$, or $3$.

**Keywords:** Consistency · Richardson extrapolation · Runge–Kutta method · Order conditions

## 1 Introduction

The different variants of classical Richardson extrapolation (RE) [2,3,5–7] are based on calculating a suitable linear combination of numerical solutions obtained by the same numerical method, but with two different step sizes. In the case of the passive RE the solutions are calculated independently on the two grids, and then these solutions are combined at the grid point of the coarse grid, while in the case of the active RE always the combined solution is used during the next time step on both grids. It has been shown by several numerical experiments that both the passive and active Richardson extrapolation increase the accuracy of the applied underlying method by one [7]. In these experiments the global error is evaluated in some norm as the time step size is successively decreased, which numerically confirms that the numerical solution is convergent to the exact solution. When the passive Richardson extrapolation is used, the

© Springer Nature Switzerland AG 2022
I. Lirkov and S. Margenov (Eds.): LSSC 2021, LNCS 13127, pp. 101–108, 2022.
https://doi.org/10.1007/978-3-030-97549-4_11

convergence of the underlying method trivially implies the convergence of the combined method, too. However, for the active Richardson extrapolation the increase of the convergence order has not yet been proven. It is known (see, e.g., [4]) that the Lipschitz property of the right-hand side function of the differential equation to be solved yields convergence of order $p$ for explicit one-step methods if the method is consistent in order $p$. Therefore, the knowledge of the consistency and its order is of vital importance to study the convergence.

In this paper we examine the consistency order of the active Richardson extrapolation when applied to any Runge–Kutta method of consistency order $p = 1, 2$, or 3.

The structure of the paper is as follows. In Section 2 the method of active Richardson extrapolation is presented. In Sect. 3 we recall the family of Runge–Kutta methods and their well-known order conditions up to order four. Finally, in Sect. 4, relying on the general order conditions we prove that the application of the active Richardson extrapolation to any Runge–Kutta method of order $p = 1, 2$, or 3 increases the order of consistency by one. Our results are illustrated with an example.

## 2  The Method of Richardson Extrapolation

Consider the Cauchy problem (1)

$$u'(t) = f(t, u(t)) \tag{1}$$
$$u(0) = u_0, \tag{2}$$

where $f : \mathbb{R} \times \mathbb{R}^d \to \mathbb{R}^d$. Assume that a numerical method is applied to solve (1), which will be referred to as underlying method. We divide $[0, T]$ into $N$ sub intervals of length $h$, and define the following two meshes on $[0, T]$:

$$\Omega_h^0 := \{t_n = nh : n = 0, 1, \ldots, N\}, \tag{3}$$

and

$$\Omega_h^1 := \{t_n = n\frac{h}{2} : n = 0, 1, \ldots, 2N\}, \tag{4}$$

The active Richardson extrapolation is defined as follows. Assume that the numerical solution has been calculated at the time level $t_{n-1}$, and denote it by $y_{n-1}$. Take a step of length $h$ by using the underlying method, and denote the resulting numerical solution by $z_n$. Take also two steps of step size $h/2$ with the underlying method, and denote the obtained numerical solution by $w_n$. The method of active Richardson extrapolation is defined as

$$y_n := \frac{2^p w_n - z_n}{2^p - 1}. \tag{5}$$

# 3 Runge–Kutta Methods and Their Order Conditions

Runge–Kutta methods have the general form

$$y_{n+1} = y_n + h \sum_{i=1}^{s} b_i k_i \tag{6}$$

with

$$k_i = f\left(t_n + c_i h, y_n + h \sum_{j=1}^{s} a_{ij} k_j\right), \quad i = 1, 2, ..., s, \tag{7}$$

where $b_i, c_i$, and $a_{ij} \in \mathbb{R}$ are given constants, and $s$ is the number of stages. Such a method is widely represented by the Butcher tableau

$$\frac{c \| A}{\| b^T} \tag{8}$$

where $c \in \mathbb{R}^{s \times 1}$, $b \in \mathbb{R}^{s \times 1}$ and the matrix $A = [a_{ij}]_{i,j=1}^{s}$ is strictly lower triangular in the case of an explicit Runge–Kutta method, and lower triangular for diagonally implicit Runge–Kutta methods. We always assume that the row-sum condition holds, i.e.,

$$c_i = \sum_{j=1}^{s} a_{ij}, i = 1, 2, \ldots, s. \tag{9}$$

Table 1 gives the conditions that are sufficient for order of consistency $p$ up to $p = 4$ in addition to the row-sum condition. (For a certain order the conditions of lower order should also be satisfied.) In this table $e$ is the vector in $\mathbb{R}^s$ all entries of which are equal to 1.

**Table 1.** Conditions for orders of consistency $p = 1, 2, 3,$ and 4 of Runge–Kutta methods.

| Order $p$ | Conditions |
|-----------|------------|
| 1 | $b^T e = 1$ |
| 2 | $b^T c = \frac{1}{2}$ |
| 3 | $b^T c^2 = \frac{1}{3}$, $b^T Ac = \frac{1}{6}$ |
| 4 | $b^T c^3 = \frac{1}{4}$, $b^T cAc = \frac{1}{8}$, $b^T Ac^2 = \frac{1}{12}$, $b^T A^2 c = \frac{1}{24}$ |

# 4 Increasing the Order by Richardson Extrapolation

The Butcher tableau of any Runge–Kutta method combined with the active Richardson extrapolation is given in Table 2 [1]. Zeroes in this tableau stand for the $s$ by $s$ null matrix, and $\rho = 1/(2^p - 1)$. Note that the combined method has three times as many stages as the underlying method.

**Table 2.** The Butcher tableau of RK+CRE.

| | $\frac{1}{2}A$ | $0$ | $0$ |
|---|---|---|---|
| $\frac{1}{2}c$ | | | |
| $\frac{1}{2}(e+c)$ | $\frac{1}{2}eb^T$ | $\frac{1}{2}A$ | $0$ |
| $c$ | $0$ | $0$ | $A$ |
| | $\left(\frac{1+\rho}{2}\right)b^T$ | $\left(\frac{1+\rho}{2}\right)b^T$ | $-\rho b^T$ |

In the following we use the notations

$$\tilde{b}^T = \left[\left(\frac{1+\rho}{2}\right)b^T, \left(\frac{1+\rho}{2}\right)b^T, -\rho b^T\right] \in \mathbb{R}^{1\times 3s},$$

$$\tilde{c} := \begin{bmatrix} \frac{1}{2}c \\ \frac{1}{2}(e+c) \\ c \end{bmatrix} \in \mathbb{R}^{3s\times 1}, \quad \tilde{A} = \begin{bmatrix} \frac{1}{2}A & 0 & 0 \\ \frac{1}{2}eb^T & \frac{1}{2}A & 0 \\ 0 & 0 & A \end{bmatrix} \in \mathbb{R}^{3s\times 3s},$$

which define the Butcher tableu of the combined method (underlying method with active RE). It is easy to see that if $b^Te = 1$, which is necessary for order 1, and the row-sum condition (9) holds for the underlying method, then the row-sum condition holds for the combined method as well. Therefore, it is not necessary to check this condition in each case separately in what follows.

### 4.1   Order Increase from 1 to 2

We introduce the notation $\tilde{e} = [1, 1, \ldots, 1]^T \in \mathbb{R}^{3s\times 1}$.

**Proposition 1.** *Assume that the underlying Runge-Kutta method (8) is first order consistent. Then its combination with the active RE yields a second order consistent method.*

*Proof.* Since $p = 1$, therefore $\rho = 1$, and the condition $b^Te = 1$ holds. We show that

$$\tilde{b}^T\tilde{e} = 1, \tag{10}$$

and

$$\tilde{b}^T\tilde{c} = \frac{1}{2}. \tag{11}$$

Checking (10):

$$\tilde{b}^T\tilde{e} = \frac{1+\rho}{2}b^Te + \frac{1+\rho}{2}b^Te - \rho b^Te = (1+\rho-\rho)b^Te = b^Te = 1.$$

Note that this holds for any value of $\rho$ (or $p$).
Checking (11):

$$\tilde{b}^T\tilde{c} = \frac{1+\rho}{2}b^T\frac{c}{2} + \frac{1+\rho}{2}b^T\frac{e+c}{2} - \rho b^Tc = \frac{1-\rho}{2}b^Tc + \frac{1+\rho}{4} = \frac{1}{2},$$

## 4.2    Order Increase from 2 to 3

**Proposition 2.** *Assume that the underlying Runge–Kutta method (8) is second order consistent. Then its combination with the active RE yields a third order consistent method.*

*Proof.* Since $p = 2$, therefore $\rho = \frac{1}{3}$. We have to verify the conditions

$$\tilde{b}^T \tilde{e} = 1, \tag{12}$$

$$\tilde{b}^T \tilde{c} = \frac{1}{2}, \tag{13}$$

$$\tilde{b}^T \tilde{c}^2 = \frac{1}{3}, \tag{14}$$

$$\tilde{b}^T \tilde{A} \tilde{c} = \frac{1}{6}. \tag{15}$$

The condition (12) holds since $\tilde{b}^T \tilde{e} = b^T e = 1$ for any order $p \geq 1$. Condition (13) is satisfied because $\tilde{b}^T \tilde{c} = \frac{1-\rho}{2} b^T c + \frac{1+\rho}{4} = \frac{1}{2}$.

The condition (14) holds since

$$\tilde{b}^T \tilde{c}^2 = \frac{1+\rho}{2} b^T \left( \frac{1}{2} c \right)^2 + \frac{1+\rho}{2} b^T \left( \frac{1}{2} (c + e) \right)^2 - \rho b^T c^2 = \frac{1}{3}.$$

Finally, (15) can be verified as

$$\tilde{b}^T \tilde{A} \tilde{c} = \left( \frac{1+\rho}{8} \right) b^T A c + \left( \frac{1+\rho}{8} \right) b^T e b^T c + \left( \frac{1+\rho}{8} \right) b^T A e$$

$$+ \left( \frac{1+\rho}{8} \right) b^T A c - \rho b^T A c = \frac{1}{6}.$$

## 4.3    Order Increase from 3 to 4

**Proposition 3.** *Assume that the underlying Runge-Kutta method (8) is third order consistent. Then its combination with the active RE yields a fourth order consistent method.*

*Proof.* Since $p = 3$, therefore $\rho = \frac{1}{7}$, and the conditions $b^T e = 1$, $b^T c = \frac{1}{2}$, $b^T c^2 = \frac{1}{3}$ and $b^T A c = \frac{1}{6}$ hold by our assumptions. We have to show that, in addition to conditions (12)–(15), the conditions

$$\tilde{b}^T \tilde{c}^3 = \frac{1}{4}, \tag{16}$$

$$\tilde{b}^T \mathrm{diag}(\tilde{c}) \tilde{A} \tilde{c} = \frac{1}{8}, \tag{17}$$

$$\tilde{b}^T \tilde{A} \tilde{c}^2 = \frac{1}{12}, \tag{18}$$

$$\tilde{b}^T \tilde{A}^2 \tilde{c} = \frac{1}{24} \tag{19}$$

are satisfied. The condition (12) $\tilde{b}^T \tilde{e} = 1$ again holds automatically. Condition (13) holds since $\tilde{b}^T \tilde{c} = \frac{1-\rho}{2} b^T c + \frac{1+\rho}{4} = \frac{1}{2}$. Also, (14) is valid since $\tilde{b}^T \tilde{c}^2 = \frac{1-3\rho}{4} b^T c^2 + \frac{1+\rho}{4} = \frac{1}{3}$. Moreover, (15) holds because $b^T Ac = \left(\frac{1-3\rho}{4}\right) b^T Ac + \frac{1+\rho}{8} = \frac{1}{6}$.

Checking (16):

$$\tilde{b}^T \tilde{c}^3 = \left(\frac{1-7\rho}{8}\right) b^T c^3 + \frac{7+7\rho}{32} = \frac{1}{4}.$$

Checking (17):

$$\tilde{b}^T \operatorname{diag}(\tilde{c}) \tilde{A} \tilde{c} =$$

$$\left[\frac{1+\rho}{2} b^T, \frac{1+\rho}{2} b^T, -\rho b^T\right] \begin{bmatrix} \operatorname{diag}\frac{c}{2} & 0 & 0 \\ 0 & \operatorname{diag}\frac{e+c}{2} & 0 \\ 0 & 0 & \operatorname{diag}c \end{bmatrix} \begin{bmatrix} \frac{A}{2} & 0 & 0 \\ \frac{1}{2} e b^T & \frac{A}{2} & 0 \\ 0 & 0 & A \end{bmatrix} \begin{bmatrix} \frac{1}{2} c \\ \frac{1}{2}(e+c) \\ c \end{bmatrix}$$

$$= \left(\frac{1-7\rho}{8}\right) (b^T \otimes c^T) Ac + \frac{21}{192}(1+\rho) = \frac{1}{8}.$$

Checking (18):

$$\tilde{b}^T \tilde{A} \tilde{c}^2 = \left(\frac{1-7\rho}{8}\right) b^T Ac^2 + \frac{7}{96}(1+\rho) = \frac{1}{12}.$$

Checking (19):

$$\tilde{b}^T \tilde{A}^2 \tilde{c} = \left(\frac{1-7\rho}{8}\right) b^T A^2 c + \frac{7}{192}(1+\rho) = \frac{1}{24}.$$

**Example:** Let us check the Butcher tableau under Table 2 by choosing the explicit Euler (EE) method as underlying method:

$$y_{n+1} = y_n + h f(t_n, y_n), \tag{20}$$

for which $A = (0), b^T = 1, c = 0$, and the method is known to be of first order. The Butcher tableau of EE + active RE according to Table 2 is given by Table 3.

**Table 3.** The Butcher tableau of EE + active RE.

$$\begin{array}{c|ccc} 0 & 0 & 0 & 0 \\ \frac{1}{2} & \frac{1}{2} & 0 & 0 \\ 0 & 0 & 0 & 0 \\ \hline & 1 & 1 & -1 \end{array}$$

This tableau corresponds to the following method:

$$k_1 = f(t_n, y_n), \quad \overline{k}_1 = f\left(t_n + \frac{1}{2}h, y_n + \frac{1}{2}hk_1\right), \quad \overline{\overline{k}}_1 = f(t_n, y_n)$$

$$y_{n+1} = y_n + h\left[b_1 k_1 + b_2 \overline{k}_1 + b_3 \overline{\overline{k}}_1\right] = y_n + h\left[k_1 + \overline{k}_1 - \overline{\overline{k}}_1\right]$$

$$= y_n + hf\left(t_n + \frac{h}{2}, \ y_n + \frac{h}{2}f(t_n, y_n)\right).$$

Note that this is exactly the explicit mid-point (MP) formula, which can also be considered as a two-stage Runge–Kutta method by the choice

$$A = \begin{pmatrix} 0 & 0 \\ \frac{1}{2} & 0 \end{pmatrix}, \ b^T = \left(0, 1\right), \ c = \begin{pmatrix} 0 \\ \frac{1}{2} \end{pmatrix}.$$

From either form it is easy to check that the conditions of second order hold.

## 5 Numerical Experiment

We considered the scalar problem

$$\begin{cases} y' = -2t \sin y, & t \in [0, 1] \\ y(0) = 1, \end{cases} \tag{21}$$

The exact solution is $y(t) = 2\cot^{-1}\left(e^{t^2}\cot(\frac{1}{2})\right)$. We solved the problem (21) using the EE method and the EE + active RE. Table 4 shows the expected increase in the order of the global error, from 1 to 2.

**Table 4.** Global errors at $t = 1$ in absolute value in Example (21)

| h | EE | Order | EE + RE | Order |
|---|---|---|---|---|
| 0.1 | 1.99e−02 | | 7.8397e−04 | |
| | | 1.0820 | | 2.1059 |
| 0.05 | 9.40e−03 | | 1.8212e−04 | |
| | | 1.0627 | | 2.0511 |
| 0.025 | 4.50e−03 | | 4.3945e−05 | |
| | | 1.0324 | | 2.0251 |
| 0.0125 | 2.20e−03 | | 1.0797e−05 | |

# 6   Conclusion

Consistency is one of the basic properties of numerical methods, associated with the local truncation error. Using the order conditions of the Runge-Kutta method for orders $p = 1, 2$, and 3, we proved that the active Richardson extrapolation increases the order of consistency by one. This result has a great significance in the study of Richardson extrapolation combined with any Runge–Kutta method since, together with the Lipschitz property of the right-hand side function $f$, it implies that the combined method will be $p$-th order convergent.

**Acknowledgements.** "Application Domain Specific Highly Reliable IT Solutions" project has been implemented with the support provided from the National Research, Development and Innovation Fund of Hungary, financed under the Thematic Excellence Programme TKP2020-NKA-06 (National Challenges Subprogramme) funding scheme. This work was completed in the ELTE Institutional Excellence Program (TKP2020-IKA-05) financed by the Hungarian Ministry of Human Capacities. The project has been supported by the European Union, and co-financed by the European Social Fund (EFOP-3.6.3-VEKOP-16-2017-00002), and further, it was supported by the Hungarian Scientific Research Fund OTKA SNN125119.

# References

1. Chan, R., Murua, A.: Extrapolation of symplectic methods for Hamiltonian problems. Appl. Numer. Math. **34**, 189–205 (2000)
2. Richardson, L.F.: The approximate arithmetical solution by finite differences of physical problems including differential equations. Philos. Trans. R. Soc. London, Ser. A **210**, 307–357 (1911)
3. Richardson, L.F.: The deferred approach to the limit, I. Single lattice. Philos. Trans. R. Soc. London, Ser. A **226**, 299–349 (1927)
4. Süli, E.: Numerical Solution of Ordinary Differential Equations. Mathematical Institute, University of Oxford (2014)
5. Zlatev, Z., Faragó, I., Havasi, Á.: Stability of the Richardson extrapolation together with the $\Theta$-method. J. Comput. Appl. Math. **235**, 507–520 (2010)
6. Zlatev, Z., Dimov, I., Faragó, I., Georgiev, K., Havasi, Á., Ostromsky, T.: Application of Richardson extrapolation for mutli-dimensional advection equations. Comput. Math. Appl. **67**(12), 2279–2293 (2014)
7. Zlatev, Z., Dimov, I., Faragó, I., Havasi, Á.: Richardson Extrapolation - Practical Aspects and Applications. De Gruyter, Berlin (2017)

# Study the Recurrence of the Dominant Pollutants in the Formation of AQI Status over the City of Sofia for the Period 2013–2020

Ivelina Georgieva$^{(\boxtimes)}$, Georgi Gadzhev, and Kostadin Ganev

National Institute of Geophysics, Geodesy and Geography – Bulgarian Academy of Sciences, Acad. G. Bonchev Street, bl. 3, 1113 Sofia, Bulgaria
iivanova@geophys.bas.bg

**Abstract.** Recently it became possible to acquire and adapt the most up-to-date models of local atmospheric dynamics - WRF, transport and transformation of air pollutants - CMAQ, and the emission model SMOKE. This gave the opportunity to conduct extensive studies of the atmospheric composition climate of the country on fully competitive modern level. Ensemble, sufficiently exhaustive and representative to make reliable conclusions for atmospheric composition - typical and extreme situations with their specific space and temporal variability was created by using of computer simulations. On this basis statistically significant ensemble of corresponding Air Quality indices (AQI) was calculated, and their climate - typical repeatability, space and temporal variability for the territory of the country was constructed. The Air Quality (AQ) impact on human health and quality of life is evaluated in the terms of AQI, which give an integrated assessment of the impact of pollutants and directly measuring the effects of AQ on human health. All the AQI evaluations are on the basis of air pollutant concentrations obtained from the numerical modelling and make it possible to revile the AQI status spatial/temporal distribution and behavior.

The presented results, allow to follow highest recurrence of the indices for the whole period and seasonally, and to analyze the possible reason for high values in the Moderate, High and Very High bands.

**Keywords:** Air pollution modeling · Air Quality indexes · AQ impact · Atmospheric composition · AQI Status

## 1 Introduction

According to the World Health Organization, air pollution severely affects the health of European citizens. There is increasing evidence of adverse effects of air pollution on the respiratory and the cardiovascular system as a result of both acute and chronic exposure [17]. There is considerable concern about impaired

© Springer Nature Switzerland AG 2022
I. Lirkov and S. Margenov (Eds.): LSSC 2021, LNCS 13127, pp. 109–116, 2022.
https://doi.org/10.1007/978-3-030-97549-4_12

and detrimental air quality conditions over many areas in Europe, especially in urbanized areas, despite 30 years of legislation and emission reductions.

A system for Chemical Weather Forecast operates in many European countries, together with the numerical weather forecast. The Bulgarian Chemical Weather Forecast and Information System (BgCWFIS) runs on five nested domains: Europe, Balkan Peninsula, Bulgaria, Sofia Municipality and Sofia City with increasing space resolution - to 1 km for the territory of Sofia City. The BgCWFIS calculate also the AQI – an integrated assessment of the impact of pollutants, directly measuring the effects of AQ on human health.

The operational results of BgCWFIS created ensemble sufficiently exhaustive and representative to allow making reliable conclusions for the atmospheric composition status and behavior. The generalisation and analysis of these results, in particular AQI is the objective of the present work.

## 2    Methods

The AQI evaluations are based on extensive computer simulations of the AQ in Sofia carried out with good resolution using up-to-date modeling tools and detailed and reliable input data. All the simulations are based on the US EPA Model-3 system: WRF v.3.2.1 - [16] Weather Research and Forecasting Model, used as meteorological pre-processor; In the System, WRF is driven by the NCEP GFS (Global Forecast System) data that can be accessed freely from [18] CMAQ v.4.6 - Community Multi-scale Air Quality model, [2,3] the Chemical Transport Model (CTM), and SMOKE - the Sparse Matrix Operator Kernel Emissions Modelling System [4] the emission pre-processor of Models-3 system. TNO inventory for 2010 [5] is exploited for the territories outside Bulgaria in the mother CMAQs domain. For the Bulgarian domains the National inventory for 2010 as provided by Bulgarian Executive Environmental Agency is used.

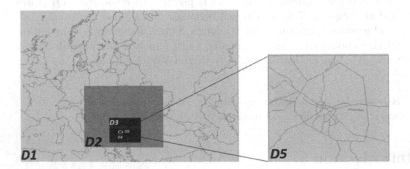

**Fig. 1.** Five computational domains (CMAQ domains are nested in WRF ones).

In BgCWFIS, climatic data is used for chemical boundary conditions following the presumption that the errors introduced by this assumption will decrease

**Table 1.** Boundaries between index points for each pollutant.

| Index | O3 Running 8 hourly mean (µg/m3) | NO2 Hourly mean (µg/m3) | SO2 15 minute mean (µg/m3) | PM10 Particles, 24 hour mean (µg/m3) | PM2.5 Particles, 24 hour mean (µg/m3) |
|---|---|---|---|---|---|
| 1 (Low) | 0-33 | 0-66 | 0-88 | 0-11 | 0-16 |
| 2 (Low) | 34-65 | 67-133 | 89-176 | 12-23 | 17-33 |
| 3 (Low) | 66-99 | 134-199 | 177-265 | 24-34 | 34-49 |
| 4 (Moderate) | 100-120 | 200-267 | 266-354 | 35-41 | 50-58 |
| 5 (Moderate) | 121-140 | 268-334 | 355-442 | 42-46 | 59-66 |
| 6 (Moderate) | 141-159 | 335-399 | 443-531 | 47-52 | 67-74 |
| 7 (High) | 160-187 | 400-467 | 530-708 | 53-58 | 75-83 |
| 8 (High) | 188-213 | 468-534 | 709-886 | 59-64 | 84-91 |
| 9 (High) | 214-239 | 535-599 | 887-1063 | 65-69 | 92-99 |
| 10 (Very High) | $\geq240$ | $\geq600$ | $\geq1064$ | $\geq70$ | $\geq100$ |

quickly to the center of the domain due to the continuous acting of the pollution sources. All other domains receive their boundary conditions from the previous domain in the hierarchy Fig. 1. BgCWFIS delivers AQ forecasts on hourly basis. According to [13] 4 main pollutants - $O_3$, $NO_2$, $SO_2$, and $PM_{10}$ are used to calculate the AQI. The further considerations in the paper are made on the basis of long term AQ simulations, which make it possible to reveal the climate of AQI spatial/temporal distribution and behavior. The AQI is defined in several segments [13], different for each considered pollutant. Different averaging periods are used for different pollutants. The breakpoints between index values are defined for each pollutant separately (Table 1). For each particular case the concentration of each pollutant falls into one of the bands, shown in Table 1. Thus the AQI for each pollutant is determined. The overall AQI, which describes the impact of the ambient pollutant mix is defined as the AQI for the pollutant with maximum value of the index. The pollutant, which determines the overall AQI for the given particular case is referred to as "dominant pollutant". Each of the AQI bands comes with advice for at-risk groups and the general population. The reference levels and Health Descriptor used are based on health-protection related limit, target or guideline values set by the EU, at national or local level or by the WHO. Detailed description of the operational performance of BgCWFIS computation of AQI is given in many papers in Bulgaria [6–9,14,15]. Results for the recurrence of all the 10 AQI, annually and by seasons, as well as for the recurrence of case when 2 pollutants – $O_3$ and $PM_{10}$ are dominant are presented in the present paper for 3 chosen hours. Such performance by indices allows to determine the dominant index (in%) during the day for the entire selected period 2013–2020.

## 3   Computational Results

Figure 2 demonstrate the spatial and diurnal variation of the annual recurrence of AQI from 2 to 10 (%) over territory of Sofia city for the whole period 2013-2020 for selected hours 05:00, 11:00 and 19:00 UTC. Here we have to mention that the Low range (formed from sum of AQI1–AQI3) the air is cleanest, so high recurrence values (100%) mean all cases with clean air (red colour) and lower recurrence values mean (blue colour), less cases with clean air or worse AQ status. AQI1 presents the cleanest situation and the recurrence there is pretty small, so the plots for AQI1 are not shown. In the other plots for the AQI4–AQI10 (Moderate, High and Very High ranges) - high recurrence values mean less favorable and respectively bad AQ status. The presented results are about AQI2 to AQI10. What can be noticed is: the recurrence of the first two indices AQI2 and AQI3 (Low range) is different during the day and it reaches 40 % over the whole city territory. The recurrence of AQI2 is about 60% over the whole city at morning hours and about 30% at noon and afternoon. For AQI3 high recurrence can be seen over the Vitosha mountain about 80% in morning and afternoon hours, and 40% at the center of the city. Higher values over the Vitosha Mountain are due to the higher concentration of $O_3$ in mountain areas and intensive $O_3$ transport from higher levels (intensive turbulence during midday). The behavior of the surface ozone is complex. The $O_3$ in Bulgaria is to a great extent due to transport from abroad [1,7,10–12]. This is the reason why the $O_3$ concentrations early in the morning are smaller (less intensive transport from higher levels), and higher at noon and afternoon (turbulence atmosphere and $O_3$ photochemistry). The second part of indices present the Moderate range are AQI4, AQI5 and AQI6. Here, as was mentioned above, the red color presents more cases with worse AQ status, and blue color is the clean air. In the plot of AQI4 in morning and afternoon hours there is high recurrence over the Vitosha mountain. At noon the cases with bad AQ reach 40% over the whole territory of the city. It should be mention that for AQI10, that present the Very High band, the plot shows recurrence of about 30% in the morning and afternoon in 3 spots of the city – the city center, the area of the airport and Kostinbrod. In the plots of other indices - AQI5, AQI6, AQI7, AQI8, and AQI9 the recurrence is zero, or so small, that cannot be seen in this scale. The next plots present the recurrence of cases in which different pollutants are the one that determine the respective AQI. Figure 3 shows the recurrence of cases for which $O_3$ determines the formation of each index from 2 to 10. Here we can say that the $O_3$ recurrence for AQI2, AQI3, and AQI4 is near 100% over the whole city except for AQI4 in morning and afternoon hours where in the city center the recurrence of cases are due to other pollutants. For AQI5 in morning hours the $O_3$ has high recurrence 100% only over the Vitosha mountain. At noon this recurrence covers the whole city except the center and Kostinbrod area, and at afternoon the $O_3$ recurrence cover already the rural area of the city. In that presentation of $O_3$ recurrence it can be judged that it determines the formation of indices for Low and Moderate range. For the other bands the $O_3$ contribution can has a very small (smaller than 10%) and so the AQI value is due to other pollutants. Figure 4 presents the recurrence of cases for which $PM_{10}$ determines the formation of the indices. In the plots for AQI2 and AQI3 the recurrence of $PM_{10}$ dominance is about 30% only in the city center during the whole day. For AQI4 the area of recurrence is wider and reaches 90% in the center and 40% in the surrounding area. The recurrence of $PM_{10}$ dominance in formation of all the other indices AQI5 to AQI10 is about 100% during the whole day and covers the center of the city, the Kostinbrod area and the Sofia airport area. This is normal having in mind that the main $PM_{10}$ sources like TPPs, manufactures and the busiest traffic network are situated in these areas.

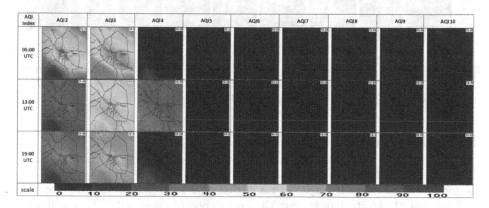

**Fig. 2.** Annual recurrence of AQI from 2 to 10 (%) over territory of Sofia city for the whole period 2013–2020 for selected hours 05:00, 11:00, and 19:00 UTC. (Color figure online)

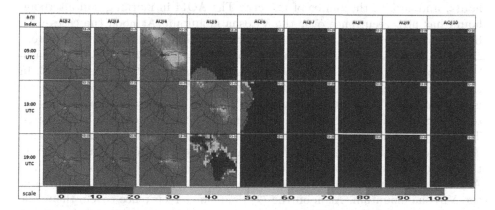

**Fig. 3.** Annual reccurence (%) of the cases in which the $O_3$ dominates in the formation of each AQI from 2 to 10 over territory of Sofia city for the whole period 2013–2020 for selected hours 05:00, 11:00, and 19:00 UTC.

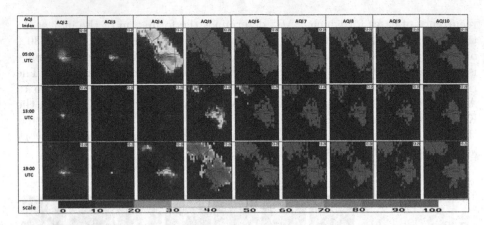

**Fig. 4.** Annual recurrence (%) of the cases in which the $PM_{10}$ dominates in the formation of each AQI from 2 to 10 over territory of Sofia city for the whole period 2013–2020 for selected hours 05:00, 11:00, and 19:00 UTC.

## 4    Discussion and Conclusion

The simulations for Sofia city show that the air quality status of Sofia is not so good (evaluated with a spatial resolution of 1 km). AQI status falls mostly in Low and Moderate bands, but the recurrence of cases with High pollution is close to 20% mostly at the city center. The recurrence of indices AQI2 and AQ3 (Low range) is different during the day and it reaches 40% over the whole city territory. The recurrence of AQI2 is about 60% over the whole city in the morning hours and about 30% at noon and afternoon. For AQI3 high recurrence can be seen over the Vitosha mountain about 80% in morning and afternoon hours, and 40% in the center of the city. The AQI4 in morning and afternoon hours has high recurrence over the Vitosha mountain. At noon the cases with bad AQ reach 40% over the whole territory of the city. The AQI10, that present the Very High band, shows recurrence about 30% in morning and afternoon in 3 spots of the city – the city center, the area of the airport and Kostinbrod. In the plots of other indices AQI5, AQI6, AQI7, AQI8, and AQI9 the recurrence is so small (less than 10%), that cannot be seen in the scale. The results for recurrence of different pollutants dominance to formation of each index shows: The $O_3$ recurrence for AQI2, AQI3, and AQI4 is near 100% over the whole city. This is not exactly the case for AQI4 in morning and afternoon hours, where in the city center the formation of the AQI is mostly due to other pollutants. For AQI5 in morning hours the $O_3$ has high dominance recurrence 100% only over the Vitosha mountain. At noon the $O_3$ dominance recurrence covers the whole city except the center and Kostinbrod area, and in afternoon the $O_3$ dominance covers already the rural area around the city. Thus for $O_3$ it can be concluded that it leads to the formation of indices from Low and Moderate range. The recurrence of $PM_{10}$ dominance to formation of AQI2 and AQI3 is about 30% only in the city center during the whole day. For AQI4 the area of recurrence is wider and about 90% in the center and 40% in the surrounding area. The dominance of $PM_{10}$ for formation of others indices AQI5 to AQI10 is about 100% during the whole day and covers the center of the city, the Kostinbrod area and the Sofia airport. This is normal having in mind that in those areas the main $PM_{10}$ sources like TPPs, manufactures and the busiest traffic network, are situated. The pollution in the city is probably due to the surface sources like road transport and also the TPPs in the city and Sofia airport. Apart from these general features the climatic behavior of the AQI probabilities is rather complex with significant spatial, seasonal and diurnal variability. The areas with slightly worse AQ status are not necessarily linked to the big pollution sources. Wide rural and even mountain regions can also have significant probability for AQI from the Moderate range. The hot spot in Sofia city, where the high value indices are with higher recurrence is in the city center. The Very High band recurrence is relatively high - about 30% in the morning and 20% in the afternoon.

**Acknowledgments.** This work has been carried out in the framework of the National Science Program "Environmental Protection and Reduction of Risks of Adverse Events and Natural Disasters", approved by the Resolution of the Council of Ministers № 577/17.08.2018 and supported by the Ministry of Education and Science (MES) of Bulgaria (Agreement № D01-363/17.12.2020). This work was supported by Contract No D01-161/28.08.2018 (Project "National Geoinformation Center (NGIC)" financed by the National Roadmap for Scientific Infrastructure 2017-2023 and by Bulgarian National Science Fund (grant DN- 14/3 from 13.12.2017). Special thanks are due to US EPA and US NCEP for providing free-of-charge data and software and to the Netherlands Organization for Applied Scientific research (TNO) for providing the high-resolution European anthropogenic emission inventory.

# References

1. Bojilova, R., Mukhtarov, P., Miloshev, N.: Climatology of the index of the biologically active ultraviolet radiation for Sofia. An empirical forecast model for predicting the UV-index. Comptes rendus de l'Academie bulgare des Sciences **73**(4), 531–538 (2020)

2. Byun, D., Ching, J.: Science Algorithms of the EPA Models-3 Community Multiscale Air Quality (CMAQ) Modeling System. EPA Report 600/R-99/030, Washington, D.C., EPA/600/R-99/030 (NTIS PB2000-100561) (1999). https://cfpub.epa.gov/si/si_public_record_report.cfm?Lab=NERL&dirEntryId=63400

3. Byun, D.: Dynamically consistent formulations in meteorological and air quality models for multiscale atmospheric studies part I: governing equations in a generalized coordinate system. J. Atmos. Sci. **56**, 3789–3807 (1999)

4. CEP: Sparse Matrix Operator Kernel Emission (SMOKE) Modeling System. University of Carolina, Carolina Environmental Programs, Research Triangle Park, North Carolina (2003)

5. Denier van der Gon, H., Visschedijk, A., van de Brugh, H., Droge, R.: A high resolution European emission data base for the year 2005. TNO-report TNO-034-UT-2010-01895 RPT-ML, Apeldoorn, The Netherlands (2010)

6. Etropolska, I., Prodanova, M., Syrakov, D., Ganev, K., Miloshev, N., Slavov, K.: Bulgarian operative system for chemical weather forecast. In: Dimov, I., Dimova, S., Kolkovska, N. (eds.) NMA 2010. LNCS, vol. 6046, pp. 141–149. Springer, Heidelberg (2011). https://doi.org/10.1007/978-3-642-18466-6_16

7. Gadzhev, G., Ganev, K., Miloshev, N., Syrakov, D., Prodanova, M.: Numerical study of the atmospheric composition in Bulgaria. Comput. Math. Appl. **65**, 402–422 (2013)

8. Gadzhev, G.: Preliminary results for the recurrence of air quality index for the city of Sofia from 2008 to 2019. In: Proceeding of 1st International Conference on Environmental Protection and Disaster RISKs, pp. 53–64 (2020). https://doi.org/10.48365/ENVR-2020.1.5

9. Ivanov, V., Georgieva, I.: Air quality index evaluations for Sofia city. In: 17th IEEE International Conference on Smart Technologies, EUROCON 2017 - Conference Proceedings, pp. 920–925 (2017). https://doi.org/10.1109/EUROCON.2017.8011246

10. Kaleyna, P., Muhtarov, P.l., Miloshev, N.: Condition of the stratospheric and mesospheric ozone layer over Bulgaria for the period 1996–2012: part 1 - total ozone content, seasonal variations. Bulg. Geophys. J. **39** 9–16 (2013)

11. Kaleyna, P., Muhtarov, P.l., Miloshev, N.: Condition of the stratospheric and mesospheric ozone layer over Bulgaria for the period 1996–2012: part 2 - total ozone content, short term variations. Bulg. Geophys. J. **39**, 17–25 (2013)

12. Kaleyna, P., Mukhtarov, P., Miloshev, N.: Seasonal variations of the total column ozone over Bulgaria in the period 1996–2012. Comptes rendus de l'Academie bulgare des Sciences **67**(7), 979–986 (2014)

13. de Leeuw, F., Mol, W.: Air quality and air quality indices: a world apart. ETC/ACC Technical Paper 2005/5 (2005). http://acm.eionet.europa.eu/docs/ETCACC_TechnPaper_2005_5_AQ_Indices.pdf

14. Syrakov, D., Etropolska, I., Prodanova, M., Ganev, K., Miloshev, N., Slavov, K.: Operational pollution forecast for the region of Bulgaria. Am. Inst. Phys. Conf. Proc. **1487**, 88–94 (2012). https://doi.org/10.1063/1.4758945

15. Syrakov, D., et al.: Downscaling of Bulgarian chemical weather forecast from Bulgaria region to Sofia city. In: American Institute of Physics Conference Proceedings, vol. 1561, pp. 120–132 (2013)

16. Shamarock, W.C., et al.: A description of the advanced research WRF version 2 (2007). http://www.dtic.mil/dtic/tr/fulltext/u2/a487419.pdf

17. World Health Organization (WHO): Health aspects of air pollution: results from the WHO project systematic review of health aspects of air pollution in Europe (2004)

18. http://www.ftp.ncep.noaa.gov/data/nccf/com/gfs/prod/

# One Solution of Task with Internal Flow in Non-uniform Fluid Using CABARET Method

Valentin A. Gushchin[1]([⊠])[ID] and Vasilii G. Kondakov[2][ID]

[1] Institute for Computer Aided Design of RAS, 19/18, 2-nd Brestskaya Street,
123056 Moscow, Russia
gushchin@icad.org.ru
[2] Nuclear Safety Institute of Russian Academy of Sciences, Moscow, Russia
kondakov@ibrae.ac.ru

**Abstract.** In this novel study the new problem of flow in density-stratified salinity liquid will be established. Additional point will be adding CABARET method in the noncompressible liquid with state equation present, that will be presented salinity and temperature dependant. Earlier, the authors solved the problem of the collapse of a homogeneously mixed spot in vertically stratified water column. Test were veried on 3d problems with following initial conditions: At time equal zero in the density-stratified liquid with nonzero salinity bounded homogeneous water column is inserted, after some time elapsed the bounds of water column are taken out. Corresponding to this test mathematical model will consist of equation of free-surface and Navier-Stocks equation for noncompressible liquid, salinity transport and thermo conductivity equations. To verify this new model test with the flow around an inverted parabola obstacle will be considered in several modes: subcritical, transcritical and supercritical. The obtained simulation results are compared with the analytical solution with shallow water approximation. A distinctive feature of such tests is the reproducibility of stable modes and the presence of a wide range of examples. To check the correctness of such for simulation of the problem an additional test example with a quasi-two-dimensional problem of collapse of homogeneously-mixed spot in a water column with vertical stratification is considered.

**Keywords:** CABARET · Shallow water · Stratification

## 1 Introduction

Study of various kinds of phenomena and processes occurring in such heterogeneous medium as the atmosphere and the ocean are both of academic and practical interest. The heterogeneity of these media linked to the effects of buoyancy – the presence of gravity. It is known that the density of sea and ocean medium depends on temperature, pressure and salinity. A number of mathematical models describing the dynamics of stratified fluids have been suggested [1,2].

© Springer Nature Switzerland AG 2022
I. Lirkov and S. Margenov (Eds.): LSSC 2021, LNCS 13127, pp. 117–123, 2022.
https://doi.org/10.1007/978-3-030-97549-4_13

In [3] using the example of the solution of the problem of the dynamics of a spot (collapse) in a stably density-stratified fluid, a numerical method is discussed which can be used to investigate the flow of an inhomogeneous incompressible viscous fluid. The possibility is foreseen of specifying the stratification of density and viscosity either by analytic formulas or by tables obtained by processing experimental data, which significantly expands the range of laminar flows under consideration. According to the model proposed in [4] the origin and development of turbulence in a stably density-stratified fluid is inseparable from internal waves and proceeds as follows. Under the action of external forces internal waves of large size arise in the stratified fluid. As a result of their nonlinear interaction and subsequent breaking up or loss of stability, regions of mixed fluid, spots, arise. These spots of mixed-up turbulent fluid evolve, gradually flattening (the collapse of turbulent spots), which in turn leads to the formation of new spots, and so on. In the evolution of a spot it is natural to consider three basic stages:

1. The initial stage: the driving force acting on the fluid particles situated inside a spot considerably exceeds the resistive forces; intense internal waves are produced by the spot.
2. The intermediate stationary stage: the motive force is mainly counterbalanced by the resistance of shape and the wave resistance due to the radiation of the internal waves; the increase of the horizontal size of the spot proceeds almost as a linear function of time, that is, the acceleration is negligibly small.
3. The concluding viscous stage: the motive force is mainly counterbalanced by viscous drag; the horizontal size of the spot changes only slightly.

Later as a result of diffusion the spot is mixed with the surrounding fluid and vanishes.

During the 40 years passed since [3] was published, the simulation of such flows has undergone a lot of changes. New physical and mathematical models have been proposed [5,6] and the quality of methods designed for solving such problems has significantly improved [7,8]. In this work, we adapt the mathematical model proposed in [9] and used for calculating the flow around a sphere and circular cylinder [6,9] to the problem of spot collapse, which was earlier solved without taking into account the diffusion of the stratifying component [3,10,11]. Consider the flat nonstationary problem about the flow occurring when a homogeneous fluid region A surrounded by a stably and continuously stratified density fluid (the stratification is assumed to be linear) collapses in the vertical direction (Fig. 1). Flow develops in the homogeneous gravity field with the acceleration due to gravity g. The undisturbed linear density distribution [9]:

$$\rho(x,y) = \rho_0 \left(1 - \frac{y}{\Lambda} + s(x,y)\right) \tag{1}$$

is characterized by the stratification scale $\Lambda = \left| \frac{1}{\rho_0} \left(\frac{\partial \rho}{\partial y}\right) \right|^{-1}$, and s is the salinity perturbation (stratifying component), which includes the salt compression ratio. We consider the plane unsteady problem of the flow which occurs when there

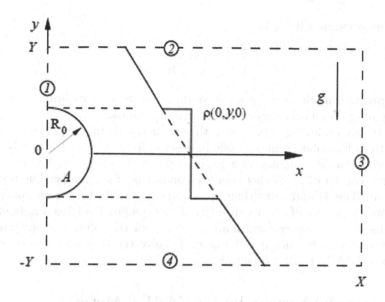

**Fig. 1.** Initial and boundary conditions for spot problem.

is a collapse of the region A of homogeneous fluid, surrounded by a stably and continuously density-stratified fluid.

The Navier-Stokes equations in the Boussinesq approximation describing the flow of this type (dynamic of collapsing spot) can be written as:

$$\frac{\partial \mathbf{v}}{\partial t} + (\mathbf{v} \cdot \nabla)\mathbf{v} = -\nabla p + \frac{1}{Re}\triangle\mathbf{v} + \frac{1}{Fr}s\frac{\mathbf{g}}{g},$$
$$\nabla \cdot \mathbf{v} = 0,$$
$$\frac{\partial s}{\partial t} + (\mathbf{v} \cdot \nabla)s = \frac{1}{Sc\cdot Re}\triangle s + \frac{v}{C},$$

(2)

where $\mathbf{v}$ is the velocity vector with components u, v, respectively, along the x and y axes of a Cartesian coordinate system selected as indicated in Fig. 1, $\rho$ is the density, p is the pressure minus hydrostatic one, s is the perturbation of salinity, the Reynolds number $Re = \rho_0 R_0^2 N/\mu$, the Froude number $Fr = R_0 N^2/g$, the Schmidt number $Sc = \mu/\rho_0 k_s$, $k_s$ is the diffusion coefficient of salts, $\mu$ is dynamic viscosity coefficient, $\mathbf{g} = (0, -g)$, g is acceleration of free fall, $\rho_0$ is the density on the level y=0, $C = \Lambda/R_0$ is scale ratio, $R_0$ is initial radius of the spot . We assume that the initial time t = 0 the system on the plane $\mathbf{R}^2$ is at rest, i.e.

$$u = 0, v = 0, (x, y) \in \mathbf{R}^2$$

the density of fluid at the spot A is

$$\rho = 1, (x, y) \in A$$

and outside of spot, i. e. in the area of $\mathbf{R}^2 \setminus A$ ,

$$\rho = 1 - \frac{y}{C} + s, (x, y) \in \mathbf{R}^2 \setminus A$$

the perturbation of salinity is

$$s = \begin{cases} \frac{y}{C}, (x,y) \in A \\ 0, (x,y) \in \mathbf{R}^2 \setminus A \end{cases}$$

As the pressure in the case of an incompressible fluid is determined with an accuracy of up to an arbitrary constant, without limiting the generality, we can select it to zero on level $y = 0$. Due to the symmetry about the plane $x = 0$, it is natural to seek a solution in only one half-plane, for example, if $x \geq 0$. Solution will search in the rectangular area $\{x, y : 0 \leq x \leq X, -Y \leq y \leq Y\}$. In the left boundary (line 1 in Fig. 1) this area are conditions of symmetry: The top (line 2), bottom (line 4) and right (line 3) boundaries should be chosen enough far away from the source of disturbance (from spots) so that setting any boundary conditions at these borders, which are necessary for the solution of the problem, does not significantly influence the flow. To solve the task we use new scheme based on CABARET [12–14] method.

## 2    Differential Scheme by CABARET Method

In [11] the authors used the CABARET method for studying the problem of spot dynamics in a fluid that is stably stratified be density. In contrast to SMIF, a new difference scheme referred to as CABARET is solved by direct calculation elliptic equations for pressure by Fast Direct Method [15]. The motion of the medium is described by the Navier-Stokes equations and the equation of salinity diffusion. The model includes also the divergence-free condition. The system of equations in dimensionless variables, where the characteristic linear size is equal to the spot radius, and the characteristic time is inversely proportional to the $Brent - V\ddot{a}is\ddot{a}l\ddot{a}$ frequency N:

$$\begin{cases} \frac{\partial u}{\partial t} + u\frac{\partial u}{\partial x} + v\frac{\partial u}{\partial y} + \frac{1}{\rho}\frac{\partial p}{\partial x} = \frac{1}{Re}\left(\frac{\partial^2 u}{\partial x^2} + \frac{\partial^2 u}{\partial y^2}\right), \\ \frac{\partial v}{\partial t} + u\frac{\partial v}{\partial x} + v\frac{\partial v}{\partial y} + \frac{1}{\rho}\frac{\partial p}{\partial y} = \frac{1}{Re}\left(\frac{\partial^2 v}{\partial x^2} + \frac{\partial^2 v}{\partial y^2}\right) - \frac{s}{Fr}, \\ \frac{\partial s}{\partial t} + u\frac{\partial s}{\partial x} + v\frac{\partial s}{\partial y} = \frac{1}{Sc\cdot Re}\left(\frac{\partial^2 s}{\partial x^2} + \frac{\partial^2 s}{\partial y^2}\right) + \frac{v}{C}, \\ \frac{\partial u}{\partial x} + \frac{\partial v}{\partial y} = 0. \end{cases} \tag{3}$$

Here the Reynolds and Froude numbers are defined as:

$$Re = \frac{\rho_0 R_0^2 N}{\mu}, Fr = \frac{R_0 N^2}{g}, Sc = \frac{\mu}{\rho_0 k_s}. \tag{4}$$

The initial conditions for the problem with a spot are as follows (Fig. 1); in a circle with a radius of $R_0$, a density of $\rho_0$ is set or unitary in a dimensionless form, and a linear distribution $\rho = 1 - y \cdot Fr$ is set around the circle, such that the densities are the same at the center of the circle $y = 0$. At time $t = 0$, the velocities are zero and the density is distributed as described above.

The system of equations (3) is reduced to a divergent form, using divergence-free condition:

$$\frac{\partial u}{\partial t} = -\left(\frac{\partial u^2}{\partial x} + \frac{\partial uv}{\partial y}\right) + \frac{1}{\rho}\frac{\partial p}{\partial x} + \frac{1}{Re}\left(\frac{\partial^2 u}{\partial x^2} + \frac{\partial^2 u}{\partial y^2}\right),$$

$$\frac{\partial v}{\partial t} = -\left(\frac{\partial uv}{\partial x} + \frac{\partial v^2}{\partial y}\right) + \frac{1}{\rho}\frac{\partial p}{\partial y} + \frac{1}{Re}\left(\frac{\partial^2 v}{\partial x^2} + \frac{\partial^2 v}{\partial y^2}\right) - \frac{s}{Fr}, \qquad (5)$$

$$\frac{\partial s}{\partial t} = -\left(\frac{\partial us}{\partial x} + \frac{\partial vs}{\partial y}\right) + \frac{1}{Sc \cdot Re}\left(\frac{\partial^2 s}{\partial x^2} + \frac{\partial^2 s}{\partial y^2}\right) + \frac{v}{C}.$$

The difference scheme for (5) can be written as follows:

$$\frac{s^{n+1/2} - s^{n-1/2}}{\tau} = -\nabla \cdot (s^n \mathbf{v}^n) + \frac{1}{Sc \cdot Re}\triangle s^{n-1/2} + \frac{v^{n-1/2}}{C},$$

$$\frac{\tilde{u} - u^{n-1/2}}{\tau} = -\nabla \cdot (u^n \mathbf{v}^n) + \frac{1}{Re}\triangle u^{n-1/2},$$

$$\frac{\tilde{v} - v^{n-1/2}}{\tau} = -\nabla \cdot (v^n \mathbf{v}^n) + \frac{1}{Re}\triangle v^{n-1/2} - \frac{s^{n-1/2}}{Fr}, \qquad (6)$$

$$\nabla \left(\frac{1}{1 - \frac{y}{C} + s^{n+1/2}}\nabla \delta p^{n+1/2}\right) = \frac{\nabla \cdot \tilde{\mathbf{v}}}{\tau},$$

$$\frac{\mathbf{v}^{n+1/2} - \tilde{\mathbf{v}}}{\tau} = -\frac{1}{1 - \frac{y}{C} + s^{n+1/2}}\nabla \delta p^{n+1/2},$$

where the concept of overpressure is introduced:

$$\delta p = p - p(y),$$

$$p(y) = \int_y^{y_0} \rho(y)g\,dy = p(y_0) - \rho_0 g\left(y - \frac{y^2}{2C}\right). \qquad (7)$$

The values with integer indices in (6) refer to the flux variables, and the variables with half-integer indices are the conservative variables. As we see, the fourth difference equation is a modified Poisson equation, where the operator takes into account density inhomogeneity. This equation is solved using the parallel conjugate gradient method with a preconditioner in the form of the usual Laplace operator with constant density. The direct solver for the inversion of the Laplace operator is obtained by using the Fourier transforms of the unknown pressure variables into two-fold decomposition and equating the corresponding component of the right-hand decomposition. Finally, the scheme ends by calculating the flow variables on the new time layer as:

$$\psi_S^{n+1} = 2\psi_C^{n+1/2} - \psi_{Sop}^n, \psi = \begin{pmatrix} s \\ u \\ v \end{pmatrix}, \qquad (8)$$

where S is the index of the face of the flux variable, C is the index of the adjacent cell, from where the flow goes in the direction of the face of S, Sop is the index of the face of the opposite face of S and belonging to cell C.

## 3   Conclusion

Our approximation assumes the division of the water column conventionally into "shallow water" and an area with stable stratification located below the first layer. In this approximation, when "stitching" the solution between the two models of the medium, by correctly taking into account the flow along the adjacent boundary of the two computational domains, we obtain a stable difference scheme. In this paper, we describe the difference schemes separately in the stable stratification approximation and in the shallow water approximation. A test case that will combine both models to find a solution will describe the decomposition of a mixed column of salt water into a seawater reservoir where there is initially a density stratification. Unfortunately, at the moment, it has not been possible to implement a numerical model to solve this problem. But despite this fact, it remains clear to us that such an approach will be effective in solving problems with the propagation of disturbances and correct allowance for internal waves. This will be the first approximation in solving the mixed problem with a stratified fluid, as opposed to the existing models where the entire region is split into Lagrangian-Eulerian cells with the solution of the problem with a free boundary.

# References

1. Turner, J.S.: Buoyancy Effects in Fluids. Cambridge University Press, Cambridge (1979)
2. Sawyer, J.S.: Environmental aerodynamics. Q. J. R. Meteorol. Soc. **104**(441), 818 (1978)
3. Gushchin, V.A.: The splitting method for problems of the dynamics of an inhomogeneous viscous incompressible fluid. USSR Comput. Math. Math. Phys. **21**(4), 190–204 (1981)
4. Barenblatt, G.I.: Dynamics of turbulent spots and intrusions in a stably stratified fluid. Izv. Atmos. Oceanogr. Phys. **14**, 139–145 (1978)
5. Gushchin, V., Matyushin, P.: Method SMIF for incompressible fluid flows modeling. In: Dimov, I., Faragó, I., Vulkov, L. (eds.) NAA 2012. LNCS, vol. 8236, pp. 311–318. Springer, Heidelberg (2013). https://doi.org/10.1007/978-3-642-41515-9_34
6. Matyushin, P., Gushchin, V.: Direct numerical simulation of the sea flows around blunt bodies. In: AIP Conference Proceedings, vol. 1690, p. 030005. AIP Publishing LLC (2015)
7. Belotserkovskii, O.M., Gushchin, V.A., Shchennikov, V.V.: Use of the splitting method to solve problems of the dynamics of a viscous incompressible fluid. USSR Comput. Math. Math. Phys. **15**(1), 190–200 (1975)
8. Gushchin, V.A.: On a one family of quasimonotone finite-difference schemes of the second order of approximation. Matematicheskoe modelirovanie **28**(2), 6–18 (2016). (in Russian)
9. Gushchin, V.A., Mitkin, V.V., Rozhdestvenskaya, T.I., Chashechkin, Yu.D.: Numerical and experimental study of the fine structure of the flow of a stratified fluid near a circular cylinder. Appl. Mech. Tech. Phys. **48**(1), 43–54 (2007). (in Russian)
10. Gushchin, V.A., Kopysov, N.A.: The dynamics of a spherical mixing zone in a stratified fluid and its acoustic radiation. Comput. Math. Math. Phys. **31**(6), 51–60 (1991). (in Russian)
11. Gushchin, V.A., Kondakov, V.G.: Mathematical modeling of free-surface flows using multiprocessor computing systems. J. Phys. Conf. Ser. **1392**, 1–6 (2019). https://doi.org/10.1088/1742-6596/1392/1/012042

12. Goloviznin, V.M., Samarskii, A.A.: Some characteristics of finite difference scheme "cabaret". Matematicheskoe modelirovanie **10**(1), 101–116 (1998). (In Russian)
13. Goloviznin, V.M., Samarskii, A.A.: Finite difference approximation of convective transport equation with space splitting time derivative. Matematicheskoe modelirovanie **10**(1), 86–100 (1998). (In Russian)
14. Goloviznin, V.M., Karabasov, S.A., Kondakov, V.G.: Generalization of the CABARET scheme to two-dimensional orthogonal computational grids. Math. Models Comput. Simul. **6**(1), 56–79 (2014). https://doi.org/10.1134/S2070048214010050
15. Kuznetsov, Y.A., Rossi, T.: Fast direct method for solving algebraic systems with separable symmetric band matrices. East-West J. Numer. Math. **4**(1), 53–68 (1996)

# Behavior and Scalability of the Regional Climate Model RegCM4 on High Performance Computing Platforms

Vladimir Ivanov[(⊠)] and Georgi Gadzhev

National Institute of Geophysics, Geodesy and Geography – Bulgarian Academy
of Sciences, Acad. G. Bonchev Street bl. 3, 1113 Sofia, Bulgaria
vivanov@geophys.bas.bg

**Abstract.** The RegCM is a regional climate model used in many studies. There are simulation runs in different domains, time periods, and regions in the world on all continents. The research works in our group are related to the historical and future climate, and its influence on the human sensation over Southeast Europe. We used the model versions 4.4 and 4.7. The main model components are the initial and boundary condition module, the physics processes parametrization module, and the dynamical core. Concerning the last one, we used the default one – the hydrostatic option corresponding to the MM5 model dynamical core. We run simulations with different combinations of parametrization schemes on the Bulgarian supercomputer Avitohol. The newer versions of the model have an additional option for using a non-hydrostatic dynamical core. The running of model simulations with different input configurations depends highly on the available computing resources. Several main factors influence the simulation times and storage requirements. They could vary much depending on the particular set of input parameters, domain area, land cover, processing cores characteristics, and their number in parallel processing simulations. The objective of that study is to analyse the RegCM model performance with hydrostatic core, and non–hydrostatic core, on the High–Performance Computing platform Avitohol.

**Keywords:** Regional climate simulation · RegCM4 · Scalability · High performance computing

## 1 Introduction

The regional climate dynamical and statistical downscaling is used to provide climate information with higher resolution than the one available from the global climate models. The Coordinated Regional Climate Downscaling Experiment [1] is a global initiative of the World Climate Research Program with communities from different continents for coordinated effort to assess and compare various regional climate downscaling techniques. One of the widely used regional climate models by the participants in CORDEX is the Regional Climate Model

© Springer Nature Switzerland AG 2022
I. Lirkov and S. Margenov (Eds.): LSSC 2021, LNCS 13127, pp. 124–131, 2022.
https://doi.org/10.1007/978-3-030-97549-4_14

[2]. Our research group uses that model to study the regional climate focusing on Bulgaria and adjacent territories, parts of the simulation domains of interest for the former project objectives. As our region is not big enough to study with global climate models, we perform high-resolution regional climate simulations, applying a dynamical downscaling technique with initial and boundary conditions as in the [5]. We used two spatial resolutions in our studies, using the Bulgarian Supercomputer "Avitohol". We studied the temperature and precipitation biases of the RegCM model version 4.4 over the Balkan Peninsula with 10 km resolution [3]. Other studies encompass a lot bigger area with 20 km resolution, covering Southeastern Europe and parts of Asia dealing with the biometeorological conditions [4] and the using of projected climate scenarios for future periods [5]. The ongoing research work involving the RegCM simulations with different spatial and time steps put the question about the model's scalability. It is well known that the increase of the spatial resolution and decreasing the time step keeping a constant domain size, leads to longer computation time. The average time for a month simulation time for the Balkan Peninsula on the Avitohol is 6 h with 16 processors [6]. RegCM simulations for the Southeast Europe domain on the 128 processors on the Avitohol show that the average run time for a month is 0.22 h. The objective of that study is to analyze the RegCM model performance for the other factors determining the scalability - the hydrostatic core, the non-hydrostatic core, and the different model physics configurations, on the High-Performance Computing platform Avitohol.

## 2    Model Configurations

For our testing purposes, we use the model RegCM version 4.7 [2]. We will use the abbreviation RegCM4 for brevity. The spatial domain includes Southeastern Europe, parts of Italy, as well as Asia Minor peninsula. We divide the study objective on two tasks. The first one is to test the dependence of the run time from the physics parameterizations schemes for hydrostatic model configurations. The simulation runs are from 29.11.1999 to 29.12.1999 with a horizontal resolution of 20 km, time step 20 s, and 18 vertical levels. The period is chosen from ERA5 reanalysis data [7], and from E-OBS observational data [8] because of the many situations with convective as well as non-convective precipitation events. That would give us a more complete picture of the model parameterization schemes performance. The second task is to test the model scalability of the hydrostatic and non-hydrostatic dynamical cores with the slowest and the fastest simulations founded from the results of the first task for 8, 16, 32, 64, and 128 processing cores. The simulations have 23 vertical levels, the time step is 20 s, and the test period is from 29.11.1999 to 14.12.1999. The horizontal resolution for the non-hydrostatic mode is 10 km and 20 km for the hydrostatic one. The RegCM has options for a hydrostatic or a non-hydrostatic one [9]. The difference between them is in the use of the vertical momentum equation. In non-hydrostatic mode, the model atmosphere is not in a hydrostatic balance, and the full vertical momentum equation apply. In that case, the model accounts for processes with notable vertical acceleration at small scales from hundreds of meters to several kilometers. The model cannot resolve the physics processes explicitly because the model resolution is too small. It is partly overcome with their simplified representation in grid size resolution by parameterization schemes. We vary the combination of these schemes for planetary-boundary layer processes (PBL scheme), microphysics atmospheric processes (MP scheme), and convection processes (CP scheme). The RegCM4 model configurations use the Biosphere–Atmosphere Transfer Scheme [10] for modelling the land-surface processes [11] and the Community Climate Model version 3 [12] for simulation of the radiative transfer processes. The tested model physics combinations and their abbreviations are given in the Fig. 1. The hydrostatic version of the model is tested with the physical parameterization schemes from the experiments in [3], and three additional ones – PBL frictionless mode, WSM5 scheme [13] for microphysical processes, and the MM5 Shallow cumulus convection scheme [15].

The calculations were implemented on the Supercomputer System "Avitohol" at the Institute of Information and Communication Technologies at the Bulgarian Academy of Sciences (IICT-BAS). The simulations for the selected domain were organized in different jobs. The tests for the model physics set-ups are run for one month. The test period for comparing the hydrostatic and non–hydrostatic cores is half for 15 days [16].

## 3   Results

Varying the PBL and microphysical schemes for each of the convective parameterizations implies the results shown in Fig. 2. The model simulations with the Grell with AS–closure show that the slowest combination is the one with UW PBL scheme and NG moisture scheme and the fastest is with frictionless PBL and SUBEX moisture scheme. The configurations with NG moisture scheme – frictionless PBL scheme, and NG moisture scheme – UW scheme are relatively slower than the other combinations. The slowest combination for the Grell–FS is the UW PBL scheme – WSM5 moisture scheme. The fastest configuration is frictionless PBL – SUBEX moisture scheme. Most of the combinations fall below 1500 s simulation time and generally have more similar run times than the Grell with AS cases. Four of the model configurations with MM5 shallow cumulus scheme have notably higher simulation times among others: frictionless PBL scheme—NG moisture scheme, Holtslag PBL scheme – NG moisture scheme, UW PBL scheme—NG moisture scheme, and UW PBL scheme – WSM5 moisture scheme. The other ones have the biggest similarity in simulation times compared to the analogic combinations for the other cumulus convection schemes.

Using the Emanuel convective scheme implies more considerable heterogeneity in the differences of the simulation run times. Although, the simulation times patterns of the slowest model configurations are the same. The case studies with the parameterization scheme for cumulus convection of Tiedtke have

| PBL | Moisture | Cumulus Convection | Notation | PBL | Moisture | Cumulus Convection | Notation |
|---|---|---|---|---|---|---|---|
| Frictionless | SUBEX | Grell AS | Hr0121 | Holtslag | Nogherotto/ Tompkins | Tiedtke | Hr1255 |
| Frictionless | SUBEX | Grell FC | Hr0122 | Holtslag | Nogherotto/ Tompkins | Kain-Fritsch | Hr1266 |
| Frictionless | SUBEX | Emanuel | Hr0144 | Holtslag | Nogherotto/ Tompkins | MM5 Shallow | Hr12n1 |
| Frictionless | SUBEX | Tiedtke | Hr0155 | Holtslag | WSM5 | Grell AS | Hr1321 |
| Frictionless | SUBEX | Kain-Fritsch | Hr0166 | Holtslag | WSM5 | Grell FC | Hr1322 |
| Frictionless | SUBEX | MM5 Shallow | Hr01n1 | Holtslag | WSM5 | Emanuel | Hr1344 |
| Frictionless | Nogherotto/ Tompkins | Grell AS | Hr0221 | Holtslag | WSM5 | Tiedtke | Hr1355 |
| Frictionless | Nogherotto/ Tompkins | Grell FC | Hr0222 | Holtslag | WSM5 | Kain-Fritsch | Hr1366 |
| Frictionless | Nogherotto/ Tompkins | Emanuel | Hr0244 | Holtslag | WSM5 | MM5 Shallow | Hr13n1 |
| Frictionless | Nogherotto/ Tompkins | Tiedtke | Hr0255 | UW | SUBEX | Grell AS | Hr2121 |
| Frictionless | Nogherotto/ Tompkins | Kain-Fritsch | Hr0266 | UW | SUBEX | Grell FC | Hr2122 |
| Frictionless | Nogherotto/ Tompkins | MM5 Shallow | Hr02n1 | UW | SUBEX | Emanuel | Hr2144 |
| Frictionless | WSM5 | Grell AS | Hr0321 | UW | SUBEX | Tiedtke | Hr2155 |
| Frictionless | WSM5 | Grell FC | Hr0322 | UW | SUBEX | Kain-Fritsch | Hr2166 |
| Frictionless | WSM5 | Emanuel | Hr0344 | UW | SUBEX | MM5 Shallow | Hr21n1 |
| Frictionless | WSM5 | Tiedtke | Hr0355 | UW | Nogherotto/ Tompkins | Grell AS | Hr2221 |
| Frictionless | WSM5 | Kain-Fritsch | Hr0366 | UW | Nogherotto/ Tompkins | Grell FC | Hr2222 |
| Frictionless | WSM5 | MM5 Shallow | Hr03n1 | UW | Nogherotto/ Tompkins | Emanuel | Hr2244 |
| Holtslag | SUBEX | Grell AS | Hr1121 | UW | Nogherotto/ Tompkins | Tiedtke | Hr2255 |
| Holtslag | SUBEX | Grell FC | Hr1122 | UW | Nogherotto/ Tompkins | Kain-Fritsch | Hr2266 |
| Holtslag | SUBEX | Emanuel | Hr1144 | UW | Nogherotto/ Tompkins | MM5 Shallow | Hr22n1 |
| Holtslag | SUBEX | Tiedtke | Hr1155 | UW | WSM5 | Grell AS | Hr2321 |
| Holtslag | SUBEX | Kain-Fritsch | Hr1166 | UW | WSM5 | Grell FC | Hr2322 |
| Holtslag | SUBEX | MM5 Shallow | Hr11n1 | UW | WSM5 | Emanuel | Hr2344 |
| Holtslag | Nogherotto/ Tompkins | Grell AS | Hr1221 | UW | WSM5 | Tiedtke | Hr2355 |
| Holtslag | Nogherotto/ Tompkins | Grell FC | Hr1222 | UW | WSM5 | Kain-Fritsch | Hr2366 |
| Holtslag | Nogherotto/ Tompkins | Emanuel | Hr1244 | UW | WSM5 | MM5 Shallow | Hr23n1 |

**Fig. 1.** Simulation cases notation for the model configurations.

similar run times distribution as in the Emanuel cases. The case studies with Kain–Fritsch cumulus convection scheme [14] have run times distribution more similar to the Grell scheme simulations. They are generally faster than the model configurations with MM5 Shallow, Emanuel, and Tiedtke schemes for cumulus

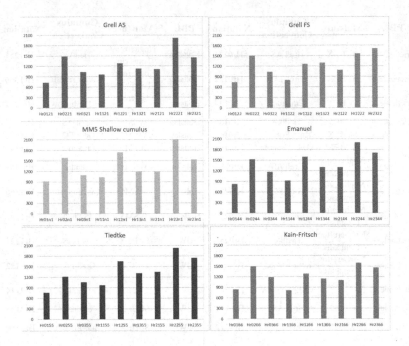

**Fig. 2.** Model simulation times for each of the convective parameterization schemes.

convection. Again, the model set-ups frictionless PBL scheme—NG moisture scheme, Holtslag PBL scheme—NG moisture scheme, UW PBL scheme—NG moisture scheme, and UW PBL scheme - WSM5 moisture scheme are the fastest ones. Generally, the slower model configuration is Hr22n1 with UW PBL scheme, and the fastest one is the Hr0121.

The variation of the all parameterization schemes over the specific moisture one reveals the differences of the model configuration from another point of view (left column of plots on the Fig. 3). The most obvious feature is the relatively faster configurations with the SUBEX moisture scheme with simulations times. It appears also, that the NG moisture scheme makes the simulation run times the biggest. The results of the model set-up variation of the PBL parameterization scheme are given in the right column of plots on the Fig. 3. The fastest simulations have frictionless PBL scheme, and the slowest ones – UW PBL scheme. The simulations with SUBEX moisture scheme for either one of the PBL groups of cases are the fastest, and the ones with the NG moisture scheme – the slowest.

The runs for the scalability comparison between the hydrostatic and non-hydrostatic cores (Fig. 4) show that the simulation time decreases linearly about several times with increasing the number of processors from 8 to 128 cores. Doubling the number of processors leads to increasing the simulation time about twice in the hydrostatic model configuration. The biggest simulation time increasing is when the number of processors changes from 32 to 64 (68%) for the Hr0121 and from 64 to 128 for the Hr22n1 (74%). The maximum run time

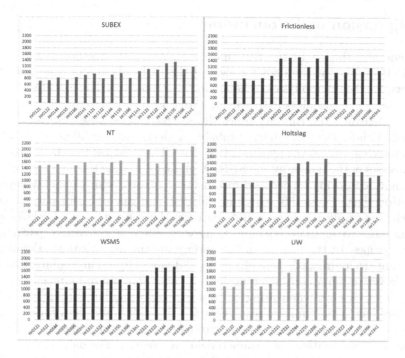

**Fig. 3.** Model simulation times for each of the microphysical parameterization schemes (left plots) and planetary boundary layer parameterization schemes (right plots).

simulation increasing for the non-hydrostatic model configurations is for core number changes from 64 to 128, and these rates of increase are lower than the corresponding ones in the hydrostatic cases. Contrary to the former set-ups, the maximum simulation time increasing is for the Hr0121 is 56%, and for the Hr22n1 is 53%. Generally, the increasing simulation times of the hydrostatic model runs are bigger than for the non-hydrostatic ones.

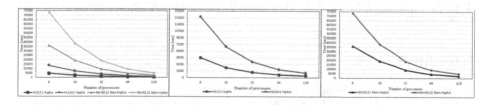

**Fig. 4.** Simulation times of the fastest and slowest schemes for the model with hydrostatic and non-hydrostatic options.

# 4    Discussion and Conclusion

The general variation in the run times for physics parameterizations is most notable if we look at the ones for each of the PBL schemes on the first place, and between the microphysical schemes in the second one. The SUBEX moisture scheme and the frictionless PBL scheme with each one of the cumulus convection schemes decreases the running times. It is expected, because the treatment of the PBL with frictionless scheme probably neglects some or all of the turbulent fluxes, which in turn saves up simulation time. The treatment of the microphysical processes in the model by the SUBEX moisture scheme is simpler, expressing in the not accounting for ice and mixed water phases, which possibly saves up simulation time. These differences in the PBL and microphysics parameterization schemes imply the biggest differences in the run times (50%–100%) when we compare two or more configurations with different convective parameterization schemes. In the process of the performing the numerical simulations, we found that the hydrostatic core of the RegCM model in 10 km resolution became unstable very often and in different periods. Therefore, we recommend using the non-hydrostatic core when using such or especially smaller spatial resolutions. These are only a preliminary results for the scalability and performance of the model in HPC—environment. Because of that, there is not a model verification for different combinations of parameterization schemes. The model performance depends on the chosen domain, time step, and resolution.

**Acknowledgments.** This work has been accomplished thanks to the computational capabilities, created in the framework of the MES Grant No. D01–221/03.12.2018 for NCDSC—part of the Bulgarian National Roadmap on RIs.

This work has been carried out in the framework of the National Science Program "Environmental Protection and Reduction of Risks of Adverse Events and Natural Disasters", approved by the Resolution of the Council of Ministers №577/17.08.2018 and supported by the Ministry of Education and Science (MES) of Bulgaria (Agreement №D01-363/17.12.2020).

Deep gratitude to the organizations and institutes (ICTP, ECMWF, ECA&D, Unidata, Copernicus Climate Data Store and all others), which provide free of charge software and data. Without their innovative data services and tools this study would not be possible.

# References

1. https://cordex.org/about/. Accessed 20 Jan 2021
2. Giorgi, F., et al.: RegCM4: model description and preliminary tests over multiple CORDEX domains. Clim. Res. **52**, 7–29 (2012). https://doi.org/10.3354/cr01018
3. Gadzhev, G., Ivanov, V., Ganev, K., Chervenkov, H.: TVRegCM numerical simulations - preliminary results. In: Lirkov, I., Margenov, S. (eds.) LSSC 2017. LNCS, vol. 10665, pp. 266–274. Springer, Cham (2018). https://doi.org/10.1007/978-3-319-73441-5_28

4. Ivanov, V., Valcheva, R., Gadzhev, G.: HPC simulations of the extreme thermal conditions in the Balkan region with RegCM4. In: Dimov, I., Fidanova, S. (eds.) HPC 2019. SCI, vol. 902, pp. 309–324. Springer, Cham (2021). https://doi.org/10.1007/978-3-030-55347-0_27

5. Gadzhev, G., Ivanov, V., Valcheva, R., Ganev, K., Chervenkov, H.: HPC simulations of the present and projected future climate of the Balkan region. In: Dimov, I., Fidanova, S. (eds.) HPC 2019. SCI, vol. 902, pp. 234–248. Springer, Cham (2021). https://doi.org/10.1007/978-3-030-55347-0_20

6. Gadzhev, G., et al.: Climate applications in a virtual research environment platform. Scalable Comput. **19**(2), 107–118 (2018). https://doi.org/10.12694/scpe.v19i2.1347

7. Muñoz Sabater, J.: ERA5-Land hourly data from 1981 to present. Copernicus Climate Change Service (C3S) Climate Data Store (CDS) (2019). https://doi.org/10.24381/cds.e2161bac. Accessed 15 Feb 2021

8. Cornes, R., van der Schrier, G., van den Besselaar, E.J.M., Jones, P.D.: An ensemble version of the E-OBS temperature and precipitation datasets. J. Geophys. Res. Atmos. **123**, 9391–9409 (2018). https://doi.org/10.1029/2017JD028200

9. Grell, A., Dudhia, J., Stauffer, D.R.: Description of the fifth generation Penn State/NCAR Mesoscale Model (MM5). Technical report TN-398+STR, NCAR, Boulder, Colorado, p. 121 (1994)

10. Dickinson, R.E., Henderson-Sellers, A., Kennedy, P.J.: Biosphere-atmosphere transfer scheme (BATS) version 1e as coupled to the NCAR community climate model. Technical report, National Center for Atmospheric Research (1993)

11. Dickinson, R.E., Kennedy, P.J., Henderson-Sellers, A., Wilson, M.: Biosphere-atmosphere transfer scheme (BATS) for the NCAR community climate model. Technical report NCARE/TN-275+STR, National Center for Atmospheric Research (1986)

12. Kiehl, J.T., et al.: Description of the NCAR community climate model (CCM3), Technical report NCAR/TN-420+STR, National Center for Atmospheric Research (1996)

13. Hong, S., Dudhia, J., Chen, S.: A revised approach to ice microphysical processes for the bulk parameterization of clouds and precipitation. Mon. Weather Rev. **132**(1), 103–120 (2004). https://doi.org/10.1175/1520-0493(2004)132⟨0103:ARATIM⟩2.0.CO;2

14. Kain, J.S., Kain, J.: The Kain - Fritsch convective parameterization: an update. J. Appl. Meteorol. **43**(1), 170–181 (2004). https://doi.org/10.1175/1520-0450(2004)043⟨0170:TKCPAU⟩2.0.CO;2

15. Bretherton, C.S., McCaa, J.R., Grenier, H.: A new parameterization for shallow cumulus convection and its application to marine subtropical cloud-topped boundary layers. part I: description and 1D results. Mon. Weather Rev. **132**(4), 864–882 (2004). https://doi.org/10.1175/1520-0493(2004)132⟨0864:ANPFSC⟩2.0.CO;2

16. Atanassov, E., Gurov, T., Karaivanova, A., Ivanovska, S., Durchova, M., Dimitrov, D.: On the parallelization approaches for intel MIC architecture. AIP Conf. Proc. **1773**, 070001 (2016). https://doi.org/10.1063/1.4964983

# Quantum Effects on 1/2[111] Edge Dislocation Motion in Hydrogen-Charged Fe from Ring-Polymer Molecular Dynamics

Ivaylo Katzarov[1,2](✉) ⓘ, Nevena Ilieva[3] ⓘ, and Ludmil Drenchev[1] ⓘ

[1] Institute of Metal Science, Equipment and Technologies with Hydro- and Aerodynamics Centre "Acad. A. Balevski" at the Bulgarian Academy of Sciences, 67, Shipchenski Prohod Street, 1574 Sofia, Bulgaria
ivaylo.katsarov@kcl.ac.uk
[2] King's College London, Strand WC2R 2LS, UK
[3] Institute of Information and Communication Technologies at the Bulgarian Academy of Sciences, Acad. G. Bonchev Street Block 25A, 1113 Sofia, Bulgaria

**Abstract.** Hydrogen influenced change of dislocation mobility is a possible cause of hydrogen embrittlement (HE) in metals and alloys. A comprehensive understanding of HE requires a more detailed description of dislocation motion in combination with the diffusion and trapping of H atoms. A serious obstacle towards the atomistic modelling of a H interstitial in Fis associated with the role nuclear quantum effects (NQEs) might play evenat room temperatures, due to the small mass of the proton. Standard molecular dynamics (MD) implementations offer a rather poor approximation for such investigations as the nuclei are considered as classical particles. Instead, we reach for *Ring-polymer MD (RPMD)*, the current state-of-the-art method to include NQEs in the calculations, which generates a quantum-mechanical ensemble of interacting particles by using MD in an extended phase space.

Here we report RPMD simulations of quantum effects on 1/2[111] edge dislocation motion in H charged Fe. The simulations results indicate that H atoms are more strongly confined to dislocation core and longer relaxation time is necessary for the edge dislocation to break away from the H atmosphere. The stronger interaction between dislocation and H atoms trapped in the core, resulting from NQEs, leads to formation of jogs along the dislocation line which reduce edge dislocation mobility in H charged Fe.

**Keywords:** Hydrogen embrittlement · Zero-point energy · Tunneling · Molecular dynamics · i-PI · EAM

## 1 Introduction

Hydrogen embrittlement and the mechanism by which it operates is a controversial scientific topic that is still under debate. One of the primary disputes

I. Lirkov and S. Margenov (Eds.): LSSC 2021, LNCS 13127, pp. 132–139, 2022.
https://doi.org/10.1007/978-3-030-97549-4_15

is the role of plasticity in hydrogen embrittlement. As the name suggests, on macroscopic scales hydrogen embrittlement results in a loss in ductility. However, there is a notable amount of plasticity associated with these failures, though not always evident outside of the microscale. The current debate is focused on the importance of this plasticity: whether it is crucial to the embrittlement process or simply a minimal and secondary result of the hydrogen-induced failure. The concept of hydrogen enhanced localized plasticity (HELP) [1,2] assumes that sufficient concentrations of hydrogen locally assists the deformation processes leading to fracture which is macroscopically brittle in appearance and behavior. Hydrogen is strongly bound to dislocation cores due to formation of trap sites at the core [3], as well as the attraction to the elastic field surrounding the dislocation. This results in the formation of an atmosphere of hydrogen [4]. In a specific temperature and strain-rate range [4,5], the hydrogen atmosphere leads to acceleration of dislocation motion in the presence of hydrogen. Two reasons are considered for this increased dislocation mobility. The first applies more to edge dislocations: the hydrogen atmosphere alters the stress field of the dislocation leading to a shielding effect and reduction of the interaction energy between dislocations and obstacles [6]. This reduction increases the dislocation mobility by allowing dislocation motion at lower stresses. The second would apply to all dislocations: hydrogen segregated to dislocations lowers the kink-pair formation energy leading to an enhancement in the kink-pair nucleation rates [7]. Higher nucleation rates would result in increased dislocation mobility when kink-pair formation is rate-limiting for dislocation motion. A comprehensive understanding of mechanisms of HE requires a detailed description of dislocation motion in combination with the diffusion and trapping of H atoms. The interactions of hydrogen with crystal defects such as dislocations in $\alpha$-Fe and their consequences are fundamentally less well understood than diffusion of hydrogen in the perfect crystal lattice, despite generally dominating the influence of hydrogen in metals. Taking into account difficulties in experimental testing and the numerous uncertainties concerning the HE and dislocation behaviour, one may see the reason behind using computer simulation methods to tackle physical problems related to this phenomena occurring at a space- and time-scales that are not available for experimental assessment. Here, we apply atomistic simulations, in particular molecular dynamics, to characterize 1/2[111] edge dislocation motion mechanisms in H charged Fe. A serious obstacle towards the atomistic modelling of a H interstitial in Fe is associated with the role of nuclear quantum effects might play even at room temperatures, due to the small mass of the proton. Standard molecular dynamics implementations are deemed inappropriate for such investigations as the nuclei are considered as classical particles, which for light nuclei such as H is often a very poor approximation. Instead, we employ Ring-polymer MD, the current state-of-the-art method to include NQEs in the calculations, which generates a quantum-mechanical ensemble of interacting particles by using MD in an extended phase space. In this work we report simulations of H nuclear quantum effects on 1/2[111] edge dislocation motion and escape from a H cloud in H charged Fe. We simulate the H effect on dislocation mobility twice, using classical Molecular Dynamics (MD) and then performing RPMD simulations, and compare the results to assess the differences. The RPMD simulations results indicate that H atoms are strongly confined to dislocation core and longer relaxation time is necessary for the edge dislocation to break away from the H atmosphere. The stronger interaction between dislocation and H atoms trapped in the core, resulting from NQEs, leads to formation of jogs along the dislocation line which reduce edge dislocation mobility in H charged Fe.

## 2   Methods

### 2.1   Ring-Polymer Molecular Dynamics

Classical MD simulations can be used to calculate a wide range of dynamical properties. However, classical MD neglects quantum-mechanical zero-point energy (ZPE) and tunneling effects in the atomic motion. In systems containing light atoms, these effects must be included to obtain the correct quantitative, and sometimes even qualitative, behavior. For example, tunneling through the reaction barrier can easily enhance the rate of a proton transfer reaction at room temperature by several orders of magnitude. One of the most successful approaches that enables the quantization of MD trajectories of complex systems containing many interacting particles is the Ring-polymer molecular dynamics [8,9]. RPMD makes use of the imaginary-time path integral formalism [10], which exploits the exact equilibrium mapping between a quantum-mechanical particle and a classical ring polymer. Using his path-integral method [10] Feynman proves that the partition function of a quantum system with a Hamiltonian of the form

$$\hat{H} = \frac{\hat{p}^2}{2m} + V(\hat{q}) \tag{1}$$

can be approximated by a purely classical partition function for a cyclic chain of n beads coupled by harmonic springs, each bead being acted by the true potential V

$$Z_n(\beta) = \frac{1}{(2\pi\hbar)^n} \int d^n\mathbf{p} \int d^n\mathbf{q} \exp\left(-\beta_n H_n(\mathbf{p}, \mathbf{q})\right) \tag{2}$$

where

$$H_n(\mathbf{p}, \mathbf{q}) = \sum_{i=1}^{n} \left[ \frac{p_i^2}{2m} + \frac{1}{2} m\omega_n^2 (q_i - q_{i-1})^2 + V(q_i) \right] \tag{3}$$

with $\beta = \frac{1}{k_B T}$, $\beta_n = \frac{\beta}{n}$ and $\omega_n^2 = \frac{1}{(\beta_n \hbar)}$. This expression becomes exact as the number of beads is infinite. The implication of this astonishing isomorphism between a single quantum particle and a cyclic chain of n classical pseudo-particles is that the techniques that have been developed for treating the statistical mechanics of classical system can be applied to quantum systems. The classical isomorphism allows us, in principle, to introduce a molecular dynamics scheme, which uses the classical dynamics generated by the Hamiltonian $H_n(\mathbf{p}, \mathbf{q})$, yielding the following equations of motion:

$$\dot{\mathbf{p}} = -\frac{\partial H_n(\mathbf{p}, \mathbf{q})}{\partial \mathbf{q}} \qquad \dot{\mathbf{q}} = \frac{\partial H_n(\mathbf{p}, \mathbf{q})}{\partial \mathbf{p}} \tag{4}$$

RPMD is formulated as a classical MD in the extended ring-polymer phase space [8]. The isomorphic Feynman beads of a quantum particle evolve in real time according to the classical dynamics generated by the classical Hamiltonian of a ring polymer consisting of n copies of the system connected by harmonic springs.

## 2.2  Molecular Dynamics Simulations of Dislocation Motion

MD simulations of the 1/2[111] edge dislocation mobility in H changed bcc Fe were performed as follows. The bcc Fe crystal was constructed of a rectangular simulation cell oriented as $X = [111]$, $Y = [1\bar{1}0]$ and $Z = [11\bar{2}]$ with dimensions $L_x = 7.43$ nm, $L_y = 6.47$ nm and $L_z = 7.00$ nm containing 28320 atoms. The simulation cell is periodic along the Z direction with a relatively large 7.0 nm periodic length. The periodicity along the Z direction effectively enforces a plane-strain boundary condition along the dislocation line direction. The atoms located within the atomic potential cut-off radius of the X and Y surfaces (the so-called inert atoms) are kept fixed during the simulations. In order to study the glide of the planar core configurations under the effect of externally applied shear stress, we started with simulation blocks containing the fully relaxed core structure of a 1/2[111] edge dislocation. The dislocation was introduced in the perfect lattice by displacing both active and inert atoms from their perfect lattice positions according to the corresponding anisotropic elastic displacement field [17]. A shear stress $\tau_{xy} = 500$ MPa was applied in the [111] direction in such a way that the dislocation was set to glide on a $(1\bar{1}0)$ plane. In practice, the shear stress is imposed by applying the appropriate homogeneous shear strain, which is evaluated with anisotropic elasticity theory. This strain was superimposed on the dislocation displacement field for all the atoms. The MD simulations are performed in the NVT ensemble with a Nosé–Hoover thermostat [11] and velocity–Verlet algorithm [12] with the integration time step being 0.5 fs. During the simulation we use the code OVITO [13] to analyze the particle coordinates and determine the dislocation core position. The applied shear stress which drives the dislocation is applied in a single pulse. After applying the stress, the simulations are run at constant stress and there is no strain rate over the simulation time. To perform RPMD simulations of the dislocation – H atmosphere interaction we used the same procedures as in the classical MD simulations. Both RPMD and classical MD simulations were performed using the open-source code i-PI [14] which contains a comprehensive array of Path-integral MD techniques. In order to run massive parallel calculations we use socket internet mode for interfacing i-PI to an embedded-atom method (EAM) potential previously developed to describe the interatomic interactions in Fe-H system [15]. We employ the same

EAM potential and i-PI code with a single pseudo-particle to run classical MD simulations. In order to achieve convergence of quantum RPMD simulations we employ 16 replicas of the entire atom system, in which each atom is represented by its 16 clones connected into a "polymer ring" (a closed Feynman path) by elastic springs. Since RPMD method comes at a great computational cost compared to a classical MD simulation with the same interatomic potential, we use a relatively small simulation block consisting of 23100 active and 5220 inert atoms. Starting from the initial simulation cell without H, H atoms are introduced by randomly inserting them into tetrahedral sites in the region of dislocation core (Fig. 1).

We performed two series of RPMD and MD simulations at a temperature of 300 K in which the amounts of H per unit length along the dislocation line direction $N_H/L_z$ are respectively 5.71 and 12.75 nm$^{-1}$.

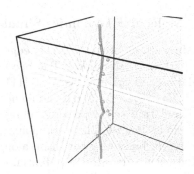

## 3   Results and Discussion

Under the application of the shear stress $\tau_{xy}$, the dislocations are driven to move on the $(1\hat{1}0)$ plane. Due to the great computational cost of RPMD compared to classical MD simulation, we impose relatively high shear stress to induce a fast dislocation reaction.

**Fig. 1.** Edge dislocation and trapped H atoms before the application of a driving force.

Both RPMD and MD simulations show that dislocation velocity induced by the applied driving force in pure Fe is 960 m/s. This velocity is high compared to H diffusivity in Fe and there is a small probability for H atoms to travel along with the dislocation. Hence, H cloud cannot keep up with the moving dislocation and after propagating for a short period in the H atmosphere the dislocation breaks away (Fig. 2). During the initial stage dislocation is surrounded by a hydrogen cloud and H atoms exert a substantial resistance to the dislocation mobility due to the solute drag.

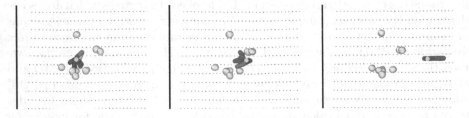

**Fig. 2.** The motion of the dislocation illustrated by a series of snapshots (RPMD simulations).

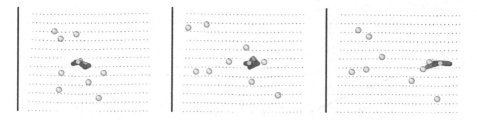

**Fig. 3.** The motion of the dislocation illustrated by a series of snapshots (MD simulations).

The dislocation velocities are nearly constant and several times smaller than the velocities of the dislocation in pure Fe. The velocity is governed by a balance between the applied driving force and the retarding force of the H cloud. The substantially lower dislocation velocity in the presence of H suggests that the dislocation motion is largely controlled by the H kinetics. Path-integral calculations show that the H binding energy at an edge dislocation is enhanced by the quantum effects [16]. The state of the H atom in the core is energetically stable and the diffusion barriers between the nearest binding sites increase. This results in a very low diffusivity compared to that in the regular lattice region. H atoms less readily travel along with the dislocation and exert a higher resistance to dislocation motion. Dislocation velocities in presence of hydrogen clouds with H per unit dislocation length 5.71 and 12.75 $nm^{-1}$ determined by RPMD are 293 and 212 m/s, respectively. The results of classical MD simulations show that the state of the H atom in the core is energetically unstable and the atom can easier leave the core for the nearest binding sites (Fig. 3). The lower classical H binding energies lead to a lower solute drag compared to the one determined by RPMD. In the classical MD simulations dislocation velocities in presence of hydrogen clouds with $H/Lz$ 5.70 and 12.75 $nm^{-1}$ are 538 and 434 m/s, respectively. The dislocation velocities before breaking away from H clouds scale inversely with the H concentration. After escaping from the H atmosphere dislocations accelerate and then move at higher velocity (1100 m/s determined by RPMD and 600 m/s calculated by classical MD).

The nuclear quantum effects arrising from H clouds also alter dislocation glide mechanism. RPMD shows that the interaction of dislocations with H atoms leads to a cross-slip and formation of jogs [17] (Fig. 4a). Jogs are steps of atomic dimension in the dislocation line which lay out of the glide plane of the dislocation. The formation of jogs has two important consequences:

a) Jogs increase the lengths of the dislocation lines and involve expenditure of additional energy;
b) Jogged dislocations will move less readily through the crystal. Since an edge dislocation can glide freely only in the plane containing its line and Burgers vector, the only way the jog can move by slip is along the screw direction. If the jog is to slip to a new position, it can only do so by a thermally activated process such as the climb [17].

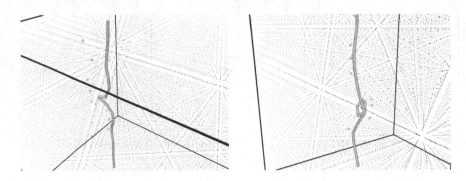

**Fig. 4.** RPMD simulations of dislocation motion (left panel); cross-slip and formation of jogs and loops (right panel).

For these reasons, jogs are almost always impediments to dislocation glide and are often sessile segments that pin the dislocation locally. The dislocation segments on both sides of the jog glide on two parallel glide planes. RPMD simulations also show that the two edge segments can pass each other – two parts of the dislocation move independently from each other forming a loop (Fig. 4b).

The intensity of jogs formation increases with increase of H concentration per dislocation length. Classical MD simulations do not predict formation of jogs.

## 4    Conclusions

The present RPMD simulations show that even at room temperature intrinsic processes in $1/2[111]$ edge dislocation motion in combination with the hydrogen diffusion are strongly influenced by H quantum-mechanical behavior. Mobility of edge dislocation in H-charged Fe is governed by a balance between the applied driving force and the retarding force of the H clouds. Due to NQEs, the H binding energy in the dislocation core and the diffusion barriers between the nearest binding sites increase. This results in a very low H diffusivity compared to that in the regular lattice region and a higher resistance to dislocation motion. The nuclear quantum effects also alter the edge dislocation glide mechanism. RPMD simulations show that the interaction of dislocations with H atoms leads to a cross-slip and formation of jogs. Jogged dislocations move less readily through the crystal and the jogs increase the dislocation length which is accompanied by expenditure of additional energy. Due to the great computational cost of RPMD method and higher activation barrier for diffusion of H in dislocation core, here we do not consider the solute drag process resulting from the H cloud travel along with the dislocation.

**Acknowledgements.** This research was supported in part by the Bulgarian Science Fund under Grant KP-06-N27/19/ 17.12.2018, the Bulgarian Ministry of Education and

Science (contract D01-205/23.11.2018) under the National Research Program "Information and Communication Technologies for a Single Digital Market in Science, Education and Security (ICTinSES)", approved by DCM # 577/17.08.2018, and the European Regional Development Fund, within the Operational Programme "Science and Education for Smart Growth 2014–2020" under the Project CoE "National center of mechatronics and clean technologies" BG05M20P001-1.001-0008-C01.

# References

1. Beachem, C.D.: A new model for hydrogen-assisted cracking (hydrogen "embrittlement"). Metall. Mater. Trans. B **3**, 441–455 (1972)
2. Robertson, I.M., Birnbaum, H.K.: An HVEM study of hydrogen effects on the deformation and fracture of nickel. Acta Metall. **34**(3), 353–366 (1986)
3. Sofronis, P., Robertson, I.M.: Transmission electron microscopy observations and micromechanical/continuum models for the effect of hydrogen on the mechanical behaviour of metals. Philos. Mag. A **82**(17–18), 3405–3413 (2002)
4. Robertson, I.M.: The effect of hydrogen on dislocation dynamics. Eng. Fract. Mech. **68**(6), 671–692 (2001)
5. Ferreira, P.J., Robertson, I.M., Birnbaum, H.K.: Hydrogen effects on the interaction between dislocations. Acta Mater. **46**(5), 1749–1757 (1998)
6. Birnbaum, H.K., Sofronis, P.: Hydrogen-enhanced localized plasticity – a mechanism for hydrogen-related fracture. Mater. Sci. Eng. A **176**(1–2), 191–202 (1994)
7. Kirchheim, R.: Revisiting hydrogen embrittlement models and hydrogen-induced homogeneous nucleation of dislocations. Scripta Mater. **62**(2), 67–70 (2010)
8. Craig, I.R., Manolopoulos, D.E.: A refined ring polymer molecular dynamics theory of chemical reaction rates. J. Chem. Phys. **123**, 034102 (2005)
9. Habershon, S., Manolopoulos, D.E., Markland, T.E., Miller, T.F., III.: Ring-polymer molecular dynamics: quantum effects in chemical dynamics from classical trajectories in an extended phase space. Annual Rev. Phys. Chem. **64**, 387–413 (2013)
10. Feynman, R.P., Hibbs, A.R.: Quntum Mechanics and Path Integral. McGraw-Hill, New York (1965)
11. Nose, S.: A unified formulation of the constant temperature molecular dynamics methods. J. Chem. Phys. **81**(1), 511–519 (1984)
12. Swope, W.C., Andersen, H.C., Berens, P.H., Wilson, K.R.: A computer simulation method for the calculation of equilibrium constants for the formation of physical clusters of molecules: application to small water clusters. J. Chem. Phys. **76**, 637 (1982)
13. Stukowski, A.: Visualization and analysis of atomistic simulation data with OVITO-the open visualization tool. Model. Simul. Mater. Sci. Eng. **18**, 015012 (2009)
14. Kapil, V., et al.: i-PI 2.0: a universal force engine for advanced molecular simulations. Comput. Phys. Commun. **236**, 214–223 (2019)
15. Ackland, G.J., Mendelev, M.I., Srolovitz, D.J., Han, S., Barashev, A.V.: Development of an interatomic potential for phosphorus impurities in $\alpha$-iron. J. Phys.: Condens. Matter **16**(27), S2629–S2642 (2004)
16. Kimizuka, H., Ogata, S.: Slow diffusion of hydrogen at a screw dislocation core in $\alpha$-iron. Phys. Rev. B **84**, 024116 (2011)
17. Hirth, J.P., Lothe, J.: Theory of Dislocations, 2nd edn. Wiley-Interscience, New York (1982)

# Degeneracy of Tetrahedral Partitions Produced by Randomly Generated Red Refinements

Sergey Korotov[1]([⊠]) [iD] and Michal Křížek[2] [iD]

[1] Western Norway University of Applied Sciences, Bergen, Norway
smkorotov@gmail.com
[2] Institute of Mathematics, Czech Academy of Sciences, Prague, Czech Republic
krizek@math.cas.cz

**Abstract.** In this paper, we survey some of our regularity results on a red refinement technique for unstructured face-to-face tetrahedral partitions. This technique subdivides each tetrahedron into 8 subtetrahedra with equal volumes. However, in contrast to triangular partitions, the red refinement strategy is not uniquely determined for the case of tetrahedral partitions which leads to the fact that randomly performed red refinements may produce degenerating tetrahedra which violate the maximum angle condition. Such tetrahedra are often undesirable in practical calculations and analysis. Thus, a special attention is needed when applying red refinements in a straightforward manner.

**Keywords:** Red refinement · Subdivision of a tetrahedron · Face-to-face refinements · Degenerating partitions · Refinement strategy

## 1 Introduction

Modern research trends and real-life needs for economical computations in complicated situations, especially in higher dimensions, require efficient and controllable adaptivity procedures in generating finite element partitions. The red refinement is one of the techniques widely used for simplicial mesh generation and adaptivity purposes in various engineering problems, see Fig. 1 for an illustration of how this refinement is performed for tetrahedral elements. However, this technique is not uniquely defined in three and higher dimensions. In particular, in the case of tetrahedral partitions, inside each tetrahedron, on each refinement level, we have three different possibilities for dividing the tetrahedron. Using some strategies (i.e. fixing rules for selection among three diagonals inside interior octahedra) one may produce regular families of tetrahedral partitions, see e.g. [4,6,7,9,10,13], choosing another strategies we may get some degenerating tetrahedral elements in the course of refinements [5,13]. Nevertheless, it is worth mentioning that for any strategy selected all produced tetrahedral partitions stay face-to-face, i.e., they are conforming.

This paper was supported by grant no. 20-01074S of the Grant Agency of the Czech Republic and RVO 67985840 of the Czech Republic.

I. Lirkov and S. Margenov (Eds.): LSSC 2021, LNCS 13127, pp. 140–147, 2022.
https://doi.org/10.1007/978-3-030-97549-4_16

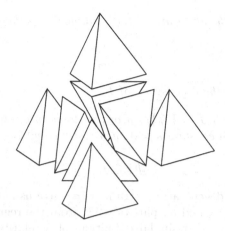

**Fig. 1.** Red refinement of a tetrahedron $T$ by midlines of its triangular faces and one interior line. The four "exterior subtetrahedra" are congruent and similar to $T$, whereas, in general, the four "interior subtehrahedra" might be different.

Non-uniqueness in the selection of diagonals and absence of similarity properties in most of the cases in three (and higher) dimensions makes an analysis of this refinement technique hard and therefore not so many (mathematical) results on this topic exist in the literature though the first results were obtained already in 1982, see [7]. In this paper we present several of our recent results on red refinements of tetrahedral partitions.

## 2   Preliminaries

Throughout the paper, we deal with face-to-face tetrahedral partitions of a bounded polyhedral domain $\Omega \subset \mathbf{R}^3$. They are denoted by $\mathcal{T}_h$, where $h$ is the so-called *discretization parameter* defined as $h = \max_{T \in \mathcal{T}_h} h_T$ with

$$h_T = \operatorname{diam} T \tag{1}$$

for a tetrahedron $T \in \mathcal{T}_h$.

In what follows, we consider a family $\mathcal{F} = \{\mathcal{T}_h\}_{h \to 0}$ of face-to-face tetrahedral partitions $\mathcal{T}_h$ of $\Omega$, see [1, p. 16]. Construction of such families can be done, for example, as in [7].

Further, the radius $r_T$ of the inscribed ball of the tetrahedron $T$ can be computed as

$$r_T = \frac{3 \operatorname{vol}_3 T}{\operatorname{vol}_2 \partial T}, \tag{2}$$

where $\operatorname{vol}_3 T$ is the three-dimensional volume of $T$, $\partial T$ denotes the boundary of $T$, $\operatorname{vol}_2 \partial T$ is the total surface area of $T$, and $r_T$ is often called the *inradius* of $T$.

**Definition 1.** *Let $T$ be an arbitrary tetrahedron. Then the ratio*

$$\sigma_T = \frac{h_T}{r_T} \tag{3}$$

*is called a measure of the degeneracy of $T$.*

**Definition 2.** *A family $\mathcal{F} = \{T_h\}_{h\to 0}$ of partitions into tetrahedra is said to be regular if there exists a constant $\kappa > 0$ such that for any $T_h \in \mathcal{F}$ and any $T \in T_h$ we have*

$$\kappa h_T \leq r_T. \tag{4}$$

It is clear that for degenerating tetrahedra $\sigma_T$ attains large values. However, $\sigma_T \leq \kappa^{-1}$ for all $T \in T_h$ and all partitions $T_h$ from the regular family $\mathcal{F}$. Note that there exist several other similar definitions of regularity and degeneracy of tetrahedral (and simplicial) elements in the literature (see e.g. [1]).

**Definition 3.** *A family $\mathcal{F}$ of partitions is said to be degenerating if for every positive integer $n \in \mathbb{N}$ there exist a partition $T_h \in \mathcal{F}$ and a tetrahedron $T \in T_h$ whose measure of degeneracy satisfies $\sigma_T > n$.*

Assume that there exists a constant $\gamma_0 < \pi$ such that for any tetrahedron $T \in T_h$ and any $T_h \in \mathcal{F}$ we have

$$\gamma_T \leq \gamma_0 \tag{5}$$

and

$$\varphi_T \leq \gamma_0, \tag{6}$$

where $\gamma_T$ is the maximum angle of all triangular faces of the tetrahedron $T$ and $\varphi_T$ is the maximum dihedral angle between faces of $T$. We say that the family $\mathcal{F}$ satisfies the *maximum angle condition* if (5)–(6) hold. According to [8], the two conditions (5) and (6) are independent.

## 3   Red Refinements and Zhang Tetrahedra

Red refinements of tetrahedra from $T_h$ proceed as follows. Each tetrahedron $T$ has four triangular faces which will be subdivided by midlines. It also has three pairs of opposite edges. Choosing one pair of these edges and connecting their midpoints by a straight line segment (called a *diagonal*) we can define a face-to-face subdivision of $T$ into 8 smaller tetrahedra, see Fig. 1. Since there are three diagonals (of the inner octahedron), we have, in general, three different red subdivisions of $T$. The formula for the lengths of these diagonals is derived in [1, p. 90].

First, let $A_0B_0C_0D_0$ be the regular tetrahedron whose edges have length 1. For $k \in \{0, 1, 2, \dots\}$ define the following midpoints

$$A_{k+1} = \frac{1}{2}(B_k + C_k), \quad B_{k+1} = \frac{1}{2}(A_k + C_k), \quad C_{k+1} = \frac{1}{2}(A_k + B_k), \quad (7)$$

and $D_{k+1}$ will be one of the midpoints of $A_kD_k$ or $B_kD_k$ or $C_kD_k$. Then the lenghts of all six edges of the tetrahedra $A_kB_kC_kD_k$ multiplied by the scaling factor $2^k$ can be divided into the following three groups:

a) If we always choose the shortest diagonal, then we get the periodic sequence

$$(1,1,1,1,1,1), \; (1,1,1,1,1,\sqrt{2}), \; (1,1,1,1,1,1), \; (1,1,1,1,1,\sqrt{2}),\dots$$

b) If we always choose the second-longest diagonal, we obtain

$$(1,1,1,1,1,1), (1,1,1,1,1,\sqrt{2}), (1,1,1,1,\sqrt{2},\sqrt{3}), (1,1,1,1,\sqrt{3},\sqrt{3}),$$

$$(1,1,1,1,\sqrt{2},\sqrt{3}),\dots$$

Since the third term is the same as the fifth term, this sequence is also periodic starting from its third term.

c) If we always choose the longest diagonal, we find (see [5] for details)

$$(1,1,1,1,1,1), (1,1,1,1,1,\sqrt{2}), (1,1,1,1,\sqrt{2},\sqrt{3}), (1,1,1,\sqrt{2},\sqrt{3},\sqrt{5}),$$

$$(1,1,1,\sqrt{3},\sqrt{5},\sqrt{7}), (1,1,1,\sqrt{5},\sqrt{7},\sqrt{11}),\dots \quad (8)$$

All terms in this sequence are different and the measure of degeneracy grows to $\infty$ as $k \to \infty$ (see Fig. 2 and Table 1). Note that this type of degeneracy is not considered in the classification of degenerated tetrahedral elements by Edelsbrunner [3]. In this case

$$D_{k+1} = \frac{1}{2}(A_k + D_k) \;\; \text{if } k \text{ is odd}, \quad \text{or} \quad D_{k+1} = \frac{1}{2}(B_k + D_k) \;\; \text{if } k \text{ is even}. \quad (9)$$

The cases a) and b) obviously lead to regular families of partitions, since the produced tetrahedra have only a finite number of different shapes. However, the case c) produces a sequence of subtetrahedra with infinite number of different shapes that does not satisfy the maximum angle condition (5)–(6) as we shall see in Sect. 4. In the case c), all $A_kB_kC_kD_k$, $k = 1, 2, \dots$, are called the *Zhang tetrahedra* in a view of the paper by Zhang [13]. They are called in this way also under any translation, rotation, reflection, and scaling.

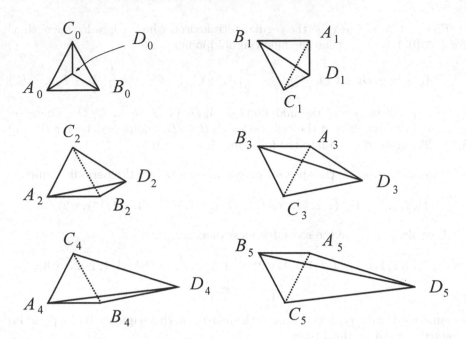

**Fig. 2.** Growth dynamics of the Zhang tetrahedra for $k = 0, 1, \ldots, 5$.

**Table 1.** Dihedral angles in degrees ° at particular edges of the Zhang tetrahedra $A_k B_k C_k D_k$, their inradii $r_k$ multiplied by the scaling factor $2^k$, and the corresponding measures of degeneracy $\sigma_k$. The maximum dihedral angles are in bold.

| Edge | $k = 0$ | $k = 1$ | $k = 2$ | $k = 3$ | $k = 4$ | $k = 5$ |
|---|---|---|---|---|---|---|
| $A_k B_k$ | **70.53** | 54.74 | 70.53 | 54.74 | 70.53 | 54.74 |
| $A_k C_k$ | **70.53** | **109.47** | 35.26 | **144.74** | 22.00 | **158.00** |
| $B_k C_k$ | **70.53** | 54.74 | **125.26** | 29.50 | **150.50** | 19.47 |
| $A_k D_k$ | **70.53** | 54.74 | 90.00 | 45.00 | 97.61 | 46.51 |
| $B_k D_k$ | **70.53** | 90.00 | 54.74 | 107.55 | 58.52 | 114.09 |
| $C_k D_k$ | **70.53** | 54.74 | 45.00 | 31.48 | 25.94 | 20.51 |
| $r_k$ | 0.204 | 0.189 | 0.171 | 0.143 | 0.127 | 0.109 |
| $\sigma_k$ | 4.899 | 7.464 | 10.156 | 15.617 | 20.840 | 29.046 |

Now let $T$ be an arbitrary tetrahedron. Consider a linear affine mapping from the regular reference tetrahedron $A_0 B_0 C_0 D_0$ to $T$. Red refinements of $T$ can be then defined via this mapping.

# 4   Main Results

First, we present several basic theorems on red refinements. Recall that the tetrahedron with vertices $(\pm 1, 0, 0)$ and $(0, \pm 1, 1)$ is called the *Sommerville tetrahedron*, see [11].

**Theorem 1.** *There exists only one type of tetrahedron $T$ (up to similarity) whose red refinement produces eight congruent subtetrahedra similar to $T$. It is the Sommerville tetrahedron.*

The proof is given in [4] (see also [7, 10, 13]).

The next result immediately follows from the fact that the four "exterior" subtetrahedra arising from the red refinement algorithm are similar to the original tetrahedron, see Fig. 1.

**Theorem 2.** *The maximum (minimum) dihedral angles between faces and also the maximum (minimum) angles in all triangular faces of all tetrahedra $T \in \mathcal{T}_h \in \mathcal{F}$ generated by the red-type refinements form nondecreasing (nonincreasing) sequences as $h \to 0$.*

The following two theorems are proved in [5].

**Theorem 3.** *For any selection of diagonals we always produce a family of partitions with $h \to 0$.*

**Theorem 4.** *The measure of degeneracy of the Zhang tetrahedra tends to $\infty$ when $k \to \infty$ and their maximum dihedral angle tends to $180°$, and the maximum angle between edges tends to $180°$ as well.*

*Remark 1.* According to [2, 12], degenerating Lagrange and Hermite finite elements lose their optimal interpolation properties in the $H^1$-Sobolev norm. Moreover, the corresponding stiffness matrices become ill-conditioned and the discrete maximum principle will be violated (see [1, p. 68]).

*Remark 2.* Let $T_0$ be the regular tetrahedron as in Sect. 3. If we choose the diagonal in the inner octahedron randomly, then we can construct a sequence of the Zhang degenerating tetrahedra $\{T_k\}_{k=0}^{\infty}$ by induction as follows. So let $T_k$ be given and denote by $s, m, \ell$ the lengths of the shortest, middle, and longest diagonal of the corresponding inner octahedron, respectively. That is

$$s \le m \le \ell,$$

where each of these inequalities may change to the equality. Now assume that we choose one of the three diagonals randomly with the same probability. If we choose the longest diagonal, we are ready. If we choose a diagonal shorter than $\ell$, then we choose as $T_{k+1}$ one of the four exterior tetrahedra which are similar to the original tetrahedron $T_k$ and we continue the refining process. After some time we have to necessarily choose the longest diagonal, since the selection process is completely random. In this way we again obtain a sequence of the Zhang tehrahedra even though some of its neighbouring entries in the sequence will be similar tetrahedra.

As a consequence of Remark 2 we get the following statement.

**Theorem 5.** *A random choice of diagonals in the red refinements produces degenerating partitions.*

*Remark 3.* Let $T_0$ be the initial regular tetrahedron and denote by $T_1, T_2, T_3,$ ... tetrahedra (up to scaling) that are generated by the red refinements with the choice of the longest diagonals, see (8). Their shapes are illustrated in Fig. 2 (cf. also Table 1). After the first red refinement step, we obtain four exterior tetrahedra with shape $T_0$ and 4 interior tetrahedra with shape $T_1$, see Fig. 1. After the second red refinement step, we get 16 tetrahedra with shape $T_0$, 32 tetrahedra with shape $T_1$, and 16 tetrahedra with shape $T_2$. Repeating this process $k$ times, we get $2^{2k}$ tetrahedra with shape $T_0$, $k2^{2k}$ tetrahedra with shape $T_1$, $k(k-1)2^{2k-1}$ tetrahedra with shape $T_2$, etc., and $2^{2k}$ tetrahedra with shape $T_k$. Hence, we see that for large $k$, less that 50% of tetrahedra remain nicely shaped (e.g. regular).

The following relevant result can be found in [4].

**Theorem 6.** *The red refinement of an acute simplex in three and higher dimensions never yields subsimplices that would be all mutually congruent.*

Its proof for tetrahedra follows immediately from Fig. 1 when we notice that for any choice of diagonal for red refinement we have four subtetrahedra sharing this diagonal. Therefore, four adjacent dihedral angles sum up $2\pi$, and at least one of them is not acute, meaning that the associated subtetrahedron is not acute, and therefore not congruent to the four corner subtetrahedra.

# References

1. Brandts, J., Korotov, S., Křížek, M.: Simplicial Partitions with Applications to the Finite Element Method. Springer, Cham (2020). https://doi.org/10.1007/978-3-030-55677-8
2. Ciarlet, P.G.: The Finite Element Method for Elliptic Problems. North-Holland, Amsterdam (1978)
3. Edelsbrunner, H.: Triangulations and meshes in computational geometry. Acta Numer **9**, 133–213 (2000)
4. Korotov, S., Křížek, M.: Red refinements of simplices into congruent subsimplices. Comput. Math. Appl. **67**, 2199–2204 (2014)
5. Korotov, S., Křížek, M.: On degenerating tetrahedra resulting from red refinements of tetrahedal partitions. Num. Anal. Appl. **14**, 335–342 (2021)
6. Korotov, S., Vatne, J.E.: On regularity of tetrahedral meshes produced by some red-type refinements. In: Pinelas, S., Graef, J.R., Hilger, S., Kloeden, P., Schinas, C. (eds.) ICDDEA 2019. SPMS, vol. 333, pp. 681–687. Springer, Cham (2020). https://doi.org/10.1007/978-3-030-56323-3_49
7. Křížek, M.: An equilibrium finite element method in three-dimensional elasticity. Apl. Mat. **27**, 46–75 (1982)
8. Křížek, M.: On the maximum angle condition for linear tetrahedral elements. SIAM J. Numer. Anal. **29**, 513–520 (1992)

9. Křížek, M., Strouboulis, T.: How to generate local refinements of unstructured tetrahedral meshes satisfying a regularity ball condition. Numer. Methods Partial Differ. Equ. **13**, 201–214 (1997)
10. Ong, M.E.G.: Uniform refinement of a tetrahedron. SIAM J. Sci. Comput. **15**, 1134–1144 (1994)
11. Sommerville, D.M.Y.: Division of space by congruent triangles and tetrahedra. Proc. Royal Soc. Edinburgh **43**, 85–116 (1923)
12. Ženíšek, A.: Convergence of the finite element method for boundary value problems of a system of elliptic equations (in Czech). Apl. Mat. **14**, 355–377 (1969)
13. Zhang, S.: Successive subdivisions of tetrahedra and multigrid methods on tetrahedral meshes. Houston J. Math. **21**, 541–556 (1995)

# Effluent Recirculation for Contaminant Removal in Constructed Wetlands Under Uncertainty: A Stochastic Numerical Approach Based on Monte Carlo Methodology

Konstantinos Liolios[1(✉)], Georgios Skodras[2], Krassimir Georgiev[1], and Ivan Georgiev[1,3]

[1] Institute of Information and Communication Technologies,
Bulgarian Academy of Sciences, Sofia, Bulgaria
kostisliolios@gmail.com, {krassimir.georgiev,ivan.georgiev}@iict.bas.bg
[2] Department of Mechanical Engineering, Faculty of Engineering,
University of Western Macedonia, Kozani, Greece
gskodras@uowm.gr
[3] Institute of Mathematics and Informatics, Bulgarian Academy of Sciences,
Sofia, Bulgaria

**Abstract.** The problem of the alternative operational technique concerning effluent re-circulation in Horizontal Subsurface Flow Constructed Wetlands (HSF CW), and the possibility of this technique to remove efficiently pollutants under uncertainty, is investigated numerically in a stochastic way. Uncertain-but-bounded input-parameters are considered as interval parameters with known upper and lower bounds. This uncertainty is treated by using the Monte Carlo method. A typical pilot case of an HSF CW concerning Biochemical Oxygen Demand (BOD) removal is presented and numerically investigated. The purpose of the present study is to compare the relevant numerical results obtained under parameter uncertainty and concerning HSF CW operation with and without the effluent recirculation technique.

## 1 Introduction

A constructed wetland (CW) is an artificial wetland which can be used for the solution of the wastewater treatment problem, especially in small and medium-sized communities, where other solutions could have negative ecological and economic effects. This is the main reason that these systems have been developed in many countries the last 30 years as reported in [1], especially in Europe, North America and Asia [2–4].

The effluent recirculation is an alternative feeding technique, which has been investigated (mostly by using experiments) in order to check whether it can increase the treatment efficiency of CWs, see e.g. [5–9]. Effluent recirculation

© Springer Nature Switzerland AG 2022
I. Lirkov and S. Margenov (Eds.): LSSC 2021, LNCS 13127, pp. 148–155, 2022.
https://doi.org/10.1007/978-3-030-97549-4_17

(or back-feeding) describes the phenomenon of taking a part of the effluent and transferring it back to the inlet of the system.

The purpose of this procedure is to investigate if the dilution of influent wastewater improves the performance of the CW. This investigation is mostly an experimental one for CWs, see e.g. [10–12]. Especially for HSF CWs, a numerical investigation for the effects of effluent recirculation has also been presented by [13], by using the Visual MODFLOW computer code [14]. The available literature shows that most investigations, which were realized about this alternative feeding technique in CWs, are related to vertical flow (VF) systems, see e.g. [15–17].

The purpose of the present study is to investigate numerically, by taking into account uncertainty concerning input parameters, whether recirculation of the effluent of wastewater can increase the removal efficiency in pilot-scale Horizontal Subsurface Flow (HSF) CWs. Emphasis is given here on the performance of these systems concerning Biochemical Oxygen Demand (BOD) removal. First, the deterministic modelling approach of the recirculation problem is briefly presented. The proposed model was calibrated in previous studies, by comparing the numerical results with available experimental ones in rectangular pilot-scale HSF CW units [18], and then was applied to evaluate the effects of recirculation [13]. Then, the deterministic problem for each set of the random input variables is solved. Finally, the probabilistic problem is treated by using Monte Carlo simulations.

## 2    Method of Analysis

### 2.1    Modelling of Recirculation

The problem of recirculation in a typical HSF CW is presented in Fig. 1 as a two-option procedure. In the first option (Fig. 1a), wastewater of flow rate $Q_{in}$ with contaminant concentration $C_{in}$ is introduced into the CW. The required Hydraulic Residence Time $HRT_1$ based on $Q_{in}$, is:

$$HRT_1 = \frac{V_p}{Q_{in}} \tag{1}$$

where $V_p$ is the pore volume of the porous medium bed.

In the second option (Fig. 1b), the effluent recirculation is presented based on the use of the recirculation factor $x = RF$. A percentage $x$ of the effluent $Q_{out,1}$ is transferred continuously back to the inlet of the system and is mixed with Qin. In this case, the inflow rate is equal with: $Q_{in,2} = Q_{in} + xQ_{out,1}$, and the inlet contaminant concentration $C_{in,2}$ is given by the following Eq. (2):

$$C_{in,2} = \frac{Q_{in}C_{in} + (xQ_{out,1})C_{out,1}}{Q_{in} + (xQ_{out,1})} \tag{2}$$

The Hydraulic Residence Time $HRT_2$ for the second option is:

$$HRT_2 = \frac{V_p}{Q_{in} + (xQ_{out,1})} = \frac{HRT_1}{1+x} \tag{3}$$

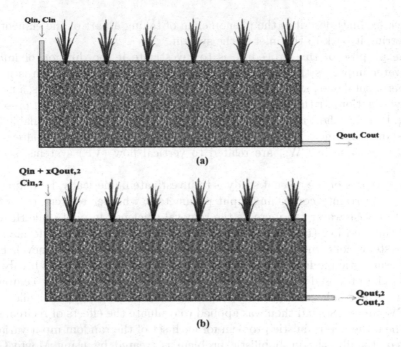

**Fig. 1.** Schematic diagram of recirculation modeling [13]: (a) First option: without recirculation; (b) Second option: with recirculation.

The usual meteorological parameters, like rainfall and evapotranspiration, are not considered; therefore, the flow remains constant (i.e., $Q_{in} = Q_{out,1}$) and Eqs. (1) and (3) can be combined to give the new reduced $HRT_2$ value, as follows:

$$HRT_2 = \frac{HRT_1}{1+x} \tag{4}$$

After the second option, at the end of the Hydraulic Residence Time $HRT_2$, the effluent wastewater of flow rate $Q_{out,2}$ has a contaminant concentration $C_{out,2}$. The question is whether recirculation improves the performance of the HSF CW, i.e., if $C_{out,2} < C_{out,1}$.

## 2.2   The Deterministic Problem

Following [18], both the above presented options of the problem of recirculation are treated in a deterministic way by solving the following system of partial differential equations.

The three-dimensional groundwater flow equation is written, using tensorial notation $(i, j = 1, 2, 3)$ [19]:

$$\frac{\partial}{\partial x_i} \left( K_{ij} \frac{\partial h}{\partial x_j} \right) = S_y \frac{\partial h}{\partial t} \tag{5}$$

$K_{ij}$: component of the hydraulic conductivity tensor; $h$: hydraulic head; $S_y$: specific yield of the porous materials.

The velocity field is computed through the Darcy relationship:

$$q_i = K_{ij}\frac{\partial h}{\partial x_j} \tag{6}$$

The partial differential equation which describes the fate and transport of a contaminant with adsorption in (3-D) transient groundwater flow systems can be written:

$$\varepsilon R_d\frac{\partial C}{\partial t} = \frac{\partial}{\partial x_i}\left(\varepsilon D_{ij}\frac{\partial C}{\partial x_j}\right) - \frac{\partial}{\partial x_i}(q_iC) + q_sC_s + \sum_n^N R_n \tag{7}$$

$q_s$: volumetric flow rate per unit area of aquifer representing fluid sources (positive) and sinks (negative); $\varepsilon$: porosity of the subsurface medium; $R_d$: retardation factor; $C$: dissolved concentration of solute; $D_{ij}$: solute hydrodynamic dispersion coefficient tensor; $C_s$: concentration of the source or sink flux; $\sum R_n$: chemical reaction term, which is given by the formula:

$$\sum R_n = -\lambda R_d C^\alpha \tag{8}$$

$\lambda, \alpha$: removal coefficients. For $\alpha = 1$, the usual linear reaction case of first-order decay is active.

The unknowns of the problem are the following five space-time functions: The hydraulic head: $h = h(xi; t)$; the three velocity components: $q_i(x_j; t)$; and the concentration: $C = C(x_i; t)$.

The numerical solution of the above deterministic problem has been obtained by using the MODFLOW code [14,20]. The input parameters concerning the various coefficients are rarely known in a safe way, which usually requires detailed in-situ measurements. Therefore, the numerical treatment of the deterministic problem is usually realized in the praxis by using a mean-value estimate of the various coefficients.

## 2.3   The Probabilistic Problem Treatment Using Monte Carlo Simulation

As well known [21–23], the Monte Carlo methodology is a broad class of computational algorithms that rely on repeated random sampling to obtain numerical results quantifying uncertainty. The underlying concept is to use randomness to solve problems that might be deterministic in principle. They are often used in physical and mathematical problems and are most useful when it is difficult or impossible to use other approaches [21,24].

In the present study, the Monte Carlo simulation is used as a repeated process of generating deterministic solutions to the above deterministic problem of Sect. 2.2. Thus, following [19,20,25], a stochastic methodology is applied, in which the various system coefficients are considered as random variables. Each solution corresponds to a set of input values. Probability density functions are used, based on lower and upper bounds for reliable estimates known either from praxis or in-situ measurements. For the quantitative estimation of the various uncertainties, a Monte Carlo simulation is applied and a statistical analysis of the obtained simulated solutions is finally performed.

## 3    Numerical Example

### 3.1    Data of the Example-Problem and Uncertain-but-Bounded Input Parameters

For the simulation of the recirculation effects by using the Monte Carlo method, the available data for a typical experimental case are used. This case concerns the pilot HSF CW orthogonal tank reported as MG-R unit in [18]. The tank has medium gravel (MG) and is planted with common reed (*Phragmites australis* reed - *R*). The dimensions of this orthogonal tank are: length, $L = 3$ m; depth, $d = 0.45$ m; and width, $w = 0.75$ m. The estimated mean values of the input parameters are: porosity, $\varepsilon = 0.36$; hydraulic conductivity coefficient, $K = 0.7640$ m/sec; pore volume, $V_p = 0.3595$ m$^3$; inflow rate, $Q_{in} = 17.119$ Litre/day; inlet contaminant concentration, $C_{in} = 389.4$ mg/Litre; and first-order removal coefficient, $\lambda = 0.15$ days$^{-1}$. The Hydraulic Residence Time (HRT) is 14 days.

In general, according relevant investigations [19], $K$ has a LOGNORMAL probability distribution, whereas $\varepsilon$ and $\lambda$ have a NORMAL probability distribution. For the investigated HSF CW case, in the Table 1 the Probability Density Functions (PDF) are shown. By COV is denoted the Coefficient of Variation.

**Table 1.** Distribution properties of uncertain input parameters

| Input parameters | Distribution $PDF$ | Mean value | $COV$ [%] |
|---|---|---|---|
| $K$ [m/sec] | LOGNORMAL | 1.538 | 25 |
| $\varepsilon$ [%] | NORMAL | 0.365 | 15 |
| $\lambda[day^{-1}]$ | NORMAL | 0.15 | 20 |

### 3.2    Representative Numerical Results and Discussion

Concerning the effects of effluent recirculation on BOD removal, the cases when the recirculation factor $x$ in Eqs. (2) and (3) has the typical values: $x = 0.0, x = 0.5$ and $x = 1.0$ are considered. So, a comparison is realized when this system operates with and without this recirculating feeding technique.

The proposed methodology based on Monte Carlo simulations is applied. For this purpose, totally 200 samples are constructed. For each sample, the deterministic problem of Sect. 2.2 is solved by using the MODFLOW code [18]. The relative optimal space discretization has resulted in 8400 cells. For each cell, values of the previously mentioned (in Sect. 2.2) five quantities, i.e., $h, C, v_i$ (where $i = 1, 2, 3$) have to be computed. Thus, in each time step, a system of $5 \times 8400 = 42000$ unknowns has to be solved. This is obtained by using the iterative SIP procedure (Strong Implicit).

After the statistical treatment of the so-obtained relevant solutions, some representative stochastic computational results concerning the mean values of removal efficiency, output concentration $C_{out}$, concentration ratios $C_{out}/C_{in}$ and COV in the considered tank (MG-R) are reported in Table 2. A relevant graphical representation is shown in Fig. 2.

**Table 2.** Mean values and COV of the BOD removal efficiency and outlet concentration, for the MG-R unit under recirculation rate x-RF = 0, 50, 100 %

| Recirculation rate $RF$ | Mean value of removal efficiency [%] | Mean value $C_{out}$ [mg/L] | Mean value $C_{out}/C_{in}$ [%] | COV [%] |
|---|---|---|---|---|
| 0.0 | 87.85 | 47.30 | 12.15 | 10.48 |
| 0.5 | 83.09 | 65.84 | 16.91 | 14.57 |
| 1.0 | 80.90 | 74.38 | 19.10 | 19.84 |

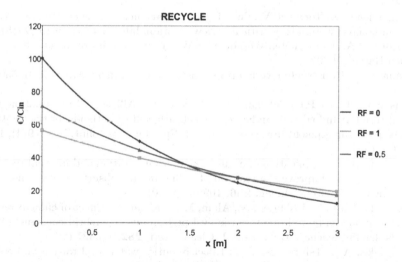

**Fig. 2.** Mean-value ratios $C/C_{in}$ of the BOD concentration along the length of the MGR unit for the values $x = RF = 0.0, 0.5 \kappa \alpha \iota 1.0$ of the recirculation factor.

As the above results show, recirculation ($RF = 0.5 \, and \, 1.0$) does not increase the performance of the CW, in comparison with the case when the wetland operates without ($RF = 0.0$) this alternative feeding technique. This consideration is in agreement with similar investigations concerning recirculation effects.

## 4    Conclusion

A stochastic approach has been developed for the problem of ground water flow and contaminant removal in constructed wetlands operating under effluent recirculation. This approach considers uncertain input parameters and is based on Monte Carlo simulations. The results of a typical numerical example showed that recirculation does not increase the performance of the CW when this system operates under this alternative feeding technique. This conclusion is in agreement with similar studies computational or experimental ones. Thus, the effluent recirculation is not suggested for the operation of pilot-scale HSF CWs, as the economical cost is not going to be covered by improvement of the performance on the removal of pollutants.

**Acknowledgements.** This work of K. Georgiev was accomplished with the support by the Grant No BG05M2OP001-1.001-0003, financed by the Science and Education for Smart Growth Operational Program (2014-2020) and co-financed by the European Union through the European structural and Investment funds.

## References

1. Kadlec, R.H., Wallace, S.: Treatment Wetlands, 2nd edn. CRC Press, Boca Raton (2009)
2. Gorgoglione, A., Torretta, V.: Sustainable management and successful application of constructed wetlands: a critical review. Sustainability **10**(11), 3910 (2018)
3. Moshiri, G.A.: Constructed Wetlands for Water Quality Improvement. CRC Press, Boca Raton (2020)
4. Vymazal, J.: Constructed wetlands for wastewater treatment. Water **2**(3), 530–549 (2010)
5. Idris, S.M., Jones, P.L., Salzman, S.A., Croatto, G., Allinson, G.: Evaluation of the giant reed (Arundo donax) in horizontal subsurface flow wetlands for the treatment of recirculating aquaculture system effluent. Environ. Sci. Pollut. Res. **19**(4), 1159–1170 (2012)
6. Lin, C.J., et al.: Application of an innovative front aeration and internal recirculation strategy to improve the removal of pollutants in subsurface flow constructed wetlands. J. Environ. Manag. **256**, 109873 (2020)
7. Saeed, T., Miah, M.J., Majed, N., Alam, M.K., Khan, T.: Effect of effluent recirculation on nutrients and organics removal performance of hybrid constructed wetlands: landfill leachate treatment. J. Clean. Prod. **282**, 125427 (2021)
8. Stefanakis, A.I., Tsihrintzis, V.A.: Effect of outlet water level raising and effluent recirculation on removal efficiency of pilot-scale, horizontal subsurface flow constructed wetlands. Desalination **248**, 961–976 (2009)
9. Torrijos, V., Gonzalo, O.G., Trueba-Santiso, A., Ruiz, I., Soto, M.: Effect of bypass and effluent recirculation on nitrogen removal in hybrid constructed wetlands for domestic and industrial wastewater treatment. Water Res. **103**, 92–100 (2016)
10. Avila, C., Pelissari, C., Sezerino, P.H., Sgroi, M., Roccaro, P., Garcia, J.: Enhancement of total nitrogen removal through effluent recirculation and fate of PPCPs in a hybrid constructed wetland system treating urban wastewater. Sci. Total Environ. **584**, 414–425 (2017)

11. Konnerup, D., Trang, N.T.D., Brix, H.: Treatment of fishpond water by recirculating horizontal and vertical flow constructed wetlands in the tropics. Aquaculture **313**(1), 57–64 (2011)

12. Sun, G., Gray, K.R., Biddlestone, A.J., Allen, S.J., Cooper, D.J.: Effect of effluent recirculation on the performance of a reed bed system treating agricultural wastewater. Process Biochem. **39**(3), 351–357 (2003)

13. Liolios, K.A., Moutsopoulos, K.N., Tsihrintzis, V.A.: Modelling alternative feeding techniques in HSF constructed wetlands. Environ. Process. **3**(1), 47–64 (2016)

14. Waterloo Hydrogeologic Inc.: Visual MODFLOW version 4.2. User's manual. U.S. Geological Survey, Virginia (2006)

15. Prost-Boucle, S., Molle, P.: Recirculation on a single stage of vertical flow constructed wetland: treatment limits and operation modes. Ecol. Eng. **43**, 81–84 (2012)

16. Zapater, M., Gross, A., Soares, M.I.M.: Capacity of an on-site recirculating vertical flow constructed wetland to withstand disturbances and highly variable influent quality. Ecol. Eng. **37**(10), 1572–1577 (2011)

17. Sklarz, M.Y., Gross, A., Soares, M.I.M., Yakirevich, A.: Mathematical model for analysis of recirculating vertical flow constructed wetlands. Water Res. **44**(6), 2010–2020 (2010)

18. Liolios, K.A., Moutsopoulos, K.N., Tsihrintzis, V.A.: Modeling of flow and BOD fate in horizontal subsurface flow constructed wetlands. Chem. Eng. J. **200–202**, 681–693 (2012)

19. Zheng, C., Bennet, G.D.: Applied Contaminant Transport Modelling, 2nd edn. Wiley, New York (2002)

20. Bear, J., Cheng, A.D.: Modeling Groundwater Flow and Contaminant Transport. Springer, Dordrecht (2010). https://doi.org/10.1007/978-1-4020-6682-5

21. Dimov, I.T.: Monte Carlo Methods for Applied Scientists. World Scientific, Singapore (2008)

22. Papadrakakis, M., Stefanou, G., Papadopoulos, V.: Computational Methods in Stochastic Dynamics. Springer, Dordrecht (2011). https://doi.org/10.1007/978-90-481-9987-7

23. Kottegoda, N.T., Rosso, R.: Statistics, Probability and Reliability for Civil and Environmental Engineers. McGraw-Hill, New York (1997)

24. Rubinstein, R.Y., Kroese, D.P.: Simulation and the Monte Carlo Method, 3rd edn. Wiley, New Jersey (2016)

25. Liolios, K., Skodras, G., Georgiev, K., Georgiev, I.: Analysis of contaminant removal in constructed wetlands under step-feeding: a Monte Carlo based stochastic treatment accounting for uncertainty. Studies in Computational Intelligence (Submitted)

# Sensitivity Study of Large-Scale Air Pollution Model Based on Modifications of the Latin Hypercube Sampling Method

Tzvetan Ostromsky[1(✉)], Venelin Todorov[1,2], Ivan Dimov[1], Rayna Georgieva[1], Zahari Zlatev[3], and Stoyan Poryazov[2]

[1] Department of Parallel Algorithms, Institute of Information and Communication Technologies, Bulgarian Academy of Sciences, Acad. G. Bonchev Street, Block 25 A, 1113 Sofia, Bulgaria
{ceco,venelin,rayna}@parallel.bas.bg, ivdimov@bas.bg

[2] Department of Information Modeling, Institute of Mathematics and Informatics, Bulgarian Academy of Sciences, Acad. Georgi Bonchev Street, Block 8, 1113 Sofia, Bulgaria
{vtodorov,stoyan}@math.bas.bg

[3] National Centre for Environment and Energy, University of Århus, Frederiksborgvej 399, P.O. Box 358, 4000 Roskilde, Denmark
zz@dmu.dk

**Abstract.** In this paper, various modifications of the Latin Hypercube Sampling algorithm have been used in order to evaluate the sensitivity of an environmental model output results for some dangerous air pollutants with respect to the emission levels and some chemical reaction rates. The environmental security importance is growing rapidly, becoming at present a significant topic of interest all over the world. Respectively, the environmental modeling has very high priority in various scientific fields. By identifying the major chemical reactions that affect the behavior of the system, specialists in various fields of application will be able to obtain valuable information about improving the model, which in turn will increase the reliability and sustainability of forecasts.

## 1 Introduction

High levels of air pollution can disrupt ecosystems and cause harm to plants, animals and humans. Therefore, it is extremely important to investigate accurately the levels of contamination [11]. It is necessary to know whether the pollution levels are below some critical levels and if so, to develop a reliable control system to keep them within these limits. Mathematical models are used to study and predict the behavior of a variety of complex systems - engineering, physical, economic, social, environmental. They determine the most important quantities that control the state and behavior of a system, as well as the quantitative regularities, that is, the mathematical laws that underlie the change of these quantities. On the other hand, it is of significant importance for developing techniques to determine the reliability, robustness, and efficiency of a mathematical model [10]. One such approach is sensitivity analysis (SA). Sensitivity studies are nowadays applied to some of the most complicated mathematical models from various intensively developing areas of application [1,4,5]. In the present study we use the

© Springer Nature Switzerland AG 2022
I. Lirkov and S. Margenov (Eds.): LSSC 2021, LNCS 13127, pp. 156–163, 2022.
https://doi.org/10.1007/978-3-030-97549-4_18

Unified Danish Eulerian Model (UNI-DEM) [12], which is a powerful large-scale air pollution model for the calculation of the concentrations of a large number of pollutants and other chemical species in the air along a certain time period. Its results can be used in various application areas (environmental protection, agriculture, health care, etc.). The large computational domain covers completely the European region and the Mediterranean.

## 2    Latin Hypercube Sampling

Latin Hypercube Sampling is a type of stratified sampling and in the case of integral approximation we must simply divide the domain $[0, 1]^d$ into $m^d$ disjoint subdomains, each of volume $\dfrac{1}{m^d}$ and to sample one point from each of them. Let this sample be $\mathbf{x}_{k,j}$, for dimensions $k = 1, \ldots, m^d$, $j = 1, \ldots, d$. LHS does not require more samples for more dimensions (variables) - it is one of the main advantages of this scheme.

Two different versions of Latin Hypercube Sampling (LHS) are compared with various seeds (seed is an integer, used for the corresponding random number generator) - the Latin Hypercube Sampling (random) [7] LHSR1 (seed = 1) and LHSR-1 (seed = −1) and the Latin Hypercube Sampling (edge) [6] LHSE1 (seed = 1) and LHSE-1 (seed = −1) algorithms have been used in our comprehensive experimental study for the first time for the particular model. The difference is simple: the Latin edge algorithm returns edge points of a Latin square, while the Latin random algorithm returns points in a Latin Random Square.

## 3    Sensitivity Studies with Respect to Emission Levels

Here we present the results for the sensitivity of UNI-DEM output (in particular, the ammonia mean monthly concentrations) with respect to the anthropogenic emissions input data variation are shown and discussed in this section. The anthropogenic emissions input consists of 4 different components: $\mathbf{E^A}$ − ammonia $(NH_3)$, $\mathbf{E^N}$ − nitrogen oxides $(NO + NO_2)$, $\mathbf{E^S}$ − sulphur dioxide $(SO_2)$, $\mathbf{E^C}$ − anthropogenic hydrocarbons.

The second degree polynomials (4 variables, 15 coefficients) fully satisfy the approximation requirements for accuracy and reliability. A detailed explanation of the approximation procedure and the reasons to choose this kind of approximation instruments is given in [3]. In [3] it is shown that the correlated sampling technique for small sensitivity indices gives reliable results for the full set of sensitivity measures. Results of the relative error (it is defined as a ratio of the absolute error (the difference in an absolute value between the estimated value and the actual (referent) value) and the corresponding referent value) estimation for the quantities $f_0$, the total variance $\mathbf{D}$, first-order $(S_i)$ and total $(S_i^{tot})$ sensitivity indices are given in Tables 1, 2, 3, 4 and 5, respectively. $f_0$ is presented by a 4-dimensional integral, while the rest of the above quantities are presented by 8-dimensional integrals, following the ideas of *correlated sampling* technique to compute sensitivity measures in a reliable way. The results show that the computational efficiency of the algorithms depends on integral's dimension and magnitude of estimated quantity. The order of relative error is different for different quantities

**Table 1.** Relative error for the evaluation of $f_0 \approx 0.048$.

| # of samples | LHSE1 | LHSE-1 | LHSR1 | LHSR-1 |
|---|---|---|---|---|
| $n$ | Relative error | Relative error | Relative error | Relative error |
| $2^{10}$ | 6.37e−05 | 2.45e−04 | 3.22e−04 | 2.43e−04 |
| $2^{12}$ | 6.18e−05 | 1.55e−04 | 2.67e−04 | 7.11e−05 |
| $2^{14}$ | 5.21e−06 | 8.44e−06 | 3.62e−05 | 8.30e−05 |
| $2^{16}$ | 1.34e−05 | 1.58e−05 | 1.62e−05 | 1.01e−05 |
| $2^{18}$ | 1.29e−05 | 5.31e−07 | 5.24e−05 | 1.52e−05 |
| $2^{20}$ | 1.53e−05 | 8.48e−07 | 8.78e−06 | 2.24e−05 |

**Table 2.** Relative error for the evaluation of the total variance $\mathbf{D} \approx 0.0002$.

| # of samples | LHSE1 | LHSE-1 | LHSR1 | LHSR-1 |
|---|---|---|---|---|
| $n$ | Relative error | Relative error | Relative error | Relative error |
| $2^{10}$ | 4.91e−02 | 2.78e−03 | 4.08e−02 | 4.68e−02 |
| $2^{12}$ | 5.10e−03 | 9.60e−03 | 1.68e−02 | 1.40e−02 |
| $2^{14}$ | 8.82e−03 | 7.55e−03 | 1.26e−02 | 1.12e−02 |
| $2^{16}$ | 8.36e−03 | 9.72e−03 | 4.43e−03 | 4.27e−03 |
| $2^{18}$ | 1.32e−04 | 7.90e−04 | 7.55e−04 | 1.14e−03 |
| $2^{20}$ | 1.03e−03 | 8.64e−04 | 3.43e−05 | 7.19e−04 |

of interest (see column *Reference value*) for the same sample size. The referent values are obtained using subroutines from Mathematica package [13] for multidimensional integration.

Table 5 is similar to Tables 3 and 4, with the only difference – the increased number of samples $n = 2^{20}$ (instead of $n = 2^{16}$ in Table 4 and $n = 2^{10}$ in Table 3). In general, this increases the accuracy of the estimated quantities. Exceptions are $S_4$, $S_4^{tot}$ and, to some extent, $S_2$ and $S_2^{tot}$, which have extremely small reference values (in terms of relative error). All LHS methods produce similar results, but for small in value sensitivity indices LHSE-1 performs better - see for example the value of $S_2^{tot}$ in Table 5.

**Table 3.** Relative error for estimation of sensitivity indices of input parameters using various Monte Carlo and quasi-Monte Carlo approaches ($n \approx 2^{10}$).

| Est. qnt. | Ref. val. | LHSE1 | LHSE-1 | LHSR1 | LHSR-1 |
|---|---|---|---|---|---|
| $S_1$ | 9e−01 | 1.13e−02 | 5.33e−03 | 4.39e−02 | 1.85e−01 |
| $S_2$ | 2e−04 | 8.53e+00 | 7.23e+00 | 1.22e+00 | 5.36e+00 |
| $S_3$ | 1e−01 | 1.48e−01 | 4.78e−02 | 4.53e−01 | 4.62e−02 |
| $S_4$ | 4e−05 | 1.85e+01 | 1.22e+01 | 1.55e+01 | 1.06e+01 |
| $S_1^{tot}$ | 9e−01 | 1.54e−02 | 4.63e−03 | 5.32e−02 | 6.91e−03 |
| $S_2^{tot}$ | 2e−04 | 1.21e+01 | 4.87e+00 | 1.15e+01 | 6.36e+00 |
| $S_3^{tot}$ | 1e−01 | 1.27e−01 | 4.53e−02 | 3.74e−01 | 3.26e−02 |
| $S_4^{tot}$ | 5e−05 | 2.25e+01 | 1.71e+01 | 1.35e+01 | 5.66e+00 |

**Table 4.** Relative error for estimation of sensitivity indices of input parameters using various Monte Carlo and quasi-Monte Carlo approaches ($n \approx 2^{16}$).

| Est. qnt. | Ref. val. | LHSE1 | LHSE-1 | LHSR1 | LHSR-1 |
|---|---|---|---|---|---|
| $S_1$ | 9e−01 | 1.19e−03 | 2.72e−03 | 2.75e−03 | 5.98e−03 |
| $S_2$ | 2e−04 | 6.16e−01 | 4.53e−01 | 1.99e−01 | 5.40e−02 |
| $S_3$ | 1e−01 | 7.19e−03 | 1.49e−02 | 1.38e−02 | 3.66e−03 |
| $S_4$ | 4e−05 | 9.65e−01 | 2.22e+00 | 4.11e+00 | 5.30e−01 |
| $S_1^{tot}$ | 9e−01 | 7.04e−04 | 1.79e−03 | 1.81e−03 | 5.15e−04 |
| $S_2^{tot}$ | 2e−04 | 9.78e−01 | 3.45e−01 | 2.36e−01 | 1.06e+00 |
| $S_3^{tot}$ | 1e−01 | 7.67e−03 | 1.76e−03 | 2.02e−02 | 6.24e−04 |
| $S_4^{tot}$ | 5e−05 | 8.66e−01 | 2.72e+00 | 1.08e+00 | 7.54e−01 |

**Table 5.** Relative error for estimation of sensitivity indices of input parameters ($n \approx 2^{20}$).

| Est. qnt. | Ref. val. | LHSE1 | LHSE-1 | LHSR1 | LHSR-1 |
|---|---|---|---|---|---|
| $S_1$ | 9e−01 | 4.66e−04 | 1.46e−04 | 1.40e−04 | 2.99e−04 |
| $S_2$ | 2e−04 | 6.38e−03 | 5.63e−02 | 1.98e−01 | 2.55e−01 |
| $S_3$ | 1e−01 | 4.01e−03 | 1.27e−03 | 4.06e−04 | 4.05e−04 |
| $S_4$ | 4e−05 | 7.09e−01 | 4.81e−01 | 5.24e−01 | 1.77e−01 |
| $S_1^{tot}$ | 9e−01 | 4.97e−04 | 1.34e−04 | 2.26e−05 | 1.31e−04 |
| $S_2^{tot}$ | 2e−04 | 5.40e−02 | 1.59e−03 | 3.09e−01 | 3.49e−01 |
| $S_3^{tot}$ | 1e−01 | 4.42e−03 | 3.09e−02 | 6.25e−04 | 1.71e−03 |
| $S_4^{tot}$ | 5e−05 | 5.85e−01 | 1.08e+00 | 3.11e−01 | 1.02e−01 |

## 4    Sensitivity Studies with Respect to Chemical Reactions Rates

In this section we will study the sensitivity of the ozone concentration values in the air over Genova with respect to the rate variation of some chemical reactions of the condensed CBM-IV scheme [12], namely: # 1, 3, 7, 22 (time-dependent) and # 27, 28 (time-independent). The simplified chemical equations of those reactions are:

$$[\#1] \quad NO_2 + h\nu \Longrightarrow NO + O;$$
$$[\#3] \quad O_3 + NO \Longrightarrow NO_2;$$
$$[\#7] \quad NO_2 + O_3 \Longrightarrow NO_3;$$

$$[\#22] \quad HO_2 + NO \Longrightarrow OH + NO_2;$$
$$[\#27] \quad HO_2 + HO_2 \Longrightarrow H_2O_2;$$
$$[\#28] \quad OH + CO \Longrightarrow HO_2.$$

The relative error estimation for the quantities $f_0$, the total variance $\mathbf{D}$ and some sensitivity indices are given in Tables 6, 7, 8, 9 and 10 respectively. The quantity $f_0$ is presented by 6-dimensional integral, while the rest of quantities are presented by 12-dimensional integrals. Table 10 is similar to Tables 9 and 8, with the only difference − the increased number of samples $n = 2^{20}$ (instead of $n = 2^{16}$ in Table 9 and $n = 2^{10}$ in

**Table 6.** Relative error for the evaluation of $f_0 \approx 0.27$.

| # of samples | LHSE1 | LHSE-1 | LHSR1 | LHSR-1 |
| $n$ | Relative error | Relative error | Relative error | Relative error |
|---|---|---|---|---|
| $2^{10}$ | 4.90e−05 | 5.86e−04 | 9.90e−05 | 1.19e−03 |
| $2^{12}$ | 7.47e−05 | 1.05e−04 | 7.21e−04 | 1.04e−04 |
| $2^{14}$ | 2.43e−04 | 1.89e−04 | 3.38e−04 | 2.26e−04 |
| $2^{16}$ | 2.99e−05 | 1.73e−05 | 2.92e−05 | 6.15e−05 |
| $2^{18}$ | 1.29e−04 | 2.09e−05 | 3.04e−05 | 3.68e−05 |
| $2^{20}$ | 2.07e−05 | 2.99e−06 | 1.31e−05 | 1.21e−05 |

**Table 7.** Relative error for the evaluation of the total variance $\mathbf{D} \approx 0.0025$.

| # of samples | LHSE1 | LHSE-1 | LHSR1 | LHSR-1 |
| $n$ | Relative error | Relative error | Relative error | Relative error |
|---|---|---|---|---|
| $2^{10}$ | 5.30e−02 | 5.01e−02 | 3.86e−03 | 1.04e−01 |
| $2^{12}$ | 1.70e−02 | 2.90e−03 | 6.63e−02 | 4.79e−02 |
| $2^{14}$ | 1.47e−02 | 1.41e−02 | 2.48e−02 | 3.25e−02 |
| $2^{16}$ | 5.81e−03 | 2.91e−03 | 7.13e−03 | 1.51e−02 |
| $2^{18}$ | 6.91e−03 | 1.73e−03 | 7.22e−04 | 4.72e−03 |
| $2^{20}$ | 2.30e−03 | 2.84e−04 | 9.33e−05 | 1.07e−03 |

**Table 8.** Relative error for estimation of sensitivity indices of input parameters ($n \approx 2^{10}$).

| Est. qnt. | Ref. val. | LHSE1 | LHSE-1 | LHSR1 | LHSR-1 |
|---|---|---|---|---|---|
| $S_1$ | 4e−01 | 8.83e−03 | 2.40e−03 | 1.35e−01 | 1.41e−01 |
| $S_2$ | 3e−01 | 1.68e−01 | 3.04e−02 | 6.03e−02 | 1.17e−01 |
| $S_3$ | 5e−02 | 1.30e−01 | 4.19e−01 | 2.52e−01 | 2.20e−02 |
| $S_4$ | 3e−01 | 1.65e−01 | 1.76e−02 | 7.06e−02 | 8.60e−02 |
| $S_5$ | 4e−07 | 3.81e+03 | 5.17e+03 | 3.10e+03 | 6.53e+03 |
| $S_6$ | 2e−02 | 5.97e−01 | 1.13e+00 | 4.29e−01 | 1.82e−01 |
| $S_1^{tot}$ | 4e−01 | 6.77e−02 | 5.56e−02 | 1.04e−02 | 1.82e−02 |
| $S_2^{tot}$ | 3e−01 | 2.39e−01 | 2.50e+00 | 1.47e−01 | 1.80e−01 |
| $S_3^{tot}$ | 5e−02 | 3.74e−02 | 1.36e−01 | 3.18e−01 | 4.92e−01 |
| $S_4^{tot}$ | 3e−01 | 2.40e−01 | 1.17e−02 | 4.71e−02 | 1.10e−01 |
| $S_5^{tot}$ | 2e−04 | 4.79e+00 | 6.65e+00 | 8.59e+00 | 2.65e+01 |
| $S_6^{tot}$ | 2e−02 | 3.72e−01 | 5.90e−01 | 7.67e−01 | 3.64e−02 |
| $S_{12}$ | 6e−03 | 2.97e+00 | 1.48e+00 | 9.64e+00 | 5.18e+00 |
| $S_{14}$ | 5e−03 | 4.97e+00 | 6.74e−01 | 2.61e+00 | 5.77e−01 |
| $S_{15}$ | 8e−06 | 8.60e+02 | 9.12e+02 | 1.00e+03 | 8.17e+02 |
| $S_{24}$ | 3e−03 | 8.40e−02 | 3.58e+00 | 3.49e+00 | 2.72e+00 |
| $S_{45}$ | 1e−05 | 1.33e+01 | 2.64e+01 | 1.12e+02 | 9.94e+01 |

**Table 9.** Relative error for estimation of sensitivity indices of input parameters using various MC approaches ($n \approx 2^{16}$).

| Est. qnt. | Ref. val. | LHSE1 | LHSE-1 | LHSR1 | LHSR-1 |
|---|---|---|---|---|---|
| $S_1$ | 4e−01 | 3.19e−03 | 5.38e−03 | 9.94e−03 | 1.49e−02 |
| $S_2$ | 3e−01 | 2.23e−02 | 2.13e−02 | 4.19e−03 | 2.34e−03 |
| $S_3$ | 5e−02 | 8.71e−02 | 6.84e−02 | 1.24e−02 | 1.58e−02 |
| $S_4$ | 3e−01 | 5.16e−03 | 5.60e−03 | 2.14e−02 | 1.62e−02 |
| $S_5$ | 4e−07 | 1.02e+03 | 1.01e+03 | 2.37e+01 | 2.29e+02 |
| $S_6$ | 2e−02 | 7.27e−02 | 4.22e−02 | 1.71e−02 | 1.03e−01 |
| $S_1^{tot}$ | 4e−01 | 7.30e−03 | 3.85e−03 | 1.88e−02 | 9.89e−03 |
| $S_2^{tot}$ | 3e−01 | 1.05e−02 | 1.73e−02 | 1.10e−02 | 4.60e−04 |
| $S_3^{tot}$ | 5e−02 | 1.00e−01 | 1.07e−01 | 2.71e−03 | 6.85e−03 |
| $S_4^{tot}$ | 3e−01 | 6.82e−03 | 1.57e−02 | 5.89e−03 | 8.41e−04 |
| $S_5^{tot}$ | 2e−04 | 5.13e−01 | 1.21e+00 | 9.24e−01 | 1.61e+00 |
| $S_6^{tot}$ | 2e−02 | 2.59e−02 | 1.01e−01 | 4.23e−02 | 1.82e−01 |
| $S_{12}$ | 6e−03 | 2.75e−02 | 1.42e−01 | 5.70e−01 | 2.68e−01 |
| $S_{14}$ | 5e−03 | 2.35e−02 | 1.10e−01 | 8.29e−01 | 1.29e+00 |
| $S_{15}$ | 8e−06 | 9.25e+02 | 9.33e+02 | 9.06e+02 | 9.25e+02 |
| $S_{24}$ | 3e−03 | 3.52e−01 | 6.26e−02 | 1.53e−01 | 5.18e−01 |
| $S_{45}$ | 1e−05 | 2.55e+00 | 3.88e+00 | 2.29e+00 | 4.13e+00 |

**Table 10.** Relative error for estimation of sensitivity indices of input parameters ($n \approx 2^{20}$).

| Est. qnt. | Ref. val. | LHSE1 | LHSE-1 | LHSR1 | LHSR-1 |
|---|---|---|---|---|---|
| $S_1$ | 4e−01 | 7.83e−04 | 7.26e−05 | 6.12e−03 | 2.73e−03 |
| $S_2$ | 3e−01 | 4.62e−03 | 3.65e−03 | 1.70e−03 | 2.91e−03 |
| $S_3$ | 5e−02 | 6.86e−03 | 5.05e−03 | 7.73e−04 | 7.33e−03 |
| $S_4$ | 3e−01 | 2.98e−04 | 2.65e−03 | 2.46e−03 | 2.21e−03 |
| $S_5$ | 4e−07 | 9.28e+00 | 1.28e+02 | 1.53e+02 | 3.06e+02 |
| $S_6$ | 2e−02 | 5.68e−03 | 6.60e−03 | 2.54e−02 | 1.09e−02 |
| $S_1^{tot}$ | 4e−01 | 1.23e−03 | 1.23e−03 | 2.39e−03 | 2.37e−03 |
| $S_2^{tot}$ | 3e−01 | 2.25e−03 | 3.68e−03 | 7.43e−03 | 4.91e−03 |
| $S_3^{tot}$ | 5e−02 | 1.39e−02 | 8.33e−03 | 1.21e−02 | 1.16e−02 |
| $S_4^{tot}$ | 3e−01 | 2.91e−04 | 3.47e−03 | 2.88e−03 | 3.23e−03 |
| $S_5^{tot}$ | 2e−04 | 4.68e−01 | 1.12e+00 | 2.68e−01 | 3.52e−01 |
| $S_6^{tot}$ | 2e−02 | 2.33e−02 | 6.06e−03 | 4.22e−02 | 9.54e−03 |
| $S_{12}$ | 6e−03 | 3.06e−01 | 1.22e−01 | 3.05e−01 | 1.39e−03 |
| $S_{14}$ | 5e−03 | 9.18e−02 | 5.75e−02 | 9.69e−02 | 6.95e−02 |
| $S_{15}$ | 8e−06 | 9.31e+02 | 9.32e+02 | 9.29e+02 | 9.27e+02 |
| $S_{24}$ | 3e−03 | 7.99e−02 | 3.96e−01 | 3.16e−02 | 2.87e−01 |
| $S_{45}$ | 1e−05 | 2.01e+00 | 4.83e−01 | 1.96e+00 | 2.41e+00 |

Table 8). In general, this increases the accuracy of the estimated quantities. Exceptions are $S_5$, $S_5^{tot}$, $S_{15}$ and $S_{45}$, which have extremely small reference values and none of the 4 methods estimates them reliably. All LHS methods produce similar results, but for small in value sensitivity indices sometimes LHSE-1 and sometimes LHSE1 give the best results - see for example the values of $S_5$ and $S_4^{tot}$ in Table 10.

## 5  Conclusion

The present study focuses on so-called environmental safety. The computational efficiency (in terms of relative error and computational time) of the Latin Hypercube Sampling Random and Edge algorithms with various seeds for multidimensional numerical integration have been studied to analyze the sensitivity of UNI-DEM air pollution model output to variations of the input anthropogenic emissions and to variations of the rates of several chemical reactions.

The various Latin Hypercube Sampling techniques have been successfully applied to compute global Sobol sensitivity measures corresponding to the influence of several input parameters on the concentrations of important air pollutants. The novelty of the proposed approaches is that the Latin Hypercube Sampling Edge algorithm with different seeds has been applied for the first time to sensitivity studies of the particular air pollution model. The numerical tests show that the presented stochastic approaches are efficient for computing the multidimensional integrals, which are used to calculate the sensitivity measures under consideration.

None of the 4 methods presented estimates reliably the quantities, which have extremely small reference values (in comparison with the largest value of the same type). In general, this natural "size effect" does not destroy the accuracy of the corresponding total sensitivity index as far as its value is much larger, so the influence of the small partial sensitivity indices is negligible (as $S_{15}$ and $S_{45}$ with respect to $S_1^{tot}$ and $S_4^{tot}$ in Tables 8, 9 and 10).

The results of our study, especially for the total sensitivity indices, are important for the decision makers and should be interpreted as follows: the larger the value of $S_i^{tot}$, the more critical the value of the $i$-th parameter with respect to the corresponding output result. For instance, analysing the results from Tables 3, 4 and 5, one can conclude that in order to reduce the ammonia mean monthly concentrations by certain restriction in the emissions, these should primarily be the emissions of $NH_3$ and $SO_2$ as their total sensitivity indices ($S_1^{tot}$ and $S_3^{tot}$) are strongly dominant, compared to the rest. On the contrary, measures to reduce the emissions of nitrogen oxides or anthropogenic hydrocarbons would not pay off for this purpose, as their corresponding total sensitivity indices ($S_2^{tot}$ and $S_4^{tot}$) are too small, so the effect of that would be negligible.

**Acknowledgments.** This work is supported by the Bulgarian National Science Fund under Project DN 12/5/2017 "Efficient Stochastic Methods and Algorithms for Large-Scale Problems". V. Todorov is also supported by the Bulgarian National Science Fund under Young Scientists Project KP-06-M32/2/2019 "Advanced Stochastic and Deterministic Approaches for Large-Scale Problems of Computational Mathematics" and by the National Scientific Program "Information and Communication Technologies for a Single Digital Market in Science, Education, and Security (ICT in SES)", contract No DO1-205/2018, financed by the Ministry of Education and Science in

Bulgaria. The work of I. Dimov is also supported by the Project KP-06-Russia/2017 "New Highly Efficient Stochastic Simulation Methods and Applications", funded by the Bulgarian National Science Fund.

# References

1. Dimitriu, G.: Global sensitivity analysis for a Chronic Myelogenous Leukemia model. In: Nikolov, G., Kolkovska, N., Georgiev, K. (eds.) NMA 2018. LNCS, vol. 11189, pp. 375–382. Springer, Cham (2019). https://doi.org/10.1007/978-3-030-10692-8_42

2. Dimov, I., Georgieva, R., Ostromsky, T., Zlatev, Z.: Variance-based sensitivity analysis of the unified Danish Eulerian model according to variations of chemical rates. In: Dimov, I., Faragó, I., Vulkov, L. (eds.) NAA 2012. LNCS, vol. 8236, pp. 247–254. Springer, Heidelberg (2013). https://doi.org/10.1007/978-3-642-41515-9_26

3. Dimov, I.T., Georgieva, R., Ostromsky, T., Zlatev, Z.: Sensitivity studies of pollutant concentrations calculated by UNI-DEM with respect to the input emissions. Central Eur. J. Math. **11**(8), 1531–1545 (2013)

4. Gocheva-Ilieva, S.G., Voynikova, D.S., Stoimenova, M.P., Ivanov, A.V., Iliev, I.P.: Regression trees modeling of time series for air pollution analysis and forecasting. Neural Comput. Appl. **31**(12), 9023–9039 (2019). https://doi.org/10.1007/s00521-019-04432-1

5. Hamdad, H., Pézerat, Ch., Gauvreau, B., Locqueteau, Ch., Denoual, Y.: Sensitivity analysis and propagation of uncertainty for the simulation of vehicle pass-by noise. Appl. Acoust. **149**, 85–98 (2019). https://doi.org/10.1016/j.apacoust.2019.01.026

6. McKay, M.D., Beckman, R.J., Conover, W.J.: A comparison of three methods for selecting values of input variables in the analysis of output from a computer code. Technometrics **21**(2), 239–245 (1979)

7. Minasny, B., McBratney, A.B.: Conditioned Latin hypercube sampling for calibrating soil sensor data to soil properties. In: Viscarra Rossel, R.A., McBratney, A.B., Minasny, B. (eds.) Proximal Soil Sensing. PSS, pp. 111–119. Springer, Dordrecht (2010). https://doi.org/10.1007/978-90-481-8859-8_9

8. Saltelli, S.: Making best use of model valuations to compute sensitivity indices. Comput. Phys. Commun. **145**, 280–297 (2002)

9. Sobol, I.M., Tarantola, S., Gatelli, D., Kucherenko, S., Mauntz, W.: Estimating the approximation error when fixing unessential factors in global sensitivity analysis. Reliab. Eng. Syst. Saf. **92**, 957–960 (2007)

10. Sobol', I.M.: Sensitivity estimates for nonlinear mathematical models. Matem. Modelirovanie **2**(1), 112–118 (1990)

11. Veleva, E., Georgiev, I.R., Zheleva, I., Filipova, M.: Markov chains modelling of particulate matter (PM10) air contamination in the city of Ruse, Bulgaria. In AIP Conference Proceedings, vol. 2302, p. 060018. AIP Publishing LLC (2020)

12. Zlatev, Z., Dimov, I.T.: Computational and Numerical Challenges in Environmental Modelling. Elsevier, Amsterdam (2006)

13. Wolfram Group: Mathematica package. http://www.wolfram.com/mathematica/

# Sensitivity Operator-Based Approach to the Interpretation of Heterogeneous Air Quality Monitoring Data

Alexey Penenko[1,2]([envelope]) [ORCID], Vladimir Penenko[1,2] [ORCID], Elena Tsvetova[1] [ORCID],
Alexander Gochakov[3] [ORCID], Elza Pyanova[1] [ORCID], and Viktoriia Konopleva[1,2]

[1] Institute of Computational Mathematics and Mathematical Geophysics SB RAS,
pr. Akademika Lavrentjeva 6, 630090 Novosibirsk, Russia
`aleks@ommgp.sscc.ru`
[2] Novosibirsk State University, Pirogova Street 1, 630090 Novosibirsk, Russia
[3] Siberian Regional Hydrometeorological Research Institute, Sovetskaya Street 30,
630099 Novosibirsk, Russia

**Abstract.** The joint use of atmospheric chemistry transport and transformation models and observational data makes it possible to solve a wide range of environment protection tasks, including pollution sources identification and reconstruction of the pollution fields in unobserved areas. Seamless usage of different measurement data types can improve the accuracy of air quality forecasting systems. The approach considered is based on sensitivity operators and adjoint equations solutions ensembles. The ensemble construction allows for the natural combination of various measurement data types in one operator equation. In the paper, we consider combining image-type, integral-type, pointwise, and time series-type measurement data for the air pollution source identification. The synergy effect is numerically illustrated in the inverse modeling scenario for the Baikal region.

**Keywords:** Air quality · Inverse modeling · Sensitivity operator · Heterogeneous measurements · Advection-diffusion-reaction model

## 1 Introduction

Air quality monitoring systems vary in their temporal and spatial coverage, the composition of the observed chemicals, and the accuracy of measurements (see the review in [7]). Measurement devices can be mounted on satellites, ground-based systems, aircraft-based, including unmanned aerial vehicles (UAVs), balloons, and rockets. Observable chemical species include ozone ($O_3$), nitrogen-dioxide ($NO_2$), formaldehyde ($HCHO$), and sulfur dioxide ($SO_2$), as well as

Supported by the grant №075-15-2020-787 in the form of a subsidy for a Major scientific project from Ministry of Science and Higher Education of Russia (project "Fundamentals, methods and technologies for digital monitoring and forecasting of the environmental situation on the Baikal natural territory").

I. Lirkov and S. Margenov (Eds.): LSSC 2021, LNCS 13127, pp. 164–171, 2022.
https://doi.org/10.1007/978-3-030-97549-4_19

aerosol optical depth. The joint use of atmospheric chemistry transport and transformation models and observational data makes it possible to solve a wide range of environment protection tasks, including pollution sources identification and reconstruction of the pollution fields in unobserved areas. Seamless usage of different data types can improve the accuracy of air quality forecasting systems.

The considered inverse modeling approach is based on sensitivity operators and adjoint equations solutions ensembles [4]. The source identification problem for the advection-diffusion-reaction model is transformed into a quasi-linear operator equation family with the sensitivity operator. The sensitivity operator is constructed of the sensitivity functions' ensemble corresponding to the measurement data elements. The ensemble structure allows for the natural combination of various measurement data types in one operator equation.

In the paper, we consider a combined use of image-type [4], integral-type, pointwise (*in situ*), and time series-type [6] measurement data for the air pollution source identification. Previously, we studied the synergy effect of using both image-type and time series-type measurements [5]. In the current work, we consider combined measurements for the expanded list of measurement types. The paper's objective is to present the sensitivity operator-based approach for the interpretation of heterogeneous measurement data.

## 2    Mathematical Background and Methods

### 2.1    Problem Statement

A convection-diffusion-reaction model for $l = 1, \ldots, N_c$ is considered in a domain $\Omega_T = \Omega \times (0, T)$, where $\Omega$ is a sufficiently smooth approximation of a bounded rectangular domain $[0, X] \times [0, Y]$ in $\mathbb{R}^2$, $T > 0$. Let $\partial \Omega_T = \partial \Omega \times [0, T]$.

$$\frac{\partial \varphi_l}{\partial t} - \nabla \cdot (\operatorname{diag}(\mu_l) \nabla \varphi_l - \mathbf{u}\varphi_l) + P_l(t, \varphi)\varphi_l = \Pi_l(t, \varphi) + f_l + r_l, \tag{1}$$
$$(\mathbf{x}, t) \in \Omega_T,$$

$$\mathbf{n} \cdot (\operatorname{diag}(\mu_l) \nabla \varphi_l) + \beta_l \varphi_l = \alpha_l, \quad (\mathbf{x}, t) \in \Gamma_{out} \subset \partial \Omega_T, \tag{2}$$
$$\varphi_l = \alpha_l, \quad (\mathbf{x}, t) \in \Gamma_{in} \subset \partial \Omega_T, \tag{3}$$
$$\varphi_l = \varphi_l^0, \quad \mathbf{x} \in \Omega, \, t = 0, \tag{4}$$

where $N_c$ is the number of considered substances, $\varphi_l = \varphi_l(x, t)$ denotes the concentration of the $l^{th}$ substance at a point $(x, t) \in \Omega_T$, $\varphi$ is the vector of $\varphi_l(\mathbf{x}, t)$ for $l = 1, \ldots, N_c$ called the state function, $L = \{1, \ldots, N_c\}$. The functions $\mu_l(\mathbf{x}, t) \in \mathbb{R}^2$ correspond to the diffusion coefficients, $\operatorname{diag}(\mathbf{a})$ is the diagonal matrix with the vector $\mathbf{a}$ on the diagonal, $\mathbf{u}(\mathbf{x}, t) \in \mathbb{R}^2$ is the underlying flow speed. $\Gamma_{in}$ and $\Gamma_{out}$ are parts of $\partial \Omega_T$ in which the vector $\mathbf{u}(\mathbf{x}, t)$ points inwards the domain $\Omega_T$ and is zero or is directed outwards the domain $\Omega_T$, correspondingly, $\mathbf{n}$ is the outer normal. Loss and production operator elements

$P_l, \Pi_l : [0, T] \times \mathbb{R}_+^{N_c} \to \mathbb{R}_+$ are defined by the transformation model. We suppose all the functions and model parameters are smooth enough for the solutions to exist and the further transformations to make sense. The functions $\alpha_l(\mathbf{x}, t)$, $\varphi_l^0(\mathbf{x})$ are boundary and initial conditions, correspondingly, $f_l(\mathbf{x}, t)$ is the *a priori* known part of the source function.

Model parameters are divided into "predefined" ones $\mathbf{v}$ and "uncertainty functions" $\mathbf{q}$ belonging to some set $Q$. Most often, the uncertainty functions in the air quality studies are unknown sources of pollution $\mathbf{q} = \{\mathbf{r}\}$ (e.g. [2]). Hence $\mathbf{v} = \{\mu, \mathbf{u}, \alpha, \varphi^0, f\}$. Let $\mathbf{r} \in Q$, where $Q$ is the set of admissible sources such that *a priori* given constraints are fulfilled. In the "Direct problem", both $\mathbf{v}$ and $\mathbf{q}$ are given and $\varphi$ has to be found from (1)–(4). In the "Inverse problem", let there be an "exact" source function $\mathbf{q}^{(*)} = \mathbf{r}^{(*)}$ to be found from (1)–(4) by given $\mathbf{v}$ and using the measurement data described in Sect. 2.3. Let $\varphi^{(*)}$ denote the solution of the direct problem with the source function $\mathbf{r}^{(*)}$.

## 2.2   Inverse Modeling with Sensitivity Operators

For the source identification problem's solution, we use the algorithm based on the ensembles of the adjoint problem solutions described in [4,6]. In the abstract form, the relation between the model state function variation and uncertainty function variation is given by the sensitivity relation

$$\left\langle S[\mathbf{q}^{(2)}, \mathbf{q}^{(1)}; h], \mathbf{q}^{(2)} - \mathbf{q}^{(1)} \right\rangle_Q = \langle h, \delta\varphi \rangle_H , \tag{5}$$

where $\langle ., . \rangle_Q$ and $\langle ., . \rangle_H$ are bi-linear forms (scalar products) on the spaces of uncertainty and state functions variations correspondingly, $S[\mathbf{q}^{(2)}, \mathbf{q}^{(1)}; h]$ denotes the sensitivity function, which is calculated by the solution of the adjoint problem (the details can be found in [4]). The solution of the adjoint problem is determined by its source function $h$, which the measurement operator defines.

If we consider a set of $\Xi \in \mathbf{N}$ functions $U = \left\{h^{(\xi)}\right\}_{\xi=1}^{\Xi}$, then combining the corresponding relations, we obtain the sensitivity operator relation

$$M_U \left[\mathbf{q}^{(2)}, \mathbf{q}^{(1)}\right] \left(\mathbf{q}^{(2)} - \mathbf{q}^{(1)}\right) = H_U\varphi\left[\mathbf{q}^{(2)}\right] - H_U\varphi\left[\mathbf{q}^{(1)}\right],$$

where

$$M_U \left[\mathbf{q}^{(2)}, \mathbf{q}^{(1)}\right] z = \sum_{\xi=1}^{\Xi} \mathbf{e}^{(\xi)} \left\langle S[\mathbf{q}^{(2)}, \mathbf{q}^{(1)}; h^{(\xi)}], z \right\rangle_Q ,$$

$$H_U\varphi = \sum_{\xi=1}^{\Xi} \mathbf{e}^{(\xi)} \left\langle h^{(\xi)}, \varphi \right\rangle_H ,$$

functions $\mathbf{e}^{(\xi)}$ are the elements of the canonical basis in $\mathbb{R}^\Xi$. The adjoint problem solutions needed to compose the sensitivity operator can be evaluated in parallel as an ensemble. Combining ensembles of solutions of adjoint equations corresponding to different measurement data types into one sensitivity operator allows us to consider heterogeneous monitoring systems.

If $\mathbf{q}^{(*)}$ is the "exact" solution of the inverse problem, $I$ is the measurement data, aggregated in the state-function form, and $\delta I$ is its perturbation, then for any $U$ and $\mathbf{q}$ the relation holds:

$$M_U \left[ \mathbf{q}^{(*)}, \mathbf{q} \right] \left( \mathbf{q}^{(*)} - \mathbf{q} \right) = H_U I + H_U \delta I - H_U \varphi \left[ \mathbf{q} \right]. \tag{6}$$

Quasi-linear operator Eq. (6) can be solved by the Newton-Kantorovich-type algorithm with the inversion, regularized by the truncated SVD [4].

### 2.3    Measurement Data in a Heterogeneous Monitoring System

Let the scalar product $\langle ., . \rangle_H$ in (5) be

$$\langle a, b \rangle_H = \sum_{l=1}^{N_c} \int_0^T \int_\Omega a_l(\mathbf{x}, t) b_l(\mathbf{x}, t) d\mathbf{x} dt.$$

We consider the following measurement types defined by their correspondence to the model state function and projection system elements $h^{(\xi)}$:

- **"Pointwise Time series"**: Time series of concentrations in specific points. In the state function terms:

$$\left\{ \varphi_{l^{(m)}}(x^{(m)}, t), t \in [0, T], \left( x^{(m)}, l^{(m)} \right) \in (\Omega \times L)_{meas} \right\},$$

projection system: $h^{(\xi)} = C(T, \theta^{(\xi)}, t)\delta(x - x^{(\xi)})\delta(l - l^{(\xi)})$;
- **"Pointwise"**: Pointwise concentration measurements at specific moments and specific points. In the state function terms:

$$\left\{ \varphi_{l^{(m)}}(x^{(m)}, t^{(m)}), \left( x^{(m)}, t^{(m)}, l^{(m)} \right) \in (\Omega_T \times L)_{meas} \right\},$$

projection system: $h^{(\xi)} = \delta(x - x^{(\xi)})\delta(t - t^{(\xi)})\delta(l - l^{(\xi)})$;
- **"Pointwise Integral"**: Integral of concentrations over the time interval in specific points. In the state function terms:

$$\left\{ \int_0^T \varphi_{l^{(m)}}(x^{(m)}, t)dt, \left( x^{(m)}, l^{(m)} \right) \in (\Omega \times L)_{meas} \right\},$$

projection system: $h^{(\xi)} = \delta(x - x^{(\xi)})\delta(l - l^{(\xi)})$;
- **"Snapshot"**: Concentration fields images at specific moments in time. In the state function terms:

$$\left\{ \varphi_{l^{(m)}}(x, t^{(m)}), \left( t^{(m)}, l^{(m)} \right) \in ([0, T] \times L)_{meas}, x \in \Omega \right\},$$

projection system: $h^{(\xi)} = C(X, \theta_x^{(\xi)}, x)C(Y, \theta_y^{(\xi)}, y)\delta(t - t^{(\xi)})\delta(l - l^{(\xi)})$.

Here $\delta$ is the appropriate delta-function and $C(X, \theta, x)$ is the element of the cosine-basis on an interval $[0, X]$:

$$C(X, \theta, x) = \frac{1}{\sqrt{X}} \begin{cases} \sqrt{2} \cos\left(\frac{\pi \theta x}{X}\right), & \theta > 0 \\ 1, & \theta = 0 \end{cases}.$$

To reflect the measurements' different reliability, we can introduce the weights for the corresponding projection system elements. In the current work, we consider all the measurements to be equally reliable, and therefore in the computations, the finite-dimensional analogs of the projection functions are normalized.

## 2.4  Inverse Modeling Scenario

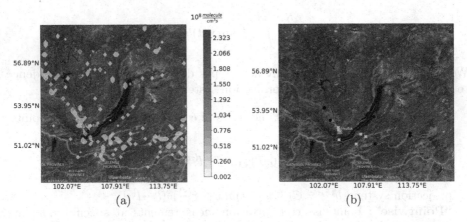

**Fig. 1.** Emission sources (a). Monitoring sites locations (b): pointwise time series (red crosses), pointwise measurements in space and time (blue circles), pointwise integrals over a time interval (magenta triangles). (Color figure online)

We used an inverse modeling scenario generated as a result of solving the direct problem for $NO$ emission sources located in the cities of the Baikal Region with the constant emission rates proportional to the city population (Fig. 1a). The meteorological data (wind fields $\mathbf{u}$) for the scenario were calculated with the COSMO model [1] for the period from 12:00 23.07.2019 to 12:00 03.08.2019. The diffusion coefficient was chosen as a constant: $\mu = 1000 \, \mathrm{m}^2/\mathrm{sec}$. As an atmospheric chemistry model, we used a photochemical nitrogen-ozone cycle model [3, p. 6], including substances $NO$, $NO_2$, $O_2$, $O_3$, $O$ involved in three chemical reactions. $O_2$ concentration is kept constant. The initial concentrations were $[NO] = 4 \cdot 10^{10}$, $[NO_2] = 1.3 \cdot 10^{10}$, $[O_2] = 5.3 \cdot 10^{18}$, $[O_3] = 1.5 \cdot 10^{12}$, $[O] = 0$ in $molecules/cm^3$.

**Table 1.** Adjoint ensemble sizes.

| Scenario | $\Xi$ | Description |
|---|---|---|
| Pointwise | 60 | 5 sites × 12 days |
| Pointwise Time series | 60 | 6 sites × 10 1D-basis elements |
| Pointwise Integral | 5 | 5 sites |
| Snapshot | 400 | 4 images × 10 1D-basis elements × 10 1D-basis elements |
| Composite | 525 | Sum of the above |

The model monitoring system measures $O_3$ concentrations. The Roshydromet monitoring sites were taken as the prototype of the measurement system (Fig. 1b). The monitoring sites were divided into three groups. In the first group, the time series of concentrations were available (red crosses in Fig. 1b). In the second group, the point measurements of concentration in space and time with a period of 24 h (blue circles in Fig. 1b) were given. In the third group of sites, the state function's integrals for the entire time interval (magenta triangles in Fig. 1b) were "measured". Images of the concentration fields were available with a period of three days.

In the numerical experiment, we estimate the synergy effect of using different measurement types. To do this, we compare the results of four scenarios for the specific measurement types from Sect. 2.3: "Pointwise", "Pointwise Time series", "Pointwise Integral", "Snapshot" to the results of the "Composite" scenario. The ensemble sizes $\Xi$ are given in Table 1. The numbers of grid points in time, space, and chemical species were $N_t = 18801$, $N_x = 60$, $N_y = 57$, $N_c = 5$. Computations were carried out on a dual Intel Xeon Gold 5220 processor with 768 GB Memory. The computation time was limited by 12 h.

## 3  Results and Discussion

The source identification algorithm convergence dynamics with respect to computation time, measured in direct problem solution times ($t_{direct} \approx 20$) s, is shown in Fig. 2. In Fig. 2a, the relative error decrease of the source identification is presented. In Fig. 2b, we compare the errors of the emitted substance concentration field reconstruction. In terms of both source and state function reconstruction, the best results are achieved in the "Composite" scenario, as expected. The worst result in terms of the source function identification is obtained in the "Snapshot" scenario, but it was comparable to the results in the other scenarios in terms of the state function reconstruction. It may be due to the "exact" solution's pointwise character, which is lost in the concentration images. Figure 3 shows the reconstruction results for different measurement system configurations. From the qualitative analysis of Fig. 3, we can conclude that in the "Composite" scenario (Fig. 3b, the reconstruction result in the unobservant areas of the "pointwise" monitoring system in Fig. 1a (to the north of Lake Baikal) is more accurate than the results obtained in the "Snapshot" scenario (Fig. 3f).

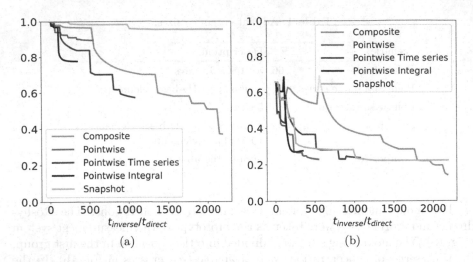

(a)

(b)

**Fig. 2.** Algorithm convergence dynamics with respect to the computation time, measured in direct problem solution times: relative source identification error (a) and emitted substance concentration field relative reconstruction error (b).

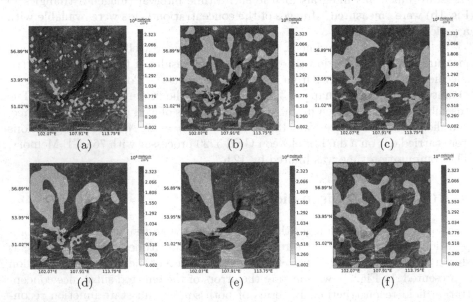

**Fig. 3.** Exact source (a); Source identification results: composite (b), pointwise (c), pointwise time series (d), pointwise integral (e), snapshot (f).

The approach based on the sensitivity operators allows both solving and analyzing the inverse problem through the sensitivity operator properties. In the future, we plan to use sensitivity operators to analyze heterogeneous monitoring systems. Another important question is the optimization of the ensemble size.

# 4    Conclusions

The sensitivity operator approach allows for seamless integration of various measurement data types in one operator equation by a simple combination of corresponding sensitivity function ensembles. The synergy effect was numerically illustrated in the inverse modeling scenario for the Baikal region.

# References

1. Baldauf, M., Seifert, A., Forstner, J., Majewski, D., Raschendorfer, M., Reinhardt, T.: Operational convective-scale numerical weather prediction with the COSMO model: description and sensitivities. Mon. Weather Rev. **139**(12), 3887–3905 (2011). https://doi.org/10.1175/mwr-d-10-05013.1
2. Elbern, H., Strunk, A., Schmidt, H., Talagrand, O.: Emission rate and chemical state estimation by 4-dimensional variational inversion. Atmos. Chem. Phys. Discuss. **7**(1), 1725–1783 (2007). https://doi.org/10.5194/acpd-7-1725-2007
3. Hundsdorfer, W., Verwer, J.G.: Numerical Solution of Time-Dependent Advection-Diffusion-Reaction Equations. Springer, Heidelberg (2013). https://doi.org/10.1007/978-3-662-09017-6
4. Penenko, A.: Convergence analysis of the adjoint ensemble method in inverse source problems for advection-diffusion-reaction models with image-type measurements. Inverse Probl. Imaging **14**(5), 757–782 (2020). https://doi.org/10.3934/ipi.2020035
5. Penenko, A.V., Gochakov, A., Antokhin, P.: Numerical study of emission sources identification algorithm with joint use of in situ and remote sensing measurement data. In: Matvienko, G.G., Romanovskii, O.A. (eds.) 26th International Symposium on Atmospheric and Ocean Optics, Atmospheric Physics. SPIE (2020). https://doi.org/10.1117/12.2575649
6. Penenko, V.V., Penenko, A.V., Tsvetova, E.A., Gochakov, A.V.: Methods for studying the sensitivity of air quality models and inverse problems of geophysical hydrothermodynamics. J. Appl. Mech. Tech. Phys. **60**(2), 392–399 (2019). https://doi.org/10.1134/s0021894419020202
7. Guide to Instruments and Methods of Observation, Volume 1-Measurement of Meteorological Variables. WMO, 2018 Edition (2018). https://library.wmo.int/index.php?lvl=notice_display&id=12407

# Using the Cauchy Criterion and the Standard Deviation to Evaluate the Sustainability of Climate Simulations

Valery Spiridonov and Hristo Chervenkov$^{(\boxtimes)}$ (iD)

National Institute of Meteorology and Hydrology, Tsarigradsko Shose blvd. 66,
1784 Sofia, Bulgaria
`hristo.tchervenkov@meteo.bg`
`http://www.meteo.bg`

**Abstract.** In simulations of climate change with climate models, the question arises as to whether the accepted 30-year period is sufficient for the model to produce sustainable results. We are looking for an answer to this question using the Cauchy criterion and the idea of saturation suggested by Lorenz, using as a sufficient condition for saturation the requirement that the standard deviation should not increase from year to year. The proposed method is illustrated by analysis of time series of observations of the temperature at 2 m and of three global climate models for the same parameter. The measured data are for the Cherni Vrah peak in Bulgaria and the modelled data, which concerns the projected future climate up to year 2100, are taken from the Climate Data Store of the Copernicus project. The example of the observations shows the instability of meteorological processes since 2000 year. All three global climate models, forced with the CMIP5 RCP4.5 scenario, show a lack of sustainability mainly in the polar regions for the period 2071–2100.

**Keywords:** Climate models · Cauchy criterion · Lorenz's saturation · Sustainability

## 1 Introduction and Problem Determination

When examining climate change using numerical models two questions arises.

1. Whether the results are independent on the initial conditions. In other words: whether climate simulation experiments are a problem without initial conditions.
2. Is a 30-year period sufficient to allow the climate model to arrive at a solution that maintains temperatures with such values that the climate norms remain stable after a fixed integration period for the remainder of the 30-year period and beyond? If so we will call it a **"sustainable solution"**.

Climate is the average weather over a relatively long period of time. The choice of a 30-year period to determine the climate norm has its history. As mentioned

© Springer Nature Switzerland AG 2022
I. Lirkov and S. Margenov (Eds.): LSSC 2021, LNCS 13127, pp. 172–179, 2022.
https://doi.org/10.1007/978-3-030-97549-4_20

in [1]: *"The concept of the 30-year climatological standard normal dates from 1935"*, In 1956, WMO recommended the use of the most recent available period of 30 years, ending in the most recent year ending with the digit 0 (which at that time meant 1921–1950). The period currently in use is 1991–2021.

Climate change is usually determined by the difference between the norms of two periods. The first is called the "reference" (or control) period and the second is the next, "future" period. Numerical climate models are used to answer the question of what climate change we can expect. The weather forecast is a series of meteorological conditions that are expected to occur. The climate is a consequence of the conditions that have occurred. The question then is what research method should be used to analyze climate change simulated by a particular climate model. Weather forecasts are made by global circulation models (GCMs) and climate change is simulated by global climate models (GCMs). The abbreviations are the same, and in fact there is no fundamental difference between the two types of models. Some studies even use the same models [3].

There are a number of studies that theoretically and practically show that only for a limited period a meaningful weather forecast can be obtained. This period is within a few months. Both types of models need initial conditions that always have some error. After a certain period of integration, the error of the weather forecast becomes comparable to the error of arbitrary value within the limits of the climatic norm. A mathematical theory of the growth of small initial errors was provided by Lorenz [4–6]. He finds that small errors initially increase and then slow down before they reach **"saturation"**. This is the moment when the root mean square error between the observed and the forecasted values reaches a limit beyond which it does not grow. Although these articles refer to long-term forecasting and "climate" is in the sense of seasonal forecasts, we will use the idea of saturation in climate simulations. GCMs are a complex system of interacting atmospheric, ocean and other models. This interaction is different for different models and it is reasonable to expect them to have different saturation periods. Saturation status should also be observed in climate models. The loss of predictability in the sense of the weather forecast is the start of climate simulation. This period often is called spin-up, It is assumed that it does not exceed 1–2 years. After this period, the climate model must begin to simulate climate. Assuming that the initial condition is one of the states of the atmosphere after reaching saturation, according to this theory the error of the weather forecast will not increase. However, this implies independence of climate norms from the initial conditions. The numerical models used to simulate climate require a different understanding of the results of this in the weather forecast. In [7], Lorentz introduced a **second type of climate predictability**: *"We shall refer to such predictions, which are not directly concerned with the chronological order in which atmospheric states occur, as climatic predictions of the second kind"*. For climate models, the term "climate simulation" (some times climate projection) is used instead of forecasting or prediction. When talking for reliability of an experiment we will consider its internal stability (hereafter sustainability), namely, whether in the chosen climatic period (say 30 years) the model reaches a state in which every eventual atmospheric conditions have occurred. The climatic norm is an average of all values during the period and therefore does not depend on the order of events. This actually is the second kind of climate predictability. If this norm reaches a sustainable value before the end of the period, it means that climate simulation gives a positive answer to the second question at the beginning.

In the present, rather demonstrative study, we will propose a method for evaluation sustainability of the climate simulation results. Whether these simulations are adequate is a different question. The proposed methodology is in fact an instrument for examining the behavior of the solution during the accepted climatic period. In this case, the period is 30 years.

## 2   Methodology

Although the definitions and expressions used herein refer to infinite series, they can be used to analyze processes that take place over a limited period of time. In the experiments with climate models, the only external factor whose change is included in the calculations is the amount of greenhouse gases. If it is constant throughout the integration period, then at some moment the model should reach some stable state. This means that the model does not fall into numerical instability and its solution is limited. Therefore, the proposed methodology requires appropriate scenarios where greenhouse gases (with attention to $CO_2$) do not grow or change much during the integration period. Some models may reach this stable state within the integration period and others may not. Insofar as climate experiments are conducted for a period of 30 years, we will examine the behavior of the model over this period. We will consider sequences of norms calculated by annual temperatures $T^p$ for different periods (denoted by $n$):

$$x_n = \frac{\sum_1^n T_n^p}{n}. \tag{1}$$

By $p$ we denote which temperatures the norm refers to. They can be from simulations for the reference (control) or future period, the difference between the future and reference periods or from observations. This period is denoted by $n$ with values from 1 to 30 (if a 30-year simulation period is accepted). We will examine the convergence of this sequence over a period of time using the Cauchy criterion. This criterion is suitable for studying convergence for each period, since it is not necessary to determine in advance the limit to which a sequence converges.

**Definition 1** *(Cauchy's criterion). The sequence $x_n$ converges to some point if and only if this holds: for every $\epsilon > 0$ there exists $K$ such that $|x_n - x_m| < \epsilon$ whenever $n, m \geq K$. This is necessary and sufficient.*

Later on Fig. 2 such convergence will be shown in the last few frames (i.e. $n, m \geq K$ when, say, $K \sim 24, 25$ depending on the models).

In order to be able to use the properties of infinite series in the case of finite sequences, we must accept a value $\epsilon$ small enough to treat the annual norms for different periods with such a small difference as indistinguishable. Since the temperatures in the meteorology are rounded to the tenth of a degree, we can set $0.05°C$ for $\epsilon$. In other words we are interested if $x_n$ defined Eq. 1 have the ability to last a long time at the same level. Below, we will use the term "sustainable" in the sense of steady behavior of the solution. A practically convenient condition for convergence of a series of annual temperatures can be found if $m$ is set to $n-1$. Although in this way we will obtain a necessary but not sufficient condition for the convergence of the Cauchy sequence 1, this condition is satisfactory for the analysis of a finite sequence with small number of elements. After some transformations we find:

$$\left| \frac{x_{n-1} - x_n}{\epsilon} \right| = \left| \frac{x_{n-1} - T_n^p}{n\epsilon} \right| < 1, \text{ for } n = \overline{2, N}. \tag{2}$$

In Eq. 2 $x_{n-1}$ is the $(n-1)$th partial sum formed as defined in Eq. 1 and $T_n^p$ can be the temperature from measurements, from the reference (control) period from future periods or from the difference between the future and the control period. It can be seen that for large $n$, condition 2 can be satisfied, although the sequence may not be convergent. That is the price to be paid for setting $m$ to $n - 1$. We will use this condition to test convergence for period of 30 years. This period is small and allows to establish possible non-convergence. This is illustrated below in Fig. 1 using observation data.

**Definition 2.** *The model is sustainable of the first kind in a given grid-point if in the Cauchy's sequence $\epsilon$ is sufficiently small.*

We will use the standard deviations $\sigma_n$ of consequence $\sigma(x_n)$ to define another criterion:

$$\sigma_n \leq \sigma_{n-1}, \text{ for } n = \overline{3, N}. \tag{3}$$

Let the minimum and maximum values of $x_n$ when $n < K$ are $x_a$ and $x_b$ respectively and they remain the same when $n \geq K$. The standard deviation has the property of decreasing or staying the same if each new term added to the sequence is within the interval $[x_a, x_b]$. If condition 3 is fulfilled for all subsequent values of $x_n$ when $n \geq K$ this will mean that the maximum and minimum values have already been reached and the process has stabilized. The standard deviation will not increase for each series of values if they are within the above interval. So, that inequality 3 is consistent with **"climatic predictability of the second kind"**, as defined by Lorenz. This is how we come to the Definition 3.

**Definition 3.** *The model is sustainable of the second kind in a given grid-point if the condition 3 is fulfilled.*

It can be proved that if the solution is sustainable of the first kind it is also sustainable of the second kind.

## 3   An Example with Observations and Three GCMs

Before looking at an example with a climate models, we will demonstrate the use of this methodology with observations from Cherni Vrah, the highest peak in Vitosha Mountain (2290 m a.s.l.). The synoptic station was established in 1935 and at that time it was the highest mountain station in Bulgaria. There has been no interruption of the observations, and the measurement method has been retained so far. There are no influences from industrial, urban changes or replacement, unlike other stations. The measurements used are for the period 1936 to 2019. Three sub-periods of 30 years were extracted from this period: 1936–1965, 1961–1990 and 1987–2019 as shown on the left plot on Fig. 1. The middle plot on Fig. 1 shows the convergence of the partial norms introduced in Eq. 1 and the right plot depicts the behavior of the series of standard deviations. For the first two periods, the partial norms converge with an error $\epsilon < 0.05$ after the 18th year from the beginning of the period, as shown in middle plot on Fig. 1. The lack of monotonically decreasing of standard deviation and convergence has been demonstrated in the third period during which temperatures rise systematically. The last period shows that the processes have not been stabilized during it. According to the above definition, process behavior is not sustainable in this period. It may be a happy coincidence that the period 1961 to 1990 is appropriate to be the reference period (at least for the Balkan Peninsula).

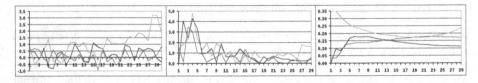

**Fig. 1.** Left: Annual temperatures (unit: °C) during the three 30-years sub-periods; Middle: Sequences according to condition 2 with a critical value 1; Right: Sequences of standard deviations (unit: °C). The sub-period 1936–1965 is shown in blue, 1961–1990 - in green and 1987–2019 - in red. (Color figure online)

We will demonstrate the proposed concept with simulations of the near surface temperature on annual basis from three GCMs. The three models are the CM3 version of the GCM of the U.S. National Oceanic and Atmospheric Administration, Geophysical Fluid Dynamics Laboratory (NOAA-GFDL), the LR-version of the model of the Max Planck Institute for Meteorology in Germany and the CM5-version of the model of CNRM (Centre National de Recherches Meteorologiques, Meteo-France, Toulouse, France) and CERFACS (Centre Europeen de Recherches et de Formation Avancee en Calcul Scientifique, Toulouse, France) (see [3] and references therein for details), noted further GFDL-CM3, MPIM-ESM and CNRM-CM5 correspondingly. The raw model output data are taken from the Climate Data Store (CDS) of the Copernicus project [2] which provides easy access to a wide range of climate datasets via a searchable catalogue. The

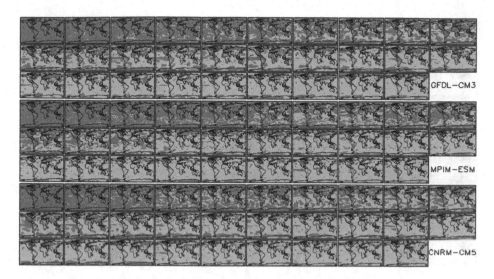

**Fig. 2.** The Cauchy sequence convergence for the considered models (Color figure online)

simulations are for the period 2071–2100 and are forced with the RCP4.5 which is probably the most suitable among all four IPCC AR5 CMIP5 scenarios for implementing the proposed methodology. In [8] for this scenario is noted: *"Representative Concentration Pathway (RCP) 4.5 is a scenario that stabilizes radiative forcing at 4.5 W/m² in the year 2100 without ever exceeding that value"*.

Figure 2 illustrate areas with slower or faster convergence of Cauchy sequences in the 30 year period for the analyzed models. The difference between the future and reference periods is analyzed. Areas where Cauchy sequences decreases or maintains its value are highlighted in green and the areas where this condition is not fulfilled - in orange. The GFDL-CM3 model shows a lack of convergence over Europe; MPIM-ESM shows a lack of convergence in the areas around the tropical zone to South America and South Africa and the CNRM-C55 - over the Middle East and Europe again. The processes have not stabilized also for all three models in many parts of the world, more systematically in the polar regions. Generally, the climate models, as a combination of ocean and atmospheric models, are very sophisticated systems. The different parameterizations and interactions between the ocean and the atmosphere lead to differences in the periods of sustainability of the modelled processes. The polar regions in which the processes of change of the snow and ice cover take place are most strongly affected by the temperature increase. These areas are most sensitive to the mechanism of interaction between the atmosphere and the ocean. The difference in this interaction is the possible reason for the different instability zones in the three models. This in turn probably implies need for longer spin-up period for these zones.

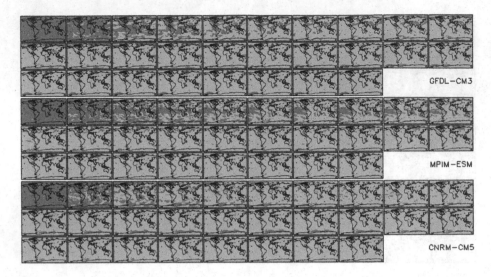

**Fig. 3.** Areas of gridcells satisfying condition 3 (green) and areas of points not satisfying it (orange) for the considered models (Color figure online)

The following Fig. 3 shows the areas in which condition 3 is or is not fulfilled for the considered models. The formal criteria for the fulfillment of this condition is $|\sigma_n - \sigma_{n-1}| < \epsilon$ for $n = \overline{3, N}$. Considering the behavior of the standard deviation we can summarize our finding concisely as follows:

- GFDL-CM3 has an area near South-East Antarctic and a very small one near the Scandinavian Peninsula.
- MPIM-ESM has such region where condition 3 is not fulfilled; It is located near the Scandinavian Peninsula.
- CNRM-CM5 has no such, at least not apparent, zones in the last four years of integration.
- Generally, areas not satisfying 3 are smaller than the areas that do not meet the Cauchy criterion 1 because condition 3 is weaker than condition 2.

## 4   Conclusion

As shown in the beginning, the 30-year period was proposed in 1935. At that time and later, modelling of climate change with numerical models was not considered. This raises the question of whether numerical climate models can produce sustainable results for such a period. The models have a stabilization period (spin-up) of 1 or more years before the true simulation of the effect of the changed amount of greenhouse gases begins. For some areas of the world, these periods may not be sufficient. Convergence may be improved if a longer adaptation period is used or if a 35-year period is used instead of the 30-year climatic period, for example. In our opinion, the lack of convergence of annual norms and standard deviation may be one of the reasons for the so-called GCM model uncertainty [9]. The proposed method for analyzing climate models allows us to see their behavior in more detail. The same methodology can be applied separately to the behavior of a model in the simulations for the reference and future periods. This would make it possible to look for the cause of one or another of the norm-change behaviors. This methodology can also be applied to changes in other meteorological elements, precipitation for example. In summary, we can say that the answer to the first question in the beginning is positive in an area if there is a "saturation" period after which a climate is formed. The proposed method shows that all three models reach a sustainable solution in large areas except the polar zones where the processes continue. In the case of scenarios with steadily increasing of greenhouse gases the proposed method can not be applied.

**Acknowledgment.** Deep gratitude to the primary model output and software vendors (CNRM, CERFACS, NOAA-GFDL, MPI-M) as well as to the CDS. This work has been co-financed by the EU through the European structural and Investment funds and funded also by the Bulgarian National Science Fund (DN14/3 13.12.2017).

# References

1. Baddour, O., Kontongomde, H. (eds.): The role of climatological normals in a changing climate. WCDMP- 61, WMO-TD 1377 (2007). https://library.wmo.int/index.php?lvl=notice_display&id=16659
2. Copernicus CDS homepage. https://cds.climate.copernicus.eu/. Accessed 1 Feb 2020
3. Global climate models included in the CDS, ECMWF site. https://confluence.ecmwf.int/display/COPSRV/Global+climate+models+included+in+the+CDS. Accessed 12 Mar 2020
4. Lorenz, E.N.: Section of planetary sciences: the predictability of hydrodynamic flow. Trans. New York Acad. Sci. **25**, 409–432 (1963). https://doi.org/10.1111/j.2164-0947.1963.tb01464.x
5. Lorenz, E.N.: A study of the predictability of a 28-variable atmospheric model. Tellus **17**, 321–333 (1965). https://doi.org/10.1111/j.2153-3490.1965.tb01424.x
6. Lorenz, E.N.: Predictability – a problem partly solved. In: Palmer, T., Hagedorn, R. (eds.) Predictability of Weather and Climate, pp. 40–58. Cambridge University Press, Cambridge (2006). https://doi.org/10.1017/CBO9780511617652.004
7. Lorenz, E.N.: Climatic predictability. In: The Physical Basis of Climate and Climate Modelling. GARP Publications Series No. 16 (1975)
8. Thomson, A.M., Calvin, K.V., Smith, S.J., et al.: RCP4.5: a pathway for stabilization of radiative forcing by 2100. Climatic Change **109**, 77 (2011). https://doi.org/10.1007/s10584-011-0151-4
9. Wootten, A., Terando, A., Reich, B.J., Boyles, R.P., Semazzi, F.: Characterizing sources of uncertainty from global climate models and downscaling techniques. J. Appl. Meteor. Climatol. **56**, 3245–3262 (2017). https://doi.org/10.1175/JAMC-D-17-0087.1

# Multidimensional Sensitivity Analysis of an Air Pollution Model Based on Modifications of the van der Corput Sequence

Venelin Todorov[1,2(✉)], Ivan Dimov[2], Rayna Georgieva[2], Tzvetan Ostromsky[2], Zahari Zlatev[3], and Stoyan Poryazov[1]

[1] Department of Information Modeling, Institute of Mathematics and Informatics, Bulgarian Academy of Sciences, Acad. Georgi Bonchev Street, Block 8, 1113 Sofia, Bulgaria
{vtodorov,stoyan}@math.bas.bg
[2] Department of Parallel Algorithms, Institute of Information and Communication Technologies, Bulgarian Academy of Sciences, Acad. G. Bonchev Street, Block 25 A, 1113 Sofia, Bulgaria
{venelin,rayna,ceco}@parallel.bas.bg, ivdimov@bas.bg
[3] National Centre for Environment and Energy, University of Århus, Frederiksborgvej 399, P.O. Box 358, 4000 Roskilde, Denmark
zz@dmu.dk

**Abstract.** An important issue when large-scale mathematical models are used to support decision makers is their reliability. Sensitivity analysis has a crucial role during the process of validating computational models to ensure their accuracy and reliability. The focus of the present work is to perform global sensitivity analysis of a large-scale mathematical model describing remote transport of air pollutants. The plain Monte Carlo approach using the well-known van der Corput sequence with various bases ($b = 3, 5, 6, 10$) has been applied for multidimensional integration to provide sensitivity studies under consideration. Sensitivity studies of the model output were performed into two directions: the sensitivity of the ammonia mean monthly concentrations with respect to the anthropogenic emissions variation, and the sensitivity of the ozone concentration values with respect to the rate variation of several chemical reactions. The numerical results show that the increase of the base leads to a higher accuracy of the estimated quantities in the most of the case studies, but the results are comparable with the results achieved using the standard van der Corput sequence with base 2.

## 1 Introduction

Mathematical models are used to study and predict the behavior of a variety of complex systems - engineering, physical, economic, social, environmental. Sensitivity studies are nowadays applied to some of the most complicated mathematical models from various intensively developing areas of application [4, 11, 12]. In a popular definition (due to A. Saltelli [7]), sensitivity analysis (SA) is the study of how uncertainty in the output of a model can be apportioned to different sources of uncertainty in the model input. The uncertainty in the model input in our case, as in many others, can be due to various reasons: inaccurate measurements or calculation, approximation, data compression, etc.

© Springer Nature Switzerland AG 2022
I. Lirkov and S. Margenov (Eds.): LSSC 2021, LNCS 13127, pp. 180–187, 2022.
https://doi.org/10.1007/978-3-030-97549-4_21

Two kinds of sensitivity analysis have been discussed in the literature: local and global [7]. Local SA studies how much some small variations of inputs around a given value can change the value of the output. Global SA takes into account all the variation range of the input parameters, and apportions the output uncertainty to the uncertainty in the input data. In the current paper the subject of our study is the global sensitivity analysis performed via the Sobol' variance-based approach [8], applied to a specific large-scale air pollution model. The Unified Danish Eulerian Model (UNI-DEM) [13] is in the focus of our study as one of the most advanced large-scale mathematical models that describes adequately all physical and chemical processes during remote transport of air pollutants.

## 2  The van der Corput Sequence

The well-known van der Corput (VDC) sequence [3, 10] is the simplest example of quasirandom number (or low-discrepancy) sequences. In general, quasirandom number sequences are less random than pseudorandom number sequences, but they are designed to reach a better uniformity in multidimensional space [6]. The star discrepancy is a commonly established measure for determining how uniformly a pointset is distributed over a region. The star discrepancy is given by:

$$D_N^* = D_N^*(\mathbf{x}_1, \dots, \mathbf{x}_N) = \sup_{\Omega \subset E^d} \left| \frac{\#\{\mathbf{x}_k \in \Omega\}}{N} - V(\Omega) \right|, \tag{1}$$

where $E^d = [0, 1)^d$, $\mathbf{x}_k = (x_k^{(1)}, x_k^{(2)}, \dots, x_k^{(s)})$ for $k = 1, 2, \dots$ [6].

The original **van der Corput** sequence is obtained when $b = 2$ [10]. The generation is quite simple. Let $n = \dots a_3(n), a_2(n), a_1(n)$ is the representation of an integer $n$ in base $b$, then the expansion of $n$ in base $b$ is generated: $n = \sum_{i=0}^{\infty} a_{i+1}(n) b^i$. The radical inverse sequence is generated by writing a decimal point followed by the digits of the expansion of $n$, in reverse order: $\phi_b(n) = \sum_{i=0}^{\infty} a_{i+1}(n) b^{-(i+1)}$. This decimal value is actually still in base $b$, so it must be properly interpreted to generate a usable value. It is fulfilled for the van der Corput sequence that $D_N^* = O\left(\frac{\log N}{N}\right)$. In the numerical experiments we will compare results obtained via van der Corput sequences with different bases - $b = 3, 5, 6, 10$ (**VDC3, VDC5, VDC6, VDC10**).

## 3  Sensitivity Studies

A description of SA procedure from a mathematical point of view is given below. Consider a scalar model output u $= f(\mathbf{x})$ corresponding to a number of non-correlated model parameters x $= (x_1, x_2, \dots, x_d)$ with a joint probability density function (p.d.f.) $p(\mathbf{x}) = p(x_1, \dots, x_d)$. In Sobol' approach [8] the parameter importance is studied via numerical integration in the terms of analysis of variance (ANOVA) model representation $f(\mathbf{x}) = f_0 + \sum_{\nu=1}^{d} \sum_{l_1 < \dots < l_\nu} f_{l_1 \dots l_\nu}(x_{l_1}, x_{l_2}, \dots, x_{l_\nu})$, where $f_0 = const$, $f(\mathbf{x})$ is a square integrable model function, $\int_0^1 f_{l_1 \dots l_\nu}(x_{l_1}, \dots, x_{l_\nu}) dx_{l_k} = 0$, $1 \le$

$k \leq \nu$, $\nu = 1, \ldots, d$, and $f_0 = \int_{E^d} f(\mathbf{x}) \mathrm{d}\mathbf{x}$. The following quantities $S_{l_1 \ldots l_\nu} = \dfrac{\mathbf{D}_{l_1 \ldots l_\nu}}{\mathbf{D}}$, $\nu \in \{1, \ldots, d\}$. are called Sobol' global sensitivity indices [8] of the corresponding order (first or higher), where $\mathbf{D}_{l_1 \ldots l_\nu}$ and $\mathbf{D}$ are the partial and the total variance of the model function, respectively. The total sensitivity index provides a measure of the total effect of a given parameter, including all the possible joint terms between that parameter and all the others. On the other hand, the total sensitivity index of a fixed parameter ($S_i^{tot}$) can be calculated with just one integral per factor in a way similar to the computation of the first-order indices following Sobol' approach.

An approach for evaluating *small* sensitivity indices (to avoid loss of accuracy because the analyzed database comes under this case) has been applied. It is a combined approach between approach of reducing of the mean value and correlated sampling suggested in [9]. Its idea is to replace the original model function with the function $\varphi(\mathbf{x}) = f(\mathbf{x}) - c$ where the constant $c \sim f_0$. Thus the partial and total variance estimations are presented in such a way:

$$\mathbf{D}_{\mathrm{y}} = \int \varphi(\mathbf{x}) \, [\varphi(\mathrm{y}, \mathrm{z}') \mathrm{d}\mathrm{x}\mathrm{d}\mathrm{z}' - \varphi(\mathbf{x}')] \mathrm{d}\mathbf{x}\mathrm{d}\mathbf{x}', \; \mathbf{D} = \int \varphi(\mathbf{x}) [\varphi(\mathbf{x}) - \varphi(\mathbf{x}')] \, \mathrm{d}\mathbf{x}\mathrm{d}\mathbf{x}'.$$

### 3.1 Plain Monte Carlo Algorithm

Plain Monte Carlo is the simplest possible stochastic approach for solving multidimensional integrals [1]. Let us consider the problem of the approximate computation of the integral $I = \int_\Omega g(\mathbf{x}) p(\mathbf{x}) \mathrm{d}\mathbf{x}$. Let $\xi$ be a random point with a p.d.f. $p(\mathbf{x})$. Introducing the random variable $\theta = g(\xi)$ such that $\mathbf{E}\theta = \int_\Omega g(\mathbf{x}) p(\mathbf{x}) \mathrm{d}\mathbf{x}$. Let the random points $\xi_1, \xi_2, \ldots, \xi_N$ be independent realizations of the random point $\xi$ with p.d.f. $p(\mathbf{x})$ and $\theta_1 = g(\xi_1), \ldots, \theta_N = g(\xi_N)$. Then an approximate value of $I$ is $\bar{\theta}_N = \frac{1}{N} \sum_{i=1}^{N} \theta_i$. The last equation defines the Plain Monte Carlo algorithm.

### 3.2 Sensitivity Studies with Respect to Emission Levels

In this section the results from sensitivity studies of UNI-DEM output (in particular, the ammonia mean monthly concentrations) with respect to the anthropogenic emissions input data variation are presented and discussed. The anthropogenic emissions input consists of 4 different components: $\mathbf{E^A}$ - ammonia ($NH_3$), $\mathbf{E^N}$ - nitrogen oxides ($NO + NO_2$), $\mathbf{E^S}$ - sulphur dioxide ($SO_2$), $\mathbf{E^C}$ - anthropogenic hydrocarbons. The results about the relative error estimation for the quantities $f_0$, the total variance $\mathbf{D}$, and the first-order ($S_i$) and the total ($S_i^{tot}$) sensitivity indices are given in Tables 1, 2, 3 and 4, respectively. The results related to the above mentioned four stochastic algorithms (**VDC3, VDC5, VDC6, VDC10**) are presented in the tables and are compared with results corresponding to the standard van der Corput sequence. The quantity $f_0$ is presented by a 4-dimensional integral, while the other quantities are presented by 8-dimensional integrals, following the ideas of *correlated sampling* technique to compute sensitivity measures in a reliable way (see [5, 8]).

The results for relative error of sensitivity indices (first- and higher order, and total) are shown for $N = 2^{10}$ and $N = 2^{20}$. Numerical tests related to a medium sample size

**Table 1.** Relative error for the evaluation of $f_0 \approx 0.048$.

| # of samples $N$ | VDC2 Relative error | VDC3 Relative error | VDC5 Relative error | VDC6 Relative error | VDC10 Relative error |
|---|---|---|---|---|---|
| $2^{10}$ | 3.72e−04 | 4.17e−03 | 1.04e−02 | 3.21e−03 | 9.91e−03 |
| $2^{12}$ | 4.39e−04 | 4.87e−03 | 1.03e−03 | 5.14e−03 | 6.50e−03 |
| $2^{14}$ | 2.15e−03 | 1.64e−03 | 8.44e−04 | 5.32e−04 | 6.50e−03 |
| $2^{16}$ | 3.11e−04 | 8.43e−04 | 4.71e−04 | 3.14e−04 | 2.99e−03 |
| $2^{18}$ | 7.66e−05 | 1.22e−03 | 8.91e−04 | 3.85e−04 | 6.24e−04 |
| $2^{20}$ | 1.48e−04 | 4.82e−04 | 1.99e−05 | 5.72e−05 | 3.92e−04 |

**Table 2.** Relative error for the evaluation of the total variance $\mathbf{D} \approx 0.0002$.

| # of samples $N$ | VDC2 Relative error | VDC3 Relative error | VDC5 Relative error | VDC6 Relative error | VDC10 Relative error |
|---|---|---|---|---|---|
| $2^{10}$ | 9.21e−02 | 1.95e−01 | 8.46e−02 | 1.75e−01 | 2.04e−01 |
| $2^{12}$ | 6.37e−02 | 4.22e−02 | 3.76e−02 | 4.33e−02 | 5.94e−02 |
| $2^{14}$ | 4.26e−02 | 1.20e−02 | 2.60e−02 | 5.24e−02 | 2.86e−02 |
| $2^{16}$ | 2.22e−03 | 4.13e−02 | 4.66e−03 | 1.46e−02 | 2.35e−02 |
| $2^{18}$ | 2.22e−03 | 1.44e−03 | 1.38e−02 | 6.87e−03 | 3.51e−04 |
| $2^{20}$ | 7.58e−03 | 3.66e−03 | 6.48e−03 | 2.61e−03 | 5.19e−03 |

**Table 3.** Relative error for the evaluation of first-order and total sensitivity indices ($N \approx 2^{10}$).

| Est. qnt. | Ref. val. | VDC3 | VDC5 | VDC6 | VDC10 |
|---|---|---|---|---|---|
| $S_1$ | 9e−01 | 3.49e−02 | 2.29e−01 | 4.36e−01 | 1.85e−01 |
| $S_2$ | 2e−04 | 2.85e+00 | 1.25e+01 | 6.74e+00 | 1.48e+01 |
| $S_3$ | 1e−01 | 3.14e−01 | 3.57e−01 | 5.91e−01 | 4.47e−01 |
| $S_4$ | 4e−05 | 2.30e+01 | 1.46e+01 | 1.30e+01 | 1.54e+01 |
| $S_1^{tot}$ | 9e−01 | 2.18e−02 | 3.45e−03 | 4.11e−03 | 3.38e−02 |
| $S_2^{tot}$ | 2e−04 | 1.97e+02 | 1.84e+02 | 7.20e+02 | 1.33e+03 |
| $S_3^{tot}$ | 1e−01 | 9.69e−01 | 1.75e−01 | 3.76e−01 | 1.38e+00 |
| $S_4^{tot}$ | 5e−05 | 3.00e+03 | 3.93e+03 | 1.26e+03 | 7.13e+03 |

**Table 4.** Relative error for the evaluation of first-order and total sensitivity indices ($N \approx 2^{20}$).

| Est. qnt. | Ref. val. | VDC2 | VDC3 | VDC5 | VDC6 | VDC10 |
|---|---|---|---|---|---|---|
| $S_1$ | 9e−01 | 3.13e−02 | 3.13e−02 | 1.15e−03 | 7.06e−03 | 1.59e−02 |
| $S_2$ | 2e−04 | 1.28e+00 | 5.02e−01 | 4.90e−01 | 2.18e−01 | 8.92e−04 |
| $S_3$ | 1e−01 | 9.13e−02 | 2.95e−02 | 9.99e−03 | 1.88e−03 | 4.60e−03 |
| $S_4$ | 4e−05 | 8.30e−01 | 3.02e−01 | 2.61e−01 | 1.32e+00 | 1.40e+00 |
| $S_1^{tot}$ | 9e−01 | 7.54e−03 | 2.65e−03 | 3.04e−04 | 1.44e−03 | 3.79e−04 |
| $S_2^{tot}$ | 2e−04 | 4.69e+01 | 1.77e+01 | 4.18e−01 | 6.69e+01 | 1.65e+01 |
| $S_3^{tot}$ | 1e−01 | 4.14e−02 | 9.19e−02 | 3.09e−02 | 9.28e−02 | 6.92e−02 |
| $S_4^{tot}$ | 5e−05 | 5.54e+02 | 3.24e+01 | 2.02e+00 | 1.04e+02 | 1.87e+01 |

($N = 2^{16}$) also have been carried out. In general, the conclusions are expected - the increase of number of samples leads to a higher accuracy of the estimated quantities. Exceptions are $S_4$ and $S_4^{tot}$ and $S_{15}$, which have extremely small reference values. One can see that all sequences produce similar results, but VDC5 has the edge for small in value sensitivity indices - see for example the estimated values of $S_2^{tot}$ and $S_4^{tot}$ in Table 4. There are some cases when the relative error corresponding to a bigger base of the van der Corput sequence is smaller than the standard sequence with base 2, but not for all cases. For example, the increase of the sample size and the base of the van der Corput sequence leads to an advantage in terms of the corresponding relative error of the estimated quantity $f_0$. It should be taken into account that the corresponding reference value of the total variance is much smaller than 0 that determines smaller order of accuracy for the same sample size in comparison of the first estimated quantity. From the results for the total variance (see Table 4) one can see that the algorithms with higher bases are characterized by a tendency of decreasing relative error.

## 3.3    Sensitivity Studies with Respect to Chemical Reactions Rates

In this section we will study the sensitivity of the ozone concentration values in the air over Genova with respect to the rate variation of some chemical reactions of the condensed CBM-IV scheme [13], namely: # 1, 3, 7, 22 (time−dependent) and # 27, 28 (time independent). The simplified chemical equations of those reactions are:

$$[\#1] \quad NO_2 + h\nu \Longrightarrow NO + O; \qquad [\#22] \quad HO_2 + NO \Longrightarrow OH + NO_2;$$
$$[\#3] \quad O_3 + NO \Longrightarrow NO_2; \qquad\qquad [\#27] \quad HO_2 + HO_2 \Longrightarrow H_2O_2;$$
$$[\#7] \quad NO_2 + O_3 \Longrightarrow NO_3; \qquad\quad [\#28] \quad OH + CO \Longrightarrow HO_2.$$

The relative error estimation for the quantities $f_0$, the total variance $\mathbf{D}$ and some sensitivity indices are given in Tables 5, 6, 7 and 8 respectively. The results related to four stochastic algorithms used for numerical integration are presented in the tables and are compared with results corresponding to the standard van der Corput sequence. The quantity $f_0$ is presented by 6-dimensional integral, while the other quantities are presented by 12-dimensional integrals. The results for relative error of sensitivity indices

**Table 5.** Relative error for the evaluation of $f_0 \approx 0.27$.

| # of samples $N$ | VDC2 Relative error | VDC3 Relative error | VDC5 Relative error | VDC6 Relative error | VDC10 Relative error |
|---|---|---|---|---|---|
| $2^{10}$ | 1.25e−02 | 9.49e−03 | 3.05e−03 | 5.12e−03 | 2.96e−04 |
| $2^{12}$ | 1.70e−03 | 1.55e−03 | 7.94e−03 | 3.98e−04 | 1.81e−03 |
| $2^{14}$ | 3.56e−03 | 1.97e−03 | 8.12e−04 | 1.36e−03 | 2.32e−04 |
| $2^{16}$ | 8.66e−04 | 5.43e−04 | 1.74e−04 | 1.46e−03 | 8.38e−04 |
| $2^{18}$ | 4.63e−04 | 2.49e−04 | 3.34e−04 | 4.75e−04 | 2.89e−04 |
| $2^{20}$ | 7.85e−05 | 1.08e−04 | 1.77e−04 | 7.72e−05 | 5.97e−05 |

**Table 6.** Relative error for the evaluation of the total variance $\mathbf{D} \approx 0.0025$.

| # of samples $N$ | VDC2 Relative error | VDC3 Relative error | VDC5 Relative error | VDC6 Relative error | VDC10 Relative error |
|---|---|---|---|---|---|
| $2^{10}$ | 2.23e−02 | 2.26e−01 | 1.44e−01 | 4.20e−02 | 4.11e−01 |
| $2^{12}$ | 2.04e−01 | 3.60e−02 | 5.86e−04 | 3.14e−02 | 4.60e−02 |
| $2^{14}$ | 3.94e−02 | 1.96e−02 | 1.65e−02 | 6.46e−02 | 1.75e−01 |
| $2^{16}$ | 1.05e−03 | 1.84e−02 | 5.96e−03 | 7.76e−02 | 3.06e−03 |
| $2^{18}$ | 1.27e−02 | 1.88e−02 | 1.80e−02 | 3.81e−03 | 2.00e−02 |
| $2^{20}$ | 1.53e−02 | 7.49e−04 | 7.72e−03 | 7.72e−03 | 5.56e−04 |

**Table 7.** Relative error for the evaluation of first-order and total sensitivity indices ($N \approx 2^{10}$).

| Est. qnt. | Ref. val. | VDC3 | VDC5 | VDC6 | VDC10 |
|---|---|---|---|---|---|
| $S_1$ | 4e−01 | 1.58e−01 | 4.40e−01 | 1.08e−03 | 4.41e−01 |
| $S_2$ | 3e−01 | 4.73e−01 | 1.04e+00 | 1.26e+00 | 2.48e−01 |
| $S_3$ | 5e−02 | 1.36e+00 | 2.06e−01 | 4.63e−01 | 2.56e+00 |
| $S_4$ | 3e−01 | 5.30e−02 | 6.78e−01 | 1.84e−01 | 5.70e−01 |
| $S_5$ | 4e−07 | 8.37e+03 | 2.20e+03 | 1.20e+03 | 2.11e+03 |
| $S_6$ | 2e−02 | 1.39e+00 | 5.64e−01 | 3.22e+00 | 3.81e+00 |
| $S_1^{tot}$ | 4e−01 | 1.13e+00 | 8.96e−02 | 6.19e−01 | 1.07e−02 |
| $S_2^{tot}$ | 3e−01 | 1.59e−01 | 2.50e+00 | 2.55e−01 | 2.20e−01 |
| $S_3^{tot}$ | 5e−02 | 4.22e+00 | 2.12e+01 | 1.69e+00 | 6.04e+00 |
| $S_4^{tot}$ | 3e−01 | 8.65e−01 | 7.06e−01 | 7.79e−01 | 9.11e−01 |
| $S_5^{tot}$ | 2e−04 | 7.52e+02 | 9.45e+02 | 6.66e+02 | 2.43e+03 |
| $S_6^{tot}$ | 2e−02 | 4.37e+00 | 2.19e+01 | 2.11e+00 | 2.34e−01 |
| $S_{12}$ | 6e−03 | 5.27e+00 | 6.02e+00 | 1.46e+00 | 3.81e+00 |
| $S_{14}$ | 5e−03 | 1.85e+00 | 8.06e+00 | 8.18e+00 | 2.05e+00 |
| $S_{15}$ | 8e−06 | 5.64e+02 | 1.05e+03 | 4.79e+02 | 5.04e+02 |
| $S_{24}$ | 3e−03 | 1.00e+01 | 2.12e+01 | 5.05e+00 | 8.59e+00 |
| $S_{45}$ | 1e−05 | 8.40e+01 | 1.66e+02 | 7.38e+01 | 1.38e+01 |

**Table 8.** Relative error for the evaluation of first-order and total sensitivity indices ($N \approx 2^{20}$).

| Est. qnt. | Ref. val. | VDC2 | VDC3 | VDC5 | VDC6 | VDC10 |
|---|---|---|---|---|---|---|
| $S_1$ | 4e−01 | 1.57e−02 | 1.52e−02 | 1.92e−02 | 1.55e−02 | 4.34e−03 |
| $S_2$ | 3e−01 | 1.39e−02 | 4.73e−03 | 1.86e−02 | 5.89e−03 | 2.71e−02 |
| $S_3$ | 5e−02 | 5.45e−02 | 2.34e−02 | 5.97e−02 | 4.91e−02 | 3.02e−02 |
| $S_4$ | 3e−01 | 7.13e−04 | 9.17e−03 | 2.29e−03 | 9.46e−06 | 8.96e−03 |
| $S_5$ | 4e−07 | 8.03e+02 | 1.33e+01 | 2.70e+01 | 1.36e+02 | 2.47e+02 |
| $S_6$ | 2e−02 | 2.09e−02 | 6.82e−02 | 1.88e−02 | 5.42e−03 | 8.76e−02 |
| $S_1^{tot}$ | 4e−01 | 1.44e−03 | 2.90e−02 | 1.70e−02 | 6.16e−03 | 5.04e−02 |
| $S_2^{tot}$ | 3e−01 | 5.82e−03 | 3.77e−03 | 2.15e−02 | 1.59e−02 | 2.60e−03 |
| $S_3^{tot}$ | 5e−02 | 1.29e−01 | 1.10e−01 | 2.39e−01 | 1.13e−01 | 1.60e−01 |
| $S_4^{tot}$ | 3e−01 | 1.75e−02 | 5.88e−03 | 6.15e−02 | 2.28e−02 | 8.15e−03 |
| $S_5^{tot}$ | 2e−04 | 6.31e+01 | 5.72e+01 | 2.72e+01 | 1.34e+01 | 9.89e−01 |
| $S_6^{tot}$ | 2e−02 | 7.22e−02 | 4.37e−02 | 4.71e−01 | 2.79e−01 | 7.63e−02 |
| $S_{12}$ | 6e−03 | 3.48e−02 | 1.17e−01 | 1.13e−01 | 1.31e−01 | 1.06e−01 |
| $S_{14}$ | 5e−03 | 2.36e−01 | 4.20e−02 | 4.96e−03 | 5.65e−02 | 2.40e−01 |
| $S_{15}$ | 8e−06 | 9.33e+02 | 9.52e+02 | 9.21e+02 | 9.34e+02 | 9.54e+02 |
| $S_{24}$ | 3e−03 | 2.04e−01 | 1.54e−01 | 3.18e−02 | 6.41e−02 | 2.69e−01 |
| $S_{45}$ | 1e−05 | 9.62e−02 | 4.47e+00 | 5.02e+00 | 5.75e−01 | 8.62e−01 |

(first- and higher order, and total) are shown for $N = 2^{10}$ and $N = 2^{20}$. Numerical tests related to a medium sample size ($N = 2^{16}$) also have been carried out. In general, the conclusions are expected - the increase of number of samples leads to a higher accuracy of the estimated quantities. Exceptions are $S_5$, $S_5^{tot}$ and $S_{15}$, which have extremely small reference values. All sequences produce similar results, but for small in value sensitivity indices VDC6 and VDC10 have the edge - see for example the estimated value of $S_{45}$ in Table 8. The trends are similar to the first sensitivity case study. One can conclude from the numerical tests that the stochastic algorithms using van der Corput sequence with different bases are efficient for the multidimensional integrals under consideration. The results for relative error for the evaluation of the estimated quantities obtained by van der Corput sequence with different bases are comparable.

## 4    Conclusion

The plain Monte Carlo method using the van der Corput sequence with various bases ($b = 3, 5, 6, 10$) has been applied for multidimensional integration ($n = 4, 8, 6, 12$). The constructed Monte Carlo algorithms have been used to provide sensitivity analysis of a large-scale mathematical model (UNI-DEM) into two directions: with respect to variation of input emissions of the anthropogenic pollutants and with respect to variation of rates of several chemical reactions. The numerical results show that the value of the base influences in a positive way on the accuracy of the estimated quantities: the increase of the base leads to smaller relative errors in the most of the case studies, but the advances are not so essential. The results for relative errors obtained by the algorithms under consideration are comparable with results achieved using the standard van der Corput sequence.

**Acknowledgment.** Venelin Todorov is supported by the Bulgarian National Science Fund under the Project KP-06-N52/5 "Efficient methods for modeling, optimization and decision making" and Project KP-06-N52/2 "Perspective Methods for Quality Prediction in the Next Generation Smart Informational Service Networks". Stoyan Apostolov is supported by the Bulgarian National Science Fund under Project KP-06-M32/2 - 17.12.2019 "Advanced Stochastic and Deterministic Approaches for Large-Scale Problems of Computational Mathematics". The work is also supported by the NSP "ICT in SES", contract No DO1-205/23.11.2018, financed by the Ministry of EU in Bulgaria and by the BNSF under Project KP-06-Russia/17 "New Highly Efficient Stochastic Simulation Methods and Applications" and Project DN 12/4-2017 "Advanced Analytical and Numerical Methods for Nonlinear Differential Equations with Applications in Finance and Environmental Pollution".

# References

1. Dimov, I.: Monte Carlo Methods for Applied Scientists. World Scientific, Singapore (2008)
2. Dimov, I., Georgieva, R., Ivanovska, S., Ostromsky, T., Zlatev, Z.: Studying the sensitivity of pollutants' concentrations caused by variations of chemical rates. J. Comput. Appl. Math. **235**(2), 391–402 (2010)
3. Faure, H., Kritzer, P., Pillichshammer, F.: From van der Corput to modern constructions of sequences for Quasi-Monte Carlo rules. Indagationes Mathematicae **26**(5), 760–822 (2015)
4. Gocheva-Ilieva, S.G., Voynikova, D.S., Stoimenova, M.P., Ivanov, A.V., Iliev, I.P.: Regression trees modeling of time series for air pollution analysis and forecasting. Neural Comput. Appl. **31**(12), 9023–9039 (2019). https://doi.org/10.1007/s00521-019-04432-1
5. Homma, T., Saltelli, A.: Importance measures in global sensitivity analysis of nonlinear models. Reliab. Eng. Syst. Saf. **52**, 1–17 (1996)
6. Karaivanova, A.: Stochastic Numerical Methods and Simulations. Demetra, Sofia (2012)
7. Saltelli, A., et al.: Global Sensitivity Analysis. The Primer. Wiley, New York (2008)
8. Sobol', I.M.: Global sensitivity indices for nonlinear mathematical models and their Monte Carlo estimates. Math. Comput. Simul. **55**(1–3), 271–280 (2001)
9. Sobol', I., Myshetskaya, E.: Monte Carlo estimators for small sensitivity indices. Monte Carlo Meth. Appl. **13**(5–6), 455–465 (2007)
10. Van der Corput, J.G.: Verteilungsfunktionen (Erste Mitteilung) (PDF), Proceedings of the Koninklijke Akademie van Wetenschappen te Amsterdam (in German), vol. 38, pp. 813–821 (1935). Zbl 0012.34705
11. Veleva, E., Georgiev, I.R.: Seasonality of the levels of particulate matter PM10 air pollutant in the city of Ruse, Bulgaria. In: AIP Conference Proceedings, vol. 2302, p. 030006. AIP Publishing LLC, December 2020
12. Veleva, E., Georgiev, I.R., Zheleva, I., Filipova, M.: Markov chains modelling of particulate matter (PM10) air contamination in the city of Ruse, Bulgaria. In: AIP Conference Proceedings, vol. 2302, p. 060018. AIP Publishing LLC, December 2020
13. Zlatev, Z., Dimov, I.T.: Computational and Numerical Challenges in Environmental Modelling. Elsevier, Amsterdam (2006)

# Running an Atmospheric Chemistry Scheme from a Large Air Pollution Model by Using Advanced Versions of the Richardson Extrapolation

Zahari Zlatev[1] , Ivan Dimov[2], István Faragó[3,4,5] , Krassimir Georgiev[2] ,
and Ágnes Havasi[3,4](✉) 

[1] Department of Environmental Science, Aarhus University, Roskilde, Denmark
`zz@envs.au.dk`
[2] Institute of Information and Communication Technologies, Bulgarian Academy of
Sciences, Sofia, Bulgaria
`ivdimov@bas.bg, georgiev@parallel.bas.bg`
[3] ELTE Eötvös Loránd University, Pázmány Péter s. 1/C, Budapest 1117, Hungary
`agnes.havasi@ttk.elte.hu`
[4] MTA-ELTE Numerical Analysis and Large Networks Research Group, Pázmány
Péter s. 1/C, Budapest 1117, Hungary
[5] Institute of Mathematics, Budapest University of Technology and Economics, Egry
J. u. 1., Budapest 1111, Hungary

**Abstract.** Atmospheric chemistry schemes, which are described mathematically by non-linear systems of ordinary differential equations (ODEs), are used in many large-scale air pollution models. These systems of ODEs are badly-scaled, extremely stiff and some components of their solution vectors vary quickly forming very sharp gradients. Therefore, it is necessary to handle the atmospheric chemical schemes by applying accurate numerical methods combined with reliable error estimators. Three well-known numerical methods that are suitable for the treatment of stiff systems of ODEs were selected and used: (a) EULERB (the classical Backward Differentiation Formula), (b) DIRK23 (a two-stage third order Diagonally Implicit Runge-Kutta Method) and (c) FIRK35 (a three-stage fifth order Fully Implicit Runge-Kutta Method). Each of these three numerical methods was applied in a combination with nine advanced versions of the Richardson Extrapolation in order to get more accurate results when that is necessary and to evaluate in a reliable way the error made at the end of each step of the computations. The code is trying at every step (A) to determine a good stepsize and (B) to apply it with a suitable version of the Richardson Extrapolation so that the error made at the end of the step will be less than an error-tolerance TOL, which is prescribed by the user in advance. The numerical experiments indicate that both the numerical stability can be preserved and sufficiently accurate results can be obtained when each of the three underlying numerical methods is correctly combined with the advanced versions of the Richardson Extrapolation.

© Springer Nature Switzerland AG 2022
I. Lirkov and S. Margenov (Eds.): LSSC 2021, LNCS 13127, pp. 188–197, 2022.
https://doi.org/10.1007/978-3-030-97549-4_22

**Keywords:** Atmospheric chemical schemes · Stiff systems of ODEs · Implicit Runge-Kutta Methods · Advanced versions of the Richardson Extrapolation

# 1   Introduction

Atmospheric chemical schemes are very important parts of many large-scale environmental models, which are used to study the long-range transport of air pollutants, [5,7], and the impact of climatic changes on some critical pollution levels, [6]. These schemes are described mathematically by non-linear systems of ordinary differential equations (ODEs), which are badly scaled, extremely stiff and some components of their solution vectors vary quickly forming very sharp gradients, [5,7]. Therefore, it is necessary to use reliable and very accurate tools in the treatment of such schemes on computers. Moreover, it is desirable to control in a reliable way the accuracy of the calculated results. The application of stable Implicit Runge-Kutta Methods, [1] and [2], combined with advanced versions of the well-known Richardson Extrapolation, [8] and [9], will be discussed in the next sections.

The selected underlying Runge-Kutta methods will be introduced in the Sect. 2. The combination of these methods with nine advanced versions of the Richardson Extrapolation will be discussed in the Sect. 3. Numerical results will be presented in the Sect. 4. Conclusions and remarks will be given in the Sect. 5.

# 2   Selection of Three Underlying Runge-Kutta Methods

The chemical schemes used in large-scale air pollution models are normally described mathematically by a non-linear system of ordinary differential equations:

$$y' = f(t,y), \ t \in [a,b], \ y \in D \subset \mathbb{R}^n \text{ and } y(a) = \eta \text{ being given.} \tag{1}$$

It is assumed that the unknown function $y(t)$ (containing the different chemical species) is sufficiently many times continuously differentiable and that any of the three numerical methods listed below can be selected and used in the numerical solution of (1). It is furthermore assumed that approximations $y_1 \approx y(t_1), y_2 \approx y(t_2), \ldots, y_N \approx y(t_N)$ of the exact solution $y(t)$ are to be calculated successively on the non-equidistant grid-points $\{t_1, t_2, \ldots, t_N\}$ starting with $t_1$ and finishing with $t_N = b$.

The first method is EULERB, the Backward Differentiation Formula (or the Implicit Euler Method). It is a first-order Runge-Kutta method, [2], p. 196, that is defined as follows:

$$y_n = y_{n-1} + hf(t_n, y_n), \ n \in \{1, 2, \ldots, N\}. \tag{2}$$

The second method is DIRK23. It is a two-stage third order Diagonally Implicit Runge-Kutta Method, [2], p. 196, defined as follows:

$$y_n = \frac{1}{2}h(k_1^n + k_2^n), \quad n \in \{1, 2, \ldots, N\}, \tag{3}$$

$$k_1^n = f\left(t_{n-1} + \frac{3+\sqrt{3}}{6}h, y_{n-1} + \frac{3+\sqrt{3}}{6}hk_1^n\right), \tag{4}$$

$$k_2^n = f\left(t_{n-1} + \frac{3-\sqrt{3}}{6}h, y_{n-1} - \frac{\sqrt{3}}{3}hk_1^n + \frac{3+\sqrt{3}}{6}hk_2^n\right). \tag{5}$$

The third method is FIRK35. It is a three-stage fifth order Fully Implicit Runge-Kutta Method, [2], p. 192, based on the following formulae:

$$y_n = \frac{16-\sqrt{6}}{36}hk_1^n + \frac{16+\sqrt{6}}{36}hk_2^n + \frac{1}{9}hk_3^n, n \in \{1, 2, \ldots, N\}. \tag{6}$$

$$k_1^n = f\left(t_{n-1} + \frac{4-\sqrt{6}}{10}h, y_{n-1} + \frac{88-7\sqrt{6}}{360}hk_1^n\right.$$
$$\left. + \frac{296-169\sqrt{6}}{1800}hk_2^n + \frac{-2+3\sqrt{6}}{225}hk_3^n\right), \tag{7}$$

$$k_2^n = f\left(t_{n-1} + \frac{4+\sqrt{6}}{10}h, y_{n-1} + \frac{296+169\sqrt{6}}{1800}hk_1^n\right.$$
$$\left. + \frac{88+7\sqrt{6}}{360}hk_2^n + \frac{-2-3\sqrt{6}}{225}hk_3^n\right), \tag{8}$$

$$k_3^n = f\left(t_{n-1} + h, y_{n-1} + \frac{16-\sqrt{6}}{36}hk_1^n + \frac{16+\sqrt{6}}{36}hk_2^n + \frac{1}{9}hk_3^n\right). \tag{9}$$

## 3   Applying Nine Versions of the Richardson Extrapolation

Any of the introduced in the previous section three numerical methods can be used in combination with the nine versions of the Richardson Extrapolation studied in [5] and [6]. The different versions, the abbreviations used and their orders of accuracy are listed in Table 1.

It is clear (see the last column in Table 1) that the accuracy can be increased very considerably when the value of parameter $q$ is increased. For example, the order of accuracy is becoming 14 when the 8TRRE is used together with FIRK35. However, it is necessary to pay something for the high accuracy: the computational work is also increased very considerably. This explains why it is worthwhile to carry out the computations with an attempt to keep both the error of the calculated solution and the number of the computational operations

**Table 1.** The nine versions of the Richardson Extrapolation, the abbreviations which will be used in the remaining part of this paper and the orders of accuracy of the different versions ($p$ being the order of accuracy of the underlying numerical method, which means that $p = 1$, $p = 3$ and $p = 5$ when EULERB, DIRK23, and FIRK35 are used. The Classical Richardson Extrapolation (CRE) was originally proposed in [3].

| $q$ | Names of the nine versions | Abbreviation | Order of accuracy |
|---|---|---|---|
| 0 | Classical Richardson Extrapolation | CRE | $p + 1$ |
| 1 | Repeated Richardson Extrapolation | RRE | $p + 2$ |
| 2 | Two Times Repeated Richardson Extrapolation | 2TRRE | $p + 3$ |
| 3 | Three Times Repeated Richardson Extrapolation | 3TRRE | $p + 4$ |
| 4 | Four Times Repeated Richardson Extrapolation | 4TRRE | $p + 5$ |
| 5 | Five Times Repeated Richardson Extrapolation | 5TRRE | $p + 6$ |
| 6 | Six Times Repeated Richardson Extrapolation | 6TRRE | $p + 7$ |
| 7 | Seven Times Repeated Richardson Extrapolation | 7TRRE | $p + 8$ |
| 8 | Eight Times Repeated Richardson Extrapolation | 8TRRE | $p + 9$ |

sufficiently small. This can be achieved by selecting, at each step of the computational process, as large as possible stepsize and a version qTRRE with as small as possible value of parameter $q$ (where $q = 0$ for CRE and $q = 1$ for RRE) under the essential requirement that the error must remain sufficiently small (according to a prescribed error tolerance TOL).

The approximate solution of (1) obtained by using the qTRRE at step $n$ is denoted by $y_n^{[q]}$ ($n = 1, 2, \ldots, N$, $q = 0, 1, \ldots, 8$). Reliable estimates of the error made in the calculation of $y_n^{[q]}$ are obtained by using $y_n^{[q]} - y_n^{[q-1]}$ if $q > 0$ and $y_n^{[0]} - y_n$ when $q = 0$. All formulae needed for the calculation of the approximations $y_n^{[q]}$ and the error estimations $y_n^{[q]} - y_n^{[q-1]}$ are listed in [8] and [9]. As an illustration, the formulae related to CRE and obtained under the assumption that two auxiliary vectors $z_n^{[0]}$ and $z_n^{[1]}$ are calculated by using the selected numerical method from Sect. 2 and by preforming one step with a stepsize $h$ and two steps with a stepsize $h/2$, starting in both cases with the available approximation $y_{n-1}^{[0]}$, are given below:

$$y_n^{[0]} = \frac{2^p z_n^{[1]} - z_n^{[0]}}{2^p - 1}, \qquad (10)$$

$$y(t_n) - z_n^{[1]} = \frac{z_n^{[1]} - z_n^{[0]}}{2^p - 1} + \mathcal{O}(h^{p+1}) \Rightarrow ERROR_n^{[0]} \approx \frac{\|z_n^{[1]} - z_n^{[0]}\|}{2^p - 1}. \qquad (11)$$

It is clear that the last relation in (11) will be true only when the stepsize $h$ is sufficiently small.

The major rules used in our Variable Stepsize Variable Formula Method, VSVFM, based on the application of any of the three numerical methods that were presented in Sect. 2 and applied together with the nine advanced versions of the Richardson Extrapolation, are sketched below.

A key parameter RATIO is calculated at every time-step by using the formula:

$$RATIO = \delta \left( \frac{TOL}{EST_n^{[q]}} \right)^{\frac{1}{p+q+1}}, \quad n = 1, 2, \ldots, N, \quad q = 0, 1, \ldots, 8, \qquad (12)$$

$$p = 1, 3, \text{ or } 5.$$

The value of the integer $q$ depends on the particular version of the Richardson Extrapolation which is used at step $n$. The integer $p$ is the order of the underlying numerical method for solving systems of ODEs. As mentioned above, the formulae used in the calculation of $EST_n^{[q]}$ are fully described in [9]. If $q = 0$, then the equality $EST_n^{[0]} = \|z_n^{[1]} - z_n^{[0]}\|/(2^p - 1)$ can be obtained by using the last relationship in (11), which implies that $EST_n^{[0]}$ will be reliable when the stepsize is sufficiently small. The coefficient $\delta$, from the right-hand-side of (12), is a precaution factor used always in the preparation of VSVFMs (see, for example, [4]); $\delta = 0.9$ is selected in all experiments.

Now, the major actions that are to be taken at the end of any step $n$ ($n = 1, 2, \ldots, N$) can be explained. These actions depend essentially on the value of parameter RATIO and are listed in Table 2. It should be emphasized here that only the most important actions are described in Table 2. Some additional rules must also be used. For example, no new increase of the stepsize is in general allowed several steps after increasing it. This rule is useful, because the theory, on which the numerical methods for solving systems of ODEs are based, is strictly speaking valid only in the case where a constant stepsize is used. Therefore, the stepsize should not be varied too often and/or with a too large amount. There is a parameter WAIT in the program, which is used to keep the same stepsize at least during two consecutive steps. However, if the check of parameter RATIO indicates that either Case 4 or Case 5 (see Table 2) has taken place, then the step is always rejected and the computations are repeated by applying a reduced stepsize.

## 4   Numerical Results

The chemical scheme was run over a time-interval of 24 h, starting at 12 o'clock at a given day and finishing at 12 o'clock on the next day. This is a very relevant and important choice, because the so chosen interval contains the periods of changes from day-time to night-time and from night-time to day-time when some chemical species vary very quickly and can cause great problems for the numerical methods. The time is measured in seconds. This means that the time-interval of 24 h, which is used in the code, is $[a, b] = [43200, 129600]$. This interval was divided into 168 sub-intervals (the length of each sub-interval being approximately 514.285 s) and the "exact" solution was calculated at the end of each sub-interval by using a constant stepsize $h = 10^{-5}$ and by running FIRK35 combined with the 8TRRE.

**Table 2.** Major rules for accepting or rejecting the approximation obtained at the current step $n$.

| Case | Ratio | Action |
|------|-------|--------|
| 1 | $0.9 \leq RATIO \leq 1.5$ | Keep the same stepsize |
| | | Increase the order of accuracy by selecting more accurate version of the Richardson Extrapolation (replacing $q$ by $q + 1$ if $q < 8$) when $RATIO < 1.0$ |
| | | Decrease the order of accuracy by selecting less accurate version of the Richardson Extrapolation (replacing $q$ by $q - 1$ if $q > 0$ when $RATIO > 1.25$). |
| | | **Continue with the calculation of** $y_{n-1}$ |
| 2 | $1.5 \leq RATIO \leq 4.0$ | Increase the stepsize by setting $h_{new} = 1.25h$ |
| | | Increase the order of accuracy by selecting a more accurate version of the Richardson Extrapolation (replacing $q$ by $q + 1$ if $q < 8$ when $RATIO > 2.0$). |
| | | **Continue with the calculation of** $y_{n-1}$. |
| 3 | $4.0 \leq RATIO$ | Increase the stepsize by setting $h_{new} = 1.5h$ |
| | | Increase the order of accuracy by selecting a more accurate version of the Richardson Extrapolation (replacing $q$ by $q + 1$ if $q < 8$ when $RATIO > 6.0$) |
| | | **Continue with the calculation of** $y_{n-1}$ |
| 4 | $0.1 \leq RATIO < 0.9$ | Decrease the stepsize by setting $h_{new} = 0.5h$ |
| | | Increase the order of accuracy by selecting a more accurate version of the Richardson Extrapolation (replacing $q$ by $q + 1$ if $q < 8$ when $RATIO < 0.25$) |
| | | **Reject the step** and repeat it by calculating a new $y_n$ |
| 5 | $RATIO < 0.1$ | Decrease the stepsize by setting $h_{new} = 0.25h$ |
| | | Increase the order of accuracy by selecting a more accurate version of the Richardson Extrapolation (replacing $q$ by $q + 1$ if $q < 8$ when $RATIO < 0.05$) |
| | | **Reject the step** and repeat it by calculating a new $y_n$ |

All three numerical methods were run with different values of the error tolerance TOL, but the code is forced to reach the end-point of each of the 168 sub-intervals. In this way, it becomes possible to check the achieved accuracy at 168 grid-points of $[a, b]$ by using the formula $\|y^{\text{exact}} - y^{\text{calculated}}\|_2 / \max(\|y^{\text{exact}}\|_2, 10^{-6})$. Some of the results are given below.

Different values of the error tolerance TOL were used with the three underlying methods. We started with DIRK23 and used $TOL \in [10^{-7}, 10^{-11}]$. The results given in Table 3 show clearly that the algorithm, the major properties of which were described in Table 2, is working very well in this situation.

Much more accurate results can be achieved, as shown in Table 4, when FIRK35 is used with $TOL \in [10^{-10}, 10^{-20}]$. Finally, if one is not interested in achieving high accuracy, then EULERB can be used with large values of the error tolerance. This is demonstrated in Table 5.

The results shown in Tables 3, 4, and 5 indicate that the number of steps is in general not changed too much when the value of the error tolerance TOL is varied. The figures presented in Table 6 explain why this is so: decreasing the value of the error tolerance leads in most of the cases not to a decrease of

**Table 3.** Results obtained by running DIRK23 combined with different versions of the Richardson Extrapolation. Five different values of the error tolerance TOL, i.e., the code is trying to keep the order of the local error estimation (which is calculated during the different runs) less than or equal to $\mathcal{O}(TOL)$.

| TOL | Successul steps | Rejected steps | Exact max. error | Estimated max. error |
|---|---|---|---|---|
| $10^{-7}$ | 342 | 0 | $9.203 \cdot 10^{-8}$ | $6.384 \cdot 10^{-8}$ |
| $10^{-8}$ | 428 | 0 | $1.906 \cdot 10^{-8}$ | $4.655 \cdot 10^{-9}$ |
| $10^{-9}$ | 362 | 0 | $3.128 \cdot 10^{-9}$ | $4.791 \cdot 10^{-10}$ |
| $10^{-10}$ | 925 | 0 | $3.932 \cdot 10^{-10}$ | $3.703 \cdot 10^{-11}$ |
| $10^{-11}$ | 3231 | 10 | $9.882 \cdot 10^{-12}$ | $3.473 \cdot 10^{-12}$ |

**Table 4.** Results obtained by running FIRK35 combined with different versions of the Richardson Extrapolation. Eleven different values of the error tolerance TOL, i.e., the code is trying to keep the order of the local error estimation (which is calculated during the different runs) less than or equal to $\mathcal{O}(TOL)$.

| TOL | Successful steps | Rejected steps | Exact max. error | Estimated max. error |
|---|---|---|---|---|
| $10^{-10}$ | 349 | 0 | 1.528E−11 | 4.770E−11 |
| $10^{-11}$ | 359 | 0 | 5.188E−12 | 7.807E−12 |
| $10^{-12}$ | 430 | 3 | 7.031E−13 | 8.479E−13 |
| $10^{-13}$ | 532 | 6 | 7.531E−14 | 8.337E−14 |
| $10^{-14}$ | 571 | 8 | 1.014E−14 | 9.979E−15 |
| $10^{-15}$ | 786 | 9 | 4.947E−15 | 9.475E−16 |
| $10^{-16}$ | 1074 | 13 | 3.184E−16 | 9.933E−17 |
| $10^{-17}$ | 619 | 9 | 2.593E−17 | 4.544E−19 |
| $10^{-18}$ | 1245 | 9 | 1.602E−17 | 6.809E−19 |
| $10^{-19}$ | 1381 | 20 | 1.059E−18 | 9.332E−20 |
| $10^{-20}$ | 1023 | 17 | 5.133E−19 | 9.507E−21 |

the stepsize, but to the selection of more accurate versions of the Richardson Extrapolation. This is important because the use of a Variable Stepsize Variable Formula Methods leads to some extra checks and calculations at each step, which are needed to decide (a) if the step is successful or it should be rejected, (b) if the stepsize should be increased and by how much it should be increased and (c) if the stepsize should be decreased and by how much it should be decreased. It is also seen that the most expensive versions of the Richardson Extrapolation are used only a few times if the error tolerance is not extremely small (see the results for $TOL = 10^{-11}$. Even if TOL is very small, 8TRRE was called only 18 times (a very small number compared with the total number of steps).

**Table 5.** Results obtained by running EULERB combined with different versions of the Richardson Extrapolation by using five different values of the error tolerance TOL, i.e., the code is trying to keep the order of the local error estimation (which is calculated during the different runs) less than or equal $\mathcal{O}(TOL)$.

| TOL | Successful steps | Rejected steps | Exact max. error | Estimated max. error |
|---|---|---|---|---|
| $10^{-2}$ | 345 | 0 | $1.054 \cdot 10^{-1}$ | $8.428 \cdot 10^{-3}$ |
| $10^{-3}$ | 348 | 0 | $4.092 \cdot 10^{-2}$ | $7.153 \cdot 10^{-4}$ |
| $10^{-4}$ | 411 | 4 | $9.614 \cdot 10^{-3}$ | $8.191 \cdot 10^{-5}$ |
| $10^{-5}$ | 434 | 10 | $2.048 \cdot 10^{-3}$ | $6.522 \cdot 10^{-6}$ |
| $10^{-6}$ | 456 | 2 | $7.588 \cdot 10^{-5}$ | $9.217 \cdot 10^{-7}$ |

**Table 6.** Numbers of calls of different versions of the Richardson Extrapolation when two values of the error tolerance TOL are used in runs where the underlying method is FIRK35.

| TOL | CRE | RRE | 2TRRE | 3TRRE | 4TRRE | 5TRRE | 6TRRE | 7TRRE | 8TRRE |
|---|---|---|---|---|---|---|---|---|---|
| $10^{-11}$ | 213 | 93 | 21 | 15 | 8 | 3 | 1 | 3 | 2 |
| $10^{-20}$ | 3 | 156 | 96 | 266 | 221 | 136 | 95 | 49 | 18 |

It should be mentioned here that all runs, results from which were presented in this section, were performed by using extended computer precision (i.e., working with about 32 significant digits of the real numbers).

# 5   Concluding Remarks

**Remark 1 – About the stability of the computational process.** The three underlying numerical methods from Sect. 2 have excellent stability properties (see [1] and [2]). However, there is a danger that the combinations of any of these methods with some of the versions of the Richardson Extrapolation may result in unstable computational process. It has been shown ([9]) that the combination of any of the underlying numerical methods with any of the nine versions of the Richardson Extrapolation is stable in extremely large stability regions in the left-hand side of the complex plane and, thus, no stability problems appeared in the runs.

**Remark 2 – About the use of other underlying numerical methods for solving ODEs.** The results presented in Sect. 4 were obtained by applying nearly the same rules for changing the stepsize and the version of the Richardson Extrapolation. This fact indicates that the strategy for varying the stepsize and the Richardson Extrapolation versions does not depend too much on the underlying methods and other underlying methods can also be selected and successfully used.

**Remark 3 – About the improvement of the performance.** The performance can be further improved in several ways. If the required accuracy is not high and/or if the solved problem is not very difficult, then it might be worthwhile to introduce a parameter $MAX_q$, so that if, for example, $MAX_q = 2$, then only the first three versions of the Richardson Extrapolation (with $q = 0$, $q = 1$ and $q = 2$) will be used in the run. Other improvements can also be made. The attempts to improve the performance are continuing.

**Acknowledgement.** "Application Domain Specific Highly Reliable IT Solutions" project has been implemented with the support provided from the National Research, Development and Innovation Fund of Hungary, financed under the Thematic Excellence Programme TKP2020-NKA-06 (National Challenges Subprogramme) funding scheme. This work was completed in the ELTE Institutional Excellence Program (TKP2020-IKA-05) financed by the Hungarian Ministry of Human Capacities. The project has been supported by the European Union, and co-financed by the European Social Fund (EFOP-3.6.3-VEKOP-16-2017-00002), and further, it was supported by the Hungarian Scientific Research Fund OTKA SNN125119. This work of K. Georgiev was accomplished with the support by the Grant No BG05M2OP001-1.001-0003, financed by the Science and Education for Smart Growth Operational Program (2014–2020) and co-financed by the European Union through the European structural and Investment funds.

# References

1. Hairer, E., Wanner, G.: Solving Ordinary Differential Equations: II Stiff and Differential-Algebraic Problems. Springer, Heidelberg (1991). https://www.springer.com/gp/book/9783540604525, https://doi.org/10.1007/978-3-642-05221-7
2. Shampine, L.F.: Tolerance proportionality in ODE codes. In: Bellen, A., Gear, C.W., Russo, E. (eds.) Numerical Methods for Ordinary Differential Equations. LNM, vol. 1386, pp. 118–136. Springer, Heidelberg (1989). https://doi.org/10.1007/BFb0089235
3. Richardson, L.F.: The deferred approach to the limit, I-single lattice. Philos. Trans. Royal Society of London Ser. A **226**, 299–349 (1927). https://doi.org/10.1098/rsta.1927.0008)
4. Zlatev, Z.: Advances in the theory of variable stepsize variable formula methods for ordinary differential equations. Appl. Math. Appl. **31**, 209–249 (1989). https://doi.org/10.1016/0096-3003(89)90120-3
5. Zlatev, Z.: Computer Treatment of Large Air Pollution Models. Kluwer Academic Publishers, Dordrecht, Boston, London (1995). (Now distributed by Springer, Berlin. https://link.springer.com/book/10.1007/978-94-011-0311-4)
6. Zlatev, Z.: Impact of future climate changes on high ozone levels in European suburban areas. Climatic Change **101**, 447–483 (2010). https://link.springer.com/article/10.1007/s10584-009-9699-7
7. Zlatev, Z., Dimov, I.: Computational and Numerical Challenges in Environmental Modelling. Elsevier, Amsterdam, Boston (2006). ISBN: 9780444522092

8. Zlatev, Z., Dimov, I., Faragó, I., Georgiev, K., Havasi, Á.: Explicit Runge-Kutta methods combined with advanced versions of the Richardson extrapolation. Comput. Meth. Appl. Math. **20**(4), 739–762 (2020). https://doi.org/10.1515/cmam-2019-0016
9. Zlatev, Z., Dimov, I., Faragó, I., Georgiev, K., Havasi, Á.: Solving stiff systems of ordinary differential equations with advanced versions of the Richardson extrapolation (2020). http://nimbus.elte.hu/~hagi/LSSC21/

# Application of Metaheuristics to Large-Scale Problems

# New Clustering Techniques of Node Embeddings Based on Metaheuristic Optimization Algorithms

Adis Alihodžić[✉], Malek Chahin, and Fikret Čunjalo

Department of Mathematics, University of Sarajevo, Zmaja od Bosne 33-35,
71000 Sarajevo, Bosnia and Herzegovina
{adis.alihodzic,malek.c,fcunjalo}@pmf.unsa.ba

**Abstract.** Node embeddings present a powerful method of embedding graph-structured data into a low dimensional space while preserving local node information. Clustering is a common preprocessing task on unsupervised data utilized to get the best insight into the input dataset. The most prominent clustering algorithm is the K-Means algorithm. In this paper, we formulate clustering as an optimization problem using different objective functions following the idea of searching for the best fit centroid-based cluster exemplars. We also apply several nature-inspired optimization algorithms since the K-Means algorithm is trapped in local optima during its execution. We demonstrate our cluster frameworks' capability on several graph clustering datasets used in node embeddings and node clustering tasks. Performance evaluation and comparison of our frameworks with the K-Means algorithm are demonstrated and discussed in detail. We end this paper with a discussion on the impact of the objective function's choice on the clustering results.

**Keywords:** Graph embedding · Node embeddings · Clustering · Community detection · Metaheuristics · K-Means clustering algorithm

## 1 Introduction

Working with graph-structured data is a growing presence in many of today's tasks. Networks can carry different types of information in nodes and their connections. Representing entities and their connections in a graph structure enriches the data with many features which a network carries in its topology. Real-world graph datasets rarely carry any labels which can be used to train machine learning models. Analysis performed on a network often include clustering of the input dataset, which allows the opportunity to group different entities into clusters based on some sense of similarity. These groups are known to call *communities*, and the task of clustering networks is often referred to as *community detection* in network sciences and its subfields. The usual approach handled to analyze large networks is to map nodes into a low dimensional space. In this way, relationships between entities can be explored by using standard

© Springer Nature Switzerland AG 2022
I. Lirkov and S. Margenov (Eds.): LSSC 2021, LNCS 13127, pp. 201–208, 2022.
https://doi.org/10.1007/978-3-030-97549-4_23

linear algebra methods. This type of data can be utilized to train many machine learning models to tackle downstream tasks. The task of mapping any node to low dimensional vector representations is known as *node embedding*. The node embedding processes usually preserves local proximity measures between nodes in the input network by encoding their neighbourhood. This process enjoys much success in solving link prediction, node classification, node clustering etc. In this paper, we are focused on node embeddings by using the *node2vec* algorithm since its primarily known for its quality of vector representations and its scalability [4]. During getting these vector representations for each entity present in the input network, the task of community detection can be accomplished by applying modern clustering algorithms. Recent successes are generated employing precisely this idea [8] which is a result, handed a reliable and scalable framework to explore groups of input entities. However, the framework can be improved by powerful metaheuristic optimization frameworks that mimic the intelligence of different types of swarms. These algorithms are also known as nature-inspired optimization algorithms and are practised to solve problems with complex solution spaces. Many optimization algorithms applied to explore these spaces tend to discover suboptimal solutions since they follow specific heuristics during their execution. The algorithms explored in this paper usually outperform state-of-the-art optimization algorithms in a wide range of problems.

## 1.1   Related Work

The incorporation of nature-inspired optimization algorithms to node embeddings' clustering does that node embeddings is still an open research field. No related research was found which are focused on the problem of clustering node embeddings using these powerful algorithms. However, there are many research papers in the literature that focused on applying nature-inspired optimization algorithms to clustering in general. Some of them are cuckoo search algorithm (CS) [13], bat algorithm (BA) [1,2,17], firefly algorithm (FA) [16], and ant colony optimization (ACO) [3]. Their goal is to create a framework capable of clustering various datasets which do not necessarily vector representations derived from an embedding algorithm. Precisely, a general clustering framework at these algorithms is being introduced and compared to the popular K-Means clustering algorithm [10]. This set of algorithms has chosen to investigate different search agents' behaviour and collective intelligence while exploring the search space. In the paper, [14], particle swarm optimization (PSO) application to K-Means is investigated in detail. It also provides an interesting choice of the objective function to direct an algorithm to find a user-defined number of groups. In the paper, [8], a framework for generating node embeddings and learning clusters, is presented on real-world datasets. The article also shows a performance measurement of various node embedding algorithms and clustering their resulting node vector representations using the K-Means algorithm. An impressive idea of this model is that it can be seen as an enhancement of other random-walk based embedding algorithms such as DeepWalk [7] and node2vec [4] to the problem of clustering of vector representations. Results of these types of algorithms are compared in this paper, as well. In the paper [6], authors presented how genetic algorithm (GA) can be implemented to solve the problem of clustering. Since evolutionary algorithms were not explored in previous research papers, this paper does include the differential evolution (DE) [9] algorithm as a representative of the family of evolutionary algorithms. In the papers, [11,12], the authors explored similar experiments on enhancing the K-Means clustering algorithm by using elephant herding optimization (EHO) [15] algorithm and bare bone fireworks algorithm (BBFA) [5]. These algorithms are a relatively new swarm intelligence algorithms applied to many different optimization problems.

### 1.2 Our Contribution

Although the problem of integrating nature-inspired optimization algorithms to the general clustering problem are given attention by some researchers; the problem of clustering node embeddings is yet to be explored. Node embedding vector representations result from an embedding model whose performance depends on the model's probabilistic choices. Local proximity measures of nodes are preserved by embedding algorithms, but each node's placement in the vector space will differ after each execution. It makes the task of clustering node embeddings more complex than clustering datasets that contain features that produce well-separated clusters. In this paper, we collect ideas found in related work and apply them to clustering node embeddings. The paper also uses real-world social network datasets, which gives readers and future researchers a sense of how our approach performs in similar network structures. This paper's original benefaction introduces a clustering framework for node embeddings based on nature-inspired algorithms, which can support as a benchmark for future researchers.

## 2 The Node Embedding Clustering Framework Based on Metaheuristic Optimization Algorithms

In this section, a framework for clustering node embeddings based on nature-inspired optimization algorithms is explained. The section is organized into three subsections. The first subsections outline possible solution representations which can be employed to model feasible solutions to the node embedding clustering problem. In the second subsection, several objective functions are described and discussed, along with the choice used in the rest of the clustering framework. The framework is illustrated in the third subsection.

### 2.1 Solution Representations

Centroid-based clustering methods such as the K-Means produce cluster exemplar points that can be handled to evaluate clustering performance. A set of solutions utilised for this framework is a two-dimensional array storing $k$ cluster exemplars. The solution $R_{ij}$ is being created as a $k \times d$ dimensional matrix, where $k$ is the predefined cluster number a user provides and $d$ is the dimension of the vector representation. Also, $|R_{ij}|$ denotes the number of vector representations belonging to cluster $R_{ij}$. Metaheuristic algorithms use these solution representations for each individual in their swarms. Details of each algorithm's notion of swarms and their behaviour are not covered in this paper. A similar approach of modelling solution representations is storing one vector of size $k \times d$. This implementation is not chosen in this research but would generate similar results. Other representations involve encoding the solution vector representation as strings. However, they would require additional implementation details and utility functions, leading to performance drops when updating solutions during the algorithm runtime.

## 2.2    Clustering as an Optimization Problem

Let us have at our disposal a set $W$ composed of $n$ vector representations $\mathbf{w}_s = (w_s^1 \ w_s^2 \cdots w_s^d) \ (s = \overline{1, n})$ of dimension $d$, where for each vector $\mathbf{w}_s$, the dimension $d$ is a user-defined parameter to the node embedding algorithm. Also, for any vector $\mathbf{w}_s$ we have exactly one node in the input network $G$, and $n = |V(G)|$. This paper does not consider the impact of parameters chosen for the embedding algorithm on the clustering task. The main aim is to group these vectors into a predefined number of clusters using some notion of similarity between each vector. There are several ways to calculate the quality of a particular clustering algorithm. The most popular is calculating the silhouette coefficient of the resulting clusters. Since the problem involves networks, we exploit the graph modularity measure expressed in the community detection algorithm [8]. This measure demonstrate the presented model's efficiency. None of these measures includes the constraint to export a user-defined number of clusters. It is an important aspect of our research because we compare our approach with the K-Means algorithm. It turns out that the number of clusters produced using the measures as mentioned earlier as objective functions is less than the ones the user-provided. However, the measure, such as the quantization error, covers an objective function that needs to include solving the problem. It can be formulated as follows

$$\min J_c = \frac{\sum_{j=1}^{N_c} [\sum_{\mathbf{z} \in R_{ij}} \frac{d(\mathbf{z}, \mathbf{m}_j)}{|R_{ij}|}]}{k} \tag{1}$$

where $i = \overline{1, p}$ is the index of the $i$-th solution from all $p$ possible solutions (or the population size), $j = \overline{1, k}$ denotes the index of the cluster exemplar of cluster $j$, and the vector $\mathbf{m}_j$ is the cluster exemplar vector of $j$-th cluster. The notation $\mathbf{z} \in R_{ij}$ is used to refer to the set of node vector representations closer to $\mathbf{m}_j$ than to any other cluster exemplar in the $i$-th solution. Finally, the notation $d(\mathbf{z}, \mathbf{m}_j)$ will be used to denote the Euclidean distance between the node vector representation and $j$-th cluster exemplar. The main reason to use the above equation is to establish the denominators' balancement, which leads to penalizing solutions with less than $k$ clusters. It is essential since reducing the number of clusters often leads to larger values in other cluster quality measures. The choice of several clusters should be a user's choice, and the algorithm should not change this without the user being aware. As stated, this type of problem will be part of future work and defined differently than the problem investigated in this research. To conclude this section, the optimization problem used to model clustering of node embeddings has presented by Eq. 1. It is intuitive since reducing the error improves clustering quality.

### 2.3   Nature-Inspired Clustering Framework for Node Embeddings

This section is closed by describing the clustering framework constructs for node embeddings based on nature-inspired optimization algorithms. Algorithm 1 gives high-level instructions on how to implement this framework using any metaheuristic optimization algorithm. These instructions do not contain implementation details, additional instructions or parameter settings the underlying algorithm carries. The mentioned algorithm can serve as a blueprint for using metaheuristic optimization algorithms to cluster node embedding vector representations obtained by a node embedding model. As stated before, experiments for this research are based on node2vec embeddings [4].

## 3   Experimental Analysis

In this section, our proposed clustering framework was compared to the traditional K-Means clustering algorithm. The framework was constructed and tested with four metaheuristic algorithms PSO, EHO, BA, and DE. The framework was implemented using Python programming language and open-source libraries such as *networkx*, *sklearn* and *numpy*. All simulations were run on an Intel Core i5-8250U CPU, 1.60 GHz with 16 GB of RAM and Ubuntu 20.04 operating system. In the embedding stage of the framework, for the node2vec algorithm, the *gensim* package implementation of the *word2vec* algorithm is used. The choice of parameters for the node2vec algorithm is as follows $d = 16$, $p = 1$, $q = 2$, $num\_walks = 20$, $walk\_length = 80$, $context\_size = 5$.

---

**Algorithm 1.** Pseudocode of the node embedding clustering framework

```
1: procedure EMBEDDINGCLUSTERING
2:     G ← input network
3:     W ← node2vec(G), vector representations
4:     P, A ← initial cluster exemplars and cluster assignments
5:     gbest ← best individual
6:     while termination condition not met do
7:         P ← cluster exemplar update step following underlying
                 metaheuristic algorithm's update rules
8:         A ← cluster reassignment step
9:         gbest ← best individual after most recent updates
       return gbest
```

---

**Table 1.** Mean quantization error and two standard deviations of the clustering obtained after 50 independent experiments applied on Facebook dataset [8].

| Tag | Dataset | K-Means | PSO | EHO | BA | DE |
|-----|---------|---------|-----|-----|-----|-----|
| $D_1$ | Government | $5.24 \pm 0.13$ | $5.33 \pm 0.24$ | $5.77 \pm 0.18$ | $5.81 \pm 0.22$ | $4.59 \pm 0.29$ |
| $D_2$ | Politicians | $4.90 \pm 0.11$ | $4.99 \pm 0.23$ | $5.47 \pm 0.17$ | $5.48 \pm 0.21$ | $4.09 \pm 0.29$ |
| $D_3$ | Athletes | $7.61 \pm 0.12$ | $7.53 \pm 0.23$ | $7.67 \pm 0.17$ | $7.33 \pm 0.16$ | $6.88 \pm 0.23$ |
| $D_4$ | Media | $8.29 \pm 0.20$ | $7.95 \pm 0.28$ | $8.05 \pm 0.19$ | $7.63 \pm 0.17$ | $7.34 \pm 0.28$ |
| $D_5$ | TV shows | $5.33 \pm 0.08$ | $4.90 \pm 0.25$ | $5.53 \pm 0.18$ | $5.44 \pm 0.24$ | $3.91 \pm 0.37$ |
| $D_6$ | Artists | $7.70 \pm 0.14$ | $7.08 \pm 0.37$ | $7.51 \pm 0.25$ | $7.12 \pm 0.37$ | $6.17 \pm 0.31$ |
| $D_7$ | Companies | $7.90 \pm 0.06$ | $7.60 \pm 0.20$ | $7.69 \pm 0.15$ | $7.35 \pm 0.15$ | $7.02 \pm 0.33$ |
| $D_8$ | Celebrities | $7.01 \pm 0.13$ | $6.76 \pm 0.25$ | $7.05 \pm 0.17$ | $6.71 \pm 0.17$ | $6.05 \pm 0.36$ |

**Table 2.** Mean execution times (in seconds) and two standard deviations produced by five clustering algorithms after 50 independent simulations done on Facebook dataset.

| Dataset | K-Means | PSO | EHO | BA | DE |
|---------|---------|-----|-----|-----|-----|
| $D_1$ | $10.9 \pm 5.7$ | $29.7 \pm 1.2$ | $45.8 \pm 4.5$ | $38.9 \pm 8.4$ | $52.3 \pm 2.7$ |
| $D_2$ | $8.7 \pm 6.9$ | $21.4 \pm 0.7$ | $33.4 \pm 0.7$ | $20.6 \pm 0.5$ | $40.5 \pm 2.2$ |
| $D_3$ | $30.4 \pm 19.4$ | $51.9 \pm 2.5$ | $80.0 \pm 5.0$ | $76.8 \pm 4.0$ | $88.1 \pm 3.9$ |
| $D_4$ | $118.3 \pm 96.9$ | $96.9 \pm 3.3$ | $137.1 \pm 1.1$ | $95.3 \pm 14.6$ | $167.1 \pm 6.1$ |
| $D_5$ | $6.0 \pm 2.3$ | $16.7 \pm 0.5$ | $25.5 \pm 0.2$ | $16.1 \pm 0.5$ | $31.2 \pm 1.9$ |
| $D_6$ | $137.3 \pm 51.3$ | $174.6 \pm 2.3$ | $277.2 \pm 6.8$ | $270.6 \pm 31.9$ | $287.0 \pm 9.8$ |
| $D_7$ | $50.7 \pm 23.3$ | $52.1 \pm 1.4$ | $80.8 \pm 5.7$ | $75.7 \pm 4.1$ | $89.4 \pm 4.1$ |
| $D_8$ | $26.0 \pm 15.2$ | $44.1 \pm 1.0$ | $63.3 \pm 1.5$ | $43.3 \pm 1.5$ | $74.5 \pm 3.1$ |

The choice of parameters used for the BA optimization algorithm is $A = 0.7$, $R = 0.7$, $Q_{min} = 0.5$, $Q_{max} = 1$, and $population\_size = 20$. For the PSO algorithm the choice of parameters is $c_1 = 1$, $c_2 = 2$, $w_{start} = 0.6$, $w_{stop} = 0.01$, and $n\_particles = 25$. For the DE algorithm values for parameters are $population\_size = 25$, $crossover\_rate = 0.3$, $differential\_weight = 0.1$. For the EHO algorithm parameters are set as follows $scale\_factor = 1$, $n\_clans = 5$, and $clan\_size = 10$.

For each algorithm the maximum number of iterations is set to 250 and the number of clusters is set to 20. All algorithms use random samples as initial solutions in the initialization step. Every algorithm was executed 50 times and their outcomes were collected and used to calculate mean errors and standard deviations as in Table 1. The mean values and standard deviations at the end of the last iteration are collected in Table 1, while the improvements of our framework compared to the K-Means algorithm are collected in Table 3. The improvements are referred to on relative percentage changes and can be expressed as $100 \times \frac{ourApproach - KMeans}{KMeans}$. The negative values indicate that the improvement

**Table 3.** The best improvements of our approaches compared to outcomes produced by the K-Means algorithm considering Facebook dataset.

| Algorithms | Facebook datasets | | | | | | | |
|---|---|---|---|---|---|---|---|---|
| | $D_1$ | $D_2$ | $D_3$ | $D_4$ | $D_5$ | $D_6$ | $D_7$ | $D_8$ |
| PSO | 1.6% | 1.9% | −1.1% | −4.1% | −8.1% | −8.1% | −3.7% | −3.5% |
| EHO | 10.1% | 11.7% | 0.8% | −2.8% | 3.7% | −2.5% | −2.6% | 0.6% |
| BA | 10.8% | 11.9% | −3.7% | −7.9% | 2.1% | −7.5% | −6.9% | −4.3% |
| DE | −12.4% | −16.4% | −9.6% | −11.5% | −26.7% | −19.8% | −11.1% | −13.7% |

is made using our approach over the K-Means. Some improvements are quite significant, emphasising the strength and capability of metaheuristic exploration to avoid falling into locally optimal solutions. Mean execution times for all of the benchmarked algorithms are collected in Table 2.

# 4   Discussion and Conclusion

In this article, a new framework for clustering node embeddings based on nature-inspired optimization algorithms is presented. Clustering node embeddings is still an open research topic which experienced multiple advances in recent years. Implementing metaheuristic optimization algorithms to solve this problem is an open research field, with this research being the first contribution. Our research is focused on developing a clustering framework suitable for clustering data in a graph structure agnostic to underlying semantics of what their entities and relationships represent. The experiments executed in this research produced promising results. Also, it has shown that metaheuristic algorithms outperform a representative of the traditional clustering algorithms. The DE algorithm exposed the most promising results for it is consistently outperforming any other of the investigated algorithm. Our research was not aimed at parameter sensitivity analysis of any of the building blocks of the presented framework, nor does it include comparing the quality of cluster results using different objective functions, as this will be part of our future work.

# References

1. Alihodzic, A., Tuba, E., Tuba, M.: An upgraded bat algorithm for tuning extreme learning machines for data classification. In: Proceedings of the Genetic and Evolutionary Computation Conference Companion, GECCO 2017, pp. 125–126. Association for Computing Machinery, New York (2017). https://doi.org/10.1145/3067695.3076088
2. Alihodzic, A., Tuba, M.: Improved bat algorithm applied to multilevel image thresholding. Sci. World J. **2014**, 16 (2014). Article ID 176718. https://doi.org/10.1155/2014/176718
3. Dorigo, M., Blum, C.: Ant colony optimization theory: a survey. Theor. Comput. Sci. **344**(2–3), 243–278 (2005). https://doi.org/10.1016/j.tcs.2005.05.020

4. Grover, A., Leskovec, J.: Node2vec: scalable feature learning for networks. In: Proceedings of the 22nd ACM SIGKDD International Conference on Knowledge Discovery and Data Mining, KDD 2016, pp. 855–864. Association for Computing Machinery, New York (2016). https://doi.org/10.1145/2939672.2939754

5. Li, J., Tan, Y.: The bare bones fireworks algorithm: a minimalist global optimizer. Appl. Soft Comput. **62**, 454–462 (2018). https://doi.org/10.1016/j.asoc.2017.10.046

6. Maulik, U., Bandyopadhyay, S.: Genetic algorithm-based clustering technique. Pattern Recognit. **33**(9), 1455–1465 (2000). https://doi.org/10.1016/S0031-3203(99)00137-5

7. Perozzi, B., Al-Rfou, R., Skiena, S.: Deepwalk: Online learning of social representations. In: Proceedings of the 20th ACM SIGKDD International Conference on Knowledge Discovery and Data Mining, KDD 2014, pp. 701–710. Association for Computing Machinery, New York (2014). https://doi.org/10.1145/2623330.2623732

8. Rozemberczki, B., Davies, R., Sarkar, R., Sutton, C.: GEMSEC: graph embedding with self clustering. In: Proceedings of the 2019 IEEE/ACM International Conference on Advances in Social Networks Analysis and Mining 2019, pp. 65–72. ACM (2019)

9. Storn, R., Price, K.: Differential evolution - a simple and efficient heuristic for global optimization over continuous spaces. J. Glob. Optim. **11**(4), 341–359 (1997). https://doi.org/10.1023/A:1008202821328

10. Tang, R., Fong, S., Yang, X., Deb, S.: Integrating nature-inspired optimization algorithms to k-means clustering. In: Seventh International Conference on Digital Information Management (ICDIM 2012), pp. 116–123 (2012). https://doi.org/10.1109/ICDIM.2012.6360145

11. Tuba, E., Dolicanin-Djekic, D., Jovanovic, R., Simian, D., Tuba, M.: Combined elephant herding optimization algorithm with k-means for data clustering. In: Satapathy, S.C., Joshi, A. (eds.) Information and Communication Technology for Intelligent Systems. SIST, vol. 107, pp. 665–673. Springer, Singapore (2019). https://doi.org/10.1007/978-981-13-1747-7_65

12. Tuba, E., Jovanovic, R., Hrosik, R.C., Alihodzic, A., Tuba, M.: Web intelligence data clustering by bare bone fireworks algorithm combined with k-means. In: Proceedings of the 8th International Conference on Web Intelligence, Mining and Semantics, WIMS 2018. Association for Computing Machinery, New York (2018). https://doi.org/10.1145/3227609.3227650

13. Tuba, M., Alihodzic, A., Bacanin, N.: Cuckoo search and bat algorithm applied to training feed-forward neural networks. In: Yang, X.-S. (ed.) Recent Advances in Swarm Intelligence and Evolutionary Computation. SCI, vol. 585, pp. 139–162. Springer, Cham (2015). https://doi.org/10.1007/978-3-319-13826-8_8

14. van der Merwe, D.W., Engelbrecht, A.P.: Data clustering using particle swarm optimization. In: The 2003 Congress on Evolutionary Computation, CEC 2003, vol. 1, pp. 215–220 (2003). https://doi.org/10.1109/CEC.2003.1299577

15. Wang, G.G., Deb, S., Gao, X.Z., Coelho, L.D.S.: A new metaheuristic optimisation algorithm motivated by elephant herding behaviour. Int. J. Bio-Inspired Comput. **8**(6), 394–409 (2017). https://doi.org/10.1504/IJBIC.2016.081335

16. Yang, X.-S.: Firefly algorithms for multimodal optimization. In: Watanabe, O., Zeugmann, T. (eds.) SAGA 2009. LNCS, vol. 5792, pp. 169–178. Springer, Heidelberg (2009). https://doi.org/10.1007/978-3-642-04944-6_14

17. Yang, X.S.: A new metaheurisitic bat-inspired algorithm. Stud. Comput. Intell. **284**, 65–74 (2010). https://doi.org/10.1007/978-3-642-12538-6_6

# A Comparison of Machine Learning Methods for Forecasting Dow Jones Stock Index

Adis Alihodžić[✉], Enes Zvorničanin, and Fikret Čunjalo

Department of Mathematics, University of Sarajevo, Zmaja od Bosne 33-35,
71000 Sarajevo, Bosnia and Herzegovina
{adis.alihodzic,fcunjalo}@pmf.unsa.ba

**Abstract.** Stock market forecasting is a challenging and attractive topic for researchers and investors, helping them test their new methods and improve stock returns. Especially in the time of financial crisis, these methods gain popularity. The algorithmic solutions based on machine learning are used widely among investors, starting from amateur ones up to leading hedge funds, improving their investment strategies. This paper made an extensive analysis and comparison of several machine learning algorithms to predict the Dow Jones stock index movement. The input features for the algorithms will be some other financial indices, commodity prices and technical indicators. The algorithms such as decision tree, logistic regression, neural networks, support vector machine, random forest, and AdaBoost have exploited for comparison purposes. The data preprocessing step used a few normalization and data transformation techniques to investigate their influence on the predictions. In the end, we presented a few ways of tuning hyperparameters by metaheuristics such as genetic algorithm, differential evolution, and immunological algorithm.

**Keywords:** Machine learning · Financial time series · Classification · Metaheuristics · Data preprocessing

## 1 Introduction

The impact of automated trading on stock exchanges is increasing, and it is estimated that about 75% of the stocks are trading on American stock exchanges using algorithmic systems. Many of them are by structure expert systems based on some manually defined rules arising from simple logic and technical analysis. Although researchers and academia are sceptical about technical analysis, it is showed that several technical indicators contribute to incremental information and may have some useful value [4]. The prediction of stock price movements is a very complex problem, and the complexity stems from the fact that several factors, including political events, market news, company earnings reports, influence its movement. Despite that, some artificial neural networks trained with technical

© Springer Nature Switzerland AG 2022
I. Lirkov and S. Margenov (Eds.): LSSC 2021, LNCS 13127, pp. 209–216, 2022.
https://doi.org/10.1007/978-3-030-97549-4_24

indicators achieved the comparable ability to recognize stock trends [5]. Unlike technical indicators, the fundamental analysis assesses stock movements based on various indicators, from macroeconomic factors such as the economy, gross domestic product, inflation rate to microeconomic indicators such as company income statements, balance sheet, P/E ratio, and others [6]. Many forms of data are used to obtain any useful stock information. By processing text from news or blogs and its sentimental analysis, it can be shown how news related to the company affects its financial returns and the volume of stock trading [7]. Some statistical methods for predicting the movement of time series are widespread in the literature and papers. Among the most common algorithms in predicting stock prices is the autoregressive integrated moving average (ARIMA) method, which assumes that a time series's future value is a linear combination of previous values and errors. This model can achieve substantial results in short-term forecasting and can be comparative with machine learning models, although it gives good results only for univariate time series [8,9]. Among modern methods of predicting prices in financial markets in the last few years, machine learning is the most common, reflected in the progressive development of its community. Some machine learning techniques, such as support vector machine (SVM) [1], random forest (RF) [2], and artificial neural networks (ANN) [3], are at the forefront [10,11]. Better predictive performance can be achieved by combining multiple methods into one whole, which is also a characteristic of ensemble methods consisting of multiple weak models [12]. Most of these algorithms can have thousands or even millions of different combinations of hyperparameters, so choosing the combination that gives the best solution is a time-consuming task. Therefore, it is a better practice to use some optimization algorithms that will, for sure, find a combination that gives an approximately optimal solution if they do not find the best combination. Among such algorithms, some metaheuristics can be mentioned, such as the genetic algorithm (GA), bat algorithm (BA), cuckoo search (CS), differential evolution method (DE), and many others [13–15]. Following an idea from the paper [16], we wanted to apply a similar set of features to another popular American stock index, Dow Jones, that tracks 30 large American companies selling on the New York Stock Exchange and the NASDAQ. The idea is to predict the index's daily movement using a set of features that consist of other indices like S&P500, NASDAQ, FTSE 100, NIKKEI 225, SSE, currencies rates such as GBP/USD, USD/CNY, USD/JNY. Additionally, commodity prices like gold and crude oil are included, and a few technical indicators from Dow Jones. The rest of the article is prepared as follows. Data preprocessing is described in Sect. 2 while scaling, and data transformation is presented in Sect. 3. In Sect. 4, we propose a model preparation with simulation results. The optimization of the support vector machine and achieved outcomes are offered in Sect. 5. In Sect. 6, experimental results are presented by the optimization of random forest. Finally, conclusions and suggestion for future work are discussed in the last section of the paper, Sect. 7.

## 2    Data Preprocessing

The entire amount of data utilised in this research includes the period from 01/01/2017 until 01/08/2020. The data were mostly complete, and those missing values are forward filled, filling gaps using the last observed value as it is usual practice for time series. As the task is to predict the daily movement of the Dow Jones index, whether it will proceed up or down the next day, it means that our problem is a binary classification. Given that, it makes sense that features will be calculated daily. Also, as they all are time series and our models take observations to be independent and usually identically distributed, it is essential to make features stationary. It that purpose, we will transform daily open prices into returns by the formula

$$r_t = \frac{O_{t+1} - O_t}{O_t} \tag{1}$$

where $r_t$ is return value at time $t$ and $O_t$ daily open price at time $t$. Note that the task is to predict the daily movement of Dow Jones or a sign of $r_t$ of Dow Jones using $r_{t-1}$ values of the mentioned features. Also, it is worth mentioning that for NIKKEI 225 and SSE return calculated from closed prices since they are trading in different time zone is used, and that way, it is closer in time but still does not cause future propagation. When it is about time series forecasting, especially stock prediction, it is important to exclude all possibilities of looking into the future. It means that training and testing have to be aligned by time, so any data shuffling or some overlap of training and test data will induce the wrong prediction. Namely, we encountered these two errors the most in the papers. In terms of technical indicator calculated from Dow Jones, we used to return and volume of the previous trading day, momentum with the lag 2 and 3 calculated as $O_t - O_{t-k}$, and price rate of change with lag 2 and 3 calculated as $(O_t - O_{t-k})/O_{t-k}$ where $k$ it the lag.

## 3    Scaling and Data Transformation

Another level employed in preparing data for machine learning algorithms is scaling or transforming all feature values according to a defined rule. Usually, in most machine learning problems, feature values in data sets vary widely, with some features reaching millions, while others do not have an absolute value greater than $10^{-6}$. This difference can make some features contribute more to computations in the algorithm, while others become negligible, which affects the algorithms' performance. On the other hand, numerical feature values can have a very curved distribution. It may be due to the presence of outliers, exponential distributions, bimodal distributions, and similar. Some algorithms, such as logistic regression (LR), assume that input values have a normal distribution, while other algorithms have been shown to give better results when the distribution is uniform. Following that, in this paper, we used data transformation methods (DTM) such as standardization (ST), min-max normalization (MM), scaling with maximum absolute value (MA), interquartile range normalization (IQR),

Cox-Box (CB) and Yeo-Johnson (YJ) transformation, inverse normal (IN) and uniform (IU) transformations.

## 4  Model Preparation

For the first experiment, we use the following algorithms:

- Logistic regression (LR);
- Support vector machine (SVM) with Gaussian RBF kernel, where $\gamma$ is reciprocal of the number of features and C is equal to 1;
- Random Forest (RF) with 100 trees, Gini impurity, without maximum depth, which means that nodes are expanded until all leaves are pure or until all leaves contain less than two samples and $\sqrt{N_f}$ features are considered while looking for the best split, where $N_f$ is the total number of features;
- Decision tree (DT) with Gini impurity and without maximum depth;
- AdaBoost with 50 decision stumps;
- Neural network (NN), feedforward neural network with one hidden layer with 100 neurons, *ReLU* activation function except for output layer where it is sigmoid, Adam optimizer with a learning rate of 0.001. A maximum number of iterations is 200 or until the loss or score is not improving by at least 0.0001 for 10 consecutive iterations.

Training and testing are used to walk forward validation with a sliding training window of one year and a test set of one month. Some mentioned methods are not deterministic and have some randomness during the building process, depending on a random seed. Because of that, we did 50 experiments for each of them and the presented results are shown in Table 1.

From the data illustrated in Table 1, it can be seen that the support vector machine achieved the best accuracy with 57.7% accuracy. Also, at least one neural network produced identical accuracy. Despite that, the best SVM model in terms of accuracy had a true positive rate of 0.891 and a true negative rate of

**Table 1.** Accuracy of Dow Jones index prediction with given classification algorithms and data transformation methods.

| DTM | LR | SVM | RF | | | DT | | | AdaBoost | | | NN | | |
|-----|-----|-----|-----|-----|-----|-----|-----|-----|-----|-----|-----|-----|-----|-----|
| | | | Max | Avg | Min | Max | Avg | Min | Max | Avg | Min | Max | Avg | Min |
| ST | .551 | **.577** | .547 | .512 | .478 | .533 | **.512** | .492 | .533 | .530 | .528 | **.577** | **.566** | .531 |
| MM | **.568** | .574 | .547 | .511 | .472 | **.536** | .511 | .492 | .533 | .530 | .528 | .570 | .555 | .530 |
| MA | .559 | .574 | .544 | .511 | .478 | .533 | **.512** | .487 | .533 | .530 | .528 | .559 | .545 | .531 |
| YJ | .564 | .554 | **.565** | .511 | .467 | .521 | .503 | .484 | .527 | .526 | .525 | .564 | .546 | .531 |
| CB | .551 | .568 | .542 | .513 | .475 | .534 | **.512** | .490 | .533 | .531 | .528 | .573 | .554 | .530 |
| IU | .548 | .574 | .550 | **.514** | **.479** | .525 | .507 | .492 | .533 | .530 | .528 | .567 | .552 | .539 |
| IN | .564 | .567 | .536 | .507 | .462 | .527 | .506 | .479 | .533 | .531 | .528 | .559 | .543 | .518 |
| IQR | .547 | .562 | .550 | .511 | .478 | .528 | .510 | .487 | .533 | .530 | .528 | .571 | .553 | **.542** |
| – | .516 | .574 | .534 | .503 | .461 | .527 | .511 | **.498** | **.544** | **.541** | **.539** | .564 | .531 | .501 |

**Table 2.** Cumulative return of Dow Jones index prediction with given classification algorithms and data transformation methods.

| DTM | LR | SVM | RF | | | DT | | | AdaBoost | | | NN | | |
|---|---|---|---|---|---|---|---|---|---|---|---|---|---|---|
| | | | Max | Avg | Min | Max | Avg | Min | Max | Avg | Min | Max | Avg | Min |
| ST | .305 | **.253** | .514 | **.054** | −.651 | .284 | −.151 | −.534 | .037 | −.008 | −.047 | **.959** | **.675** | .230 |
| MM | .249 | .149 | .393 | .028 | −.604 | .173 | −.144 | −.715 | .045 | −.004 | −.047 | .583 | .291 | −.106 |
| MA | .262 | .149 | .564 | −.002 | −.595 | .237 | −.159 | −.404 | .045 | −.02 | −.047 | .744 | .460 | .264 |
| YJ | .365 | .012 | **.793** | −.001 | −.663 | .179 | −.197 | −.514 | .192 | .154 | .108 | .900 | .595 | **.322** |
| CB | .431 | −.049 | .531 | .048 | **−.408** | .183 | −.166 | .584 | .045 | .001 | −.047 | .949 | .634 | .276 |
| IU | .500 | .149 | .580 | .039 | −.579 | .161 | −.179 | −.544 | .034 | −.017 | −.047 | .888 | .507 | .296 |
| IN | **.761** | .139 | .466 | .015 | −.614 | .127 | −.221 | −.695 | .034 | −.013 | −.057 | .739 | .344 | −.016 |
| IQR | .244 | −.033 | .483 | .030 | −.580 | .142 | −.180 | −.527 | .045 | .001 | −.036 | .918 | .551 | .256 |
| − | −.028 | .149 | .572 | −.005 | −.912 | **.390** | **.071** | −.310 | **.390** | **.336** | **.290** | .784 | .064 | −.365 |

0.155, indicating that class 1 (up) predictions dominate. Regarding that, accuracy is not the best metric for the evaluation, and instead of accuracy, we will use more indicative metric, such as cumulative return. Cumulative return is calculated as the sum of daily returns, where the following formula calculates the daily return $(d_r)$ as

$$
d_r = \begin{cases} |r_t|, & \text{if the movement of the index is accurately predicted} \\ -|r_t|, & \text{otherwise} \end{cases} \tag{2}
$$

Thus, if the same amount of money was invested every day, the cumulative return is the percentage of that amount, ultimately realized as profit. This metric is more important from a financial perspective, and besides it, the Sharpe ratio and volatility may be helpful. The simulation results by the cumulative return are presented in Table 2.

Based on outcomes in Table 2, it can be inferred that the neural network gains the highest cumulative return of 0.959, and its total accuracy was 56.2% with a real positive rate of 71.7% and a real negative rate of 35.3%. By analyzing all the results, the influence of individual transformers on them is seen, so it can be seen that standardization (ST) and Cox-Box transformation (CB) are superior to others. It is also interesting that models without transformation have worse results than others, confirming the importance of transformations in machine learning problems.

## 5 Optimization of Support Vector Machine

From the results presented in the previous section, it can be witnessed that the SVM did not produce satisfying outcomes. The reason for this may be that a specified set of hyperparameters is not the best tuned for a given problem. Therefore, in this section, we tried to optimize the parameters maximizing cumulative return. As in the paper [14], the values of kernel coefficient $\gamma$ and penalty parameter $C$ are selected with an exponential step, i.e. $\gamma \in \{2^{-15}, 2^{-13}, ..., 2^3\}$ and

**Table 3.** Top 10 models of support vector machine by cumulative return.

| Kernel | DTM | $C$ | $\gamma$ | Accuracy | TPR | TNR | Cumul. return |
|--------|-----|-----|----------|----------|-----|-----|---------------|
| RBF  | MM | 8192  | 0.125 | **0.587** | 0.677 | 0.464 | **1.192** |
| POLY | IN | 32768 | 0.002 | 0.579 | **0.883** | 0.169 | 1.068 |
| POLY | IN | 64    | 0.016 | 0.579 | **0.883** | 0.169 | 1.068 |
| POLY | IN | 4096  | 0.004 | 0.579 | **0.883** | 0.169 | 1.068 |
| POLY | IN | 0.125 | 0.125 | 0.579 | **0.883** | 0.169 | 1.068 |
| POLY | IN | 512   | 0.008 | 0.579 | **0.883** | 0.169 | 1.068 |
| POLY | IN | 8     | 0.031 | 0.579 | **0.883** | 0.169 | 1.068 |
| POLY | IN | 1     | 0.063 | 0.579 | **0.883** | 0.169 | 1.068 |
| RBF  | MM | 32768 | 0.063 | 0.579 | 0.667 | 0.460 | 1.055 |
| RBF  | YJ | 16384 | 0.002 | 0.579 | 0.675 | **0.468** | 1.044 |

$C \in \{2^{-5}, 2^{-3}, ..., 2^{15}\}$, respectively. Also, the features' scale values are from $[-1, 1]$ or $[0, 1]$, which some data transformations will certainly do. Thus, in the experiment, a grid search method will be applied for the SVM algorithm with hyperparameters, including all transformations as in the previous section and various types of kernel functions such as polynomial (POLY), Gauss (Radial basis function or RBF) and sigmoid functions.

## 5.1  Results

This section has used the Grid search method as one of the simplest techniques for optimizing SVM parameters $C$ and $\gamma$. Grid search yielded almost 12,000 models of the SVM with different combinations of hyperparameters, where the top 10 outcomes in terms of cumulative return were shown in Table 3.

The highest result in cumulative return is 1.192, which is 0.233 more than the previous best result achieved by the neural network, as shown in Table 2. Also, this method achieved a higher accuracy of 0.586 compared to the previous one who had 0.577.

## 6  Optimization of Random Forest

In this section, we will employ three metaheuristics to adjust the hyperparameters of the random forest. Compared to other methods, the random forest is a valuable technique since it has a broader hyperparameter set. The following hyperparameters of random forest are being used for optimization:

- Number of trees $\in \{2, 4, .., 10, 20, 30, ..., 140\}$;
- Impurity $\in \{\text{Gini, Entropy}\}$;
- Maximal tree depth $\in \{2, 4, ..., 10, 15, 20, 25, ..., 50, \text{unlimited}\}$;

**Table 4.** Accuracy of metaheuristics and comparison between first and last iteration.

| Method | Best | | Worst | | Average | | Error | |
|--------|-------|-------|-------|-------|---------|-------|--------|--------|
|        | First | Last  | First | Last  | First   | Last  | First  | Last   |
| GA     | 0.543 | 0.572 | 0.532 | 0.554 | 0.539   | 0.565 | 0.0031 | 0.0050 |
| IA     | **0.564** | **0.582** | **0.555** | **0.577** | **0.560** | **0.580** | 0.0026 | **0.0015** |
| DE     | 0.553 | 0.566 | 0.547 | 0.557 | 0.550   | 0.559 | **0.0020** | 0.0027 |

- The minimum number of samples required to split a node∈ {2, 4, ..., 10, 15, 20, 25, 30};
- The number of features ($n$) to consider when looking for the best split ∈ {$\sqrt{n}$, $\log_2 n$, n};
- all mentioned transformations;
- training set in months ∈ {12, 13, ..., 23};
- test set in days ∈ {20, 21, 22, ..., 29}.

For the specified set of hyperparameters exist, more than 12 million different combinations. Still, this set could be even more extensive, but due to limited resources, so that more experiments could be done within a reasonable time, we restricted it. Also, for the same reason, the prediction period is limited to only one year, and it is 2018. The optimization metric is accuracy as it is more stable over a shorter period. Through empirical analysis of the problem, specific kinds of features are used as well as metaheuristics such as genetic algorithm (GA), differential evolution method (DE), and immunological algorithm (IA). For each of the metaheuristics, ten independent experiments are conducted. Also, the number of agents for all metaheuristics is limited to 49 and the number of iterations to 50. The obtained results are drawn in Table 4.

In Table 4, it can be seen that the best result in almost all indicators was achieved by the IA. The GA showed the largest standard deviation among the ten experiments by far, while the DE produced the worst average score and smallest improvement through iterations.

## 7    Conclusion and Future Work

This paper presented a systematic approach for solving the Dow Jones index's daily forecasting and described exact methodological steps, from data prepro- cessing to optimizing hyperparameters by metaheuristics. Based on simulation outcomes, it can be concluded that transformation techniques for preprocessing data can boost machine learning algorithms' performance. Some of the presented methods achieved good outcomes in terms of cumulative return. Also, the appli- cation of metaheuristics has produced promising results. Therefore, in general, we think that metaheuristics are not sufficiently represented in the industry, and their potential is neglected. This paper used all presented features together, but we think feature selection could be achieved higher results. In future work, we will employ several new metaheuristics for feature selection.

# References

1. Cortes, C., Vapnik, V.: Support-vector networks. Mach. Learn. **20**(3), 273–297 (1995)
2. Breiman, L.: Random forests. Mach. Learn. **45**, 5–32 (2001)
3. McCulloch, W.S., Pitts, W.: A logical calculus of the ideas immanent in nervous activity. Bull. Math. Biophys. **5**(4), 115–133 (1943)
4. Lo, A.W., Mamaysky, H., Wang, J.: Foundations of technical analysis: computational algorithms, statistical inference, and empirical implementation. J. Finance **55**(4), 1705–1765 (2000)
5. Ticknor, J.L.: A Bayesian regularized artificial neural network for stock market forecasting. Expert Syst. Appl. **40**(14), 5501–5506 (2013)
6. Kotu, V., Deshpande, B.: Data Science: Concepts and Practice, 2nd edn. Elsevier, Amsterdam (2019)
7. Zhang, W., Skiena, S.: Trading strategies to exploit blog and news sentiment. In: Proceedings of the Fourth International AAAI Conference on Weblogs and Social Media (ICWSM). AAAI, pp. 375–378, 2010
8. Ariyo, A. A., Adewumi, A. O., Ayo, C. K.: Stock price prediction using the ARIMA model. In: 2014 UKSim-AMSS 16th International Conference on Computer Modelling and Simulation, pp. 106–112. IEEE (2014)
9. Adebiyi, A. A., Adewumi, A. O., Ayo, C. K.: Comparison of ARIMA and artificial neural networks models for stock price prediction. J. Appl. Math. **2014**, Article ID 614342 (2014). https://doi.org/10.1155/2014/614342
10. Patel, J., Shah, S., Thakkar, P., Kotecha, K.: Predicting stock and stock price index movement using trend deterministic data preparation and machine learning techniques. Expert Syst. Appl. **42**(1), 259–268 (2015)
11. Kim, S., Ku, S., Chang, W., Song, J.W.: Predicting the direction of us stock prices using effective transfer entropy and machine learning techniques. IEEE Access **8**, 111660–111682 (2020)
12. Ballings, M., Van den Poel, D., Hespeels, N., Gryp, R.: Evaluating multiple classifiers for stock price direction prediction. Expert Syst. Appl. **42**(20), 7046–7056 (2015)
13. Tuba, M., Alihodzic, A., Bacanin, N.: Cuckoo search and bat algorithm applied to training feed-forward neural networks. In: Yang, X.-S. (ed.) Recent Advances in Swarm Intelligence and Evolutionary Computation. SCI, vol. 585, pp. 139–162. Springer, Cham (2015). https://doi.org/10.1007/978-3-319-13826-8_8
14. Tuba, E., Capor Hrosik, R., Alihodzic, A., Jovanovic, R., Tuba, M.: Support vector machine optimized by fireworks algorithm for handwritten digit recognition. In: Simian, D., Stoica, L.F. (eds.) MDIS 2019. CCIS, vol. 1126, pp. 187–199. Springer, Cham (2020). https://doi.org/10.1007/978-3-030-39237-6_13
15. Kazem, A., Sharifi, E., Hussain, F.K., Saberi, M., Hussain, O.K.: Support vector regression with chaos-based firefly algorithm for stock market price forecasting. Appl. Soft Comput. **13**(2), 947–958 (2013)
16. Liu, C., Wang, J., Xiao, D., Liang, Q.: Forecasting SP 500 stock index using statistical learning models. Open J. Stat. **06**, 1067–1075 (2016)

# Optimal Knockout Tournaments: Definition and Computation

Amelia Bădică[1], Costin Bădică[1(✉)], Ion Buligiu[1], Liviu Ion Ciora[1], and Doina Logofătu[2]

[1] University of Craiova, Craiova, Romania
cbadica@software.ucv.ro
[2] Frankfurt University of Applied Sciences, Frankfurt am Main, Germany

**Abstract.** We study competitions structured as hierarchically shaped single elimination tournaments. We define optimal tournaments by maximizing attractiveness such that topmost players will have the chance to meet in higher stages of the tournament. We propose a dynamic programming algorithm for computing optimal tournaments and we provide its sound complexity analysis.

**Keywords:** Optimization · Knockout tournament · Dynamic programming · Algorithm complexity

## 1 Introduction

In this paper we propose a formal definition of competitions that have the shape of hierarchical single-elimination tournaments, also known as *knockout tournaments*. We introduce methods to quantitatively evaluate the attractiveness and competitiveness of a given tournament. We consider that a tournament is more attractive if competition is encouraged in higher stages, i.e. topmost players will have the chance to meet in higher stages of the tournament, thus increasing the stake of their matches.

In knockout tournaments, the result of each match is always a win of one of the two players, i.e. draws are not possible. A knockout tournament is hierarchically structured as a complete binary tree such that each leaf represents one player or team that is enrolled in the tournament, while each internal node represents a game of the tournament.

Our main achievements can be summarized as follows:

1. An exact formula for counting the total number of knockout tournaments, showing that the number of tournaments grows very large with the number of players.
2. A tournament cost function based on players' quota that assigns a higher cost to those tournaments where highly ranked players tend to meet in higher stages, thus making the tournament more attractive and competitive.
3. A dynamic programming algorithm for computing optimal tournaments.
4. The complexity analysis of the proposed algorithm.

© Springer Nature Switzerland AG 2022
I. Lirkov and S. Margenov (Eds.): LSSC 2021, LNCS 13127, pp. 217–225, 2022.
https://doi.org/10.1007/978-3-030-97549-4_25

Knockout tournaments attracted researches in operations research, combinatorics and statistics. The problem is also related to intelligent planning and scheduling [6]. A method for augmenting a tournament with probabilistic information based on tournament results was proposed in [4]. The effectiveness of tournament plans based on dominance graphs is studied in [5]. Tournament problem was also a source of inspiration for programming competitions [2]. A more recent work addressing competitiveness development and ranking precision of tournaments is [1].

Knockout tournaments have applications in sports (e.g. tennis tournaments) and online games. While in the former case there is a relatively low number of players, thus not raising special computational challenges, the number of players in massive multiplayer online games can grow such that computing the optimal tournament is a more difficult problem.

## 2   Knockout Tournaments

We consider hierarchically structured knockout tournaments such that the result of each match is always a win of one of the two players, i.e. draws are not possible.

**Definition 1.** *(2-partition) Let us consider a finite nonempty set $\Sigma$ with an even number of elements. A 2-partition of $\Sigma$ is a partition $(\Sigma', \Sigma'')$ of $\Sigma$ such that $|\Sigma'| = |\Sigma''|$.*

An $n$-stage hierarchical tournament is structured as a complete binary tree with $2^n$ leaves representing players or teams that are enrolled in the tournament. The idea is that the tournament has $n$ stages such that in stage 1 there are $2^{n-1}$ matches, ..., in stage $k$ there are $2^{n-k}$ matches, and in stage $n$ there is a single match – the tournament final.

If $n = 0$ there is match so the singleton participant is the winner of the tournament.

If $n \geq 1$ then the tournament has $n$ stages numbered $1, 2, \ldots, n$. Each $i = 1, 2, \ldots, n$ consists of $2^{n-i}$ matches transforming the current $(n - i + 1)$-stage tournament into an $(n - i)$-stage tournament. For $i = n$, the last 0-stage defines the winner of the tournament.

**Definition 2.** *(Set of tournaments) Let us consider a finite nonempty set $\Sigma$ with $2^n$ elements. The set of $n$-stage tournaments $\mathcal{T}_n(\Sigma)$ is defined as follows:*

- *if $n = 0$ then $\mathcal{T}_0(\Sigma) = \Sigma$.*
- *if $n \geq 1$ then $\mathcal{T}_n(\Sigma) = \{\{t', t''\}|$ there is a 2-partition $(\Sigma', \Sigma'')$ of $\Sigma$ such that $t' \in \mathcal{T}_{n-1}(\Sigma'), t'' \in \mathcal{T}_{n-1}(\Sigma'')\}$.*

*Example 1.*

$$
\begin{aligned}
T_0(\{1\}) &= \{1\} \\
T_1(\{1, 2\}) &= \{\{1, 2\}\} \\
T_2(\{1, 2, 3, 4\}) &= \{\{\{1, 3\}, \{2, 4\}\}, \{\{1, 4\}, \{2, 3\}\}, \{\{1, 2\}, \{3, 4\}\}\}
\end{aligned}
$$

Let us now build a tournament $t \in \mathcal{T}_3(\{1,2,3,4,5,6,7,8\})$. We first consider a 2-partition of $\Sigma = \{1,2,3,4,5,6,7,8\}$ defined by $\Sigma' = \{1,2,4,7\}$ and $\Sigma'' = \{3,5,6,8\}$. Then we chose two tournaments $t' \in \mathcal{T}_2(\Sigma')$ and $t'' \in \mathcal{T}_2(\Sigma'')$ defined by:

$$t' = \{\{1,4\},\{2,7\}\}$$
$$t'' = \{\{3,8\},\{5,6\}\}$$

We can now define $t$ as:

$$t = \{\{\{1,4\},\{2,7\}\},\{\{3,8\},\{5,6\}\}\}$$

**Proposition 1.** *(Counting tournaments) The set $\mathcal{T}_n(\Sigma)$ contains $\frac{(2^n)!}{2^{2^n-1}}$ tournaments.*

*Proof.* An $n$-stage tournament is a complete binary tree with $2^n$ leaves and $2^n-1$ internal nodes. The sequence of leaves is a permutation of the set $\Sigma$. Each such sequence defines a tree. But, for each internal node, exchanging its left and right sub-tree is a tournament invariant. There are $(2^n)!$ permutations of leaves and there are $2^n - 1$ independent ways of exchanging left and right sub-tree of the tree that defines each permutation, producing the same tournament. So the total number of $n$-stage tournaments is $\frac{(2^n)!}{2^{2^n-1}}$.

*Example 2.* The number of 2-stage tournaments is $\frac{(2^2)!}{2^{2^2-1}} = \frac{4!}{8} = 3$.

Observe that applying the formula, we obtain 315 3-stage tournaments. Let us obtain this result using a different reasoning. Let us count the number of 2-partitions of a set with 8 elements. There are $\binom{8}{4}/2 = 35$ possibilities, as we consider the 4-combinations of 8 elements, and we divide by 2 as the order of the subsets of a partition does not matter. Now, for each set of the 2-partition there are 3 2-stage tournaments, so multiplying we get a total of 9 possibilities. So the number of 3-stage tournaments is $9 \cdot 35 = 315$.

## 3   Pairings

Each stage of a tournament defines possible pairings between players. Each possible pairing defines a possible match.

**Definition 3.** *(Pairings) Let $t \in \mathcal{T}_n(\Sigma)$ an $n$-stage tournament for $n \geq 1$. A pairing is a set $\{x,y\} \subseteq \Sigma$ representing a possible match that can occur at stage $n$ of $t$. Let $t = \{\{t',t''\}|t' \in \mathcal{T}_{n-1}(\Sigma'), t'' \in \mathcal{T}_{n-1}(\Sigma'')\}$. The set $p(t)$ of pairings associated with tournament $t$ is defined as $p(t) = \{\{x,y\}|x \in \Sigma', y \in \Sigma''\}$.*

Observe that the number of pairings associated with tournament $t \in \mathcal{T}_n(\Sigma)$ is $2^{2n-2}$. Note also that every match is possible in a given $n$-stage tournament, at one and only one of its stages $k = 1,2,\ldots,n$. Stage 1 represents the leaves of the tree, while stage $n$ represents the final of the tournament. Stage $k$ contains $2^{n-k}$ sets of pairings, each set containing $2^{2k-2}$ pairings. So the total number of possible pairings is:

$$\sum_{k=1}^{n} 2^{n-k} \cdot 2^{2k-2} = 2^{n-2} \cdot \sum_{k=1}^{n} 2^k = 2^{n-2} \cdot (2^{n+1} - 2) = \frac{2^n(2^n - 1)}{2} \qquad (1)$$

As there are $|\Sigma| = 2^n$ players, Eq. (1) shows that the total number of pairings is the same as the total number of possible matches.

## 4   Tournament Cost

Let $s_{i,j}$ be the stage of a tournament where players $i, j$ can meet. There is a unique stage $k = 1, 2, \ldots, n$ of the tournament that contains pairing $\{i, j\}$ so $s_{i,j} = k$ is well defined.

Intuitively, the higher are quotations of players $i$ and $j$, the better is to let them meet in a higher stage of the tournament in order to increase the stake of their match.

We assume in what follows that a quotation $q_i \in (0, +\infty)$ is available for each player $i \in \Sigma$. Quotations can be obtained from the players' current ranking or by other means.

**Table 1.** Players' ranking and quotation for $n = 4$.

| Player $i$ | Rank $r_i$ | Quota $q_i = n + 1 - r_i$ |
|---|---|---|
| 1 | 1 | 4 |
| 4 | 2 | 3 |
| 3 | 3 | 2 |
| 2 | 4 | 1 |

**Table 2.** Games playing stages for each tournament and tournament costs.

| $s^{t_1}$ | 1 | 2 | 3 | 4 | $s^{t_2}$ | 1 | 2 | 3 | 4 | $s^{t_3}$ | 1 | 2 | 3 | 4 |
|---|---|---|---|---|---|---|---|---|---|---|---|---|---|---|
| 1 | | 2 | 1 | 2 | 1 | | 2 | 2 | 1 | 1 | | 1 | 2 | 2 |
| 2 | | | 2 | 1 | 2 | | | 1 | 2 | 2 | | | 2 | 2 |
| 3 | | | | 2 | 3 | | | | 2 | 3 | | | | 1 |
| Cost | | 59 | | | | | 56 | | | | | 60 | | |

**Definition 4.** *(Tournament cost) Let $t \in \mathcal{T}_n(\Sigma)$ be an $n$-stage tournament and let $s_{i,j}^t \in \{1, 2, \ldots, n\}$ be the stage of $t$ where players $i, j$ can meet. Let $q_i > 0$ be the quotations of players for all $i \in \Sigma$. The cost of $t$ is defined as:*

$$Cost(t) = \sum_{i,j \in \Sigma, i<j} q_i q_j s_{ij}^t \qquad (2)$$

Obviously, better ranked players have a higher quotation. We assume that if player $i$ has rank $r_i$ then its quota is $q_i$ such that whenever $r_i < r_j$ we have $q_i > q_j$. For example, if there are $N = 2^n$ players then we can choose $q_i = N + 1 - r_i$ for all $i = 1, \ldots, N$.

**Definition 5.** *(Optimal tournament) A tournament such that its cost computed with Eq. (2) is maximal is called* optimal tournament *and it is defined by:*

$$OptC(\Sigma) = \max_{t \in \mathcal{T}_n(\Sigma)} Cost(t)$$
$$t^* = argmax_{t \in \mathcal{T}_n(\Sigma)} Cost(t) \tag{3}$$

*Example 3.* Let us consider 4 players (see Table 1). We assume that each player has a unique rank from 1 to 4. Now, if we choose $q_i = 5 - r_i$ then, using this approach for defining players' quota, player 2 with rank 4 is assigned quotation $q_2 = 1$.

We consider the three tournaments $t_1, t_2, t_3 \in T_2(\{1, 2, 3, 4\})$ from Example 1. According to Eq. (2), the cost of a tournament $t \in T_2(\{1, 2, 3, 4\})$ is:

$$Cost(t) = \sum_{1 \le i < j \le 4}^{4} s_{ij}^t q_i q_j$$
$$Cost(t) = s_{12}^t \cdot 4 \cdot 1 + s_{13}^t \cdot 4 \cdot 2 + s_{14}^t \cdot 4 \cdot 3 + s_{23}^t \cdot 1 \cdot 2 + s_{24}^t \cdot 1 \cdot 3 + s_{34}^t \cdot 2 \cdot 3 \tag{4}$$

Substituting stage values $s_{ij}^t$ for each tournament from Table 2 into Eq. (4) we obtain the tournaments' cost values from Table 2. We observe that in this case the best tournament is $t_3$. Actually it can be easily checked that the best tournament is $t_3$ for whatever values of the quota that are decreasingly ordered according to the ranks.

## 5    A Dynamic Programming Algorithm for Computing Optimal Tournaments

In what follows, for any subset of players $\Sigma$ we denote by $q_\Sigma$ the following sum:

$$q_\Sigma = \sum_{i \in \Sigma} q_i \tag{5}$$

**Proposition 2.** *(Recurrence for tournament cost) Let $t \in \mathcal{T}_n(\Sigma)$ be an n-stage tournament such that $\Sigma$ is a finite nonempty set with $2^n$ elements. Then:*

$$Cost(t) = \begin{cases} 0 & n = 0 \\ Cost(t') + Cost(t'') + nq_{\Sigma'}q_{\Sigma \setminus \Sigma'} & n \ge 1, t = \{t', t''\} \\ & t' \in \mathcal{T}_{n-1}(\Sigma'), t'' \in \mathcal{T}_{n-1}(\Sigma \setminus \Sigma') \end{cases} \tag{6}$$

*Proof.* If $n = 0$ then $\Sigma$ has a single player so the result is obvious, as no games are played to determine the winner of the tournament.

If $n \ge 1$ then $t = \{t', t''\}$. If $i \in \Sigma'$ and $j \in \Sigma \setminus \Sigma'$ then $s_{ij}^t = n$. So Equation (2) gives:

$$Cost(t) = \sum_{i,j \in \Sigma', i<j} q_i q_j s_{ij}^{t'} + \sum_{i,j \in \Sigma \setminus \Sigma', i<j} q_i q_j s_{ij}^{t''} + \sum_{i \in \Sigma', j \in \Sigma \setminus \Sigma'} q_i q_j s_{ij}^t =$$
$$Cost(t') + Cost(t'') + n \sum_{i \in \Sigma', j \in \Sigma \setminus \Sigma'} q_i q_j = Cost(t') + Cost(t'') + nq_{\Sigma'}q_{\Sigma \setminus \Sigma'} \tag{7}$$

**Proposition 3.** *(Recurrences for optimal tournaments)*

a. *The optimal tournament cost OptC introduced by Eq. (3) can be defined recursively as follows:*

$$
OptC(\Sigma) = \begin{cases} 0 & n = 0, |\Sigma| = 1 \\ \max_{\Sigma' \subseteq \Sigma, |\Sigma'| = 2^{n-1}} [OptC(\Sigma') + \\ \quad OptC(\Sigma \setminus \Sigma') + nq_{\Sigma'}q_{\Sigma \setminus \Sigma'}] & n \geq 1, |\Sigma| = 2^n \end{cases}
$$

(8)

b. *The optimal tournament can be determined by recording the subsets $\Sigma'$ that maximize OptC in Equation (8) for $n \geq 1$, $|\Sigma| = 2^n$ as follows:*

$$
OptS(\Sigma) = \operatorname*{argmax}_{\Sigma' \subseteq \Sigma, |\Sigma'| = 2^{n-1}} OptC(\Sigma') + OptC(\Sigma \setminus \Sigma') + nq_{\Sigma'}q_{\Sigma \setminus \Sigma'} \quad (9)
$$

*Proof.* The proof follows by applying the maximization operation in Eq. (6) and observing that the term $nq_{\Sigma'}q_{\Sigma \setminus \Sigma'}$ does not depend on $t = \{t', t''\}$.

Moreover, the sets $OptS(\Sigma)$ can be used to construct an optimal tournament. Let $|\Sigma| = 2^n$ and $\Sigma_n = \Sigma$. We define: $\Sigma_{n-1} = OptS(\Sigma_n)$, ..., $\Sigma_0 = OptS(\Sigma_1)$. Then the optimal tournament $t^*$ can be defined recursively as follows:

$$
t^* = t_n(\Sigma_n)
$$

$$
t_i(\Sigma_i) = \begin{cases} \{t_{i-1}(\Sigma_{i-1}), t_{i-1}(\Sigma_i \setminus \Sigma_{i-1})\} & i \geq 1 \\ j & i = 0, \Sigma_0 = \{j\} \end{cases} \quad (10)
$$

Proposition 3 (Eq. (8) in particular) can be used to design a dynamic programming algorithm for computing the optimal tournament and its cost. The cost of the optimal tournament is computed with the help of $OptC$ vector that is indexed by all the subsets of $\Sigma$ of cardinal: $2^0, 2^1, \ldots 2^n$. Note that the size of this vector is:

$$
S_n = \sum_{i=0}^{n} \binom{2^n}{2^i} \quad (11)
$$

Additionally we must save in vector $OptS$ of size $S_n$ the subsets $\Sigma'$ determined using Eq. (9), such that we can reuse them to construct the optimal tournament using Eq. (10). Our proposed algorithm is presented as Algorithm 1.

---

**Algorithm 1.** $OptTourCost(\Sigma, N = 2^n, q)$ algorithm for computing the cost of the optimal tournament

---

**Input:** $N = 2^n, n \in \mathbb{N}$. $N$ represents the number of players.
$\qquad$ $q$. Vector of size $N$ representing the players' quota.
$\qquad$ $\Sigma$ such that $|\Sigma| = 2^n$. $\Sigma$ represents the set of players.
**Output:** $OptC$. Vector of costs of the optimal sub-tournaments.
$\qquad$ $OptS$. Vector of sets to construct the optimal tournament.
1: **for** $i = 1, N$ **do**
2: $\quad$ $OptC[\{i\}] \leftarrow 0$
3: **end for**
4: **for** $k = 1, n$ **do**
5: $\quad$ **for** $\Sigma_1 \subseteq \Sigma$ s.t. $|\Sigma_1| = 2^k$ **do**
6: $\quad\quad$ $Cmax \leftarrow -\infty$
7: $\quad\quad$ **for** $\Sigma' \subseteq \Sigma_1$ s.t. $|\Sigma'| = 2^{k-1}$ **do**
8: $\quad\quad\quad$ $C \leftarrow OptC(\Sigma') + OptC(\Sigma \setminus \Sigma')$
9: $\quad\quad\quad$ $ql \leftarrow 0$
10: $\quad\quad\quad$ **for** $i \in \Sigma'$ **do**
11: $\quad\quad\quad\quad$ $ql \leftarrow ql + q_i$
12: $\quad\quad\quad$ **end for**
13: $\quad\quad\quad$ $qr \leftarrow 0$
14: $\quad\quad\quad$ **for** $i \in \Sigma_1 \setminus \Sigma'$ **do**
15: $\quad\quad\quad\quad$ $qr \leftarrow qr + q_i$
16: $\quad\quad\quad$ **end for**
17: $\quad\quad\quad$ $C \leftarrow C + k * ql * qr$
18: $\quad\quad\quad$ **if** $C > Cmax$ **then**
19: $\quad\quad\quad\quad$ $Cmax \leftarrow C$
20: $\quad\quad\quad\quad$ $Smax \leftarrow \Sigma'$
21: $\quad\quad\quad$ **end if**
22: $\quad\quad$ **end for**
23: $\quad\quad$ $OptC[\Sigma_1] \leftarrow Cmax$
24: $\quad\quad$ $OptS[\Sigma_1] \leftarrow Smax$
25: $\quad$ **end for**
26: **end for**

---

**Algorithm 2.** $OptTour(\Sigma, n, OptS)$ algorithm for computing the optimal tournament

---

**Input:** $n \in \mathbb{N}$. $n$ represents the number of stages.
$\qquad$ $OptS$. Vector of size $S_n$ (see Equation (11)) determined by Algorithm 1.
$\qquad$ $\Sigma$ such that $|\Sigma| = 2^n$. $\Sigma$ represents the set of players.
**Output:** Returns the optimal tournament.
1: **if** $n = 0$ **then**
2: $\quad$ Let $\Sigma = \{j\}$
3: $\quad$ **return** $j$
4: **end if**
5: $\Sigma' \leftarrow OptS(\Sigma)$
6: $t' \leftarrow OptTour(\Sigma', n - 1, OptS)$
7: $t'' \leftarrow OptTour(\Sigma \setminus \Sigma', n - 1, OptS)$
8: **return** $\{t', t''\}$

The algorithm for computing an optimal tournament based on the values saved in vector $OptS$ by Algorithm 1 is presented as Algorithm 2.

**Proposition 4.** *(Correctness of Algorithms 1 and 2)*

a. *The value $OptC[\Sigma]$ computed by Algorithm 1 represents the cost of the optimal tournament.*
b. *The tournament determined by Algorithm 2 is the optimal tournament.*

*Proof.* **Proof of a.** Algorithm 1 computes the values of $OptC$ and $OptS$ in bottom-up fashion starting with subsets of cardinal $2^0$, then with $2^1$, ..., and finally with $2^n$. The computation follows Eqs. (8) and (9). Therefore the correctness of this point follows from Proposition 3.

**Proof of b.** Algorithm 2 computes the optimal tournament using Eqs. (10). As values of $OptS$ are correctly determined according to point "a", it follows that the tournament computed by Algorithm 2 is the optimal tournament.

**Proposition 5.** *(Complexity of Algorithms 1 and 2)   Let us consider tournaments with $n \geq 1$ stages and $N = 2^n$ players.*

a. *Space complexity of Algorithms 1 and 2 is $\Theta\left(\frac{2^N}{\sqrt{N}}\right)$.*
b. *Time complexity of Algorithm 1 is $\Theta((2\sqrt{2})^N)$.*
c. *Time complexity of Algorithm 2 is $\Theta(N)$.*

*Proof.* The proof is using the Stirling approximation of the factorial, written in inequality form [3], in fact showing that $p! = \Theta\left(\sqrt{p}\left(\frac{p}{e}\right)^p\right)$:

$$\sqrt{2\pi} \leq \frac{p!}{\sqrt{p}\left(\frac{p}{e}\right)^p} \leq e \tag{12}$$

Using this observation it is not difficult to prove that:

$$\binom{2p}{p} = \Theta\left(\frac{2^{2p}}{\sqrt{p}}\right) \tag{13}$$

**Proof of a.** The space complexity of Algorithms 1 and 2 is given by the size of vectors $OptC$ and $OptS$ (see Eq. (11)). But the asymptotically dominant term of this summation is $\binom{2^n}{2^{n-1}}$. Then the result follows using Eq. (13) for $p = 2^{n-1}$.

**Proof of b.** Algorithm 1 contains one "for" loop (lines 4–26) including other three nested "for" loops. The first inner "for" loop (lines 5–25) is executed $\binom{N}{2^k}$ times. The second inner "for" loop (lines 7–22) is executed $\binom{2^k}{2^{k-1}}$ times. The third inner "for" loop (lines 10–12 and 14–16) is executed $2^{k-1}$ times. The total number of steps is given by:

$$\sum_{k=1}^{n} \binom{N}{2^k}\binom{2^k}{2^{k-1}} 2^{k-1} \tag{14}$$

Observe that the asymptotically dominant term of this summation is obtained for $k = n - 1$ and it can be transformed using Equation (13), thus concluding the proof:

$$\binom{N}{2^{n-1}}\binom{2^{n-1}}{2^{n-2}}2^{n-2} = \Theta\left(\frac{2^{2^n}}{\sqrt{2^{n-1}}}\frac{2^{2^{n-1}}}{\sqrt{2^{n-2}}}2^{n-2}\right) = \Theta\left(\frac{1}{\sqrt{2}}2^N 2^{N/2}\right) = \Theta((2\sqrt{2})^N) \quad (15)$$

**Proof of c.** Observe that for $n \geq 1$ time complexity of Algorithm 2 satisfies the equation $T(n) = 2T(n-1)$. This shows that the time complexity is $\Theta(2^n) = \Theta(N)$.

# 6 Conclusions

We introduced optimal knockout tournaments and we theoretically analyzed a dynamic programming algorithm for their computation. As future work we plan to carry out an experimental evaluation of the proposed algorithm. Moreover, we would like to augment our model by incorporating the uncertainty of matches' winners.

# References

1. Bao, N.P.H., Xiong, S., Iida, H.: Reaper tournament system. In: Chisik, Y., Holopainen, J., Khaled, R., Luis Silva, J., Alexandra Silva, P. (eds.) INTETAIN 2017. LNICST, vol. 215, pp. 16–33. Springer, Cham (2018). https://doi.org/10.1007/978-3-319-73062-2_2
2. CodeChef: Tennis tournament (2012). https://www.codechef.com/COOK27/problems/TOURNAM. Accessed 21 March 2021
3. Dutka, J.: The early history of the factorial function. Arch. Hist. Exact Sci. **43**(3), 225–249 (1991). https://doi.org/10.1007/BF00389433
4. Hartigan, J.A.: Probabilistic completion of a knockout tournament. Ann. Math. Stat. **37**(2), 495–503 (1966). https://doi.org/10.1214/aoms/1177699533
5. Maurer, W.: On most effective tournament plans with fewer games than competitors. Ann. Stat. **3**(3), 717–727 (1975). https://doi.org/10.1214/aos/1176343135
6. Vu, T., Shoham, Y.: Fair seeding in knockout tournaments. ACM Trans. Intell. Syst. Technol. **3**(1), 9:1–9:17 (2011). https://doi.org/10.1145/2036264.2036273

# Risk Registry Platform for Optimizations in Cases of CBRN and Critical Infrastructure Attacks

Nina Dobrinkova[1]([✉]), Evangelos Katsaros[2], and Ilias Gkotsis[3]

[1] Institute of Information and Communication Technologies, Bulgarian Academy of Sciences, Sofia, Bulgaria
nina.dobrinkova@iict.bas.bg, ninabox2002@gmail.com
[2] European University Cyprus, Engomi, Cyprus
[3] Center for Security Studies (KEMEA), Athens, Greece

**Abstract.** Nowadays the world faces a wide range of complex challenges and threats to its security. The origination of modern threats, among others takes into consideration the proliferation of weapons of mass destruction (WMD) and their delivery systems. Rapid advances in science and technology, have proven to be misused by terrorist groups who develop the necessary knowledge and capacity to turn them into Chemical-biological-radiological and nuclear (CBRN) threats against the civil population. Adequate response and cross-sectoral cooperation in case of emergency situation is the base for low number of casualties and fast localization of the threat sources. Risk Registry platforms used to optimize the CBRN and critical infrastructure attacks are important crisis management capacity tools which need to be further developed using the nowadays information and communication technologies (ICT). Identifying and formulating registry algorithms containing documented cases of executed threats in these thematic areas, including the technological side of attacks, is very important for tactical planning and optimizations. In our paper we will describe a system implemented as a Risk Registry platform used for table top or field test exercises in cases of CBRN response by teams in Bulgaria, Greece, and Cyprus.

## 1 Introduction

The Confrontation of CBRN-Terrorism Threats project (COBRA project) has been designed in the framework of ISF-Police Action Grant (Call: ISFP-2018-AG-CT). The project main aim was to address the prevention and preparedness measures against the increasing wide range of complex attacks to the European country's security. CBRN threats so far are considered as a low probability but with high impact risks. Even at a small scale, a CBRN attack may have a considerable impact on the societies and economies against which they are used, resulting in significant and lasting disruption, widespread fear and uncertainty. Two of the main objectives of the EC Action Plan against CBRN (2017) are related to enhance preparedness against chemical, biological, radiological and nuclear security risks with main focus on:

© Springer Nature Switzerland AG 2022
I. Lirkov and S. Margenov (Eds.): LSSC 2021, LNCS 13127, pp. 226–233, 2022.
https://doi.org/10.1007/978-3-030-97549-4_26

- Ensuring a more robust preparedness for and response to CBRN security incidents;
- Enhancing our knowledge of CBRN risks.

In this direction, stress tests with optimization algorithms for reaction in such cases will allow the first responders to protect better soft targets (e.g. a Public Space) and/or a Critical Infrastructure (e.g. Trans Adriatic Pipeline paths) or cross-border zones where immigrant groups stay in one place unprotected from such attacks. The E.C., the Parliament and the Council have made many landmark decisions to establish trust and security for the European Citizens, being threatened by terrorist attacks and disruptive other actions. Decisions and actions such as the restrictions to the access to firearms, the combat against terrorist financing, better border and screening equipment and closing the information gaps in the large-scale information systems are examples of tangible actions that are being done. Next to the legislative and policy actions to protect the E.U. citizens, the Commission implements the Action plan to improve the protection of public places. The European Parliament and the Council Presidency launched a Roadmap for a More United, Stronger and More Democratic Union, providing guidance and support to Member States at national, regional and local levels. While there can never be 'zero risk', these operational measures will support Member States in detecting threats, reducing the vulnerability of public spaces, mitigating the consequences of a terrorist attack and improving cooperation. Based on the EU Counter-Terrorism Strategy, which was adopted in 2005, COBRA project main aims are to address and enhance the following two strands:

- PROTECT citizens and critical infrastructure by reducing vulnerabilities against attacks;
- RESPOND in a coordinated way by preparing for addressing terrorist attacks and minimizing the consequences of CBRN issues, improving capabilities to deal with the aftermath and taking into account the needs of victims.

Our paper has three main sections. The first one focus on a possible application of a methodology for multicriteria analyses quantification of risk and a post-countermeasure of the risk suggested by Linkov et al. in 2012 [1]. The second define the basic CBRN threats, which can be used as case studies. The last section covers the potential implementation of a Risk registry platform tested as demo in Greece, Bulgaria, and Cyprus.

## 2  Methodology

The theoretical bases of CBRN Multi-Criteria Decision Analysis (MCDA) applied to terrorism are described in details by Ezell et al. in 2008 and Bier et al. in 2009 [2,3]. The Linkov et all in 2012 [1], propose Risk Informed Decision Framework (RIDF), which integrates three independent methodologies - Life Cycle Analysis (LCA), Risk Analysis (RA) and Multi-Criteria Decision Analysis (MCDA).

COBRA project research phase was focused on the above-mentioned methodologies in order to evaluate what data was possible to be collected for tests in the three demo countries.

In the RIDF approach the central piece is dedicated to the Multi-Criteria Decision Analysis (MCDA). MCDA facilitates the formulation of the objective function of the decision problem discussed in details from Keeney in 1992, Linkov and Moberg in 2011, and Linkov et al. in 2006 [4–6]. Most applications of MCDA in the field of CBRN threats rely on other tools such as decision trees for the ultimate decision recommendations, and are limited to evaluating consequences of a terrorist attack only. The second element of RIDF that had a vital part of the evaluation and research phase in COBRA project was the Life Cycle Analysis (LCA). In most cases LCA has been used to establish a full account of environmental impacts from anthropogenic activities. The suggested approach by Linkov et all in 2012 [1], formalizes LCA in a way the descriptions of the different stages in cases of a terrorist threats. The mathematical background connecting the life cycle elements to specific threat consequences proposed by Linkov et all in 2012 [1], is based on the standard risk equation definition:

$$Risk = Threat \times Vulnerability \times Consequences \qquad (1)$$

With suggested modification by Linkov et all in 2012 [1] to Eq. (1):

$$Risk = P(Attack) \times P(Success|Attack) \times \qquad (2)$$
$$\{Mortality, Morbidity, DirectEconomicLoss, IndirectEconomicLoss\}$$

the quantitative value of each of the components in Eq. (2) is suggesting a guided expert elicitation exercise. In Eq. (2), the difference between a baseline quantification of risk and a post-countermeasure quantification of risk is a natural measure of countermeasure benefit. If divided by the costs of the countermeasure, this difference gives a useful measure of the countermeasure cost effectiveness. Implicit in this definition is the life cycle nature of the risk of terrorism threats. Terrorism threats are processed as a sequence of events which mirror the order of the right side of the risk Eq. (2). These common stages of terrorist attacks are defined more specifically by Mohan & von Winterfeldt (2010) [7] and Linkov et all in 2012 [1] as six separate ones:

1. Plan - Terrorist cells (TC) develop a strategic course of action;
2. Collect Resources - An orchestrated effort to collect primary resources;
3. Develop Attack Capability - TC engage in developing the attack mode;
4. Successful Intrusion and Use - Security breach and attack carried out;
5. Direct Impact - Immediate human and economic loss from the attack;
6. Indirect Impact - Attack effects are felt over time and space.

The descriptions given in [1,7] state that stages 1 through 3 form the first term of the risk equation, P(Attack), stage 4 is the term P(Success—Attack) and stages 5 & 6 cover the last term of consequences Mortality, Morbidity, Direct Economic Loss, Indirect Economic Loss.

The COBRA project primarily focusses on data collection for identification, examination, analysis, and systematic categorization of the risks related to CBRN threats against soft targets as well as terrorism targeting European Critical Infrastructures. This data is geo-localized for Greece, Cyprus, and Bulgaria in order to determine the relative range of threats.

## 3 Data Evaluation for CBRN Risks at National, Regional, and Cross Border Levels for Greece, Bulgaria, Cyprus

CBRN incidents according to the Centre for the protection of Natural Infrastructure (more details can be found in the official link: https://www.cpni.gov.uk/chemical-biological-radiological-and-nuclear-cbrn-threats-0, the CBRN threats and hazards are categorized to the following types: (based on https://www.europarl.europa.eu/RegData/etudes/STUD/2018/604960/IPOL_STU(2018)604960_EN.pdf):

1. Chemical - Poisoning or injury caused by chemical substances, including traditional (military) chemical warfare agents, harmful industrial or household chemicals;
2. Biological - Illnesses caused by the deliberate release of dangerous bacteria or viruses or by biological toxins (e.g. ricin, found in castor oil beans);
3. Radiological - Illness caused by exposure to harmful radioactive materials;
4. Nuclear - Life-threatening health effects caused by exposure to harmful radiation, thermal or blast effects arising from a nuclear detonation.

Often E-explosives are included which could be used in combination with the above categories. Given this classification, Chemical, Biological, Radiological and Nuclear (CBRN) threats concern a variety of events deriving from naturally occurring incidents, accidents or malicious actions, their origin could be:

1. Naturally occurring incidents that may be triggered from natural hazards such as fire, flood, storm near hazardous infrastructures of materials;
2. Accidents concerning critical infrastructures' due to failure of technological factors and/or r human actions, errors and negligence;
3. Malicious actions concern the deliberate malicious use of CBRN materials and knowledge in order to cause casualties (terrorist attacks).

Threats can also be divided based on their occurrence:

1. Regular threats, occurring frequently, as a result the mitigation system has general knowledge of how to respond to them. For such threats it is economically feasible to have emergency preparedness plans and equipment;
2. Irregular threats refer to one kind event. For them it is practically impossible to provide a standard response as they are unexpected, although they may be imagined in a scenario;

3. Unexampled events are basically events that are virtually impossible to imagine and are included or not included in preparatory scenarios. They can be standalone events, or a series of cascading events incorporated in a way not previously known.

Provided the above, it should be taken in consideration that, in order to address CBRN threats, their risk must be identified on a national level and then processed according the methodological stages. The best way to address observed shortcomings is by simulation of CBRN scenarios describing hypothetical threats, including scale, severity and general impact, casualties, (and possible) number of neighboring countries that could be affected.

Such scenarios are developed for COBRA project within the scope of the three participating countries—Greece, Bulgaria, and Cyprus. The selected scenarios describe hypothetical but possible situations with low, medium or high probability to occur and their impact has a range from limited to tremendous consequences. The scenarios in terms of occurrence include regular and irregular threats. The assessed case studies cover a list of known CBRN threats as defined in the research phase of the project. Specific questionnaire has been formulated, combined with semi-structured discussions with experts on CBRN in the three countries national stakeholders, aiming to produce a general assessment of CBRN threats in terms of likelihood and impact for the evaluated threats.

The case studies for validation and demonstration are uploaded to the Risk Registry platform of the project. The COBRA training activities organized as computer assisted table top exercises will be fulfilled until the end of its lifetime.

## 4   Risk Registry Platform and Its Tools

Based on open-source GIS tools we built a Risk Registry Platform suitable for a decision support tool of first responder teams. The application is based on open-source GIS software using: Geoserver, Qgis, Web App Builder, Boundless WEBSDK, OpenLayers. The Geoserver allows the user to display spatial information to the operation room teams. The QGIS base is a professional GIS (Geographic Information System) cross-platform application that is Free and Open Source Software (FOSS) which reduce the support cost of the system. The Web App Builder is a plugin for QGIS that allows easy creation of web applications. The Boundless WEBSDK provides tools for easy-to-build JavaScript-based web mapping applications. The GeoServer used allows processing of maps and data from a variety of sources. The GeoServer comes with a browser-based management interface and connects to multiple data sources at the back end. The final visualization is based on OpenLayers software that display data into the web browser. The OpenLayers software makes displaying of dynamic and field team localization inputs as a vector data and markers loaded on the spot.

This risk registry developed in COBRA project is having three levels of treat scenarios:

1. Regular threats - they can occur often and the system learns how to respond, by adding every time economical feasible response with emergency preparedness plans and equipment. Example cases but not limited to these options, are added to the system are gas lines or trucks transporting fuels that can cause mass casualties after a wrong work with fire nearby;
2. Irregular threats - they occur as a one of a kind events and thus the examples chosen are from the historical events described well in the CBRNe literature. Examples of such events are:
   - London July 7th 2005, 3 subways, 1 bus, 52 dead and 700+ wounded people. Coordinated terrorist attack that terrified London city that day.
   - Madrid March 11th 2004, 192 dead and 1500+ wounded people. Ten bombs were exploded at four different places in the same city at the same time. Many affected lives and authorities updating contingency plans with grades of unexpected measures for the future.
3. Unexampled events - that are virtually impossible to imagine and yet has happened. Example for such event is the New York September 11th 2001, series of four coordinated terrorist attacks by the Islamist terrorist group Al-Qaeda against the United States. 2,977 people were killed, 19 hijackers committed murder-suicide, and 6,000+ others were injured. Two of the planes, crashed into the North and South towers of the World Trade Center complex in Lower Manhattan. A third flight, was hijacked over Ohio and was crashed into the west side of the Pentagon, which led to a partial collapse of its west side. The fourth and final plane, crashed into a field near Shanksville, Pennsylvania after a struggle between passengers and hijackers. Investigators were unable to determine the exact target, but concluded the plane was planned to crash into the White House or the Capitol Building.

These scenarios are used as basic options which every team from Bulgaria, Greece, and Cyprus could play with and make according to the exercise needs more difficult or just use the basic parameters during the training.

Description of all details that will allow comparison with real CBRN events to extract useful conclusions from past ones are the main functionalities included to the system design. The goal is to identify and formulate a registry containing documented cases of realized threats in the CBRN thematic area including also data on the technological side of attacks. Regarding Critical Infrastructures and the exploitation of their vulnerabilities as a result of terrorist attacks, the Risk registry platform provides a corresponding layer with possible scenarios for reactions where the basic examples are used or modified by adding local specificities of the three countries.

The risk registry users view screens are available to Fig. 1 and Fig. 2. On Fig. 1 the general map is available where all project countries are located with the predefined scenarios adapted according to the specific local case study scenarios.

The risk registry initial screen with the availability to add scenarios per country is available to Fig. 2. Operational team can add a new attack (real or virtual) in the tested scenario in an easy user-friendly panel with automatic geo-localization window.

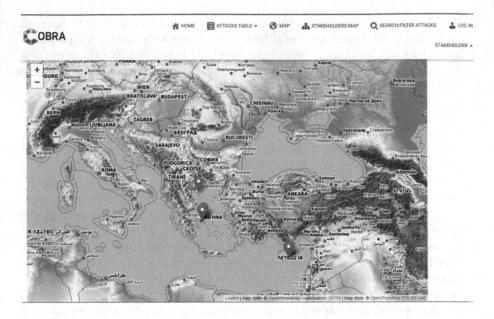

**Fig. 1.** Risk registry platform general map with scenarios per project country

**Fig. 2.** Adding a new attack in the scenario panel - gives map for location and panel to add parameters, representing the specific real case

## 5   Conclusion

The COBRA project Risk Registry platform is born by the need of CBRN events analysis tool. The responsible authorities optimizations in cases of a CBRN strategies and action plans (wherever exist), are general and only after a real event they are updated. The civil protection prevention, preparedness and management policies based on predefined scenarios and stakeholder's opinion can bring to the CBRN response phase faster and more adequate reactions with low number of casualties. The risk registry platform can be used for analysis of CBRN events from the past that can produce useful information and conclusions exploited after analyses of the operational team. The platform can be implemented after the project end as a training tool in CBRN or CBRNe exercises that can test operational team's decision-making processes, having in mind lessons learned from the past events.

**Acknowledgment.** This work has been partially supported by the CBRN-Terrorism Threats project (Acronym: COBRA) funded under ISF-Police Action Grant: 861789 and the Bulgarian National Scientific Fund project number DFNI DN12/5 "Efficient Stochastic Methods and Algorithms for Large-Scale Problems".

# References

1. Linkov I., Tkachuk A., Canis L., Mohan M., Keisler J., Risk informed decision framework for integrated evaluation of countermeasures against CBRN threats. J. Homel. Secur. Emerg. Manag. **9**(1), Article no. 17 (2012). https://doi.org/10.1515/1547-7355.1926
2. Ezell, B., von Winterfeldt, D.: Probabilistic risk analysis and terrorism. Biosecur. Bioterrorism **7**, 108–110 (2009)
3. Bier, V.M., Cox Jr., L.A., Azaiez, M.N.: Why both game theory and reliability theory are important in defending infrastructure against intelligent attacks. In: Bier, V.M.M., Azaiez, M.N. (eds.) Game Theoretic Risk Analysis of Security Threats. ISORMS, vol. 128, pp. 1–11. Springer, Boston (2009). https://doi.org/10.1007/978-0-387-87767-9_1
4. Keeney, R.L.: Value Focused Thinking: A Path to Creative Decision Making. Harvard University Press, Cambridge (1992)
5. Linkov, I., Moberg, E.: Multi-Criteria Decision Analysis: Environmental Applications and Case Studies. CRC Press, Boca Raton (2011)
6. Linkov, I., Satterstrom, K., Kiker Batchelor, C.G., Bridges, T.: From comparative risk assessment to multi-criteria decision analysis and adaptive management: recent developments and applications. Environ. Int. **32**, 1072–1093 (2006)
7. Mohan M., von Winterfeldt D.: Using risk assessment, economic assessment, and risk management to improve preparedness for terrorist attacks and natural disasters: a methodology for a qualitative assessment of target capabilities of the department of homeland security, technical report to the National Center for Risk and Economic Analysis of Terrorism Events (CREATE), University of Southern California (2010)

# Influence of the ACO Evaporation Parameter for Unstructured Workforce Planning Problem

Stefka Fidanova[1(✉)] and Olympia Roeva[2]

[1] Institute of Information and Communication Technologies,
Bulgarian Academy of Sciences, Sofia, Bulgaria
stefka@parallel.bas.bg
[2] Institute of Biophysics and Biomedical Engineering,
Bulgarian Academy of Sciences, Sofia, Bulgaria
olympia@biomed.bas.bg

**Abstract.** Optimization of the production process is important for every factory or organization. The better organization can be done by optimization of the workforce planing. The main goal is decreasing the assignment cost of the workers with the help of which, the work will be done. The problem is NP-hard, therefore it can be solved with algorithms coming from artificial intelligence. The problem is to select employers and to assign them to the jobs to be performed. The constraints of this problem are very strong and for the algorithms is difficult to find feasible solutions especially when the problem is unstructured. We apply Ant Colony Optimization Algorithm to solve the problem. We investigate the algorithm performance according evaporation parameter. The aim is to find the best parameter setting.

## 1 Introduction

One of the main decision making problem in industry is workforce planing. It is very important for human resource management. It is a hard optimization problem (NP-hard), which includes multiple level of complexity. Its two main parts are selection and assignment. The first part is selection of employers from the set of available workers. After that is assignment of the selected workers to jobs, which the worker will perform. The aim is to minimize assignment cost in the framework of work requirements.

The problem is a hard optimization problem and exact methods and traditional numerical methods can not solve it for a reasonable time. These methods can be useful only for small simplified variants of the problem. A deterministic version of workforce planing problem is studied in [13,19]. In [13] the workforce planning is reformulated as mixed integer programming. It is shown that the mixed integer program is much easier to solve the problem than the non-linear program. In [19] the model includes workers differences and the possibility of

© Springer Nature Switzerland AG 2022
I. Lirkov and S. Margenov (Eds.): LSSC 2021, LNCS 13127, pp. 234–241, 2022.
https://doi.org/10.1007/978-3-030-97549-4_27

workers training and upgrading. A variant with random demands of the problem is considered in [4, 20]. Two stage program of scheduling and allocating with random demands is proposed in [4]. Other variant of the problem is to include uncertainty [14, 16, 18, 25, 26]. A lot of authors skip some of the constraints to simplify the problem. Mixed linear programming is applied in [6] and in [20] is utilized decomposition method, but for the more complex non-linear workforce planning problems, the convex methods are not applicable.

Metaheuristic methods are more and more applied on hard optimization problems, because they can find close to optimal solutions even for large-scale difficult problems [2, 17, 21, 23, 24]. Different metaheuristics are used for solving workforce planing problem. Between them are genetic algorithm [1, 15], memetic algorithm [22], scatter search [1] etc.

Ant Colony Optimization (ACO) algorithm is proved to be very effective to solve different classes of complex optimization problems [7, 12]. In our previous works [8–10] we propose ACO algorithm for workforce planning. We have considered the variant of the workforce planning problem proposed by Alba in [1]. Current paper is the continuation of [8] and [10] and further develops the ideas behind [8] and [10]. The values of the algorithm parameters is very important for algorithm management. We investigate the influence of evaporation parameter on algorithm performance. The aim is to find the best parameter setting.

The rest of the paper is organized as follows. In Sect. 2 the mathematical description of the problem is presented. Section 3 describes ACO algorithm for workforce planing problem. Section 4 discusses computational results. A conclusion and possibilities for future work are done in Sect. 5.

## 2   Definition of Workforce Planning Problem

The workforce planning problem proposed in [1] and [11] is explored in this paper. The set of jobs $J - \{1, \ldots, m\}$ need to be completed during a fixed period of time. The job $j$ requires $d_j$ hours to be finished. $I = \{1, \ldots, n\}$ is the set of candidates workers to be assigned. Every worker must perform every of assigned to him job minimum $h_{min}$ hours to work in efficient way. The worker $i$ is available $s_i$ hours. One worker can be assigned to maximum $j_{max}$ jobs. The set $A_i$ shows on which jobs, the worker $i$ is qualified. Maximum $t$ workers can be assigned during the all period, or at most $t$ workers may be selected from the set of workers $I$. The selected workers need to be capable to complete all the jobs they are assigned. The goal is to find feasible solution, that optimizes the objective function.

The cost of assigning the worker $i$ to the job $j$ is $c_{ij}$. The description of the mathematical model of the workforce planning problem is as follows:

$$x_{ij} = \begin{cases} 1 \text{ if the worker } i \text{ is assigned to job } j \\ 0 \text{ otherwise} \end{cases}$$

$$y_i = \begin{cases} 1 \text{ if worker } i \text{ is selected} \\ 0 \text{ otherwise} \end{cases}$$

$$z_{ij} = \text{number of hours that worker } i$$
$$\text{is assigned to perform job } j$$

$$Q_j = \text{set of workers qualified to perform job } j$$

$$\text{Minimize} \sum_{i \in I} \sum_{j \in A_i} c_{ij}.x_{ij} \tag{1}$$

Subject to

$$\sum_{j \in A_i} z_{ij} \leq s_i.y_i \quad i \in I \tag{2}$$

$$\sum_{i \in Q_j} z_{ij} \geq d_j \quad j \in J \tag{3}$$

$$\sum_{j \in A_i} x_{ij} \leq j_{max}.y_j \quad i \in I \tag{4}$$

$$h_{min}.x_{ij} \leq z_{ij} \leq s_i.x_{ij} \quad i \in I, j \in A_i \tag{5}$$

$$\sum_{i \in I} y_i \leq t \tag{6}$$

$$x_{ij} \in \{0,1\} \quad i \in I, j \in A_i$$
$$y_i \in \{0,1\} \quad i \in I$$
$$z_{ij} \geq 0 \qquad i \in I, j \in A_i$$

The objective function is the minimization of the total assignment cost. The number of hours for each selected worker is limited (inequality 2). The work must be done in full (inequality 3). The number of the jobs, that every worker can perform is limited (inequality 4). There is minimal number of hours that every job must be performed by every assigned worker to can work efficiently (inequality 5). The number of assigned workers is limited (inequality 6).

This mathematical model can be used with other objectives too. If $\tilde{c}_{ij}$ is the cost the worker $i$ to performs the job $j$ for one hour, then the objective function can minimize the cost of the all jobs to be finished.

$$f(x) = \text{Min} \sum_{i \in I} \sum_{j \in A_i} \tilde{c}_{ij}.x_{ij} \tag{7}$$

## 3    Application of ACO on Workforce Planning

Nature inspired methods become more and more popular last decades. Between them is ACO. It is metaheuristic method, inspired by real ants behavior when ants look for a food. Real ants use pheromone, a chemical substance which they synthesis, to mark their way and can return back. An ant moves in random and detecting previously laid pheromone it decides whether to follow it and reinforce it with a new added pheromone. Thus the more ants follow a trail, the more attractive that trail is. The pheromone evaporates during the time and the pheromone level of not used and less used paths decreases and they become less desirable. Thus the nature prevents the ants to follow wrong and useless path. The ants can find a shorter path between the source of the food and the nest only by their collective intelligence.

## 3.1   Basic ACO Algorithm

The NP-hard problem can not be solved with exact or traditional methods using reasonable computational resources like time and memory. For large hard problems it is more reasonable to be applied some metaheuristic method. The goal is to be found near optimal solution in fast way [5].

Marco Dorigo is the first who applied ant behavior for solving hard optimization problems [3]. Later various variants are proposed. The main difference between them is the pheromone updating [5]. The ACO simulates ant behavior when the member of the colony search for a food. The problem is represented by graph. The solutions are represented by paths in a graph and the aim is to find shorter path in a frame of given constraints. Transition probability rule is used to specify how to include new nodes in the partial solution.

The transition probability $P_{i,j}$, is a product of the heuristic information $\eta_{i,j}$ and the pheromone trail level $\tau_{i,j}$ related to the move from node $i$ to the node $j$, where $i, j = 1, \ldots, n$.

$$P_{i,j} = \frac{\tau_{i,j}^a \eta_{i,j}^b}{\sum\limits_{k \in Unused} \tau_{i,k}^a \eta_{i,k}^b}, \tag{8}$$

where $Unused$ is the set of unused nodes of the graph.

The initial pheromone level is the same for all elements of the graph and is set to a positive constant value $\tau_0$, $0 < \tau_0 < 1$. After that at the end of the current iteration the ants update the pheromone level [5]. A node becomes more desirable if it accumulates more pheromone.

The main update rule for the pheromone is:

$$\tau_{i,j} \leftarrow \rho\tau_{i,j} + \Delta\tau_{i,j}, \tag{9}$$

where $\rho$ decreases the value of the pheromone, which mimics evaporation in a nature. $\Delta\tau_{i,j}$ is a new added pheromone, which is proportional to the quality of the solution. The value of the objective function shows the quality of the solution.

The first node of the solution is randomly chosen. Random start allows a few ants to be used, much less than other population based metaheuristics. It is a diversification of the search in a search space. The heuristic information represents the prior knowledge of the problem, which is used to better manage the algorithm performance. The pheromone is a global history of the ants to find optimal solution, their global intelligence. It is a tool for concentration of the search around best so far solutions.

## 3.2  ACO for Workforce Planing

One of the main things for the ACO successes is representation of the problem by graph. We decide to represent workforce planning problem by 3 dimensional graph where the node $(i, j, z)$ corresponds to worker $i$ to be assigned to the job $j$ for time $z$. Three random numbers are generated for every ant: the first random number is from the interval $[0, \ldots, n]$ and corresponds to the worker we assign; the second random number is from the interval $[0, \ldots, m]$ and shows the job which this worker will perform. The third random number is from the interval $[h_{min}, \min\{d_j, s_i\}]$ and shows number of hours worker $i$ is assigned to performs job $j$. They correspond to the start node of the ant. Next nodes are included in the solution like in the traditional ACO algorithm, applying transition probability rule. We include new nodes in the solution till the solution is completed.

The following heuristic information is applied:

$$\eta_{ijl} = \begin{cases} l/c_{ij} & l = z_{ij} \\ 0 & otherwise \end{cases} \tag{10}$$

The meaning of this heuristic information is assignment of the cheapest worker as longer as possible. We chose to include the node with a highest probability in the partial solution. When there are more than one possibilities with the same probability, the next node is chosen in a random way between them.

The constraints are taken in to account, when a new node is included: how many workers are assigned till now; how many time slots every worker is assigned till now; how many time slots are assigned per job till now. If some of the constraints is violated, then the probability of this move is set to 0. When no more possibilities to include new node (the transition probability is 0 for all unused nodes) it is an indicator that the solution is constructed. When the constructed solution is feasible the value of the objective function is the sum of the assignment cost of the assigned workers. In other case the value of the objective function is set to be equal to $-1$.

In our algorithm only elements of feasible solutions receive new pheromone and it is proportional to the 1 over the value of the objective function.

$$\Delta \tau_{i,j} = \frac{\rho - 1}{f(x)} \tag{11}$$

Thus more pheromone will be accumulated by the elements of better solutions and they will become more attractive in the next iteration. The iteration best solution is compared with the global best solution and if on the current iteration the some of the ants achieve better solution it becomes the new global best. The number of iterations is the end condition.

# 4    Computational Results and Discussion

In this section we tested the performance of our algorithm and the influence of the evaporation parameter on 10 unstructured problems. We prepare a software, written in C, which realizes our algorithm. It is run on Pentium desktop computer at 2.8 GHz with 4 GB of memory. An artificially generated unstructured problem, considered in [1], is used for the tests.

In our previous work [8] it is shown that our ACO algorithm outperforms the genetic and scatter search algorithms proposed in [1]. In [10] we learn the influence of the evaporation parameter on structured problems. A problem is structured when the time to finish a job is proportional to the minimal time, which a job must be performed by a worker. The algorithm finds feasible solutions more often for structured problems than for unstructured.

We perform 30 independent runs with every one of the five values of the evaporation parameter, because the algorithm is stochastic and to guarantee the robustness of the average results. We apply ANOVA test for statistical analysis to guarantee the significance of the difference between the average results. We compare the average number of iterations needed to find the best result for every test problem. The needed number of iterations for every test problem can be very different, because the specificity of the tests. Therefore for comparison we use ranking as more representative. The algorithm with some fixed value for evaporation is on the first place, if it achieves the best solution with less average number of iterations over 30 runs, according other values and we assign to it 1, we assign 2 to the value on the second place, 3 to the value on the third place, 4 to the value of the forth place and 5 to the value with most number of iterations. On some cases can be assigned same numbers if the number of iterations to find the best solution is the same. We sum the ranking of the cases over all 10 test problems all 5 values of cvaporation parameter, to find final ranking of the different values of the evaporation parameter.

**Table 1.** Evaporation parameter ranking

|              | $\rho = 0.1$ | $\rho = 0.3$ | $\rho = 0.5$ | $\rho = 0.7$ | $\rho = 0.9$ |
|--------------|---------|---------|---------|---------|---------|
| First place  | 3 times | 0 times | 1 times | 2 times | 4 times |
| Second place | 2 times | 2 times | 1 times | 3 times | 2 times |
| Third place  | 3 times | 4 times | 2 times | 0 imes  | 2 times |
| Forth place  | 1 times | 3 times | 3 times | 3 times | 1 times |
| Fifth plase  | 1 times | 1 times | 3 times | 2 times | 1 times |
| Ranking      | 25      | 33      | 36      | 30      | 23      |

Table 1 shows the ranking of the evaporation parameter. When the ranking is a less number, the meaning is better performance with this evaporation parameter. Least number of iterations is needed when the evaporation parameter is equal to 0.9. In this case the algorithm is on the first place four times,

on the second place is 2 times, on the third place 2 times, and on the fourth
and fifth places respectively one time. The ranking is equal to 23. The worst
results are achieved when the evaporation parameter is equal to 0.5. With evaporation parameter equal to 0.1 are achieved results a little bit worse when the
evaporation parameter is equal to 0.9. If we compare with the influence of the
evaporation parameter on structured problems, we will see a difference. On structured problems the best ranking is when the evaporation parameter is 0.1 and
when the evaporation parameter increases the ranking becomes worsens. We can
explain this result with the structure of the data.

## 5    Concluding Remarks

In this paper we learn the influence of the evaporation parameter of ACO algorithm when it is applied on unstructured workforce problem, the number of iterations needed to find best solution. The achieved results show that the best results
are achieved with a minimal and maximal value of the evaporation parameter
and the worst performance is when the evaporation parameter is in the middle
of the interval of the parameter values $(0, 1)$. In a future we will continue our
work on other aspects of workforce planning problem.

**Acknowledgment.** This work is partially supported by the Projects: KP-06-N22/1
"Theoretical Research and Applications of InterCriteria Analysis" and by the Bulgarian
Scientific Fund by the grant DN 12/5.

## References

1. Alba E., Luque G., Luna F., Parallel metaheuristics for workforce planning. J.
   Math. Model. Algorithms **6**(3), 509–528 (2007). https://doi.org/10.1007/s10852-
   007-9058-5
2. Albayrak, G., Özdemir, İ: A state of art review on metaheuristic methods in time-
   cost trade-off problems. Int. J. Struct. Civ. Eng. Res. **6**(1), 30–34 (2017)
3. Bonabeau, E., Dorigo, M., Theraulaz, G.: Swarm Intelligence: From Natural to
   Artificial Systems. Oxford University Press, New York (1999)
4. Campbell, G.: A two-stage stochastic program for scheduling and allocating cross-
   trained workers. J. Oper. Res. Soc. **62**(6), 1038–1047 (2011)
5. Dorigo M, Stutzle T.: Ant Colony Optimization. MIT Press, Cambridge (2004)
6. Easton, F.: Service completion estimates for cross-trained workforce schedules
   under uncertain attendance and demand. Prod. Oper. Manag. **23**(4), 660–675
   (2014)
7. Fidanova, S., Roeva, O., Paprzycki, M., Gepner, P.: InterCriteria analysis of ACO
   start startegies. In: Proceedings of the 2016 Federated Conference on Computer
   Science and Information Systems, pp. 547–550 (2016)
8. Fidanova, S., Luque, G., Roeva, O., Paprzycki, M., Gepner, P.: Ant colony optimization algorithm for workforce planning. In: FedCSIS 2017, IEEE Xplorer, IEEE
   Catalog Number CFP1585N-ART, pp. 415–419 (2017)
9. Roeva, O., Fidanova, S., Luque, G., Paprzycki, M., Gepner, P.: Hybrid ant colony
   optimization algorithm for workforce planning. In: FedCSIS 2018, IEEE Xplorer,
   pp. 233–236 (2018)

10. Fidanova, S., Luque, G., Roeva, O., Ganzha, M.: Ant colony optimization algorithm for workforce planning: influence of the evaporation parameter. In: Proceedings of the 2019 Federated Conference on Computer Science and Information Systems, Annals of Computer Science and Information Systems, pp. 181–185 (2019). ISSN: 2300-5963

11. Glover, F., Kochenberger, G., Laguna, M., Wubbena, T.: Selection and assignment of a skilled workforce to meet job requirements in a fixed planning period. In: MAEB 2004, pp. 636–641 (2004)

12. Grzybowska, K., Kovács, G.: Sustainable supply chain - supporting tools. In: Proceedings of the 2014 Federated Conference on Computer Science and Information Systems, vol. 2, pp. 1321–1329 (2014)

13. Hewitt, M., Chacosky, A., Grasman, S., Thomas, B.: Integer programming techniques for solving non-linear workforce planning models with learning. Eur. J. Oper. Res. **242**(3), 942–950 (2015)

14. Hu, K., Zhang, X., Gen, M., Jo, J.: A new model for single machine scheduling with uncertain processing time. J. Intell. Manuf. **28**(3), 717–725 (2015). https://doi.org/10.1007/s10845-015-1033-9

15. Li, G., Jiang, H., He, T.: A genetic algorithm-based decomposition approach to solve an integrated equipment-workforce-service planning problem. Omega **50**, 1–17 (2015)

16. Li, R., Liu, G.: An uncertain goal programming model for machine scheduling problem. J. Intell. Manuf. **28**(3), 689–694 (2014). https://doi.org/10.1007/s10845-014-0982-8

17. Mucherino, A., Fidanova, S., Ganzha, M.: Introducing the environment in ant colony optimization. In: Fidanova, S. (ed.) Recent Advances in Computational Optimization. SCI, vol. 655, pp. 147–158. Springer, Cham (2016). https://doi.org/10.1007/978-3-319-40132-4_9

18. Ning Y., Liu J., Yan L., Uncertain aggregate production planning. Soft Comput. **17**(4), 617–624 (2013)

19. Othman, M., Bhuiyan, N., Gouw, G.: Integrating workers' differences into workforce planning. Comput. Ind. Eng. **63**(4), 1096–1106 (2012)

20. Parisio, A., Jones, C.N.: A two-stage stochastic programming approach to employee scheduling in retail outlets with uncertain demand. Omega **53**, 97–103 (2015)

21. Roeva, O., Atanassova, V.: Cuckoo search algorithm for model parameter identification. Int. J. Bioautom. **20**(4), 483–492 (2016)

22. Soukour, A., Devendeville, L., Lucet, C., Moukrim, A.: A Memetic algorithm for staff scheduling problem in airport security service. Expert Syst. Appl. **40**(18), 7504–7512 (2013)

23. Tilahun, S.L., Ngnotchouye, J.M.T.: Firefly algorithm for discrete optimization problems: a survey. J. Civ. Eng. **21**(2), 535–545 (2017)

24. Toimil, D., Gómes, A.: Review of metaheuristics applied to heat exchanger network design. Int. Trans. Oper. Res. **24**(1–2), 7–26 (2017)

25. Yang, G., Tang, W., Zhao, R.: An uncertain workforce planning problem with job satisfaction. Int. J. Mach. Learn. Cybern. **8**(5), 1681–1693 (2016). https://doi.org/10.1007/s13042-016-0539-6. http://rd.springer.com/article/10.1007/s13042-016-0539-6

26. Zhou, C., Tang, W., Zhao, R.: An uncertain search model for recruitment problem with enterprise performance. J. Intell. Manuf. **28**(3), 695–704 (2014). https://doi.org/10.1007/s10845-014-0997-1

# BINMETA: A New Java Package for Meta-heuristic Searches

Antonio Mucherino[✉]

IRISA, University of Rennes 1, Rennes, France
antonio.mucherino@irisa.fr

**Abstract.** We present a new Java package, named BINMETA, for the development and the study of meta-heuristic searches for global optimization. The solution space for our optimization problems is based on a discrete representation, but it does not restrict to combinatorial problems, for every representation on computer machines finally reduces to a sequence of bits. We focus on general purpose meta-heuristics, which are not tailored to any specific subclass of problems. Although we are aware that this is not the first attempt to develop one unique tool implementing more than one meta-heuristic search, we are motivated by the following three main research lines on meta-heuristics. First, we plan to collect several implementations of meta-heuristic searches, developed by several programmers under the common interface of the package, where a particular attention is given to the common components of the various meta-heuristics. Second, the discrete representation for the solutions that we employ allows the user to perform a preliminary study on the degrees of freedom that is likely to give a positive impact on the performance of the meta-heuristic searches. Third, the choice of Java as a programming language is motivated by its flexibility and the use of a high-level objective-oriented paradigm. Finally, an important point in the development of BINMETA is that a meta-heuristic search implemented in the package can also be seen as an optimization problem, where its parameters play the role of decision variables.

## 1 Introduction

Meta-heuristic searches are general purpose methods for global optimization [10,14], which are generally inspired by the observation of the nature, or of animal behavior, including the social behavior of human beings [6]. Among the most "famous" meta-heuristics we can cite the family of Generic Algorithms [13], and swarm intelligence approaches such as Ant Colony Optimization [4]. The BINMETA package that we present for the first time in this article is intended for collecting the implementations of several of the meta-heuristic searches that have been proposed in the scientific literature in recent years.

We plan our BINMETA package to be the instrument for a wide comparison among different methods and strategies implemented in various meta-heuristic searches, that can potentially be complementary and complete each other under the common interface provided by the package. One main idea is to keep the implementations rather "simple", while remaining effective, and very modular, so that the involvement of other

© Springer Nature Switzerland AG 2022
I. Lirkov and S. Margenov (Eds.): LSSC 2021, LNCS 13127, pp. 242–249, 2022.
https://doi.org/10.1007/978-3-030-97549-4_28

researchers in the development of the package is encouraged. In other words, the main approach for the development of this Java package is the so-called incremental build approach [11]. The initial versions of the package have already been deposited on the GitHub[1], in a public repository. We make reference in the following to the $13^{th}$ commit (code: 2007aa2) to this repository.

We make the choice to represent the solution space of all considered optimization problems in a combinatorial space, so that the implemented meta-heuristic searches can work in a discrete space. A discrete representation is naturally accommodated for combinatorial problems, but we do not aim to limit, with this choice for the representation of the solutions, our package to solely combinatorial problems. In fact, the solutions to every optimization problem need to admit a suitable discrete representation (in practice, a representation as a bit string) in order to be treated by digital machines. Moreover, we pay particular attention to having *optimal* discrete representations for the solutions of our optimization problems, where we intend optimality in terms of the degrees of freedom for the solutions.

BINMETA is developed in Java. The choice of a high-level and object-oriented programming language comes first of all for encouraging a wide collaboration in its development (as already mentioned above). Moreover, Java allows us to develop the different parts of the package (optimization problems, meta-heuristic searches) under a common interface, where every part can be easily related to the others. One initial advantage of this flexibility is given by the fact that meta-heuristic searches are implemented so that they are compatible with the interface describing the objective functions of the optimization problems. The selection of the "optimal" parameters for a given meta-heuristic search for the solution of a given optimization problem can in this way be seen as another optimization problem, where the parameters of the meta-heuristic search play the role of decision variables. The solution to this parameter tuning problem can therefore be attempted with our Java package.

We point out that this is not the first software project on meta-heuristic searches. There exist in fact several implementations of single (or of particular subgroups of) meta-heuristic searches; the more one meta-heuristic search is studied and used in the scientific community, the more there are implementations available. Other software tools may not focus on meta-heuristics but they may contain some of their implementations for tackling some specific problems (examples can be found in [14]). Finally, other software projects collecting several meta-heuristics include the software tool PARADISEO [2] (written in C++), as well as the software tool JMETAL [5], written in Java and mainly focused on multi-objective optimization. To the best of our knowledge, apart from the main features already pointed out above (data representation and optimization, organization of class interfaces), BINMETA is the first software project whose development strictly follows an incremental programming approach, and it encourages the participation of several developers.

The rest of the paper is organized as follows. In Sect. 2, we will give the details of the binary representations that will be employed for the possible solutions to our optimization problems. Section 3 will present some of the optimization problems that are currently available in the BINMETA package, while Sect. 4 will describe the very

---

[1] https://github.com/mucherino/binMeta.

first meta-heuristic searches that have been implemented in the package. Finally, Sect. 5 will present some preliminary experiments, and Sect. 6 will conclude the paper.

## 2    Representation of the Solutions

Some meta-heuristic searches were originally conceived and developed to admit a discrete representation of the solutions. For example, in Ant Colony Optimization (ACO) [1], the solutions are vertices of a given graph, where the involved *ants* move by stepping from one vertex to another where an edge exists between the two. The search space is therefore discrete and admits representations of the entire set of potential solutions in a combinatorial space. Other meta-heuristics, instead, are strongly based on continuous representations of the solutions: one example is the recent Spiral Optimization meta-heuristics [15], where the constructed *spirals* admit continuous representations in Euclidean spaces. Even if initially conceived for a discrete or continuous representation of the solutions, the implementation adaptations of meta-heuristic searches from one to the other representation have been attempted in previous works (see for example [3]).

In our Java package, the choice to represent the solutions to all optimization problems in a combinatorial space comes from the observation that all suitable representations on a computer machine *need* to be discrete (and finite). Even continuous problems with real-valued variables, for example, need to have their variables represented on computer machines in a binary format, which essentially corresponds to a well-formatted "bit string" (the *floating-point* representation is for example the proper format for the real-valued variables). Instead of using these generally provided standard data types, BINMETA gives the user the possibility to develop ad-hoc and specific binary representation for the involved variables. If a given variable can take up to four different values, for example, then 2 bits are sufficient to represent it. If solutions of the problem at hand are sets of variables that can each take up to four values, these solutions can be represented in binary format as a *unique* bit string concatenating the 2 bits necessary for the representation of each variable.

We point out that the use of one unique bit string for the representation of the solutions is much more efficient than considering arrays of single variables. In fact, in the previous example, we would have an array of variables of type byte, each covering 8 bits in the computer memory. However, only four values can be taken by each variable stored in the byte, and therefore 6 out of the 8 bits are actually not used. Moreover, if they were used, they would correspond to values that are not feasible for these variables, so that constraints need to be included and verified for a correct use of the array of bytes. These two issues are immediately overcome when employing one unique bit string representation.

## 3    The First Optimization Problems

We give in this section some examples of optimization problems that are currently implemented in our Java package BINMETA. Some of them are simple problems, which were included in the package with the only aim to performing simple tests on the meta-heuristic searches, before attempting the solution of harder problems. Some of our optimization problems are classical problems arising in the field of Operational Research [18]. Our current package version includes the SUBSET SUM PROBLEM, the NUMBER PARTITION PROBLEM and the KNAPSACK PROBLEM.

In the following sections, we will focus instead on two other optimization problems that we have included in the package mainly for testing and presentation purposes (Sects. 3.1 and 3.2). Finally, in Sect. 3.3, we will briefly discuss the possibility to solve optimization problems where the objective function provides, for a given set of parameters' values, the performance of the implemented meta-heuristic searches (that we will discuss in Sect. 4).

## 3.1 The Pi Objective

This objective is related to an optimization problem that can be considered as an *easy* problem: we decided to describe it in details for giving an example of simple problem with a set of real-valued variables that can be encoded as a unique bit string having fixed length.

The basic idea is to find the positions on a unit circle (lying in the two-dimensional Euclidean space) of $n$ points $x_i$ such that the objective

$$Pi(\{x_1, x_2, \ldots, x_n\}) = \sum_{i=1}^{n} ||x_i - x_{i+1}||$$

is maximized, where it is supposed that $x_{n+1}$ coincides with $x_1$, and where $|| \cdot ||$ represents the Euclidean norm. Finding the set of $n$ points that maximizes the value of the Pi function is equivalent to improving the approximation of the number $\pi$ that is obtained.

As mentioned above, this is not a hard problem: if every point $x_i$ is constrained to take positions in a portion of circle that is in size proportional to $1/n$, then the optimal solution can be simply identified by local optimization. However, we did not impose this constraint in the implementation of the objective in our BINMETA package.

The precision of the point positions, in terms of number of bits used their representation, can naturally have an important impact on the obtained approximation of $\pi$. Since every $x_i$ belongs to the unit circle, we can represent it with only one real value (a vector angle); a discrete set of possible values for this angle can then be represented by a bit sub-string having a predefined length (which will correspond to the precision of the representation). Finally, one possible solution to the problem can be represented by the bit string concatenating all sub-strings related to the points $x_i$.

## 3.2 The Fermat Objective

Fermat's last theorem states that no solutions for the equation $x^n + y^n = z^n$ exist when $x$, $y$, and $z$ are positive integers larger than 1, when the value of $n$ is fixed and is greater than 2. This theorem was proved in 1995 by Sir. Andrew Wiles [17], about 350 years after the formulation of the theorem by Pierre de Fermat. We consider the following objective:

$$F(x, y, z) = |z^n - x^n - y^n|,$$

which basically measures the violation of the Fermat equation for a fixed value of $n$, and for three possible integers $x$, $y$ and $z$. If we attempt the minimization of $F$ when $n = 2$, we know that there exist solutions where the value of the function can be zero. However, if we could find a triplet $(\hat{x}, \hat{y}, \hat{z})$ for which the value of this objective is zero when $n > 2$, this would correspond to find a counterexample to Fermat's last theorem.

Differently from the problem in the previous section, optimizing the Fermat objective is NP-hard. This can be easily proved by noticing that the problem is equivalent to a SUBSET SUM instance where, in the set of all positive integers of the type $x^n$ smaller than a given upper bound, and which also contains their opposites $-x^n$, it is necessary to verify whether there exist a subset whose element sum is zero, with the additional constraint that the cardinality of the subset must be 3. Two copies of $x^n$ may be included in the initial set to consider equations where $x = y$.

This objective allows us to point out the importance of the choice of the binary representation (the final bit string) for the possible solutions to the problem. We can remark, first of all, that by inverting $x$ and $y$ in the original equation, the solution does not change. Therefore, half of the possible solutions (which are symmetric w.r.t. other solutions) can be removed by imposing the constraint $x \leq y$. Moreover, we can also remark that the values that $z$ can take are also constrained by the values assigned to the other two integers.

Once the number of bits for the integer representations is selected (equivalent to giving an upper bound on the values of the integers), the final bit string representing a possible solution to the Fermat problem is the concatenation of the three sub-strings encoding the three integers $x$, $y$ and $z$. However, for the reasons given above, only $x$ is represented in absolute value, while the difference $y - x$ is encoded by the second sub-string, and the difference $z - y$ is encoded by the third sub-string.

### 3.3 Meta-heuristic Searches as Objective Functions

Recent works have been focusing on the problem of setting up automatically the parameters involved in meta-heuristic searches [12]. In our Java package, a special class of optimization problems is given by the set of implemented meta-heuristic searches, where the objective functions provide a measure of their performance when invoked to solve a certain problem. The decision variables are in this case the set of parameters that are used to invoke the meta-heuristic search, represented as a bit string in our implementations. The main idea is to find the parameters that allow the search to achieve the best performance when dealing with a specific problem (or a particular class of problems). In BINMETA, the meta-heuristic search A may be used for optimizing the parameters of the meta-heuristic search B in the attempt to solve the optimization problem C. The situation where A = B is also feasible from a technical point of view in our Java package.

## 4   The First Meta-heuristic Searches

We focus in this section on one of the first meta-heuristic searches that have been implemented in our Java package BINMETA, the Wolf Search meta-heuristics (see Sect. 4.1).

Notice however that a Random Walk meta-heuristics has also been implemented, as well as a local optimization procedure inspired by the gradient descent method but acting directly on the binary representations of our solutions.

## 4.1  Wolf Search

Wolves can hunt individually or in a group. These groups of wolves are however local, they do not comprise the entire set of wolves in a given area. A wolf moving individually in search for food would alternate its efforts in catching a new prey by itself, and in verifying whether a near wolf has been luckier in its searches and eventually join it. The risk for a threat, given by the presence of other predators, can potentially make the wolf decide to leave its current search, in order to escape in a safer area. We implemented the Wolf Search as described in [16], because well adapted to the spirit of our Java package; we remark that a more sophisticated meta-heuristic search, that is also based on the hunting behavior of wolves, was subsequently proposed in [7].

The Wolf Search meta-heuristics is based on the following three main steps. First of all, every wolf in the group attempts to perform an individual move in the search space, with the aim of finding better hunting conditions (more food, more preys, represented by a solution with better objective function values). The new solution is accepted only if it does improve the objective function value, and only if it was not recently considered (for this second criterium, we suppose that the wolves are equipped with an ephemeral memory containing previously explored solutions). In our implementation, the wolf initially looks randomly in its range of vision, and then tries to improve its current solution by performing a local search.

If the individual search does not lead to any new better solutions, then the wolf looks around itself to verify whether other wolves had been luckier in the search. To simulate locality, only the wolves positioned in solutions that are close in distance to the current wolf's solution are taken into consideration. The current wolf can therefore decide to join another wolf when this other one is in its range of vision, and if this other wolf is currently carrying a better solution. Joining another wolf corresponds to standing in the solution it is carrying, or in a very near solution.

Finally, every wolf has a given risk probability for a threat, which will imply a random movement in the search space, but limited to the wolf's range of vision. This movement is completely independent from the current value of the objective function; since the wolf behavior may make the wolves escape from good quality solutions, these solutions are not included in the ephemeral memory when escaping, in a way that these wolves may come back exploring the same part of the search space when the threat has disappeared.

## 5  Preliminary Experiments

We propose some preliminary computational experiments, where the Random Walk and Wolf Search meta-heuristics (see Sect. 4.1) are employed for solving the optimization problems represented by the objective function Pi (see Sect. 3.1) and Fermat (see Sect. 3.2).

**Table 1.** Every meta-heuristic run for 1 s, on a standard computer laptop. Since our meta-heuristics are implemented to solve minimization problems, a negative sign is assigned to every evaluation of the Pi objective.

| | $n$ | #digits | #bits | Random Walk | Wolf Search | | $n$ | #digits | #bits | Random Walk | Wolf Search |
|---|---|---|---|---|---|---|---|---|---|---|---|
| *Pi* | 10 | 3 | 30 | −3.0880 | −3.0901 | *Fermat* | 2 | 10 | 30 | 1.0 | 0.0 |
| | 25 | 4 | 100 | −3.1192 | −3.1325 | | 3 | 12 | 36 | 16353.0 | 1.0 |
| | 50 | 5 | 250 | −3.1202 | −3.1321 | | 3 | 15 | 45 | 1849940.0 | 2.0 |

On the left-hand side of Table 1, we report some solutions found when optimizing the Pi objective. The value of $n$ indicates the number of points on the unit circle, while *#digits* indicates the number of bits used for the representation of their corresponding angle. The total number of bits (*#bits*) is a consequence of the previous two values. The experiments show that better results can be obtained with larger values for both $n$ and *#digits*. As expected, Wolf Search performs better than Random Walk.

On the right-hand side of Table 1, some experiments concerning the Fermat objective are proposed. In this case, $n$ is the exponent in the Fermat equation, while *#digits* is the number of bits used in the representation of each integer $x$, $y$, and $z$ (and therefore the total number of bits corresponds to 3 times this value). When $n = 2$, Wolf Search is able to find a solution for which the objective is zero: in this solution, $x = 3$, $y = 5$, and $z = 10$. In the other two experiments, solutions to the equation are not supposed to exist.

## 6    Conclusions

We introduced the new BINMETA package, a Java package for the implementation and the study, under a common interface, of several meta-heuristic searches for global optimization. Future works will consist in extending the number of objectives, as well as the number of meta-heuristic searches, that are implemented, so that BINMETA can become one of the main references for the study of meta-heuristic searches. We also plan to introduce the concept of meta-heuristic *environment* in our package, as described in [9]. Finally, we point out again the importance in choosing the binary representation for the solutions: this is an important point that was studied for example in [8] for the distance geometry problem, which we plan to include in the near future in our package.

**Acknowledgments.** Throughout the entire article, the reader may have noticed that the plural form is employed even if there is only one author. This author actually needs to thank the collaboration of some Master students that worked on this software package in the framework of course projects. The identity of the students that gave the most important contributions appear (in different forms) in the source files (see GitHub repository).

This work is partially supported by the international project MULTIBIOSTRUCT funded by the ANR French funding agency (ANR-19-CE45-0019).

# References

1. Atanassova, V., Fidanova, S., Popchev, I., Chountas, P.: Generalized nets, ACO-algorithms and genetic algorithm. In: Sabelfeld, K.K., Dimov, I. (eds.) Monte Carlo Methods and Applications, pp. 39–46. De Gruyter (2012)
2. Cahon, S., Melab, N., Talbi, E.-G.: ParadisEO: a framework for the reusable design of parallel and distributed metaheuristics. J. Heuristics **10**, 357–380 (2004)
3. Crawford, B., Soto, R., Astorga, G., García, J., Castro, C., Paredes, F.: Putting continuous metaheuristics to work in binary search spaces. Complexity **2017** (2017). Article ID 8404231, 19 p.
4. Dorigo, M., Birattari, M.: Ant colony optimization. In: Sammut, C., Webb, G.I. (eds.) Encyclopedia of Machine Learning, pp. 36–39. Springer, Boston (2010). https://doi.org/10.1007/978-0-387-30164-8
5. Durillo, J.J., Nebro, A.J.: jMetal: a Java framework for multi-objective optimization. Adv. Eng. Softw. **42**, 760–771 (2011)
6. Fister Jr, I., Yang, X.-S., Fister, I., Brest, J., Fister, D.: A brief review of nature-inspired algorithms for optimization. Elektrotehniski Vestnik **80**(3), 1–7 (2013)
7. Mirjalili, S., Mirjalili, S.M., Lewis, A.: Grey wolf optimizer. Adv. Eng. Softw. **69**, 46–61 (2014)
8. Mucherino, A.: An analysis on the degrees of freedom of binary representations for solutions to discretizable distance geometry problems. In: Fidanova, S. (eds) WCO 2020. SCI, vol. 986, pp. 251–255. Springer, Cham (2022). https://doi.org/10.1007/978-3-030-82397-9_13
9. Mucherino, A., Fidanova, S., Ganzha, M.: Ant colony optimization with environment changes: an application to GPS surveying. In: IEEE Conference Proceedings, Federated Conference on Computer Science and Information Systems (FedCSIS15), Workshop on Computational Optimization (WCO15), Lodz, Poland, pp. 495–500 (2015)
10. Mucherino, A., Seref, O.: Modeling and solving real-life global optimization problems with meta-heuristic methods. In: Pardalos, P.M., Papajorgji , P.J. (eds.) Advances in Modeling Agricultural Systems. SOIA, vol. 25, pp. 403–419. Springer, Boston (2009). https://doi.org/10.1007/978-0-387-75181-8_19
11. Pressman, R.S., Maxim, B.R.: Software Engineering: A Practitioner's Approach, 9th edn. McGraw-Hill Education (2019). 704 p.
12. Pukhkaiev, D., Semendiak, Y., Götz, S., Aßmann, U.: Combined selection and parameter control of meta-heuristics. In: IEEE Conference Proceedings, Symposium Series on Computational Intelligence (SSCI20), Canberra, Australia, pp. 3125–3132 (2020)
13. Sivanandam, S., Deepa, S.: Introduction to Genetic Algorithms. Springer, Heidelberg (2008). 442 p. https://doi.org/10.1007/978-3-540-73190-0
14. Sörensen, K., Glover, F.: Metaheuristics, encyclopedia of operations research and management. Science **62**, 960–970 (2013)
15. Tamura, K., Yasuda, K.: Spiral optimization algorithm using periodic descent directions. SICE J. Control Meas. Syst. Integr. **9**(3), 134–143 (2016)
16. Tang, R., Fong, S., Yang, X.S., Deb, S.: Wolf search algorithm with ephemeral memory. In: IEEE Proceedings, 7th International Conference on Digital Information Management (ICDIM 2012), Macau, pp. 165–172 (2012)
17. Wiles, A.: Modular elliptic curves and Fermat's last theorem. Ann. Math. **141**(3), 443–551 (1995)
18. Woeginger, G.J.: Exact algorithms for NP-hard problems: a survey. In: Jünger, M., Reinelt, G., Rinaldi, G. (eds.) Combinatorial Optimization — Eureka, You Shrink! LNCS, vol. 2570, pp. 185–207. Springer, Heidelberg (2003). https://doi.org/10.1007/3-540-36478-1_17

# Synergy Between Convergence and Divergence—Review of Concepts and Methods

Kalin Penev[✉]

Faculty of Business, Law and Digital Technologies, Solent University,
East Park Terrace, Southampton SO14 0YN, UK
Kalin.Penev@solent.ac.uk

**Abstract.** Modern Industry 4.0 technologies face a challenge in dealing with billions of connected devices, petabyte-scale of generated data, and exponentially growing internet traffic. Artificial Intelligence and Evolutionary algorithms can resolve variety of large optimisation problems. Many methods employed in search for solutions often fall in stagnation or in unacceptable results, which reminds for classical dilemma exploration versus exploitations closely related with convergence and diversity of the explored solutions. This article reviews convergence and divergence centred algorithms and discuses synergy between convergence and divergence in adaptive heuristics.

**Keywords:** Convergence · Divergence · Synergy · Adaptive heuristic algorithms · Tabu search · Particle swarm optimisation · Genetic algorithms · Differential evolution · Scatter search · Novelty search · Surprise search · Free search

## 1 Introduction

Exponentially growing number of connected devices, generated data and internet traffic demands fast and efficient methods which can provide optimal solutions in technology and science. The concept for natural selection and generation of new species applied to computational optimisation could help to improve optimisation process for large scale task. In this context Evolutionary computation (EC) as a branch of computational intelligence, included into the broad framework of bioinspired heuristics needs more research efforts.

According to [11], in contrast to the natural evolution artificial evolution in Evolutionary Computation suffers an endemic lack of diversity during evolutionary optimisation processes and all candidate solutions frequently homologize. This situation is usually described as stagnation or premature convergence to a local suboptimum.

To cope with stagnations many approaches are proposed [2,3,7,9,13]. This article discusses synergy between convergence and divergence in adaptive heuristic algorithms, aims to highlight and compare strengths and limitations of existing algorithms.

© Springer Nature Switzerland AG 2022
I. Lirkov and S. Margenov (Eds.): LSSC 2021, LNCS 13127, pp. 250–256, 2022.
https://doi.org/10.1007/978-3-030-97549-4_29

# 2    Methodologies Review

This section reviews classical and novel algorithms with specific original concepts such as Tabu search [5], Particle Swarm Optimisation [2], Genetic algorithms [3,7], Differential Evolution [13], Scatter search [6], Novelty search [8], Surprise search [4], Free Search [9]. Most of these techniques are effective on of limited types of landscapes and implemented concepts can lead to improvement of further methods. Some of them are dominated by convergence process, other intentionally try to diverge over the search space to generate appropriate diversity or to achieve other objectives.

## 2.1    Convergence Centred Methods

This section reviews several methods base on modification strategy, which aims to converge to the optimal solution and using specific techniques and approaches to prevent and avoid premature convergence to suboptimal locations.

**Tabu Search.** The Tabu Search idea is—"Tabu search is a metastrategy for guiding known heuristics to overcome local optimality" [5]. Tabu Search is based on the temporary prohibition of moves in order to avoid cycles in the search trajectory [1]. Some open problems of Tabu Search are: - the determination of an appropriate "prohibition" for a given task, the adoption of minimal computational complexity algorithms for using memory, the robustness of the technique for a wide range of different problems [1]. Using the concept of Tabu Search, a local search can be guided to go beyond local optima through metaheuristic methods that use the information obtained in the previous part of the run. In the Reactive Tabu Search (RTS) algorithm a simple feedback scheme influences the value of the prohibition parameter in the Tabu Search, so that a balance of exploration versus exploitation is obtained that is appropriate for the local characteristics of the task [1]. The concept of Tabu Search aims to facilitated convergence but limits divergence across the whole space.

**Particle Swarm Optimisation.** Particle Swarm Optimisation (PSO) is motivated by social behaviour of organisms such as bird flocking and fish schooling. PSO is a method for optimisation of continuous nonlinear functions [2]. PSO algorithm is not only a tool for optimisation, but also a tool for representing socio-cognition of human and artificial agents, based on principles of social psychology. Some scientists suggest that knowledge, which is optimised by social interaction and thinking is not only private but also interpersonal. In a PSO system, particles fly around in a multidimensional search space. During flight, each particle adjusts its position according to its own experience, and according to the experience of a neighbouring particle, making use of the best position encountered by itself and its neighbour. PSO combines local search methods with global search methods, attempting to balance exploration and exploitation [2]. Aim of the adjustment based on best positions is to converge to the optimum. This concept to certain extent limits divergence over the whole search space.

**Genetic Algorithms.** Genetic algorithms (GA) are a family of computational models inspired by Darwin's theory about evolution [7,14]. An algorithm is started with a set of solutions (represented by chromosomes) called "population". Solutions from one population are taken and used to form a new population. This is motivated by a hope, that the new population will be better than the old one. Solutions, which are selected to form new solutions (offspring), are chosen according to their fitness - the more suitable they are the more chances they have, to reproduce. This is repeated until some condition (for example expiration of the period for search or improvement of the best solution) is satisfied. The GA has wide range of modifications and variations according to representation of the variables, recombination strategies, modification strategies and replacement strategies. An original idea explored in real-value The Genetic Algorithm is implemented and coding BLX-a modification [3]. The real coded BLX-a GA also selects two parents from the current population, makes a crossover between them and produces an offspring (new solution). The offspring can mutate with small probability. Then the GA incorporates the offspring in the new generation if it is better than an individual from current generation. Aim of the crossover is to converge to the optimal solution, while the aim of mutation is to diverge over the search space. Balance, between these processes, relays on random generation. For large scale tasks this leads to the need for large number of iterations, computational resources, and time.

**Differential Evolution.** Differential Evolution (DE) [13] is based on the idea for generating trial parameter vectors. It is an optimiser for multivariate functions. Starting from an initial random population of points (interpreted as vectors) the procedure iteratively updates the population using two basic operations: mutation and recombination. DE is applicable to functions that are non-differentiable, non-linear, or otherwise resistant to traditional approximation techniques. A similar technique used by neural researchers to deduce network functions from data samples, simulated annealing, could also model non-linear/non-differentiable functions, but DE is faster and requires fewer restarts [13]. Aim of recombination in DE is to converge to the optimal solution, and the aim of mutation is to diverge over the search space. Balance, between these processes can be controlled by different modification strategies [10].

## 2.2   Divergence Centred Methods

This section focuses on methods which utilises divergence and applies specific techniques for identification of appropriate solutions and termination of the process.

**Scatter Searchs.** According to the literature "The Scatter Search process, building on the principles that underlie the surrogate constraint design, is organized to capture information not constrained separately in the original vectors, and to take advantage of auxiliary heuristic methods both for' selecting the elements to be combined and for generating new vectors." [6]. The original form of scatter search [6] can be presented in the following 3 steps: (1) Generate a starting set of solution vectors by heuristic processes designed for the problem; (2) Create new points consisting of linear combinations of subsets of the current reference solutions. The linear combinations are chosen to produce points both inside and outside the convex regions spanned by the reference solutions, modified by generalized rounding processes to yield integer values for integer-constrained vector components. (3) Extract a collection of the best solutions generated in Step 2 to be used as starting points for a new application of the heuristic processes of Step 1. Repeat these steps until reaching a specified iteration limit. Three features of Scatter Search deserve attention: - First the linear combinations are structured according to the goal of generating weighted centres of selected subregions, allowing for no convex combinations that project these centres into regions external to the original reference solutions. Second, the strategies for selecting particular subsets of solutions to combine in Step 2 are designed to make use of clustering, which allows different types of strategic variation by generating new solutions 'within clusters' and 'across clusters'. Third, the method is organized to use supporting heuristics that can start from infeasible solutions, and which removes the restriction that solutions selected as starting points for reapplying the heuristic processes must be feasible [6]. The concept of Scatter Search can be classified as an example which aims to utilise predominantly divergence.

**Novelty Search.** Novelty Search algorithm [8] is based on the concept for openended evolutionary system, which continually produces novel forms [12]. This is different from other evolutionary algorithms, which aim to achieve optimal solution. The Novelty Search utilises the divergence over the search space, which relays on the hope that the groped optimal value can be achieved accidentally. It is acknowledged that: "it is likely more efficient to take the most promising results from novelty search and further optimize them based on an objective function. This idea exploits the strengths of both approaches: Novelty Search effectively finds approximate solutions, while objective optimization is good for tuning approximate solutions. Alternatively, novelty search could be applied when a traditional evolutionary algorithm converges, to replenish diversity in the population." [8] Proposed modification strategy uses distance between the solutions and factors based on previous experiments to generate divergence from current locations.

**Surprise Search.** Surprise Search as stated in [4] can also be classified as a divergence centred search algorithm which has demonstrated acceptable performance on robot morphology evolution, maze navigation, and other twodimensional tasks. According to the literature [4] it modifies Novelty Search. Divergence is based on a definition of surprise as "deviation from the expected." Surprise Search modification strategy inherits modification strategy concept based on distance. The distance is generated on past difference and expectations for further behaviour. This, two steps, approach favours individuals that diverge from predicted future trends. With certain probability the predictions could be infeasible or unreachable by conventional search when applied to real tasks such as points outside a maze that should be traversed. Proposed in Surprise Search divergence strategy brings improvements appropriate for certain applications. [4]

## 3   Synergy Between Convergence and Divergence

This section illustrates the Free Search concept published earlier. Free Search is adaptive heuristic method [9] for real coded optimisation. This algorithm harmonises divergence and convergence using highly unlimited modification strategy and natural relations between stochastically generated values [9].

Optimisation process starts with exploration over the search space implemented as individual journeys within their neighbour space [9]. In the beginning algorithm has no knowledge about the search space and exploration is highly random. This is achieved by multiplication of several stochastic variables. The first exploration generates knowledge stored in a form of qualitative indicators related with evaluated locations. These indicators facilitate individuals' sensibility for orientation within the search space for further search. [9]

On initial stage locations' quality and individuals' sensibility are uniformly distributed among low, medium, and high levels. Individuals with low level of sensibility can select for start position any marked location. The individuals with high sensibility can select for start position marked locations with high quality and will ignore locations with low quality. When marked locations quality highly differs and stochastically generated sensibility produces accidentally high values only, then the individuals will search around the area of the highest quality solutions. Such situations appear naturally. In this manner process converges to high quality locations. [9]

Other situation which naturally appears is when marked locations qualities are very similar and randomly generated sensibility is low. In this case individuals can select low quality marked locations with high probability, which indirectly will decrease the probability for selection of high quality marked locations. Individuals with low sensibility can select to explore around locations marked with low quality.

As far as locations quality is independent on their position within the search space, similar quality locations could be remotely distributed. This facilitates divergence across the entire search space. Sensibility varies across all the individuals and during the optimisation process. In this manner convergence and divergence are harmonised in the search of optimal solution. It helps to minimise generated new solutions according to the achieved quality. Modification strategy is independent on the distance, which prevent generation of many location if applied to large scale tasks. [9]

# 4 Discussion

Modern Industry 4.0 technologies face several challenges such as: - maximising connected devices; optimising use of infrastructure; maximising processed data; minimising use of resources; maximising internet traffic; minimising traffic delays; maximising number of users and minimising energy use, and waste. All of these are optimisation problems, at large and growing scale. In this context reasonable hope relies on Artificial & Computational Intelligence and Evolutionary & Heuristic algorithms, which can resolve variety of large optimisation problems. Often methods employed in large scale search converge prematurely and fall in stagnation. A promising approach, which can help to escape from or prevent premature convergence is synergy between convergence and divergence.

# 5 Conclusion

This article reviews several classical and novel convergence centred and divergence centred algorithms. Discussion on synergy between convergence and divergence highlights the role of evolutionary and heuristic optimisation methods in optimisation, of demanded by Industry 4.0 large scale tasks. Presented review of classical and novel algorithms and concepts suggests a good potential for synergy between convergence & divergence. More further research efforts in this area should be done.

# References

1. Battiti R.: Reactive search: toward self-tuning heuristics. In: Rayward-Smith V.J., (ed.), Modern Heuristic Search Methods, vol. 4, pp. 61–83. John Wiley and Sons Ltd (1996)
2. Eberhart R., Kennedy J.: Particle swarm optimisation. In: Proceedings of the 1995 IEEE International Conference on Neural Networks, vol. 4, pp. 1942–1948. IEEE Press (1995)
3. Eshelman L.J., Schaffer J.D.: Real-coded genetic algorithms and interval-schemata. In: Foundations of Genetic Algorithms 2, Morgan Kaufman Publishers, San Mateo, pp. 187–202 (1993)
4. Gravina, D., Liapis, A. Yannakakis, G. N.: Surprise search: beyond objectives and novelty. In: Proceeding Genetic Evolution Computer Conference, pp. 677–684 (2016)
5. Glover F.: Tabu search - part II. ORSA J. Comput. 2, 4–32 (1990)
6. Glover F.: Scatter search and path relinking, Chapter Nineteen In: Corne, D., Dorigo, M., Glover, F. (eds.), New Ideas in Optimisation. ISBN 007 7095065, McGraw-Hill International (UK) Limited, pp. (294-316) (1999)
7. Holland, J.: Adaptation in natural and artificial systems. University of Michigan Press (1975)
8. Lehman, J. Stanley, K.O.: Exploiting open-endedness to solve problems through the search for novelty. In: International Conference on Artificial Life (ALIFE XI), pp. 329–336. MIT Press (2008)

9. Penev K.: Adaptive computing in support of traffic management. In: Parmee, I., (ed.) Adaptive Computing in Design and Manufacturing 2004, Bristol, UK, pp. 295–306 (2004)
10. Price, K., Storn R.: Differential evolution, Dr, Dobb's J. **22**(4), pp. 18–24 (1997)
11. Squillero, G., Tonda, A.T.: Divergence of character and premature convergence: a survey of methodologies for promoting diversity in evolutionary optimization. Inf. Sci. **329**, 782–799 (2016)
12. Standish, R.: Open-ended artificial evolution. Int. J. Comput. Intell. Appl. **3**(167) (2003)
13. Storn, R., Price, K.: Differential evolution - a simple and efficient adaptive scheme for global optimisation over continuous space, ICSI, TR-95-012 (1995)
14. Whitley, D.: A genetic algorithm tutorial, Comput. Sci. Dept. Colorado State Univ. Technical Report CS-93-103 (1993)

# Advanced Stochastic Approaches Based on Optimization of Lattice Sequences for Large-Scale Finance Problems

Venelin Todorov[1,2(✉)], Ivan Dimov[2], Rayna Georgieva[2], Stoyan Apostolov[3], and Stoyan Poryazov[1]

[1] Department of Information Modeling, Bulgarian Academy of Sciences, Institute of Mathematics and Informatics, Academy Georgi Bonchev Street, Block 8, 1113 Sofia, Bulgaria
{vtodorov,stoyan}@math.bas.bg
[2] Department of Parallel Algorithms, Bulgarian Academy of Sciences, Institute of Information and Communication Technologies, Academy G. Bonchev Street, Block 25 A, 1113 Sofia, Bulgaria
{venelin,rayna}@parallel.bas.bg, ivdimov@bas.bg
[3] Faculty of Mathematics and Informatics, Sofia University, Sofia, Bulgaria

**Abstract.** In this work we study advanced stochastic methods for solving a specific multidimensional problems related to computation of European style options in computational finance. Recently, stochastic methods have become a very important tool for high-performance computing of very high-dimensional problems in computational finance. Here, a different kind of optimal generating vectors have been applied for the first time to a specific problem in computational finance. Numerical tests show that they give superior results to the stochastic approaches used up to now. The advantages and disadvantages of various highly efficient stochastic approaches for multidimensional integrals related to evaluation of European style options have been analyzed.

## 1 Introduction

The pricing of options is a very important in financial markets today [3,5] and especially difficult when the dimension of the problem goes higher [4]. Monte Carlo (MC) and quasi-Monte Carlo (QMC) methods are appropriate for solving multidimensional problems [2,10], since their computational complexity increases polynomially, but not exponentially with the dimensionality. MC methods are used not only for option pricing, but also in other problems in computational finance. The QMC methods using special deterministic sequences achieve higher accuracy and computational efficiency compared to the MC methods. Options have been widely traded since the creation of the organized exchange in 1973 [12]. The famous Black-Scholes model provides explicit closed form solutions for the values of the European style call and put options. Nowadays much of the focus is on high-dimensional problems that are more interesting and inspired by practical cases. However, some algorithms are not easily adaptable to multidimensional case representing a significant challenge from a computational viewpoint [3]. That is

I. Lirkov and S. Margenov (Eds.): LSSC 2021, LNCS 13127, pp. 257–265, 2022.
https://doi.org/10.1007/978-3-030-97549-4_30

why we need advanced approaches that are efficient enough in high-dimensional cases in terms of accuracy and computational time. The basic definitions and terminology used in the paper can be found in [9].

## 2   Problem Settings

The value of an asset $S$ is described by the stochastic differential equation [4]:

$$dS = \mu S dt + \sigma S dX, \tag{1}$$

where $\sigma$ is the volatility of the asset, characterizing the fluctuations in the price, $\mu$ is the drift rate [9]. MC and QMC methods are suitable when the solution is described as the expectation of a random variable, which lies in the definition of the risk-neutral evaluation formula for the European options [9]:

$$V(S, t) = E\left(e^{-r(T-t)}h(S(T)) \mid S(t) = S, \mu = r\right), \tag{2}$$

where $E$ is the expectation operator and $h(S)$ is the payoff function. We follow the ideas of Lai and Spanier [9] and to avoid plagiarism all the definitions can be found there.

Consider an European call option [9] whose payoff depends on $k > 1$ assets with prices $S_i, i = 1, ..., k$. Each asset is modeled by a random walk (1)

$$dS_i = \mu_i S_i dt + \sigma_i S_i dX_i,$$

where $\sigma_i$ is the annualized standard deviation for the $i$-th asset and $dX_i$ is an increment of a Brownian motion. The Brownian motion is a special type Markov stochastic process, which has the property $dX \sim N(0, \sqrt{dt})$, where $N(\mu, \sigma)$ is the normal distribution with mean $\mu$ and variance $\sigma^2$. Following [9] we assume that at expiry time $T$, and risk-free interest rate $r$, the payoff is given by $h(S'_1, \ldots, S'_k)$, where $S'$ denotes the value of the $i$-th asset at expiry. Then the value of the option satisfies:

$$V = e^{-r(T-t)}(2\pi(T-t))^{-k/2}(\det \Sigma)^{-1/2}(\sigma_1 \ldots \sigma_k)^{-1}$$

$$\int_0^\infty \cdots \int_0^\infty \frac{h(S'_1, \ldots, S'_k)}{S'_1 \ldots S'_k}$$

$$\exp\left(-0.5\alpha^\top \Sigma^{-1}\alpha\right) dS'_1 \ldots dS'_k,$$

$$\alpha_i = \left(\sigma_i(T-t)^{1/2}\right)^{-1} \left(\ln(S'_i/S_i) - (r - \sigma_i^2/2)(T-t)\right),$$

$r$ is the risk-free interest rate and $\Sigma$ is the covariance matrix, whose entry with index $(i, j)$ is the covariance $dX_i$ and $dX_j$ of the $k$ assets. According to [9] the most important case in recent models is when the payoff function is the exponent function.

We will now give a brief explanation which demonstrates the strength of the MC and QMC approach [5]. According to [5] we will choose 100 nodes on the each of the coordinate axes in the $s$-dimensional cube $G = E^s$ and we have to evaluate about $10^3$ values of the function $f(x)$. Assume a time of $10^{-7}s$ is necessary for calculating one value of the function [5]. So, a time of order $10^{23}$ s will be necessary for computation of the integral, and 1 year has $31536 \times 10^3$ s.

Now MC approach consists of generating N pseudorandom values (points) (PRV) in $G$; in evaluating the values of $f(x)$ at these points; and averaging the computed values of the function. For each uniformly distributed random (UDR) point in $G$ we have to generate 100 UDR numbers in $[0, 1]$. Assume that the expression in front of $h^{-6}$ is of order 1 [5]. Here $h = 0.1$, and we have $N \approx 10^6$; so, it will be necessary to generate $30 \times 10^6 = 3 \times 10^7$ PRV. Usually, 2 operations are sufficient to generate a single PRV. According to [5] the time required to generate one PRV is the same as that for computation the value of $f(x)$. So, in order to solve the task with the same accuracy, a time of

$$3 \times 10^7 \times 2 \times 10^{-7} \approx 6 \ s$$

will be necessary. We summarize that in the case of 30-dimensional integral it is about $1.5 \times 10^{22}$ times faster than the deterministic one. That motivates our study on the new highly efficient stochastic approaches for the problem under consideration.

## 3    Highly Efficient Stochastic Approaches Based on Lattice Rules

We will use this rank-1 lattice sequence [11]:

$$\mathbf{x}_k = \left\{ \frac{k}{N} \mathbf{z} \right\}, \ k = 1, \ldots, N, \tag{3}$$

where $N$ is an integer, $N \geq 2$, $\mathbf{z} = (z_1, z_2, \ldots z_s)$ is the generating vector and $\{z\}$ denotes the fractional part of $z$. For the definition of the $E_s^\alpha(c)$ and $P_\alpha(z, N)$ see [11] and for more details, see also [1].

In 1959 Bahvalov proved that [1] there exists an optimal choice of the generating vector $\mathbf{z}$:

$$\left| \frac{1}{N} \sum_{k=0}^{N-1} f\left(\left\{\frac{k}{N}\mathbf{z}\right\}\right) - \int_{[0,1)^s} f(u)du \right| \leq cd(s, \alpha) \frac{(\log N)^{\beta(s,\alpha)}}{N^\alpha}, \tag{4}$$

for the function $f \in E_s^\alpha(c)$, $\alpha > 1$ and $d(s, \alpha), \beta(s, \alpha)$ do not depend on $N$.

The generating vector $\mathbf{z}$ which satisfies (4), is an optimal generating vector [11]. While the theoretical result establish the existence of optimal generating vectors the main bottleneck lies in the creation of the optimal vectors, especially for very high dimensions [9].

The first generating vector in our study is the generalized Fibonacci numbers of the corresponding dimension:

$$\mathbf{z} = (1, F_n^{(s)}(2), \ldots, F_n^{(s)}(s)), \tag{5}$$

where we use that $F_n^{(s)}(j) := F_{n+j-1}^{(s)} - \sum\limits_{i=0}^{j-2} F_{n+i}^{(s)}$ and

$F_{n+l}^{(s)}$ $(l = 0, \ldots, j-1, j$ is an integer, $2 \le j \le s)$ is the term of the $s$-dimensional Fibonacci sequence [11].

If we change the generating vector to be optimal in the way described in [8] we have improved the lattice sequence. This is a 200-dimensional base-2 generating vector of prime numbers for up to $2^{20} = 1048576$ points, constructed recently by Dirk Nuyens [8]. The special choice of this optimal generating vector is definitely more efficient than the Fibonacci generating vector, which is only optimal for the two dimensional case [11]. For this improved lattice rule is satisfied [8]:

$$D_N^* = \mathcal{O}\left(\frac{\log^s N}{N}\right).$$

## 4  Numerical Examples and Results

The numerical study includes high-performance computing of the following multidimensional integrals:

Example 1 (s=3):

$$\int\limits_{[0,1]^3} e^{x_1 x_2 x_3} \approx 1,14649907. \tag{6}$$

Example 2 (s=4):

$$\int\limits_{[0,1]^4} x_1 x_2^2 e^{x_1 x_2} \sin(x_3) \cos(x_4) \approx 0,1089748630. \tag{7}$$

Example 3 (s=5):

$$\int\limits_{[0,1]^5} e^{\sum\limits_{i=1}^{5} 0.5 a_i x_i^2 (2 + \sin \sum\limits_{j=1, j \ne i}^{5} x_j)} \approx 2,923651, \tag{8}$$

where $a_i = (1; 0, 5; 0, 2; 0, 2; 0, 2)$.

Example 4 (s=7):

$$\int\limits_{[0,1]^7} e^{1 - \sum\limits_{i=1}^{3} \sin(\frac{\pi}{2} \cdot x_i)} . arcsin \left( \sin(1) + \frac{\sum\limits_{j=1}^{7} x_j}{200} \right) \approx 0,75151101. \tag{9}$$

Example 5 (s=10):

$$\int\limits_{[0,1]^{10}} \frac{4 x_1 x_3^2 e^{2 x_1 x_3}}{(1 + x_2 + x_4)^2} e^{x_5 + \cdots + x_{10}} \approx 14,808435. \tag{10}$$

**Table 1.** Relative error for the 3-dimensional integral.

| N | HAL | FIBO | CRUDE | LAT |
|---|---|---|---|---|
| $2^{10}$ | 1.13e-03 | 1.06e-03 | 2.16e-03 | **5.23e-05** |
| $2^{12}$ | 4.60e-04 | 3.82e-03 | 2.82e-03 | **3.06e-04** |
| $2^{14}$ | **9.03e-05** | 4.09e-04 | 7.97e-04 | 1.75e-04 |
| $2^{16}$ | 2.62e-05 | **4.77e-06** | 7.28e-04 | 1.25e-04 |
| $2^{18}$ | 8.11e-06 | 4.95e-06 | 1.04e-04 | **7.18e-07** |
| $2^{20}$ | 2.45e-06 | **6.85e-07** | 6.70e-05 | 7.46e-07 |

**Table 2.** Relative error for the 4-dimensional integral.

| N | HAL | FIBO | CRUDE | LAT |
|---|---|---|---|---|
| $2^{10}$ | 9.53e-03 | 9.28e-03 | 4.36e-02 | **6.29e-03** |
| $2^{12}$ | 4.06e-03 | 3.94e-03 | 1.27e-02 | **1.14e-03** |
| $2^{14}$ | 6.72e-04 | **3.56e-04** | 2.65e-02 | 1.55e-03 |
| $2^{16}$ | 2.11e-04 | **1.01e-04** | 3.47e-03 | 1.63e-03 |
| $2^{18}$ | **5.09e-05** | 1.34e-04 | 3.98e-03 | 7.23e-05 |
| $2^{20}$ | **1.78e-05** | 2.68e-05 | 5.62e-04 | 9.18e-05 |

Example 6 (s=20):

$$\int\limits_{[0,1]^{20}} e^{\prod\limits_{i=1}^{20} x_i} \approx 1,00000949634. \tag{11}$$

Example 7 (s=25):

$$\int\limits_{[0,1]^{25}} \frac{4x_1 x_3^2 e^{2x_1 x_3}}{(1+x_2+x_4)^2} e^{x_5+\cdots+x_{20}} x_{21} \ldots x_{25} \approx 108,808. \tag{12}$$

Example 8 (s=30):

$$\int\limits_{[0,1]^{30}} \frac{4x_1 x_3^2 e^{2x_1 x_3}}{(1+x_2+x_4)^2} e^{x_5+\cdots+x_{20}} x_{21} \ldots x_{30} \approx 3,244540. \tag{13}$$

The results are presented in tables including the relative error (RE) of the MC and QMC method that has been used, the CPU-time (T) in seconds and the number of realizations of the random variable (N). We will make a high-performance computation, including the Optimized lattice rule (LAT), the Fibonacci based rule (FIBO), the simplest plain MC algorithm (CRUDE) [5] and the Halton quasi-random sequence (HAL) [6,7].

**Table 3.** Relative error for the 5-dimensional integral.

| N | HAL | FIBO | CRUDE | LAT |
|---|---|---|---|---|
| $2^{10}$ | 3.07e-03 | 8.47e-04 | 1.87e-04 | **1.06e-04** |
| $2^{12}$ | 8.62e-04 | 1.11e-03 | 2.33e-01 | **6.45e-04** |
| $2^{14}$ | 1.79e-04 | 6.40e-04 | 1.24e-03 | **1.75e-04** |
| $2^{16}$ | **5.70e-05** | 5.91e-05 | 1.41e-04 | 3.29e-04 |
| $2^{18}$ | **1.74e-05** | 2.49e-05 | 9.05e-04 | 3.35e-05 |
| $2^{20}$ | **4.85e-06** | 6.96e-06 | 5.49e-04 | 3.60e-05 |

**Table 4.** Relative error for the 7-dimensional integral.

| N | HAL | FIBO | CRUDE | LAT |
|---|---|---|---|---|
| $2^{10}$ | **2.05e-03** | 3.43e-02 | 2.84e-03 | 2.68e-03 |
| $2^{12}$ | 1.28e-03 | 3.16e-03 | 5.85e-03 | **5.48e-04** |
| $2^{14}$ | 4.91e-04 | 3.89e-04 | 2.56e-03 | **3.17e-04** |
| $2^{16}$ | 3.51e-04 | **7.68e-05** | 4.27e-03 | 2.82e-04 |
| $2^{18}$ | 3.08e-04 | **2.78e-04** | 1.51e-03 | 3.01e-04 |
| $2^{20}$ | 2.96e-04 | **2.87e-04** | 2.54e-04 | 2.92e-04 |

**Table 5.** Relative error for the 10-dimensional integral.

| N | HAL | FIBO | CRUDE | LAT |
|---|---|---|---|---|
| $2^{10}$ | 5.96e-02 | **2.02e-03** | 1.05e-01 | 2.75e-03 |
| $2^{12}$ | **1.12e-02** | 2.76e-02 | 3.09e-02 | 2.92e-02 |
| $2^{14}$ | **1.55e-03** | 4.58e-02 | 2.72e-02 | 2.82e-03 |
| $2^{16}$ | **7.45e-04** | 1.03e-01 | 4.40e-03 | 6.31e-03 |
| $2^{18}$ | **2.48e-04** | 4.46e-02 | 5.41e-03 | 1.18e-03 |
| $2^{20}$ | **1.33e-04** | 1.38e-02 | 7.53e-04 | 2.09e-03 |

**Table 6.** Relative error for the 20-dimensional integral.

| N | HAL | FIBO | CRUDE | LAT |
|---|---|---|---|---|
| $2^{10}$ | 5.27e-07 | 2.32e-06 | 4.46e-07 | **1.51e-07** |
| $2^{12}$ | 2.35e-07 | 4.76e-07 | 2.06e-07 | **1.80e-07** |
| $2^{14}$ | 1.71e-07 | 9.54e-07 | 9.49e-08 | **8.52e-08** |
| $2^{16}$ | 8.41e-08 | 9.54e-07 | 8.79e-08 | **6.77e-08** |
| $2^{18}$ | **2.58e-08** | 9.54e-07 | 6.38e-08 | 8.75e-08 |
| $2^{20}$ | 1.33e-08 | 2.84e-06 | 1.32e-08 | **5.15e-10** |

**Table 7.** Relative error for the 25-dimensional integral.

| N | HAL | FIBO | CRUDE | LAT |
|---|---|---|---|---|
| $2^{10}$ | 5.44e-01 | 1.00e+00 | **2.42e-01** | 3.33e-01 |
| $2^{12}$ | 1.29e-01 | 1.00e+00 | 1.39e-02 | **1.88e-03** |
| $2^{14}$ | 1.34e-01 | 7.53e-01 | 1.01e-01 | **2.08e-02** |
| $2^{16}$ | 5.37e-02 | 4.58e-01 | 6.99e-02 | **1.36e-02** |
| $2^{18}$ | 1.13e-02 | 3.45e-01 | 1.34e-02 | **2.99e-03** |
| $2^{20}$ | 8.36e-03 | 2.88e-01 | 1.22e-02 | **6.43e-03** |

**Table 8.** Relative error for the 30-dimensional integral.

| N | HAL | FIBO | CRUDE | LAT |
|---|---|---|---|---|
| $2^{10}$ | 6.86e-01 | 1.00e+00 | 3.09e-01 | **2.09e-01** |
| $2^{12}$ | 3.40e-01 | 1.00e+00 | **2.12e-02** | 2.93e-01 |
| $2^{14}$ | 2.25e-01 | 1.00e+00 | **3.90e-02** | 1.25e-01 |
| $2^{16}$ | 1.29e-01 | 8.58e-01 | 6.87e-02 | **3.50e-02** |
| $2^{18}$ | 2.14e-02 | 8.11e-01 | **1.45e-02** | 5.60e-02 |
| $2^{20}$ | 1.84e-02 | 4.45e-01 | 1.36e-02 | **9.57e-03** |

For the 3-dimensional integral, for a number of samples $N = 2^{20}$, Fibonacci based lattice rule and optimized lattice rule produce result of the same order with 2 orders better than the other two algorithms - see Table 1. For the 4-dimensional integral, for a number of samples $N = 2^{20}$, Fibonacci based lattice rule and Halton sequence produce better results than the optimized lattice rule - see Table 2. For the 5-dimensional integral for $N = 2^{20}$ again the best approach is Halton sequence, followed by the Fibonacci based lattice rule, with 1 order better than the optimized lattice rule and with 2 orders better than the Crude MC algorithm - see see Table 3. For the 7-dimensional integral $N = 2^{20}$ all the methods produce relative errors of the same order - see Table 4. For the 10-dimensional integral for $N = 2^{20}$ the best relative error is obtained by the Halton sequence, followed by the Crude MC algorithm and optimized lattice rule, and here one can observe that the Fibonacci based lattice rule started to give worse relative errors with increasing the dimensionality of the integral - see Table 5. For the 20-dimensional integral $N = 2^{20}$ the optimized lattice rule has shown its strength giving relative error two order better than both Halton sequence and Crude MC - see Table 6. For the 25-dimensional integral $N = 2^{20}$ the optimized lattice rule and Halton sequence produce results of the same order with 2 order better than Fibonacci based lattice rule - see Table 7. For the 30-dimensional integral $N = 2^{20}$ the optimized lattice rule gives the best relative error one order better than both Halton sequence and Crude MC and Fibonacci based lattice rule gives the worst results - see Table 8. To summarize FIBO accuracy is lower than the accuracy of the optimized lattice rules LAT with increasing the dimensionality of the integral, which is the case in many financial applications.

## 5    Conclusion

A comprehensive experimental study of optimized lattice rule, Fibonacci lattice sets, Halton sequence and Crude Monte Carlo algorithm has been done for the first time on some case test functions related to option pricing. Optimized lattice rule described here is not only one of the best available algorithms for high dimensional integrals but also one of the few possible methods, because in this work we show that the deterministic algorithms need a huge amount of time for the evaluation of the multidimensional integral, as it was discussed in this paper. The numerical tests show that the improved lattice rule is efficient for multidimensional integration and especially for computing multidimensional integrals of a very high dimensions, where Fibonacci based lattice rule gives very bad accuracy. It is an important element since this may be crucial in order to achieve a more reliable interpretation of the results in European style options which is foundational in computational finance. Since the performance of the lattice rule depends on the choice of the generator vectors, the presented lattice rule with an optimal generating vector is an optimization over the Fibonacci generalized vector.

**Acknowledgment.** Venelin Todorov is supported by the Bulgarian National Science Fund under the Project KP-06-N52/5 "Efficient methods for modeling, optimization and decision making" and Project KP-06-N52/2 "Perspective Methods for Quality Prediction in the Next Generation Smart Informational Service Networks". Stoyan Apostolov is supported by the Bulgarian National Science Fund under Project KP-06-M32/2 - 17.12.2019 "Advanced Stochastic and Deterministic Approaches for Large-Scale Problems of Computational Mathematics". The work is also supported by the NSP "ICT in SES", contract No DO1-205/23.11.2018, financed by the Ministry of EU in Bulgaria and by the BNSF under Project KP-06-Russia/17 "New Highly Efficient Stochastic Simulation Methods and Applications" and Project DN 12/4-2017 "Advanced Analytical and Numerical Methods for Nonlinear Differential Equations with Applications in Finance and Environmental Pollution".

## References

1. Bakhvalov, N.: On the approximate calculation of multiple integrals. J. Complexity **31**(4), 502–516 (2015)
2. Asenov, A., Pencheva, V., Georgiev, I.: Modelling passenger service rate at a transport hub serviced by a single urban bus route as a queueing system. In: 2019 IOP Conference Series: Materials Science and Engineering, pp. 664 012034
3. Boyle, P.P., Lai, Y., Tan, K.: Using lattice rules to value low-dimensional derivative contracts. In: 9th International AFIR Colloquium Proceedings, vol. 2, pp. 111–134 (2001)
4. Caflisch, R.E., Morokoff, W., Owen, A.: Valuation of mortgage-backed securities using Brownian bridges to reduce effective dimension. J. Comp. Finance **1**(1), 27–46 (1997)
5. Dimov, I.: Monte carlo methods for applied scientists. London, Singapore, World Scientific, New Jersey, p. 291 (2008)
6. Halton, J.: On the efficiency of certain quasi-random sequences of points in evaluating multi-dimensional integrals. Numerische Mathematik **2**, 84–90 (1960)
7. Halton, J., Smith, G.B.: Algorithm 247: Radical-inverse Quasi-Random point sequence. Commun. ACM **7**, 701–702 (1964)

8. Kuo, F.Y., Nuyens, D.: Application of quasi-Monte Carlo methods to elliptic PDEs with random diffusion coefficients - a survey of analysis and implementation. Foundations Comput. Math. **16**(6), 1631–1696 (2016)

9. Lai, Y., Spanier, J.: Applications of Monte Carlo/Quasi-Monte Carlo methods in finance: option pricing. In: Niederreiter, H., Spanier, J. (eds.) Monte-Carlo and Quasi-Monte Carlo Methods 1998. Springer (2000). https://doi.org/10.1007/978-3-642-59657-5_19

10. Pencheva, V., Asenov, A., Georgiev, I., Sładkowski, A.: Research on the state of urban passenger mobility in Bulgaria and prospects for using low carbon energy for transport. In: Sładkowski, A. (eds.) Ecology in Transport: Problems and Solutions. LNCS. vol. 124, pp. 441–504. Springer, Cham (2020). https://doi.org/10.1007/978-3-030-42323-0_8

11. An historical overview of lattice point sets. In: Fang, K.T., Niederreiter, H., Hickernell, F.J. (eds.) Monte Carlo and Quasi-Monte Carlo Methods 2000. Springer, Heidelberg, pp. 158–167 (2000)

12. Wilmott, P., Dewynne, J., Howison, S., Option pricing: mathematical models and computation, Oxford University Press (1995)

# Intuitionistic Fuzzy Approach
# for Outsourcing Provider Selection
# in a Refinery

Velichka Traneva(✉) ⓘ and Stoyan Tranev ⓘ

"Prof. Asen Zlatarov" University, "Prof. Yakimov" Blvd, 8000 Bourgas, Bulgaria
tranev@abv.bg
http://www.btu.bg

**Abstract.** Outsourcing is a new approach to the transfer of a business process that is traditionally carried out by the organization to an independent external service provider. Outsourcing is a good strategy for companies that need to reduce operating costs and improve competitiveness. It is important that companies select out the most eligible outsourcing providers.

In this study, an intuitionistic fuzzy multi-criteria decision making approach is used for selecting the most appropriate outsourcing service provider for an oil refining enterprise on the Balkan peninsula. An optimal outsourcing problem is formulated and an algorithm for selection the most eligible outsourcing service provider is proposed using the concept of index matrices (IMs), where the evaluations of outsourcing candidates against criteria formulated by several experts are intuitionistic fuzzy pairs. The proposed decision-making model takes into account the ratings of the experts and the weighting factors of the evaluation criteria according to their priorities for the outsourcing service. Due to complexity of outsourcing process, the real numbers are not enough to characterize evaluation objects. Fuzzy sets (FSs) of Zadeh use the single membership function to express the degree to which the element belongs to the fuzzy set. As a result, the FSs is unable to express non-membership and hesitation degrees. Since intuitionistic fuzzy sets (IFSs) of Atanassov consider the membership and non-membership degrees simultaneously, they are more flexible than the FSs in modeling with uncertainty. The originality of the paper comes from the proposed decision-making model and its application in an optimal intuitionistic fuzzy (IF) outsourcing problem of a refinery. The presented approach for selection the most suitable outsourcing service provider can be applied to problems with imprecise parameters and can be extended in order to obtain an optimal solution for other types of multidimensional outsoursing problems by using $n$-dimensional index matrices.

Work on Sect. 1 and Sect. 2 is supported by the Asen Zlatarov University through project Ref. No. NIX-449/2021 "Modern methods for making management decisions", the work on Sect. 3 and Sect. 4 is supported by the Ministry of Education and Science under the Programme "Young scientists and postdoctoral students", approved by DCM # 577/17.08.2018.

I. Lirkov and S. Margenov (Eds.): LSSC 2021, LNCS 13127, pp. 266–274, 2022.
https://doi.org/10.1007/978-3-030-97549-4_31

**Keywords:** Decision making · Index matrix · Intuitionistic fuzzy
logic · Outsourcing · Refinery

# 1 Introduction

Outsourcing is a transfer of a business process that has been traditionally operated and managed internally to an external service provider [12]. In Group Decision-Making (GDM) [4] a set of experts in a given field is involved in a decision process concerning the selection of the best alternative among a set of predefined ones. The uncertainty in the GDM-problem may be caused by the characteristics of the candidates for the outsourcing service providers, which can be unavailable or indeterminate. It may also be a consequence of the inability of the experts to formulate a precise evaluation [27]. IFSs (see [2,4]) are an extension of FSs of Zadeh [28] and provides the tools to describe the imprecision in data more accurately, by allowing a hesitancy degree. In the following we will make a brief literature review of methods for solving outsourcing problems:

Decision making methods about outsourcing are multi-attribute decision making methods [15,24], such as AHP [25], ANP [21], PROMETHEE [11], balanced score card [21], and TOPSIS [11]. Wang and Yang [25] have proposed a method by using AHP and PROMETHEE in making information system outsourcing decisions. Modak et al. [19] have proposed a new approach to outsourcing problem of a coal mining organisation based on a balanced scorecard and fuzzy AHP. Wang et al. [26] have applied a TODIM approach based on multi-hesitant fuzzy linguistic information to evaluate logistics outsourcing providers. Bottani and Rizzi [11] have proposed a fuzzy TOPSIS methodology based on a structured framework for the selection of the most appropriate service provider. Kahraman et al. [14] have developed a novel approach using hesitant linguistic term sets for supplier selection problem. An outsourcing model for the evaluation was given by Araz et al. [1], in which first evaluates the alternative using the PROMETHEE method and then fuzzy goal programming is used to select the most suitable alternative. Liu and Wang [16] have developed a fuzzy linear programming approach for the selection of providers by integrating fuzzy Delphi method. Hsu et al. [13] have proposed a hybrid decision model for selection of outsourcing companies, combining the DEMATEL, ANP, and grey relation methods. In [24] some disadvantages of the methods are presented as follows:

1) The crisp real numbers (see [21,25]) or FSs [16,28] are used for the studies to represent the information. IFSs [4] take into account the hesitancy degree and are more flexible and practical than FSs in dealing with uncertainty.
2) In the methods [16,18,25] a single decision maker (expert) is included.
3) The fuzzy linear programming method [16] has used fuzzy data.

The IF outsourcing provider selection problems are formulated in [24] and IF linear programming method is proposed for solving such problems. In [15], an interval-valued IF AHP and TOPSIS -based methodology is proposed to select of the best alternative among alternative outsource manufacturers.

In this study, an IF optimal autsoursing problem is formulated and an algorithm for selection the most appropriate outsourcing service provider is proposed applying the toolkit of IMs [3], where the evaluations of the outsourcing candidates against criteria are intuitionistic fuzzy pairs (IFPs, [9]). The proposed decision-making model takes into account the ratings of the experts and the weighting factors of the evaluation criteria according to their priorities for the outsourcing service. A numerical case study containing five criteria and four alternatives for an outsourcing service provider in a refinery is also provided.

The rest of this paper is structured as follows: Sect. 2 describes the basic definitions of the theories of IMs and IFPs. Section 3 formulates an IF multicriteria decision making problem for the selection of outsourcing service provider and proposes an index-marix algorithm for its solution. Finally, in Sect. 4, the proposed approach is demonstrated on the real IF problem of selecting an outsourcing supplier of construction and installation works in a refinery. Section 5 concludes the paper and gives aspects for future research.

## 2    Preliminaries

### 2.1    Short Remarks on IFPs

The **IFP** has the form of an ordered pair $\langle a, b \rangle = \langle \mu(p), \nu(p) \rangle$, where $a, b \in [0, 1]$ and $a + b \leq 1$, that is used as an evaluation of a proposition $p$ [9]. Let us have two IFPs $x = \langle a, b \rangle$ and $y = \langle c, d \rangle$. We recall some basic operations [4,9]:

$$\neg x = \langle b, a \rangle; \qquad\qquad x \wedge_1 y = \langle \min(a, c), \max(b, d) \rangle;$$
$$x \vee_1 y = \langle \max(a, c), \min(b, d) \rangle; \ x \wedge_2 y = x + y = \langle a + c - a.c, b.d \rangle;$$
$$x@y = \langle \tfrac{a+c}{2}, \tfrac{b+d}{2} \rangle; \qquad\qquad x \vee_2 y = x.y = \langle a.c, b + d - b.d \rangle;$$

and relations with IFPs [4,9,20]

$$
\begin{array}{llll}
x \geq y & \text{iff} & a \geq c \text{ and } b \leq d; & x \geq_\square y \quad \text{iff} \quad a \geq c; \\
x \geq_\circ y & \text{iff} & b \leq d; & x \geq_R y \quad \text{iff} \quad R_{\langle a,b \rangle} \leq R_{\langle c,d \rangle}, \qquad (1) \\
\multicolumn{4}{l}{\text{where} \quad R_{\langle a,b \rangle} = 0.5(2 - a - b)(1 - a).}
\end{array}
$$

### 2.2    Definition and Operations Over 3-D Intuitionistic Fuzzy Index Matrices (3-D IFIM)

Let $\mathcal{I}$ be a set. By 3-D IFIM $[K, L, H, \{\langle \mu_{k_i, l_j, h_g}, \nu_{k_i, l_j, h_g} \rangle\}]$, $(K, L, H \subset \mathcal{I})$ and elements, which are IFPs, we denote the object [5,22]:

| $h_g \in H$ | $l_1$ | $\cdots$ | $l_j$ | $\cdots$ | $l_n$ |
|---|---|---|---|---|---|
| $k_1$ | $\langle \mu_{k_1,l_1,h_g}, \nu_{k_1,l_1,h_g} \rangle$ | $\cdots$ | $\langle \mu_{k_1,l_j,h_g}, \nu_{k_1,l_j,h_g} \rangle$ | $\cdots$ | $\langle \mu_{k_1,l_n,h_g}, \nu_{k_1,l_n.h_g} \rangle$ |
| $\vdots$ | $\vdots$ | $\cdots$ | $\vdots$ | $\cdots$ | $\vdots$ |
| $k_m$ | $\langle \mu_{k_m,l_1,h_g}, \nu_{k_m,l_1,h_g} \rangle$ | $\cdots$ | $\langle \mu_{k_m,l_j,h_g}, \nu_{k_m,l_j,h_g} \rangle$ | $\cdots$ | $\langle \mu_{k_m,l_n,h_g}, \nu_{k_m,l_n,h_g} \rangle$ |

Let $\mathcal{X}, \mathcal{Y}, \mathcal{Z}, \mathcal{U}$ be fixed sets. Let operations "$*$" and "$\circ$" be defined so that: $* : \mathcal{X} \times \mathcal{Y} \to \mathcal{Z}$ and $\circ : \mathcal{Z} \times \mathcal{Z} \to \mathcal{U}$. Following [5,22], we recall some operations over IMs. Let us have two 3D-IFIMs $A = [K, L, H, \{\langle \mu_{k_i,l_j,h_g}, \nu_{k_i,l_j,h_g} \rangle\}]$ and $B = [P, Q, R, \{\langle \rho_{p_r,q_s,r_d}, \sigma_{p_r,q_s,r_d} \rangle\}]$.

**Addition-(max,min)** $A \oplus_{(\max,\min)} B = [K \cup P, L \cup Q, H \cup R, \{\langle \phi_{t_u,v_w,x_y}, \psi_{t_u,v_w,x_y} \rangle\}]$, where $\langle \phi_{t_u,v_w,x_y}, \psi_{t_u,v_w,x_y} \rangle$ is defined in [5].

**Transposition:** $A^T$ is the transposed IM of $A$.

**Multiplication with a constant** $\alpha A = [K, L, H\{\alpha \langle \mu_{k_i,l_j,h_g}, \nu_{k_i,l_j,h_g} \rangle\}]$

   **Multiplication** $A \odot_{(\circ,*)} B = [K \cup (P - L), Q \cup (L - P), H \cup R, \{\langle \phi_{t_u,v_w,x_y}, \psi_{t_u,v_w,x_y} \rangle\}]$, where $\langle \phi_{t_u,v_w,x_y}, \psi_{t_u,v_w,x_y} \rangle$

$$
= \begin{cases}
\langle \mu_{k_i,l_j,h_g}, \nu_{k_i,l_j,h_g} \rangle, & \text{if } t_u = k_i \in K \\
& \& \ v_w = l_j \in L - P - Q \ \& \ x_y = h_g \in H \\
& \text{or } t_u = k_i \in K - P - Q \\
& \& \ v_w = l_j \in L \ \& \ x_y = h_g \in H; \\
\langle \rho_{p_r,q_s,r_d}, \sigma_{p_r,q_s,r_d} \rangle, & \text{if } t_u = p_r \in P \\
& \& \ v_w = q_s \in Q - K - L \ \& \ x_y = r_d \in R \\
& \text{or } t_u = p_r \in P - L - K \\
& \& \ v_w = q_s \in Q \ \& \ x_y = r_d \in R; \\
\langle \underset{l_j = p_r \in L \cap P}{\circ} (*(\mu_{k_i,l_j,h_g}, \rho_{p_r,q_s,r_d})), & \text{if } t_u = k_i \in K \ \& \ v_w = q_s \in Q \\
\underset{l_j = p_r \in L \cap P}{*} (\circ(\nu_{k_i,l_j,h_g}, \sigma_{p_r,q_s,r_d}))\rangle, & \& \ x_y = h_g = r_d \in H \cap R; \\
\langle 0, 1 \rangle, & \text{otherwise.}
\end{cases}
$$

where $\langle \circ, * \rangle \in \{\langle \max, \min \rangle, \langle \min, \max \rangle\}$.

**Aggregation operation by one dimension** [23]

$$
\alpha_{K,\#_q}(A, k_0) = \quad
\begin{array}{c|ccc}
 & l_1 & \cdots & l_n \\
\hline
k_0 & \overset{m}{\underset{i=1}{\#_q}} \langle \mu_{k_i,l_1}, \nu_{k_i,l_1} \rangle & \cdots & \overset{m}{\underset{i=1}{\#_q}} \langle \mu_{k_i,l_n}, \nu_{k_i,l_n} \rangle
\end{array}, (1 \le q \le 10) \quad (2)
$$

If we use $\#_1^*$ in the operation (2) we obtain super pessimistic aggregation operation, etc., $\#_{10}^*$ - super optimistic aggregation operation.

**Substitution.** Local substitution over $A$ is defined for the pair of indices $(p, k_i)$ by $\left[ \frac{p}{k_i}; \perp; \perp \right] A = [(K - \{k_i\}) \cup \{p\}, L, H, \{a_{k_i,l,h}\}]$.

**Projection.** Let $M \subseteq K$, $N \subseteq L$ and $U \subseteq H$. Then, $pr_{M,N,U} A = [M, N, U, \{b_{k_i,l_j,h_g}\}]$, and for each $k_i \in M$, $l_j \in N$ and $h_g \in U$, $b_{k_i,l_j,h_g} = a_{k_i,l_j,h_g}$.

# 3   An Index-Matrix Approach to Intuitionistic Fuzzy Problem for Selection an Outsourcing Provider

Here, we formulate an optimal IF autsoursing problem and propose an index-matrix algorithm for selection the most eligible outsourcing service provider.

Let us assume that a company has decided to outsource an activity to an external service provider in order to reduce its costs. The company has an evaluation system of the applicants $\{k_1, \ldots, k_i, \ldots, k_m\}$ (for $i = 1, \ldots, m$) for a providing outsourcing service $v_e (1 \le e \le u)$, containing criteria $\{c_1, \ldots, c_j, \ldots, c_n\}$ (for $j = 1, \ldots, n$). Experts $\{d_1, \ldots, d_s, \ldots, d_D\}$ (for $s = 1, \ldots, D$) in the organization assess the weighting coefficients under the form of IFPs of the assessment criteria $c_j$ (for $j = 1, \ldots, n$) according to their priority for this service $v_e$ - $pk_{c_j, v_e}$ (for $j = 1, \ldots, n$). Each expert belong to $\{r_1, \ldots, r_s, \ldots, r_D\}$ has an IFP rating $r_s = \langle \delta_s, \epsilon_s \rangle$ is an IFP ($1 \le s \le D$). At the moment of applying for outsourcing service provider, all candidates are evaluated by the experts according to the criteria for the outsourcing service assignment $\dot{v_e}$, and their evaluations $ev_{k_i, c_j, d_s}$ (for $1 \le i \le m, 1 \le j \le n, 1 \le s \le D$) are IFPs. The aim of the problem is to optimally allocate outsourcing service provider among its candidates.

To find an optimal solution we propose the following index-matrix algorithm, which is named IFIMOA, described with mathematical notation and pseudocode:

**Step 1.** 3-D evaluation IFIM $EV[K, C, E, \{ev_{k_i, c_j, d_s}\}]$ is formed in accordance with the above problem, where $K = \{k_1, k_2, \ldots, k_m\}$, $C = \{c_1, c_2, \ldots, c_n\}$, $E = \{d_1, d_2, \ldots, d_D\}$ and the element $\{ev_{k_i, c_j, d_s}\} = \langle \mu_{k_i, c_j, d_s}, \nu_{k_i, c_j, d_s} \rangle$ (for $1 \le i \le m, 1 \le j \le n, 1 \le s \le D$) is the estimate of the $d_s$-th expert for the $k_i$-th candidate by the $c_j$-th criterion. The expert is not sure about for its evaluation according to a given criterion due to changes in some uncontrollable factors. The evaluations are under the form of IFPs. Let us have the set of intervals of the expert evaluations by all criteria for all candidates $[p^1_{k_i, c_j, d_s}; p^2_{k_i, c_j, d_s}]$ and let

$$A_{min, i, j, s} = \min_{1 \le i \le m, 1 \le j \le n, 1 \le s \le D} p^1_{k_i, c_j, d_s} < \max_{1 \le i \le m, 1 \le j \le n, 1 \le s \le D} p^2_{k_i, c_j, d_s} = A_{max, i, j, s}.$$

For the interval $[p^1_{k_i, c_j, d_s}; p^2_{k_i, c_j, d_s}]$ we construct IFPs [4] as follows:

$$\mu_{k_i, c_j, d_s} = \frac{p^1_{k_i, c_j, d_s} - A_{min, i, j, s}}{A_{max, i, j, s} - A_{min, i, j, s}}, \nu_{k_i, c_j, d_s} = \frac{A_{max, i, j, s} - p^2_{k_i, c_j, d_s}}{A_{max, i, j, s} - A_{min, i, j, s}}. \text{ Go to } Step \; 2.$$

**Step 2.** Let the score (rating) $r_s$ of the $s$-th expert ($s \in E$) be specified by an IFP $\langle \delta_s, \epsilon_s \rangle$. $\delta_s$ and $\epsilon_s$ are interpreted respectively as his degree of competence and of incompetence. Then we create $EV^*[K, C, E, \{ev^*_{k_i, c_j, d_s}\}]$ $= r_1 pr_{K, C, d_1} EV \oplus_{(max, min)} r_2 pr_{K, C, d_2} EV \ldots \oplus_{(max, min)} r_D pr_{K, C, d_D} EV$. $EV := EV^*(ev_{k_i, l_j, d_s} = ev^*_{k_i, l_j, d_s}, \forall k_i \in K, \forall l_j \in L, \forall d_s \in E).$

$\alpha_E$-th aggregation operation is applied to find the aggregated value of the $k_i$-th candidate against the $c_j$-th criterion in a present moment $h_f \notin E$:

$$R = \alpha_{E,\#_q}(EV, h_f) = \left\{ \begin{array}{c|c} c_j & h_f \\ \hline k_1 & \overset{D}{\underset{s=1}{\#_q}} \langle \mu_{k_1,c_j,d_s}, \nu_{k_1,c_j,d_s} \rangle \\ \vdots & \vdots \\ k_m & \overset{D}{\underset{s=1}{\#_q}} \langle \mu_{k_m,c_j,d_s}, \nu_{k_m,c_j,d_s} \rangle \end{array} \middle| c_j \in C \right\}, (1 \leq q \leq 10), \text{ go to } Step \ 3.$$

**Step 3.** Let us define the 3-D IFIM $PK$ of the weight coefficients of the assessment criterion according to its priority to the service $v_e$ $(1 \leq e \leq u)$:

$$PK[C, v_e, h_f\{pk_{c_j,v_e,h_f}\}] = \begin{array}{c|c} h_f & v_e \\ \hline c_1 & pk_{c_1,v_e,h_f} \\ \vdots & \vdots \\ c_j & pk_{c_j,v_e,h_f} \\ \vdots & \vdots \\ c_n & pk_{c_n,v_e,h_f} \end{array},$$

where $C = \{c_1, c_2, \ldots, c_n\}$, and for $1 \leq j \leq n : pk_{c_j,v_e,h_f}$ are IFPs. Let us calculate $R^T[K, C, h_f]$ and 3-D IFIM $B[K, v_e, h_f\{b_{k_i,v_e,h_f}\}] := R^T \odot_{(\circ,*)} PK$, which contains the cumulative estimates of the $k_i$-th candidate (for $1 \leq i \leq m$) for the $v_e$-th outsourcing service and $\langle \circ, * \rangle$ is an operation from (2). Go to *Step 4*.

**Step 4.** The aggregation operation $\alpha_{K,\#_q}(B, k_0)$ is applied by the dimension $K$ to find the most suitable candidate for the vacant position $v_e$.

$$al_{K,\#_q}(B, k_0) = \begin{array}{c|c} & v_e \\ \hline k_0 & \overset{m}{\underset{i=1}{\#_q}} \langle \mu_{k_i,v_e}, \nu_{k_i,v_e} \rangle \end{array}, \text{ where } k_0 \notin K, 1 \leq q \leq 10. \text{ Go to } Step \ 5.$$

**Step 5.** After finishing of the procedure, we need to determine whether there are correlations between some of these criteria. The procedure in IFS-case, based on the intercriteria analysis is (ICrA, see [8]) is discussed in [10]. Let $\alpha, \beta \in [0, 1]$ be given, so that $\alpha + \beta \leq 1$. The criteria $C_k$ and $C_l$ are in $(\alpha, \beta)$-positive consonance, if $\mu_{C_k,C_l} > \alpha$ and $\nu_{C_k,C_l} < \beta$; $(\alpha, \beta)$-negative consonance, if $\mu_{C_k,C_l} < \beta$ and $\nu_{C_k,C_l} > \alpha$; $(\alpha, \beta)$-dissonance, otherwise. After application of the ICrA over IFIM $R$ we determine which criteria are in a consonance. Let two criteria be in consonance. Then, we can evaluate their complexity, and the criterion with a higher complexity can be eliminated from the future decision making process. If $O = \{O_1, ..., O_V\}$ is the set of criteria that can be omitted, then reduce the IM $R$ by $R* = R_{(O, \perp)}$. The complexity of the algorithm without step 5 is $O(Dmn)$ (the complexity of the ICrA in this step is $O(m^2 n^2)$ [7]).

## 4 Real-World Application

In this section, the proposed IM approach from the Sect. 3 is applied to a real case study. A refinery is engaged in the decision about selection of its autosourcing refinery service provider for construction and installation works related to the replacement of obsolete and physically worn out equipment and technical re-equipment of the objects of an oil refining enterprise. For this purpose, the oil-refinery invites three experts $d_1, d_2$ and $d_3$ to evaluate the candidates for the autosourcing refinery service. There are four potential logistics providers for

further evaluating $k_i$ (for $1 \leq i \leq 4$). The real evaluation system of optimal out-soursing service provider selection is based on the basis of 5 criteria as follows: $C_1$ - Compliance of the outsourcing service provider with its corporate culture; $C_2$ - Understanding of the outsourcing service by the provider; $C_3$ - Necessary resources of the outsourcing provider for the implementation of the outsourcing service; $C_4$ - Price of the provided outsourcing service; $C_5$ - Opportunity for strategic development of the outsourcing service together with the outsourcing-assignor. The weighting coefficients under the form of IFPs of the assessment criteria $c_j$ (for $j = 1, ..., 5$) according to their priority for the outsourcing service $v_e$ - $pk_{c_j,v_e}$ (for $j = 1, ..., 5$) and the ratings of the experts $\{r_1, r_2, r_3\}$ be given. The aim of the problem is to optimally select autosourcing service provider.
Solution of the problem: **Step 1.** 3-D evaluation IFIM $EV[K, C, E, \{es_{k_i,c_j,d_s}\}]$ is formed in accordance with the above problem:

$$= \left\{ \begin{array}{c|ccccc|ccccc|ccccc}
d_1 & c_1 & c_2 & c_3 & c_4 & c_5 & d_2 & c_1 & c_2 & c_3 & c_4 & c_5 & d_3 & c_1 & c_2 & c_3 & c_4 & c_5 \\
\hline
k_1 & \langle 0.4,0.3\rangle & \langle 0.7,0.1\rangle & \langle 0.3,0.2\rangle & \langle 0.5,0.1\rangle & \langle 0.6,0.0\rangle & k_1 & \langle 0.3,0.2\rangle & \langle 0.9,0.1\rangle & \langle 0.4,0.3\rangle & \langle 0.4,0.0\rangle & \langle 0.5,0.1\rangle & k_1 & \langle 0.5,0.3\rangle & \langle 0.6,0.1\rangle & \langle 0.2,0.3\rangle & \langle 0.2,0.5\rangle & \langle 0.6,0.1\rangle \\
k_2 & \langle 0.7,0.1\rangle & \langle 0.5,0.1\rangle & \langle 0.5,0.2\rangle & \langle 0.3,0.4\rangle & \langle 0.7,0.0\rangle & k_2 & \langle 0.6,0.1\rangle & \langle 0.4,0.2\rangle & \langle 0.4,0.0\rangle & \langle 0.4,0.3\rangle & \langle 0.6,0.0\rangle & k_2 & \langle 0.7,0.0\rangle & \langle 0.6,0.1\rangle & \langle 0.8,0.0\rangle & \langle 0.1,0.4\rangle & \langle 0.4,0.0\rangle \\
k_3 & \langle 0.5,0.2\rangle & \langle 0.6,0.1\rangle & \langle 0.6,0.2\rangle & \langle 0.2,0.4\rangle & \langle 0.6,0.1\rangle & k_3 & \langle 0.4,0.3\rangle & \langle 0.7,0.0\rangle & \langle 0.5,0.2\rangle & \langle 0.2,0.5\rangle & \langle 0.7,0.1\rangle & k_3 & \langle 0.6,0.1\rangle & \langle 0.7,0.1\rangle & \langle 0.7,0.0\rangle & \langle 0.2,0.3\rangle & \langle 0.7,0.1\rangle \\
k_4 & \langle 0.6,0.0\rangle & \langle 0.7,0.1\rangle & \langle 0.8,0.0\rangle & \langle 0.4,0.3\rangle & \langle 0.5,0.2\rangle & k_4 & \langle 0.5,0.0\rangle & \langle 0.6,0.1\rangle & \langle 0.7,0.0\rangle & \langle 0.5,0.1\rangle & \langle 0.3,0.2\rangle & k_4 & \langle 0.8,0.0\rangle & \langle 0.4,0.3\rangle & \langle 0.6,0.0\rangle & \langle 0.6,0.1\rangle & \langle 0.4,0.2\rangle
\end{array} \right\},$$

where the IFP $\{ev_{k_i,c_j,d_s}\}$ (for $1 \leq i \leq 4, 1 \leq j \leq 5, 1 \leq s \leq 3$) is the estimate of the $d_s$-th expert for the $k_i$-th candidate by the $c_j$-th criterion.

**Step 2.** Let the experts have the following rating coefficients respectively $\{r_1, r_2, r_3\} = \{\langle 0.7, 0.0\rangle, \langle 0.6, 0.0\rangle, \langle 0.8, 0.0\rangle\}$. We create $EV^*[K, C, E, \{ev^*\}]$ $= r_1 pr_{K,C,d_1} EV \oplus_{(max,min)} r_2 pr_{K,C,d_2} EV \oplus_{(max,min)} r_3 pr_{K,C,d_3} EV$.
$EV := EV^*$. Let us apply the optimistic aggregation operation $\alpha_{E,(max,min)}$ $(EV, h_f) = R[K, h_f, C]$ to find the aggregate value of the $k_i$-th candidate against the $c_j$-th criterion in a current time-moment $h_f \notin D$.
**Step 3.** Let us create a 3-D IFIMs $PK$ of the weight coefficients of the criteria

and $B = R^T \odot_{(\circ, *)} PK = \begin{array}{c|c} h_f & v_e \\ \hline k_1 & \langle 0.8, 0.0002\rangle \\ k_2 & \langle 0.852, 0\rangle \\ k_3 & \langle 0.846, 0\rangle \\ k_4 & \langle 0.89, 0\rangle \end{array}$ .

**Step 4.** $k_4$ is the optimal service outsoursing provider with the maximum degree of acceptance $\mu = 0.89$ and the minimum degree of rejection $\nu = 0$.
**Step 5.** After application of the ICrA with $\alpha = 0.85$ and $\beta = 0.10$ over IM $R$ we determine that there are not criteria in a consonance. The evaluation system for selecting an outsourcing service provider in the refinery is optimized.

## 5    Conclusion

We presented a new approach to decision making in the process of outsoursing by integrating incomplete information from the experts, based on the concepts of IFSs and IMs. This method is used on real life data, associated with a refinery. The outlined approach for selection the most suitable outsourcing service provider can be applied to both problems with clear and imprecise parameters, and can be extended in order to obtain the optimal solution for other types of multidimensional outsoursing problems by $n$-dimensional index matrices [6].

# References

1. Araz, C., Ozfirat, P., Ozkarahan, I.: An integrated multicriteria decision-making methodology for outsourcing management. Comput. Oper. Res. **34**, 3738–3756 (2007)
2. Atanassov, K.T.: Intuitionistic fuzzy sets. VII ITKR Session, Sofia, 20–23 June 1983 (Deposed in Centr. Sci.-Techn. Library of the Bulg. Acad. of Sci., 1697/84)
3. Atanassov, K.: Generalized index matrices. Comptes rendus de l'Academie Bulgare des Sciences **40**(11), 15–18 (1987)
4. Atanassov, K.: On Intuitionistic Fuzzy Sets Theory. STUDFUZZ, vol. 283. Springer, Heidelberg (2012). https://doi.org/10.1007/978-3-642-29127-2
5. Index Matrices: Towards an Augmented Matrix Calculus. SCI, vol. 573. Springer, Cham (2014). https://doi.org/10.1007/978-3-319-10945-9
6. Atanassov, K.: n-Dimensional extended index matrices Part 1. Adv. Stud. Contemp. Math. **28**(2), 245–259 (2018)
7. Atanassova, V., Roeva, O.: Computational complexity and influence of numerical precision on the results of intercriteria analysis in the decision making process. Notes Intuitionistic Fuzzy Sets **24**(3), 53–63 (2018)
8. Atanassov, K., Mavrov, D., Atanassova, V.: Intercriteria decision making: a new approach for multicriteria decision making, based on index matrices and intuitionistic fuzzy sets. In: Issues in IF Sets and Generalized Nets, vol. 11, pp. 1–8 (2014)
9. Atanassov, K., Szmidt, E., Kacprzyk, J.: On intuitionistic fuzzy pairs. Notes Intuitionistic Fuzzy Sets **19**(3), 1–13 (2013)
10. Atanassov, K., Szmidt, E., Kacprzyk, J., Atanassova, V.: An approach to a constructive simplification of multiagent multicriteria decision making problems via ICrA. Comptes rendus de lAcademie bulgare des Sciences **70**(8), 1147–1156 (2017)
11. Bottani, E., Rizzi, A.: A fuzzy TOPSIS methodology to support outsourcing of logistics services. Supply Chain Manage. Int. J. **11**, 294–308 (2006)
12. Handley, S.M.: The evaluation. Analysis and management of the business outsourcing process. Ph.D. thesis, Graduate School of The Ohio State University (2008)
13. Hsu, C., Liou, J., Chuang, Y.: Integrating DANP and modified grey relation theory for the selection of an outsourcing provider. Expert Syst. Appl. **40**, 2297–2304 (2013)
14. Kahraman, C., Oztaysi, B., Cevik Onar, S.: A multicriteria supplier selection model using hesitant fuzzy linguistic term sets. J. Multiple-Val. Logic Soft Comp. **26**, 315–333 (2016)
15. Kahraman, C., Öztayşi, B., Çevik Onar, S.: An integrated intuitionistic fuzzy AHP and TOPSIS approach to evaluation of outsource manufacturers. J. Intell. Syst. **29**(1), 283–297 (2020). https://doi.org/10.1515/jisys-2017-0363
16. Liu, H., Wang, W.: An integrated fuzzy approach for provider evaluation and selection in third-party logistics. Expert Syst. Appl. **36**, 4387–4398 (2009)
17. Ljubojević, S., Pamučar, D., Jovanović, D., Vešović, V.: Outsourcing transport service: a fuzzy multi-criteria methodology for provider selection based on comparison of the real and ideal parameters of providers. Oper. Res. **19**(2), 399–433 (2017). https://doi.org/10.1007/s12351-017-0293-x
18. Min, H., DeMond, S., Joo, J.: Evaluating the comparative managerial efficiency of leading third party logistics providers in North America. Benchmarking Int. J. **20**, 62–78 (2013)
19. Modak, M., Pathak, K., Ghosh, K.: Performance evaluation of outsourcing decision using a BSC and fuzzy AHP approach: a case of the Indian coal mining organization. Resour. Policy **52**, 181–191 (2017)

20. Szmidt, E., Kacprzyk, J.: Amount of information and its reliability in the ranking of Atanassov intuitionistic fuzzy alternatives. In: Rakus-Andersson, E. (eds.) Recent Advances in Decision Making, SCI, vol. 222, pp. 7–19. Springer, Heidelberg (2009). https://doi.org/10.1007/978-3-642-02187-9_2

21. Tjader, Y., May, J., Shang, J., Vargas, L., Gao, N.: Firm-level outsourcing decision making: a balanced scorecard-based analytic network process model. Int. J. Prod. Econ. **147**, 614–623 (2014)

22. Traneva, V., Tranev, S.: Index Matrices as a Tool for Managerial Decision Making. Publ. House of the Union of Scientists, Bulgaria (2017). (in Bulgarian)

23. Traneva, V., Tranev, S., Stoenchev, M., Atanassov, K.: Scaled aggregation operations over two- and three-dimensional index matrices. Soft. Comput. **22**(15), 5115–5120 (2018). https://doi.org/10.1007/s00500-018-3315-6

24. Wan, S.-P., Wang, F., Lin, L.-L., Dong, J.-Y.: An IF linear programming method for logistics outsourcing provider selection. Knowl.-Based Syst. **82**, 80–94 (2015)

25. Wang, J., Yang, D.: Using a hybrid multi-criteria decision aid method for information system outsourcing. Compute. Oper. Res. **34**, 3691–3700 (2007)

26. Wang, J., Wang, J., Zhang, H.: A likelihood-based TODIM approach based on multi-hesitant fuzzy linguistic information for evaluation in logistics outsourcing. Comput. Industr. Eng. **99**, 287–299 (2016)

27. Yager, R.: Non-numeric multi-criteria multi-person decision making. Int. J. Group Decis. Making Negot. **2**, 81–93 (1993)

28. Zadeh, L.: Fuzzy sets. Inf. Control **8**(3), 338–353 (1965)

# Quantitative Relationship Between Particulate Matter and Morbidity

Petar Zhivkov[1]([⊠])(iD) and Alexandar Simidchiev[2]([⊠])(iD)

[1] Institute of Information and Communication Technologies - Bulgarian Academy of Sciences (IICT-BAS), acad. Georgi Bonchev Street Bl. 2, 1113 Sofia, Bulgaria
pzhivkov@iit.bas.bg
[2] Medical Institute of Ministry of Interior, Gen Skobelev blvd. 79, 1606 Sofia, Bulgaria
alex@simidchiev.net

**Abstract.** Air pollution is a major environmental health problem affecting everyone. According to the World Health Organization (WHO), there is a close relationship between small particles (PM10 and PM2.5) and increased morbidity or mortality, both daily and over time. We investigated this quantitative relationship in Sofia by comparing levels of particulate matter with a baseline number of hospital, emergency department visits, asthma prevalence, and other morbidity outcomes from 4 local health sources. The methods for this comparison model are linear correlation and non-parametric correlation analysis of a time series study conducted in Sofia from 1 January 2017 to 31 May 2019. We introduce in this study an optimized spatial and time coverage of air quality by including data from a network of citizen stations. These benefits are weighed against limitations, such as model performance, the precision of the data in days with high humidity, and the appropriateness of which will depend on epidemiological study design. The final results that will be presented can be used for optimizing healthcare and pharmaceutical planning by justifying what acute morbidities are mostly affected by higher concentrations of PM10 and PM2.5.

**Keywords:** Morbidity · Particulate matter · Air pollution

## 1 Introduction

The human health effects of exposure to outdoor air pollutants is considered a global health concern [1]. The links between urban Air Pollution (AP) and human health are consistently and clearly established by many researchers [2,3], especially for short-term effects such as cardiovascular events [4], neurovascular [5,6] and asthma [7]. The cost of human lives is significant as a recent refined modelling suggests that there are nearly 9 million deaths per year from AP [8]. Around 25% of premature deaths associated with AP are respiratory by nature [9]. Significant literature of epidemiologic studies suggests a correlation between

© Springer Nature Switzerland AG 2022
I. Lirkov and S. Margenov (Eds.): LSSC 2021, LNCS 13127, pp. 275–283, 2022.
https://doi.org/10.1007/978-3-030-97549-4_32

acute morbidity and exposure to air pollution from Particulate Matter (PM) [10]. Most of these data comes from time series analyses [11] comparing the variations in hospitalization with the average particulate matter variations [12]. PM is responsible for airway inflammations [13] and given that respiratory system is a common portal for entry, minimizing airway exposure also minimizes the cardiovascular effects [14–16]. Also, multicity studies exist such as the European Air Pollution and Health: a European Approach (APHEA) project [17] and the American National Morbidity, Mortality, and Air Pollution Study (NMMAPS) [18] that both provide a consistent evidence for association health and air pollutants for multiple cities by covering a large geographic area. The consequences of air pollution can be seen as an enhancement to a person's risk of illness or injury or as an additional general well-being risk acquired by a population [19]. The aim of air quality management is to control or avoid adverse impacts on air pollution to public health. Therefore, it is important to define such effects that are deemed "adverse," and to distinguish them from those effects that are not deemed adverse, thus concentrating protection efforts on the contaminants that cause the most extreme health impacts.

## 1.1   Air Quality Control - Difference in Norms and Legislation

The WHO guidelines for air pollution can be looked more like recommendations, they do not have a mandatory character. While the EU Air Quality Directive can be looked as a legislation that every EU member has to follow. The most noticeable difference is that there is no limit for daily limits for PM2.5 in the EU Air Quality directive. For the purpose of this research we will use the WHO guidelines as they are advisable worldwide and having purely health aspects in the consideration, while the EU Directive includes some also politico-economic aspects in recital.

## 1.2   Case Study

Sofia is the only European capital situated in a valley and is characterized by high quantity of anthropogenic emissions and by frequent occurrence of stagnant meteorological conditions. The city has a population of 1.2 million people [19] and is situated in the Sofia valley. The area is recognized as a problematic location where especially during winter there are numerous exceedances despite the European legislation that aims at air pollution control. The combination of cold winters and its situation predisposes to temperature inversions that last several days to a week. Many urban areas located in a valley without adequate air exchange encounter significant air pollution problems that are linked with the local atmospheric peculiarities [20]. A strong inversion and light precipitation and/or wind were the major causes for trapping pollutants in the air mainly during winter time. The air quality in Sofia deteriorates significantly during winter compared to summer. Hence it will be a good model to test our hypothesis of a significant link between air pollution and health consequences. In addition, due to burning fossil fuels for household heating, mainly wood, charcoal but also sometimes rubbish, the levels of PM rise significantly. The literature abounds with models that can be used to assess the pollution/health hypothesis. A correlation between temperature and mortality around Sofia is made by using models with linear and non-linear terms [21]. Other research that examines the genotoxicity of ambient air in 3 European cities, including Sofia shows that winter air pollution is six- to 10-fold higher in comparison with summer air [22].

## 2    Aim and Methods

This will be the first such study with real data from official and civil sources of information on particulate air pollution, comprehensive data from the activities of the center for emergency medical care, data from two of the largest hospitals related to the access to emergency care in Sofia. The results can be compared and contrasted with other international studies with local data. The aim of this study we formulated as follows.

### 2.1    Aim

To find which and with how much acute morbidity is increasing in days that PM levels do not meet the WHO guidelines using time series analysis.

The aim can be further reframed in the following four tasks:

(1) Compare data about air quality with health data originating from hospitals and ambulance services.
(2) Analyze the comparison and highlight the key problem areas.
(3) Test the hypothesis that data from low-cost sensors can be useful in such epidemiological research.
(4) Identify future prospects and summarize the key areas where further research is needed to improve model performance.

### 2.2    Data and Modeling

To complete the task of this research, firstly we will discover which days within the observed period concentrations of particulate matter are exceeding the WHO limits. Thereafter we will compare the health data for the days that air quality is within the WHO guidelines with the days that are not.

**Air Pollution Data.** Air quality data in a form of hourly average concentrations of PM2.5 and PM10 is obtained for 2018 and 2019 as independent series for each element from official sensors provided from (a) the Executive Environment Agency (EEA) and (b) low-cost stations from the air.bg network (a contributors driven global sensor network that creates Open Environmental Data).

PM10 is particulate matter 10 micrometers or less in diameter, PM2.5 is particulate matter $2.5\,\mu m$ or less in diameter. The data is collected from 5 official monitoring sites (Druzhba, Nadezhda, Hipodruma, Pavlovo, and Mladost). Beta attenuation method is used measuring PM10 and PM2.5 according to the European Directive 2008/50/EC (Directive, 2008). The use of air quality data from low-cost stations aims to increase the potential benefits from traditional monitoring networks with additional geographic and temporal measurement resolution [23]. Furthermore, low-cost stations can contribute to data where limited air quality stations exist [24]. Low-cost sensors were considered fit for many specific purposes including expand the conversations with communities [25]. In Sofia, at the moment of the research there are with additional geographic and temporal measurement resolution, adding over 300 citizen science (low-cost) monitors that can deliver at-the-moment information.

**Health Data.** Due to the voluntary nature of data provision, there are different time characteristics and formats from different sources:

- Summarized data on the activity of the Center for Emergency Medical Aid Sofia by diagnoses of the center—from 01.01.2017 to 14.03.2019
- Information on the activity of diagnoses from DCC Tokuda—02.01.2018 to 31.12.2018
- Information on the activity of diagnoses from Pirogov outpatient clinic— 01/01/2018 to 31/05/2019
- Information on the activity of hospitalized diagnoses from Pirogov— 01/01/2018 to 31/12/2018

The research uses the International Classification of Diseases (ICD) and more specifically ICD-10 for segmentation of the diseases and the identified morbidity. ICD-10 is the 10th revision of the ICD, a medical classification list by the World Health Organization (WHO) that is used by the time this study is conducted. It contains codes for diseases, signs and symptoms, abnormal findings, complaints, social circumstances, and external causes of injury or diseases [26,27].

## 2.3   Methods

For this research we use a time series analysis with correlation methods for analyzing the air quality and health data. The statistical methods in the research fall in the two categorizations: parametric and nonparametric. Parametric comparisons are based on the premise that the variable is continuous and normally distributed. Nonparametric approaches are used where data is continuous with non-normal distribution or any other form of data other than continuous variables. The parametric Pearson correlation test (1) is used for comparing the two sources of air quality data. It provides a measure of the linear association between the two continuous variables (usually just referred to as correlation coefficient). Correlation coefficients for each (x, y) pair are determined to carry out the evaluation, and the values of x and y are consequently replaced by their ranks. Applying the test findings to a coefficient of correlation ranging from -1 to 1.

$$r = \frac{\sum_{i=1}^{n}(x_i - \bar{x})(y_i - \bar{y})}{\sqrt{\sum_{i=1}^{n}(x_i - \bar{x})^2}\sqrt{\sum_{i=1}^{n}(y_i - \bar{y})^2}} \tag{1}$$

Parametric methods are better ways to measure the difference between the groups relative to their equivalent nonparametric methods, but due to certain strict criteria, including data normality and sample size, we cannot use parametric tests in any situation and instead use their alternate nonparametric methods:

- For correlation analysis between pairs of variables is also used the non-parametric method Spearman's rho.
- Intergroup comparison for two unpaired groups is used the non-parametric Mann - Whitney U test between means.

**Model Design.** We develop an air quality and health indicator coupling model based on data about particulate matter and registrations in hospitals and EMA. As exposure requires time to develop some health effects [28], we introduce a lag-shift analysis of 1, 2, and 3 days. In this way the research will show not only which hospital admissions and EMA entries change most due to air quality, but we can see how they develop in time. In Fig. 1 the exposure-effect method is illustrated where dose is the exposure multiplied the time. Some health effects need more time to show symptoms than others.

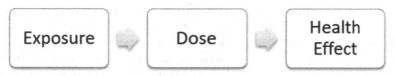

**Fig. 1.** Air quality and health indicator coupling model based on the exposure-effect method

## 3 Results

### 3.1 Aggregation of Air Quality Data

A recent study suggests that low-cost sensors can offer coarse details on the air quality detected, but are not ready for applications involving high precision [23]. We decided to test the hypothesis that low-cost air quality sensors can be a good supplement to official air quality stations from EEA and see if we clean and aggregate the data we can have a correlation that will be useful for considering some coarse details. In the comparison we have cleaned the data where humidity has been above 70%. These conditions do not necessarily mean that the data is incorrect, but under such conditions the manufacturer does not guarantee the predicted accuracy of 10% [29].

### 3.2 Accute Morbidity

In the data of hospitalized patients with more serious condition, we observe the following correlations (Tables 1 and 2). An increase (relative to background levels) of respiratory and thoracic diseases from the 1st to the 3rd day after excess of PM10 by 120%, as well as heart failure in the same time period by 19%. Excess PM2.5 was associated with a 59% increase in pulmonary embolism rate on days 2 and 3.

The results show an increase in EMA contacts with 11% on days with exceedances of PM10 and 13.5% on days with exceedances of PM2.5, for a period of at least 3 days. Mean increase (relative to background levels) in neuroses on day 2 after PM10 exceedance by 1%, on heart attacks on day 3 after excess by 8%, on strokes immediately after excess by 9%, lasting up to day 3 after excess. The same applies to hypertensive conditions, which increase by 5% immediately around the exceedance and until the third day following the event.

**Table 1.** Comparison of non-parametric data from hospitalized patients in serious conditions with WHO norms for average daily concentration of PM10.

| Morbidity | ICD-10 | Without LAG | | LAG 1 day | | LAG 2 days | | LAG 3 days | |
|---|---|---|---|---|---|---|---|---|---|
| | | Z | p | Z | p | Z | p | Z | p |
| Respiratory system | C30-C39 | −1,504 | 0,133 | −2,211 | 0,027 | −2,870 | 0,004 | −2,862 | 0,004 |
| Heart failure | I50 | −1,729 | 0,084 | −3,656 | 0,001 | −3,475 | 0,001 | −2,821 | 0,005 |

**Table 2.** Comparison of non-parametric data from hospitalized patients in serious conditions with WHO norms for average daily concentration of PM2.5.

| Morbidity | ICD-10 | Without LAG | | LAG 1 day | | LAG 2 days | | LAG 3 days | |
|---|---|---|---|---|---|---|---|---|---|
| | | Z | p | Z | p | Z | p | Z | p |
| Respiratory system | C30-C39 | −0,361 | 0,718 | −1,270 | 0,204 | −1,265 | 0,206 | −2,119 | 0,034 |
| Pulmonary embolism | I26 | −1,601 | 0,109 | −2,302 | 0,021 | −2,894 | 0,004 | −1,852 | 0,064 |
| Heart failure | I50 | −1,586 | 0,113 | −2,831 | 0,005 | −3,195 | 0,001 | −2,799 | 0,005 |

Further, acute infections of the upper respiratory tract increase by 47% and pneumonia by 60%, respectively. Chronic Obstructive Pulmonary Disease (COPD) after a day is 36% more. Regarding asthma, the allergic asthma records raise more in days with increased pollution in comparison to non-allergic asthma.

**Conclusive Comments.** In conclusion, acute morbidity increased in the days where PM concentrations did not meet WHO Air Quality Guidelines target.

**Key Findings.** While the respiratory illnesses mediated a higher and more rapid uptake when air quality goes above healthy limits, other connected to air pollution diseases as cardiovascular and neurovascular also increased.

**Implications.** This novelty study will be useful to medical physicists, environmental scientists, and policy makers. Beside improving air quality, policy makers should focus on mechanisms of informing citizens for upcoming days with high air pollution and keeping a medical surveillance of people at risk groups.

## 4    Conclusion and Future Research

The exposure-effect method is complicated, and precise measurements are very difficult to achieve even in an exposure chamber. However, it becomes apparent with the methods of epidemiological research and population statistics that air pollution contributes to expected changes in certain disease indicators. In conclusion, similar to data from published studies in cities in Europe, America and Asia, increased levels of air pollution are associated with higher levels of diagnosing health consequences.

**Future Research.** Further understanding of this morbidity-particulate matter relationship will be important to exploit this study into a novel risk and recommendation delivery system for these specific illnesses and to improve public health. With few exceptions, short-term health effects are measured by using averaged citywide air pollution concentrations for exposure indicators. This can lead to exposure misclassification and thus to bias [33]. Measurements of the relations between health effects and air pollution can be improved with more geographically precise exposure measurements and a method to calculate personal exposure during the day, and compare it to the health condition of each individual.

**Acknowledgements.** The authors would like to thank the anonymous reviewers for their helpful comments, and also the hospitals Pirogov, Tokuda, the Emergency Medical Aid Sofia, Faculty of Public Health at MU Sofia, Sofia Municipality, Air for Health, Air Solutions, and air.bg for the cooperating with data for this research. In addition, the project Grant No BG05M2OP001-1.001-0003, financed by the Science and Education for Smart Growth Operational Program and co-financed by the European Union through the European structural and Investment funds.

# References

1. Thurston, G.D., et al.: A joint ERS/ATS policy statement: what constitutes an adverse health effect of air pollution? An analytical framework. Eur. Respiratory J. **49**(1) (2017)
2. Samet, J., Krewski, D.: Health effects associated with exposure to ambient air pollution. J. Toxicol. Environ. Health Part A **70**(3–4), 227–242 (2007)
3. Li, P., et al.: The acute effects of fine particles on respiratory mortality and morbidity in Beijing, 2004–2009. Environ. Sci. Pollut. Res. **20**(9), 6433–6444 (2013)
4. Vermylen, J., et al.: Ambient air pollution and acute myocardial infarction. J. Thrombosis Haemostasis **3**(9), 1955–1961 (2005)
5. Tallon, L.A., et al.: Cognitive impacts of ambient air pollution in the National Social Health and Aging Project (NSHAP) cohort. Environ. Int. **104**, 102–109 (2017)
6. Cherrie, J.W., et al.: Effectiveness of face masks used to protect Beijing residents against particulate air pollution. Occupat. Environ. Med. **75**(6), 446–452 (2018)
7. Traboulsi, H., et al.: Inhaled pollutants: the molecular scene behind respiratory and systemic diseases associated with ultrafine particulate matter. Int. J. Molecular Sci. **18**(2), 243 (2017)
8. Pope, C.A., Dockery, D.W., Schwartz, J.: Review of epidemiological evidence of health effects of particulate air pollution. Inhalation Toxicol. **7**(1), 1–18 (1995)

9. Bell, M.L., Samet, J.M., Dominici, F.: Time-series studies of particulate matter. Annu. Rev. Public Health **25**, 247–280 (2004)
10. Dominici, F., Sheppard, L., Clyde, M.: Health effects of air pollution: a statistical review. Int. Statist. Rev. **71**(2), 243–276 (2003)
11. Guan, T., et al.: The effects of facemasks on airway inflammation and endothelial dysfunction in healthy young adults: a double-blind, randomized, controlled crossover study. Particle Fibre Toxicol. **15**(1), 30 (2018)
12. Hedley, A.J., et al.: Cardiorespiratory and all-cause mortality after restrictions on Sulphur content of fuel in Hong Kong: an intervention study. Lancet **360**(9346), 1646–1652 (2002)
13. Clancy, L., et al.: Effect of air-pollution control on death rates in Dublin, Ireland: an intervention study. Lancet **360**(9341), 1210–1214 (2002)
14. Shi, J., et al.: Cardiovascular benefits of wearing particulate-filtering respirators: a randomized crossover trial. Environ. Health Perspect. **125**(2), 175–180 (2017)
15. Katsouyanni, K., et al.: Short term effects of air pollution on health: a European approach using epidemiologic time series data: the APHEA protocol. J. Epidemiol. Community Health **50**(Suppl 1), S12–S18 (1996)
16. Samet, J.M., et al.: The national morbidity, mortality, and air pollution study. Part II: Morbidity Mortality Air Pollution United States Res. Rep. Health Eff. Inst .**94**(pt 2), 5–79 (2000)
17. Gochfeld, M., Burger, J.: Disproportionate exposures in environmental justice and other populations: the importance of outliers. Am. J. Public Health **101**(S1), S53–S63 (2011)
18. NSI: Population by cities (2019)
19. Rendón, A.M., et al.: Effects of urbanization on the temperature inversion breakup in a mountain valley with implications for air quality. J. Appl. Meteorol. Climatol. **53**(4), 840–858 (2014)
20. Pattenden, S., Nikiforov, B., Armstrong, B.: Mortality and temperature in Sofia and London. J. Epidemiol. Community Health **57**(8), 628–633 (2003)
21. Gábelová, A., et al.: Genotoxicity of environmental air pollution in three European cities: Prague, Košice and Sofia. Mutat. Res. Genetic Toxicol. Environ. Mutagenesis **563**(1), 49–59 (2004)
22. Castell, N., et al.: Can commercial low-cost sensor platforms contribute to air quality monitoring and exposure estimates? Environ. Int. **99**, 293–302 (2017)
23. Zimmerman, N., et al.: Closing the gap on lower cost air quality monitoring: machine learning calibration models to improve low-cost sensor performance. Atmos. Meas. Tech. Discuss **2017**, 1–36 (2017)
24. Morawska, L., et al.: Applications of low-cost sensing technologies for air quality monitoring and exposure assessment: how far have they gone? Environ. Int. **116**, 286–299 (2018)
25. World Health Organization: International classification of diseases: [9th] ninth revision, basic tabulation list with alphabetic index : World Health Organization (1978)
26. World Health Organization: International Classification of Diseases (ICD) Information Sheet (2018)
27. Kampa, M., Castanas, E.: Human health effects of air pollution. Environ. Pollut. **151**(2), 362–367 (2008)
28. Jayaratne, R., et al.: The influence of humidity on the performance of a low-cost air particle mass sensor and the effect of atmospheric fog. Atmospheric Measure. Tech. **11**(8), 4883–4890 (2018)

29. Wang, W., et al.: Particulate air pollution and ischemic stroke hospitalization: how the associations vary by constituents in Shanghai, China. Sci. Total Environ. **695**, 133780 (2019)

30. Shin, S., et al.: Ambient air pollution and the risk of atrial fibrillation and stroke: a population-based cohort study. Environ. Health Perspect. **127**(8), 87009 (2019)

31. Zhang, R., et al.: Acute effects of particulate air pollution on ischemic stroke and hemorrhagic stroke mortality. Front. Neurol. **9**, 827 (2018)

32. Laurent, O., et al.: Air pollution, asthma attacks, and socioeconomic deprivation: a small-area case-crossover study. Am. J. Epidemiol. **168**(1), 58–65 (2008)

# Advanced Discretizations and Solvers for Coupled Systems of Partial Differential Equations

# Decoupling Methods for Systems of Parabolic Equations

Petr N. Vabishchevich[1,2]($\boxtimes$) [ID]

[1] Nuclear Safety Institute, Russian Academy of Sciences, Moscow, Russia
[2] Academy of Sciences of the Republic of Sakha (Yakutia), Yakutsk, Russia

**Abstract.** We consider the decoupling methods of the Cauchy problem's numerical solution for a system of multidimensional parabolic equations. Of most significant interest for computational practice is the case when the equations are related to each other. Splitting schemes are constructed for such vector problems when the transition to a new level in time is provided by solving common scalar problems for individual components of the solution. Two main classes of decoupling methods are distinguished by allocating the diagonal part of the problem's matrix operator and its lower and upper triangular parts. An increase in the approximate solution of explicit-implicit schemes is achieved by using some three-level approximations in time. Special attention is paid to when the time derivatives of the solution components are related to each other.

**Keywords:** System of parabolic equations · Decoupling methods · Explicit-implicit schemes

## 1 Introduction

Many applied problems are associated with the approximate solution of initial-boundary value problems for systems of nonstationary partial differential equations when the vector of unknowns' individual components are tied together. Component splitting schemes are used to obtain separate problems for finding components on a new time level.

Numerical methods for solving the corresponding nonstationary vector boundary value problems are based on approximations in time used for scalar equations. In the approximate solution of problems for first-order evolution equations, two- and three-level time approximations are used. Stability conditions are given by the theory of stability (correctness) of operator-difference schemes, and they are formulated as operator inequalities in finite-dimensional Hilbert spaces [4,5].

When studying systems of equations, special attention should be paid to particular explicit-implicit approximations [1], which provide a transition to a new level in time by solving a sequence of separate simple (mono physical) problems. Here we consider such splitting schemes [2,6] for the approximate solution of parabolic systems of equations. They are formulated as vector evolution equations on the direct sum of Hilbert spaces. Two- and three-level schemes are

© Springer Nature Switzerland AG 2022
I. Lirkov and S. Margenov (Eds.): LSSC 2021, LNCS 13127, pp. 287–294, 2022.
https://doi.org/10.1007/978-3-030-97549-4_33

considered for the additive representation of the problem operator with selecting the diagonal part of the operator. The second class of methods is based on a triangular two-component representation of the problem operator. Similar splitting schemes for general evolutionary equations were proposed in the work [9]. More general problems are characterized by the connection with each other of the solution components' derivatives in time.

## 2    Formulation of the Problem

Let $H_\alpha$, $\alpha = 1, 2, \ldots, m$ be finite-dimensional real Hilbert (Euclidean) spaces in which the scalar product and norm are $(\cdot, \cdot)_\alpha$ and $\| \cdot \|_\alpha$, $\alpha = 1, 2, \ldots, m$ respectively. Individual components of the solution will be denoted by $u_\alpha(t)$, $\alpha = 1, 2, \ldots, m$ for each $t$ $(0 \le t \le T, T > 0)$. A solution is sought for a system of first-order evolution equations:

$$\frac{du_\alpha}{dt} + \sum_{\beta=1}^{p} A_{\alpha\beta} u_\beta = f_\alpha, \quad \alpha = 1, 2, \ldots, m. \tag{1}$$

Here $f_\alpha(t)$, $\alpha = 1, 2, \ldots, m$ are given, and $A_{\alpha\beta}$ are linear constants (independent of $t$) operators from $H_\beta$ to $H_\alpha$ $(A_{\alpha\beta} : H_\beta \to H_\alpha)$ for all $\alpha = 1, 2, \ldots, m$. The system of Eq. (1) is supplemented with the initial conditions

$$u_\alpha(0) = u_\alpha^0, \quad \alpha = 1, 2, \ldots, m. \tag{2}$$

We will interpret the system of Eq. (1) as one evolutionary equation for vector $\boldsymbol{u} = \{u_1, u_2, \ldots, u_m\}$:

$$\frac{d\boldsymbol{u}}{dt} + \boldsymbol{A}\boldsymbol{u} = \boldsymbol{f}(t), \quad 0 < t \le T, \tag{3}$$

where $\boldsymbol{f} = \{f_1, f_2, \ldots, f_m\}$, and for elements operator matrix $\boldsymbol{A}$ we have the representation

$$\boldsymbol{A} = \{A_{\alpha\beta}\}, \quad \alpha, \beta = 1, 2, \ldots, m.$$

On the direct sum of spaces $\boldsymbol{H} = H_1 \oplus H_2 \oplus \cdots \oplus H_m$ put

$$(\boldsymbol{u}, \boldsymbol{v}) = \sum_{\alpha=1}^{m} (u_\alpha, v_\alpha)_\alpha, \quad \|\boldsymbol{u}\|^2 = \sum_{\alpha=1}^{m} \|u_\alpha\|_\alpha^2.$$

Taking into account (2), we have

$$\boldsymbol{u}(0) = \boldsymbol{u}^0, \tag{4}$$

where $\boldsymbol{u}^0 = \{u_1^0, u_2^0, \ldots, u_m^0\}$.

We will consider the Cauchy problem (3), (4) provided that the operator $\boldsymbol{A}$ is self-adjoint and positive definite in $\boldsymbol{H}$:

$$\boldsymbol{A} = \boldsymbol{A}^* \ge \delta \boldsymbol{I}, \quad \delta > 0, \tag{5}$$

where $I$ is the identity operator in $H$. The self-adjoint property is associated with the fulfillment of the equalities

$$A_{\alpha\beta} = A_{\beta\alpha}^*, \quad \alpha, \beta = 1, 2, \ldots, m$$

for the operators of the original system of Eq. (1).

For $D = D^* > 0$ through $H_D$ denote the space $H$ equipped with scalar product $(y, w)_D = (Dy, w)$ and the norm $\|y\|_D = (Dy, y)^{1/2}$. The simplest a priori estimate for the solution of the Cauchy problem (3), (4), which we will be guided by when studying the corresponding operator-difference schemes, has the form

$$\|u(t)\|_A^2 \leq \|u^0\|_A^2 + \frac{1}{2} \int_0^t \|f(\theta)\|^2 d\theta. \tag{6}$$

It provides the stability of solutions of the problem (3), (4) to the initial data and the right-hand side.

We arrive at systems of evolutionary equations of type (1) when approximating in space a number of applied problems. As an example, we note a system of coupled parabolic equations of the second order, which is typical for diffusion problems in multicomponent media, when a solution to the system is sought in the domain $\Omega$

$$\frac{\partial u_\beta}{\partial t} - \sum_{\beta=1}^m \operatorname{div}(d_{\alpha\beta}(x)\operatorname{grad} u_\beta) = f_\alpha(x, t), \ x \in \Omega, \ 0 < t \leq T, \ \alpha = 1, 2, \ldots, m.$$

The coefficients $d_{\alpha\beta}$ describe diffusion processes: self-diffusion—for $\alpha = \beta$ and cross diffusion—for $\alpha \neq \beta$.

For an approximate solution of the differential-operator problem (3), (4), we will use the usual weighted schemes. When using a two-level scheme, Eq. (3) is approximated by the difference equation

$$\frac{y^{n+1} - y^n}{\tau} + A(\sigma y^{n+1} + (1 - \sigma)y^n) = \varphi^n, \tag{7}$$

where $\sigma$ is a numeric parameter (weight), which is usually $0 \leq \sigma \leq 1$ and for example $\varphi^n = f(\sigma t^{n+1} + (1 - \sigma)t^n)$. For simplicity, we restrict ourselves to the case of the same weight for all equations of the system (1). Taking into account (4), we supplement (7) with the initial condition

$$y^0 = u^0. \tag{8}$$

If $\sigma \geq 0.5$, then the operator-difference scheme (7) is unconditionally stable in $H_A$ and the difference solution satisfies the level-by-level estimate

$$\|y^{n+1}\|_A^2 \leq \|y^n\|_A^2 + \frac{\tau}{2}\|\varphi^n\|^2. \tag{9}$$

The estimate (9) acts as a grid analogue of the estimate (6) and ensures the unconditional stability of the difference scheme with weights (7), (8) under the natural constraint $\sigma \geq 0.5$. Considering the corresponding problem for the error, we make sure that the solution of the operator-difference problem (7), (8) converges to the solution of differential-difference problem (3), (4) in $H_A$ for $\sigma \geq 0.5$ with $\mathcal{O}((2\sigma - 1)\tau + \tau^2)$. For $\sigma = 0.5$, we have the second order of convergence in $\tau$.

The transition to a new time level in the scheme (7) is associated with solving the problem

$$(I + \sigma\tau A)y^{n+1} = \psi^n.$$

In relation to the original problem (1), (2), it is necessary to solve the system of coupled equations

$$y_\alpha^{n+1} + \sum_{\beta=1}^m \sigma\tau A_{\alpha\beta} y_\beta^{n+1} = \psi_\alpha^n, \quad \alpha = 1, 2, \ldots, p.$$

For an approximate solution of such a problem, one or another iterative method can be used, in particular, of the block type [3].

The second possibility more completely takes into account the specifics of the nonstationary problems under consideration and is associated with the construction of splitting schemes, when the transition to a new time level is associated with the solution of simpler problems. For problems like (1), (2), it is natural to focus on splitting schemes, when the transition to a new time level is provided by solving problems like

$$y_\alpha^{n+1} + \sigma\tau A_{\alpha\alpha} y_\alpha^{n+1} = \widetilde{\psi}_\alpha^n, \quad \alpha = 1, 2, \ldots, m.$$

This corresponds to the fact that the computational implementation is provided by inverting only the diagonal part of the operator matrix $I + \sigma\tau A$.

## 3    Component Splitting Schemes

Let us recall the stability conditions for the two-level scheme for the problem (3), (4), which we write in the form

$$B\frac{y^{n+1} - y^n}{\tau} + Ay^n = \varphi^n. \tag{10}$$

The main result [4,5] is formulated as follows.

**Lemma 1.** *Let $A$ in (10) be a self-adjoint positive operator, and the operator $B$ satisfy the condition*

$$B \geq I + \frac{\tau}{2}A. \tag{11}$$

*Then the scheme (8), (10) is stable in $H_A$ and the a priori estimate (9) is true.*

It is most natural to construct schemes of component-wise splitting on the basis of moving the diagonal part of the operator matrix $\boldsymbol{A}$ to the upper level in time. We use the notation

$$\boldsymbol{D} = \mathrm{diag}\{A_{11}, A_{22}, \ldots, A_{mm}\}.$$

We assume that there exists $\gamma > 0$ such that

$$\boldsymbol{A} \leq \gamma \boldsymbol{D}. \tag{12}$$

An approximate solution is found from the equation

$$\frac{y^{n+1} - y^n}{\tau} + \boldsymbol{D}(\sigma y^{n+1} + (1 - \sigma)y^n) + (\boldsymbol{A} - \boldsymbol{D})y^n = \varphi^n. \tag{13}$$

**Theorem 1.** *Explicit-implicit scheme (8), (13) under the inequality (12) is stable in $\boldsymbol{H}_{\boldsymbol{A}}$ for*

$$2\sigma \geq \gamma, \tag{14}$$

*wherein the a priori estimate (9) is correct.*

*Proof.* It is enough to check the inequality (11). In case (12) for scheme (13) we have

$$\boldsymbol{B} = \boldsymbol{I} + \sigma \tau \boldsymbol{D} \geq \boldsymbol{I} + \frac{1}{\gamma} \sigma \tau \boldsymbol{A}.$$

When choosing the weight parameter $\sigma$ taking into account (12), inequality (11) obviously holds.                                                                                          □

The second class of component-wise splitting schemes is based on the triangular splitting of the operator matrix $\boldsymbol{A}$:

$$\boldsymbol{A} = \boldsymbol{A}_1 + \boldsymbol{A}_2, \quad \boldsymbol{A}_1^* = \boldsymbol{A}_2. \tag{15}$$

Additive presentation (15) corresponds to the choice

$$\boldsymbol{A}_1 = \begin{pmatrix} \frac{1}{2}A_{11} & 0 & \cdots & 0 \\ A_{21} & \frac{1}{2}A_{22} & \cdots & 0 \\ \cdots & \cdots & \cdots & 0 \\ A_{m1} & A_{m2} & \cdots & \frac{1}{2}A_{mm} \end{pmatrix}, \quad \boldsymbol{A}_2 = \begin{pmatrix} \frac{1}{2}A_{11} & A_{12} & \cdots & A_{1m} \\ 0 & \frac{1}{2}A_{22} & \cdots & A_{2m} \\ \cdots & \cdots & \cdots & A_{m-1\,m} \\ 0 & 0 & \cdots & \frac{1}{2}A_{mm} \end{pmatrix}.$$

Instead of (13) we will use the scheme (10), in which

$$\boldsymbol{B} = \boldsymbol{I} + \sigma \tau \boldsymbol{A}_1. \tag{16}$$

An increase in the accuracy of the approximate solution with the choice of $\sigma = 0.5$ is achieved in schemes of the alternating triangular method [4,6], when the operator $\boldsymbol{B}$ is represented in the following factorized form

$$\boldsymbol{B} = (\boldsymbol{I} + \sigma \tau \boldsymbol{A}_1)(\boldsymbol{I} + \sigma \tau \boldsymbol{A}_2). \tag{17}$$

The following statement is proved similarly to Theorem 1.

**Theorem 2.** *With $\sigma \geq 1$ the scheme (8), (10), (15), (16) is unconditionally stable in $H_A$, and the factorized scheme (8), (10), (15), (17)—for $\sigma \geq 0.5$. The difference solution satisfies the level-by-level estimate (9).*

*Proof.* In the case (15), (16) we have

$$(By, y) = (y, y) + \frac{1}{2}\sigma\tau(Ay, y)$$

and inequality (11) holds for $\sigma \geq 1$. For (15), (17)

$$B = I + \sigma\tau A + \sigma^2\tau^2 A_1 A_2, \quad B = B^* \geq I + \sigma\tau A.$$

Thus, the inequality (11) holds for $\sigma \geq 0.5$. Taking this into account, the required statement follows from Lemma 1.                                                                    $\square$

The computational realization of schemes based on triangular splitting (15) is associated with the sequential determination of individual components of the approximate solution. For example, when using the factorized scheme (8), (10), (15), (16) an approximate solution to the problem (1), (2) can be carried out on the basis of a sequential solution of the following simpler problems:

$$y_\alpha^{n+1/2} + \frac{1}{2}\sigma\tau A_{\alpha\alpha}y_\alpha^{n+1/2} = \check{\psi}_\alpha^n,$$

$$y_\alpha^{n+1} + \frac{1}{2}\sigma\tau A_{\alpha\alpha}y_\alpha^{n+1} = \hat{\psi}_\alpha^n, \quad \alpha = 1, 2, \ldots, m.$$

Like the weighted scheme (7), (8), the factorized scheme has the second order of convergence for $\sigma = 0.5$ and the first for other values of the weight.

## 4    Three-Level Schemes of Component Splitting

An increase in the accuracy of component-wise splitting schemes for vector evolutionary equations can be provided using three-level schemes. The three-level scheme for the approximate solution of the problem (3), (4), can be written in the form

$$B\frac{y^{n+1} - y^{n-1}}{2\tau} + R(y^{n+1} - 2y^n + y^{n-1}) + Ay^n = \varphi^n, \tag{18}$$

$$n = 1, 2, \ldots, N - 1,$$

for given

$$y^0 = u^0, \quad y^1 = v^0. \tag{19}$$

Let us formulate the stability conditions for scheme (18) in the form of the following statement [4, 5].

**Lemma 2.** *Let in (18) the operators $\boldsymbol{R}$ and $\boldsymbol{A}$ be constant self-adjoint operators. When conditions*

$$\boldsymbol{B} \geq \boldsymbol{I}, \quad \boldsymbol{A} > 0, \quad \boldsymbol{R} > \frac{1}{4}\boldsymbol{A} \tag{20}$$

*are true, the scheme (18), (19) is stable and the a priori estimate*

$$\mathcal{E}^{n+1} \leq \mathcal{E}^1 + \frac{1}{2}\sum_{k=1}^{n}\tau\|\varphi^k\|^2 \tag{21}$$

*holds, in which*

$$\mathcal{E}^{n+1} = \frac{1}{4}(\boldsymbol{A}(\boldsymbol{y}^{n+1} + \boldsymbol{y}^n), \boldsymbol{y}^{n+1} + \boldsymbol{y}^n)$$
$$+ (\boldsymbol{R}(\boldsymbol{y}^{n+1} - \boldsymbol{y}^n), \boldsymbol{y}^{n+1} - \boldsymbol{y}^n) - \frac{1}{4}(\boldsymbol{A}(\boldsymbol{y}^{n+1} - \boldsymbol{y}^n), \boldsymbol{y}^{n+1} - \boldsymbol{y}^n).$$

When performing (12) similarly to (13), we will use an explicit-implicit three-level scheme of the second order of accuracy

$$\frac{\boldsymbol{y}^{n+1} - \boldsymbol{y}^{n-1}}{2\tau} + \boldsymbol{D}(\sigma\boldsymbol{y}^{n+1} + (1 - 2\sigma)\boldsymbol{y}^n + \sigma\boldsymbol{y}^{n-1}) + (\boldsymbol{A} - \boldsymbol{D})\boldsymbol{y}^n = \varphi^n. \tag{22}$$

**Theorem 3.** *Explicit-implicit three-level scheme (19), (22) with constraints (12) is stable for*

$$4\sigma > \gamma, \tag{23}$$

*moreover, the a priori estimate (21) is true.*

*Proof.* Stability is established based on lemma 2. In case (22) we have

$$\boldsymbol{B} = \boldsymbol{I}, \quad \boldsymbol{R} = \sigma\boldsymbol{D}$$

in (18). Taking into account (12), the last inequality (20) will be satisfied under the constraints (23) on the weight $\sigma$. Thus, the conditions of the lemma 2 are satisfied.    □

We also note the possibility of increasing the accuracy of the alternating-triangular method schemes due to the transition from a two-level scheme to a three-level one. Such variants of multilevel schemes of the alternating triangular method for scalar problems were discussed by us earlier in [7,8].

When splitting (15), instead of the scheme (10), (17) we will use

$$(\boldsymbol{I} + \sigma\tau\boldsymbol{A})\frac{\boldsymbol{y}^{n+1} - \boldsymbol{y}^n}{\tau} + \sigma^2\tau\boldsymbol{A}_1\boldsymbol{A}_2(\boldsymbol{y}^{n+1} - 2\boldsymbol{y}^n + \boldsymbol{y}^{n-1}) + \boldsymbol{A}\boldsymbol{y}^n = \varphi^n. \tag{24}$$

It corresponds to the regularization of a two-level scheme with weights due to an additional term proportional to $\boldsymbol{A}_1\boldsymbol{A}_2$. The computational implementation of (24) is similar to the scheme (10), (17).

Taking into account that

$$\frac{y^{n+1} - y^n}{\tau} = \frac{y^{n+1} - y^{n-1}}{2\tau} + \frac{y^{n+1} - 2y^n + y^{n-1}}{2\tau},$$

we write the scheme (24) in the form (18) for

$$B = I + \sigma\tau A, \quad R = \frac{1}{2\tau}(I + \sigma\tau A) + \sigma^2\tau A_1 A_2.$$

For $\sigma \geq 0.5$, conditions (20) are satisfied. Lemma 2 allows us to state the following statement.

**Theorem 4.** *For $\sigma \geq 0.5$, the scheme (19), (24) is unconditionally stable, and the a priori estimate (21) is valid for the approximate solution.*

# References

1. Hundsdorfer, W.H., Verwer, J.G.: Numerical Solution of Time-Dependent Advection-Diffusion-Reaction Equations. Springer, Cham (2003). https://doi.org/10.1007/978-3-662-09017-6
2. Marchuk, G.I.: Splitting and alternating direction methods. In: Ciarlet, P.G., Lions, J.L. (eds.) Handbook of Numerical Analysis, vol. I, pp. 197–462. North-Holland (1990)
3. Saad, Y.: Iterative methods for sparse linear systems. SIAM (2003)
4. Samarskii, A.A.: The Theory of Difference Schemes. Dekker, New York (2001)
5. Samarskii, A.A., Matus, P.P., Vabishchevich, P.N.: Difference schemes with operator factors. Kluwer (2002)
6. Vabishchevich, P.N.: Additive Operator-Difference Schemes: Splitting Schemes. de Gruyter, Berlin (2013)
7. Vabishchevich, P.N.: Three level schemes of the alternating triangular method. Comput. Math. Math. Phys. **54**(6), 953–962 (2014)
8. Vabishchevich, P.N.: Explicit schemes for parabolic and hyperbolic equations. Appl. Math. Comput. **250**, 424–431 (2015)
9. Vabishchevich, P.N.: Explicit-implicit schemes for first-order evolution equations. Differential Equat. **56**(7), 882–889 (2020)

# Optimal Control of ODEs, PDEs and Applications

# Random Lifting of Set-Valued Maps

Rossana Capuani$^{(\boxtimes)}$ , Antonio Marigonda , and Marta Mogentale

Department of Computer Sciences, University of Verona,
Strada Le Grazie 15, 37134 Verona, Italy
{rossana.capuani,antonio.marigonda}@univr.it

**Abstract.** In this paper we discuss the properties of particular set-valued maps in the space of probability measures on a finite-dimensional space that are constructed by mean of a suitable *lift* of set-valued map in the underlying space. In particular, we are interested to establish under which conditions some good regularity properties of the original set-valued map are inherited by the lifted one. The main motivation for the study is represented by multi-agent systems, i.e., finite-dimensional systems where the number of (microscopic) agents is so large that only macroscopical description are actually available. The macroscopical behaviour is thus expressed by the superposition of the behaviours of the microscopic agents. Using the common description of the state of a multi-agent system by mean of a time-dependent probability measure, expressing the fraction of agents contained in a region at a given time moment, the results of this paper yield regularity results for the macroscopical behaviour of the system.

**Keywords:** Set-valued map · Multi-agent systems · Statistical description

## 1 Introduction

In the last decade, the mathematical analysis of *complex systems* attracted a renewed interest from the applied mathematics community in view of its capability to model many real-life phenomena with a good degree of accuracy. In particular, the field of application of such models ranges from social dynamics (e.g., pedestrian dynamics, social network models, opinion formation, infrastructure planning) to financial markets, from big data analysis to life sciences (e.g. flocking).

All those systems are characterized by the presence of a *large* number of individuals, called *agents*, usually moving in a finite-dimensional space $\mathbb{R}^d$ under the effect of a *global field* which is possibly affected also by the current agent configuration. In its simplest setting, each agent moves along the steepest descent direction of a functional (which is the *same* for all the agents) which can also take into account interaction effects between them. The interaction between the agents may range from the simplest, e.g., avoiding collision, or attraction/repulsion effects, to more complex ones, involving also penalization of overcrowding/dispersion, or further state constraints on the density of the agents.

I. Lirkov and S. Margenov (Eds.): LSSC 2021, LNCS 13127, pp. 297–305, 2022.
https://doi.org/10.1007/978-3-030-97549-4_34

Due to the huge number of agents, a description of the motion of each agent becomes impossible. Therefore, using the simplifying assumptions that all the agents of the collective are *indistinguishable*, only a *macroscopical* (statistical) description of the system is feasible. In this sense, at each instant $t$ of time the state of the system is described by a time-depending positive Borel measure $\mu_t$ whose meaning is the following: given a region $A \subseteq \mathbb{R}^d$, the quotient $\dfrac{\mu_t(A)}{\mu_t(\mathbb{R}^d)}$ represents the fraction of agents that at time $t$ are present in the region $A$. If we suppose that the total amount of the agents does not change in time, we can normalize the quotient by taking $\mu_t(\mathbb{R}^d) \equiv 1$, i.e., the evolution of the system can be represented by a family of probability measures indexed by the time parameter.

Under reasonable assumptions on the agents' trajectories, the family of measures describing the evolution of the system obeys to the *continuity equation*

$$\partial_t \mu_t + \operatorname{div}(v_t \mu_t) = 0, \tag{1}$$

coupled with an initial data $\mu_0$, representing the initial state of the system. The equation must be understood in the sense of distribution. The time-depending vector field $(t, x) \mapsto v_t(x)$ represents the macroscopical vector field along which the mass flows, and $v_t \mu_t$ is the flux. This leads to an absolutely continuous curve in the space of probability measures, endowed with the Wasserstein distance.

The link between the trajectories of the microscopic agents and the macroscopical evolution of the system is given by the *superposition principle* (see [1] and [7] in the context of differential inclusions): namely, every solution $t \mapsto \mu_t$ of (1) can be represented by the pushforward $e_t \sharp \eta$ of a probability measure $\eta \in \mathscr{P}(\mathbb{R}^d \times C^0([0, T]))$ concentrated on pairs $(x, \gamma)$, where $\gamma$ is any integral solution of $\dot{\gamma}(t) = v_t(\gamma(t))$ satisfying $\gamma(0) = x$, and $e_t(x, \gamma) = \gamma(t)$ is the evaluation operator. Conversely, given a Borel family of absolutely continuous curves in $[0, T]$, any probability measure $\eta$ concentrated on the pairs $(\gamma(0), \gamma)$ defines by the pushforward $\mu_t = e_t \sharp \eta$ an absolutely continuous curve $t \mapsto \mu_t$, which solves (1) for a vector field $v_t(\cdot)$ representing the weighted average of the speeds of the trajectories of the agents concurring at point $x$ at time $t$. We recall that in this case $v_t(\cdot)$ is an *average* and therefore it may happens that *no agent* is following the integral curves of the vector field, even if the macroscopical effects will be a displacement along it.

The problem discussed in this paper is the following. We suppose to have a set-valued map $S$ associating to every point $x \in \mathbb{R}^d$ a set of curves in $\mathbb{R}^d$ representing the allowed trajectories of the agent which at initial time $t = 0$ is at $x$. Our goal is to study the corresponding properties of the set-valued map describing the family of macroscopical trajectories in the space of probability measures. It turns out that this amounts to study the properties of a new set-valued map associating to each probability measure $\mu$ the set of probability measure concentrated on the graph of $S(\cdot)$ and whose first marginal is equal to $\mu$.

The main example is when the set $S(x)$ describes the trajectories of a differential inclusion with initial data $x$, and we want to derive regularity properties of the macroscopical evolutions from the properties of the set-valued map defining the differential inclusion. The regularity properties of the solution map is crucial in the study of many optimization problems and to generalize mean field models, like e.g. in [4–6].

The paper is structured as follows: in Sect. 2 we give some preliminaries and basic definitions, in Sect. 3 we prove the main results, while in Sect. 4 we explore possible extensions and further developments.

## 2  Preliminaries

Let $(X, d_X)$ be a separable metric space. We denote by $\mathscr{P}(X)$ the set of Borel probability measures on $X$ endowed with the weak* topology induced by the duality with the Banach space $C_b^0(X)$ of the real-valued continuous bounded functions on $X$ with the uniform convergence norm. For any $p \geq 1$, we set the space of Borel probability measures with finite $p$-*moment* as

$$\mathscr{P}_p(X) = \left\{ \mu \in \mathscr{P}(X) : \int_X d_X^p(x, \bar{x}) \, d\mu(x) < +\infty \text{ for some } \bar{x} \in X \right\}.$$

Given complete separable metric spaces $(X, d_X)$, $(Y, d_Y)$, for any Borel map $r : X \to Y$ and $\mu \in \mathscr{P}(X)$, we define the *push forward measure* $r\sharp\mu \in \mathscr{P}(Y)$ by setting $r\sharp\mu(B) = \mu(r^{-1}(B))$ for any Borel set $B$ of $Y$.

**Definition 1 (Transport plans and Wasserstein distance).** *Let $X$ be a complete separable metric space, $\mu_1, \mu_2 \in \mathscr{P}(X)$. We define the set of* admissible transport plans *between $\mu_1$ and $\mu_2$ by setting $\Pi(\mu_1, \mu_2) = \{\pi \in \mathscr{P}(X \times X) : \mathrm{pr}_i\sharp\pi = \mu_i, \ i = 1, 2\}$, where for $i = 1, 2$, we defined $\mathrm{pr}_i : X \times X \to X$ by $\mathrm{pr}_i(x_1, x_2) = x_i$. The inverse $\pi^{-1}$ of a transport plan $\pi \in \Pi(\mu, \nu)$ is defind by $\pi^{-1} = i\sharp\pi \in \Pi(\nu, \mu)$, where $i(x, y) = (y, x)$ for all $x, y \in X$. The $p$-Wasserstein distance between $\mu_1$ and $\mu_2$ is*

$$W_p^p(\mu_1, \mu_2) = \inf_{\pi \in \Pi(\mu_1, \mu_2)} \int_{X \times X} d_X^p(x_1, x_2) \, d\pi(x_1, x_2).$$

*If $\mu_1, \mu_2 \in \mathscr{P}_p(X)$ then the above infimum is actually a minimum, and we define*

$$\Pi_o^p(\mu_1, \mu_2) = \left\{ \pi \in \Pi(\mu_1, \mu_2) : W_p^p(\mu_1, \mu_2) = \int_{X \times X} d_X^p(x_1, x_2) \, d\pi(x_1, x_2) \right\}.$$

*The space $\mathscr{P}_p(X)$ endowed with the $W_p$-Wasserstein distance is a complete separable metric space, moreover for all $\mu \in \mathscr{P}_p(X)$ there exists a sequence $\{\mu^N\}_{N \in \mathbb{N}} \subseteq \mathrm{co}\{\delta_x : x \in \mathrm{supp}\,\mu\}$ such that $W_p(\mu^N, \mu) \to 0$ as $N \to +\infty$.*

*Remark 1.* Recalling formula (5.2.12) in [1], when $X$ is a separable Banach space we have $W_p(\delta_0, \mu) = \mathrm{m}_p^{1/p}(\mu) = \left( \int_{\mathbb{R}^d} \|x\|_X^p \, d\mu(x) \right)^{1/p}$, for all $\mu \in \mathscr{P}_p(X), p \geq 1$. In particular, if $t \mapsto \mu_t$ is $W_p$-continuous, then $t \mapsto \mathrm{m}_p^{1/p}(\mu_t)$ is continuous.

**Definition 2 (Set-valued maps).** *Let $X, Y$ be sets. A set-valued map $F$ from $X$ to $Y$ is a map associating to each $x \in X$ a (possible empty) subset $F(x)$ of $Y$. We will write $F : X \rightrightarrows Y$ to denote a set-valued map from $X$ to $Y$. The graph of a set-valued map $F$ is $\operatorname{graph} F := \{(x, y) \in X \times Y : y \in F(x)\} \subseteq X \times Y$, while the domain of $F$ is $\operatorname{dom} F := \{x \in X : F(x) \neq \emptyset\} \subseteq X$. Given $A \subseteq X$, we set $\operatorname{graph}(F_{|A}) := \operatorname{graph} F \cap (A \times Y) = \{(x, y) \in A \times Y : y \in F(x)\}$. A selection of $F$ is a map $f : \operatorname{dom} F \rightarrow Y$ such that $f(x) \in F(x)$ for all $x \in \operatorname{dom} F$.*

The following Lemma is a direct consequence of [8, Theorem 7.1].

**Lemma 1 (Borel selection of the metric projection).** *Let $X, Y$ be complete separable metric spaces. Assume that $S : X \rightrightarrows Y$ is a continuous set-valued map with compact nonempty images. Then there exists a Borel map $g : X \times Y \rightarrow Y$ such that $g(x, y) \in S(x)$ and $d_Y(y, g(x, y)) = d_{S(x)}(y)$ for all $(x, y) \in X \times Y$, i.e., for every $x \in X$ and $y \in Y$ we have that $g(x, y)$ is a metric projection of $y$ on $S(x)$.*

## 3   Results

In this section, we will introduce the notion of random lift of set-valued maps and its main properties.

**Definition 3 (Random lift of set-valued maps).** *Let $X, Y$ be complete separable metric spaces, $S : X \rightrightarrows Y$ be a set valued map. Define the set-valued map $\mathscr{P}(S) : \mathscr{P}(X) \rightrightarrows \mathscr{P}(X \times Y)$ as follows*

$$\mathscr{P}(S)(\mu) := \left\{ \boldsymbol{\eta} \in \mathscr{P}(X \times Y) : \operatorname{graph} S \supseteq \operatorname{supp} \boldsymbol{\eta}, \ and \ \operatorname{pr}^{(1)} \sharp \boldsymbol{\eta} = \mu \right\},$$

*where $\mu \in P(X)$ and $\operatorname{pr}^{(1)}(x, y) = x$ for all $(x, y) \in X \times Y$. The set-valued map $\mathscr{P}(S)(\cdot)$ will be called the random lift of $S(\cdot)$.*

*Remark 2.* Directly from the definition, we have that

1. $\mathscr{P}(S)$ has convex images (even if $S$ has not convex images): indeed for all $\boldsymbol{\eta}_i \in \mathscr{P}(S)(\mu)$, $i = 1, 2$, and $\lambda \in [0, 1]$, set $\boldsymbol{\eta}_\lambda = \lambda \boldsymbol{\eta}_1 + (1 - \lambda) \boldsymbol{\eta}_2$, and notice that $\operatorname{supp} \boldsymbol{\eta}_\lambda \subseteq \operatorname{supp} \boldsymbol{\eta}_1 \cup \operatorname{supp} \boldsymbol{\eta}_2 \subseteq \operatorname{graph} S$, and for all Borel set $A \subseteq X$

$$\boldsymbol{\eta}_\lambda((\operatorname{pr}^{(1)})^{-1}(A)) = \lambda \boldsymbol{\eta}_1((\operatorname{pr}^{(1)})^{-1}(A)) + (1 - \lambda) \boldsymbol{\eta}_2((\operatorname{pr}^{(1)})^{-1}(A))$$
$$= \lambda \mu(A) + (1 - \lambda)\mu(A) = \mu(A),$$

   and so $\operatorname{pr}^{(1)} \sharp \boldsymbol{\eta}_\lambda = \mu$.

2. Given a Borel set $A \subseteq X$, $\mu \in \mathscr{P}(X)$, $\boldsymbol{\eta} \in \mathscr{P}(S)(\mu)$, we have

$$\boldsymbol{\eta}\Big((X \times Y) \backslash (\operatorname{graph}(S_{|A}))\Big) = \boldsymbol{\eta}\Big[\big((X \times Y) \backslash \operatorname{graph} S\big) \cup \big((X \times Y) \backslash (A \times Y)\big)\Big]$$
$$= \boldsymbol{\eta}\big((X \times Y) \backslash (A \times Y))\big) = \boldsymbol{\eta}\big((X \backslash A) \times Y\big) = \mu(X \backslash A),$$

   recalling that $\operatorname{supp} \boldsymbol{\eta} \subseteq \operatorname{graph} S$ and that $\operatorname{pr}^{(1)} \sharp \boldsymbol{\eta} = \mu$.

**Lemma 2 (Closure of the graph of the lift).** *$S$ has closed graph if and only if $\mathscr{P}(S)$ has closed graph.*

*Proof.* Suppose that $S$ has closed graph. Indeed, let $\{(\mu_n, \eta_n)\}_{n \in \mathbb{N}} \subseteq$ graph $\mathscr{P}(S)$ be a sequence converging to $(\mu, \eta) \in \mathscr{P}(X) \times \mathscr{P}(X \times Y)$. Since $\mathrm{pr}^{(1)}$ is continuous, we have that $\{\mathrm{pr}^{(1)} \sharp \eta_n\}_{n \in \mathbb{N}}$ narrowly converges to $\mathrm{pr}^{(1)} \sharp \eta$, and therefore, since $\mathrm{pr}^{(1)} \sharp \eta_n = \mu_n$, by passing to the limit we get $\mathrm{pr}^{(1)} \sharp \eta = \mu$.

On the other hand, let $(x, y) \in \mathrm{supp}\,\eta$. Then by [1, Proposition 5.1.8], there is a sequence $\{(x_n, y_n)\}_{n \in \mathbb{N}}$ such that $(x_n, y_n) \in \mathrm{supp}\,\eta_n$ and $(x_n, y_n) \to (x, y)$ in $Y$. By assumption, $(x_n, y_n) \in \mathrm{graph}\,S$, and, since $\mathrm{graph}\,S$ is closed, we have $(x, y) \in \mathrm{graph}\,S$. Thus $\mathrm{supp}\,\eta \subseteq \mathrm{graph}\,S$. We conclude that $\eta \in \mathscr{P}(S)(\mu)$. Suppose now that $\mathscr{P}(S)$ has closed graph. Let $\{(x_n, y_n)\}_{n \in \mathbb{N}}$ be a sequence in $\mathrm{graph}(S)$ converging to $(x, y)$. In particular, we have that $\{(\delta_{x_n}, \delta_{x_n} \otimes \delta_{y_n})\}_{n \in \mathbb{N}}$ is a sequence in $\mathrm{graph}(\mathscr{P}(S))$ converging to $(\delta_x, \delta_x \otimes \delta_y)$, which therefore belongs to $\mathrm{graph}(\mathscr{P}(S))$ by assumption. Thus $(x, y) \in \mathrm{graph}(S)$. $\square$

**Lemma 3 (Narrow compactness).** *Suppose that for every compact $K \subseteq X$ the set $\mathrm{graph}\,(S_{|K}) := \mathrm{graph}\,S \cap (K \times Y)$ is compact in $X \times Y$. Then, endowed $\mathscr{P}(X)$ and $\mathscr{P}(X \times Y)$ with the narrow topology, we have that for every relative compact $\mathscr{K} \subseteq \mathscr{P}(X)$, the set $\mathscr{P}(S)(\mathscr{K}) := \bigcup_{\mu \in \mathscr{K}} \mathscr{P}(S)(\mu)$ is relatively compact. Furthermore, if $S$ has closed graph, then $\mathscr{P}(S)$ has compact images.*

*Proof.* The first part of the statement is a direct consequence of our assumption, Theorem 5.1.3 in [1] and Remark 2. To prove the second part, we take $\mathscr{K} = \{\mu\}$ obtaining that $\mathscr{P}(S)(\mu)$ is relatively compact. Since $S$ has closed graph, by Lemma 2 we have that $\mathscr{P}(S)$ has closed graph, in particular it has closed images. Therefore $\mathscr{P}(S)(\mu)$ is compact. $\square$

**Lemma 4 (Uniform integrability).** *Suppose that $S$ has at most linear growth, i.e., there exists $C, D > 0$ and $(\bar{x}, \bar{y}) \in X \times Y$ such that $d_Y(y, \bar{y}) \leq C d_X(x, \bar{x}) + D$ for all $(x, y) \in \mathrm{graph}\,S$. Then if $\mathscr{K} \subseteq \mathscr{P}(X \times Y)$ is a set with uniformly integrable 2-moments, we have that $\mathscr{P}(S)(\mathscr{K})$ has uniformly integrable 2-moments.*

*Proof.* By the linear growth, we have in particular that for every bounded $K \subseteq X$ the set $\mathrm{graph}\,(S_{|K}) := \mathrm{graph}\,S \cap (K \times Y)$ is bounded in $X \times Y$: indeed, let $R > 0$ such that $K \subseteq B(\bar{x}, R)$. Then by the linear growth we have

$$d^2_{X \times Y}((x, y), (\bar{x}, \bar{y})) = d^2_X(x, \bar{x}) + d^2_Y(y, \bar{y}) \leq R^2 + (CR + D)^2,$$

for all $(x, y) \in \mathrm{graph}\,S \cap (K \times Y)$. Thus $\mathrm{graph}\,(S_{|K}) \subseteq B((\bar{x}, \bar{y}), \sqrt{R^2 + (CR + D)^2})$.

Fix $\varepsilon > 0$. Since $\mathscr{K}$ has uniformly integrable 2-moments, we have that there exists $r_\varepsilon > 1$ such that for all $\mu \in \mathscr{K}$ it holds $\int_{X \backslash B(\bar{x}, r_\varepsilon)} d^2_X(x, \bar{x})\, d\mu(x) \leq \varepsilon$. In particular, for all $\mu \in \mathscr{K}$ it holds $\int_{X \backslash B(\bar{x}, r_\varepsilon)} \left[ d^2_X(x, \bar{x}) + (C d_X(x, \bar{x}) + D)^2 \right] d\mu(x)$ $\leq (1 + C + D)\varepsilon$, since on $X \backslash B(\bar{x}, r_\varepsilon)$ we have $d_X(x, \bar{x}) \geq 1$ by the choice of $r_\varepsilon$.

By assumption, the set graph $\left(S_{|B(\bar{x},r_\varepsilon)}\right)$ is bounded, and so there exists $k_\varepsilon > 0$ such that $B_{X \times Y}((\bar{x},\bar{y}),k) \supseteq \text{graph}\left(S_{|B(\bar{x},r_\varepsilon)}\right)$ for all $k \geq k_\varepsilon$. Recalling that all $\eta \in \mathscr{P}(S)(\mathscr{K})$ are supported on graph $S$, we have

$$
\int_{(X \times Y) \setminus B_{X \times Y}((\bar{x},\bar{y}),k)} d_{X \times Y}^2((x,y),(\bar{x},\bar{y}))\, d\eta(x,y) =
$$

$$
= \int_{\text{graph } S \setminus \text{graph}\left(S_{|B(\bar{x},r_\varepsilon)}\right)} d_X^2(x,\bar{x}) + d_Y^2(y,\bar{y})\, d\eta(x,y)
$$

$$
\leq \int_{[X \setminus B(\bar{x},r_\varepsilon)] \times Y} \left[d_X^2(x,\bar{x}) + (Cd_X(x,\bar{x}) + D)^2\right] d\eta(x,y)
$$

$$
= \int_{X \setminus B(\bar{x},r_\varepsilon)} \left[d_X^2(x,\bar{x}) + (Cd_X(x,\bar{x}) + D)^2\right] d\mu(x) \leq (1 + C + D)\varepsilon.
$$

Therefore $\mathscr{P}(S)(\mathscr{K})$ has uniformly integrable 2-moments. $\qquad\blacksquare$

**Corollary 1.** *Suppose that for every compact $K \subseteq X$ the set graph $\left(S_{|K}\right) := \text{graph } S \cap (K \times Y)$ is compact in $X \times Y$ and that $S$ has at most linear growth, i.e., there exists $C, D > 0$ and $(\bar{x},\bar{y}) \in X \times Y$ such that $d_Y(y,\bar{y}) \leq Cd_X(x,\bar{x}) + D$ for all $(x,y) \in \text{graph } S$. Then the following holds true.*

1. *For every relatively compact $\mathscr{K} \subseteq \mathscr{P}_2(X)$, the set $\mathscr{P}(S)(\mathscr{K})$ is relatively compact in $\mathscr{P}_2(X \times Y)$. In particular, the restriction $\mathscr{P}(S)_{|\mathscr{P}_2(X)}$ of $\mathscr{P}(S)$ to $\mathscr{P}_2(X)$ takes values in $\mathscr{P}_2(X \times Y)$.*
2. *If furthermore $S$ has closed graph, then $\mathscr{P}(S)_{|\mathscr{P}_2(X)}$ has closed graph in $\mathscr{P}_2(X \times Y)$, and so compact images in $\mathscr{P}_2(X \times Y)$.*

*Proof.* We prove (1). By Proposition 7.1.5 in [1], $\mathscr{K}$ has uniformly integrable 2-moments and it is tight. According to Lemma 3 and Lemma 4, we have that $\mathscr{P}(S)(\mathscr{K})$ has uniformly integrable 2-moments and it is tight. Again by Proposition 7.1.5 in [1], we conclude that $\mathscr{P}(S)(\mathscr{K})$ is relatively compact in $\mathscr{P}_2(X \times Y)$.

To prove (2), notice that $W_2$-convergence implies narrow convergence. Thus every $W_2$-converging sequence $\{(\mu_n, \boldsymbol{\eta}_n)\}_{n \in \mathbb{N}} \subseteq \text{graph } \mathscr{P}(S) \cap \mathscr{P}_2(X \times Y)$ is narrowly converging to the same limit, say $(\mu, \boldsymbol{\eta})$. Since $\mathscr{P}(S)$ has closed graph in $\mathscr{P}(X \times Y)$, we have that the limit $(\mu, \boldsymbol{\eta})$ belongs to graph $\mathscr{P}(S)$, and since $\mu \in \mathscr{P}_2(X)$, by item (1) we have that $(\mu, \boldsymbol{\eta}) \in \text{graph } \mathscr{P}(S)_{|\mathscr{P}_2(X \times Y)}$. $\qquad\blacksquare$

**Theorem 1 (Lipschitz continuity).** *Suppose that $S$ is Lipschitz continuous with compact images and for every compact $K \subseteq X$ the set graph $\left(S_{|K}\right) := \text{graph } S \cap (K \times Y)$ is compact in $X \times Y$. Then*

1. $\mathscr{P}(S)(\mathscr{P}_2(X)) \subseteq \mathscr{P}_2(X \times Y)$;
2. *for any $\mathscr{K}$ relatively compact in $\mathscr{P}_2(X)$ we have that $\mathscr{P}(S)(\mathscr{K})$ is relatively compact in $\mathscr{P}_2(X \times Y)$. In particular, the restriction of $\mathscr{P}(S)$ to $\mathscr{P}_2(X)$ has compact images in $\mathscr{P}_2(X \times Y)$;*
3. $\mathscr{P}(S)_{|\mathscr{P}_2(X)} : \mathscr{P}_2(X) \rightrightarrows \mathscr{P}_2(X \times Y)$ *is Lipschitz continuous with* $\text{Lip } \mathscr{P}(S) \leq \sqrt{1 + (\text{Lip } S)^2}$.

*Proof.* According to the previous results, to prove (1–2) it is sufficient to show that $S$ has at most linear growth. Fix $\bar{x} \in X$ and $\bar{y} \in S(\bar{x})$. Let $y \in S(x)$, and $y' \in S(\bar{x})$ be such that $d_{S(\bar{x})}(y) = d_Y(y, y')$. The existence of such $y'$ follows from the compactness of $S(\bar{x})$. By Lipschitz continuity, we have

$$d_Y(y, \bar{y}) \leq d_Y(y, y') + d_Y(y', \bar{y}) = d_{S(\bar{x})}(y) + \sup_{y_1, y_2 \in S(\bar{x})} d_Y(y_1, y_2)$$

$$\leq \operatorname{Lip} S \cdot d_X(x, \bar{x}) + \sup_{y_1, y_2 \in S(\bar{x})} d_Y(y_1, y_2).$$

The compactness of $S(\bar{x})$ yields the boundedness of $D := \sup_{y_1, y_2 \in S(\bar{x})} d_Y(y_1, y_2)$, the linear growth follows by taking $C = \operatorname{Lip} S$.

We prove now (3). According to Lemma 1, there exists a Borel map $g : X \times Y \to Y$ such that $d_Y(y, g(x, y)) = d_{S(x)}(y)$ and $g(x, y) \in S(x)$. Let $\mu^{(1)}, \mu^{(2)} \in \mathscr{P}_2(X)$, $\pi \in \Pi_o^2(\mu^{(1)}, \mu^{(2)})$. Take $\eta^{(1)} \in \mathscr{P}(S)(\mu^{(1)})$ and disintegrate it w.r.t. $\operatorname{pr}^{(1)}$, i.e., $\eta^{(1)} = \mu^{(1)} \otimes \eta_x$, where $\{\eta_x\}_{x \in X}$ is family of Borel probability measures on $Y$, uniquely defined for $\mu^{(1)}$-a.e. $x \in X$. Define the Borel map $T : X \times Y \times X \to (X \times Y) \times (X \times Y)$ by $T(x_1, x_2, y_1) = ((x_1, y_1), (x_2, g(x_2, y_1)))$, and the measure $\theta \in \mathscr{P}(X \times X \times Y)$ by

$$\int_{X \times X \times Y} \varphi(x_1, x_2, y_1) \, d\theta(x_1, x_2, y_1) = \int_{X \times X} \int_Y \varphi(x_1, x_2, y_1) \, d\eta_{x_1}(y_1) \, d\pi(x_1, x_2).$$

The measure $T\sharp\theta$ belongs to $\mathscr{P}((X \times Y) \times (X \times Y))$. Defined $\operatorname{pr}_{X \times Y}^{(i)} : (X \times Y) \times (X \times Y) \to X \times Y$ as $\operatorname{pr}_{X \times Y}^{(i)}((x_1, y_2), (x_2, y_2)) = (x_i, y_i)$ for $i = 1, 2$, we obtain $\operatorname{pr}_{X \times Y}^{(1)} \sharp(T\sharp\theta) = \eta^{(1)}$.

Define $\eta^{(2)} := \operatorname{pr}_{X \times Y}^{(2)} \sharp(T\sharp\theta)$. Notice that $\operatorname{pr}^{(1)} \sharp \eta^{(2)} = \operatorname{pr}^{(2)} \sharp \pi = \mu^{(2)}$ by construction. We prove that $\operatorname{supp} \eta^{(2)} \subseteq \operatorname{graph} S$. Indeed, let $A$ be an open set disjoint from $\operatorname{graph} S$. Then

$$\eta^{(2)}(A) = \int_{X \times Y} \chi_A(x, y) \, d\eta^{(2)}(x, y) =$$

$$\iint_{(X \times Y)^2} \chi_A(x_2, g(x_2, y_1)) \, d\theta(x_1, x_2, y_1) = 0,$$

since $(x_2, g(x_2, y_1)) \in \operatorname{graph} S$ for all $y_1 \in Y$. We obtain that $\eta^{(2)} \in \mathscr{P}(S)(\mu^{(2)})$ and $T\sharp\theta \in \Pi(\eta^{(1)}, \eta^{(2)})$. Thus

$$W_2^2(\eta^{(1)}, \eta^{(2)}) \leq \iiint_{X \times X \times Y} [d_X^2(x_1, x_2) + d_Y^2(y_1, g(x_2, y_1))] \, d\eta_{x_1}(y_1) \, d\pi(x_1, x_2)$$

$$= W_2^2(\mu^{(1)}, \mu^{(2)}) + \iiint_{X \times X \times Y} d_{S(x_2)}^2(y_1) \, d\eta_{x_1}(y_1) \, d\pi(x_1, x_2)$$

$$\leq W_2^2(\mu^{(1)}, \mu^{(2)}) + \iint_{X \times X} (\operatorname{Lip} S)^2 \cdot d_X^2(x_1, x_2) \, d\pi(x_1, x_2)$$

$$= [1 + (\operatorname{Lip} S)^2] \cdot W_2^2(\mu^{(1)}, \mu^{(2)}),$$

the Lipschitz continuity estimate follows.

# 4   Applications and Extensions

The main application of the above result is the following one.

**Definition 4 (Solution set-valued map).** *Let $X$ be a finite-dimensional real space, $F : [a,b] \times X \rightrightarrows X$ be a Lipschitz set-valued map with compact convex nonempty values, $I = [a,b] \subseteq \mathbb{R}$ be a compact interval, $\theta = \{\theta_t\}_{t \in I} \in C^0(I; \mathscr{P}_2(X))$, and $\mu \in \mathscr{P}_2(X)$. We define the set-valued maps $S_I^F : X \rightrightarrows C^0(I; X)$, $\Xi_I^F : \mathscr{P}_2(\mu) \rightrightarrows \mathscr{P}_2(X \times C^0(I; X))$ and $\Upsilon_I^F : \mathscr{P}_2(\mu) \rightrightarrows C^0(I; \mathscr{P}_2(X))$ by*

$$S_I^F(x) := \{\zeta(\cdot) \in C^0(I; X) : \dot{\zeta}(t) \in F(t, \zeta(t)) \text{ for a.e. } t \in I, \ \zeta(a) = x\},$$
$$\Xi_I^F(\mu) := \{\eta \in \mathscr{P}(X \times C^0(I; X)) : \operatorname{supp} \eta \subseteq \operatorname{graph} S_I^F, \ e_a \sharp \eta = \mu\},$$
$$\Upsilon_I^F(\mu) := \{e_I \sharp \eta : \eta \in \Xi_I^F(\mu)\}.$$

**Proposition 1.** *Let $X$ be a finite-dimensional space, $I = [a,b]$ a compact interval of $\mathbb{R}$, $F : I \times \mathscr{P}_2(X) \times X \rightrightarrows X$ be a Lipschitz set-valued map with nonempty compact and convex values. Given a Lipschitz continuous curve $\theta = \{\theta_t\}_{t \in I} \subseteq \mathscr{P}_2(X)$, we have that the set-valued map $\Xi_I^F : \mathscr{P}_2(X) \rightrightarrows \mathscr{P}_2(X \times C^0(I; X))$ (defined as in Definition 4) enjoys the following properties:*

1. *$\Xi_I^F$ has nonempty compact convex images;*
2. *$\operatorname{Lip} \Xi_I^F \leq \sqrt{1 + e^{2(b-a)\operatorname{Lip} F \cdot (1+b-a)}}$;*
3. *for every relatively compact $\mathscr{K} \subseteq \mathscr{P}_2(X)$ we have that $\Xi_I^F(\mathscr{K})$ is relatively compact in $\mathscr{P}_2(X \times Y)$.*

*Proof.* Standard result in differential inclusion theory (see e.g. from Theorem 1 and Corollary 1 in Sect. 2, Chap. 2 of [2], and Filippov's Theorem, see e.g. Theorem 10.4.1 in [3]) yields all the properties needed on $S_I^F(\cdot)$ to have that its random lift $\Xi_I^F$ enjoys the requested properties.

The notion introduced in the previous section allows to transfer informations from a Lipschitz set-valued map between complete metric separable spaces to its natural lift in the space of probability measures. It is possible, in this setting, also to add a Lipschitz dependence of $F$ on the current state of the system. In this case the existence of trajectories in the space of probability measures follows from a straightforward application of Banach contraction principle to the set-valued map $\Upsilon_I^F$.

# References

1. Ambrosio, L., Gigli, N., Savare, G.: Gradient flows in metric spaces and in the space of probability measures, 2nd edn. Lectures in Mathematics ETH Zürich. Birkhäuser Verlag, Basel (2008)
2. Aubin, J.-P., Cellina, A.: Differential inclusions, Grundlehren der Mathematischen Wissenschaften [Fundamental Principles of Mathematical Sciences], vol. 264. Springer, Cham (1984)

3. Aubin, J.-P., Frankowska, H.: Set-Valued Analysis, Modern Birkhäuser Classics, Reprint of the 1990 edition [MR1048347]. Birkhäuser Boston Inc., Boston (2009)
4. Cannarsa, P., Capuani, R.: Existence and uniqueness for mean field games with state constraints. In: Cardaliaguet, P., Porretta, A., Salvarani, F. (eds.) PDE Models for Multi-Agent Phenomena. SIS, vol. 28, pp. 49–71. Springer, Cham (2018). https://doi.org/10.1007/978-3-030-01947-1_3
5. Cannarsa, P., Capuani, R., Cardaliaguet, P.: $C^{1,1}$-smoothness of constrained solutions in the calculus of variations with application to mean field games. Math. Eng. **1**(1), 174–203 (2018). https://doi.org/10.3934/Mine.2018.1.174
6. Cannarsa, P., Capuani, R., Cardaliaguet, P.: Mean field games with state constraints: from mild to pointwise solutions of the PDE system. Calculus Variat. Partial Different. Equat. **60**(3), 1–33 (2021). https://doi.org/10.1007/s00526-021-01936-4
7. Lirkov, I., Margenov, S. (eds.): LSSC 2017. LNCS, vol. 10665. Springer, Cham (2018). https://doi.org/10.1007/978-3-319-73441-5
8. Himmelberg, C.J.: Measurable relations. Fund. Math. **87**(1), 53–72 (1975)

# Hölder Regularity in Bang-Bang Type Affine Optimal Control Problems

Alberto Domínguez Corella$^{(\boxtimes)}$ and Vladimir M. Veliov

Institute of Statistics and Mathematical Methods in Economics,
Vienna University of Technology, Vienna, Austria
alberto.corella@tuwien.ac.at

**Abstract.** This paper revisits the issue of Hölder Strong Metric sub-Regularity (HSMs-R) of the optimality system associated with ODE optimal control problems that are affine with respect to the control. The main contributions are as follows. First, the metric in the control space, introduced in this paper, differs from the ones used so far in the literature in that it allows to take into consideration the bang-bang structure of the optimal control functions. This is especially important in the analysis of Model Predictive Control algorithms. Second, the obtained sufficient conditions for HSMs-R extend the known ones in a way which makes them applicable to some problems which are non-linear in the state variable and the Hölder exponent is smaller than one (that is, the regularity is not Lipschitz).

**Keywords:** Optimal control · Affine problems · Hölder metric sub-regularity

## 1 Introduction

Consider the following affine optimal control problem

$$\min_{u \in \mathcal{U}} \left\{ l(x(T)) + \int_0^T \left[ w(t, x(t)) + \langle s(t, x(t)), u(t) \rangle \right] dt \right\}, \tag{1}$$

subject to

$$\dot{x}(t) = a(t, x(t)) + B(t, x(t))u(t), \quad x(0) = x_0. \tag{2}$$

Here the state vector $x(t)$ belongs to $\mathbb{R}^n$ and the control function belongs to the set $\mathcal{U}$ of all Lebesgue measurable functions $u : [0, T] \to U$, where $U \subset \mathbb{R}^m$. Correspondingly, $l : \mathbb{R}^n \to \mathbb{R}$ and $w : \mathbb{R} \times \mathbb{R}^n \to \mathbb{R}$ are real-valued functions, $s : \mathbb{R} \times \mathbb{R}^n \to \mathbb{R}^m$ and $a : \mathbb{R} \times \mathbb{R}^n \to \mathbb{R}^n$ are vector-valued functions, and $B$ is an $(n \times m)$- matrix-valued function taking values in $\mathbb{R} \times \mathbb{R}^n$. The initial state $x_0 \in \mathbb{R}^n$ and the final time $T > 0$ are fixed.

We make the following basic assumption.

Supported by the Austrian Science Foundation (FWF) under grant No P31400-N32.

I. Lirkov and S. Margenov (Eds.): LSSC 2021, LNCS 13127, pp. 306–313, 2022.
https://doi.org/10.1007/978-3-030-97549-4_35

**Assumption 1.** *The set $U$ is a convex compact polyhedron. The functions $f$ :* $\mathbb{R} \times \mathbb{R}^n \times \mathbb{R}^m \to \mathbb{R}^n$ *and* $g : \mathbb{R} \times \mathbb{R}^n \times \mathbb{R}^m \to \mathbb{R}^m$ *given by*

$$f(t, x, u) := a(t, x) + B(t, x)u, \quad g(t, x, u) := w(t, x) + \langle s(t, x), u \rangle,$$

*and $l : \mathbb{R}^n \to \mathbb{R}$ are measurable and bounded in $t$, locally uniformly in $(x, u)$, and differentiable in $x$. Moreover, these functions and their first derivatives in $x$ are Lipschitz continuous in $x$, uniformly in $(t, u) \in [0, T] \times U$.*

With the usual definition of the Hamiltonian

$$H(t, x, u, p) := g(t, x, u) + \langle p, f(t, x, u) \rangle,$$

the local form of the Pontryagin principle for problem (1)–(2) can be represented by the following *optimality system* for $x$, $u$ and an absolutely continuous function $p : [0, T] \to \mathbb{R}^n$ : for almost every $t \in [0, T]$

$$0 = \dot{x}(t) - f(t, x(t), u(t)), \tag{3}$$

$$0 = x(0) - x_0, \tag{4}$$

$$0 = \dot{p}(t) + \nabla_x H(t, x(t), p(t), u(t)), \tag{5}$$

$$0 = p(T) - \nabla l(x(T)), \tag{6}$$

$$0 \in \nabla_u H(t, x(t), p(t), u(t)) + N_U(u(t)), \tag{7}$$

where $N_U(u)$ is the usual normal cone to the convex set $U$ at $u \in \mathbb{R}^m$. The optimality system can be recast as a generalized equation

$$0 \in \Phi(x, p, u), \tag{8}$$

where the *optimality map* $\Phi$ is defined as

$$\Phi(x, p, u) := \begin{pmatrix} -\dot{x} + f(\cdot, x, u) \\ x(0) - x_0 \\ \dot{p} + \nabla_x H(\cdot, x, p, u) \\ p(T) - \nabla l(x(T)) \\ \nabla_u H(\cdot, x, p) + N_U(u) \end{pmatrix}. \tag{9}$$

We remind the general definition of the property of Hölder Strong Metric sub-Regularity (HSMs-R) of a map, introduced under this name in [6] and appearing earlier in [4] (see the recent paper [2] for a comprehensive analysis of this property).

**Definition 1.** *Let $(\mathcal{Y}, d_{\mathcal{Y}})$ and $(\mathcal{Z}, d_{\mathcal{Z}})$ be metric spaces. A set-valued map $\Phi$ : $\mathcal{Y} \to \mathcal{Z}$ is strongly Hölder sub-regular at $\hat{y} \in \mathcal{Y}$ for $\hat{z} \in Z$ with exponent $\theta > 0$ if $\hat{z} \in \Phi(\hat{y})$ and there exist positive numbers $a, b$ and $\kappa$ such that if $y \in \mathcal{Y}$ and $z \in \mathcal{Z}$ satisfy*

$$i) \quad z \in \Phi(y) \qquad ii) \quad d_{\mathcal{Y}}(y, \hat{y}) \le a, \qquad iii) \quad d_{\mathcal{Z}}(z, \hat{z}) \le b,$$

*then*

$$d_{\mathcal{Y}}(y, \hat{y}) \le \kappa d_{\mathcal{Z}}(z, \hat{z})^\theta.$$

*We call $a, b$ and $\kappa$ parameters of strong Hölder sub-regularity. If $\theta = 1$, then the property is called SMs-R.*

In this paper, we reconsider this property for the optimality map $\Phi$, with an appropriate definition of the metric space where $y = (x, p, u)$ takes values and of the image space. It is well known that the HSMs-R property of the optimality map plays an important role in the analysis of stability of the solutions and of approximation methods in optimization, in general. We refer to [2] for general references, and to [10], where more bibliography on the utilization of the HSMs-R property in the error analysis of optimal control problems is provided. We mention that a sufficient condition for SMs-R follows from the fundamental paper [5], but it does not apply to affine problems.

The paper contains two main contributions.

(i) Usually in the investigations of regularity of the optimality map for affine problems (see [9] and the bibliography therein) the metric in the control space is related to the $L^1$-norm, which does not give information about the structure of the control function even if the optimal control is of bang-bang type, as assumed later in this paper. The metric in $\mathcal{U}$ that we define in the present paper captures some structural similarities of the controls, thus the regularity property in this metric is closer to (but weaker then) the so called *structural stability*, investigated in e.g. [7,8]. The SMs-R or HSMs-R properties of the optimality map $\Phi$ in this metric is especially important in the analysis of Model Predictive Control algorithms.

(ii) The obtained sufficient conditions for HSMs-R extend the known ones (e.g. [1,9]) in a way which makes them applicable to some problems which are non-linear in the state variable and the Hölder exponent $\theta$ is smaller than one.

## 2   The Main Result

First of all we define the metric spaces $\mathcal{Y}$ and $\mathcal{Z}$ of definition and images of the set-valued map $\Phi$ in (8), (9). For that we introduce some notations.

Using geometric (rather than analytic) terminology, we denote by $V$ the set of vertices of $U$, and by $E$ the set of all unit vectors $e \in \mathbb{R}^m$ that are parallel to some edge of $U$. Let $Z$ be a fixed non-empty subset of $[0, T]$. For $\varepsilon \ge 0$ and for $u_1, u_2 \in \mathcal{U}$ denote $\Sigma(\varepsilon) := [0, T] \setminus (Z + [-\varepsilon, \varepsilon])$, and for $u_1, u_2 \in \mathcal{U}$ define

$$d^*(u_1, u_2) := \inf \{\varepsilon > 0 : u_1(t) = u_2(t) \text{ for a.e. } t \in \Sigma(\varepsilon)\}.$$

For $Z = \emptyset$ we formally define $d^*(u_1, u_2) = 0$ if $u_1 = u_2$ a.e., and $d^*(u_1, u_2) = T$ else. It is easy to check that $d^*$ is a shift-invariant metric in $\mathcal{U}$. For a shift-invariant metric $d$ in any metric space we shorten the notation $d(y_1, y_2)$ as $d(y_1 - y_2)$. Then we define the spaces

$$\mathcal{Y} := W^{1,1}([0, T]; \mathbb{R}^n) \times W^{1,1}([0, T]; \mathbb{R}^n) \times \mathcal{U},$$
$$\mathcal{Z} := L^1([0, T]; \mathbb{R}^n) \times \mathbb{R}^n \times L^1([0, T]; \mathbb{R}^n) \times \mathbb{R}^n \times L^\infty([0, T]; \mathbb{R}^n)$$

with the metrics

$$dy(x, p, u) := \|x\|_{1,1} + \|p\|_{1,1} + d^*(u),$$

$$d_{\mathcal{Z}}(\xi, \eta, \pi, \zeta, \rho) := \|\xi\|_1 + |\eta| + \|\pi\|_1 + |\zeta| + \|\rho\|_\infty.$$

The particular set $Z$ in the definition of $d^*$ will be defined in the next lines. The map $\Phi$ defined in (9) is now considered as a map acting on $\mathcal{Y}$ with images in $\mathcal{Z}$. The normal cone $N_{\mathcal{U}}(u)$ to the closed convex set $\mathcal{U} \subset L^1([0, T]; \mathbb{R}^m)$ that appears in (9) is a subset of the dual space $L^\infty([0, T]; \mathbb{R}^m)$, which can be equivalently defined as

$$N_{\mathcal{U}}(u) := \begin{cases} \emptyset & \text{if } u \notin \mathcal{U} \\ \{v \in L^\infty([0, T]; \mathbb{R}^m) : v(t) \in N_U(u(t)) \text{ for a.e. } t \in [\tau, T]\} & \text{if } u \in \mathcal{U}. \end{cases}$$

By a standard argument, problem (1)–(2) has a solution, hence system (8) has a solution, as well. Let $\hat{y} = (\hat{x}, \hat{p}, \hat{u}) \in \mathcal{Y}$ be a reference solution of the optimality system (8). Denote by

$$\hat{\sigma} := \nabla_u H(\cdot, \hat{x}, \hat{p}) = B(\cdot, \hat{x})^\top \hat{p} + s(\cdot, \hat{x})$$

the so-called *switching function* corresponding to the triple $(\hat{x}, \hat{p}, \hat{u})$. We extend the definition of the switching function in the following way. For any $u \in \mathcal{U}$ define the function

$$[0, T] \ni t \mapsto \sigma[u](t) := B(t, x(t))^\top p(t) + s(t, x(t)),$$

where $(x, p)$ solves the system (3)–(6) for the given $u$.

**Assumption 2.** *There exist numbers $\gamma_0 > 0$, $\alpha_0 > 0$ and $\nu \geq 1$ such that*

$$\int_0^T \langle \sigma[u](t), u(t) - \hat{u}(t) \rangle \, dt \geq \gamma_0 \|u - \hat{u}\|_1^{\nu+1}$$

*for all $u \in \mathcal{U}$ with $\|u - \hat{u}\|_1 \leq \alpha_0$.*

The following assumption is standard in the literature on affine optimal control problems, see e.g. [3, 7, 9].

**Assumption 3.** *There exist numbers $\tau > 0$ and $\mu > 0$ such that if $s \in [0, T]$ is a zero of $\langle \hat{\sigma}, e \rangle$ for some $e \in E$, then*

$$|\langle \hat{\sigma}(t), e \rangle| \geq \mu |t - s|^\nu,$$

*for all $t \in [s - \tau, s + \tau] \cap [0, T]$. Here $\nu$ is the number from Assumption 2.*

Assumption 1 implies, in particular, that the set

$$\hat{Z} := \{s \in [0, T] : \langle \hat{\sigma}(s), e \rangle = 0 \text{ for some } e \in E\}$$

is finite. In what follows the set $Z$ in the definition of the metric $d^*$ will be fixed as $Z = \hat{Z}$.

The following result is well-known for a box-like set $U$; under the present assumptions it is proved in [9, Proposition 4.1].

**Lemma 1.** *Under Assumptions 1 and 3, $\hat{u}$ is (equivalent to) a piecewise constant function with values in the set $V$ of vertices of $U$. Moreover, there exists a number $\gamma > 0$ such that*

$$\int_0^T \langle \hat{\sigma}(t), u(t) - \hat{u}(t) \rangle \, dt \geq \gamma \|u - \hat{u}\|_1^{\nu+1}$$

*for all $u \in \mathcal{U}$.*

As a consequence of the lemma, Assumption 2 is implied by Assumption 3, provided that there exists $\gamma_1 < \gamma$ such that

$$\int_0^T \langle \sigma[\hat{u} + v](t) - \sigma[\hat{u}](t), v(t) \rangle \, dt \geq -\gamma_1 \|v\|_1^{\nu+1} \tag{10}$$

for all $v \in \mathcal{U} - \hat{u}$ with $\|v\|_1 \leq \alpha_0$. Notice that in the case of a linear-quadratic problem, condition (10) reduces to the one in [9, Corollary 3.1].

The main result in this paper follows.

**Theorem 1.** *Let Assumption 1–3 be fulfilled. There exist positive numbers $a, b$ and $\kappa$ such that if $y = (x, p, u) \in \mathcal{Y}$ and $z \in \mathcal{Z}$ satisfy*

$$i) \quad z \in \Phi(x, p, u) \qquad ii) \quad \|u - \hat{u}\|_1 \leq a, \qquad iii) \quad d_{\mathcal{Z}}(z, \hat{z}) \leq b,$$

*then*

$$d_{\mathcal{Y}}(y, \hat{y}) \leq \kappa d_{\mathcal{Z}}(z, \hat{z})^{-\nu^2}.$$

*In particular, the optimality map $\Phi : \mathcal{Y} \to \mathcal{Z}$ is Hölder strongly metrically subregular at $\hat{y} = (\hat{x}, \hat{p}, \hat{u})$ for zero with exponent $1/\nu^2$.*

## 3     Proof of the Theorem

We begin with two lemmas.

**Lemma 2.** *There exists a positive number $\kappa_0$ such that for every $\varepsilon \in (0, T)$, $t \in \Sigma(\varepsilon)$, and $e \in E$ it holds that*

$$|\langle \hat{\sigma}(t), e \rangle| \geq \kappa_0 \varepsilon^{\nu}.$$

*Proof.* For brevity we use the notations

$$\hat{\sigma}_e(t) := \langle \hat{\sigma}(t), e \rangle, \qquad e \in E,$$

$$\delta := \inf\{|\hat{\sigma}_e(t)| : e \in E, t \in \Sigma(\tau)\} > 0.$$

Let $t \in \Sigma(\tau)$. Then

$$|\hat{\sigma}_e(t)| \geq \delta = \frac{\delta}{\varepsilon^{\nu}}\varepsilon^{\nu} \geq \frac{\delta}{T^{\nu}}\varepsilon^{\nu}.$$

Now let $t \in \Sigma(\varepsilon) \setminus \Sigma(\tau)$. This implies, in particular, that the set $Z = \hat{Z}$ is non-empty, since $\Sigma(\tau) = [0, T]$ if $\hat{Z} = \emptyset$. Then for some $s \in \hat{Z}$ and $e \in E$ with $\varepsilon \leq |t - s| \leq \tau$ it is fulfilled that

$$|\hat{\sigma}_e(t)| \geq \mu|t - s|^{\nu} \geq \mu\varepsilon^{\nu}.$$

This implies the claim of the lemma with $\kappa_0 = \min\{\mu, \delta/T^{\nu}\}$.     $\square$

**Lemma 3.** *There exist positive numbers $\kappa_1$ and $\rho_1$ such that for every functions $\sigma \in L^\infty$ with $\|\hat\sigma - \sigma\|_\infty \leq \rho_0$ and $u \in \mathcal{U}$ with $\sigma(t) + N_U(u(t)) \ni 0$ for a.e. $t \in [0,T]$ it holds that*

$$u(t) = \hat u(t) \quad \text{for a.e. } t \in \Sigma\big(\kappa_1\|\sigma - \hat\sigma\|_\infty^{\frac{1}{\nu}}\big).$$

*Proof.* Consider the case $\|\sigma - \hat\sigma\|_\infty > 0$ and $\hat Z \neq 0$, the other cases are similar. For a vertex $v \in V$ we denote

$$E(v) = \Big\{ \frac{v' - v}{|v' - v|} : v' \text{ is a neighboring vertex to } v \Big\} \subset E.$$

From (7) applied to $(\hat x, \hat p, \hat u)$, we obtain that $\langle \hat\sigma(t), v - \hat u(t)\rangle \geq 0$ for a.e. $t$ and for every $v \in V$. From Lemma 1 we know that $\hat u(t) \in V$ for a.e. $t$. This implies that for a.e. $t$ it holds that $\hat\sigma_e(t) \geq 0$ for all $e \in E(\hat u(t))$. Let us fix such a $t$ which, moreover, belongs to $\Sigma\big(\kappa_1\|\sigma - \hat\sigma\|_\infty^{\frac{1}{\nu}}\big)$; the number $\kappa_1$ will be defined in the next lines. Then according to Lemma 2 we have that

$$\hat\sigma_e(t) \geq \kappa_0\Big(\kappa_1\|\sigma - \hat\sigma\|_\infty^{\frac{1}{\nu}}\Big)^\nu = \kappa_0(\kappa_1)^\nu \|\sigma - \hat\sigma\|_\infty.$$

Let us choose $\kappa_1$ and $\rho_1$ such that $\kappa_0\kappa_1^\nu > 1$ and $\rho_1 < (T/\kappa_1)^\nu$. Then

$$\sigma_e(t) := \langle \sigma(t), e\rangle = \hat\sigma_e(t) + (\sigma_e(t) - \hat\sigma_e(t)) > \|\sigma - \hat\sigma\|_\infty - \|\sigma - \hat\sigma\|_\infty = 0.$$

Thus we obtain that

$$\langle \sigma(t), v - \hat u(t)\rangle > 0 \quad \text{for every } v \in V \setminus \{\hat u(t)\}.$$

This implies that $\hat u(t)$ is the unique solution of $\sigma(t) + N_U(u) \ni 0$, hence $u(t) = \hat u(t)$. $\qquad\square$

*Proof of Theorem 1.* In the proof we use the constants involved in the assumptions and in the lemmas above. Let $z = (\xi, \eta, \pi, \upsilon, \rho) \in \mathcal{Z}$ and $y = (\tilde x, \tilde\lambda, \tilde u) \in \mathcal{Y}$ such that $z \in \Phi(y)$. Denote $\Delta := \rho + [\sigma_\tau[\tilde u] - \nabla_u H(\bar p, \tilde x, \tilde\lambda)]$. Using the Grönwall's inequality, we can find constants $c_1$ and $c_2$ (indepent of $y$ and $z$) such that $\|\Delta\|_\infty \leq c_1\|z\|_{\mathcal{Z}}$ and $\|y - \hat y\|_{\mathcal{Y}} \leq c_2\|\tilde u - \hat u\|_1$. Let $a := \alpha_0$, since $\Delta - \sigma[\tilde u] = \rho - \nabla_u H(\bar p, \tilde x, \tilde\lambda) \in N_{\mathcal{U}}(\tilde u)$, we have

$$\int_0^T \langle \Delta - \sigma[\tilde u], \hat u - \tilde u\rangle \leq 0.$$

Now, by Assumption 2

$$0 \geq \int_0^T \langle \Delta - \sigma[\tilde u], \hat u - \tilde u\rangle = \int_0^T \langle \sigma[\tilde u], \tilde u - \hat u\rangle + \int_0^T \langle \Delta, \hat u - \tilde u\rangle$$

$$\geq \gamma_0 \Big(\int_0^T |\tilde u - \hat u|\Big)^{\nu+1} - \|\Delta\|_\infty \int_0^T |\tilde u - \hat u|.$$

Hence,

$$\|\tilde{u} - \hat{u}\|_1 \leq \frac{1}{\gamma_0^{\frac{1}{\nu}}}\|\Delta\|_\infty^{\frac{1}{\nu}} \leq \frac{c_1^{\frac{1}{\nu}}}{\gamma_0^{\frac{1}{\nu}}}\|z\|_{\mathcal{Z}_\tau}^{\frac{1}{\nu}}.$$

With $\kappa' := c_2 c_1^{\frac{1}{\nu}} \gamma_0^{-\frac{1}{\nu}}$ we obtain that

$$\|y - \hat{y}\|_{\mathcal{Y}} \leq \kappa' \|z\|_{\mathcal{Z}}^{\frac{1}{\nu}}.$$

There exists a constant $c_3 > 0$ such that $\|\sigma[\tilde{u}] - \sigma[\hat{u}]\|_\infty \leq c_3\|\tilde{u} - \hat{u}\|_1$, hence

$$\|\sigma[\tilde{u}] - \rho - \sigma[\hat{u}]\|_\infty \leq c_3\kappa'\|z\|_{\mathcal{Z}}^{\frac{1}{\nu}} + \|\rho\|_\infty \leq (c_3\kappa' + 1)\|z\|_{\mathcal{Z}}^{\frac{1}{\nu}}.$$

Let $b$ be small enough so Lemma 3 holds with $\sigma = \sigma[\tilde{u}] - \rho$. We get $d^*(\tilde{u}, \hat{u}) \leq$
$\left[(c_3\kappa' + 1)\|z\|_{\mathcal{Z}}^{\frac{1}{\nu}}\right]^{\frac{1}{\nu}} = (c_3\kappa' + 1)^{\frac{1}{\nu}}\|z\|_{\mathcal{Z}}^{\frac{1}{\nu^2}}$. Finally,

$$d_{\mathcal{Y}}(y, \hat{y}) \leq (c_3\kappa' + 1)^{\frac{1}{\nu}}\|z\|_{\mathcal{Z}}^{\frac{1}{\nu^2}} + k'\|z\|_{\mathcal{Z}}^{\frac{1}{\nu}} \leq \kappa\|z\|_{\mathcal{Z}}^{\frac{1}{\nu^2}},$$

where $\kappa := (c_3\kappa' + 1) + \kappa'$.                                            □

## 4    An Example

Let $\alpha : [0, +\infty) \to \mathbb{R}$ be a differentiable function with Lipschitz derivative such that $\alpha$ attains its minimum at zero and $\alpha'(x) \geq 0$ for all $x \in [0, +\infty)$, e.g., $\alpha(x) = x^2$ or $\alpha(x) = 1 - e^{-x^2}$. Moreover, let $T$ be a positive number and let $\beta : [0, T] \to \mathbb{R}$ be $\nu$-times differentiable $(\nu \geq 1)$ and satisfy $\beta(t) > 0$ for $t > 0$, $\beta(0) = \ldots = \beta^{(\nu-1)}(0) = 0$, $\beta^{(\nu)}(0) \neq 0$. Consider the following optimal control problem

$$\min_{u \in \mathcal{U}} \left\{ \int_0^T \left[\alpha(x(t)) + \beta(t)u(t)\right] dt \right\}, \tag{11}$$

subject to

$$\dot{x}(t) = u(t), \quad x(0) = 0, \quad u(t) \in [0, 1] \quad \text{a.e. in } [0, T]. \tag{12}$$

The optimality system of problem (11)–(12) is given by

$$0 = \dot{x}(t) - u(t), \quad x(0) = 0, \tag{13}$$
$$0 = \dot{p}(t) + \alpha'(x(t)), \quad p(T) = 0, \tag{14}$$
$$0 \in \beta(t) + p(t) + N_{[0,1]}(u(t)). \tag{15}$$

Hence the switching function corresponding to each control $u \in \mathcal{U}$ is given by $\sigma[u](t) := \beta(t) + p[u](t)$, where $p[u](t) = \int_t^T \alpha'(x[u](s)) \, ds$ and $x[u](t) =$

$\int_0^t u(s)\,\mathrm{d}s$, $t \in [0,T]$. It is clear that the unique minimizer of problem (11)–(12) is $(\hat{x}, \hat{p}, \hat{u}) = (0,0,0)$, and consequently its switching function is given by $\hat{\sigma}(t) = \beta(t)$. Since $\hat{\sigma}$ satisfies $\hat{\sigma}(0) = \ldots = \hat{\sigma}^{(\nu-1)}(0) = 0$ and $\hat{\sigma}^{(\nu)}(0) \neq 0$, we have that Assumption 3 is satisfied with the same number $\nu$ but not with $\nu - 1$. Now, observe that for $v \in \mathcal{U} - \hat{u}$ we have

$$\int_0^T \langle \sigma[\hat{u} + v](t) - \sigma[\hat{u}](t), v(t) \rangle \, \mathrm{d}t = \int_0^T \langle p[\hat{u} + v](t), v(t) \rangle \, \mathrm{d}t \geq 0.$$

Thus, in accordance with (10), Assumption 2 is satisfied.

# References

1. Alt, W., Schneider, C., Seydenschwanz, M.: Regularization and implicit Euler discretization of linear-quadratic optimal control problems with bang-bang solutions. Appl. Math. Comput. **287/288**, 104–124 (2016)
2. Cibulka, R., Dontchev, A.L., Kruger, A.Y.: Strong metric subregularity of mappings in variational analysis and optimization. J. Math. Anal. Appl. **457**, 1247–1282 (2018)
3. Domínguez Corella, A., Quincampoix, M., Veliov, V.M.: Strong bi-metric regularity in affine optimal control problems. Pure Appl. Funct. Anal. (to appear). https://orcos.tuwien.ac.at/research/
4. Dontchev, A.L.: Characterizations of Lipschitz stability in optimization. In: Recent Developments in Well-Posed Variational Problems, pp. 95–115. Kluwer (1995)
5. Dontchev, A.L., Hager, W.W.: Lipschitzian stability in nonlinear control and optimization. SIAM J. Control Optim. **31**, 569–603 (1993)
6. Dontchev, A.L., Rockafellar, R.T.: Regularity and conditioning of solution mappings in variational analysis. Set Valued Anal. **12**, 7–109 (2004)
7. Felgenhauer, U.: On stability of bang-bang type controls. SIAM J. Control Optim. **41**(6), 1843–1867 (2003)
8. Felgenhauer, U., Poggolini, L., Stefani, G.: Optimality and stability result for bang-bang optimal controls with simple and double switch behavior. Control Cybern. **38**(4B), 1305–1325 (2009)
9. Osmolovskii, N.P., Veliov, V.M.: Metric sub-regularity in optimal control of affine problems with free end state. ESAIM Control Optim. Calc. Var. **26** (2020). Paper No. 47
10. Preininger, J., Scarinci, T., Veliov, V.M.: Metric regularity properties in bang-bang type linear-quadratic optimal control problems. Set Valued Var. Anal. **27**(2), 381–404 (2019)

# Simultaneous Space-Time Finite Element Methods for Parabolic Optimal Control Problems

Ulrich Langer[1]([⊠]) and Andreas Schafelner[2]

[1] Institute for Computational Mathematics, Johannes Kepler University Linz,
Altenbergerstr. 69, 4040 Linz, Austria
ulanger@numa.uni-linz.ac.at
[2] Doctoral Program "Computational Mathematics", Johannes Kepler University
Linz, Altenbergerstr. 69, 4040 Linz, Austria
andreas.schafelner@dk-compmath.jku.at
http://www.numa.uni-linz.ac.at, https://www.dk-compmath.jku.at

**Abstract.** This work presents, analyzes and tests stabilized space-time finite element methods on fully unstructured simplicial space-time meshes for the numerical solution of space-time tracking parabolic optimal control problems with the standard $L_2$-regularization.

**Keywords:** Parabolic optimal control problems · $L_2$-regularization · Space-time finite element methods

## 1 Introduction

Let us consider the following space-time tracking optimal control problem: For a given target function $y_d \in L_2(Q)$ (desired state) and for some appropriately chosen regularization parameter $\varrho > 0$, find the state $y \in Y_0 = \{v \in L^2(0,T; H_0^1(\Omega)) : \partial_t v \in L^2(0,T; H^{-1}(\Omega)), v = 0 \text{ on } \Sigma_0\}$ and the control $u \in U = L_2(0,T; L_2(\Omega)) = L_2(Q)$ minimizing the cost functional

$$J(y,u) = \frac{1}{2} \int_Q |y - y_d|^2 \, dQ + \frac{\varrho}{2} \|u\|_{L_2(Q)}^2 \tag{1}$$

subject to the linear parabolic initial-boundary value problem (IBVP)

$$\partial_t y - \text{div}_x(\nu \nabla_x y) = u \text{ in } Q, \quad y = 0 \text{ on } \Sigma, \quad y = 0 \text{ on } \Sigma_0, \tag{2}$$

where $Q := \Omega \times (0,T)$, $\Sigma := \partial\Omega \times (0,T)$, $\Sigma_0 := \Omega \times \{0\}$, $T > 0$ is the final time, $\partial_t$ denotes the partial time derivative, $\text{div}_x$ is the spatial divergence operator, $\nabla_x$ is the spatial gradient, and the source term $u$ on the right-hand side of the parabolic PDE serves as control. The spatial domain $\Omega \subset \mathbb{R}^d$, $d = 1, 2, 3$, is

Supported by the Austrian Science Fund under the grant W1214, project DK4.

I. Lirkov and S. Margenov (Eds.): LSSC 2021, LNCS 13127, pp. 314–321, 2022.
https://doi.org/10.1007/978-3-030-97549-4_36

supposed to be bounded and Lipschitz. We assumed that $0 < \nu_1 \leq \nu(x,t) \leq \nu_2$ for almost all $(x,t) \in Q$ with positive constants $\nu_1$ and $\nu_2$.

This standard setting was already investigated in the famous book by J.L. Lions [6], but it is a fundamental model problem for constructing efficient numerical discretization techniques and solvers. Since the state equation (2) has a unique solution $y \in Y_0$, one can conclude the existence of a unique control $u \in U$ minimizing the quadratic cost functional $J(S(u), u)$, where $S$ is the solution operator mapping $u \in U$ to the unique solution $y \in Y_0$ of (2); see, e.g., [6] and [9]. There is a huge number of publications devoted to the numerical solution of the optimal control problem (1)–(2) with the standard $L_2(Q)$ regularization; see, e.g., [9]. The overwhelming majority of the publications uses some time-stepping or discontinuous Galerkin method for the time discretization in combination with some space-discretization method like the finite element method; see, e.g., [9]. The unique solvability of the optimal control problem can also be established by showing that the optimality system has a unique solution. In [5], the Banach-Nečas-Babuška theorem was applied to the optimality system to show its well-posedness. Furthermore, the discrete inf-sup condition, which does not follow from the inf-sup condition in the infinite-dimensional setting, was established for continuous space-time finite element discretization on fully unstructured simplicial space-time meshes. The discrete inf-sup condition implies stability of the discretization and a priori discretization error estimates. Distributed controls $u$ from the space $U = L_2(0,T; H^{-1}(\Omega))$ together with energy regularization were investigated in [4], where one can also find a comparison of the energy regularization with the $L_2(Q)$ and the sparse regularizations.

In this paper, we make use of the maximal parabolic regularity of the reduced optimality system in the case of the $L_2(Q)$ regularization and under additional assumptions imposed on the coefficient $\nu$. Then we can derive a stabilized finite element discretization of the reduced optimality system in the same way as it was done for the state equation in our preceding paper [3]. The properties of the finite element scheme lead to a priori discretization error estimates that are confirmed by the numerical experiments.

## 2   Space-Time Finite Element Discretization

Eliminating the control $u$ from the optimality system by means of the gradient equation $p + \varrho u = 0$, we arrive at the reduced optimality system the weak form of which reads as follows: Find the state $y \in Y_0$ and the adjoint state $p \in P_T$ such that, for all $v, q \in V = L_2(0,T; H_0^1(\Omega))$, it holds

$$
\begin{aligned}
\varrho \int_Q \left[ \partial_t y\, v + \nu\, \nabla_x y \cdot \nabla_x v \right] dQ + \int_Q p\, v\, dQ &= 0, \\
- \int_Q y\, q\, dQ + \int_Q \left[ -\partial_t p\, q + \nu\, \nabla_x p \cdot \nabla_x q \right] dQ &= - \int_Q y_d\, q\, dQ,
\end{aligned}
\tag{3}
$$

where $P_T := \{ p \in L^2(0,T; H_0^1(\Omega)) : \partial_t p \in L^2(0,T; H^{-1}(\Omega)), p = 0 \text{ on } \Sigma_T \}$. The variational reduced optimality system (3) is well-posed; see [5, Theorem 3.3].

Moreover, we additionally assume that the coefficient $\nu(x,t)$ is of bounded variation in $t$ for almost all $x \in \Omega$. Then $\partial_t u$ and $Lu := -\mathrm{div}_x(\nu\nabla_x u)$ as well as $\partial_t p$ and $Lp := -\mathrm{div}_x(\nu\nabla_x p)$ belong to $L_2(Q)$; see [2]. This property is called maximal parabolic regularity. In this case, the parabolic partial differential equations involved in the reduced optimality system (3) hold in $L_2(Q)$. Therefore, the solution of the reduced optimality system (3) is equivalent to the solution of the following system of coupled forward and backward parabolic PDEs: Find $y \in Y_0 \cap H^{L,1}(Q)$ and $p \in P_T \cap H^{L,1}(Q)$ such that the coupled PDE optimality system

$$\begin{aligned}
\varrho\big[\partial_t y - \mathrm{div}_x(\nu\nabla_x y)\big] &= -p \quad \text{in } L_2(Q), \\
-\partial_t p - \mathrm{div}_x(\nu\nabla_x p) &= y - y_d \quad \text{in } L_2(Q)
\end{aligned} \tag{4}$$

hold, where $H^{L,1}(Q) = \{v \in H^1(Q) : Lv := -\mathrm{div}_x(\nu\nabla_x v) \in L_2(Q)\}$. The coupled PDE optimality system (4) is now the starting point for the construction of our coercive and consistent space-time finite element scheme.

Let $\mathcal{T}_h$ be a regular decomposition of the space-time cylinder $Q$ into simplicial elements, i.e., $\overline{Q} = \bigcup_{K\in\mathcal{T}_h} \overline{K}$, and $K \cap K' = \emptyset$ for all $K$ and $K'$ from $\mathcal{T}_h$ with $K \neq K'$; see, e.g., [1] for more details. On the basis of the triangulation $\mathcal{T}_h$, we define the space-time finite element spaces

$$Y_{0h} = \{y_h \in C(\overline{Q}) : y_h(x_K(\cdot)) \in \mathbb{P}_k(\hat{K}), \forall K \in \mathcal{T}_h, y_h = 0 \text{ on } \overline{\Sigma}\cup\overline{\Sigma}_0\}, \tag{5}$$

$$P_{Th} = \{p_h \in C(\overline{Q}) : p_h(x_K(\cdot)) \in \mathbb{P}_k(\hat{K}), \forall K \in \mathcal{T}_h, p_h = 0 \text{ on } \overline{\Sigma}\cup\overline{\Sigma}_T\}, \tag{6}$$

where $x_K(\cdot)$ denotes the map from the reference element $\hat{K}$ to the finite element $K \in \mathcal{T}_h$, and $\mathbb{P}_k(\hat{K})$ is the space of polynomials of the degree $k$ on the reference element $\hat{K}$.

For brevity of the presentation, we set $\nu$ to 1. The same derivation can be done for $\nu$ that fulfill the condition $\mathrm{div}_x(\nu\nabla_x w_h)|_K \in L_2(K)$ for all $w_h$ from $Y_{0h}$ or $P_{Th}$ and for all $K \in \mathcal{T}_h$ (i.e., piecewise smooth) in addition to the conditions imposed above. Multiplying the first PDE in (4) by $v_h + \lambda\partial_t v_h$ with $v_h \in Y_{0h}$, and the second one by $q_h - \lambda\partial_t q_h$ with $q_h \in P_{Th}$, integrating over $K$, integrating by parts in the elliptic parts where the positive scaling parameter $\lambda$ does not appear, and summing over all $K \in \mathcal{T}_h$, we arrive at the variational consistency identity

$$a_h(y, p; v_h, q_h) = \ell_h(v_h, q_h) \quad \forall(v_h, q_h) \in Y_{0h} \times P_{Th}, \tag{7}$$

with the combined bilinear and linear forms

$$\begin{aligned}
a_h(y, p; v, q) = \sum_{K\in\mathcal{T}_h} \int_K \Big[&\varrho\big(\partial_t y\, v + \lambda\partial_t y\partial_t v + \nabla_x y \cdot \nabla_x v - \lambda\Delta_x y\, \partial_t v\big) \\
&+ p(v + \lambda\partial_t v) - \partial_t p\, q + \lambda\partial_t p\partial_t q + \nabla_x p \cdot \nabla_x q \\
&+ \lambda\Delta_x p\, \partial_t q - y(q - \lambda\partial_t q)\Big]\, \mathrm{d}K \quad \text{and}
\end{aligned} \tag{8}$$

$$\ell_h(v, q) = -\sum_{K\in\mathcal{T}_h} \int_K y_d(q - \lambda\partial_t q)\, \mathrm{d}K, \tag{9}$$

respectively. Now, the corresponding consistent finite element scheme reads as follows: Find $(y_h, p_h) \in Y_{0h} \times P_{Th}$ such that

$$a_h(y_h, p_h; v_h, q_h) = \ell_h(v_h, q_h) \quad \forall (v_h, q_h) \in Y_{0h} \times P_{Th}. \tag{10}$$

Subtracting (10) from (7), we immediately get the Galerkin orthogonality relation

$$a_h(y - y_h, p - p_h; v_h, q_h) = 0 \quad \forall (v_h, q_h) \in Y_{0h} \times P_{Th}, \tag{11}$$

which is crucial for deriving discretization error estimates.

## 3  Discretization Error Estimates

We first show that the bilinear form $a_h$ is coercive on $Y_{0h} \times P_{Th}$ with respect to the norm

$$\|(v, q)\|_h^2 = \varrho \|v\|_{h,T}^2 + \|q\|_{h,0}^2 = \varrho \left( \|v(\cdot, T)\|_{L_2(\Omega)}^2 + \|\nabla_x v\|_{L_2(Q)}^2 + \lambda \|\partial_t v\|_{L_2(Q)}^2 \right)$$
$$+ \|q(\cdot, 0)\|_{L_2(\Omega)}^2 + \|\nabla_x q\|_{L_2(Q)}^2 + \lambda \|\partial_t q\|_{L_2(Q)}^2.$$

Indeed, for all $(v_h, q_h) \in Y_{0h} \times P_{Th}$, we get the estimate

$$a_h(v_h, q_h; v_h, q_h) = \sum_{K \in \mathcal{T}_h} \int_K \left[ \varrho(\partial_t v_h \, v_h + \lambda |\partial_t v_h|^2 + |\nabla_x v_h|^2 - \lambda \Delta_x v_h \, \partial_t v_h) \right.$$
$$+ q_h(v_h + \lambda \partial_t v_h) - \partial_t q_h \, q_h + \lambda |\partial_t q_h|^2 + |\nabla_x q_h|^2$$
$$\left. + \lambda \Delta_x q_h \, \partial_t q_h - v_h(q_h - \lambda \partial_t q_h) \right] dK$$
$$\geq \mu_c \|(v_h, q_h)\|_h^2, \tag{12}$$

with $\mu_c = 1/2$ provided that $\lambda \leq c_{inv}^{-2} h^2$, where $c_{inv}$ denotes the constant in the inverse inequality $\|\text{div}_x(\nabla_x w_h)\|_{L_2(K)} \leq c_{inv} h^{-1} \|\nabla_x w_h\|_{L_2(K)}$ that holds for all $w_h \in Y_{0h}$ or $w_h \in P_{Th}$. For $k = 1$, the terms $\Delta_x v_h$ and $\Delta_x q_h$ are zero, and we do not need the inverse inequality, but $\lambda$ should be also $\mathcal{O}(h^2)$ in order to get an optimal convergence rate estimate. The coercivity of the bilinear form $a_h$ immediately implies uniqueness and existence of the finite element solution $(y_h, p_h) \in Y_{0h} \times P_{Th}$ of (10). In order to prove discretization error estimates, we need the boundedness of the bilinear form

$$|a_h(y, p; v_h, q_h)| \leq \mu_b \|(y, p)\|_{h,*} \|(v_h, q_h)\|_h \quad \forall (v_h, q_h) \in Y_{0h} \times P_{Th}, \tag{13}$$

and for all $y \in Y_{0h} + Y_0 \cap H^{L,1}(Q)$ and $p \in P_{Th} + P_T \cap H^{L,1}(Q)$, where

$$\|(y, p)\|_{h,*}^2 = \|(y, p)\|_h^2 + \varrho \sum_{K \in \mathcal{T}_h} \lambda \|\Delta_x y\|_{L_2(K)}^2 + [(\varrho + 1)\lambda^{-1} + \lambda] \|y\|_{L_2(Q)}^2$$
$$+ \sum_{K \in \mathcal{T}_h} \lambda \|\Delta_x p\|_{L_2(K)}^2 + [2\lambda^{-1} + \lambda] \|p\|_{L_2(Q)}^2.$$

Indeed, using Cauchy's inequalities and the Friedrichs inequality $\|w\|_{L_2(Q)} \le c_{F\Omega}\|\nabla_x w\|_{L_2(Q)}$ that holds for all $w \in Y_0$ or $w \in P_T$, we can easily prove (13) with $\mu_b = (\max\{4, 1 + \lambda c_{F\Omega}^2, 3 + \varrho^{-1}, 1 + \lambda c_{F\Omega}^2 \varrho^{-1}\})^{1/2}$. Now, (11), (12), and (13) immediately lead to the following Céa-like estimate of the discretization error by some best-approximation error.

**Theorem 1.** *Let $y_d \in L_2(Q)$ be a given target, and let $\nu \in L_\infty(Q)$ fulfill the assumptions imposed above. Furthermore, we assume that the regularization (cost) parameter $\varrho \in \mathbb{R}_+$ is fixed. Then the Céa-like estimate*

$$\|(y - y_h, p - p_h)\|_h \le \inf_{v_h \in Y_{0h}, q_h \in P_{Th}} \left( \|(y - v_h, p - q_h)\|_h + \frac{\mu_b}{\mu_c}\|(y - v_h, p - q_h)\|_{h,*} \right)$$

*holds, where $(y, p)$ and $(y_h, p_h)$ are the solutions of (3) and (10), respectively.*

This Céa-like estimate immediately yields convergence rate estimates of the form

$$\|(y - y_h, p - p_h)\|_h \le c(u, p)h^s \tag{14}$$

with $s = \min\{k, l\}$ provided that $y \in Y_0 \cap H^{L,1}(Q) \cap H^{l+1}(Q)$ and $p \in P_T \cap H^{L,1}(Q) \cap H^{l+1}(Q)$, where $l$ is some positive real number defining the regularity of the solution; see [3] for corresponding convergence rate estimates for the state equation only.

## 4  Numerical Results

Let $\{\phi^{(j)} : j = 1, \ldots, N_h\}$ be a nodal finite element basis for $Y_{0h}$, and let $\{\psi^{(m)} : m = 1, \ldots, M_h\}$ be a nodal finite element basis for $P_{Th}$. Then we can express each finite element function $y_h \in Y_{0h}$ and $p_h \in P_{Th}$ via the finite element basis, i.e. $y_h = \sum_{j=1}^{N_h} y_j \phi^{(j)}$ and $p_h = \sum_{m=1}^{M_h} p_m \psi^{(m)}$, respectively. We insert this ansatz into (10), test with basis functions $\phi^{(i)}$ and $\psi^{(n)}$, and obtain the system

$$\mathbf{K}_h \begin{pmatrix} \mathbf{y}_h \\ \mathbf{p}_h \end{pmatrix} = \begin{pmatrix} \mathbf{0} \\ \mathbf{f}_h \end{pmatrix}$$

with $\mathbf{K}_h = (a_h(\phi^{(j)}, \psi^{(m)}; \phi^{(i)}, \psi^{(n)}))_{i,j=1,\ldots,N_h}^{m,n=1,\ldots,M_h}$, $\mathbf{f}_h = (\ell_h(0, \psi^{(n)}))_{n=1,\ldots,M_h}$, $\mathbf{y}_h = (y_j)_{j=1,\ldots,N_h}$ and $\mathbf{p}_h = (p_m)_{m=1,\ldots,M_h}$. The (block)-matrix $\mathbf{K}_h$ is non-symmetric, but positive definite due to (12). Hence the linear system is solved by means of the flexible General Minimal Residual (GMRES) method, preconditioned by a block-diagonal algebraic multigrid (AMG) method, i.e., we apply an AMG preconditioner to each of the diagonal blocks of $\mathbf{K}_h$. Note that we only need to solve once in order to obtain a numerical solution of the space-time tracking optimal control problem (1)–(2), consisting of state and adjoint state. The control can then be recovered from the gradient equation $p + \varrho u = 0$.

The space-time finite element method is implemented by means of the C++ library MFEM [7]. We use *BoomerAMG*, provided by the linear solver library *hypre*, to realize the preconditioner. The linear solver is stopped once the initial residual is reduced by a factor of $10^{-8}$. We are interested in convergence rates with respect to the mesh size $h$ for a fixed regularization (cost) parameter $\varrho$.

## 4.1 Smooth Target

For our first example, we consider the space-time cylinder $Q = (0,1)^3$, i.e., $d = 2$, the manufactured state

$$y(x,t) = \sin(x_1 \pi) \sin(x_2 \pi) \left( a\, t^2 + b\, t \right),$$

as well as the corresponding adjoint state

$$p(x,t) = -\varrho \sin(x_1 \pi) \sin(x_2 \pi) \left( 2\, \pi^2 a\, t^2 + (2\, \pi^2 b + 2\, a)t + b \right),$$

with $a = \frac{2\pi^2+1}{2\pi^2+2}$ and $b = 1$. The desired state $y_d$ and the optimal control $u$ are then computed accordingly, and we fix the regularization parameter $\varrho = 0.01$. This problem is very smooth and devoid of any local features or singularities. Hence, we expect optimal convergence rates. Indeed, as we can observe in Fig. 1, the error in the $\|(\cdot,\cdot)\|_h$-norm decreases with a rate of $\mathcal{O}(h^k)$, where $k$ is the polynomial degree of the finite element basis functions.

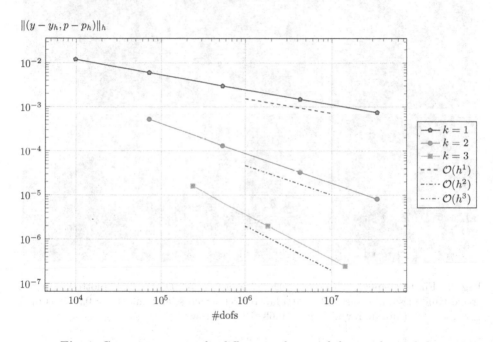

**Fig. 1.** Convergence rates for different polynomial degrees $k = 1, 2, 3$.

## 4.2 Discontinuous Target

For the second example, we consider once more the space-time cylinder $Q = (0,1)^3$, and specify the target state

$$y_d(x,t) = \begin{cases} 1, & \sqrt{(x_1 - 0.5)^2 + (x_2 - 0.5)^2 + (t - 0.5)^2} \leq 0.25, \\ 0, & \text{else,} \end{cases}$$

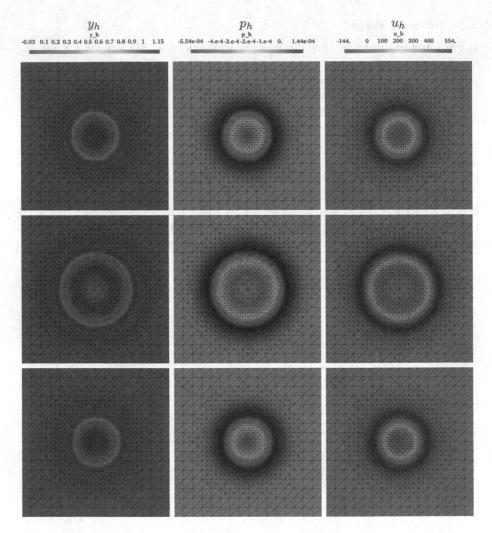

**Fig. 2.** Finite element solutions, with $J(y_h, u_h) = 3.5095 \times 10^{-3}$, plotted over the space-time mesh, obtained after 20 adaptive refinements, and cut at $t = 0.3125$ (upper row), $t = 0.5$ (middle row), and $t = 0.6875$ (lower row).

as an expanding and shrinking circle that is nothing but a *fixed* ball in the space-time cylinder $Q$. We use the fixed regularization parameter $\varrho = 10^{-6}$. Here, we do not know the exact solutions for the state or the optimal control. Thus, we cannot compute any convergence rates for the discretization error. However, the discontinuous target state may introduce local features at the (hyper-)surface of discontinuity. Hence, it might be beneficial to use adaptive mesh refinements driven by an a posteriori error indicator. In particular, we use the residual based indicator proposed by Steinbach and Yang [8], applied to the residuals of the reduced optimality system (4). The final indicator is then the sum of the squares of both parts.

In Fig. 2, we present the finite element functions $y_h$, $p_h$, and $u_h$, plotted over cuts of the space-time mesh $\mathcal{T}_h$ at different times $t$. We can observe that the mesh refinements are mostly concentrated in annuli centered at $(0.5, 0.5)$, e.g. for $t = 0.5$, the outer and inner radii are $\sim \frac{7}{36} \pm \frac{1}{36}$, respectively; see Fig. 2 (middle row).

## 5  Conclusions

We proposed a stable, fully unstructured, space-time simplicial finite element discretization of the reduced optimality system of the standard space-time tracking parabolic optimal control problem with $L_2$-regularization. We derived a priori discretization error estimates. We presented numerical results for two benchmarks. We observed optimal rates for the example with smooth solutions as predicted by the a priori estimates. In the case of a discontinuous target, we used full space-time adaptivity. In order to get the full space-time solution $(y_h, p_h, u_h)$, we have to solve one large-scale system of space-time finite element equations. We used a parallel version of the flexible GMRES preconditioned by AMG. It is clear that this approach can be generalized to more advanced optimal control problems including non-linear state equations and control constraints.

## References

1. Ciarlet, P.G.: The Finite Element Method for Elliptic Problems. North-Holland Publishing Co., Amsterdam (1978)
2. Dier, D.: Non-autonomous maximal regularity for forms of bounded variation. J. Math. Anal. Appl. **425**, 33–54 (2015)
3. Langer, U., Neumüller, M., Schafelner, A.: Space-time finite element methods for parabolic evolution problems with variable coefficients. In: Apel, T., Langer, U., Meyer, A., Steinbach, O. (eds.) FEM 2017. LNCSE, vol. 128, pp. 247–275. Springer, Cham (2019). https://doi.org/10.1007/978-3-030-14244-5_13
4. Langer, U., Steinbach, O., Tröltzsch, F., Yang, H.: Space-time finite element discretization of parabolic optimal control problems with energy regularization. SIAM J. Numer. Anal. **59**, 675–695 (2021)
5. Langer, U., Steinbach, O., Tröltzsch, F., Yang, H.: Unstructured space-time finite element methods for optimal control of parabolic equations. SIAM J. Sci. Comput. **43**, A744–A771 (2021)
6. Lions, J.L.: Optimal control of systems governed by partial differential equations, vol. 170. Springer, Berlin (1971). https://doi.org/10.1007/978-3-642-65024-6
7. MFEM: Modular finite element methods library. mfem.org
8. Steinbach, O., Yang, H.: Comparison of algebraic multigrid methods for an adaptive space-time finite-element discretization of the heat equation in 3D and 4D. Numer. Linear Algebra Appl. **25**(3), e2143 nla.2143 (2018)
9. Tröltzsch, F.: Optimal Control of Partial Differential Equations: Theory, Methods and Applications, Graduate Studies in Mathematics, vol. 112. American Mathematical Society, Providence (2010)

# A New Algorithm for the LQR Problem with Partially Unknown Dynamics

Agnese Pacifico<sup>(✉)</sup> ⓘ, Andrea Pesare ⓘ, and Maurizio Falcone ⓘ

Dipartimento di Matematica, Sapienza Università di Roma, Rome, Italy
pacifico.1699761@studenti.uniroma1.it,
{pesare,falcone}@mat.uniroma1.it

**Abstract.** We consider an LQR optimal control problem with partially unknown dynamics. We propose a new model-based online algorithm to obtain an approximation of the dynamics *and* the control at the same time during a single simulation. The iterative algorithm is based on a mixture of Reinforcement Learning and optimal control techniques. In particular, we use Gaussian distributions to represent model uncertainty and the probabilistic model is updated at each iteration using Bayesian regression formulas. On the other hand, the control is obtained in feedback form via a Riccati differential equation. We present some numerical tests showing that the algorithm can efficiently bring the system towards the origin.

**Keywords:** Reinforcement learning · LQR problem · Numerical methods

## 1 Introduction

The Linear-Quadratic Regulator (LQR) optimal control problem [1,3] is a classical problem in control theory with a wide range of applications. When the dynamics is known, the optimal control is obtained in feedback form solving a backward Riccati differential equation.

Some Reinforcement Learning (RL) problems can be seen as LQR problems where the dynamics is partially or completely unknown. Some RL algorithms try to learn a model for the dynamics and use this model to find the optimal policy. These are called *model-based algorithms*. Others recover the optimal control directly, without reconstructing the dynamics, and these are called *model-free algorithms*. For a broad overview on RL, we recommend [14].

The connection between RL and optimal control theory was already identified in the past [15]. Recently, Palladino and co-authors tried to clarify this relationship via some rigorous proofs, identifying some RL tasks as real optimal control problems with unknown dynamics [8,10,11]. In this context, we propose a *new model-based algorithm for LQR problems* where the dynamics is partially

This research has been partially supported by the INdAM Research group GNCS.

I. Lirkov and S. Margenov (Eds.): LSSC 2021, LNCS 13127, pp. 322–330, 2022.
https://doi.org/10.1007/978-3-030-97549-4_37

unknown, which takes contributions from both fields. In fact, our algorithm can be considered as a case of Bayesian RL, a class of model-based algorithms where the controller builds a stochastic model of the dynamics and updates it according to Bayesian statistics [6]. On the other hand, we borrow the LQR solution from optimal control theory, to get the synthesis of a suitable control [1].

In particular, our algorithm is similar to PILCO [2], from which it takes the use of Gaussian distributions and the whole process of Bayesian update. However, we propose here some novelties. The first is that in PILCO the optimal control is chosen through a gradient descent algorithm in a class of controls, whereas we solve a Riccati differential equation to identify the minimizer. The second is that PILCO needs several trials to reconstruct the dynamics and stabilise the system. Our algorithm is designed to approximate the dynamics *and* to find a suitable control in a single run but can be applied only to linear dynamical systems. Finally, let us mention that other works have already dealt with LQR problems with unknown dynamics (see i.e. [4,5,7,9] and references therein), but they all need several trials to converge, whereas our method works with just one.

The paper is structured as follows. In Sect. 2 we recall the LQR problem. In Sect. 3 we present our algorithm and discuss some implementation details. Finally, in Sect. 4 we show and discuss some numerical tests.

## 2   The Classical LQR Problem

The Linear-Quadratic Regulator (LQR) problem [1,3] is an optimal control problem with linear dynamics and quadratic cost. In the finite horizon case, the state of the system $x(t) \in \mathbb{R}^n$ evolves according to the following controlled dynamics

$$\begin{cases} \dot{x}(t) = \widehat{A}x(t) + Bu(t), & t \in [0, T] \\ x(0) = x_0. \end{cases} \tag{1}$$

We will denote by $\mathbb{R}^{m \times n}$ the space of matrices with $m$ rows and $n$ columns; $\mathbb{I}_n$ and $\mathbf{0}_n$ will be respectively the identity and zero matrices in $\mathbb{R}^{n \times n}$. Here $\widehat{A} \in \mathbb{R}^{n \times n}$ and $B \in \mathbb{R}^{n \times m}$. The control function $u(t) \in \mathbb{R}^m$ must be chosen among the admissible controls $\mathcal{U} := \{u : [0, T] \to \mathbb{R}^m \text{ Lebesgue measurable}\}$ to minimize the quadratic cost functional

$$J_{x_0}[u] := \frac{1}{2} \left( \int_0^T \left( x(t)^T Q x(t) + u(t)^T R u(t) \right) dt + x(T)^T Q_f x(T) \right). \tag{2}$$

The *main assumptions* on the cost matrices are the following:

- $Q, Q_f \in \mathbb{R}^{n \times n}$ are symmetric and positive semi-definite;
- $R \in \mathbb{R}^{m \times m}$ is symmetric and positive definite.

For any $x_0 \in \mathbb{R}^n$, we will call $\bar{u}$ an optimal control starting from $x_0$ if and only if

$$J_{x_0}[\bar{u}] \leq J_{x_0}[u] \quad \forall u \in \mathcal{U}. \tag{3}$$

When the dynamics is fully known, the optimal control can be obtained in feedback form [1]. Indeed, if $P(t)$ is the unique symmetric solution of the Riccati differential equation

$$\begin{cases} -\dot{P}(t) = \widehat{A}^T P(t) + P(t)\widehat{A} - P(t)BR^{-1}B^T P(t) + Q, & t \in [0,T] \\ P(T) = Q_f, \end{cases} \tag{4}$$

the optimal control is given by $\bar{u}(t) = -R^{-1}B^T P(t)x(t)$.

We want to investigate: what happens if the dynamics is partially unknown? How can one find a suitable control? We considered the following framework:

**Setting.** *We assume that the matrices $B \in \mathbb{R}^{n \times m}$, $Q, Q_f \in \mathbb{R}^{n \times n}$ and $R \in \mathbb{R}^{m \times m}$ are given, whereas the state matrix $\widehat{A} \in \mathbb{R}^{n \times n}$ is unknown.*

## 3   An Online Algorithm for the LQR Problem

In this section, we describe our model-based algorithm able to solve the LQR problem without knowing the matrix $\widehat{A}$. The algorithm is online, meaning that it doesn't need any previous simulation or computation. Our goal is twofold: first, we look for a good estimate for the unknown dynamics matrix $\widehat{A}$; and secondly, we want to choose a control that can steer the trajectory towards the origin. Furthermore, we assume that we can run the experiment just once, so the system must be controlled while the dynamics is still uncertain. To this end, we use a technique to get an estimate of the matrix $\widehat{A}$ and to update the control at the same time.

We divide the interval $[0,T]$ into equal time steps of length $\Delta t$, globally we will have $N$ time steps and we group them into rounds of $S$ steps each. The $i$-th round will be denoted by $[t_{i-1}, t_i]$ and a superscript index will indicate a single time step:

$$t_i^j = t_{i-1} + j\Delta t, \quad j = 0, \dots, S.$$

The current knowledge of the dynamics matrix is represented as a probability distribution over matrices. For each round, two major operations are carried out:

1. At the beginning of each round, a probability $\pi_{i-1}$ is given from the previous round. We use the mean $\bar{A}_{i-1}$ of this distribution and compute a feedback control for the round solving a Riccati equation;
2. At the end of each round, the current probability distribution is updated using Bayesian formulas, according to the trajectory observed during the round. The output is a new probability distribution $\pi_i$.

The whole algorithm is summarised below as Algorithm 1. In the following, we will give more technical details.

**Prior Distribution.** The algorithm requires the choice of a prior distribution $\pi_0$. We fix some $m_0 \in \mathbb{R}^n$ and $\Sigma_0 \in \mathbb{R}^{n \times n}$ and consider a random matrix $A$ such

---

**Algorithm 1** An online algorithm for the LQR problem

Divide $[0, T]$ into $M$ intervals of length $\Delta t$, with $\Delta t = \frac{T}{M}$;
Group the intervals in *rounds*, each containing $S$ intervals;
Choose a prior distribution $\pi_0$ over matrices;
**for** $i$ from 1 to the *rounds* number **do**
    Find a feedback control $u_i^*$ solving a Riccati eq. with the mean matrix $\bar{A}_{i-1}$;
    Use $u_i^*$ as control for all the steps in the $i$-th *round*;
    Observe the actual trajectory;
    Update $\pi_i$ according to the data from the observed trajectory
**end for**

---

that each of its rows $r_k$ is distributed as an independent Gaussian vector with mean $m_0$ and covariance matrix $\Sigma_0$

$$A = \begin{bmatrix} r_1^T \\ \vdots \\ r_n^T \end{bmatrix} \qquad r_k \sim \mathcal{N}(m_0, \Sigma_0) \qquad \forall k = 1, \ldots, n$$

A typical choice for the prior distribution, when no information about the true matrix $\widehat{A}$ is available, is $m_0 = (0, \ldots, 0)^T$ and $\Sigma_0 = nm\,\mathbb{I}_n$.

**Feedback Control.** At the beginning of the round $[t_{i-1}, t_i]$ our knowledge of the matrix $\widehat{A}$ is described by the distribution $\pi_{i-1}$. In order to find the control to apply, we solve the evolutive *Riccati* equation associated with the matrix $\bar{A}_{i-1}$, where $\bar{A}_{i-1}$ is the mean of the distribution $\pi_{i-1}$. The Riccati equation reads

$$\begin{cases} -\dot{P}(t) = \bar{A}_{i-1}^T P(t) + P(t)\bar{A}_{i-1} - P(t)BR^{-1}B^T P(t) + Q & t \in [t_{i-1}, T] \\ P(T) = Q_f. \end{cases} \tag{5}$$

If we denote by $P_i(t)$ the solution of (5), our control will be the feedback control given by

$$u_i^*(t_i^j) = -R^{-1}B^T P_i(t_i^j)x_i^j \qquad j = 0, \ldots S - 1, \tag{6}$$

where $x_i^j = x(t_i^j)$ is the observed state of the system. Since the control must be defined for all instants $t \in [t_{i-1}, t_i]$, we choose a piecewise constant control: $u_i^*(t) = u_i^*(t_i^j)$ for $t \in [t_i^j, t_i^{j+1}]$. Note that we need real-time observations of the system state to compute $u_i^*$, since it depends on $x$.

**Distribution Update.** At the end of each round we can update the probability distribution using *Bayesian regression formulas*. More precisely, we know that during the time step $[t_i^j, t_i^{j+1}]$ the system evolves according to (1) where we plug-in the chosen piecewise constant control. We can easily get an approximation of the state derivative with a finite difference scheme using the state observations $x_i^j = x(t_i^j)$. By a first-order scheme, rearranging the terms, we get

$$\widehat{A}x_i^j \simeq \frac{x_i^{j+1} - x_i^j}{\Delta t} - Bu^*(t_i^j) \qquad j = 0, \ldots, S - 1. \tag{7}$$

We interpret these data as observations of the dynamics with a Gaussian noise due to the error in the derivative approximation and the measurements. We treat each row $\widehat{r}_k$ of $\widehat{A}$ separately. Denoting by $y^j$ the right hand side in (7), we can write its $k$-th component as

$$y^j_{(k)} = \widehat{r}_k x^j_i + \varepsilon \qquad \text{with } \varepsilon \sim \mathcal{N}(0, \sigma^2),$$

where $\sigma$ is a parameter chosen according to the order of the derivative approximation.

For each row $k$ we have a prior distribution $r_k \sim \mathcal{N}(m_0, \Sigma_0)$ given by $\pi_{i-1}$. We define $X$ as the matrix whose columns are $x^0_i, \ldots, x^{S-1}_i$ and $y$ as the column vector $y^0_{(k)}, \ldots, y^{S-1}_{(k)}$. Therefore we obtain a posterior distribution $r_k|_{y,X} \sim \mathcal{N}(m, \Sigma)$ where $\Sigma^{-1} = \frac{1}{\sigma^2} X X^T + \Sigma_0^{-1}$ and $m = \Sigma(Xy/\sigma^2 + \Sigma_0^{-1} m_0)$. For more details about Bayesian Linear Regression see for instance the extensive monograph by Rasmussen and Williams [12].

**Higher-Order Schemes.** The derivative in (7) can be approximated with higher order finite difference schemes [13]. Note that trajectory regularity is required in the interval containing the nodes used in the approximation. While first-order approximation uses only two nodes, higher-order approximations use more nodes, thus the control cannot jump at each time step and we have to keep it constant for more steps.

*Remark 1 (Heuristic argument for convergence).* The algorithm cannot find the optimal control for the problem, since at the beginning the matrix $\widehat{A}$ is unknown and it needs at least some steps to have a good estimate for $\widehat{A}$. However, from Bayesian Regression theory (see [12]) we know that the more data we observe, the more precise our distribution $\pi_i$ becomes, eventually converging to the Dirac delta $\delta_{\widehat{A}}$.

Furthermore, let us note that since $\pi_i$ is converging to $\delta_{\widehat{A}}$, then also the mean matrix $\bar{A}_i$ of $\pi_i$ is converging to $\widehat{A}$. Thus, after few rounds, the feedback control computed by the algorithm (which is using $\bar{A}_i$) in the interval $[t_i, T]$ should be close to the optimal control of a trajectory of the real dynamics which starts from the same point $x^S_{i-1}$.

Finally, when $\Delta t \to 0$, the algorithm reaches earlier a good estimate of the dynamics matrix. This means that also the computed control is closer to the optimal one. All these heuristics are confirmed by the numerical simulations in the next section.

## 4    Some Numerical Tests

The following numerical tests for the algorithm described above were performed in MATLAB and took few seconds to run.

**Test 1.** We first consider a dynamical system where the state lies in $\mathbb{R}^2$ and the control is 1-dimensional, i.e. $n = 2$ and $m = 1$. The LQR problem is defined by the following matrices:

$$\widehat{A} = \begin{pmatrix} 0 & 1 \\ -1 & 0 \end{pmatrix} \quad B = \begin{pmatrix} 0 \\ 1 \end{pmatrix} \quad Q = \begin{pmatrix} 1 & 0 \\ 0 & 1 \end{pmatrix} \quad R = 0.1 \quad Q_f = \begin{pmatrix} 0 & 0 \\ 0 & 0 \end{pmatrix}.$$

The time horizon is set to $T = 5$ and the starting point is $x_0 = (0, 1)^T$. We assume that the matrix $\widehat{A}$ is unknown to the algorithm, though we use it to simulate the dynamics. We choose the prior distribution as recommended in Sect. 3, using $m_0 = (0, 0)^T$ and $\Sigma_0 = 2I_2$; for all the tests we set $\sigma = \sqrt{10\Delta t^p}$ and $S = 2p$, where $p$ is the order of the scheme used in the derivative approximation. Figure 1 shows the *piecewise constant* controls chosen by the algorithm with $p = 1$ for different values of $\Delta t$ and the corresponding trajectories. Recall that the matrix $\widehat{A}$ is completely unknown at the beginning, so the control we apply in the first steps depends only on the prior distribution we have chosen and clearly is not accurate. This causes the trajectory to deviate from the optimal solution. However, after few steps the matrix $\widehat{A}$ is well approximated and the algorithm manages to steer the state towards the origin anyway. In Table 1a we have reported the cost of the solution found by the algorithm for different choices of $\Delta t$. When $\Delta t$ is smaller, the algorithm recovers the matrix $\widehat{A}$ quickly and thus the deviation from the optimal solution is smaller. This confirms the heuristics of Remark 1. We can also observe a numerical order of convergence equal to 1.

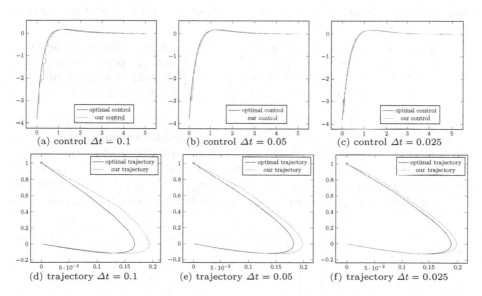

(a) control $\Delta t = 0.1$     (b) control $\Delta t = 0.05$     (c) control $\Delta t = 0.025$

(d) trajectory $\Delta t = 0.1$     (e) trajectory $\Delta t = 0.05$     (f) trajectory $\Delta t = 0.025$

**Fig. 1.** Simulations of Test 1 for different values of $\Delta t$. The first row shows the control chosen by the algorithm as a function of time (in red) compared with the optimal control (in blue), computed knowing the matrix $\widehat{A}$; the second row shows the trajectories in $\mathbb{R}^2$. The red dot is the trajectory starting point. (Color figure online)

**Table 1.** Numerical results for Test 1. (a): Cost of the trajectory for different values of $\Delta t$. The cost is compared with the optimal cost computed knowing $\widehat{A}$ ($C^*$ last row). (b): Error for $\widehat{A}$ after the simulation using different $\Delta t$ and different finite difference schemes for the derivatives approximation; $p$ is the scheme order.

(a)

| $\Delta t$ | Cost | Error | Order |
|---|---|---|---|
| 0.1 | 0.3897 | 0.0063 | - |
| 0.05 | 0.3866 | 0.0032 | 0.98 |
| 0.025 | 0.3849 | 0.0015 | 1.09 |
| $C^*$ | 0.3834 | - | - |

(b)

| | $p = 1$ | | $p = 2$ | | $p = 4$ | |
|---|---|---|---|---|---|---|
| $\Delta t$ | Error | Order | Error | Order | Error | Order |
| 0.1 | 0.167 | - | 9.74e-3 | - | 1.63e-4 | - |
| 0.05 | 0.087 | 0.94 | 1.91e-3 | 2.3 | 3.00e-6 | 5.8 |
| 0.025 | 0.045 | 0.95 | 4.14e-4 | 2.2 | 1.03e-7 | 4.8 |
| 0.01 | 0.018 | 1.00 | 6.30e-5 | 2.1 | 1.70e-9 | 4.5 |

We tried different finite difference schemes for the approximation of the state derivatives (see "*Higher-order schemes*" in Sect. 3). Table 1b shows the error in the approximation of $\widehat{A}$ at the end of the simulation, when using schemes of order $p = 1, 2, 4$ and for different values of $\Delta t$. As expected, when we consider more accurate approximations of the gradient, we get better estimations of the matrix $\widehat{A}$. Unfortunately, the same does not hold for the solution costs, which are not significantly improved if compared with the ones found by the first-order approximation.

**Test 2.** For the second test we choose $n = 4$, $m = 3$ and $T = 10$. Our matrices are

$$\widehat{A} = \begin{pmatrix} -0.0215 & -0.7776 & -0.1922 & 0.9123 \\ -0.3246 & 0.5605 & -0.8071 & 0.1504 \\ 0.8001 & -0.2205 & -0.7360 & -0.8804 \\ -0.2615 & -0.5166 & 0.8841 & -0.5304 \end{pmatrix}, \quad B = \begin{pmatrix} -0.2937 & -0.6620 & -0.0982 \\ 0.6424 & 0.2982 & 0.0940 \\ -0.9692 & 0.4634 & -0.4074 \\ -0.9140 & 0.2955 & 0.4894 \end{pmatrix},$$

$Q = \frac{1}{4}\mathbb{I}_4$, $R = \frac{1}{3}\mathbb{I}_3$ and $Q_f = \mathbb{I}_4$, and the starting point is $x_0 = (1, 1, 1, 1)^T$. We set $\Delta t = 0.025$, $S = 4$ and use a second order approximation for the derivatives. Fig. 2 shows the control found by the algorithm and Fig. 3 the corresponding trajectory. The behaviour observed in Test 1 is even more visible here: in the first steps the control is not accurate since we do not know $\widehat{A}$, but after few steps the algorithm learns more on the matrix and manages to bring the state

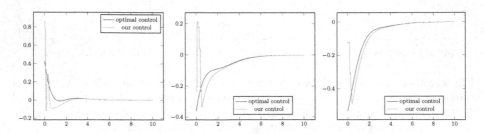

**Fig. 2.** The three components of the control in Test 2

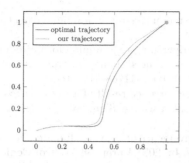 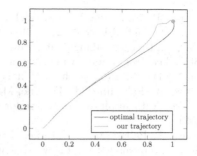

**Fig. 3.** The trajectory of Test 2 in $\mathbb{R}^4$, represented by projecting components in couples: $(x_1, x_2)$ on the left and $(x_3, x_4)$ on the right. The red dots indicate the starting point of the trajectory. (Color figure online)

to the origin. The cost of the control found by the algorithm is 1.111, whereas the optimal cost computed using $\widehat{A}$ is 1.056.

## 5   Conclusions

We proposed a new algorithm designed to deal with LQR problems when the dynamics is partially unknown. Numerical tests presented in Sect. 4 showed how it manages to approximate the dynamics *and* to find a suitable control that brings the system towards the origin in a single simulation. Future works include the convergence analysis of the algorithm and possible extensions to the nonlinear case.

## References

1. Anderson, B.D., Moore, J.B.: Optimal Control: Linear Quadratic Methods. Courier Corporation (2007)
2. Deisenroth, M., Rasmussen, C.E.: PILCO: a model-based and data-efficient approach to policy search. In: Proceedings ICML 2011, pp. 465–472 (2011)
3. Fleming, W.H., Rishel, R.W.: Deterministic and Stochastic Optimal Control, vol. 1. Springer, New York (2012). https://doi.org/10.1007/978-1-4612-6380-7
4. Fong, J., Tan, Y., Crocher, V., Oetomo, D., Mareels, I.: Dual-loop iterative optimal control for the finite horizon LQR problem with unknown dynamics. Syst. Control Lett. **111**, 49–57 (2018)
5. Frueh, J.A., Phan, M.Q.: Linear quadratic optimal learning control (LQL). Int. J. Control **73**(10), 832–839 (2000)
6. Ghavamzadeh, M., Mannor, S., Pineau, J., Tamar, A., et al.: Bayesian reinforcement learning: a survey. Foundations Trends® Mach. Learn. **8**(5–6), 359–483 (2015)
7. Li, N., Kolmanovsky, I., Girard, A.: LQ control of unknown discrete-time linear systems-a novel approach and a comparison study. Optimal Control Appl. Methods **40**(2), 265–291 (2019)

8. Murray, R., Palladino, M.: A model for system uncertainty in reinforcement learning. Syst. Control Lett. **122**, 24–31 (2018)
9. Pang, B., Bian, T., Jiang, Z.-P.: Adaptive dynamic programming for finite-horizon optimal control of linear time-varying discrete-time systems. Control Theory Technol. **17**(1), 73–84 (2019). https://doi.org/10.1007/s11768-019-8168-8
10. Pesare, A., Palladino, M., Falcone, M.: Convergence results for an averaged LQR problem with applications to reinforcement learning. Sign. Syst. Math. Control (2021)
11. Pesare, A., Palladino, M., Falcone, M.: Convergence of the value function in optimal control problems with unknown dynamics. In: 2021 European Control Conference (ECC). IEEE (forthcoming)
12. Rasmussen, C., Williams, C.: Gaussian Processes for Machine Learning. Adaptive Computation and Machine Learning. MIT Press, Cambridge (2006)
13. Strikwerda, J.C.: Finite Difference Schemes and Partial Differential Equations. SIAM (2004)
14. Sutton, R.S., Barto, A.G.: Reinforcement Learning: An Introduction. MIT Press, Cambridge (2018)
15. Sutton, R.S., Barto, A.G., Williams, R.J.: Reinforcement learning is direct adaptive optimal control. IEEE Control. Syst. **12**(2), 19–22 (1992)

# Tensor and Matrix Factorization
# for Big-Data Analysis

# Solving Systems of Polynomial Equations—A Tensor Approach

Mariya Ishteva[1]([✉])[ID] and Philippe Dreesen[2][ID]

[1] Department Computer Science, ADVISE-NUMA, KU Leuven, Geel, Belgium
`mariya.ishteva@kuleuven.be`
[2] Department Electrical Engineering, ESAT-STADIUS, KU Leuven, Leuven, Belgium
`philippe.dreesen@kuleuven.be`

**Abstract.** Polynomial relations are at the heart of mathematics. The fundamental problem of solving polynomial equations shows up in a wide variety of (applied) mathematics, science and engineering problems. Although different approaches have been considered in the literature, the problem remains difficult and requires further study.

We propose a solution based on tensor techniques. In particular, we build a partially symmetric tensor from the coefficients of the polynomials and compute its canonical polyadic decomposition. Due to the partial symmetry, a structured canonical polyadic decomposition is needed. The factors of the decomposition can then be used for building systems of linear equations, from which we find the solutions of the original system.

This paper introduces our approach and illustrates it with a detailed example. Although it cannot solve any system of polynomial equations, it is applicable to a large class of sub-problems. Future work includes comparisons with existing methods and extending the class of problems, for which the method can be applied.

**Keywords:** Systems of polynomial equations · Tensors · Canonical polyadic decomposition · Partial symmetry

## 1 Introduction

Solving systems of multivariate polynomial equations is a fundamental problem in mathematics, having a multitude of scientific and engineering applications. This task typically involves square systems (as many equations as unknowns), which generically have a solution set consisting of isolated points. Computational methods for solving polynomial systems are largely dominated by the symbolic

This research was supported by KU Leuven Research Fund; KU Leuven start-up-grant STG/19/036 ZKD7924; FWO (EOS Project 30468160, SBO project S005319N, Infrastructure project I013218N, TBM Project T001919N, Research projects G028015N, G090117N, PhD grants SB/1SA1319N, SB/1S93918, SB/151622); ERC AdG "Back to the Roots" (885682); Flemish Government (AI Research Program); PD is affiliated to Leuven. AI - KU Leuven institute for AI. Part of this work was done while the authors were with Vrije Universiteit Brussel (VUB-ELEC).

© Springer Nature Switzerland AG 2022
I. Lirkov and S. Margenov (Eds.): LSSC 2021, LNCS 13127, pp. 333–341, 2022.
https://doi.org/10.1007/978-3-030-97549-4_38

Groebner basis approach [6], although other approaches exist, such as homotopy continuation methods.

From the (numerical) linear algebra viewpoint, there is a less known connection between polynomial systems and eigenvalue problems. In the 1980s, the work of Stetter [16], among others, demonstrated that eigenvalue problems are at the core of polynomial systems. The eigenvalue problem formulation, returning all solutions of the system, can be obtained from a Groebner basis, or from resultant-based approaches [6]. Although the number of solutions grows quickly with system size and degree (it is equal to the product of the equations degrees), the computational bottleneck in these approaches is at the steps preceding the eigenvalue problem formulation, such as computing a Groebner basis, or manipulating large resultant-based matrices.

In this article, we look for the hidden eigenvalue problem in another way, by exploring the connection between polynomials and tensors (multiway arrays). In particular, we build a partially symmetric tensor from the coefficients of the polynomials and decompose this tensor with a canonical polyadic decomposition, which can often be reformulated as an eigenvalue problem [7]. Due to partial symmetry, a structured version of the canonical polyadic decomposition is needed. The factors of the decomposition can then be used for building systems of linear equations, from which we find the solutions of the polynomial system.

The remainder of this paper is organized as follows. Section 2 introduces the basic background material. Section 3 first presents the main idea in the case of bivariate polynomial equations of degree two and then discusses generalizations to more equations (more variables) and higher degree polynomials. Section 4 summarizes the main conclusions and discusses future work.

## 2    Background Material

This section introduces our notation, the canonical polyadic decompositions of tensors, and the link between (symmetric) tensors and polynomials.

### 2.1    Notation

We use lowercase ($a$), boldface lowercase ($\mathbf{a}$), uppercase boldface ($\mathbf{A}$), and calligraphic font ($\mathcal{A}$) for scalars, vectors, matrices, and tensors (a $d$th-order tensor is a multiway array with $d$ indices), respectively. The elements of the vectors, matrices, and tensors are accessed as $a_i$, $A_{ij}$ and $\mathcal{A}_{i_1 \ldots i_d}$, respectively.

$\mathbf{A}^\top$ denotes the transpose of the matrix $\mathbf{A}$ and the symbols $\circ$, $\otimes$ and $\odot$ stand for the outer, the Kronecker, and the Khatri-Rao product, respectively.

The elements of the $n$-mode product $\mathcal{A} \bullet_n \mathbf{x} \in \mathbb{R}^{I_1 \times \cdots \times I_{n-1} \times I_{n+1} \times \cdots \times I_d}$, where $1 \leq n \leq d$, of a tensor $\mathcal{A} \in \mathbb{R}^{I_1 \times \cdots \times I_d}$ and a vector $\mathbf{x} \in \mathbb{R}^{I_n}$ are defined as

$$(\mathcal{A} \bullet_n \mathbf{x})_{i_1 \ldots i_{n-1} i_{n+1} \ldots i_d} = \sum_{i_n=1}^{I_n} \mathcal{A}_{i_1 \ldots i_d} x_{i_n}.$$

Thus, if $\mathcal{A}$ is a third-order tensor, the products $\mathcal{A} \bullet_n \mathbf{x}$ are matrices whose elements are the scalar products of the vector $\mathbf{x}$ and the mode-$n$ vectors of $\mathcal{A}$. The product $\mathcal{A} \bullet_1 \mathbf{x} \bullet_2 \mathbf{x}$ is a vector and the product $\mathcal{A} \bullet_1 \mathbf{x} \bullet_2 \mathbf{x} \bullet_3 \mathbf{x}$ is a scalar.

## 2.2 The Canonical Polyadic Decomposition of Tensors

A tensor $\mathcal{A}$ has rank equal to one, if it can be written as an outer product of vectors. For example, for a third-order tensor $\mathcal{A}$ of rank one, we have $\mathcal{A} = \mathbf{a} \circ \mathbf{b} \circ \mathbf{c}$, that is, $\mathcal{A}_{ijk} = a_i b_j c_k$. Every tensor can be expressed as a sum of rank-one tensors, i.e., for a third-order tensor $\mathcal{A}$ we have $\mathcal{A} = \sum_{i=1}^{r} \mathbf{a}_i \circ \mathbf{b}_i \circ \mathbf{c}_i$. We denote this by $\mathcal{A} = [\![\mathbf{A}, \mathbf{B}, \mathbf{C}]\!]$, where $\mathbf{a}_i$, $\mathbf{b}_i$, and $\mathbf{c}_i$ are the $r$ columns of $\mathbf{A}$, $\mathbf{B}$, and $\mathbf{C}$, respectively. If $r$ is minimal, we call this the canonical polyadic decomposition (CPD) of $\mathcal{A}$ [2,10] (Fig. 1) and $r$ is called the rank of tensor $\mathcal{A}$.

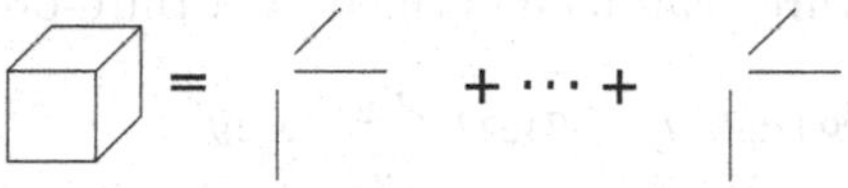

**Fig. 1.** The canonical polyadic decomposition decomposes a third-order tensor into a minimal sum of rank-one terms (outer products of vectors). The number of terms is called the rank of the tensor.

For symmetric tensors (tensors invariant under permutations of the indices), symmetric decompositions are considered, i.e., decompositions with identical factors ($\mathcal{A} = [\![\mathbf{A}, \mathbf{A}, \mathbf{A}]\!]$). We deal with partially symmetric tensors (symmetric with respect to some of the modes), so we seek a partially symmetric decomposition (where some of the factors are identical), for example $\mathcal{A} = [\![\mathbf{A}, \mathbf{A}, \mathbf{C}]\!]$.

Computing the CPD of a tensor is a difficult problem in general but has been studied extensively in the literature. This problem can often be reduced to an eigenvalue problem and our preference goes for this option [7]. In some cases, iterative algorithms are required, for which a number of implementations are available, for example Tensorlab [17]. For further details on tensor decompositions and their applications, we refer to the overview papers [3,5,8,11,12,14], books [4,9,13,15], and references therein.

## 2.3 A Link Between Polynomials and Symmetric Tensors

We next discuss the link between polynomials and symmetric tensors, first in the case of polynomials of degree two and then for higher-degree polynomials.

**Polynomials of Degree Two.** Every polynomial of degree two can be associated with a symmetric coefficient matrix $\mathbf{C}$. A bivariate polynomial in $x$ and $y$ can be written as $\begin{bmatrix} x & y & 1 \end{bmatrix} \mathbf{C} \begin{bmatrix} x \\ y \\ 1 \end{bmatrix}$. For example, consider the polynomials $p$ and $q$,

$$p(x,y) = -x^2 + 2xy + 8y^2 - 12x = \begin{bmatrix} x & y & 1 \end{bmatrix} \begin{bmatrix} -1 & 1 & -6 \\ 1 & 8 & 0 \\ -6 & 0 & 0 \end{bmatrix} \begin{bmatrix} x \\ y \\ 1 \end{bmatrix},$$

$$(1)$$

$$q(x,y) = 2x^2 + 8xy + \tfrac{7}{2}y^2 + 8x - 2y - 2 = \begin{bmatrix} x & y & 1 \end{bmatrix} \begin{bmatrix} 2 & 4 & 4 \\ 4 & \tfrac{7}{2} & -1 \\ 4 & -1 & -2 \end{bmatrix} \begin{bmatrix} x \\ y \\ 1 \end{bmatrix}.$$

We denote $\begin{bmatrix} x \\ y \\ 1 \end{bmatrix}$ by $\mathbf{u}$. In case of more variables, the length of $\mathbf{u}$ increases, but any polynomial can still be written as $\mathbf{u}^\top \mathbf{C} \mathbf{u}$, with some symmetric matrix $\mathbf{C}$.

**Polynomials of Higher Degree.** Every polynomial of higher degree can be associated with a higher-order tensor (instead of with a matrix). For example, a polynomial of degree three can be described by a third-order tensor $\mathcal{C}$:

$$a_{111}x^3 + 3a_{112}x^2 y + 3a_{122}xy^2 + a_{222}y^3$$
$$+ 3a_{110}x^2 + 6a_{120}xy + 3a_{220}y^2 + 3a_{100}x + 3a_{200}y + a_{000}$$
$$= \mathcal{C} \bullet_1 \mathbf{u} \bullet_2 \mathbf{u} \bullet_3 \mathbf{u},$$

where $\mathcal{C} \in \mathbb{R}^{3 \times 3 \times 3}$ has the following frontal slices

$$\begin{bmatrix} a_{111} & a_{112} & a_{110} \\ a_{112} & a_{122} & a_{120} \\ a_{110} & a_{120} & a_{100} \end{bmatrix}, \quad \begin{bmatrix} a_{112} & a_{122} & a_{120} \\ a_{122} & a_{222} & a_{220} \\ a_{120} & a_{220} & a_{200} \end{bmatrix}, \quad \begin{bmatrix} a_{110} & a_{120} & a_{100} \\ a_{120} & a_{220} & a_{200} \\ a_{100} & a_{200} & a_{000} \end{bmatrix}.$$

In case of more variables, the length of $\mathbf{u}$ increases, but the polynomial can still be written as $\mathcal{C} \bullet_1 \mathbf{u} \bullet_2 \mathbf{u} \bullet_3 \mathbf{u}$, with some symmetric tensor $\mathcal{C}$.

## 3    Solving Polynomial Systems by Tensor Decompositions

This section first presents the main idea in the case of two polynomials (in two variables) of degree two and then discusses possible extensions to the cases of more equations (and variables) and higher degree polynomials.

### 3.1    Solving Systems of Two Polynomial Equations of Degree Two

For systems of two bivariate polynomial equations of degree two, our approach consist of four steps: building a partially symmetric tensor from the coefficients of the polynomials (Step 1), computing its partially symmetric CPD (Step 2) and using the factors of the decomposition to build systems of linear equations, from which we find the solutions $(x, y)$ (Steps 3–4). We use the polynomials from (1) to illustrate the approach, see also Fig. 2.

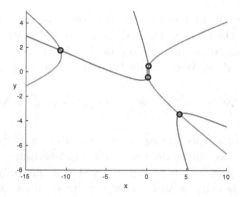

**Fig. 2.** The equations from the running example are visualized as the blue and red lines. The solutions of the system are the four points of intersection, circled in black. (Color figure online)

**Step 1.** We associate one matrix with each equation (as in (1)) and stack them behind each other in a partially symmetric third-order tensor $\mathcal{T} \in \mathbb{R}^{3 \times 3 \times 2}$. The system then becomes

$$\mathcal{T} \bullet_1 \mathbf{u} \bullet_2 \mathbf{u} = \begin{bmatrix} 0 \\ 0 \end{bmatrix}, \quad \text{with } \mathbf{u} = \begin{bmatrix} x \\ y \\ 1 \end{bmatrix}. \tag{2}$$

**Step 2.** We next decompose $\mathcal{T}$ in (partially symmetric) rank-one terms,

$$\mathcal{T} = [\![\mathbf{V}, \mathbf{V}, \mathbf{W}]\!],$$

with $\mathbf{V} \in \mathbb{R}^{3 \times r}$ and $\mathbf{W} \in \mathbb{R}^{2 \times r}$, where $r$ is the rank of $\mathcal{T}$. The typical ranks of a $3 \times 3 \times 2$ tensor are three and four over $\mathbb{R}$. However, if we allow decompositions with complex numbers, the typical rank is only three. In the following, we thus consider the rank to be three and will work with complex numbers, if necessary. For our example the rank over $\mathbb{R}$ is three. We obtain[1]

$$\mathbf{V} = \begin{bmatrix} 2 & 1 & 1 \\ 1 & -1 & 2 \\ 2 & -2 & 0 \end{bmatrix}, \quad \mathbf{W} = \begin{bmatrix} -1 & 1 & 2 \\ \frac{1}{2} & -1 & 1 \end{bmatrix}.$$

The system of polynomial equations (2) can now be re-written as

$$\mathcal{T} \bullet_1 \mathbf{u} \bullet_2 \mathbf{u} = [\![\mathbf{V}, \mathbf{V}, \mathbf{W}]\!] \bullet_1 \mathbf{u} \bullet_2 \mathbf{u} = [\![\mathbf{u}^\top \mathbf{V}, \mathbf{u}^\top \mathbf{V}, \mathbf{W}]\!] = \begin{bmatrix} 0 \\ 0 \end{bmatrix}. \tag{3}$$

---

[1] The CPD is invariant under some scaling and permutation of the columns of the factors. Here we have chosen to re-scale the solution obtained from the eigenvalue algorithm, in order to simplify the numbers in the example.

In the following, we will first ignore the fact that the last element of $\mathbf{u}$ equals one. The system (3) then has an intrinsic scaling indeterminacy. We will resolve this issue by re-scaling $\mathbf{u}$ in the last step of the algorithm.

**Step 3.** Let $\mathbf{z} = \mathbf{V}^\top \mathbf{u}$. We re-write Eq. (3) and solve it for $\mathbf{z}$:

$$[\![\mathbf{u}^\top \mathbf{V}, \mathbf{u}^\top \mathbf{V}, \mathbf{W}]\!] = [\![\mathbf{z}^\top, \mathbf{z}^\top, \mathbf{W}]\!] = \mathbf{W}(\mathbf{z}^\top \odot \mathbf{z}^\top)^\top = \mathbf{W}\mathbf{z}.^2 = \begin{bmatrix} 0 \\ 0 \end{bmatrix}, \quad (4)$$

where the elements of $\mathbf{z}.^2$ are the squared elements of $\mathbf{z}$. This is a linear system of equations for $\mathbf{z}.^2$ (we disregard the scaling indeterminacy for now and will resolve it in Step 4.). $\mathbf{z}.^2$ leads to eight possible $\mathbf{z}$ but only four of them are essentially different because $\mathbf{z}$ and $-\mathbf{z}$ eventually produce the same $(x, y)$, due to the scaling indeterminacy of $\mathbf{u}$.

For our example,

$$\mathbf{z}.^2 = \begin{bmatrix} 0.8242 \\ 0.5494 \\ 0.1374 \end{bmatrix},$$

so the four (essentially different) solutions are

$$\mathbf{z}^{(1)} = \begin{bmatrix} 0.9078 \\ 0.7412 \\ 0.3706 \end{bmatrix}, \mathbf{z}^{(2)} = \begin{bmatrix} -0.9078 \\ 0.7412 \\ 0.3706 \end{bmatrix}, \mathbf{z}^{(3)} = \begin{bmatrix} 0.9078 \\ -0.7412 \\ 0.3706 \end{bmatrix}, \mathbf{z}^{(4)} = \begin{bmatrix} 0.9078 \\ 0.7412 \\ -0.3706 \end{bmatrix}. \quad (5)$$

**Step 4.** We find four solutions for $\mathbf{u}$ from $\mathbf{z}^\top = \mathbf{u}^\top \mathbf{V}$ by solving the four systems of linear equations (one for each $\mathbf{z}^{(i)}$ from (5))

$$\mathbf{V}^\top \mathbf{u} = \mathbf{z}.$$

Finally, we rescale each of the four solutions for $\mathbf{u}$, so that the last element becomes one. The first two elements are then $x$ and $y$.

For our example we obtain

$$\mathbf{u}^{(1)} = \begin{bmatrix} 0.5497 \\ -0.0895 \\ -0.0510 \end{bmatrix} = -0.0510 \begin{bmatrix} -10.7766 \\ 1.7553 \\ 1 \end{bmatrix}, \mathbf{u}^{(2)} = \begin{bmatrix} -0.0555 \\ 0.2131 \\ -0.5049 \end{bmatrix} = -0.5049 \begin{bmatrix} 0.1100 \\ -0.4220 \\ 1 \end{bmatrix},$$

$$\mathbf{u}^{(3)} = \begin{bmatrix} 0.0555 \\ 0.1575 \\ 0.3196 \end{bmatrix} = 0.3196 \begin{bmatrix} 0.1737 \\ 0.4929 \\ 1 \end{bmatrix}, \mathbf{u}^{(4)} = \begin{bmatrix} 0.5497 \\ -0.4602 \\ 0.1343 \end{bmatrix} = 0.1343 \begin{bmatrix} 4.0929 \\ -3.4263 \\ 1 \end{bmatrix}.$$

The solutions are

$$(x^{(1)}, y^{(1)}) = (-10.7766, 1.7553), \quad (x^{(2)}, y^{(2)}) = (0.1100, -0.4220),$$
$$(x^{(3)}, y^{(3)}) = (0.1737, 0.4929), \quad (x^{(4)}, y^{(4)}) = (4.0929, -3.4263).$$

The procedure is summarized as Algorithm 1.

**Remark.** In our example we have four real roots. It is also possible for two bivariate equations of degree two to have four complex roots or two real and two complex roots. The proposed algorithm can deal with these cases as well.

---

**Algorithm 1.** Solving a polynomial system of equations via tensor decomposition

---

**Input:** A system of 2 polynomial equations of degree 2 (and 2 variables)
**Output:** The solutions $(x^{(i)}, y^{(i)})$, $i = 1, \ldots, 4$ of the system

---

1: Reformulate the problem as $\mathcal{T} \bullet_1 \mathbf{u} \bullet_2 \mathbf{u} = \begin{bmatrix} 0 \\ 0 \end{bmatrix}$.

2: Decompose the tensor $\mathcal{T}$ in (partially symmetric) rank-one terms, $\mathcal{T} = [\![ \mathbf{V}, \mathbf{V}, \mathbf{W} ]\!]$.

3: Solve the linear system $\mathbf{W}(\mathbf{z}.^2) = \begin{bmatrix} 0 \\ 0 \end{bmatrix}$ for $\mathbf{z}.^2$.
   Find the 4 (essentially different) solutions $\mathbf{z}^{(i)}$, $i = 1, \ldots, 4$.

4: Solve the linear systems $\mathbf{V}^\top \mathbf{u}^{(i)} = \mathbf{z}^{(i)}$ for $\mathbf{u}^{(i)}$, $i = 1, \ldots, 4$.
   Normalize $\mathbf{u}^{(i)}$, $i = 1, \ldots, 4$ so that the last elements become 1.
   Extract the solutions $(x^{(i)}, y^{(i)})$, $i = 1, \ldots, 4$ by removing the last element of each $\mathbf{u}^{(i)}$.

---

A polynomial system of equations can also have roots at infinity. These roots correspond to solutions of the homogenized version of the system, where a third variable is introduced that multiplies each monomial (zero, one or two times) to complete it to degree two. Solutions at infinity are the solutions, for which the additional variable is zero. Our algorithm can find such roots as well. In this case, the last entry of $\mathbf{u}$ becomes zero.

### 3.2 Systems with More Variables or Higher Degree Polynomials

We now briefly discuss two generalizations of the main problem (2): the cases of larger number of variables or higher degree polynomials.

**More Variables.** In case of more variables, the length of $\mathbf{u}$ and the number of slices of the system's tensor will increase, but we could proceed in a similar way if the rank of the tensor is small enough. If the rank is very large (even if we allow complex factors), a modification of the main algorithm will be necessary. A possible direction to consider here would be to reformulate the problem as

$$\mathbf{W}(\mathbf{V} \odot \mathbf{V})^\top (\mathbf{u} \otimes \mathbf{u}) = \mathbf{0}$$

and solve for $\mathbf{u} \otimes \mathbf{u}$ as in [1]. This line of research is left for future work.

**Polynomials of Higher Degree.** In case of polynomials of higher degree, the associated tensors are of higher order. For example, if we have bivariate polynomials of degree three, we associate a third-order tensor (instead of a matrix) with each equation and stack these tensors in a fourth-order tensor $\mathcal{T} \in \mathbb{R}^{3 \times 3 \times 3 \times 2}$. The decomposition of $\mathcal{T}$ in rank-one terms contains then one additional factor $\mathbf{V}$,

$$\mathcal{T} = [\![ \mathbf{V}, \mathbf{V}, \mathbf{V}, \mathbf{W} ]\!].$$

We can proceed in a similar way as in Sect. 3.1, except that now in the first system we solve for $\mathbf{z}.^3$. Unfortunately, the rank of $\mathcal{T}$ could increase as well.

# 4    Conclusions and Perspectives

We proposed a new procedure for solving bivariate polynomial equations, exclusively using tools from numerical (multi-)linear algebra, such as the eigenvalue decomposition and solving linear systems. Although our approach is currently not general enough for solving an arbitrary system of polynomials, it is applicable to a large class of sub-problems. The core computational steps of the procedure are i) a CPD of the coefficient tensor, and ii) solving linear systems involving the CPD factors. In many cases the CPD is known to be an eigenvalue problem in disguise. For this reason, our new approach is particularly interesting as said eigenvalue problem is phrased 'directly' in the equations' coefficients, as opposed to existing methods in which the eigenvalue formulation follows after several computationally intensive steps.

Future work will focus on comparing and establishing connections to existing eigenvalue-based approaches for polynomial system solving. We aim to generalize the method to deal with systems in more variables and of larger degrees, which involves higher-order coefficient tensors, instead of matrices. In this context it remains to be seen what is the (complex) rank of the resulting coefficient tensor, and how to generalize all the steps of the proposed algorithm.

# References

1. Boussé, M., Vervliet, N., Domanov, I., Debals, O., De Lathauwer, L.: Linear systems with a canonical polyadic decomposition constrained solution: algorithms and applications. Numer. Linear Algebra Appl. **25**(6), e2190 (2018)
2. Carroll, J., Chang, J.: Analysis of individual differences in multidimensional scaling via an N-way generalization of "Eckart-Young" decomposition. Psychometrika **35**(3), 283–319 (1970)
3. Cichocki, A., et al.: Tensor decompositions for signal processing applications. From two-way to multiway component analysis. IEEE Sig. Process. Mag. **32**(2), 145–163 (2015)
4. Cichocki, A., Zdunek, R., Phan, A., Amari, S.: Nonnegative Matrix and Tensor Factorizations. Wiley (2009)
5. Comon, P.: Tensors: a brief introduction. IEEE Sig. Process. Mag. **31**(3), 44–53 (2014)
6. Cox, D.A., Little, J., O'Shea, D.: Ideals, Varieties, and Algorithms. UTM, Springer, Cham (2015). https://doi.org/10.1007/978-3-319-16721-3
7. Domanov, I., De Lathauwer, L.: Canonical polyadic decomposition of third-order tensors: relaxed uniqueness conditions and algebraic algorithm. Linear Algebra Appl. **513**, 342–375 (2017)
8. Grasedyck, L., Kressner, D., Tobler, C.: A literature survey of low-rank tensor approximation techniques. GAMM-Mitteilungen **36**(1), 53–78 (2013)
9. Hackbusch, W.: Tensor Spaces and Numerical Tensor Calculus, Springer Series in Computational Mathematics, vol. 42. Springer, Heidelberg (2012). https://doi.org/10.1007/978-3-642-28027-6
10. Harshman, R.A.: Foundations of the PARAFAC procedure: model and conditions for an "explanatory" multi-mode factor analysis. In: UCLA Working Papers in Phonetics, vol. 16, no. (1), pp. 1–84 (1970)

11. Khoromskij, B.N.: Tensors-structured numerical methods in scientific computing: survey on recent advances. Chemom. Intell. Lab. Syst. **110**(1), 1–19 (2012)
12. Kolda, T.G., Bader, B.W.: Tensor decompositions and applications. SIAM Rev. **51**(3), 455–500 (2009)
13. Kroonenberg, P.M.: Applied Multiway Data Analysis. Wiley (2008)
14. Sidiropoulos, N.D., De Lathauwer, L., Fu, X., Huang, K., Papalexakis, E.E., Faloutsos, C.: Tensor decomposition for signal processing and machine learning. IEEE Trans. Signal Process. **65**(13), 3551–3582 (2017)
15. Smilde, A., Bro, R., Geladi, P.: Multi-Way Analysis. Applications in the Chemical Sciences. Wiley, Chichester (2004)
16. Stetter, H.J.: Numerical Polynomial Algebra. SIAM (2004)
17. Vervliet, N., Debals, O., Sorber, L., Van Barel, M., De Lathauwer, L.: Tensorlab 3.0, March 2016. https://www.tensorlab.net/

# Nonnegative Tensor-Train Low-Rank Approximations of the Smoluchowski Coagulation Equation

Gianmarco Manzini[1]($^{(\boxtimes)}$), Erik Skau[2], Duc P. Truong[2], and Raviteja Vangara[1]

[1] Theoretical Division, Los Alamos National Laboratory, Los Alamos, NM, USA
gmanzini@lanl.gov
[2] Computer, Computational and Statistics Division, Los Alamos National Laboratory, Los Alamos, NM, USA

**Abstract.** We present a finite difference approximation of the nonnegative solutions of the two dimensional Smoluchowski equation by a nonnegative low-order tensor factorization. Two different implementations are compared. The first one is based on a full tensor representation of the numerical solution and the coagulation kernel. The second one is based on a tensor-train decomposition of solution and kernel. The convergence of the numerical solution to the analytical one is investigated for the Smoluchowski problem with the constant kernel and the influence of the nonnegative decomposition on the solution accuracy is investigated.

**Keywords:** Smoluchowski equation · Multidimensional problem · Nonnegative tensor factorization · Low-order tensor decomposition · Tensor-train method

## 1 Introduction

The multidimensional Smoluchowski equation [17,23] for the particle distribution field $n(v, t)$, i.e., the number of particles with size $v$ at time $t$, reads as

$$\frac{\partial n}{\partial t} = L^1(n) - nL^2(n) \qquad \text{in} \quad \Omega \times [0, T], \tag{1}$$

$$n(v, t = 0) = n_o(v), \tag{2}$$

where $v \in \Omega = \mathbb{R}_+^d$, $T > 0$, and the integrals in the right-hand side are given by:

$$L^1(n) = \frac{1}{2} \int_0^v K(v - u, u) \, n(v - u, t) \, n(u, t) \, du \tag{3}$$

$$L^2(n) = \int_0^\infty K(v, u) \, n(u, t) \, du. \tag{4}$$

Los Alamos National Laboratory.

I. Lirkov and S. Margenov (Eds.): LSSC 2021, LNCS 13127, pp. 342–350, 2022.
https://doi.org/10.1007/978-3-030-97549-4_39

The initial value problem (IVP) (1)–(2) describes a system of particles that are uniformly distributed and chaotically moving in space, and that can combine through pairwise inelastic collisions to form other particles with bigger size (binary aggregation). The process is governed by the so-called *coagulation kernel* $K(u, v)$, which, from physical consideration, we assume to be symmetric and positive, i.e., $K(u, v) = K(v, u) \geq 0$. The first integral in (3) provides the rate at which two particles with size $u$ and $v - u$ may collide and form a new particle with size $v$. The second integral in (4) provides the rate at which particles of size $v$ are removed from the particle distribution field as they collide with a particle of any size $u$ and form a new particle with size $v + u$.

In the multidimensional setting, the particles are composed by (at most) $d$ substances and the independent vector-valued variable $v = (v^{(1)}, v^{(2)}, \ldots, v^{(d)}) \in \mathbb{R}^d$, the *particle size* or *particle volume*, collects the volumetric fraction of the substances in a given particle. We assume that every $v_i$, $i = 1, \ldots, d$, vary in the range $[V_{min}, V_{max}]$, where $V_{min}$ may be related to some dissipation mechanism of sufficiently small particles and $V_{max}$ to the natural precipitation of sufficiently big particles (e.g., because of the gravitation). The integral operators in (4) must be interpreted in the multidimensional setting, so that for a given integrable scalar function $q(u)$ they read as

$$\int_0^v q(u)\, du = \int_0^{v^{(1)}} \int_0^{v^{(2)}} \ldots \int_0^{v^{(d)}} q(u^{(1)}, u^{(2)}, \ldots, u^{(d)})\, du^{(1)}\, du^{(2)} \ldots du^{(d)}.$$

Although the well-posedness of the Smoluchowski equation has been theoretically proven [16,24], analytical solutions of the Smoluchowski equations are not known except for the very few special cases of constant, additive and multiplicative kernels [4,10,11,13]. So, in practical applications, we often need to resort to numerical discretizations providing an approximation to $n(v, t)$. A commonly adopted but not very accurate method for the numerical resolution of the multidimensional Smoluchowski IVP relies on Monte-Carlo simulations, cf. [9]. More accurate methods could be provided by classical finite-difference schemes. Unfortunately, a straightforward application of these schemes is impractical because their computational complexity grows exponentially with respect to the dimensionality. This bad scaling with respect to $d$ is a well-known effect called the *curse of dimensionality*. A breakthrough is found in the series of papers of References [2,14,15], where it is proposed to modify the finite difference schemes by using low-rank tensor-train representations [19,20] of the unknown distribution field $n(v, t)$ and the Smoluchowski kernel $K(u, v)$ on a multidimensional grid and fast convolution algorithms [7] for the approximation of the integrals in (4).

Although this approach has led to the design of a very efficient method for numerically solving the Smoluchowski equation, several issues still remain open. One of these open issues is surely related to the nonnegativity of the numerical solution. It is obvious from its definition that the particle distribution function $n(v, t)$ must be nonnegative. However, nonnegativity is not automatically inherited by a numerical approximation if the numerical method is not specifically designed to ensure that. A different approach based on reformulating the

Smoluchowski equation as a conservation law ensures the nonnegativity under a specific Courant-Friedrichs-Lax (CFL) constraint relating the time and space step size [1,5,6,12,21]. However, no such result is available at the moment for the approach that we consider in this work. An important question here, which has been raised in [22], is if a nonnegative tensor decomposition may help or even be necessary to ensure such nonnegativity property.

In this work, we present some very preliminary results that investigate this topic on a two-dimensional version of the scheme proposed in [15]. To this end, we use the nonnegative matrix factorization (NMF). The hierarchical alternating least squares (HALS) algorithm is used due to its ability to achieve high order of accuracy [3,8]. This is important to maintain the accuracy order of the solver. Our very preliminary results show that using NMF-HALS may imply some loss of accuracy of the solution that can be compensated by accepting an increasing computational complexity. The outline of the paper is as follows. In Sect. 2, we briefly review the tensor-based finite difference approximation of the Cauchy-Smoluchowski equation. In Sect. 3, we assess the performance of the method on a representative problem. In Sect. 4, we offer final remarks and conclusion and discuss possible future work.

## 2 Finite Difference Approximation of the Smoluchowski Equation

To solve the Smoluchowski equation (1), we introduce a multidimensional grid by partitioning the domain $[V_{min}, V_{max}]$ in each direction into $N$ equispaced subintervals with step size $h = (V_{max} - V_{min})/N$. We label the nodes of this grid by using the multi-index $i = (i^{(1)}, i^{(2)} \ldots, i^{(d)})$, with $i^{(\ell)}$ being the mesh partitioning index in the $\ell$-th direction. The grid nodes are denoted by $v_i = \left(v_{i^{(1)}}^{(1)}, v_{i^{(2)}}^{(2)} \ldots, v_{i^{(d)}}^{(d)}\right)$. The $i$-th cell of the grid is given by $I_i = \left[v_{i^{(1)}}^{(1)}, v_{i^{(1)}+1}^{(1)}\right] \times \left[v_{i^{(2)}}^{(2)}, v_{i^{(2)}+1}^{(2)}\right] \times \ldots \times \left[v_{i^{(d)}}^{(d)}, v_{i^{(d)}+1}^{(d)}\right]$. We let $T$ denote the final time of the evolution of $n(v,t)$ and we partition the time interval $[0,T]$ by $M$ equally sized time subintervals with time step $\tau = T/M$. We denote the time instants $t^k = k\tau$, $\tau = 0, \ldots, M$, so that $t^0 = 0$ and $t^M = T$. We use the abbreviation $n^k(v)$ to denote the solution field $n(v, t^k)$ at the time step $t^k = k\tau$. The time step $\tau$ and space step $h$ are normally chosen to have a sufficiently small approximation error.

The predictor-corrector scheme for the one-dimensional version of the Smoluchowski equation (1) reads as:

$$n^{k+\frac{1}{2}}(v_i) = \frac{\tau}{2}\left(L_i^1(n^k) - n^k(v_i)L_i^2(n^k)\right) + n^k(v_i), \qquad i = 0, \ldots, N,$$

$$n^{k+1}(v_i) = \tau\left(L_i^1(n^{k+\frac{1}{2}}) - n^{k+\frac{1}{2}}(v_i)L_i^2(n^{k+\frac{1}{2}})\right) + n^k(v_i), \qquad i = 0, \ldots, N, \quad (5)$$

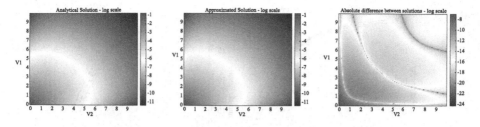

**Fig. 1.** Analytical solution $n(v,t)$ (left panel), numerical approximation (middle panel), and absolute difference between the two (left panel) at the final integration time $T = 1$. The numerical approximation is obtained by using the TT-based implementation. The color palette is in logarithmic scale and the labels are the exponents of 10.

where $L_i^1$ and $L_i^2$ are the discrete counterparts of the integral operators in (3)–(4). These two discrete operators are built by applying the trapezoidal rule. Specifically, $L_0^1(n) = 0$ $(i = 0)$ and

$$L_i^1(n) = \frac{h^d}{2} \sum_{i_1=1}^{i} \Big( K(v_{i-i_1}, v_{i_1}) n(v_i) n(v_{i-i_1}) + K(v_{i-i_1+1}, v_{i_1-1}) n(v_{i_1-1}) n(v_{i-i_1+1}) \Big)$$

for $i > 0$, and

$$L_i^2(n) = \frac{h^d}{2} K(v_i, v_0) n(v_0) + h \sum_{i_1=1}^{N-1} \Big( K(v_i, v_i) n(v_{i_1}) + \frac{h}{2} K(v_i, v_N) n(v_N) \Big)$$

for $i = 0, \ldots, N$. The summations in the previous formulas must be interpreted in the multidimensional setting as follows:

$$\sum_{i_1=1}^{i} [\ldots]_{i_1} = \sum_{i_1^{(1)}=1}^{i^{(1)}} \sum_{i_1^{(2)}=1}^{i^{(2)}} \cdots \sum_{i_1^{(d)}=1}^{i^{(d)}} [\ldots]_{i_1^{(1)}, i_1^{(2)}, \ldots i_1^{(d)}}. \tag{6}$$

For the implementation of the second order Runge-Kutta predictor-corrector scheme, we use the fast algorithms proposed in [15].

## 2.1   Tensor-train Representation

A $d$-dimensional function $f = f(v^{(1)}, \ldots, v^{(d)})$ is represented in the tensor train (TT) format through a multi-linear combination of 3-dimensional functions as follows [2]:

$$f(v^{(1)}, \ldots, v^{(d)}) = \sum_{\alpha_0=1}^{r_0} \cdots \sum_{\alpha_d=1}^{r_d} f^{(1)}(\alpha_0, v_1, \alpha_1) \ldots f^{(d)}(\alpha_{d-1}, v_d, \alpha_d).$$

Here, $v^{(1)} \ldots, v^{(d)}$ are the "physical" variables, $\alpha_k = 1, \ldots, r_k$ are the inner indices of the decomposition and $r_k$ are the *TT-ranks* of the decomposition.

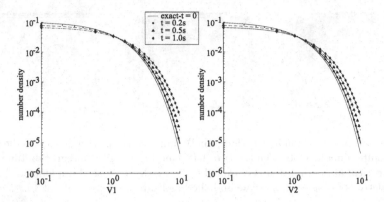

**Fig. 2.** Comparison between the total exact and numerical distribution functions as dependent on the volumetric fraction $v^{(1)}$ (left) and $v^{(2)}$ (right), which are obtained by integrating on the other independent variable. The distribution curves from the exact solution are shown at time $t = 0$ (solid line), $t = 0.2$ s (dotted line), $t = 0.5$ s (dashed line), and $t = 1$ s (dashed-dotted line). The distribution curves from the numerical solution are shown by using only symbols corresponding to the nodal values at time $t = 0.2$ (dots), $t = 0.5$ s (triangles), $t = 1$ s (diamonds). The numerical results are obtained using the TT-based implementation of the solver. The comparison shows a good agreement between the numerical solution and the exact solution.

We conventionally assume that $r_0 = r_d = 1$ to close the expression above. The three-dimensional functions $f^{(k)}(\alpha_{k-1}, v_k, \alpha_k)$ are called the *TT-cores* of the decomposition. The function $f(v^{(1)}, \ldots, v^{(d)})$ can be sampled at the $N^d$ nodes of the multidimensional grid introduced in the previous section. Such straightforward representation can be stored in a multidimensional tensor $\mathcal{F} = \left( f_{i^{(1)}, i^{(2)}, \ldots i^{(d)}} \right)$ where the entry is the function value at the node labeled by $i = \left( i^{(1)}, i^{(2)}, \ldots, i^{(d)} \right)$. This tensor clearly has $N^d$ entries. Instead, the TT-representation requires only $\mathcal{O}(dNr^2)$ to be stored, where we set $r = \max_k r_k$ and each value $f_{i^{(1)}, i^{(2)}, \ldots i^{(d)}}$ can be computed in $\mathcal{O}(dr^2)$ operations as it only needs $d$ multiplications of $r \times r$-sized matrices by a vector of size $r$. The TT-approximation of a given multidimensional tensor can be found with an assigned tolerance or ranks by using the *TT-Cross* algorithm. In particular, the TT-Cross algorithm can provide a TT-approximation of $f$ with maximal TT-rank $r = \max_k r_k$ in $\mathcal{O}(dNr^3)$ operations if a routine that evaluates $f$ at any grid point is available. More details can be found in References [19,20].

Using the TT-format to represent functions and coagulation kernels makes the calculation of the integrals $L_i^1(n)$ and $L_i^2(n)$ practically feasible. In fact, a direct application of the trapezoidal rule would require the evaluation of $L_i^1(n)$ at all the grid points with the complexity of $\mathcal{O}(N^{2d})$ operations. Instead, such integral operator is a lower-triangular convolution and if the integral arguments are given in TT-format, such integral can be evaluated in just $\mathcal{O}(d^2 r^4 N \log N)$ operations by the algorithm proposed in [7]. Similarly, if the integrands of $L_i^1(n)$

are given in TT-format, such integral operator can be evaluated by using fast TT-arithmetic algorithms in $\mathcal{O}(dNr^4)$ operations, see [7,19,20].

Given that the coagulation kernel $K(u,v)$, the particle distribution field $n(v,t)$, and the trapezoidal weights for computing the integrals $L^1$ and $L^2$ are inherently nonnegative, ones might be interested in a nonnegative tensor train (NTT) representation. This is a TT representation whose cores are constrained to be nonnegative. For solving the two-dimensional Smoluchowski equations, the NTT reduces to the special case of NMF.

## 3   Numerical Results

We solve the Smoluchowski IVP (1)–(2) for $d = 2$ by taking $K(u,v) = 1$ and the initial solution:

$$n(v_1, v_2, t = 0) = n_0(v_1, v_2) = \gamma \exp\left(-v_1 - v_2\right) \tag{7}$$

where we recall that $v = (v_1, v_2)$ and $\gamma$ is a normalization factor so that the integral of the density function $n_0(v_1, v_2)$ over $[V_{min}, V_{max}]^d$ is equal to 1. The analytical solution at any time $t > 0$ is given by [13, Eq. (58)]:

$$
\begin{aligned}
n(v_1, v_2, t) &= \frac{\gamma}{(t+1)^2} \exp\left(-v_1 - v_2\right) I_0\left(2\left(\frac{v_1 v_2 t}{t+1}\right)^{\frac{1}{2}}\right) \\
&= \frac{n_0(v_1, x_2)}{(t+1)^2} I_0\left(2\left(\frac{v_1 v_2 t}{t+1}\right)^{\frac{1}{2}}\right),
\end{aligned}
\tag{8}
$$

where $I_0$ is the zero-th order modified Bessel function.

We solve the Smoluchowski IVP assuming that $V_{min} = 0$, $V_{max} = 10$, $T = 1$, and considering all the combinations for the pairs $(N, \tau)$ with $N \in \{25, 50, 100, 150\}$ and $\tau \in \{1e\text{--}1, 5e\text{--}2, 2.5e\text{--}2, 1.25e\text{--}2\}$. The predictor-corrector scheme and the trapezoidal rule are expected to provide a convergence rate proportional to $\mathcal{O}(h^2 + \tau^2)$, where $h = V_{max}/(N-1)$. Figure 1 shows the exact solution, its numerical approximation and the difference between them at the final integration time $t = 1$. To assess the accuracy of the method and the impact of the TT decomposition on the global accuracy of the method, we consider the $L^2$ approximation error at the final integration time $T = 1$ that is obtained from two different implementations of the solver. The first implementation uses a full representation of the numerical solution and the kernel, the second implementation uses the TT representation. Both solvers are implemented in Python3 and the TT-format one is based on the TT Toolbox library [18]. The numerical solutions are comparable as can be seen from Table 1, which reports the approximation errors with respect to the exact solution. In particular, we see the TT-based solver reproduces the same behavior of the solver based on the full representation and that the expected second-order convergence can be seen for both solvers on the first three entries along the diagonal. This evidence makes us to conclude that the TT-decomposition introduces an error that is smaller

**Table 1.** Performance comparison between the implementations using the full tensor format and the TT format. The errors are comparable and shows a second-order convergence along the diagonal (simultaneous refinement in time and space) on the first three entries.

| $\tau/N$ | Full format | | | | TT format | | | |
|---|---|---|---|---|---|---|---|---|
| | 25 | 50 | 100 | 150 | 25 | 50 | 100 | 150 |
| 1.e−1 | 7.98e−3 | 1.12e−3 | 1.63e−3 | 1.87e−3 | 7.98e−3 | 1.12e−3 | 1.63e−3 | 1.87e−3 |
| 5.e−2 | 9.17e−3 | 1.61e−3 | 3.63e−4 | 6.33e−4 | 9.17e−3 | 1.61e−3 | 3.63e−4 | 6.33e−4 |
| 2.5e−2 | 9.48e−3 | 1.86e−3 | 9.47e−5 | 3.48e−4 | 9.48e−3 | 1.86e−3 | 9.50e−5 | 3.49e−4 |
| 1.25e−2 | 9.56e−3 | 1.92e−3 | 8.23e−5 | 2.82e−4 | 9.55e−3 | 1.92e−3 | 8.11e−5 | 2.85e−4 |

**Table 2.** Performance of NMF-HALS scheme with $r = 10$ and $r = 20$

| $\tau/N$ | $r = 10$ | | | | $r = 20$ | | | |
|---|---|---|---|---|---|---|---|---|
| | 25 | 50 | 100 | 150 | 25 | 50 | 100 | 150 |
| 1e−1 | 8.38e−3 | 2.47e−3 | 2.63e−3 | 2.86e−3 | 8.12e−3 | 1.96e−3 | 2.4e−3 | 2.91e−3 |
| 5.e−2 | 9.62e−3 | 2.81e−3 | 2.09e−3 | 2.24e−3 | 9.38e−3 | 2.35e−3 | 1.91e−3 | 2.26e−3 |
| 2.5e−2 | 9.91e−3 | 2.97e−3 | 2.07e−3 | 2.17e−3 | 9.69e−3 | 2.54e−3 | 1.90e−3 | 2.19e−3 |
| 1.25e−2 | 9.97e−3 | 3.00e−3 | 2.05e−3 | 2.15e−3 | 9.76e−3 | 2.59e−3 | 1.90e−3 | 2.17e−3 |

than the other errors of the approximation, which are originated by the time and space discretization. Figure 2 shows the time evolution of the particle distribution function $n(\cdot, v)$ versus the first volumetric fraction $v^{(1)}$ (left panel) and the second volumetric fraction $v^{(2)}$ (right panel). To total distribution field with respect to $v^{(\ell)}$, $\ell = 1, 2$, is obtained by integrating on the other variable. The curves of the exact distribution field at the four times $t = 0$, $t = 0.2\,\mathrm{s}$, $t = 0.5\,\mathrm{s}$ and $t = 1\,\mathrm{s}$ are plotted as solid, dotted, dashed, and dashed-dotted lines. The results from the numerical simulation using the TT-based solver ($\tau = 1e - 1\,\mathrm{s}$, rank $r = 10$ and $N = 20$) are plotted against these curves by using symbols, e.g., solid dots at $t = 0.2\,\mathrm{s}$, triangles at $t = 0.5\,\mathrm{s}$, and diamonds at $t = 1\,\mathrm{s}$. This figure shows an excellent agreement between the numerical solution and the exact solution.

Then, we use NMF-HALS to investigate how the method performs when we use a nonnegative factorization of the solution and the kernel. The experiments were run with $r = 10$ and $r = 20$. In particular, at any time step (*i*) the current solution is approximated by NMF-HALS; (*ii*) we compute the integrals $L_i^1$ and $L_i^2$, and update the solution at the predictor step still using NMF-HALS approximation; (*iii*) we finally update the solution at the corrector step. We report these results in Table 2 and show that the NMF scheme can reach the error level of the full-format and TT-format when the space and time step sizes are large ($\tau \in \{1.e-1, 5e-2\}$ and $N \in \{25, 50\}$). On the most refined space and time steps, the error may not decrease due to numerical effects such as ill-conditioning or a not enough accurate approximations of the integrals in (3)–(4). Moreover, we see that increasing the rank of the decomposition from $r = 10$

to $r = 20$ is beneficial in some cases, as for example for $(\tau, N) = (1.25e-1, 100)$, but not for all $(\tau, N)$. This drawback may also depend on the interplay with the numerical effects mentioned above and could require the design of a more accurate algorithm for extracting nonnegative factors. This issue will be investigated in our future work.

## 4   Conclusions

In this work, we compared the accuracy of two implementations of a finite difference approximation for the numerical resolution of the Smoluchowski equation. The first implementation is a standard finite difference scheme, while the second one was based on a TT-decomposition, that can be optionally nonnegative. Our preliminary results confirm that using a regular TT-decomposition does not impact significantly in terms of accuracy of the numerical approximation, although providing a great reduction of the computational complexity from $\mathcal{O}(N^2)$ to $\mathcal{O}(rN \log N)$ operations. On the other hand, the convergence of the method is more affected when using the nonnegative TT representation of the unknown field and coagulation kernel.

**Acknowledgments.** The authors were partially supported by the LDRD program of Los Alamos National Laboratory under project numbers 20210485ER and 20190020DR. Los Alamos National Laboratory is operated by Triad National Security, LLC, for the National Nuclear Security Administration of U.S. Department of Energy (Contract No. 89233218CNA000001).

## References

1. Bourgade, J.P., Filbet, F.: Convergence of a finite volume scheme for coagulation-fragmentation equations. Math. Comput. **77**, 851–883 (2008)
2. Chaudhury, A., Oseledets, I., Ramachandran, R.: A computationally efficient technique for the solution of multi-dimensional PBMs of granulation via tensor decomposition. Comput. Chem. Eng. **61**, 234–244 (2014)
3. Cichocki, A., Zdunek, R., Amari, S.: Hierarchical ALS algorithms for nonnegative matrix and 3D tensor factorization. In: Davies, M.E., James, C.J., Abdallah, S.A., Plumbley, M.D. (eds.) ICA 2007. LNCS, vol. 4666, pp. 169–176. Springer, Heidelberg (2007). https://doi.org/10.1007/978-3-540-74494-8_22
4. Fernández-Díaz, J.M., Gómez-García, G.J.: Exact solution of Smoluchowski's continuous multi-component equation with an additive kernel. Europhys. Lett. (EPL) **78**(5), 56002 (2007)
5. Filbet, F., Laurençot, P.: Mass-conserving solutions and non-conservative approximation to the Smoluchowski coagulation equation. Arch. Math. **83**, 558–567 (2004)
6. Filbet, F., Laurençot, P.: Numerical simulation of the Smoluchowski coagulation equation. SIAM J. Sci. Comput. **25**(6), 2004–2028 (2004)
7. Kazeev, V.A., Khoromskij, B.N., Tyrtyshnikov, E.E.: Multilevel Toeplitz matrices generated by tensor-structured vectors and convolution with logarithmic complexity. SIAM J. Sci. Comput. **35**, A1511–A1536 (2013)

8. Kim, J., He, Y., Park, H.: Algorithms for nonnegative matrix and tensor factorizations: a unified view based on block coordinate descent framework. J. Global Optim. **58**(2), 285–319 (2014)
9. Kruis, F.E., Maisels, A., Fissan, H.: Direct simulation Monte Carlo method for particle coagulation and aggregation. AIChE J. **46**(9), 1735–1742 (2006)
10. Leyvraz, F.: Large-time behavior of the Smoluchowski equations of coagulation. Phys. Rev. A **29**, 854–858 (1984)
11. Leyvraz, F., Tschudi, H.R.: Singularities in the kinetics of coagulation processes. J. Phys. A Math. Gen. Phys. **14**, 3389–3405 (1981)
12. Liu, H., Gröpler, R., Warnecke, G.: A high order positivity preserving DG method for coagulation-fragmentation equations. SIAM J. Sci. Comput. **41**(3), B448–B465 (2019)
13. Lushnikov, A.A.: Evolution of coagulating systems: III. Coagulating mixtures. J. Colloid Interface Sci. **54**, 94–101 (1976)
14. Matveev, S.A., Smirnov, A.P., Tyrtyshnikov, E.E.: A fast numerical method for the Cauchy problem for the Smoluchowski equation. J. Comput. Phys. **282**, 23–32 (2015)
15. Matveev, S.A., Zheltkov, D.A., Tyrtyshnikov, E.E., Smirnov, A.P.: Tensor train versus Monte Carlo for the multicomponent Smoluchowski coagulation equation. J. Comput. Phys. **316**, 164–179 (2016)
16. Melzak, Z.A.: A scalar transport equation. Trans. Am. Math. Soc. **85**, 547–560 (1957)
17. Müller, H.: Zur allgemeinen Theorie ser raschen Koagulation. Fortschrittsberichte über Kolloide und Polymere Kolloidchemische Beihefte **27**, 223–250 (1928)
18. Oseledets, I.: Python implementation of the TT-Toolbox. https://github.com/oseledets/ttpy
19. Oseledets, I., Tyrtyshnikov, E.: TT-cross approximation for multidimensional arrays. Linear Algebra Appl. **432**, 70–88 (2010)
20. Oseledets, I.V.: Tensor-train decomposition. SIAM J. Sci. Comput. **33**, 2295–2317 (2011)
21. Qamar, S., Warnecke, G.: Solving population balance equations for two-component aggregation by a finite volume scheme. Chem. Eng. Sci. **62**, 679–693 (2007)
22. Shcherbakova, E., Tyrtyshnikov, E.: Nonnegative tensor train factorizations and some applications. In: Lirkov, I., Margenov, S. (eds.) LSSC 2019. LNCS, vol. 11958, pp. 156–164. Springer, Cham (2020). https://doi.org/10.1007/978-3-030-41032-2_17
23. Smoluchowski, M.: Versuch einer mathematischen Theorie der Koagulationskinetik kolloider Lösungen. Z. Phys. Chem. **92U**(1), 129–168 (1918)
24. White, W.H.: A global existence theorem for Smoluchowski's coagulation equations. Proc. Am. Math. Soc. **80**, 273–276 (1980)

# Boolean Hierarchical Tucker Networks on Quantum Annealers

Elijah Pelofske[1], Georg Hahn[2(✉)], Daniel O'Malley[3], Hristo N. Djidjev[5], and Boian S. Alexandrov[4]

[1] CCS-3, Los Alamos National Laboratory, Los Alamos, NM 87545, USA
[2] Harvard University T.H. Chan School of Public Health, Boston, MA 02115, USA
ghahn@hsph.harvard.edu
[3] EES-16, Los Alamos National Laboratory, Los Alamos, NM 87545, USA
[4] T-1, Los Alamos National Laboratory, Los Alamos, NM 87545, USA
[5] Institute of Information and Communication Technologies,
Bulgarian Academy of Sciences, Sofia, Bulgaria

**Abstract.** Quantum annealing is an emerging technology with the potential to solve some of the computational challenges that remain unresolved as we approach an era beyond Moore's Law. In this work, we investigate the capabilities of the quantum annealers of D-Wave Systems, Inc., for computing a certain type of Boolean tensor decomposition called Boolean Hierarchical Tucker Network (BHTN). Boolean tensor decomposition problems ask for finding a decomposition of a high-dimensional tensor with categorical, [true, false], values, as a product of smaller Boolean core tensors. As the BHTN decompositions are usually not exact, we aim to approximate an input high-dimensional tensor by a product of lower-dimensional tensors such that the difference between both is minimized in some norm. We show that BHTN can be calculated as a sequence of optimization problems suitable for the D-Wave 2000Q quantum annealer. Although current technology is still fairly restricted in the problems they can address, we show that a complex problem such as BHTN can be solved efficiently and accurately.

## 1 Introduction

One of the most powerful tools for extracting latent (hidden) features from data is factor analysis [8]. Traditionally, factor analysis approximates some data-matrix $X \in \mathbb{R}^{n \times m}$ by a product of two factor matrices, $X \approx AB$, where $A \in \mathbb{R}^{n \times k}$, $B \in \mathbb{R}^{k \times m}$, and $k \ll n, m$. Various factorizations can be obtained by imposing different constraints. Assuming the input data is a matrix $X \in \mathbb{B}^{n \times m}$, where $\mathbb{B} = \{0, 1\}$, Boolean factorization is looking for two binary matrices $A \in \mathbb{B}^{n \times k}$, $B \in \mathbb{B}^{k \times m}$ such that $X = AB$, where $x_{ij} = \vee_{l=1}^{k} a_{il} y_{lj} \in \mathbb{B}$ and $\vee$ is the logical "or" operation $(1 + 1 = 1)$. The *Boolean rank* of $X$ is the smallest $k$ for which such an exact representation, $X = AB$, exists. Interestingly, in contrast to the non-negative rank, the Boolean rank can be not only bigger or equal, but also much smaller than the logarithm of the real rank [3,6]. Hence,

I. Lirkov and S. Margenov (Eds.): LSSC 2021, LNCS 13127, pp. 351–358, 2022.
https://doi.org/10.1007/978-3-030-97549-4_40

low-rank factorization for Boolean matrices is of particular interest. Past works on utilizing quantum annealing for matrix factorization were focused mostly on non-negative matrix factorizations [4,7].

Instead of using matrix representation, many contemporary datasets are high-dimensional and need to be represented as *tensors*, that is, as multidimensional arrays. Tensor factorization is the high-dimensional generalization of matrix factorization [5]. In this work, we present an algorithm for Boolean Hierarchical Tucker Tensor Network computation using the D-Wave 2000Q annealer. *Boolean Hierarchical Tucker Tensor Network (BHTN)* is a special type of tensor factorization, in which tensor product operations are structured in a tree-like fashion. We start by reshaping/unfolding a high-dimensional tensor into an equivalent matrix representation. Afterwards, this matrix representation is decomposed recursively by reshaping and splitting using Boolean matrix factorization. We show that the factorization problem of Boolean matrices, which is the most computationally challenging step, can be reformulated as a *quadratic unconstrained binary optimization (QUBO)* problem, which is a type of problem suitable for solving on D-Wave's quantum annealer. Problems previously solved on D-Wave include maximum clique, minimum vertex cover, graph partitioning, and many other NP-hard optimization problems. The D-Wave 2000Q device situated at Los Alamos National Laboratory has over 2000 qubits, but because of the limited connections (couplers) between those qubits, it is usually limited to solving optimization problems of sizes (number of variables) not higher than 65. Nevertheless, due to the special structure of our algorithm, we are able to compute BHTNs of much larger tensors. The largest one we tried was of order 8, containing a total of 65536 elements.

Our contribution is threefold: First, we present a recursive algorithm that calculates a BHTN. Second, we reformulate the Boolean matrix factorization as a higher-order binary optimization (HUBO) problem. Third, after having converted the HUBO to a QUBO, suitable for processing via quantum annealing, we use D-Wave 2000Q to compute BHTNs for input Boolean tensors of various orders, ranks, and sizes. No quantum algorithms for computing BHTNs have been previously published.

The article is structured as follows: Section 2 starts by presenting a high-level overview of Hierarchical Tucker Networks, which allow us to recursively decompose tensors into a series of lower-order tensors. We show that on each level of the recursion, we are faced with factorizations of Boolean matrices, a problem which is equivalent to minimizing a higher-order binary optimization problem that can be solved on D-Wave. We evaluate our algorithm on different input tensors in Sect. 3. The article concludes with a discussion in Sect. 4.

## 2  Methods

This section describes all components of the Boolean Hierarchical Tucker Network leading to a higher-order binary optimization problem, and the subsequent solving step on D-Wave.

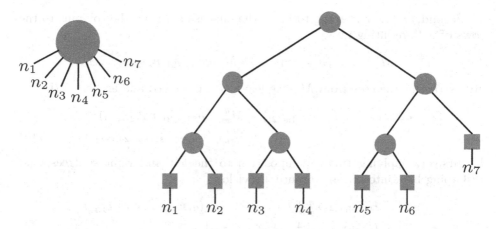

**Fig. 1.** An illustration of the Hierarchical Tucker Network. An input order-7 tensor (left) and its decomposition tree $HT$ (right). Each circle node of $HT$ (except the root) is an order-3 "core" tensor, and each square is a matrix. An edge connecting two nodes denotes a contraction operation.

## 2.1 Hierarchical Tucker Decomposition

We assume we are given an order-$d$ input tensor $\mathcal{T}$ to be transformed into a BHTN $\mathcal{HT}$. A visualization of the structure of a BHTN is given in Fig. 1. We use the same notation $\mathcal{HT}$ both for the BHTN as well as its associated decomposition tree. We describe a recursive algorithm, and assume that the tensor at the current level of recursion is $T(n_1, \ldots, n_s, q)$, where $n_i$ is the size in the $i$-th dimension and $q$ is the dimension to connect $T$, through a contraction operation, to the higher-level factor in the decomposition tree. We assume that $q = 1$ corresponds to level 0 of the decomposition tree (i.e., we create a dummy dimension of the input tensor, for uniformity of representation), and for all other levels we have $q > 1$. Let $i$ denote the current iteration level. The algorithm consists of a sequence of reshaping and splitting operations, and the output is a subtree $HT$ of $\mathcal{HT}$ corresponding to a factorization of $T$.

Initially, we reshape $T$ into a matrix $M = M(n_1, \ldots, n_{s/2}, n_{s/2+1}, \ldots, n_s q)$. Assuming $s > 3$, the following recursion is admissible.

First, we split $M$ into two matrices of specified dimensions, using a Boolean matrix factorization algorithm, leading to

$$M \to M'(n_1, \ldots, n_{s/2}, r_{(1,s/2)}) * M''(r_{(1,s/2)}, n_{s/2+1}, \ldots, n_s q). \quad (1)$$

We will use $M'$ to compute (by recursion) the left branch (left subsubtree) of the decomposition subtree $HT$ corresponding to $T$, while $M''$ will be used to define the right branch of $HT$, as well as one order-3 tensor, called *core*, which will be the root of $HT$ and which connects those two (left and right) branches, and also connects $HT$ to its parent 3-d tensor in $\mathcal{HT}$.

Second, we use reshaping to move the dimension $q$ from the columns to the rows of $M''$, resulting in

$$(1) \to M'(n_1, \ldots, n_{s/2}, r_{(1,s/2)}) * M''(qr_{(1,s/2)}, n_{s/2+1}, \ldots, n_s). \qquad (2)$$

By extracting the core from $M''$, we leave $M'$ unchanged and are left with

$$(2) \to M'(n_1, \ldots, n_{s/2}, r_{(1,s/2)}) * M''_{\text{core}}(qr_{(1,s/2)}, r_{(s/2+1,s)})$$
$$* M''_{\text{right}}(r_{(s/2+1,s)}, n_{s/2+1}, \ldots, n_s). \qquad (3)$$

Recursively applying this decomposition to the left and right subtrees, and reshaping back into tensors, eventually yields

$$(3) \to HT_{\text{left}}([1, s/2], r_{(1,s/2)}) \times T_{\text{core}}(q, r_{(1,s/2)}, r_{(s/2+1,s)})$$
$$\times HT_{\text{right}}([s/2+1, s], r_{(s/2+1,s)}) \qquad (4)$$
$$= HT([1, s], q), \qquad (5)$$

where we use $[k_1, k_2]$ to denote the sequence $k_1, k_1 + 1, \ldots, k_2$.

The decomposition outlined here allows us to decompose a tensor which has been flattened out as a matrix so long as $s > 3$. For $s \leq 3$, the decomposition is constructed explicitly.

As can be seen in the algorithm's description, there are two types of operations: reshaping, which can be implemented by reordering the elements of the tensor/matrix in a straightforward manner, and factorization of a Boolean matrix into a product of two Boolean matrices, which is a more complex problem and will be considered in the following subsection.

## 2.2    Iterative Boolean Matrix Factorization

Suppose we are given a Boolean matrix $X \in \mathbb{B}^{n \times m}$, which we aim to represent as a product, $X = AB$, where $A$ and $B$ are also Boolean matrices, and we will reduce the latter problem into a series of problems suitable for solving on D-Wave. We can approximately solve this problem by using an iterative minimization procedure,

$$A = \arg \min_Y d(X, YB), \qquad (6)$$

$$B = \arg \min_Y d(X, AY), \qquad (7)$$

where $d(\cdot, \cdot)$ denotes the Hamming distance. Note that when iteratively solving Eq. (6)–(7) both minimizations can be performed with one subroutine by taking transposes, hence we focus only on solving Eq. (7). This problem can be further reduced into a single-column Boolean factorization problem, i.e.,

$$B_i = \arg \min_y d(X_i, Ay) \text{ for } i \in \{1, \ldots, m\}, \qquad (8)$$

for $y = Y_i$, since, for $A$ fixed, values of the $i$-th column of $X = AY$ depend only on values of the $i$-th column of $Y$. Let $T_i = \{j : X_{ji} = 1\}$ be the set of all indices with entry *true* in column $X_i$, and likewise let $F_i = \{j : X_{ji} = 0\}$ be the set of all indices with entry *false* in column $X_i$. Then we can write

$$d(X_i, Ay) = C - \sum_{j \in T_i} f((A^\top)_j \cdot y) + \sum_{j \in F_i} f((A^\top)_j \cdot y), \tag{9}$$

where $f(x_1, \ldots, x_n) = 1 - \prod_{i=1}^{n}(1 - x_i)$ and $C$ is equal to the number of non-zero entries in column $X_i$. Moreover, note that $f((A^\top)_j \cdot y)$ can be reformulated as $f(y_{j_1}, y_{j_2}, \ldots, y_{j_{r'}})$, where $\{j_1, j_2, \ldots, j_{r'}\} = \{j : A_{ji} = 1\}$. This makes Eq. (9) a so-called higher order binary optimization (HUBO) problem in a (possibly complete) subset of the variables $\{y_1, y_2, \ldots, y_r\}$, where $r$ is the rank of the factorization.

To solve Eq. (9) on D-Wave, we first need to convert it into a quadratic unconstrained binary optimization (QUBO) problem. A QUBO in $n$ binary variables is given by a minimization of the following function,

$$\sum_{i=1}^{n} a_i x_i + \sum_{i<j} a_{ij} x_i x_j, \tag{10}$$

where $x_i \in \{0, 1\}$ for $i \in \{1, \ldots, n\}$ are the variables, and $a_i, a_{ij} \in \mathbb{R}$, for $i, j \in \{1, \ldots, n\}$, are chosen by the user to define the problem under investigation. Note that while the objective function of a QUBO is a polynomial of degree two, the degree of a HUBO can be arbitrarily large, e.g., a HUBO can include terms such as $a_{ijk} x_i x_j x_k$.

Conversions from higher order problems to QUBO are part of the D-Wave API [1,2]. The penalty factor (called "strength" in the D-Wave documentation) for the HUBO to QUBO conversion was always chosen as the maximum of the absolute value of any HUBO coefficient. In our experiments this choice preserved the minimum ground state solution while not causing the coefficients to become too large.

To solve the QUBO obtained after conversion, its coefficients are mapped onto the quantum chip, and a number of parameters such as the annealing time, the number of anneals, and other embedding-related ones are set by the user. Valid annealing times for D-Wave 2000Q are between 1 and 2000 microseconds, and the number of anneals specifies how many times the annealing process is repeated in a single annealer call. Since current quantum technology is noisy, and the outcome of quantum measuring is not deterministic, hundreds or thousands of anneals are usually done in a single call to the annealer, and the best of the proposed solutions is chosen.

## 3 Experiments

This section presents our experimental results for our BHTN implementation. All results are computed with the algorithms of Sect. 2.

Since these algorithms call the D-Wave 2000Q, which allows for the specification of several tuning parameters, we first calibrate a variety of D-Wave

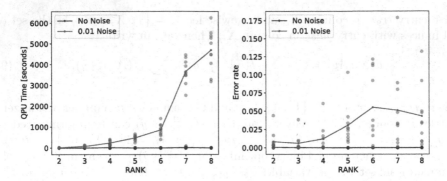

**Fig. 2.** QPU times (left) and error rates (right) as a function of the rank while keeping the order 4 and the size 8 fixed. Tensor without added noise in blue, and tensor with added noise in red. Solid lines connect the means of the scatter plot values. (Color figure online)

parameters (in particular, annealing time, an embedding parameter called chain strength, and the number of anneals). The details of calibration results are omitted here for brevity of the exposition.

We are interested in investigating how our algorithm performs if we vary certain properties of the tensor being factored (characterized by the order and the size of the tensor) and if the factorization model is different (i.e., if we vary the rank). In the following, we always fix two of these three parameters and vary the third. We measure the error rates and the QPU (Quantum Processing Unit) times.

Additionally, we perform each experiment twice, once without adding noise to each tensor, and once with noise. In the latter case, we randomly apply independent bit flips with a probability of 0.01 to the tensor under investigation. We generate the factors of the BHTN of our test tensors as Boolean matrices with Bernoulli entries (entry 1 chosen with probability $p$, and entry 0 chosen with probability $1 - p$), which are then reshaped as tensors. By multiplying them into the final input tensor using the BHTN format, we make sure that each of our test tensors (before adding noise) has at least one exact BHTN representation. Also, we ensure that each tensor we generate (either without added noise, or with added noise) does not solely contain only entries 0 (false) or entries 1 (true).

After having generated a Boolean test tensor $\mathcal{T}$, we apply the algorithm of Sect. 2 to it and record the BHTN. We then compare $\mathcal{T}$ and the tensor obtained by multiplying the tensors of the computed BHTN by counting the number of elements that are dissimilar between both tensors, and divide that count by the number of elements in $\mathcal{T}$.

Figure 2 shows the results as we vary the rank of the tensor while keeping the order, 4, and the size, 8, fixed. The QPU time and error rate both show little or no dependence on the rank for the scenario without added noise (blue curves). With noise, we observe that the QPU time and the error rates seem to

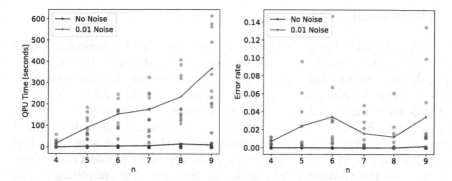

**Fig. 3.** QPU times (left) and error rates (right) as a function of the size while keeping the rank 4 and the order 4 fixed. Tensor without added noise in blue, and tensor with added noise in red. Solid lines connect the means of the scatter plot values. (Color figure online)

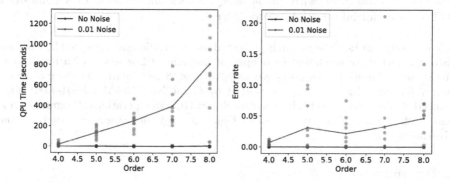

**Fig. 4.** QPU times (left) and error rates (right) as a function of the order while keeping the rank 4 and the size 4 fixed. Tensor without added noise in blue, and tensor with added noise in red. Solid lines connect the means of the scatter plot values. (Color figure online)

be dependent on the rank. Note that the QPU time goes up with the rank due to the larger number of anneals needed to achieve the same error rate.

Figure 3 shows the results for the scenario in which we fix both the rank and the order at 4, and vary the size of the tensor. Without added noise, we observe very low QPU times and error rates. With added noise, we observe a roughly linear increase in QPU time as a function of the size. Error rates for the scenario with added noise fluctuate without significantly changing with $n$, increasing from size 4 to 6, going down from 6 to 8, and up again at size 9.

Finally, we look at the behavior of the algorithm of Sect. 2 as we vary the order of the tensor (Fig. 4). We observe that without noise both the error rate and the QPU time remain fairly constant and close to zero. With noise, we observe a linearly increasing trend for the QPU time measurements. The error rate plot for tensors with noise has a less clear trend as a function of order, although we also observe an increase in error rates as the order increases.

# 4   Discussion

This article considers computing Boolean Hierarchical Tucker Networks (BHTNs) on quantum annealers. To this end, we introduce an algorithm that allows us to break down an input tensor as a product of smaller-order Boolean tensors. The latter can again be factored in a recursive fashion to produce a BHTN.

On a lower implementation level, we consider a Boolean matrix factorization problem, for which we present an iterative factorization algorithm that reformulates the Boolean matrix factorization as the minimization of a (higher order) unconstrained binary optimization problem. This problem can be solved efficiently on the D-Wave 2000Q quantum annealer after converting it into a quadratic unconstrained binary optimization (QUBO) problem, the type of function the D-Wave 2000Q quantum annealer is designed to minimize.

An experimental section considers the factorization of synthetic Boolean tensors of varying ranks, sizes, and orders. We show that the D-Wave 2000Q annealer, in connection with our strategy, allows one to accurately compute a BHTN for an initial tensor of up to order 8.

**Acknowledgments.** The research presented in this article was supported by the Laboratory Directed Research and Development program of Los Alamos National Laboratory under the project numbers 20190020DR and 20190065DR. The work of Hristo Djidjev has been also partially supported by Grant No BG05M2OP001-1.001-0003, financed by the Science and Education for Smart Growth Operational Program (2014–2020) and co-financed by the European Union through the European structural and Investment funds.

# References

1. D-Wave Systems Inc.: Create a binary quadratic model from a higher order polynomial (2020). https://docs.ocean.dwavesys.com/projects/dimod/en/latest/reference/generated/dimod.higherorder.utils.make_quadratic.html#dimod.higherorder.utils.make_quadratic
2. D-Wave Systems Inc.: Higher-order models (2020). https://docs.ocean.dwavesys.com/projects/dimod/en/latest/reference/higherorder.html
3. DeSantis, D., Skau, E., Alexandrov, B.: Factorizations of binary matrices-rank relations and the uniqueness of Boolean decompositions. arXiv preprint arXiv:2012.10496 (2020)
4. Golden, J., O'Malley, D.: Reverse annealing for nonnegative/binary matrix factorization. Plos One **16**(1), e0244026 (2021)
5. Kolda, T.G., Bader, B.W.: Tensor decompositions and applications. SIAM Rev. **51**(3), 455–500 (2009)
6. Monson, S.D., Pullman, N.J., Rees, R.: A survey of clique and biclique coverings and factorizations of (0,1)-matrices. Bull. ICA **14**, 17–86 (1995)
7. O'Malley, D., Vesselinov, V., Alexandrov, B., Alexandrov, L.: Nonnegative/binary matrix factorization with a D-wave quantum annealer. PloS One **13**(12), e0206653 (2018)
8. Spearman, C.: "General intelligence," objectively determined and measured. Am. J. Psychol. **15** (1961)

# Topic Analysis of Superconductivity Literature by Semantic Non-negative Matrix Factorization

Valentin Stanev[1(✉)], Erik Skau[2], Ichiro Takeuchi[1], and Boian S. Alexandrov[3]

[1] Department of Materials Science and Engineering,
University of Maryland, College Park, MD 20742, USA
vstanev@umd.edu
[2] Computer, Computational and Statistical Sciences Division,
Los Alamos National Laboratory, Los Alamos, NM 87545, USA
[3] Theoretical Division, Los Alamos National Laboratory,
Los Alamos, NM 87545, USA

**Abstract.** We analyze a corpus consisting of more than 17,000 abstracts in the general field of superconductivity, extracted from the arXiv – an online repository of scientific articles. We utilize a recently developed topic modeling method called SeNMFk, extending the standard Non-negative Matrix Factorization (NMF) methods by incorporating the semantic structure of the text, and adding a robust system for determining the number of topics. With SeNMFk, we were able to extract coherent topics validated by human experts. From these topics, a few are relatively general and cover broad concepts, while the majority can be precisely mapped to particular scientific effects or measurement techniques. The topics also differ by ubiquity, with only three topics prevalent in almost 40% of the abstract, while each specific topic tends to dominate a small subset of the abstracts. These results demonstrate the ability of SeNMFk to produce a layered and nuanced analysis of large scientific corpora.

## 1 Introduction

Robust scientific activity is vital for economic and technological progress, and the ability to deal with various existing and emerging challenges. However, the current explosion of research work and publications is creating its own challenges. The total global research output already exceeds 2.6 million articles annually, and has grown at an average rate of 4% each year over the last decade[1]. This makes it impossible for individual scientists to keep up with important developments in their fields [11] and can overload entire journals[2]. Even the very recent research surge addressing the COVID-19 pandemic created a flood of papers, making it difficult to distinguish the important from trivial findings and thus impeding

---

[1] https://ncses.nsf.gov/pubs/nsb20206/.
[2] In 2018, the academic journal *The Review of Higher Education* had to suspend accepting submissions due to a two-year backlog.

© Springer Nature Switzerland AG 2022
I. Lirkov and S. Margenov (Eds.): LSSC 2021, LNCS 13127, pp. 359–366, 2022.
https://doi.org/10.1007/978-3-030-97549-4_41

the efforts of scientists, health workers, and policymakers to quickly discover the most pertinent information[3].

This underscores the need for automated methods capable of organizing and analyzing the vast amounts of scientific publications appearing every day [14]. In the last two decades, Machine Learning (ML) methods for Natural Language Processing (NLP) have demonstrated a robust ability to model text data [22]. ML systems that can help to reduce the burden of analyzing existing literature will become an indispensable research tool. However, many outstanding problems still have to be addressed before such systems become a reality.

In every collection of documents, each document is a combination of some recurring themes or hidden (latent) concepts called topics. Extracting these topics and representing each document as their combination – known as *topic modeling* – is one of the most important and common tasks in the analysis of text corpora, and is a key in the efforts to reduce unorganized text corpora into actionable information. There are a number of classical topic modeling methods: Latent Semantic Analysis (LSA) [5], Probabilistic LSA (PLSA) [8], Latent Dirichlet Allocation (LDA) [3], Non-negative Matrix Factorization (NMF) [21]. These have been used extensively to analyze large text corpora in a variety of applications (see, for example, Refs. [2,6,9,15,20]).

Recently, we reported a new method, SeNMFk, which showed a superior performance in identifying the number of topics in several benchmark text corpora when compared to other state-of-the-art techniques [19]. SeNMFk extends the standard NMF methods in two key directions. Standard NMF models the topics by decomposing the term frequency-inverse document frequency (TF-IDF) representation of the corpus, a matrix $X$, into a product of two low-rank non-negative matrices: $W$, which represents the topics, and $H$, which represents the coordinates of each of the documents in the topic-space. SeNMFk utilizes a coupled minimization of the TF-IDF matrix, $X$, and the word-context matrix, $M$, to account for the semantic structure of the texts. Each cell of $M$ represents the number of times two words (i, j) occur in a predetermined window (for example, +/- 5 words). The words in this window are the context words. Importantly, it has been demonstrated that word and context embeddings can be obtained by factorizing the normalized word-context matrix [12]. Second, SeNMFk includes a robust system for determining the latent dimension via random sampling and a subsequent custom clustering of the topic vectors. The custom clustering is used to determine the number of semantic-enhanced topics based on their stability.

Here, we extend the work of Ref. [19] by presenting a study of the performance of SeNMFk on a corpus of scientific publications: an important step in testing the ability of the model to extract valid coherent topics from a real-world data. We applied SeNMFk to a corpus of scientific abstracts from arXiv – an online repository of scientific articles[4]. We focused on abstracts from the field of superconductivity, which has been a subject of intense research ever since its

---

[3] "Scientists are drowning in COVID-19 papers. Can new tools keep them afloat?", J. Brainard, Science (2020).

[4] https://arxiv.org.

discovery more than a century ago. This effort has led to many major discoveries and resulted in five Nobel prizes. Despite its long history, superconductivity remains a vibrant scientific field [7].

Using SeNMFk, we analyzed a corpus consisting of more than 17,000 abstracts. We were able to extract 29 stable topics which form 21 clusters. The topics were validated by human experts, and also exhibit better coherence than pure NMF-extracted topics. The topics show a significant range in their specificity: a few relatively general topics cover broad concepts, while the majority of the topics can be precisely mapped to particular scientific effects, measurement techniques or applications. The topics also differ by ubiquity, with only three topics prevalent in almost 40% of the abstracts. Specific topics tend to dominate a small subset of the corpus – 1%–4% of the abstracts. These results show the ability of SeNMFk to produce a nuanced analysis of large scientific corpora, and underscore the enormous potential of this method as a research tool.

## 2   Methods: SeNMFk

NMF is an unsupervised learning method that approximates a non-negative matrix, $X \in \mathbb{R}_+^{F \times N}$, by a product of two factor matrices, $W \in \mathbb{R}_+^{F \times k}$, and $H \in \mathbb{R}_+^{k \times N}$, such that $X_{ij} \approx \sum_{s=1}^{k} W_{is} H_{sj}$ [16]. $W$ and $H$ are both non-negative and have one small dimension $k$. When NMF is applied to a text corpus for text mining, the basis patterns, represented by the columns of the matrix $W$, are the *topics* whose linear combinations span the entire corpus, and the total number of topics is equal to $k$. Each document is represented as a linear combination of the extracted topics with coefficients given by the columns of $H$.

There are a lot of notable topic modeling efforts within the NMF family of methods (see, for example, Refs. [4,10]). However, a serious limitation of the classical NMF for topic modeling tasks is its inability to integrate the semantics of the words via, for example, word embeddings [13], successfully used in deep learning. Several recent works have presented different approaches for incorporating semantics information in NMF (see Refs. [1,17,18]) All these semantic-assisted NMF methods demonstrate an excellent coherence of the extracted topics and high-quality document clusters, but they require the number of latent topics, $k$, as a prior. In fact, accurately identifying the number of latent topics is a challenging task for all topic models. Yet, determining the correct latent dimension is imperative for a meaningful analysis of a corpus, as underestimating the number of topics results in poor topic separation, *under-fitting*, while overestimation leads to noisy and superfluous topics, *over-fitting*.

To address this issue, recently we developed a robust topic modeling method called SeNMFk. SeNMFk is an NMF method that incorporates: (a) the semantic structure through term-context relations with a coupled minimization of the TF-IDF representation of the corpus, $X$, and Shifted Positive Point-wise Mutual Information (SPPMI) [12] matrix, $M$, and (b) determination of the latent dimension via random sampling of pairs of $X$ and $M$ matrices and a subsequent custom clustering of the topic vectors in $W$. This custom clustering allows us to find the latent dimension of the semantic-enhanced topics based on their stability. Here, we utilize SeNMFk, by solving the optimization problem,

$$\underset{W \in \mathbb{R}_+^{F \times k}, H \in \mathbb{R}_+^{k \times N}, G \in \mathbb{R}_+^{k \times F}}{\text{minimize}} \frac{1}{2}||X - WH||^2 + \frac{\lambda}{2}||M - WG||^2,$$

which we solve efficiently by concatenating the TF-IDF matrix $X$ with SPPMI matrix, $M$, and applying pyDNMFk[5] on the concatenation $[X|M]$. SeNMFk showed an excellent performance in identifying the number of topics in several benchmark text corpora, thus demonstrating its ability to avoid both under- and over-fitting [19].

## 3   Data

The corpus we analyzed consists of abstracts extracted from the arXiv – an open access repository of electronic preprints and postprints, currently containing more than 1.8 millions scientific articles, predominantly from the fields of physics and other natural sciences. We collected the abstracts from the meta-data repository for arXiv, made available through OAI-PMH (Open Archives Initiative Protocol for Metadata Harvesting). The abstracts were filtered by category and only the ones with "cond-mat.supr-con" ("Condensed Matter - Superconductivity") in their list of the categories were kept. There were more than 35,000 abstracts in this particular group. To ensure we are analyzing abstracts of a research screened by a robust peer-review process, we only considered abstracts which include DOI pointing to a published version of the article. To reduce the number of documents, we have focused on a span of twelve years – from 2007 to 2018. These years contain some significant developments in the field, including the discovery of an entire new high-temperature superconducting family and the emergence of topological superconductors as a major research topic. After applying these filters, there were 17,394 documents in the corpus, representing roughly half of all superconductivity abstracts in arXiv.

The raw abstract texts were preprocessed using standard NLP practices, including removal of the punctuation and standard English stop words, as well as the discovery of bigrams and trigrams. After the preprocessing, we converted the cleaned corpus to a TF-IDF representation. The vocabulary consists of 9433 tokens. It has to be noted that one significant factor complicating the analysis is the presence of chemical formulas. These obviously carry a lot of information but are not properly tokenized by standard NLP packages. Using specialized tools for chemical notation tokenization is an important future research direction.

---

[5] The software is publicly available at https://github.com/lanl/pyDNMFk.

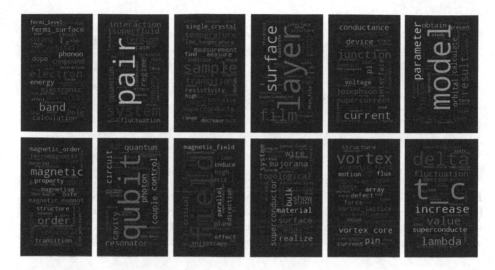

**Fig. 1.** Some of the topics discussed in the text, presented by their word clouds. The topics are ordered by their importance, from the most important at the top left to the least important at the bottom right: "Normal state", "Superconducting state", "Experiment – general", "Experiment – thin-film", "Experiment - Josephson junction", "Theory – general", "Magetism", "Quantum computing", "Experiment - magnetic field", "Topological superconductivity", "Vortex", "$T_c$".

## 4   Discussion

SeNMFk extracted 29 topics; examining the top words in the topics, we can extract the themes they represent; the word cloud diagrams of a few selected topics are shown in Fig. 1[6].

The first important observation is the topic specificity varies significantly, with some covering broad concepts, while others clearly being very thematically focused, reflecting, for example, a particular experimental technique or physical phenomenon. The second – connected – observation is that the relative importance of the topics also varies in a wide range: three of them are the leading topics in the respective 14%, 13%, and 12% of the abstracts (thus dominating a combined 39% of all documents). Another three topics form the next tier, leading in 8%, 8%, and 7%, respectively, of the abstracts (see Fig. 2a). None of the remaining 23 topics is leading in more than 5% of the abstracts, and eight topics are not leading in any of the abstracts.

Not surprisingly, the three top topics are fairly broad. One of them can be associated with a general description of the properties and origin of the superconducting phase, with top words such as "superconducting pair", "superfluid", "phase diagram". Another one obviously stems from descriptions of the normal state electronic structure (band structure) underlying the superconducting

---

[6] All results are available at https://github.com/vstanev1/-NLP_arxiv_supercon.

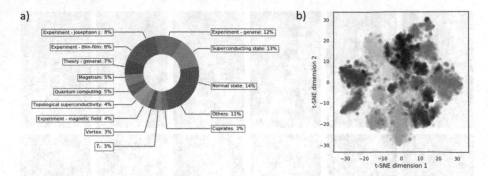

**Fig. 2.** a) The distribution of topics by the percentage of abstracts each topic dominates (note that category "Others" represents the combined contribution of the eight topics which dominate in less than 2% of the abstracts). b) The t-SNE projection of the abstracts using their coordinates in topic space, given by the columns of the $H$ matrix.

phase, with words like "electron", "band", "fermi_surface", "phonon" (electron-phonon interaction are the cause of superconductivity in many materials). The third of these broad topics is related to the experimental nature of many of the abstracts, with top words such as "sample", "measure", and "single_crystal".

The topics in the next tier are more specific. One contains words such as "layer", "surface", "film" and reflects the significant emphasis in the field on thin-film experiments. Another one clearly originates in work on Josephson junctions physics, with words like "junction", "josephson", "current", "conductance". The third topic in this tier is dominated by the words "result", "model", "calculate", "parameter", and "theory", and reflects the theoretical work in the field. The remaining topics are typically very specific and rarely dominate in any given abstract. Two of the topics with appreciable weight (with top words "qubit" and "topological", and dominating 5% and 4% of the abstracts, respectively) reflect some of the most exciting recent developments in the field – the emergence of the idea of topologically nontrivial superconductors, and the development of superconducting devices for quantum computing. Other topics describe particular experimental techniques: one is associated with pressure studies (top words: "pressure", "gpa" = giga pascal) and another with applying external magnetic field (top words: "field", "magnetic_field"). A few topics can be mapped to uses of superconductors for applications such as cables and digital circuits.

Interestingly, only two topics can be directly matched to a particular materials group. One topic (with top word "pseudogap") is based on studies of cuprates – the first high-temperature superconducting family. Another one is based on the second – iron-based – high-temperature superconducting family discovered in 2008 (one of the top words is "feas_layer" – many of the materials in this group indeed contain FeAs layers). This surprising relative scarcity (all experimental work is done on concrete materials) is at least partially explained by the problems associated with tokenizing chemical names.

Extracting and analyzing the topics is only the first steps in the corpus analysis. To obtain a more nuanced general picture, we cluster the column-vectors of the $H$ matrix obtained by the optimization procedure of SeNMFk (see Sect. 2). This yields 21 well-separated clusters of abstracts formed around a single dominant topic, but containing appreciable contributions from other topics as well. For example, there is a cluster which is a superposition of several theoretically-focused topics: the abstracts in this clusters with high probability present theoretical work. Another cluster is formed around the cuprate topic and contains almost equal components of the three most general topics ("superconducting state", "normal state", "experimental studies"), and noticeable contributions from few other specific topics – this reflects the diversity of approaches and techniques used in the field.

Using the results of SeNMFk, we can also visualize the entire corpus by projecting the coordinates of each abstract in topic-space using a dimensionality reduction technique. t-SNE was used in Fig. 2b, and the colors show the cluster index of the abstracts.

**Acknowledgments.** This research was funded by DOE National Nuclear Security Administration (NNSA) - Office of Defense Nuclear Nonproliferation R&D (NA-22), and supported by the LANL LDRD grant 20190020DR and DOE BES STTR DE-SC0021599.

# References

1. Ailem, M., Salah, A., Nadif, M.: Non-negative matrix factorization meets word embedding. In: Proceedings of the 40th International ACM SIGIR Conference on Research and Development in Information Retrieval, pp. 1081–1084 (2017)
2. Bisgin, H., Liu, Z., Kelly, R., Fang, H., Xu, X., Tong, W.: Investigating drug repositioning opportunities in FDA drug labels through topic modeling. BMC Bioinf. **13**(S15) (2012)
3. Blei, D.M., Ng, A.Y., Jordan, M.I.: Latent Dirichlet allocation. J. Mach. Learn. Res. **3**, 993–1022 (2003)
4. Choo, J., Lee, C., Reddy, C.K., Park, H.: UTOPIAN: user-driven topic modeling based on interactive nonnegative matrix factorization. IEEE Trans. Visual Comput. Graphics **19**(12), 1992–2001 (2013)
5. Deerwester, S., Dumais, S.T., Furnas, G.W., Landauer, T.K., Harshman, R.: Indexing by latent semantic analysis. J. Am. Soc. Inf. Sci. **41**(6), 391–407 (1990)
6. Greene, D., Cross, J.P.: Unveiling the political agenda of the European parliament plenary: a topical analysis. In: WebSci 2015. ACM (2015)
7. Hirsch, J., Maple, M., Marsiglio, F.: Superconducting materials classes: introduction and overview. Physica C: Supercond. Appl. **514**, 1–8 (2015). https://doi.org/10.1016/j.physc.2015.03.002. https://www.sciencedirect.com/science/article/pii/S0921453415000933
8. Hofmann, T.: Probabilistic latent semantic analysis. arXiv preprint arXiv:1301.6705 (2013)
9. Hong, L., Davison, B.D.: Empirical study of topic modeling in twitter. In: SOMA 2010. ACM (2010)

10. Kim, H., Choo, J., Kim, J., Reddy, C.K., Park, H.: Simultaneous discovery of common and discriminative topics via joint nonnegative matrix factorization. In: Proceedings of the 21th ACM SIGKDD International Conference on Knowledge Discovery and Data Mining, pp. 567–576 (2015)

11. Landhuis, E.: Scientific literature: information overload. Nature 535(7612), 457–458 (2016). https://doi.org/10.1038/nj7612-457a

12. Levy, O., Goldberg, Y.: Neural word embedding as implicit matrix factorization. In: Advances in Neural Information Processing Systems, pp. 2177–2185 (2014)

13. Mikolov, T., Sutskever, I., Chen, K., Corrado, G.S., Dean, J.: Distributed representations of words and phrases and their compositionality. In: Advances in Neural Information Processing Systems, pp. 3111–3119 (2013)

14. Gold in the text? Nature 483(7388), 124–124 (2012). https://doi.org/10.1038/483124a

15. Nguyen, T.H., Shirai, K.: Topic modeling based sentiment analysis on social media for stock market prediction. In: Proceedings of the 53rd Annual Meeting of the Association for Computational Linguistics and the 7th International Joint Conference on Natural Language Processing (Volume 1: Long Papers), pp. 1354–1364. ACL (2015)

16. Paatero, P., Tapper, U.: Positive matrix factorization: a non-negative factor model with optimal utilization of error estimates of data values. Environmetrics 5(2), 111–126 (1994)

17. Salah, A., Ailem, M., Nadif, M.: Word co-occurrence regularized non-negative matrix tri-factorization for text data co-clustering. In: Thirty-Second AAAI Conference on Artificial Intelligence (2018)

18. Shi, T., Kang, K., Choo, J., Reddy, C.K.: Short-text topic modeling via nonnegative matrix factorization enriched with local word-context correlations. In: Proceedings of the 2018 World Wide Web Conference, pp. 1105–1114 (2018)

19. Vangara, R., et al.: Semantic nonnegative matrix factorization with automatic model determination for topic modeling. In: 2020 19th IEEE International Conference On Machine Learning And Applications (ICMLA). IEEE (2020)

20. Wang, C., Blei, D.M.: Collaborative topic modeling for recommending scientific articles. In: Proceedings of the 17th ACM SIGKDD International Conference on Knowledge Discovery and Data Mining. ACM (2011)

21. Xu, W., Liu, X., Gong, Y.: Document clustering based on non-negative matrix factorization. In: Proceedings of the 26th Annual International ACM SIGIR Conference on Research and Development in Informaion Retrieval, pp. 267–273 (2003)

22. Young, T., Hazarika, D., Poria, S., Cambria, E.: Recent trends in deep learning based natural language processing [Review Article]. IEEE Comput. Intell. Mag. 13(3), 55–75 (2018). https://doi.org/10.1109/MCI.2018.2840738

# Machine Learning and Model Order Reduction for Large Scale Predictive Simulations

# Deep Neural Networks and Adaptive Quadrature for Solving Variational Problems

Daria Fokina[1,2]([⊠]), Oleg Iliev[1,2,3], and Ivan Oseledets[4,5]

[1] Fraunhofer ITWM, Kaiserslautern, Germany
`daria.fokina@itwm.fraunhofer.de`
[2] Technical University Kaiserslautern, Kaiserslautern, Germany
[3] Institute of Mathematics and Informatics, Bulgarian Academy of Sciences,
Sofia, Bulgaria
[4] Skolkovo Institute of Science and Technology, Moscow, Russia
[5] Marchuk Institute of Numerical Mathematics RAS, Moscow, Russia

**Abstract.** The great success of deep neural networks (DNNs) in such areas as image processing, natural language processing has motivated also their usage in many other areas. It has been shown that in particular cases they provide very good approximation to different classes of functions. The aim of this work is to explore the usage of deep learning methods for approximation of functions, which are solutions of boundary value problems for particular differential equations. More specific, the class of methods known as physics-informed neural network will be explored. Components of the DNN algorithms, such as the definition of loss function and the choice of the minimization method will be discussed while presenting results from the computational experiments.

**Keywords:** Physics informed neural networks · Variational problem · Adaptive quadrature

## 1 Introduction

In the past few years machine learning (ML) has become quite a common tool in many areas. For example, the development of deep neural network architectures and the availability and diversity of deep learning frameworks has made it widely used in computer vision. The further development of new models, such as transformers, has allowed to achieve a great success in natural language processing. However, when it comes to physical problems, machine learning only starts being actively used. For instance, while there is a certain success in geophysical modeling, machine learning struggles to solve initial-boundary-value problems. Here for the most cases the ML-approaches have not yet achieved neither the accuracy of classical numerical methods, nor their speed. However, it was shown

Supported by BMBF project 05M2020-ML-MORE.

I. Lirkov and S. Margenov (Eds.): LSSC 2021, LNCS 13127, pp. 369–377, 2022.
https://doi.org/10.1007/978-3-030-97549-4_42

that machine learning algorithms can be beneficial when combined with standard methods [2,4].

Recently a class of algorithms called physics-informed machine learning has gained a lot of attention. It should be noted that there exist a lot of terms expressing the same idea, e.g. physics-aware, physics-guided, physics-driven, etc. machine learning. The key principle behind this idea is to use the physical knowledge about the problem. In particular, an interesting approach called physics-informed neural networks (PINN) was proposed in [6]. The authors are solving partial differential equations using deep neural networks. In order to do that they introduce a new loss function, which is basically the squared $l_2$-norm of the residual of the considered PDE. We present a similar approach, but instead of minimizing the $l_2$-norm of the residual, we use the natural variational formulation and minimize the energy functional. We also note that such approach does not require computation of second-order derivatives. We demonstrate the applicability of the proposed method for an elliptic problem with zero-Dirichlet and Neumann boundary conditions for one- and two-dimensional test cases.

The loss function computation requires the integral estimation. We suggest to use for that an adaptive quadrature rule, for which we predefine the error of the estimation. We compare its performance with a standard method - Monte Carlo sampling. The advantage of the adaptive method is that it requires much less points to reach the desired accuracy and allows to adjust the density of the points arrangement depending on the complexity of the integrand.

We continue with the description of the proposed method in Sect. 2. Section 3 presents the results from the computational experiments, and short conclusions are drawn in Sect. 4.

## 2   Method Description

As a starting point we suppose that we have a mapping $u : \mathbb{R}^n \to \mathbb{R}$ that we would like to approximate. In all our experiments we use a simple fully-connected neural network of depth $d$, i.e. at each layer $i$ of the network we have the linear transform, followed by an activation function $g$:

$$l_i(x) = g(W_i^T x + b_i), \, i = 1, \ldots, d, \tag{1}$$

where $W_i \in \mathbb{R}^{h_{i-1} \times h_i}$, $b_i \in \mathbb{R}^{h_i}$, $h_0 = n, h_d = 1$. We denote the set of all the parameters of the neural network as $\theta := \{(W_i, b_i)\}_{i=1}^d$. The parameters are found via minimization of a certain functional, called *loss* function. In the standard algorithms for function approximation, to find $u$ we would need its values in a number of points $\{(x_i, y_i)\}_{i=1}^N$ in order to compute the loss. For our setting we don't need the values of $u$, but we use other information about the function. Likewise it was done in [6], the governing differential equation for $u$ has to be defined.

## 2.1   Loss Function Definition

Now we add a constraint on $u$ that it is the solution of the PDE with Dirichlet or Neumann boundary conditions in a form:

$$\Delta u(x) = f(x), \ x \in \Omega$$

$$\text{S.t. } u(x) = 0 \text{ (Dirichlet) or } \frac{\partial u(x)}{\partial \mathbf{n}(x)} = 0 \text{ (Neumann)}, \ x \in \partial\Omega, \tag{2}$$

where $\mathbf{n}(x)$ is a normal to $\partial\Omega$ at point $x$. In the standard way we parametrize $\hat{u}(x)$ - approximation of $u$ as a neural network and denote it as $\hat{u}(x; \theta)$. We define a loss function as energy functional and minimize it with respect to $\theta$:

$$\mathcal{L}(\theta) = \int_{\Omega} ((\nabla\hat{u}(x; \theta), \nabla\hat{u}(x; \theta)) + f(x)\hat{u}(x; \theta))\, dx \to \min_{\theta \in \Theta}, \tag{3}$$

where $\Theta$ is the set of all admissible parameters, it includes all such $\theta$ that $\hat{u}(x; \theta)$ satisfies the introduced boundary conditions.

For Dirichlet boundary conditions, the formulation is the following:

$$\mathcal{L}(\theta) = \int_{\Omega} ((\nabla\hat{u}(x; \theta), \nabla\hat{u}(x; \theta)) + f(x)\hat{u}(x))dx \to \min_{\theta : \hat{u}(x;\theta)|_{\partial\Omega}=0}. \tag{4}$$

In this case we need the approximation $\hat{u}$ a priori satisfy boundary conditions. We do it by introducing soft constraint via a $L_2$-penalty:

$$\mathcal{L}(\theta) = \int_{\Omega} ((\nabla\hat{u}(x; \theta), \nabla\hat{u}(x; \theta)) + f(x)\hat{u}(x; \theta))dx + \lambda \int_{\partial\Omega} \|\hat{u}(x; \theta)\|_2^2 dx \to \min_{\theta}. \tag{5}$$

In the case of Neumann boundary conditions for the uniqueness of the solution an additional constraint is introduced:

$$\int_{\Omega} u(x)dx = 0. \tag{6}$$

The solution to this problem can be found via minimization of the following functional:

$$\int_{\Omega} \left( (\nabla\hat{u}(x; \theta), \nabla\hat{u}(x; \theta)) + f(x) \left( \hat{u}(x; \theta) - \int_{\Omega} \hat{u}(x; \theta)dx \right) \right) dx \to \min_{\theta}. \tag{7}$$

Equation 3 shows the considered physics-informed loss function. For this formulation we need to compute the integral. Standard Monte Carlo (MC) way to do it is to sample some points from the domain uniformly and get the average of the integrand, i.e. if $\{x_i\}_{i=1}^{N}$ is the set of sampled points, the loss value will be:

$$\mathcal{L}(\theta) = \frac{1}{N} \sum_{i=1}^{N} \left( (\nabla \hat{u}(x_i; \theta), \nabla \hat{u}(x_i; \theta)) + f(x_i) \left( \hat{u}(x_i; \theta) - \left( \frac{1}{N} \sum_{i=1}^{N} \hat{u}(x_i; \theta) \right) \right) \right) \tag{8}$$

However, this estimate has an error of order of $1/\sqrt{N}$, where $N$ is the number of sampled points. There exist different other methods for quadrature computation, like quasi-random methods. We propose to use an adaptive quadrature scheme.

## 2.2   Adaptive Quadrature Scheme

The basic idea of adaptive quadrature is to split the computational domain into smaller subdomains and recompute the quadrature, until the desired error value is achieved. In order to estimate the integral value, a base quadrature algorithm is required. In our case it is just a weighted sum of the function values in a number of points, so the algorithm needs only to define the weights and the set of points (Algorithm 1). When we compute the integral on the domain different from the reference domain of the base algorithm, we need to transform the points so, that the new points fit the new domain (function TRANSFORM in the algorithm). After that function values are calculated in the new points.

---

**Algorithm 1.** Base quadrature

---

**Require:** $P = \{p_i\}_{i=1}^{K}$, $p_i \in \mathbb{R}^n$ - set of points for integration, $w \in \mathbb{R}^K$ - weights, $f$ - integrand, $\mathcal{I}$ - domain for integration
  **function** QUADRATURE($P, w, f, \mathcal{I}$)
    $\hat{P} \leftarrow$ TRANSFORM($P, \mathcal{I}$)
    **return** Vol($\mathcal{I}$) $\cdot \sum_i w_i f(\hat{p}_i)$
  **end function**

---

Using this algorithm we recompute the quadrature by adaptively refining the intervals. The schematic idea of adaptive quadrature computation is provided in Algorithm 2. Depending on the dimensionality of the problem we use different types of domains. For 1D case line segments are used, for 2D case—triangles. At each step the domain is split into $k$ equal non-overlapping subdomains, which are similar to the initial domain ($k = 2$ for 1D, $k = 4$ for 2D). This approach allows us to distribute the considered points unevenly. This can be quite useful for complex functions with oscillations of high frequencies in a part of the domain.

## 3   Computational Experiments

The code was written using JAX [1]. The algorithm for quadrature estimation, including the points and weights values, was taken from quadpy package [5]. We minimize the functional (3) using Adam optimizer with learning rate scheduling. The used activation function is $\tanh(x)$.

---

**Algorithm 2.** Adaptive quadrature

---

**Require:** $f$ - integrand, $\mathcal{I}s$ - set of domains on which the quadrature is computed, QUADRATURE$_{base}$ - algorithm for quadrature estimation (Alg. 1), QUADRATURE$_{comp}$ - the second quadrature algorithm for comparison with base algorithm.
   **function** ADAPTIVE_QUADRATURE($f, \mathcal{I}s$)
      $I \leftarrow 0$
      **for** $\mathcal{I}$ in $\mathcal{I}s$ **do**
         $I_1 \leftarrow$ QUADRATURE$_{base}(f, \mathcal{I})$
         $I_2 \leftarrow$ QUADRATURE$_{comp}(f, \mathcal{I})$
         del $\mathcal{I}$
         **if** $|I_1 - I_2| < \epsilon_{abs}$ **then**
            $I \leftarrow I + I_1$
         **else**
            $\mathcal{I}' \leftarrow$ SPLIT_DOMAIN($\mathcal{I}$)
            $\mathcal{I}s$.append($\mathcal{I}'$)
         **end if**
      **end for**
      **return** $I$
   **end function**

---

### 3.1    1D Case

The following examples are considered for the one-dimensional case:

1. $f(x) = 1$, $x \in [0, 1]$ and zero-Dirichlet boundary conditions
2. $f(x) = \cos(\pi x)$, $x \in [0, 1]$ and zero-Neumann boundary conditions
3. $f(x) = \tanh''(\alpha x)$, $x \in [0, 1]$ and zero-Neumann boundary conditions

In all the cases we can analytically find the solution and estimate the error between the prediction and the reference solution. As basic architecture a fully connected neural network with hidden size ($h_i$) 10 and depth ($d$) 10 is taken. For the adaptive algorithm the optimizer's initial learning rate $10^{-2}$ was multiplied by 0.1 after iterations 2000 and 3000. For the Monte Carlo estimate a constant learning rate $10^{-3}$ is considered in case of Neumann BC, and for Dirichlet BC the learning rate has value 0.01 and is mulitiplied by 0.1 after iterations 500 and 3000. Gauss-Kronrod scheme of degree 7 is used for the adaptive integration. We use 10000 samples for MC estimates. For the validation of the method the relative $L_2$ error is computed via adaptive algorithm. For the test relative errors the $L_2$-norms are approximated using 1000 random points. The error values are presented in Table 1.

Figure 1 shows the result for the first case with Dirichlet boundary conditions for both approaches for computing integral. The chosen parameter $\lambda$ was set to 50 for Monte Carlo and 100 for adaptive quadrature. It can be observed that although the curves on the plots look very similar, boundary conditions are not satisfied, i.e. the function plot is a little displaced by some $\epsilon$, and the resulting relative error is $\sim 0.05$ (Fig. 1c). This issue has to be investigated more thoroughly and this is a task for the future work.

For the case of Neumann boundary conditions the result is much better (Fig. 2). In this case we also provide the relative error for a larger number of samples. One can notice from Fig. 2c that for Monte Carlo method the final error for 1000000 samples had order only of $10^{-2}$. Adaptive quadrature used only 45 points and has reached the error of order $10^{-3}$.

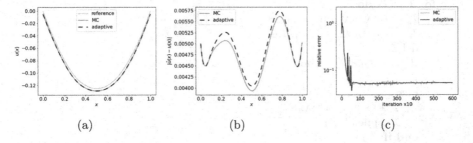

(a)                              (b)                              (c)

**Fig. 1.** Result for $f(\mathbf{x}) = 1$ and zero-Dirichlet boundary conditions: Figure 1a is the prediction visualization, Fig. 1b shows the prediction error $|\hat{u}(x) - u(x)|$, Fig. 1c shows relative errors during the training.

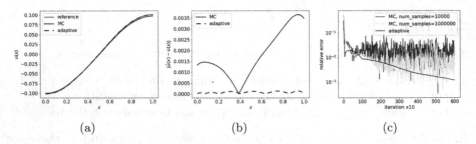

(a)                              (b)                              (c)

**Fig. 2.** Result for $f(\mathbf{x}) = \cos(\pi x)$ and zero-Neumann boundary conditions: Figure 2a is the prediction visualization, Fig. 2b shows the prediction error $|\hat{u}(x) - u(x)|$, Fig. 2c shows relative errors during the training.

In the third case (Fig. 3). the gradient descent presumably gets stuck in the local minimum. Here we experiment with different training approaches.

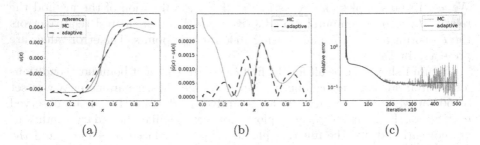

(a)                              (b)                              (c)

**Fig. 3.** Result for $f(\mathbf{x}) = \tanh''(15x)$ and zero-Neumann boundary conditions: Figure 3a is the prediction visualization, Fig. 3b shows the prediction error $|\hat{u}(x) - u(x)|$, Fig. 3c shows relative errors during the training.

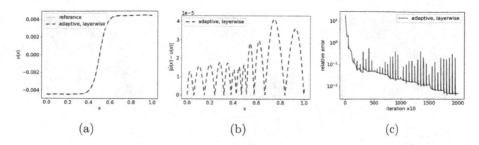

**Fig. 4.** Result of layerwise training for $f(x) = \tanh''(15x)$: Figure 4a is the prediction visualization, Fig. 4b shows the prediction error $|\hat{u}(x) - u(x)|$, Fig. 4c shows relative errors during the training.

*Layerwise Training.* We try to train each layer independently. The process is the following: fix all the layers, except first one, optimize its parameters for $k$ iterations, then do the same for the second layer, the third, etc. And repeat the procedure $N$ times. This approach has allowed us to reach error of 0.0035 in comparison with 0.09 for the standard case. Figure 4 visualize the result, we show the relative error only for first 20000 iterations for a clearer view. The peaks on Fig. 4c correspond to the switch between the layers in the optimization procedure. The training continued for 50000 iterations, the layer was changed each 500 iterations.

## 3.2   2D Case

Further we consider $\mathbf{x} \in [0,1] \times [0,1]$ and $f(\mathbf{x}) = \cos(\pi x_1) + \cos(\pi x_2)$. The used neural network architecture is fully-connected as in the one-dimensional case. The depth and the hidden size are set to 4 and 10 correspondingly. For Monte Carlo quadrature the learning rate 0.01 is taken for the optimizer, it is multiplied by 0.1 after 300, 800 and 1500 iterations. In case of the adaptive estimate the starting learning rate is 0.01, the update is performed after 300 iterations. The quadrature is computed via Dunavant scheme [3].

For a such simple right-hand side the algorithms shows quite a good result (Fig. 5). And compared to Monte Carlo integration it requires less points as it was in one-dimensional case. The comparison in terms of achieved test error and number of points is provided in Table 1. We use 10000 uniformly sampled points to estimate the error.

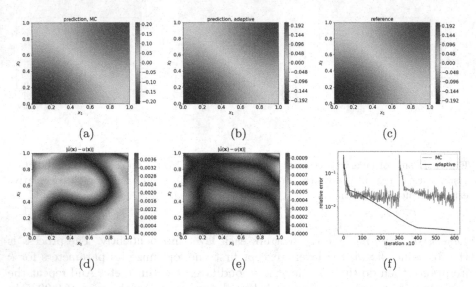

**Fig. 5.** Result for $f(\mathbf{x}) = \cos(\pi x_1) + \cos(\pi x_2)$: Fig. 5a is for the Monte Carlo estimate, Fig. 5b is for the adaptive quadrature, Fig. 5c is for the reference solution, Fig. 5d shows the prediction error $|\hat{u}(\mathbf{x}) - u(\mathbf{x})|$ for Monte Carlo, Fig. 5e—for the adaptive algorithm, 5f shows relative errors during the training.

**Table 1.** Comparison of adaptive and Monte Carlo methods

| Right-hand side | BC type | Method | Number of iterations | Relative $L_2$-error | Number of points |
|---|---|---|---|---|---|
| $f(x) = 1$ | Dirichlet | MC | 4000 | 0.0524 | 10000 |
| | | Adaptive | 4000 | 0.0535 | 45 |
| $f(x) = \cos(\pi x)$ | Neumann | MC | 6000 | 0.0286 | 10000 |
| | | Adaptive | 6000 | 0.0011 | 45 |
| $f(x) = \tanh''(15(x - 0.5))$ | Neumann | MC | 5000 | 0.3002 | 10000 |
| | | Adaptive | 5000 | 0.2025 | 105 |
| | | Layerwise | 50000 | 0.0047 | 147 |
| $f(\mathbf{x}) = \cos(\pi x_1) + \cos(\pi x_2)$ | Neumann | MC | 6000 | 0.0144 | 10000 |
| | | Adaptive | 8000 | 0.0016 | 266 |

## 4    Conclusion

The proposed in the paper pipeline including loss function based on energy norm of the residual of the considered PDE gives very promising results. Here it should be noted that adaptive quadrature gives a better result than Monte Carlo in terms of the achieved error estimate in all the cases. It also requires much less points for the estimation. However, because of the recursive structure of the procedure, it takes longer to compute. Monte Carlo integration requires more points for the more accurate result, and when the number of samples reaches 1000000, the computation times become comparable.

In case of Neumann boundary conditions for adaptive quadrature with an error in quadrature estimation $\sim 10^{-4}$ we achieve relative $L_2$ error of $\sim 10^{-3}$ for the approximated function. For the case of function with steep gradient, training only one layer of the network at once has proved itself to be useful.

In case of Dirichlet boundary conditions we couldn't achieve relative error lower than $\sim 0.05$, this may be due to the used penalty in the functional. There are different ways to overcome this problem—reconsider the choice of loss, architecture, optimizer, penalty coefficient $\lambda$. These are the issues for the further work.

# References

1. Bradbury, J., et al.: JAX: composable transformations of Python+NumPy programs (2018). http://github.com/google/jax
2. Dal Santo, N., Deparis, S., Pegolotti, L.: Data driven approximation of parametrized PDEs by reduced basis and neural networks. J. Comput. Phys. **416**, 109550 (2020). https://doi.org/10.1016/j.jcp.2020.109550. https://www.sciencedirect.com/science/article/pii/S0021999120303247
3. Dunavant, D.A.: High degree efficient symmetrical gaussian quadrature rules for the triangle. Int. J. Numer. Methods Eng. **21**(6), 1129–1148 (1985). https://doi.org/10.1002/nme.1620210612. https://onlinelibrary.wiley.com/doi/abs/10.1002/nme.1620210612
4. Kochkov, D., Smith, J.A., Alieva, A., Wang, Q., Brenner, M.P., Hoyer, S.: Machine learning accelerated computational fluid dynamics (2021). https://arxiv.org/pdf/2102.01010.pdf
5. Quadpy. https://github.com/nschloe/quadpy
6. Raissi, M., Perdikaris, P., Karniadakis, G.E.: Physics informed deep learning (part I): data-driven solutions of nonlinear partial differential equations (2017). https://arxiv.org/abs/1711.10561

# A Full Order, Reduced Order and Machine Learning Model Pipeline for Efficient Prediction of Reactive Flows

Pavel Gavrilenko[3], Bernard Haasdonk[2], Oleg Iliev[3], Mario Ohlberger[1], Felix Schindler[1(✉)], Pavel Toktaliev[3], Tizian Wenzel[2], and Maha Youssef[2]

[1] Mathematics Münster, Westfälische Wilhelms-Universität Münster, Einsteinstr. 62, 48149 Münster, Germany
{mario.ohlberger,felix.schindler}@uni-muenster.de
[2] Institute of Applied Analysis and Numerical Simulation, Pfaffenwaldring 57, 70569 Stuttgart, Germany
{haasdonk,tizian.wenzel,maha.youssef-ismail}@mathematik.uni-stuttgart.de
[3] Fraunhofer-Institut für Techno und Wirtschaftsmathematik, Fraunhofer-Platz 1, 67663 Kaiserslautern, Germany
{oleg.iliev,pavel.gavrilenko,pavel.toktaliev}@itwm.fraunhofer.de

**Abstract.** We present an integrated approach for the use of simulated data from full order discretization as well as projection-based Reduced Basis reduced order models for the training of machine learning approaches, in particular Kernel Methods, in order to achieve fast, reliable predictive models for the chemical conversion rate in reactive flows with varying transport regimes.

**Keywords:** Reactive flow · Model order reduction · Machine learning

## 1 Introduction

Reactive mass transport in porous media with catalytic reactions is the basis for many industrial processes and systems, such as fuel cells, photovoltaic cells, catalytic filters for exhaust gases and catalytic burners. The usual way of designing and testing prototypes of such devices is expensive and time-consuming. While modeling and simulation of the processes at the pore scale of the porous media can help to optimize the design of device catalytic components, it is currently limited by the fact that such simulations lead to large amounts of data (each simulation may consist of hundreds of TB). Moreover, processes under consideration depend on a large number of parameters. As a consequence, the development of

Funded by BMBF under contracts 05M20PMA, 05M20VSA, 05M20AMD. Funded by the Deutsche Forschungsgemeinschaft (DFG, German Research Foundation) under Germany's Excellence Strategy EXC 2044-390685587, Mathematics Münster: Dynamics - Geometry - Structure. Funded by the Deutsche Forschungsgemeinschaft (DFG, German Research Foundation) under Germany's Excellence Strategy EXC 2075-390740016. We acknowledge the support by the Stuttgart Center for Simulation Science (SimTech).

© Springer Nature Switzerland AG 2022
I. Lirkov and S. Margenov (Eds.): LSSC 2021, LNCS 13127, pp. 378–386, 2022.
https://doi.org/10.1007/978-3-030-97549-4_43

approaches to solve these problems with large amounts of data, as well as for the prediction of the chemical conversion rate using modern data-based methods is essential for fast, reliable predictive models.

The purpose of this paper is to demonstrate a computational pipeline which combines direct computational tools and different data-based methods for a simple model problem with industry relevant aspects. As a basic underlying model for reactive flow we consider the following scalar convection-diffusion-reaction (CDR) model problem for a concentration $c_\mu$, i.e.

$$\partial_t c_\mu - \Delta c_\mu + \mu_2 \nabla \cdot (u c_\mu) + \mu_1 c_\mu = 0, \tag{1}$$

posed on a unit domain with prescribed velocity $u$ of unit magnitude, where we consider the Damköhler and Peclét numbers $Da, Pe \in \mathbb{R}^+$ as parameters, i.e. $\mu = (\mu_1, \mu_2)^\top = (Da, Pe)^\top$. The problem will be complemented with suitable initial and boundary data, as well as with a suitable quantity of interest that needs to be evaluated.

In Sect. 2, we detail the problem formulation and involved concepts constituting the full approximation pipeline, including

- a full order model (**FOM**),
- a reduced order model (**ROM**) based on FOM data and
- a machine learning (**ML**) based model based on FOM and ROM data.

Section 3 describes the associated software pipeline while numerical experiments in Sect. 4 give a proof of concept of our approach for a simple, but industrially relevant example.

## 2 Approximation and Data Based Learning

Given a bounded set of input parameters $\mathcal{P} \subset \mathbb{R}^p$, $p \subset \mathbb{N}$ and an end time $T > 0$, we seek to efficiently and accurately approximate the evaluation of a function $f \in L^2(\mathcal{P}; L^2([0, T]))$ at a fixed set of finitely many time points for varying inputs $\mu \in \mathcal{P}$. The function $f$, modeling a quantity of interest (QoI), is implicitly given by $f(\mu; t) := s_\mu(c_\mu(t))$, where for any input parameter $\mu \in \mathcal{P}$, $s_\mu \in V'$ is a linear functional and the state $c_\mu \in L^2(0, T; V)$ with $\partial_t c_\mu \in L^2(0, T; V')$ is the unique weak solution of a parabolic partial differential equation (PDE)

$$\langle \partial_t c_\mu, v \rangle + a_\mu(c_\mu, v) = l_\mu(v) \quad \text{for all } v \in V, \quad \text{and } c_\mu(0) = c_0, \tag{2}$$

given initial data $c_0 \in V$, a Gelfand triple of suitable Hilbert spaces $V \subset H^1(\Omega) \subset L^2(\Omega) \subset V'$ associated with a spatial domain $\Omega$, a continuous linear functional $l_\mu \in V'$ and a continuous and coercive bilinear form $a_\mu : V \times V \to \mathbb{R}$, for each parameter $\mu \in \mathcal{P}$.

Since we consider stationary non-homogeneous Dirichlet boundary data $g_D \in H^{1/2}(\Gamma_D)$ on the Dirichlet boundary $\Gamma_D \subset \partial\Omega$ and Neumann data $g_N \in L^2(\Gamma_N)$ on the Neumann boundary $\Gamma_N := \partial\Omega \backslash \Gamma_D$, we select an extension of the Dirichlet data $\tilde{g}_D \in H^1(\Omega)$, such that $\tilde{g}_D|_{\Gamma_D} = g_D$ in the sense of

traces, and consider the shifted solution trajectory $c_{0,\mu} := c_\mu - \tilde{g}_D \in V :=$
$H^1_{\Gamma_D}(\Omega) := \{v \in H^1(\Omega) \mid v|_{\Gamma_D} = 0 \text{ in the sense of traces}\}$. We then obtain the
weak formulation of (1) by noting $\partial_t \tilde{g}_D = 0$, letting $l_{0,\mu}(v) := \int_{\Gamma_N} g_N v \, ds$,
$a_\mu(c, v) := \int_\Omega (\nabla c - \mu_2 u c) \cdot \nabla v \, dx + \int_\Omega \mu_1 c v \, dx + \int_{\Gamma_N} (\mu_2 c u) \cdot n v \, ds$, and
$l_\mu(v) := l_{0,\mu}(v) - a_\mu(\tilde{g}_D, v)$ in (2), recalling $\mu = (\mu_1, \mu_2)^\top = (Da, Pe)^\top$. Thus,
(2) describes the time evolution of the shifted solution and we obtain the original
solution by $c_\mu := c_{0,\mu} + \tilde{g}_D$.

## 2.1 Approximation by a Full Order Model (FOM)

Since we may not evaluate $f$ exactly, given any $N_T \in \mathbb{N}$, we consider a so-called
full order model (FOM) approximation,

$$f_h : \mathcal{P} \to \mathbb{R}^{N_T}, \quad \mu \mapsto \big(f_h(\mu; t_1), \ldots, f_h(\mu; t_{N_T})\big)^\top, \quad f_h(\mu; t) := s_\mu(c_{h,\mu}(t)), \quad (3)$$

stemming for simplicity from an implicit Euler discretization on an equidistant
temporal grid (yielding $N_T$ points $\{t_n\}_{1 \le n \le N_T} \subset [0, T]$), and a conforming spatial discretization space $V_h \subset V$ of fixed dimension $N_h \in \mathbb{N}$, yielding for each
$1 < n \le N_T$ the unique solution $c_{h,\mu}(t_n) \in V_h$ of

$$\frac{1}{\Delta t}\big(c_{h,\mu}(t_{n+1}) - c_{h,\mu}(t_n), v\big)_{L^2(\Omega)} + a_\mu(c_{h,\mu}(t_{n+1}), v) = l_\mu(v) \quad \forall v \in V_h, \quad (4)$$

assuming $c_{h,\mu}(t_1) := c_0 \in V_h$, for simplicity. Depending on $N_h$ and $N_T$, the
computation of $c_{h,\mu}$ (and thus the evaluation of $f_h$) in the context of parameter studies, optimal design, uncertainty quantification or related applications
involving $f_h$, may easily be prohibitively costly.

In the non-parametric setting, i.e. when considering (2) for a single input
tuple $\mu = (Da, Pe)^\top$, adaptive Finite Element methods as well as adaptive time
stepping schemes are the methods of choice, in particular when considering long-
time integration. While these are also applicable in the context of model order
reduction, we restrict ourselves to fixed equidistant temporal and spatial grids.
As approximation space $V_h$ in (4) we use standard conforming piecewise linear
finite elements, assuming for simplicity $\tilde{g}_D \in V_h$.

Note that other suitable choices include stabilized FEM or (upwind) Finite
Volume (FV) schemes (in particular for large Péclet numbers which might
induce steep spatial gradients or oscillations) and interior penalty discontinuous Galerkin (DG) schemes (in particular higher order variants to benefit from
smooth spatial parts of the state).

## 2.2 Reduced Basis Reduced Order Model (ROM)

Employing machine-learning (ML) techniques such as artificial neural networks
or kernel methods, one could directly utilize $f_h$ to learn an approximation $f_{ml}$ :
$\mathcal{P} \to \mathbb{R}^{N_T}$ to efficiently provide cheap approximations of $f_h$. In particular, ML-
based models usually do not require the computation of a (reduced) state and
may even compute all $N_T$ values at once without time-integration - or even

provide predictions for continuous times $t$ instead of for $N_T$ discretized time values. However, to ensure good approximation properties of such a ML-based model, the computation of the training and validation data involving $f_h$ may still be too demanding. So, a method for rapidly generating additional training data for data augmentation is required.

As an intermediate step, we thus employ a structure preserving Reduced Basis (RB) reduced order model (ROM), which we obtain by Galerkin-projection of (4) onto a carefully crafted low-dimensional RB subspace $V_{rb} \subset V_h$ (see [1]). While the construction of the RB space also involves solving (4), it then allows the computation of a reduced state $c_{rb,\mu}(t_n) \in V_{rb}$, given for $1 \leq n \leq N_T$ as the unique solution of

$$\frac{1}{\Delta t}\big(c_{rb,\mu}(t_{n+1}) - c_{rb,\mu}(t_n), v\big)_{L^2(\Omega)} + a_\mu(c_{rb,\mu}(t_{n+1}), v) = l_\mu(v) \quad \forall v \in V_{rb}, \quad (5)$$

and thus the evaluation of an RB QoI $f_{rb} : \mathcal{P} \to \mathbb{R}^{N_T}$,

$$\mu \mapsto \big(f_{rb}(\mu; t_1), \ldots, f_{rb}(\mu; t_{N_T})\big)^\top \quad \text{with} \quad f_{rb}(\mu; t) := s_\mu(c_{rb,\mu}(t)), \quad (6)$$

with a computational complexity only depending on $N_{rb} := \dim V_{rb}$, not $N_h$, owing to a pre-computation of all $V_h$-dependent quantities arising in (5) under mild assumptions on the parametrization of $a_\mu$ and $l_\mu$.

Since the RB-ROM (5) simply arises as the Galerkin projection of (2) onto $V_{rb}$, it is fully defined once we specify the RB space, the construction of which is a delicate matter: it should be as low-dimensional as possible, to ensure a good online-performance of the resulting RB model; it should be as rich as possible, to ensure good approximation properties of the resulting RB model; however, at the same time, its construction greatly impacts the overall performance of the scheme and should be as cheap as possible. For simplicity we employ the method of snapshots: we collect a series of state trajectories $\{c_{h,\mu}\}_{\mu \in \mathcal{P}_{POD}}$ for a set of a priori specified training parameters $\mathcal{P}_{POD} \subset \mathcal{P}$ and simply obtain $V_{rb}$ by proper orthogonal decomposition (POD) of the resulting snapshot Gramian.

Note that the approximation quality of the resulting model can only be assessed in a post-processing step involving the computational complexity of the FOM again. An alternative adaptive construction of $V_{rb}$ can be achieved by means of an iterative POD-greedy algorithm steered by a posteriori error control, which at the same time allows an efficient quantification of the induced model reduction error independent of the FOM complexity. However, the computational complexity of the required offline-computations may be significant (in particular for problems with long time-integration).

In particular for spatially highly resolved FOMs with $N_h \gg 1$, evaluating $f_{rb}$ might be orders of magnitude faster than $f_h$. However, the solution of (5) still requires time integration which is why for temporally highly resolved FOMs with $N_T \gg 1$, we employ advanced machine learning ROMs using greedy kernel methods to directly learn the mapping $f_{ml} : \mathcal{P} \to \mathbb{R}^{N_T}$, thus skipping the time integration (detailed in the following section).

## 2.3   Approximation by Machine Learning: Kernel Methods

We consider $\mathcal{P} \subset \mathbb{R}^2$ and for our problem complexity shallow instead of deep learning architectures are sufficient. We apply kernel methods which in our case rely on strictly positive definite kernels, which are in the scalar case symmetric functions $k : \mathcal{P} \times \mathcal{P} \to \mathbb{R}$, such that the so called kernel matrix $(A_n)_{i,j} = k(\mu_i, \mu_j)$, $i, j = 1, .., n$ is positive definite for any set $\{\mu_1, .., \mu_n\} \subset \mathcal{P}, n \in \mathbb{N}$ of pairwise distinct inputs (see [7]). The well known Gaussian kernel $k(x, y) = \exp(-\|x - y\|_2^2)$ is an example of a strictly positive definite kernel.

Associated to a strictly positive definite kernel on a domain $\mathcal{P}$, there is a space of continuous function, the so called Reproducing Kernel Hilbert Space (RKHS), $\mathcal{H}_k(\mathcal{P})$. For given input data $\{\mu_1, .., \mu_N\} \subset \mathcal{P}$ and corresponding target data $\{y_1, .., y_N\} \subset \mathbb{R}^{d_{\text{out}}}$, learning with kernels refers to a minimization of a loss-functional over the RKHS. Using a mean-squared error loss with a standard norm regularization, a kernel representer theorem states that a solution

$$f_{\text{ml}} := \arg\min_{f \in \mathcal{H}_k(\mathcal{P})} \mathcal{L}(f), \quad \mathcal{L}(f) = \frac{1}{N} \sum_{i=1}^{N} \|y_i - f(\mu_i)\|_{\mathbb{R}^{d_{\text{out}}}}^2 + \lambda \cdot \|f\|_{\mathcal{H}_k(\mathcal{P})}^2 \quad (7)$$

can be found in the finite dimensional subspace spanned by the data, i.e. there exists $f_{\text{ml}} \in \text{span}\{k(\cdot, \mu_i), i = 1, .., N\}$. In order to learn a sparse and fast surrogate $f_{\text{ml}}$, one strives not to use all the training data, but only a meaningful subset. While a global optimization is combinatorial infeasible, we use greedy methods in order to select a subset, as implemented in the vectorial kernel orthogonal greedy algorithm (VKOGA) [6]. Algorithms of this type start with an empty set $X_0 := \{\}$, and incrementally select the next point $X_{n+1} := X_n \cup \{x_{n+1}\}$, $x_{n+1} \in \{\mu_1, .., \mu_N\}$, for instance according to the $f$-greedy selection criterion, $x_{n+1} := \arg\max_{\mu_i, i=1,..,N} \|(y_i - s_n)(\mu_i)\|_{\mathbb{R}^{d_{\text{out}}}}$, whereby $s_n$ is the kernel approximant based on the input data $X_n$ and corresponding target data. These greedy kernel approximation algorithms have been studied thoroughly in terms of approximation speed, stability and generalization, for instance the unregularized case, $\lambda = 0$ within (7), in [8].

## 3   Software Environment

We aim for a flexible, user-friendly and performant software environment to meet the requirements arising from Sect. 2, with the Python-based and freely available model reduction library pyMOR[1] [5] at its core:

**Interchangable FOM:** As argued in Sect. 2.1, FV schemes or conforming or DG FEM schemes might be required for the problem at hand, the choice of which is not always clear a priori. Thus, while pyMOR's built-in numpy/scipy-based discretization might be desirable for quick prototyping, more advanced multi-purpose discretization libraries such as deal.II, dune-gdt[2], FEniCS or

---

[1] Available at https://pymor.org, including references to other software libraries.
[2] Available at https://github.com/dune-community/dune-gdt.

`NGSolve` (all freely available) are often required for more advanced problems. Some applications, for instance for multi species reactive porous media transport with non-linear source terms on computer tomography-based geometries, require more specialized libraries such as Fraunhofers in house library `PoreChem` [3].

Convenient `pyMOR` wrappers are available for all of the above libraries with a well-defined API, to allow a unified handling of the resulting FOM, regardless of its origin. For instance, given the model problem described in Sect. 2 with scalar Dammköhler- and Péclet-number as input parameters, we can call

```
c_h = fom.solve({'Da': 1, 'Pe': 1e-3})    # compute c_{h,μ} from (4)
f_h = fom.output({'Da': 1, 'Pe': 1e-3})   # compute f_h(μ) from (3)
```

on any FOM obtained from one of the above libraries[3].

**Interchangable MOR Algorithms:** As argued in Sect. 2.2, there exist several established algorithms for the generation of a reduced basis, most prominently (for instationary problems) the POD (or method of snapshots) and the POD-greedy algorithm, where the applicability of each algorithm in terms of accuracy and performance is often not clear a priori. Once wrapped as `pyMOR` models, all FOMs stemming from the above libraries expose access to (the application of) their operators (e.g. those induced by $a_\mu$ and $l_\mu$, products such as the $L^2$- or $H^1$-semi product) and all vector arrays (containing state snapshots such as $c_\mu$) of these FOMs fulfill the same API. `pyMOR` thus ships a large variety of generic algorithms applicable such as a stabilized `gram_schmidt`, an `rb_greedy` and `rb_adaptive_weak_greedy` and `pod` algorithm, as well as the distributed or incremental `hapod` algorithm (see below). For instance, calling

```
snapshots = fom.solution_space.empty()
for mu in parameter_space.sample_randomly(10):
    snapshots.append(fom.solve(mu))
pod_basis, _ = pod(snapshots, product=fom.h1_0_product)
```

computes a POD basis for any FOM stemming from one of the above libraries. Similarly, the Galerkin projection of FOMs (i.e. its operators, products and outputs) is provided generically, yielding a structurally similar ROM as in:

```
pod_reductor = InstationaryRBReductor(
    fom, RB=pod_basis, product=fom.h1_0_product)
pod_rom = pod_reductor.reduce()
```

**Expandable and User-Friendly Environment:** The high-level interactive nature of the Python programming language allows quick prototyping for beginners and experts alike, while its rich and free ecosystem allows access to high-performance libraries such as the above mentioned ones or `pytorch`[4], often used in the context of machine learning. In addition, `pyMOR`s API and generic algorithms allows for flexible high-level user code: since all models behave similarly, and all outputs and reduced data structures are `numpy`-based, a call like

---

[3] The code examples throughout this section contain actual (shortened) Python usercode, usually encountered in `pyMOR`-based applications.

[4] https://pytorch.org/

```
f_rb = pod_rom.output({'Da': 1, 'Pe': 1e-3})      # f_{rb}(μ) from (6)
abs_linf_output_err = np.max(np.abs(f_h - f_rb)) # ‖f_h(μ) − f_{rb}(μ)‖_{L^∞}
```

works for any combination of FOM and generated ROM. It is thus easily possible to prototype and evaluate new algorithms (such as advanced MOR and ML algorithms) as well as create custom applications, such as the workflow from Sect. 2, involving the greedy kernel methods from the VKOGA library [9].

# 4  Numerical Experiments: Reactive Flow

To demonstrate the approach detailed in Sect. 2, we consider the one dimensional CDR equation for dimensionless molar concentration variable, $c_\mu(x,t) \in \mathbb{R}^+$, i.e. (1) posed on the unit domain $\Omega := (0,1)$ with $u = 1$, $c_0 := 0$ initial values, Dirichlet boundary $\Gamma_D := \{0\}$, Dirichlet data $g_D = 1$, Neumann boundary $\Gamma_N := \{1\}$ and Neumann data $g_N = 0$. Here we choose diffusion time as a characteristic time and $T = 3$ to ensure a near-stationary QoI, namely the break through curve $s_\mu(t) := \int_{\Gamma_N} c_\mu(t)\, ds$, at the end of the simulation. This model is widely used as a basis in chemical engineering and industry plug-flow/perfectly stirred reactors models, and as a consequence part of real designing processes in the industry. It is an excellent compromise between the complexity of real industrial models of the catalytic process, and a simple mathematical formulation providing main features of transport processes. For this choice of initial- and boundary data, an analytical solution of (1) is available owing to [2].

We consider the diffusion dominated regime, i.e. $\mu = (Da, Pe)^\top \in \mathcal{P} := [10^{-3}, 1]^2$ and are interested in an overall relative approximation error w.r.t. the target QoI $f$ in $L^\infty(\mathcal{P}_{\text{test}}; L^2([0,T]))$ of less than one percent, measured over a finite test set $\mathcal{P}_{\text{test}} \subset \mathcal{P}$. We thus require a relative error of $10^{-4}$ from all approximation components, and use the analytical solution from [2] to calibrate the FOM to compute the reference QoI $f_h$ with a relative error less than $10^{-4}$, yielding $h = 2^{-6}$ and $\Delta t = 2^{-13}$ (thus $N_T = 24576$ time steps) as sufficient for the diffusion dominated regime. We also use a finer spatial grid in a second experiment to additionally represent the state $c_\mu$ accurately. As spatial product over $V$ we chose the full $H^1$-product, which is also used for orthonormalization. For the discretization, we use the preliminary Python bindings of dune-gdt[5] to provide a pyMOR-compatible FOM, as detailed in Sect. 3.

For the method of snapshots, we select the four outermost points of $\mathcal{P}$ as training parameters $\mathcal{P}_{\text{POD}}$ and use pyMOR's implementation of the incremental Hierarchical approximate POD (HAPOD) from [4], inc_vectorarray_hapod, with a tolerance of $10^{-4}$. Handling a dense snapshot Gramian of size $(4 \cdot 24576)^2$, as in the classical POD, would otherwise be infeasible. We use a VKOGA implementation[6] with default parameters and train it using the four inputs $\mu \in \mathcal{P}_{\text{POD}}$ and already computed FOM outputs $f_h(\mu)$ used to build the POD basis, as well

---

[5] Similar to https://github.com/ftschindler-work/dune-gdt-python-bindings.
[6] Available at https://github.com/GabrieleSantin/VKOGA.

**Table 1.** Accuracy and runtime (in seconds) of the FOM, RB-ROM and VKOGA models (and respective approximations of $f$) from the proposed pipeline for the experiment from Sect. 4 for an "output-accurate" FOM (first row) and a "state-and-output-accurate" FOM (second row). Offline time comprises all parts to build the model (FOM: grid + FEM assembly; RB-ROM: FOM + snapshots + HAPOD + Galerkin projection onto $V_{rb}$; VKOGA: RB-ROM + snapshots + fitting). Online time denotes average time to evaluate the respective $f_*$ (FOM: four inputs; RB-ROM: ten inputs; VKOGA: 1000 inputs). "rel. err." denote respective relative errors (RB-ROM: $\|f_h(\mu) - f_{rb}(\mu)\|_{l_2}/\|f_h(\mu)\|_{l_2}$ over $\mathcal{P}_{POD}$ + five random inputs; VKOGA: $\|f_{rb}(\mu) - f_{ml}(\mu)\|_{l_2}/\|f_{rb}(\mu)\|_{l_2}$ over 50 random inputs). "p. o." denotes pay-off of the full pipeline after this many queries of $f_{ml}$ compared to $f_h$ (respective offline + online).

| FOM ($f_h$) | | | RB-ROM ($f_{rb}$) | | | | VKOGA model ($f_{ml}$) | | | | |
|---|---|---|---|---|---|---|---|---|---|---|---|
| $N_h$ | offline | online | offline | $N_{rb}$ | online | rel. err. | offline | points | online | rel. err. | p. o. |
| 65 | 3.62e-2 | 6.99e0 | 1.56e2 | 12 | 4.15e0 | 7.81e-6 | 9.76e2 | 51 | 4.12e-4 | 3.31e-6 | 140 |
| 65537 | 1.63e0 | 3.39e3 | 2.47e4 | 14 | 4.21e0 | 3.31e-6 | 2.55e4 | 51 | 4.15e-4 | 3.19e-6 | 8 |

as 196 randomly selected inputs $\mu \in \mathcal{P}$ and corresponding RB-ROM approximations $f_{rb}(\mu)$.

The performance and approximation properties of the resulting approximate models[7] are given in Table 1.

## 5   Conclusion

We propose a pipeline of consecutively built full order, reduced order and machine-learned models to approximate the evaluation of a QoI depending on a parabolic PDE, and numerically analyze its performance for an industrially relevant one-dimensional problem (see Table 1). While the similar dimensions of the RB-ROM, $N_{rb}$, and the number of selected VKOGA points for both runs indicate, that the low spatial resolution for the state is sufficient to approximate the QoI in this example, the proposed pipeline pays off after 140 queries, compared to only using the FOM. The second run demonstrates an even more pronounced pay off after 8 queries for higher spatial resolutions (as in higher dimensions).

## References

1. Benner, P., Cohen, A., Ohlberger, M., Willcox, K. (eds.): Model Reduction and Approximation: Theory and Algorithms. Computational Science and Engineering, vol. 15. SIAM, Philadelphia (2017)
2. van Genuchten, M., Alves, W.: Analytical solutions of the one-dimensional convective-dispersive solute transport equation. Technical Bulletin - United States Department of Agriculture, vol. 1661, p. 49 (1982)

---

[7] Computed on a dual socket compute server equipped with two Intel Xeon E5-2698 v3 CPUs with 16 cores running at 2.30 GHz each and 256 GB of memory available.

3. Greiner, R., Prill, T., Iliev, O., van Setten, B.A., Votsmeier, M.: Tomography based simulation of reactive flow at the micro-scale: particulate filters with wall integrated catalyst. Chem. Eng. J. **378**, 121919 (2019)
4. Himpe, C., Leibner, T., Rave, S.: Hierarchical approximate proper orthogonal decomposition. SIAM J. Sci. Comput. **40**(5), A3267–A3292 (2018)
5. Milk, R., Rave, S., Schindler, F.: pyMOR - generic algorithms and interfaces for model order reduction. SIAM J. Sci. Comput. **38**(5), S194–S216 (2016)
6. Santin, G., Haasdonk, B.: Kernel methods for surrogate modeling. arXiv arXiv:1907:10556 (2019)
7. Wendland, H.: Scattered Data Approximation. Cambridge Monographs on Applied and Computational Mathematics, vol. 17. Cambridge University Press, Cambridge (2005)
8. Wenzel, T., Santin, G., Haasdonk, B.: Analysis of data dependent greedy kernel algorithms: convergence rates for $f$-, $f \cdot P$- and $f/P$-greedy (2021). https://arxiv.org/abs/2105.07411
9. Wirtz, D., Haasdonk, B.: A vectorial kernel orthogonal greedy algorithm. Dolomites Res. Notes Approx. **6**, 83–100 (2013)

# A Multiscale Fatigue Model
# for the Degradation of Fiber-Reinforced
# Materials

N. Magino[1], J. Köbler[1], H. Andrä[1]([⊠]), F. Welschinger[2], R. Müller[3], and M. Schneider[4]

[1] Fraunhofer Institute for Indsutrial Mathematics ITWM, Kaiserslautern, Germany
{nicola.magino,heiko.andrae}@itwm.fraunhofer.de
[2] Robert Bosch GmbH, Corporate Sector Research and Advance Engineering, Renningen, Germany
[3] University of Kaiserslautern, Kaiserslautern, Germany
[4] Karlsruhe Institute of Technology (KIT), Karlsruhe, Germany

**Abstract.** Short-fiber reinforced materials show material degradation under fatigue loading prior to failure. To investigate these effects, we model the constituents by an isotropic fatigue damage model for the matrix material and isotropic linear-elastic material model for the fibers. On the microscale we compute the overall material response for cell problems with different fiber orientation states with FFT-based methods. We discuss a concept to model order reduction, that enables us to apply the model efficiently on component scale.

**Keywords:** Fatigue damage · Short-fiber reinforced · Multiscale

## 1 Introduction

Thermoplastics such as polyether ether ketone (PEEK) [1], polypropylene [2] or polyethylene (PE) [3] show a material degradation prior to failure.

In recent years, fiber reinforced thermoplastics have become widely popular in engineering applications. As thermoplastics serve as the matrix for these materials, the effective mechanical properties of these composite materials are strongly affected by those of the thermoplastic. The degradation of elastic properties in short-fiber reinforced thermoplastics was observed in PBT-PET-based [4], HDPE [5] and polyamid-based [6] materials.

We are interested in the material degradation of short-fiber reinforced thermoplastics. Since local fiber orientations in engineering components strongly depend on the injection molding process, an infinite number of fiber orientations has to be considered. Consequently, experimental procedures are time and cost

MS acknowledges financial support of the German Research Foundation (DFG) within the International Research Training Group "Integrated engineering of continuous-discontinuous long fiber reinforced polymer structures" (GRK 2078).

I. Lirkov and S. Margenov (Eds.): LSSC 2021, LNCS 13127, pp. 387–392, 2022.
https://doi.org/10.1007/978-3-030-97549-4_44

intensive. Thus, we aim at developing complementary simulation methods. We use a multiscale approach for modeling fatigue in short-fiber reinforced thermoplastics [7] which is able to account for the geometric properties of the microscale. The authors recently proposed a fatigue damage model in the context of multiscale simulations [8]. With the help of precomputations on the microscale we extract macroscopic material equations making use of nonuniform transformation field analysis [9,10].

In Sect. 2 we discuss the matrix material and the approach to model order reduction in detail. Subsequently, we demonstrate the capabilities of the model order reduction approach on fiber orientation structures in Sect. 3.

## 2   Modeling and Theoretical Background

Following a recently proposed approach published by the authors [8], we briefly discuss the fatigue damage model of the matrix.

A material model framed as a generalized standard materials (GSM) [11] is used. We chose to formulate the material model directly in logarithmic cycle space and denote the continuous time scale variable by $\overline{N} = \log_{10}(N)$ in terms of the cycle $N$. The derivative w. r. t. the cycle-scale variable $\overline{N}$ is denoted by $(\cdot)' = \mathrm{d}(\cdot)/\mathrm{d}\overline{N}$. We define the free energy $w$ and the dissipation potential $\phi$ as

$$w\left(\varepsilon, d\right) = \frac{1}{2\left(1+d\right)}\, \varepsilon : \mathbb{C} : \varepsilon, \qquad \phi\left(d'\right) = \frac{1}{2\alpha}\left(d'\right)^{2} \tag{1}$$

in terms of the strain amplitude tensor $\varepsilon$, the scalar damage variable $d \geq 0$, the stiffness tensor $\mathbb{C}$ and a damage velocity parameter $\alpha$.

On the cell $Y$ and after discretization in cycle space via an implicit Euler scheme we consider the Ortiz-Stainier [12,13] functional

$$F\left(\bar{\varepsilon}, u, d\right) = \left\langle \frac{1}{2\left(1+d\right)}\, \left(\bar{\varepsilon} + \nabla^{s}u\right) : \mathbb{C} : \left(\bar{\varepsilon} + \nabla^{s}u\right) + \frac{1}{2\alpha\triangle\overline{N}}\left(d - d^{n}\right)^{2} \right\rangle_{Y} - \bar{\varepsilon} : \Sigma \tag{2}$$

where $\langle . \rangle_{Y}$ denotes averaging over $Y$ and $\Sigma$ the macroscopic stress amplitude. The microscale computations are based on minimizing the functional in Eq. (2). However, to derive macroscopic evolution equations within the framework of nonuniform transformation field analysis [9,10], we reformulate the above functional in terms of the stress

$$\sigma = \frac{\partial w(\varepsilon, d)}{\partial \varepsilon} \tag{3}$$

and obtain the functional

$$S\left(\sigma, d\right) = \left\langle -\frac{\left(1+d\right)}{2}\, \sigma : \mathbb{S} : \sigma + \frac{1}{2\alpha\triangle\overline{N}}\left(d - d^{n}\right)^{2} \right\rangle_{Y}. \tag{4}$$

In contrast to the variational problem (2), the functional in the mixed formulation (4) is a polynomial in the variables $(\sigma, d)$. This enables precomputing of all relevant data in a nonuniform transformation analysis.

We are thus concerned with the mixed variational problem

$$\max_{\substack{\operatorname{div} \sigma = 0 \\ \langle \sigma \rangle_Y = \Sigma}} \min_d \left( S\left(\sigma, d\right) \right). \tag{5}$$

Upon Galerkin discretization, it remains to study existence and uniqueness of saddle points of this problem. Thus we define the operator

$$\mathcal{A}\left(\vec{\sigma}, \vec{d}\right) = \left( -\frac{\partial S}{\partial \vec{\sigma}}, \frac{\partial S}{\partial \vec{d}} \right), \tag{6}$$

whose roots are precisely critical points of the mixed variational problem (4). Under the condition

$$d \geq d_- \tag{7}$$

for some $d_- > -1$, $\mathcal{A}$ can be shown to be strongly monotone. Thus, as long as the constraint $d \geq d_- > -1$ is satisfied, classical monotone operator theory [14] implies that there is a unique root of $\mathcal{A}$.

The discretized operator $\mathcal{A}_{M_d}$ on $M_d$ damage modes

$$d = \sum_{a=1}^{M_d} d_a \delta_a \tag{8}$$

inherits the mathematical structure of the continuous problem. To proceed, let us discuss a scenario how to incorporate the constraint $d \geq d_-$ in the discretized system. Suppose that the damage modes $\delta_a$ are non-negative point-wise, i.e., the non-negativity condition

$$\delta_a\left(x\right) \geq 0 \quad \text{holds for all} \quad x \in Y \quad \text{and} \quad a = 1, \dots, M_d.$$

Then, the the condition $d\left(x\right) \geq 0$ is satisfied if all damage coefficients $d_a \geq 0$ are non-negative. In particular, a solution $\left(\vec{\sigma}, \vec{d}, \vec{\mu}\right) \in \mathbb{R}^{M_\sigma} \times \mathbb{R}^{M_d} \times \mathbb{R}^{M_d}$ of the KKT conditions

$$\mathcal{A}_M\left(\vec{\sigma}, \vec{d}, \vec{\mu}\right) = \left(0, \vec{\mu}\right),$$
$$\mu_a \geq 0, \quad \delta_a \geq 0, \quad \mu_a \delta_a = 0, \quad a = 1, \dots, M_d \tag{9}$$

may be sought.

In the following section we discuss an approach to extract positive damage modes and investigate the capability of the method to approximate the full field solution.

# 3   Computational Results

We consider the three microstructures shown in Fig. 1. These structures were generated by the sequential addition and migration algorithm [15] and represent

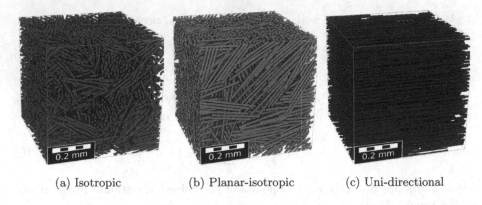

(a) Isotropic          (b) Planar-isotropic          (c) Uni-directional

**Fig. 1.** Structures

isotropic, planar-isotropic and uni-dircetional fiber orientations. Upon spatial discretization on a staggered grid [17] and prior to model order reduction, we apply uniaxial extension at stress amplitudes of 100 MPa in the $xx$-, $yy$- and $zz$-components as well as shear stress amplitudes of 100 MPa in the $xy$-, $xz$- and $yz$-components to the microstructures and solve the equations by a nonlinear conjugate gradient method [16].

As discussed in Sect. 2, monotonicity of the stress-based operator used in the reduced order model will be ensured if the damage field is point-wise bounded from below by a constant $d_-$ strictly larger than $-1$. Thus we start by restricting the damage modes to be non-negative. For extracting meaningful positive damage modes we use the modes obtained from proper orthogonal decomposition (POD) and split them into positive and negative parts. More precisely, for a given mode $d$, we define two associated modes

$$d_1^+ = \langle d \rangle_+ \quad \text{and} \quad d_2^+ = -\langle -d \rangle_+, \tag{10}$$

where $\langle q \rangle_+ = q$ for $q \geq 0$ and $\langle q \rangle_+ = 0$ otherwise. These modes, referred to as the POD$^+$-modes, are collected in a mode set $\{\delta_a\}$.

In the reduced order model, constraining the damage field coefficients $d_a$ to be larger or equal to zero ensures that the reconstructed damage fields $d^{\text{mor}}$

$$d^{\text{mor}} = \sum_{a=1}^{M^d} d_a \delta_a \tag{11}$$

are non-negative everywhere. The non-negativity constraints on the damage coefficients $d_a$ are enforced by a primal-dual active-set strategy [18]. Due to finite arithmetic precision, we slightly relax the non-negativity constraint to $\delta_a > -10^{-10}$.

On the left hand of Fig. 2, the accuracy of the resulting reduced order model is shown for the reference structures shown in Fig. 1, quantified by the relative strain amplitude error

$$e^{\text{rom}} = \max_i \left( \frac{\|\bar{\varepsilon}_{\max}(\overline{N}_i) - \bar{\varepsilon}_{\max}^{\text{rom}}(\overline{N}_i)\|}{\|\bar{\varepsilon}_{\max}(\overline{N}_i)\|} \right). \tag{12}$$

The number of modes refers to the number of stress modes $M_\sigma$ incorporated to the reduced order model. The number of associated damage modes $M_d$ computes as

$$M_d = 1 + 2(M_\sigma - 1). \tag{13}$$

Indeed, due to the POD procedure, all but the first damage mode have vanishing mean. In particular, the first damage mode is strictly positive, and all further modes have non-trivial positive and negative parts. Thus, this choice of the number of damage modes arises naturally from precomputing an equal number of POD-modes for stress and damage.

As shown in Fig. 1, the error measure (12) decreases with increasing number of modes incorporated into the reduced order model for all shown load cases. The load cases shown are uniaxial tension in $xx$-, $yy$- and $zz$-direction as well as the shear load cases in $xy$-, $xz$- and $yz$-direction, each at a constant load amplitude of 50 MPa. Load cases with equal macroscopic properties e.g., tension in $xx$-, $yy$- and $zz$-direction on the isotropic structure, show a similar approximation behavior. The reduced order model is able to approximate the full-field predictions quite accurately. More precisely, the maximum relative strain error is encountered for all considered load cases at nine stress modes is 1.08%.

The model is thus ready for application on engineering components.

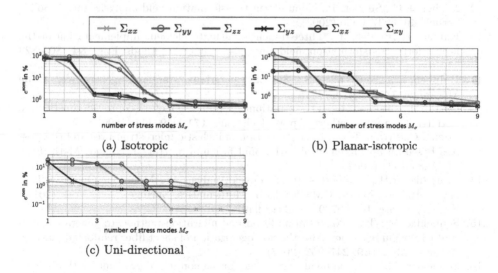

(a) Isotropic

(b) Planar-isotropic

(c) Uni-directional

**Fig. 2.** Accuracy study for POD$^+$ modes

# References

1. Shresthaa, R., Simsiriwongb, J., Shamsaeia, N., Moser, R.: Cyclic deformation and fatigue behavior of polyether ether ketone, viscoplasticity and ductile damage. Int. J. Fatigue **60**, 411–427 (2016)
2. Drozdov, A.D.: Cyclic strengthening of polypropylene under strain-controlled loading, viscoplasticity and ductile damage. Mater. Sci. Eng. A **528**, 8781–8789 (2011)
3. Avanzini, A.: Mechanical characterization and finite element modelling of cyclic stress-strain behaviour of ultra high molecular weight polyethylene, viscoplasticity and ductile damage. Mater. Des. **29**, 330–343 (2007)
4. Klimkeit, B., et al.: Fatigue damage mechanisms in short fiber reinforced PBT+PET GF30. Mater. Sci. Eng. A **528**, 1577–1588 (2011)
5. Krairi, A.: Multiscale modeling of the damage and failure of homogeneous and shortfiber reinforced thermoplastics under monotonic and fatigue loadings, Prom.: Doghri, I., Université catholique de Louvain (2015)
6. Chebbi, E., Mars, J., Wali, M., Dammak, F.: Fatigue behavior of short glass fiber reinforced polyamide 66: experimental study and fatigue damage modelling. Period. Polytech. Mech. Eng. **60**(4), 247–255 (2016)
7. Köbler, J., Magino, N., Andrä, H., Welschinger, F., Müller, R., Schneider, M.: A computational multi-scale model for the stiffness degradation of short-fiber reinforced plastics subjected to fatigue loading. Comput. Methods Appl. Mech. Eng. **373**, 113522 (2021)
8. Magino, N., Köbler, J., Andrä, H., Welschinger, F., Müller, R., Schneider, M.: A multiscale high-cycle fatigue-damage model for the stiffness degradation of fiber-reinforced materials based on a mixed variational framework. Comput. Methods Appl. Mech. Eng. **388**, 114198 (2021)
9. Michel, J.-C., Suquet, P.: Nonuniform transformation field analysis. Int. J. Solids Struct. **40**, 6937–6955 (2003)
10. Fritzen, F., Böhlke, T.: Three-dimensional finite element implementation of the nonuniform transformation field analysis. Numer. Methods Eng. **84**, 803–829 (2010)
11. Halphen, B., Nguyen, Q.S.: Sur les matériaux standards generalisés. J. de Mécanique **14**, 508–520 (1975)
12. Ortiz, M., Stainier, L.: The variational formulation of viscoplastic constitutive updates. Comput. Methods Appl. Mech. Eng. **171**, 419–444 (1999)
13. Miehe, C.: Strain-driven homogenization of inelastic microstructures and composites based on an incremental variational formulation. Int. J. Numer. Meth. Eng. **55**, 1285–1322 (2002)
14. Bauschke, H.H., Combettes, P.L.: Convex Analysis and Monotone Operator Theory in Hilbert Spaces. CMS Books in Mathematics, Springer, New York (2017). https://doi.org/10.1007/978-1-4419-9467-7
15. Schneider, M.: The sequential addition and migration method to generate representative volume elements for the homogenization of short fiber reinforced plastics. Comput. Mech. **59**, 247–263 (2017)
16. Schneider, M.: A dynamical view of nonlinear conjugate gradient methods with applications to FFT-based computational micromechanics. Comput. Mech. **66**(1), 239–257 (2020). https://doi.org/10.1007/s00466-020-01849-7
17. Schneider, M., Ospald, F., Kabel, M.: Computational homogenization of elasticity on a staggered grid. Int. J. Numer. Meth. Eng. **105**(9), 693–720 (2016)
18. Hintermüller, M., Ito, K., Kunisch, K.: The primal-dual active set method as a semismooth Newton method. SIAM J. Optim. **13**, 865–888 (2003)

# A Classification Algorithm for Anomaly Detection in Terahertz Tomography

Clemens Meiser[1]([✉])[iD], Thomas Schuster[1][iD], and Anne Wald[2][iD]

[1] Department of Mathematics, Saarland University, Saarbrücken, Germany
{meiser,thomas.schuster}@num.uni-sb.de
[2] Institute of Numerical and Applied Mathematics, University of Göttingen, Göttingen, Germany
a.wald@math.uni-goettingen.de

**Abstract.** Terahertz tomography represents an emerging field in the area of nondestructive testing. Detecting outliers in measurements that are caused by defects is the main challenge in inline process monitoring. An efficient inline control enables to intervene directly during the manufacturing process and, consequently, to reduce product discard. We focus on plastics and ceramics and propose a density-based technique to automatically detect anomalies in the measured data of the radiation. The algorithm relies on a classification method based on machine learning. For a verification, supervised data are generated by a measuring system that approximates an inline process. The experimental results show that the use of terahertz radiation, combined with the classification algorithm, has great potential for a real inline manufacturing process.

**Keywords:** Anomaly detection · Inline monitoring · Terahertz tomography

## 1 Introduction

Terahertz (THz) radiation is a part of the electromagnetic spectrum. The corresponding frequencies from 0.1 to 10 THz are located between microwaves and infrared radiation. Due to this special position in the electromagnetic spectrum, THz radiation is characterized by ray and wave character. It is possible to obtain information about the amplitude from measurements of the absorption of the radiation whereas the phase can be identified using time-of-flight measurements. The radiation is non-ionizing. It can penetrate many materials, especially non-conductive ones such as many ceramics, and it does not require a medium to couple with [5]. Furthermore, the radiation achieves a better resolution compared to microwaves because of its shorter wavelength [10]. In spite of this wide range of advantages the so called *THz gap*, referring to a lack of effective transducers and detectors [15,16], prevented an extensive application. This gap has only

Supported by the German Plastics Center (SKZ) and partially funded by German Federation of Industrial Research Associations (AiF) under 19948 N.

I. Lirkov and S. Margenov (Eds.): LSSC 2021, LNCS 13127, pp. 393–401, 2022.
https://doi.org/10.1007/978-3-030-97549-4_45

recently been closed. The field of THz inspection has expanded rapidly and has nowadays the chance to compete with X-ray, ultrasound and microwaves. THz radiation has become an interesting tool for many applications. It is used, for example, for security detection [13], the control of car paints, and in the pharmaceutical industry [15]. In particular, the radiation receives increasing attention in the field of nondestructive testing (NDT), where many techniques have been adopted and adapted from competing technologies like computerized tomography (CT) or ultrasound [11,14,16]. While we observe a fast progress in the offline control of NDT, THz systems are currently too slow for inline inspection. Recommendations indicate that the systems have to treble their acquisition speed [3]. Especially in the inline manufacturing, there is a great need for applications because defects, such as cracks, voids and inclusions are mostly produced during the process [15]. To avoid short-cycle products and to intervene directly, a fast and valid method is necessary. First investigations of a contactless and nondestructive inline control with THz radiation, for instance, were shown in [7]. An overview of THz tomography techniques is presented by Guillet et al. [6].

In our context, we evaluate inline measurements of THz radiation with a machine learning (ML) technique called *anomaly detection* (AD) in order to test its application in inline monitoring, and, more precisely, the detection of defects in the product. The algorithm is based on the learning of a multivariate Gaussian distribution. Considering the definition of an anomaly as a significant variation from typical values [9], their detection is perfectly suited for the underlying application. We use training data from a real-time measurement generated by a system that approximates an inline process. We obtain a data set encoding intensity, refraction, reflection and temporal information. These supervised data are used for learning whether an inline measurement lies inside a certain norm and, subsequently, for detecting deviations from this norm. To complete our investigations, we test our AD algorithm on an unknown object.

## 2   A Classification Algorithm for Inline Monitoring

The idea of AD to identify measurements that deviate from an expected behavior is a typical ML task. Applications can be found in fields like fraud detection, insurance, health care and cyber security [1], or, as in our case, in the monitoring of an inline manufacturing process. The starting point is a set of training data $\{x^{(1)}, x^{(2)}, ..., x^{(m)}\} \subset \mathbb{R}^d$. We assume that the training set contains only measurements from an intact object. Each single data point consists of $d$ attributes, called features, which are represented by real numbers. Assuming that the data $x^{(i)}$ are realizations of a real-valued random variable with probability density function $p(x)$, it is appropriate to identify typical data from intact objects with large values $p(x^{(i)})$, whereas anomalies can be characterized by small values $p(x^{(i)})$. For a given training data set, we first estimate a probability density function $p : \mathbb{R}^d \to \mathbb{R}$. Subsequently, we decide, depending on a threshold parameter $\epsilon^*$, whether a new data point $x_{\text{test}}$ is an anomaly or not. The threshold parameter is computed via a learning algorithm. For this purpose, we use a cross validation set and a decision function. The algorithm is inspired by [8,12].

In order to estimate $p$, we assume that the data and, more precisely, its features follow a Gaussian distribution, which is on the one hand motivated by our own measured data, see Fig. 3, on the other hand it is a common procedure to describe the scattering of measurements as normally distributed, see [4]. In case of a multivariate set of data $x^{(i)} \in \mathbb{R}^d$, $i = 1, 2, ..., m$, as realizations of an $\mathcal{N}(\mu, \Sigma)$-distributed random variable $X$ we compute the expected value $\mu \in \mathbb{R}^d$ and the covariance matrix $\Sigma \in \mathbb{R}^{d \times d}$, obtaining

$$p(x; \mu, \Sigma) = \frac{1}{(2\pi)^{\frac{d}{2}} |\Sigma|^{\frac{1}{2}}} e^{-\frac{1}{2}(x-\mu)^T \Sigma^{-1}(x-\mu)} \tag{1}$$

with

$$\mu = \frac{1}{m} \sum_{i=1}^m x^{(i)} \quad \text{and} \quad \Sigma = \frac{1}{m} \sum_{i=1}^m (x^{(i)} - \mu)(x^{(i)} - \mu)^T, \tag{2}$$

where $|\Sigma|$ denotes the determinant of $\Sigma$.

In a second step, we learn the threshold $\epsilon^*$. To this end we need a labeled cross validation set

$$\{(x_{CV}^{(1)}, y_{CV}^{(1)}), (x_{CV}^{(2)}, y_{CV}^{(2)}), ..., (x_{CV}^{(l)}, y_{CV}^{(l)})\} \subset \mathbb{R}^d \times \{0, 1\}, \quad l \in \mathbb{N},$$

with labels $y_{CV}^{(i)} \in \{0, 1\}$, where $y_{CV}^{(i)} = 1$ means that $x_{CV}^{(i)}$ is anomalous, whereas $y_{CV}^{(i)} = 0$ indicates a typical measurement $x_{CV}^{(i)}$. For any $\epsilon \geq 0$ we compute the *decision function* $f$ by

$$f(x_{CV}^{(i)}, \epsilon) = \begin{cases} 1, & \text{if } p(x_{CV}^{(i)}; \mu, \Sigma) < \epsilon \\ 0, & \text{otherwise,} \end{cases} \quad i = 1, ..., l. \tag{3}$$

By means of $f$ we compute the *confusion matrix* $\mathbf{C} \in \mathbb{N}^{2 \times 2}$,

$$\mathbf{C} = \begin{pmatrix} TP & FP \\ FN & TN \end{pmatrix},$$

for a fixed threshold $\epsilon$, where the entries represent the number of data points correctly labeled as positive (TP), data points falsely labeled as positive (FP), data points correctly labeled as negative (TN), and data points incorrectly labeled as negative (FN) (cf. [2]). The confusion matrix $\mathbf{C}$ characterizes the quality of the classification given $\epsilon$ and ideally resembles a diagonal matrix. From its entries we deduce the two values prec (*precision*) and rec (*recall*) by

$$\text{prec} = \frac{TP}{TP + FP} \quad \text{and} \quad \text{rec} = \frac{TP}{TP + FN}.$$

If the classifier works correctly, we have prec = rec = 1, and it performs poorly if both values are close to zero. Finally we compute the $F_1$-score $F_1(\epsilon)$ as the harmonic mean of prec and rec,

$$F_1(\epsilon) = 2 \frac{\text{prec} \cdot \text{rec}}{\text{prec} + \text{rec}}.$$

The threshold $\epsilon^*$ is then determined as the value maximizing the $F_1$-Score,

$$\epsilon^* := \arg\max_{\epsilon \in [0, p_{\max}]} F_1(\epsilon). \tag{4}$$

Here, the parameter $p_{\max}$ represents the maximum value of the probability density function $p$. Finally we evaluate the algorithm by means of a test set

$$\{(x_{\text{test}}^{(1)}, y_{\text{test}}^{(1)}), (x_{\text{test}}^{(2)}, y_{\text{test}}^{(2)}), ..., (x_{\text{test}}^{(l)}, y_{\text{test}}^{(l)})\} \subset \mathbb{R}^d \times \{0, 1\}.$$

The test set usually consists of measured and/or simulated data and contains as many normal and anomalous data as the cross validation set. If the evaluation fails, then the set of training data should be enhanced. After the parameters have been learned, a classification can be used to indicate whether an irregularity exists for an unknown data set: If the value of the probability density function falls below the optimal threshold parameter $\epsilon^*$, then the inline process should be intervened. A summary of the density-based AD and classification algorithm is given by Algorithm 1.

---

**Algorithm 1.** Density-based Anomaly Detection

**INPUT:**

- Training set $\{x^{(1)}, x^{(2)}, ..., x^{(m)}\} \subset \mathbb{R}^d$
- Cross Validation set $\{(x_{CV}^{(1)}, y_{CV}^{(1)}), (x_{CV}^{(2)}, y_{CV}^{(2)}), ..., (x_{CV}^{(l)}, y_{CV}^{(l)})\} \subset \mathbb{R}^d \times \{0, 1\}$
- Test set $\{(x_{\text{test}}^{(1)}, y_{\text{test}}^{(1)}), (x_{\text{test}}^{(2)}, y_{\text{test}}^{(2)}), ..., (x_{\text{test}}^{(l)}, y_{\text{test}}^{(l)})\} \subset \mathbb{R}^d \times \{0, 1\}$
- Measured data $\{x^{(1)}, x^{(2)}, ..., x^{(J)}\} \subset \mathbb{R}^d$

**STEP 1:** Consider the training set as realizations of an $\mathcal{N}(\mu, \Sigma)$-distributed random variable, $\mu \in \mathbb{R}^d$, $\Sigma \in \mathbb{R}^{d \times d}$, and estimate the probability density function (1) by (2).

**STEP 2:** Use the cross validation set and the decision function (3) to construct the confusion matrix $\mathbf{C}$ and to find an optimal threshold parameter $\epsilon^* \geq 0$ by maximizing the $F_1$-Score (4).

**STEP 3:** Evaluate the algorithm by using the test set. If the evaluation fails, then enhance the training data set.

**CLASSIFICATION:**
for $j = 1 : J$

     if $p(x^{(j)}, \mu, \Sigma) < \epsilon^*$ then set $y^{(j)} = 1$ (i.e., an outlier has been detected)
     else $y^{(j)} = 0$ (no outlier/defect detected)

---

# 3    The THz Measuring System

The THz measuring system simulates the procedure of an inline monitoring process. The emitter and the receivers are placed on a turntable which rotates around the object under investigation. It is possible to shift the turntable vertically, while at the same time the observed object is fixed. The emitter sends electromagnetic radiation of a frequency between 0.12 and 0.17 THz and, simultaneously, measures reflection data. One receiver is located opposite the emitter to register deviations in the transmission process. A second one is placed close to the first one to collect information on the refraction of the radiation. Figure 1 illustrates the setup of the THz tomograph, while Fig. 2 shows the real system. We receive two dimensional data $x^{(i[k,z])}$, $k = 1, ..., K$, $z = 1, ..., Z$, since the measuring system is shifted in $K$ steps, where in each step a complete 360° rotation in $Z$ steps of the measuring system is performed (i.e., a slice of the object is scanned). During an real inline process, however, it is different: for example, in an extrusion process, the material moves through a horizontally and vertically fixed measuring system that rotates around the object. Furthermore, the inline monitoring is typically a continuous process and the object is scanned in a helical shape. In this case, we thus acquire 3D data, but since the object is not shifted but moved along, there are only few data points per slice. For our investigations we used solid pipes made of polyethylene with a diameter of 10 cm and various lengths. The material has a refractive index of about $n = 1.53$ and an absorption coefficient of about $\alpha = 0.06$ cm$^{-1}$. After scanning the pipes without defects, we manufactured horizontal and vertical holes in some pipes to generate synthetic defects. Furthermore, we filled some holes with materials like oil and metal. This way we obtain a data set consisting of 220400 measurements from intact samples and 105965 anomalous data points from defect samples. We split it into three subsets: a training set, a cross validation set and a test set. The cross validation set and the test set each are composed of 50% of the anomalous data and 20% of the typical data, while the training set just includes 60% of the unaffected elements. One single data point $x^{(i)} = x^{(i[k,z])}$ is composed of five features: in each position $[k, z]$, where $k = 1, ..., K$ refers to the shift and $z = 1, ..., Z$ to the angle

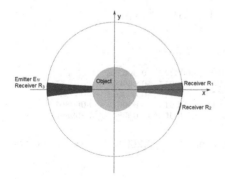

**Fig. 1.** Schematic THz tomograph          **Fig. 2.** Original THz tomography system

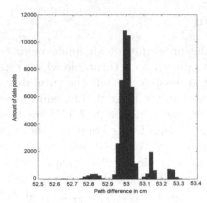 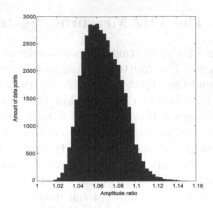

**Fig. 3.** Distribution of the measured transmission data from receiver $R_1$

position, the receivers $R_1$ and $R_3$ measure absorption and phase information, while receiver $R_2$ only registers the absorption information since no reference signal is available that is required for the phase information.

Figure 3 shows the distribution of a typical set of data measured by receiver $R_1$. We see that, indeed, the measurements resemble a Gaussian distribution concerning both the absorption data as well as the phase shifts.

## 4    Numerical Results

In this section, we present the computational results of our investigations. By using the data set described in Sect. 3, we evaluate the algorithm and, more generally, determine whether the application of terahertz radiation for the inline monitoring of plastics is suitable. At the end, we investigate an unknown pipe with the learned algorithm to localize the locations of the defects.

By including the measured values of receivers $R_1$, $R_2$ and $R_3$, we integrate information about transmission, reflection and refraction, respectively, of the terahertz radiation in our setting. The multivariate Gaussian distribution is estimated as

$$p(x; \mu, \Sigma) = \frac{1}{(2\pi)^{\frac{5}{2}} |\Sigma|^{\frac{1}{2}}} e^{-\frac{1}{2}(x-\mu)^T \Sigma^{-1}(x-\mu)}$$

with

$$\mu = \begin{pmatrix} 53.601002 \\ 1.015691 \\ 0.139608 \\ 0.010417 \\ 83.159275 \end{pmatrix}, \Sigma = \begin{pmatrix} 0.258068 & -0.012679 & 0.003472 & 0.000218 & -0.355831 \\ -0.012679 & 0.001625 & -0.000057 & -0.000001 & -0.012545 \\ 0.003472 & -0.000057 & 0.002129 & -0.000013 & -0.034956 \\ 0.000218 & -0.000001 & -0.000013 & 0.000016 & -0.000867 \\ -0.355831 & -0.012545 & -0.034956 & -0.000867 & 3.299866 \end{pmatrix}.$$

We obtain the learned threshold parameter $\epsilon^* = 2.260130$ and the corresponding confusion matrix

$$\mathbf{C} = \begin{pmatrix} 52982 & 434 \\ 0 & 43266 \end{pmatrix}.$$

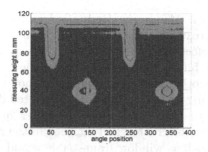

**Fig. 4.** Anomaly detection of an unknown pipe

The respective $F_1$-score approaches the ideal value of 1 and is given by $F_1(\epsilon^*) = 0.995921$, which is an impressive result. Only 434 out of 96682 data points are false positive and all anomalous data points are found.

We finally apply the AD process to investigate an unknown solid pipe that potentially contains defects. We use scanning data with the above mentioned five features to calculate values of the probability density function with estimated expected values and covariance matrix. Figure 4 visualizes the results $y = \left(y^{(i[k,z])}\right)_{k,z}$ according to Algorithm 1: The anomalous data with $y^{(i[k,z])} = 1$ are marked yellow. Two defects are detected by our algorithm, which each appears twice in the plot since they are scanned in intervals of 180° when the system is rotated. We read the plot from top to bottom: The first horizontal yellow lines represent the transition between air and pipe. They are followed by an area of about 40 mm which includes a defect. After a small section with no defects, a second damage of about 10 mm follows. The last measurements are unaffected. By comparing our results with the exact dimension of the pipe, we note that again promising results are achieved: The solid pipe was built with two damaged areas, a vertical hole of 4 cm from above and a lateral hole with a diameter of 8 mm. Note that the aim of our investigation was not to determine the exact dimensions of the defects and to characterize them, but to localize the approximate anomalous areas which was completely achieved.

Considering the computational time of Algorithm 1, the second step is the most expensive one, since an optimization problem is solved. The total time depends on the amount of data and, in our case (Intel Core i7-8565U processor), it is about three seconds. Further, the performance of the algorithm increases by the number of correct measurements that are used for the training process, whereas the partition of the correct data to the training set, cross validation set and test set does not influence the generalizability significantly.

## 5 Discussion and Conclusion

In this article we evaluated THz tomographic measurements with an AD algorithm to investigate its use for inline process monitoring of plastics. We introduced the algorithm and tested it on a real data set measured at the 'German

Plastics Center' (SKZ) in Würzburg, Germany. The results show that our presented technique has great potential for applying it in a real-time inline system. A good detection of defects and anomalous data was demonstrated.

Nevertheless, it is a future challenge to transfer the results to further materials and productions. Another task is to find physical models that can simulate the features of the measuring system to integrate simulated data. The main advantage of a simulation is that we can extend the number and types of defects. Not every single defect has to be created and included in the real material. For complex inline products, such as window frames, this would have huge advantages. One possibility for simulating the intensity is given in Tepe et al. [11]. A possibility to simulate the time-of-flight measurement is given by the eikonal equation. With regard to the use of our technique in industry, it would make sense not just to detect the defects, but also to classify them in order to intervene directly during the production process.

# References

1. Chandola, V., Banerjee, A., Kumar, V.: Anomaly detection: a survey. ACM Comput. Surv. (CSUR) **41**(3), 1–58 (2009)
2. Davis, J., Goadrich, M.: The relationship between precision-recall and ROC curves. In: Proceedings of the 23rd International Conference on Machine Learning, pp. 233–240 (2006)
3. Dhillon, S., et al.: The 2017 terahertz science and technology roadmap. J. Phys. D Appl. Phys. **50**(4), 043001 (2017)
4. Eden, K., Gebhard, H.: Dokumentation in der Mess-und Prüftechnik. Springer, Wiesbaden (2014). https://doi.org/10.1007/978-3-658-06114-2
5. Ferguson, B., Zhang, X.C.: Materials for terahertz science and technology. Nat. Mater. **1**(1), 26–33 (2002)
6. Guillet, J.P., et al.: Review of terahertz tomography techniques. J. Infrared Millim. Terahertz Waves **35**(4), 382–411 (2014)
7. Krumbholz, N., et al.: Monitoring polymeric compounding processes inline with THz time-domain spectroscopy. Polym. Test. **28**(1), 30–35 (2009)
8. Limthong, K.: Real-time computer network anomaly detection using machine learning techniques. J. Adv. Comput. Netw. **1**(1), 126–133 (2013)
9. Mehrotra, K., Mohan, C., Huang, H.: Anomaly Detection Principles and Algorithms. Springer, Cham (2017). https://doi.org/10.1007/978-3-319-67526-8
10. Nüßler, D., Jonuscheit, J.: Terahertz based non-destructive testing (NDT): making the invisible visible. tm-Technisches Messen **1**(ahead-of-print) (2020)
11. Tepe, J., Schuster, T., Littau, B.: A modified algebraic reconstruction technique taking refraction into account with an application in terahertz tomography. Inverse Probl. Sci. Eng. **25**(10), 1448–1473 (2017)
12. Tharwat, A.: Classification assessment methods. Appl. Comput. Inform. **17**(6), 168–192 (2020)
13. Tzydynzhapov, G., Gusikhin, P., Muravev, V., Dremin, A., Nefyodov, Y., Kukushkin, I.: New real-time sub-terahertz security body scanner. J. Infrared Millim. Terahertz Waves **41**, 1–10 (2020)

14. Wald, A., Schuster, T.: Terahertz tomographic imaging using sequential subspace optimization. In: Hofmann, B., Leitao, A., Zubelli, J.P. (eds.) New Trends in Parameter Identification for Mathematical Models. Trends in Mathematics, Birkhäuser Basel (2018)
15. Zhong, S.: Progress in terahertz nondestructive testing: a review. Front. Mech. Eng. **14**(3), 273–281 (2018). https://doi.org/10.1007/s11465-018-0495-9
16. Zouaghi, W., Thomson, M., Rabia, K., Hahn, R., Blank, V., Roskos, H.: Broadband terahertz spectroscopy: principles, fundamental research and potential for industrial applications. Eur. J. Phys. **34**(6), 179–199 (2013)

# Reduced Basis Methods for Efficient Simulation of a Rigid Robot Hand Interacting with Soft Tissue

Shahnewaz Shuva[1]([✉]), Patrick Buchfink[1], Oliver Röhrle[2],
and Bernard Haasdonk[1]

[1] Institute of Applied Analysis and Numerical Simulation,
University of Stuttgart, Stuttgart, Germany
{shuvasz,buchfipk,haasdonk}@mathematik.uni-stuttgart.de
[2] Institute for Modeling and Simulation of Biomechanical Systems,
University of Stuttgart, Stuttgart, Germany
roehrle@simtech.uni-stuttgart.de

**Abstract.** We present efficient reduced basis (RB) methods for the simulation of a coupled problem consisting of a rigid robot hand interacting with soft tissue material. The soft tissue is modeled by the linear elasticity equation and discretized with the Finite Element Method. We look at two different scenarios: (i) the forward simulation and (ii) a feedback control formulation of the model. In both cases, large-scale systems of equations appear, which need to be solved in real-time. This is essential in practice for the implementation in a real robot. For the feedback-scenario, we encounter a high-dimensional Algebraic Riccati Equation (ARE) in the context of the linear quadratic regulator. To overcome the real-time constraint by significantly reducing the computational complexity, we use several structure-preserving and non-structure-preserving reduction methods. These include reduced basis techniques based on the Proper Orthogonal Decomposition. For the ARE, we compute a low-rank-factor and hence solve a low-dimensional ARE instead of solving a full dimensional problem. Numerical examples for both cases (i) and (ii) are provided. These illustrate the approximation quality of the reduced solution and speedup factors of the different reduction approaches.

**Keywords:** Model order reduction · Soft tissue · Robotics

## 1 Introduction

The ability to manipulate deformable objects using robots has diverse applications with enormous economical benefits. The applications include food industry,

The research leading to this publication has received funding from the German Research Foundation (DFG) as part of the International Research Training Group "Soft Tissue Robotics" (GRK 2198/1) and under Germany's Excellence Strategy - EXC 2075 – 390740016. We acknowledge the support by the Stuttgart Center for Simulation Science (SimTech).

I. Lirkov and S. Margenov (Eds.): LSSC 2021, LNCS 13127, pp. 402–409, 2022.
https://doi.org/10.1007/978-3-030-97549-4_46

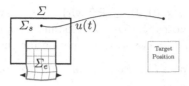

**Fig. 1.** Schematic view of gripper soft tissue system with target position.

medical sectors, automobile industry, soft material processing industry and many more. Although grasping and manipulation of rigid objects by robots is a mature field in robotics, with over three decades of works, the study of deformable objects has not been as extensive in the robotics community [4]. Here, we model the elastic object using existing linear elastic theory with two characterization parameters, namely, first and second Lamé parameter. The two most challenging and frequently studied manipulation tasks on planar deformable objects are grasping and controlling its deformations [5]. Grasping an object consists of positioning the end effectors of a robot hand on the object to lift and hold it in the air, which involves the challenges of slipping off and non-linear contact mechanics. For the sake of simplicity, we assume the soft tissue material to be attached to the rigid robot's end effectors. We are interested in controlling the object using a feedback controller and transfer it to a target position. Here, we focus on two different scenarios, (i) the forward simulation of the coupled problem, where the robot hand along with the soft tissue material follows a prescribed trajectory, and (ii) a feedback control such that the robot hand along with the soft tissue material cost-optimally reaches a target position and then stabilizes. In Fig. 1, a schematic view is provided. After the discretization of both problems, large-scale systems of equations appear. Another challenging issue besides modeling and controlling is that the simulation of the model has to be finished in real-time because in a modern robotics software, like Franka Control Interface [8], all real-time loops have to be finished within 1 ms. A feasible solution of this is to simulate a reduced model instead of a full order model. We developed and applied different structure-preserving and non-structure-preserving methods to maintain specific structures, e.g. block structure in the reduced system. The paper is organized in the following way. In Sect. 2, we discuss the modeling based on a spatially continuous formulation, which is followed by a large-scale spatially discrete problem in the context of the forward and LQR control problem. In Sect. 3, we discuss different reduced basis methods for both cases. Next in Sect. 4, we apply the MOR techniques in two numerical examples. Finally, we compare the approximation quality of the reduced solutions and provide a comparison of the execution time in contrast to the full order model.

## 2   Model Problems

We model the soft tissue material by the following time-dependent linear elasticity partial differential equation (PDE). We assume a two-dimensional spatial domain $\Omega_e := (0,1) \times (0,2) \subset \mathbb{R}^2$ with boundary, $\Gamma = \Gamma_D \cup \Gamma_N$ where $\Gamma_D = \{0,1\} \times [1.5, 2]$ denotes the Dirichlet boundary and $\Gamma_N = \Gamma \setminus \Gamma_D$ denotes the Neumann boundary. The governing equation for the displacement field is

$$\nabla \cdot \boldsymbol{\sigma} + \boldsymbol{F} = \rho \ddot{\boldsymbol{q}}^c \quad \text{in} \quad \Omega_e \times (0,T)$$

with suitable initial and boundary conditions where $\boldsymbol{q}^c$ is the displacement vector and $\boldsymbol{\sigma}$ the Cauchy stress tensor. $\boldsymbol{F}$ is the external body force per unit volume and $\rho$ is the mass density. The constitutive equation is $\boldsymbol{\sigma} = C : \boldsymbol{\epsilon}$ according to Hooke's law for elastic material, where $C$ is a fourth-order tensor and : is the contraction operator. For isotropic and homogeneous media holds $\boldsymbol{\sigma} = 2\mu\boldsymbol{\epsilon} + \lambda tr(\boldsymbol{\epsilon})\boldsymbol{I}$ where $\lambda$, $\mu$ are the Lamé parameters. The strain is expressed in terms of the gradient of the displacement with $\boldsymbol{\epsilon} = \frac{1}{2}(\nabla \boldsymbol{q}^c + (\nabla \boldsymbol{q}^c)^T)$.

## 2.1   Forward Model

After discretization using the Finite Element Method (FEM), we get the following second order system of differential equations,

$$M(\rho)\ddot{q}(t) + K(\mu, \lambda)q(t) = f(t) + B_u u(t) \tag{1}$$

with the parameters $\rho, \lambda, \mu \in \mathbb{R}$ and the state vector $q \in \mathbb{R}^n$ which is decomposed into the displacement vector $q_s \in \mathbb{R}^{n_s}$ of the solid robot hand and the displacement vector $q_e \in \mathbb{R}^{n_e}$ of the soft tissue material, i.e. $q := [q_s, q_e]^T \in \mathbb{R}^n$, $n := n_s + n_e$. The mass and stiffness matrix $M, K \in \mathbb{R}^{n \times n}$, the influence matrix $B_u \in \mathbb{R}^{n \times m}$, the body force $f(t) = \left[f_s^T(t) \ f_e^T(t)\right]^T \in \mathbb{R}^n$ and input vector $u(t) = \left[u_s^T(t) \ u_e^T(t)\right]^T \in \mathbb{R}^n$ in the above equations are as follows: $M = \left[M_{ss} \ M_{se}; M_{es} \ M_{ee}\right]$, $K = \left[K_{ss} \ K_{se}; K_{es} \ K_{ee}\right]$, $B_u = \text{blkdiag}(B_{us}, B_{ue})$. Here the block matrices with the subscript $se$ or $es$ refer to the coupled matrix coefficient between the vectors $q_s$ and $q_e$ and the ; indicates a new block row. With all non-zero block matrices, the problem can be interpreted as a two-way coupled problem [7]. For simplification, we assume that the mass of the solid system is large enough compared to the mass of the elastic system, such that the influence of the motion of the elastic system on the solid body is negligible but vice versa is relevant, i.e. $M_{se}$ is chosen as a zero matrix and $M_{es}$ is a non-zero matrix. Since we consider the robot hand as a rigid body, $K_{ss}$ is set to zero. The contact zone follows the motion of the solid body, so we ignore $K_{se}$. The values of the displacement vectors on the Dirichlet boundary of the elastic body are determined by the displacement vectors of the solid body $q_s$, i.e. $K_{es}$ is a non-zero matrix. With these assumptions, we obtain the following one-way coupled model

$$\begin{bmatrix} M_{ss} & 0 \\ M_{es} & M_{ee} \end{bmatrix} \begin{bmatrix} \ddot{q}_s(t) \\ \ddot{q}_e(t) \end{bmatrix} + \begin{bmatrix} 0 & 0 \\ K_{es} & K_{ee} \end{bmatrix} \begin{bmatrix} q_s(t) \\ q_e(t) \end{bmatrix} = \begin{bmatrix} f_s(t) \\ f_e(t) \end{bmatrix} + \begin{bmatrix} B_{us} \\ 0 \end{bmatrix} \begin{bmatrix} u_s(t) \\ u_e(t) \end{bmatrix} \tag{2}$$

Here, the elasticity part of the state vector is coupled with the acceleration and displacement of the solid state vector. For solving Eq. 2, we transformed it into system of first order differential equations, where the state vector is composed of the displacement $q(t)$ and the velocity $v(t) = \dot{q}(t)$.

## 2.2   Linear Quadratic Regulator (LQR)

For feedback control, we formulate a LQR problem and assume the weighing matrices $Q \in \mathbb{R}^{p \times p}$, $p \leq 2n$ to be symmetric positive semi-definite and $R \in \mathbb{R}^{m \times m}$ to be symmetric positive definite. We define the quadratic cost functional

$$J(u) = \int_0^\infty [x(t)^T C^T Q^{1/2} Q^{1/2} C x(t) + u(t)^T R u(t)] dt$$

with $E\dot{x}(t) = Ax(t) + Bu(t)$, $y(t) = Cx(t)$, $x(0) = x_0$, time $t \in \mathbb{R}_+$, the state $x(t) = \begin{bmatrix} q_s^T(t) & \dot{q}_s^T(t) & q_e^T(t) & \dot{q}_e^T(t) \end{bmatrix}^T \in \mathbb{R}^{2n}$, the input $u(t) \in \mathbb{R}^m$, the output $y(t) \in \mathbb{R}^p$ and system matrices $E, A \in \mathbb{R}^{2n \times 2n}, B \in \mathbb{R}^{2n \times m}, C \in \mathbb{R}^{p \times 2n}$ with $E =$

$$\begin{bmatrix} E_{ss} & 0 \\ E_{es} & E_{ee} \end{bmatrix}, A = \begin{bmatrix} A_{ss} & 0 \\ A_{es} & A_{ee} \end{bmatrix}, B = \begin{bmatrix} 0 \\ B_{us} \\ 0 \end{bmatrix}, C = \begin{bmatrix} 0\,0\,0\,0\,1\,0\ldots0\,0\,0\ldots0 \\ 0\,0\,0\,0\,0\,0\ldots0\,1\,0\ldots0 \end{bmatrix}.$$

If we assume $(E, A, B)$ to be stabilizable and $(E, A, Q^{1/2}C)$ to be detectable, the optimal control problem (OCP) possesses a unique solution $u(t) = K_f x(t)$ with the feedback gain matrix $K_f = -R^{-1}B^T PE \in \mathbb{R}^{m \times 2n}$. Here, the symmetric positive semi-definite matrix $P \in \mathbb{R}^{2n \times 2n}$ is the unique stabilizing solution of the Generalized Algebraic Riccati Equation (ARE) [1]

$$E^T PA + A^T PE - E^T PBR^{-1}B^T PE + C^T QC = 0 \tag{3}$$

Note that this is a large-scale problem with $4n^2$ unknowns and therefore of quadratic complexity.

## 3   Reduced Basis Methods

The RB-method [9] approximates the solution in a low-dimensional subspace that is constructed from solutions of the large-scale problems (Eq. 1 and 3). To construct a reduced basis matrix $V := \text{span}\{\psi_1, \ldots \psi_{N_V}\} \in \mathbb{R}^{2n \times N_V}$ with $N_V$ basis functions, we define a snapshot matrix $X := [x_1, \ldots, x_{n_{sn}}] \in \mathbb{R}^{2n \times n_{sn}}$ with $1 \leq i \leq n_{sn}$ so-called snapshots $x_i$, which are acquired by solving the corresponding large-scale problem. For different reduction techniques, we use different splittings of the snapshot matrix: a split $X_e \in \mathbb{R}^{2n_e \times n_{sn}}$, $X_s \in \mathbb{R}^{2n_s \times n_{sn}}$ in elastic and solid part, a split $X_q, X_v \in \mathbb{R}^{n \times n_{sn}}$ in velocity and displacement and a combination of both splits $X_{s_q}, X_{s_v} \in \mathbb{R}^{n_s \times n_{sn}}, X_{e_q}, X_{e_v} \in \mathbb{R}^{n_e \times n_{sn}}$. The reduction techniques are based on the so-called Proper Orthogonal Decomposition (POD). It chooses the reduced basis as the first $k$ left-singular vectors of the snapshot matrix $X$ which we denote with $\text{POD}_k(X)$. In practice, the POD basis is computed via the truncated Singular Value Decomposition of $X$. In the following, the methods are classified as non-structure-preserving and structure-preserving techniques.

## 3.1    MOR for Forward Problem

For the model reduction, we apply Galerkin projection with a suitable basis matrix $V$. The non-structure-preserving methods are the following:

- **Global POD**: Here, we implement a global POD, i.e. perform a singular value decomposition (SVD) of the snapshot matrix, $V = \text{POD}_k(X)$.
- **Componentwise POD**: In this technique, we apply the POD algorithm for displacement and velocity snapshots separately, i.e. $V_1 = \text{POD}_{k_1}(X_q)$, $V_2 = \text{POD}_{k_2}(X_v)$ and then construct $V = \text{blkdiag}(V_1, V_2)$.

As structure-preserving techniques, we apply the following methods:

- **POD with fixed solid modes**: The reduced basis space is constructed in the following way: We apply POD only in the elastic part of the snapshots, i.e. $V_e = \text{POD}_{k_e}(X_e)$ and hence construct $V = \text{blkdiag}(I_{2n_s}, V_e)$ where $I_{2n_s} \in \mathbb{R}^{2n_s \times 2n_s}$ is the identity matrix. The unreduced solid modes are still computationally efficient since $n_s \ll n_e$.
- **Componentwise POD with fixed solid modes**: Here we construct $V = \text{blkdiag}(I_{2n_s}, V_{e_q}, V_{e_v})$, with $V_{e_q} = \text{POD}_{k_{e_q}}(X_{e_q})$, $V_{e_v} = \text{POD}_{k_{e_v}}(X_{e_v})$.
- **Global Proper Symplectic Decomposition(GPSD)**: As our system for $u = 0$ is a Hamiltonian system, we construct a so-called orthosymplectic basis matrix. As basis generation technique, we use the so-called PSD Complex SVD based on the SVD of a modified snapshot matrix $Y = [X\ JX]$, where $J$ is the corresponding Poisson matrix. For more details, we refer to [6].
- **PSD with fixed solid modes**: Here, we apply the PSD only on the elastic part.

## 3.2    RB-ARE for LQR

In practice, the dimension of the state space and hence of $P$ is typically very high. The dimension $m$ of the input $u(t)$ and $p$ of the output $y(t)$ is much smaller than the number of states $2n$. The computation of $P$ is often very expensive or even impossible [2]. Instead of determining $P$, we compute $ZZ^T \approx P$ with low-rank-factor $Z \in \mathbb{R}^{2n \times N}$ of low-rank $N \ll 2n$. The methods without fixing the solid modes include:

- **POD of ARE solution $P$**: We construct the reduced basis matrix using the POD of the solution of ARE, i.e. $V = \text{POD}_k(P)$.
- **Weighted POD**: We assume a weighting matrix $W = \frac{1}{2}(E + E^T)$ and apply a weighted POD [3], i.e. $V := W^{-\frac{1}{2}}\text{POD}_k(W^{\frac{1}{2}}P)$.

  The method with fixed solid modes are the following:

- **POD with fixed solid modes**: We decompose the solution $P$ of the ARE into $P_{ss}$, $P_{se}$, $P_{es}$ and $P_{ee}$ and apply $V_e := \text{POD}_k([P_{es}, P_{ee}])$ and construct $V := \text{blkdiag}(I_{2n_s}, V_e)$.

- **Componentwise POD**: We decompose blocks of $P$ in displacement and velocity and use $\text{POD}_{k_1}([P_{es_{11}}, P_{ee_{11}}, P_{ee_{12}}])$ and $\text{POD}_{k_2}([P_{es_{21}}, P_{ee_{21}}, P_{ee_{22}}])$ for displacement and velocity part separately to construct V.
- **POD of decomposed** $P$: We use $V_1 := \text{POD}_{k_1}([P_{ee_{11}}, P_{ee_{12}}, P_{ee_{21}}, P_{ee_{22}}])$ where $P_{ee_{11}}, P_{ee_{12}}, P_{ee_{21}}, P_{ee_{22}} \in \mathbb{R}^{n_e \times n_e}$ with $V := \text{blkdiag}(I_{2n_s}, V_1, V_1)$.

We introduce the reduced basis approximation $\hat{P} = VP_N V^T$, where $N \ll 2n$ and $P_N \in \mathbb{R}^{N \times N}$ is the solution of the reduced ARE

$$E_N^T P_N A_N + A_N^T P_N E_N - E_N^T P_N B_N R^{-1} B_N^T P_N E_N + C_N^T Q C_N = 0.$$

$V$ is constructed by one of the above mentioned methods (one parameter setting), $A_N = V^T A V$, $E_N = V^T E V$, $B_N = V^T B$ and $C_N = C V$.

## 4   Numerical Examples

We present our numerical experiments for both, the forward and the LQR problem. As a time integrator, we use the implicit mid-point rule with $n_t = 600$ time steps. We choose ranges of the first and second Lamé parameter, $\lambda \in [30, 500]$ and $\mu \in [20, 500]$. We consider one trajectory for a particular value of the parameters ($\lambda = 50$, $\mu = 50$) which results in $n_{sn} = n_t$ snapshots. The initial state is $x(0) = 0 \in \mathbb{R}^{2n}$. As a target position we consider $\bar{x}(T) = [5\,5\ldots3\,3\ldots]^T$ with the final time $T = 3$ and 300 for the forward and the LQR problem, respectively. For the simulation and reduction of the model, we use the software package RBmatlab [9].

### 4.1   Forward Problem

We determine $N_V$ to include 99.9% energy of the POD functional. An example of a reduced forward problem is illustrated in Fig. 2 using the PSD method. To measure the quality of the reduced solution we compute the relative error $\frac{\|X - \hat{X}\|_F}{\|X\|_F}$ (see Fig. 3), where $X \in \mathbb{R}^{2n \times n_{sn}}$ is the snapshot matrix introduced above, $\hat{X} := VX_r$ is the reconstructed solution, $X_r \in \mathbb{R}^{N_V \times n_{sn}}$ is the matrix gathering the reduced solution time-instances as columns and $\|.\|_F$ is the Frobenius norm. It shows that the structure-preserving methods are better than the non-structure-preserving methods. The best technique is the PSD without fixed solid modes which results in an error of $\approx 10^{-3}$ for the reduced system when using at least $N_V = 14$ basis functions. The execution time for the different methods is between $0.097\,\text{s}$ and $0.101\,\text{s}$ depending on the number of basis functions. The PSD with $N_V = 14$ requires $0.097\,\text{s}$ which is much less compared to the execution time $16.22\,\text{s}$ for the high-dimensional solution ($2n = 1916$). For a size of the reduced models larger than 25, we do not get any improvements in accuracy because the singular values do not show a sharp decay after that.

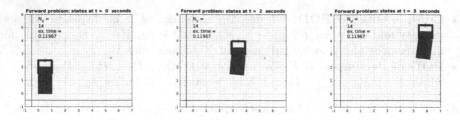

**Fig. 2.** Forward simulation of the reduced coupled problem

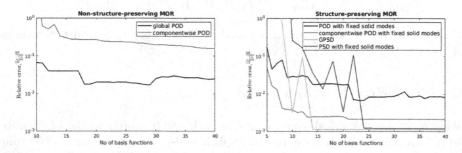

**Fig. 3.** Approximation quality of reduced solutions.

## 4.2  LQR Problem

We first provide the simulation using the reduced controller in Fig. 4 using $V$ from the global POD method. It shows that the reduced solution is able to compute a reconstructed stabilizing solution $\hat{P}$ which drives the soft tissue to its target position. We applied the MOR methods listed in Sect. 3.2. The relative error of the reduced ARE solution is shown in Fig. 5 which indicates that the methods without fixed solid modes provide better results for smaller numbers of basis functions. We also see that the computation time for Eq. 3 ($2n = 880$) requires 84.53 s which we can reduce to a runtime between 0.051 s and 0.057 s depending on the number of basis functions, e.g. 0.053 s for the global POD with $N_V = 8$.

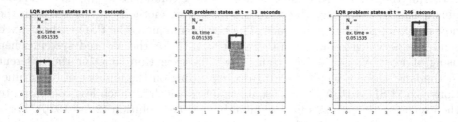

**Fig. 4.** Simulation of the reduced LQR problem

 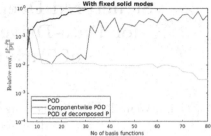

**Fig. 5.** Relative error for the controller-scenario. Errors above 100% are not depicted.

## 5 Conclusion and Outlook

We compared the quality and speedup of different basis generation techniques for a large-scale soft tissue model in the context of a forward problem and a feedback control problem. In the forward problem, we see that the PSD-based methods provide better results in terms of accuracy than the POD-based methods. For two example reduced models with a good accuracy, we are able to achieve speedup factors of 167 and 1600, respectively, which motivates us to apply these methods in parametric problems for multi-query scenarios. The future work includes adding an obstacle using state constraints in order to look at a more general class of control problems, greedy basis generation for the multi-query case and modeling using a non-linear elastic material law.

## References

1. Schmidt, A., Haasdonk, B.: Reduced basis approximation of large scale parametric algebraic Riccati equations. ESAIM Control Optim. Calc. Var. **24**, 129–151 (2018)
2. Schmidt, A.: Feedback control for parametric partial differential equations using reduced basis surrogate models. Ph.D. thesis, University of Stuttgart (2018)
3. Volkwein, S.: Proper Orthogonal Decomposition: Theory and Reduced-Order Modeling. POD Lecture Note (2013, preprint)
4. Arriola-Rios, V.E., Guler, P., Ficuciello, F., Kragic, D., Siciliano, B., Wyatt, J.L.: Modeling of deformable objects for robotic manipulation: a tutorial and review. Front. Robot. AI **7**, 82 (2020)
5. Jia, Y.B., Guo, F., Lin, H.: Grasping deformable planar objects: squeeze, stick/slip analysis, and energy-based optimalities. Int. J. Robot. Res. **33**(6), 866–897 (2014)
6. Buchfink, P., Bhatt, A., Haasdonk, B.: Symplectic model order reduction with non-orthonormal bases. Math. Comput. Appl. **24**(2), 43 (2019)
7. Benner, P., Feng, L.: Model order reduction for coupled problems. Appl. Comput. Math. **V14**(1), 3–22 (2015)
8. Franka Control Interface Document (2017). http://frankaemika.github.io/docs/
9. Haasdonk, B.: Reduced basis methods for parametrized PDEs - a tutorial introduction for stationary and instationary problems. In: Benner, P., Cohen, A., Ohlberger, M., Willcox, K. (eds.) Model Reduction and Approximation: Theory and Algorithms, Philadelphia. SIAM (2016)

# Structured Deep Kernel Networks for Data-Driven Closure Terms of Turbulent Flows

Tizian Wenzel[1]([⊠]), Marius Kurz[2], Andrea Beck[3], Gabriele Santin[4][iD], and Bernard Haasdonk[1]

[1] Institute for Applied Analysis and Numerical Simulation, University of Stuttgart, Stuttgart, Germany
{tizian.wenzel,bernard.haasdonk}@mathematik.uni-stuttgart.de
[2] Institute of Aerodynamics und Gas Dynamics, University of Stuttgart, Stuttgart, Germany
marius.kurz@iag.uni-stuttgart.de
[3] Laboratory of Fluid Dynamics and Technical Flows, Otto-von-Guericke-Universität Magdeburg, Magdeburg, Germany
andrea.beck@ovgu.de
[4] Digital Society Center, Bruno Kessler Foundation, Trento, Italy
gsantin@fbk.eu

**Abstract.** Standard kernel methods for machine learning usually struggle when dealing with large datasets. We review a recently introduced Structured Deep Kernel Network (SDKN) approach that is capable of dealing with high-dimensional and huge datasets - and enjoys typical standard machine learning approximation properties.

We extend the SDKN to combine it with standard machine learning modules and compare it with Neural Networks on the scientific challenge of data-driven prediction of closure terms of turbulent flows. We show experimentally that the SDKNs are capable of dealing with large datasets and achieve near-perfect accuracy on the given application.

**Keywords:** Machine learning · Structured deep kernel networks · Closure terms · Turbulent flows

## 1 Introduction

In modern science and engineering there is an increasing demand for efficient and reliable data-based techniques. On the one hand, an unprecedented amount of data is nowadays available from various sources, notably complex computer simulations. On the other hand, traditional model-based methods are often required to integrate the additional information acquired through measurements.

A particularly interesting situation is the case of surrogate modeling [8], where the underlying ground-truth is some engineering model represented by a computational simulation, which is accessible but expensive to execute. In this

© Springer Nature Switzerland AG 2022
I. Lirkov and S. Margenov (Eds.): LSSC 2021, LNCS 13127, pp. 410–418, 2022.
https://doi.org/10.1007/978-3-030-97549-4_47

setting the accurate simulation provides a discrete set of input-output pairs, and the resulting data-based model, or surrogate, can be used to replace to some extent the full simulation in order to simplify its understanding and analysis.

In this paper we use a recently proposed Structured Deep Kernel Network, which is build on a representer theorem for deep kernel learning [3], and extend and apply it to multivariate regression using time-series data. The technique is applied to the prediction of closure terms for turbulent flows, where a comparison with standard neural network techniques is drawn. The results show the full flexibility of the proposed setup while archiving near-perfect accuracy.

The paper is organized as follows. To begin with, we recall in Sect. 1.1 and 1.2 some background information about machine learning and Artificial Neural Networks (ANNs). In Sect. 2 the novel Structured Deep Kernel Network (SDKN) is reviewed and combined with standard machine learning modules. Section 3 provides background information on our application setting and the need for machine learning techniques. The subsequent Sect. 4 explains the numerical experiments and their results as well as the practicability of the SDKN. Section 5 summarizes the results and provides an outlook.

## 1.1  Regression in Machine Learning

Machine learning for regression tasks is usually posed as an optimization problem. For given data $\mathcal{D} := (x_i, y_i)_{i=1}^n$ with $x_i \subset \mathbb{R}^{d_{in}}, y_i \subset \mathbb{R}^{d_{out}}$, the goal is to find a function $f$ which approximates $f(x_i) = y_i$. Mathematically speaking, this refers to minimizing a loss functional $\mathcal{L}_D$. In case of a mean-squared error (MSE) loss, this means searching for

$$f^* := \arg \min_{f \in \mathcal{H}} \mathcal{L}_D(f), \quad \mathcal{L}_D(f) = \frac{1}{n} \sum_{i=1}^n \|y_i - f(x_i)\|_2^2 + \lambda \cdot \mathcal{R}(f) \qquad (1)$$

over a suitable space of functions $\mathcal{H}$, whereby $\mathcal{R}(f)$ is a function-dependent regularization term with regularization parameter $\lambda \in \mathbb{R}$. The space of functions $\mathcal{H}$ is usually parametrized, i.e. we have $f(x) = f(x, \theta)$ for some parameters $\theta$.

We consider here two very popular techniques that define $f$, namely Artificial Neural Networks (ANNs) and kernel methods. ANNs enjoy favorable properties for high dimensional data approximation due to representation learning. The other approach is given by kernel methods, which have a sound theoretical background, however struggle when dealing with large and high-dimensional datasets.

## 1.2  Neural Networks

ANNs (see e.g. [5]) are a prevalent class of ansatz functions in machine learning. The common feedforward ANN architecture consists of $L$ consecutive layers, whereby layer $l$ transforms its input $x \in \mathbb{R}^{d_{l-1}}$ according to

$$f_l(x) = \sigma \left( W_l x + b_l \right), \qquad (2)$$

with a weight matrix $W_l \in \mathbb{R}^{d_l \times d_{l-1}}$ and a bias vector $b \in \mathbb{R}^{d_l}$. The weight matrices $\{W_l\}_{l=1,..,L}$ and bias vectors $\{b_l\}_{l=1,..,L}$ for all the layers are parameters of the ANN, which are obtained by solving the optimization task in Eq. (1). A common choice for the non-linear activation function $\sigma(\cdot)$ is the rectified linear unit (ReLU, $\sigma(\cdot) = \max(\cdot, 0)$), which is applied element-wise. The output of a layer is then passed as input to the succeeding layer. The feedforward ANN can thus be written as a concatenation of the individual layers $f(x) = f_L \circ .. \circ f_1(x)$.

In contrast to standard kernel methods which will be recalled in the next section, the basis of the ANN is not determined a priori, but depends on its weights and biases. ANNs can thus learn a suitable basis automatically from the data they are trained on. Moreover, the concatenation of non-linear transformations allows the ANN to recombine the non-linear basis functions of the preceeding layers to increasingly complex basis functions in the succeeding layers, which renders ANNs highly suitable for high-dimensional data.

Numerous variants of the feedforward ANN have been proposed for particular kinds of data and applications. For sequential data, recurrent neural networks (RNNs) have established themselves as state of the art. A common representative of RNNs is the GRU architecture proposed in [4].

## 2   Structured Deep Kernel Networks

Here, we present a brief summary of kernel methods and the architecture presented in [10]. For a thorough discussion, we refer the reader to that reference, as our goal here is to present an extension of the orignal SDKN to sequential data.

Standard kernel methods rely on the use of possibly strictly positive definite kernels like the Gaussian kernel

$$k(x, z) = \exp(-\|x - z\|_2^2). \tag{3}$$

In this framework, a standard Representer Theorem [6] simplifies the loss minimization of Eq. (1) and states that a loss-minimizing function can be found in a subspace $V_n := \mathrm{span}\{k(\cdot, x_i), i = 1, .., n\}$ spanned by the data, i.e.

$$f^*(\cdot) = \sum_{j=1}^n \alpha_j k(\cdot, x_j). \tag{4}$$

The corresponding coefficients $\alpha_j \in \mathbb{R}^{d_{out}}$, $j = 1, .., n$ can then be found by solving a finite dimensional and convex optimization problem. This is the basis to provide both efficient solution algorithms and approximation-theoretical results. However the choice of a fixed kernel $k(x, y)$ restricts these standard kernel methods, as the feature representation of the data is implicitly determined by the choice of $k$. Furthermore, assembling and solving the linear system to determine the coefficients $\alpha_j$ poses further problems in the big data regime.

In order to overcome these shortcoming, a deep kernel representer theorem [3] has been leveraged in [10] to introduce Strucutured Deep Kernel Networks,

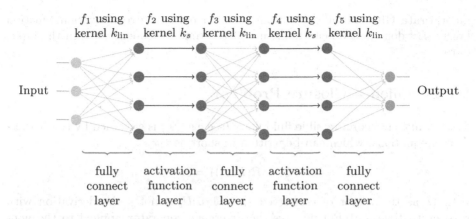

**Fig. 1.** Visualization of the Structured Deep Kernel Network. Gray arrows refer to layers using the linear kernel, while black arrows refer to layers using the single dimensional kernel layer. The braces below the layers indicate similarities to neural networks.

which alleviate these obstacles by putting kernel methods into a structured multilayer setup: The deep kernel representer theorem considers a concatenation of $L$ functions as $f = f_L \circ .. \circ f_1$ and states that the problem can be restricted to

$$f_l^*(\cdot) = \sum_{j=1}^n \alpha_{lj} k_l(\cdot, f_{l-1}^* \circ .. \circ f_1^*(x_j)), \qquad l = 2, .., L \qquad (5)$$

with $\alpha_{lj} \in \mathbb{R}^{d_l}$. This is a generalization of the standard representation of Eq. (4) to the multilayer setting. The paper [10] proceeds by choosing non-standard kernels:

1. For $l$ odd, vector-valued linear kernels are picked: $k_{\mathrm{lin}}(x, z) = \langle x, z \rangle_{\mathbb{R}^{d_l-1}} \cdot I_{d_l}$, whereby $I_{d_l} \in \mathbb{R}^{d_l \times d_l}$ is the identity matrix.
2. For $l$ even, single-dimensional kernels are used, that use single components $x^{(i)}, z^{(i)}$ of the vectors $x, z \in \mathbb{R}^{d_l-1}$: $k_s(x, z) = \mathrm{diag}(k(x^{(1)}, z^{(1)}), .., k(x^{(d)}, z^{(d)}))$ for some standard kernel $k : \mathbb{R} \times \mathbb{R} \to \mathbb{R}$, e.g. the Gaussian from Eq. (3).

This choice of kernels introduces a structure which is depicted in Fig. 1, which motivates the naming *Structured* Deep Kernel Networks and allows for comparison with neural networks: The linear kernels of the odd layers give rise to fully connected layers (without biases), while even layers with their single-dimensional kernels can be viewed as *optimizable activation function layers*. In order to obtain a sparse model and alleviate possible overfitting issues, we only use an expansion size of e.g. $M = 5 \ll n$ within the sum of Eq. (5), which amounts to fix the remaining coefficients to zero. Even for this sparse representation and special choice of kernels, it is proven in [10] that the SDKNs satisfy universal approximation properties, legitimizing their use for machine learning tasks.

The proposed setup is sufficiently flexible such that it can be combined with standard neural network modules: For the application described in Sect. 3 we incorporate GRU-modules by using them in between two activation function layers. By doing so, we can make use of the time-dependence within the input data.

## 3    Turbulence Closure Problem

The evolution of compressible fluid flows (see e.g. [9]) is governed by the Navier-Stokes equations, which can be written in short as

$$U_t + R(F(U)) = 0, \tag{6}$$

with $U$ as the vector of conserved variables. $U_t$ denotes the derivation with respect to time and $R()$ denotes the divergence operator applied to the non-linear fluxes $F(U)$. Most engineering flows of interest exhibit turbulent behavior. Such turbulent flows are inherently chaotic dynamical systems with a multiscale character. The direct numerical simulation (DNS) of turbulence is thus only feasible for simple geometries and low Reynolds numbers, which correspond to a small range of active flow scales. The framework of large eddy simulation (LES) addresses these restrictions by resolving only the large energy-containing scales by applying a low-pass filter $\overline{(\cdot)}$ to the underlying Navier-Stokes equations. However, the filtered flux term $\overline{R(F(U))}$ is generally unknown, since it depends on the full solution $U$. To this end, the coarse-scale solution is typically advanced in time by some appropriate numerical discretization $\tilde{R}(\overline{U})$, which then yields

$$\overline{U}_t + \tilde{R}(\overline{U}) = \underbrace{\tilde{R}(\overline{U}) - \overline{R(F(U))}}_{\text{perfect LES closure}}. \tag{7}$$

Solving these filtered equations becomes feasible in terms of computational cost, but the right-hand side of Eq. (7) exhibits the unknown *closure terms*, which describe the effects of the unresolved small-scale dynamics on the resolved scales.

The task of turbulence modeling can thus be stated as finding some model $M$ which recovers the closure terms solely from the filtered flow field:

$$\left( \tilde{R}(\overline{U}) - \overline{R(F(U))} \right) \approx M(\overline{U}). \tag{8}$$

A myriad of different models have been proposed in literature over the last decades to derive the mapping $M$ based on mathematical and physical reasoning. While the accuracy of the closure model is crucial for obtaining a reliable description of the flow field, no universal and generally *best* model has been identified to date. Therefore, increasing focus is laid upon finding this mapping $M$ from data by leveraging the recent advances in machine learning. See [2] for an extensive review. In that reference, machine learning is used to directly recover the unknown flux term $\overline{R(F(U))} = f(\overline{U})$ without positing any prior assumptions on the functional form of the underlying mapping $f(\cdot)$.

# 4 Numerical Application

In the present work, the dataset described in [1,7] (to which we refer for further details on the following configurations) is used as training set for the machine learning algorithms. This dataset is based on high-fidelity DNS of decaying homogeneous isotropic turbulence (DHIT). The coarse-scale quantities according to Eq. (7) are obtained by applying three different LES filters to the DNS solution: A global Fourier cutoff filter ("Fourier"), a local $L_2$-projection filter onto piecewise polynomials ("projection") and a local Top-hat filter, as shown in Fig. 2. The latter two are based on the typical representations of the solution in discontinuous Galerkin (DG) and Finite-Volume (FV) schemes, respectively.

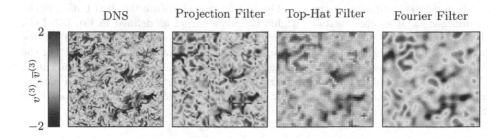

**Fig. 2.** Two-dimensional slices of the three-dimensional z-velocity field $v^{(3)}$ for the high-fidelity DNS and for the corresponding filtered velocity fields $\overline{v}^{(3)}$ of the three investigated LES filters. Each slice of the filtered flow field contains $48^2$ solution points.

For each filter, the corresponding dataset comprises nearly 30 million samples and a separate DHIT simulation is used as blind testing set. Since turbulence is a time-dependent phenomenon, the input features for each training sample are chosen as a time series of the filtered three-dimensional velocity vector $\overline{v}^{(i)}$, $i = 1, 2, 3$ at a given point in space. The target quantity is the three-dimensional closure term $\overline{R(F(U))}^{(i)}$, $i = 1, 2, 3$ in the last timestep of the given series. To investigate the influence of temporal resolution, a variety of different sampling strategies are examined, which are given in Table 1. Before training, the input features of the training dataset were normalized to zero mean and unit variance.

**Table 1.** Different time series for training. $N_{seq}$ denotes the number of time instances per sample and $\Delta t_{seq}$ is the time increment between two successive time instances.

|                | GRU1      | GRU2      | GRU3      |
| -------------- | --------- | --------- | --------- |
| $N_{seq}$      | 3         | 10        | 21        |
| $\Delta t_{seq}$ | $10^{-3}$ | $10^{-4}$ | $10^{-4}$ |

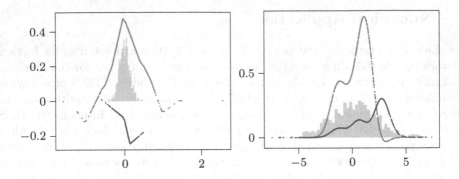

**Fig. 3.** Visualization of exemplary single-dimensional kernel function mappings before (red) and after optimization (green). The histogramms indicate the distribution of the training data after optimization. Right: Gaussian kernel as defined in Eq. (3). Left: Wendland kernel of order 0, which is defined as $k(x,y) = \max(1 - \|x - y\|_2, 0)$. These mappings can be interpreted as optimizable activation functions, see Sect. 2: On the left, the final activation function looks simpler than before (less kinks), indicating an improved generalization. On the right, the final activation function has a smaller magnitude with a similar shape, indicating a smaller relevance in the prediction of the SDKN. (Color figure online)

The ANN architecture from [7] was used as baseline model, which incorporates a GRU layer to leverage the temporal information in the data. Such a GRU layer was also introduced into the SDKN framework, which clearly demonstrates its modularity. In order to obtain a fair comparison of both the SDKN and ANN approach, both models exhibit mostly the same setup: The structure of the networks is given via their input dimension and output dimension of 3 each and the dimensions of the inner layers are chosen as $32, 64, 48, 24$. This amounts to a total of 31400 optimizable parameters for the ANN and 32240 for the SDKN. The slight difference is related to the additional parameters $\alpha_{lj}$ within the single-dimensional kernel layers, see Eq. (5) for $l$ even, and the unused bias-parameters within the SDKN. Concerning the the results in Table 2, the Gaussian kernel from Eq. (3) was used for the single-dimensional kernels within the SDKN, but we remark that the use of other kernels yields qualitatively the same results (Fig. 3). We further remark that both models are within the underparametrized regime, as the number of parameters is significantly below the number of training samples.

The Adam optimizer was used for training with an initial learning rate of $10^{-3}$, which was then halved after each 10 epochs for the NN and after each 5 epochs for the SDKN. The NN was optimized for 50 epochs and the SDKN for 25 epochs. This distinction keeps the overall optimization time comparable, since the SDKN optimization is about a factor of 2 more time-consuming per epoch. This stems from the optimization of the additional parameters (e.g. the $\alpha_{lj}$ for $l$ even in Eq. (5)) related to its setup. For training, the MSE loss from Eq. (1) was used without any regularization, since the use of a relatively small batch size (128) is likely to provide a sufficient regularization effect. We remark that both

**Table 2.** Overview of the results on the test set after optimization of the Neural Network (NN) and the Structured Deep Kernel Network (SDKN): Cross-correlation (left) and Loss (right).

| Cross-Correlation | | GRU1 | GRU2 | GRU3 |
|---|---|---|---|---|
| Projection | ANN | 0.9989 | 0.8163 | 0.9989 |
| | SDKN | 0.9989 | 0.9989 | 0.9988 |
| Top-Hat | ANN | 0.9992 | 0.9992 | 0.9992 |
| | SDKN | 0.9991 | 0.9992 | 0.9992 |
| Fourier | ANN | 0.9992 | 0.9993 | 0.9993 |
| | SDKN | 0.9992 | 0.9993 | 0.9993 |

| MSE-Loss | | GRU1 | GRU2 | GRU3 |
|---|---|---|---|---|
| Projection | ANN | 3.235e-01 | 4.996e+01 | 3.233e-01 |
| | SDKN | 3.253e-01 | 3.261e-01 | 3.368e-01 |
| Top-Hat | ANN | 3.155e-02 | 2.917e-02 | 2.888e-02 |
| | SDKN | 3.222e-02 | 2.989e-02 | 2.893e-02 |
| Fourier | ANN | 1.179e-02 | 9.737e-03 | 9.452e-03 |
| | SDKN | 1.177e-02 | 1.007e-02 | 9.587e-03 |

architecture and training are hyperparameter-optimized for the ANN setting, as this was the model of choice in [7]. The experiments were run on an Nvidia GTX1070 using PyTorch implementations of the models. The optimization took about 12 GPU-hours each.

Table 2 summarizes the results for the different datasets and cases: Both the final MSE-loss as well as the cross-correlation is given, with the latter as a common similarity measure for turbulence. The reported cross-correlations match the ones reported in [7], thus validating their results: Even in our runs, the ANN using the GRU2 setup performed badly on the DG dataset with only a final cross-correlation of 0.8163, which is way worse than the other listed cross-correlations. The SDKN reached at least the same test accuracies, even without any further hyperparameter optimization. Especially there is no drop in accuracy for the prediction related to the DG dataset in conjunction with the GRU2 case, as it can be observed in Table 2: The SDKN reaches a cross-correlation of 0.9989 which is en par with the other GRU-setups, in contrast to the performance of the ANN setup. This might indicate that the SDKN is less likely to get stuck in a local minima compared to the ANN.

## 5 Conclusion and Outlook

In this paper an extension of a recently proposed Structured Deep Kernel Network (SDKN) setup to time-series data was introduced. The proposed SDKN model was compared to standard Neural Network models on the challenging task of data-driven prediction of closure terms for turbulence modeling. With help of machine learning models, significant speed ups in the simulation of turbulent flows can be achieved. It was shown numerically that the SDKN can reach the same near-perfect accuracy as hyper-parameter optimized ANNs for several variants of the task at the same training cost.

The mathematical background which the SDKN was built on seems promising for further theoretical analysis. From the application point of view, the goal is to optimize the SDKN only for a few epochs and then use the optimized kernel in conjunction with standard shallow kernel models, for which efficient and reliable methods exists. This is expected to significantly speed up the training phase of the surrogate model.

**Ackowledgements.** The authors acknowledge the funding of the project by the Deutsche Forschungsgemeinschaft (DFG, German Research Foundation) under Germany's Excellence Strategy - EXC 2075 - 390740016 and funding by the BMBF project ML-MORE.

# References

1. Beck, A., Flad, D., Munz, C.D.: Deep neural networks for data-driven LES closure models. J. Comput. Phys. **398**, 108910 (2019)
2. Beck, A., Kurz, M.: A perspective on machine learning methods in turbulence modeling. GAMM-Mitteilungen **44**(1), e202100002 (2021)
3. Bohn, B., Rieger, C., Griebel, M.: A representer theorem for deep kernel learning. J. Mach. Learn. Res. **20**, 2302–2333 (2019)
4. Cho, K., van Merrienboer, B., Bahdanau, D., Bengio, Y.: On the properties of neural machine translation: encoder-decoder approaches. In: Proceedings of SSST-8, Eighth Workshop on Syntax, Semantics and Structure in Statistical Translation, pp. 103–111. Association for Computational Linguistics, Stroudsburg (2014)
5. Goodfellow, I., Bengio, Y., Courville, A.: Deep Learning. MIT Press, Cambridge (2016)
6. Kimeldorf, G.S., Wahba, G.: A correspondence between Bayesian estimation on stochastic processes and smoothing by splines. Ann. Math. Statist. **41**(2), 495–502 (1970)
7. Kurz, M., Beck, A.: A machine learning framework for LES closure terms. Electron. Trans. Numer. Anal. **56**, 117–137 (2022). https://doi.org/10.1553/etna_vol56s117
8. Santin, G., Haasdonk, B.: Kernel methods for surrogate modeling. In: Benner, P., Schilders, W., Grivet-Talocia, S., Quarteroni, A., Rozza, G., Silveira, L.M. (eds.) Model Order Reduction, Volume 1: System- and Data-Driven Methods and Algorithms, pp. 311–354. De Gruyter (2021). https://doi.org/10.1515/9783110498967-009
9. Serrin, J.: Mathematical principles of classical fluid mechanics. In: Truesdell, C. (ed.) Fluid Dynamics I/Strömungsmechanik I, pp. 125–263. Springer, Heidelberg (1959). https://doi.org/10.1007/978-3-642-45914-6_2
10. Wenzel, T., Santin, G., Haasdonk, B.: Universality and optimality of structured deep kernel networks (2021). https://arxiv.org/abs/2105.07228

# HPC and Big Data: Algorithms and Applications

# On the Use of Low-discrepancy Sequences in the Training of Neural Networks

E. Atanassov$^{(\boxtimes)}$, T. Gurov, D. Georgiev, and S. Ivanovska

Institute of Information and Communication Technologies, Bulgarian Academy of Sciences, Acad. G. Bonchev str., Block 25A, 1113 Sofia, Bulgaria
{emanouil,gurov,dobromir,sofia}@parallel.bas.bg

**Abstract.** The quasi-Monte Carlo methods use specially designed deterministic sequences with improved uniformity properties compared with random numbers, in order to achieve higher rates of convergence. Usually certain measures like the discrepancy are used in order to quantify these uniformity properties. The usefulness of certain families of sequences with low discrepancy, like the Sobol and Halton sequences, has been established in problems with high practical value as in Mathematical Finance. Multiple studies have been done about applying these sequences also in the domains of optimisation and machine learning. Currently many types of neural networks are used extensively to achieve break-through results in Machine Learning and Artificial Intelligence. The process of training these networks requires substantial computational resources, usually provided by using powerful GPUs or specially designed hardware.

In this work we study different approaches to employ efficiently low-discrepancy sequences at various places in the training process where their uniformity properties can speed-up or improve the training process. We demonstrate the advantage of using Sobol low-discrepancy sequences in benchmark problems and we discuss various practical issues that arise in order to achieve acceptable performance in real-life problems.

**Keywords:** Quasi-Monte Carlo algorithms · Neural networks · Stochastic gradient descent

## 1  Introduction

Machine learning has found many real-life applications, in many cases using huge amounts of data in order to train models to classify or predict unknown quantities. Because of the increasing amount of data and complexity of the tasks there is growing need to use powerful HPC techniques in order to achieve the desired accuracy in acceptable time. The use of GPGPU computing is very popular between practitioners and has led to the adoption of many standardized software packages, which encapsulate powerful Machine Learning/Artificial Intelligence concepts in an easy-to-use APIs and software patterns. Especially popular are the Artificial Neural Networks (ANNs), which found many areas

© Springer Nature Switzerland AG 2022
I. Lirkov and S. Margenov (Eds.): LSSC 2021, LNCS 13127, pp. 421–430, 2022.
https://doi.org/10.1007/978-3-030-97549-4_48

of application. Despite the success of these techniques, the training of ANNs is compute intensive and can be a time-consuming and expensive process. The number of parameters in a Neural Network can be in order of millions, even billions. Training an ANN can be formulated as an optimisation problem, but due to the high dimension of the implicit space the problem is usually solved through combining statistical and deterministic methods. Perhaps the most popular method is the so-called Stochastic Gradient Descent (SGD), frequently combined with techniques using Momentum, batching, etc. More about these can be read at [4]. Since using these methods involves random choices, they can be thought of as some kind of Monte Carlo (MC) methods. This view raises the issue of whether successful quasi-Monte Carlo (QMC) methods can be developed in order to achieve faster convergence. We remind that a basic Monte Carlo method consists of approximating an integral

$$\int_{E^s} f(x)dx$$

with the finite sum

$$\frac{1}{N} \sum_{i=1}^{N} f(x_i),$$

where $x_i$ are sampled from uniform distribution in the $s$-dimensional unit cube $E^s$. The corresponding quasi-Monte Carlo method in this case would be obtained by using an $s$-dimensional sequence that is uniformly distributed in the unit cube $E^s$. These specially constructed sequences are usually defined in a way so that their discrepancy is as low as possible. More about the uniform distribution of sequences and quasi-Monte Carlo methods can be read at [9].

The quasi-Monte Carlo methods easily outperform typical Monte Carlo methods when the dimension of the function space under consideration is low and when the sub-integral functions are smooth. Cases when the dimension is very high (thousands in some cases) and the quasi-Monte Carlo methods still outperform the MC, are also studied, most notably in financial mathematics (see, e.g., [6]). When considering the process of training a neural network as a Monte Carlo algorithm and trying to design a superior quasi-Monte Carlo method, one has to avoid constructions that lead to extremely high dimensions and also strive to increase the number of points, $N$, in order to achieve practically usable designs. In this work we study two quasi-Monte Carlo algorithms with wide applicability and we demonstrate how they are easily implemented in popular software packages for Machine Learning/training of ANNs and also how they can be superior than the usually used SGD variants in series of tests. We show that from point of view of computational time they do not increase the required time noticeably. We discuss the domain of applicability of the algorithms, some practical issues when dealing with bigger datasets and we provide directions for further research on the topic.

## 2    Setting of the Problem

The training process of the ANNs usually consists in updating the parameters of the network based on the available data, organized in epochs, where one epoch means going over the whole dataset. One item from the dataset can be an image, speech or text fragment, etc., and the desired output can be a real number or a class designation for classification problems. In order to speed-up processing, usually a mini-batch containing several items is fed to the updating process and the parameters are updating based on the desired outputs for the whole batch. When we consider certain loss function $f$ and its gradient, $\nabla f$, the SGD method updates the network based on the gradient of the loss function, but computed not over all examples but only over the chosen examples from the mini-batch. We can think of the SGD method as approximating the real gradient with the gradient from the randomly chosen mini-batch. However, the situation is more complex than that because the use of the full gradient may not be better than SGD because of possibility to reach a local minimum, as well as because of the popularity of momentum strategies like ADAM, Nesterov, etc. Regardless of these ramifications, we can consider the main step of the regular gradient descent method to be expressed as

$$w := w - \frac{\eta}{N} \sum_{i=1}^{N} \nabla Q_i(w),$$

where the objective function is expressed as the sum of errors over all samples:

$$Q(w) = \frac{1}{N} \sum_{i=1}^{N} Q_i(w), \tag{1}$$

potentially including some regularization term. The parameters $w$ are mainly weights and biases. In such a setting one step of SGD consists in

$$w := w - \eta \nabla Q_i(w)$$

where the sample $i$ is chosen randomly from all samples and $\eta$ is the learning rate. In such a setting one epoch of training consists in performing the above step one time for each of the available samples. Usually the training process will involve several epochs, although in a situation with a huge dataset this may not be the case. As the computation of the gradient even for one sample may be computationally heavy and since the regularisation term may not be dependent on the samples, there is some saving in computing the gradient for several samples, let us say $m$ samples, at once as compared to computing $m$ times the gradient for one sample. This can be especially true for computations on a GPU. That is why it is popular to use a mini-batch of samples for one step of SGD, i.e., performing the step like this:

$$w := w - \frac{\eta}{m} \sum_{i=1}^{m} \nabla Q_{s_i}(w), \tag{2}$$

where the samples $s_1, \ldots, s_m$ form a mini-batch of size $m$. In such case one epoch of training will require $N/m$ steps. Apart from efficiency, there are other considerations that support the use of mini-batches, but there are also reasons to limit the size of the mini-batch to a relatively low number like 128 or 256. One can look at Eq. (2) as an approximation of (1) via Monte Carlo method. This was our motivation to look into the possibility of applying a quasi-Monte Carlo method aiming at achieving better accuracy.

## 3   Related Work

There have been many attempts to use low-discrepancy sequences in Machine Learning and in particular in the training of Neural Networks. For example, they have been used in the problem of the initial settings of weights - see [13] for a general overview and then [8] for the QMC approach. The data augmentation process where the training samples are randomly transformed in order to obtain more data to train the network is also amenable to QMC treatment. If few random numbers used to decide how to transform a particular sample, then we can hope that a low-dimensional sequence with low-discrepancy will cover the inherent space of the possible samples better. In our test setting we also tried this technique, but we were not able to obtain visible improvement, so we are not going to discuss this further.

## 4   QMC Algorithms for the Stochastic Gradient Descent Setting

Our aim in designing QMC algorithms for the setting of training neural networks using SGD with mini-batch was to make use of the better distribution of the low-discrepancy sequences in order to obtain better approximations to the true gradient when using the mini-batch and to have balanced distribution of the samples from the different classes in the mini-batches. We assume that the number of epochs in training is limited in advance. This is not an important limitation, because a large dataset naturally does not allow high number of epochs. If this number of epochs defines the dimension of the sequence $s$ and the total number samples in the training dataset is $N$, we use the first $N$ terms of the Sobol sequence in dimension $s$ [12]. For each fixed epoch $k$, we order the corresponding coordinates of the Sobol sequence:

$$\left\{ x_i^{(k)} \right\}_0^{N-1}$$

and thus obtain a permutation $\tau_k$ of the numbers from 0 to $N-1$, so that $x_{\tau_k(0)}^{(k)}$ is the smallest coordinate. In our first QMC algorithm the samples in the epoch $k$ are ordered as $\tau_k(0), \ldots, \tau_k(N-1)$, while in the second QMC algorithm the inverse permutation is used, so that the samples are ordered starting with $\tau_k^{-1}(0)$. While doing this, we assume that the training dataset is ordered so that samples from the same class are consecutive, since in our tests random shuffling of the dataset before doing these re-orderings negates any positive effect from them. If the neural network is supposed to predict certain real-valued quantity, then we would expect the training samples to be ordered according to this quantity. In a more general setting we would like the samples to be grouped in a way that puts similar samples closer, where the measure of similarity can be arbitrarily defined. Once the order of the samples in the epoch is fixed, the consecutive samples are grouped in mini-batches of size $m$ and fed to be SGD algorithm or its variant. Since the low-discrepancy sequence provides low correlation between different coordinates, we expect that in this way we use the variations between samples from the same class in an optimal way. Since for the Sobol sequence it is preferable to use number of terms that is power-of-two, in our tests we always limit the training dataset to have power-of-two elements. If for some reason $m$ does not divide $N$, then we should discard the last incomplete mini-batch and start the next epoch as described above. Although the QMC Algorithm 1 seems to be more natural, we found out that QMC Algorithm 2 is competitive and even more promising in some tests, which was the reason we describe it here as another option of how to use QMC for the SGD training.

## 5    Implementation of the Algorithms

The QMC algorithms 1 and 2 only impact the ordering of samples and their organization into mini-batches. Our practical implementation of the QMC algorithms consisted in integrating the Sobol sequence generation code from [3] into a publicly available python code [10] which was developed by its author in order to replicate the performance of the networks described in a well known paper [7] when recognize classes of images from the ImageNet dataset. More precisely, our test problem consists in training a well-known Neural Network (ResNet V2, [11]) over images from the benchmark ImageNet dataset [5]. To train the network, we modified the data loading code in [10] in order to achieve the desired ordering of the samples, as required by QMC algorithms 1 and 2. Since the original code involves data augmentation through application of various random transformations to the image that is being loaded, the same image is transformed in a different way for each new epoch. Regardless of that, we make sure to our ordering, defined by the Sobol sequence, instead of the random shuffling that was in the code, retaining the augmentation step. For the testing images the data augmentation procedure is not applied, as is the situation in the original code.

## 6    Numerical Results

The test problem consists in training a well-known Neural Network (ResNet V2, [11]) to recognize the classes of images from the well-known ImageNet dataset [5]. We used publicly available code (in Python, [10]) which performed well in the popular benchmark problem in its default settings. Therefore we changed only the loading of the dataset in way as to ensure that the order of using the images was as it was prescribed in the algorithm. Since the original code involves data augmentation through application of various random transformations to the image that is being loaded, we never process the exact same image twice. For the testing images the data augmentation procedure is not applied, as is the situation in the original code. We kept this part of the code unchanged in order to obtain fair comparison. The test were run on our HPC system, consisting of servers Fujitsu Primergy RX 2540 M4, NVIDIA Tesla V100 32 GB, 128 GB RAM, CPU 2x Intel Xeon Gold 5118 2.30 GHz 24 cores, 2x800 GB SSD. The code uses Tensorflow and Keras with GPU computing enabled. The mini-batch size was left at its default setting of 128, which was also a good choice for us. For a low-discrepancy sequence in the algorithm we used the Sobol sequence [2]. As this sequence is well known to perform better when the number of terms is equal to a power-of-two, we adjusted the dataset to include number of images that are powers of two. In order to see the effects of the size of the dataset on the performance of QMC vs MC training, We devised subsets of the whole ImageNet dataset with three different sizes -

- small size - $2^{14}$ images in training, 13 categories, 650 images in testing,
- medium size - $2^{17}$ images in training, 104 categories, 5200 images in testing,
- big size - 1281167 images in training, 1000 categories, 50000 images in testing.

In the next figures we show the performance comparison, where we display the percentage correct classifications vs the number of epochs in training. One can see in Fig. 1 and 2 that both QMC algorithms outperform the regular (Monte Carlo) SGD algorithm for the small-size dataset. The results in Table 1 show the accuracy of both QMC algorithms when using different scrambling variants of the Sobol sequence. The Owen-type scrambling is used in Seq. 1, while the less computationally intensive Matousek scrambling is used for Seq. 2. We expected

**Table 1.** Accuracy of both QMC algorithms, medium size dataset

| Algorithm | Epoch | | | | | | | | | |
|---|---|---|---|---|---|---|---|---|---|---|
| | 1 | 2 | 3 | 4 | 5 | 7 | 8 | 10 | 20 | 30 |
| Testing | | | | | | | | | | |
| Monte Carlo | 0.123 | 0.241 | 0.369 | 0.483 | 0.542 | 0.581 | 0.632 | 0.695 | 0.762 | 0.807 |
| QMC Alg. 1 Seq. 1 | 0.123 | 0.318 | 0.470 | **0.526** | **0.579** | 0.612 | 0.636 | **0.703** | 0.766 | 0.801 |
| QMC Alg. 2 Seq. 1 | **0.147** | 0.329 | 0.378 | 0.378 | 0.563 | 0.603 | 0.661 | 0.685 | 0.765 | 0.810 |
| QMC Alg. 1 Seq. 2 | 0.121 | **0.339** | 0.439 | 0.515 | 0.537 | 0.633 | **0.680** | 0.681 | 0.761 | **0.813** |
| QMC Alg. 2 Seq. 2 | 0.143 | 0.335 | **0.481** | 0.481 | 0.557 | **0.647** | 0.641 | 0.679 | **0.780** | 0.811 |
| Training | | | | | | | | | | |
| Monte Carlo | 0.061 | 0.192 | 0.316 | 0.408 | 0.472 | 0.556 | 0.584 | 0.622 | 0.732 | 0.803 |
| QMC Alg. 1 Seq. 1 | 0.074 | 0.239 | 0.379 | 0.466 | 0.520 | 0.592 | 0.617 | 0.655 | 0.751 | 0.823 |
| QMC Alg. 2 Seq. 1 | 0.076 | 0.236 | 0.373 | 0.460 | 0.519 | 0.591 | 0.613 | 0.653 | 0.752 | 0.824 |
| QMC Alg. 1 Seq. 2 | **0.078** | **0.249** | **0.386** | **0.473** | **0.529** | **0.596** | **0.622** | **0.659** | **0.759** | **0.832** |
| QMC Alg. 2 Seq. 2 | 0.074 | 0.241 | 0.380 | 0.465 | 0.520 | 0.592 | 0.615 | 0.652 | 0.752 | 0.822 |

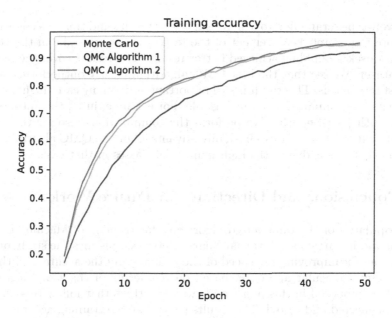

**Fig. 1.** Training accuracy with increase in epochs, small-size dataset

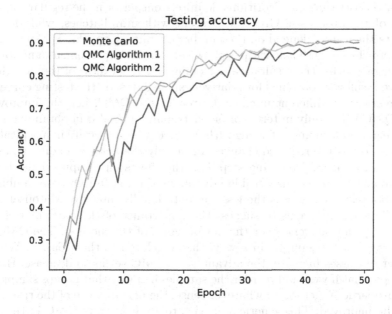

**Fig. 2.** Testing accuracy with increase in epochs, small-size dataset

the Owen-type scrambling to achieve superiour results and this was confirmed in the more important measurement of the testing accuracy, while in the training the Matousek scrambling achieved better results. The results cover the medium-size dataset. We see that the best QMC algorithm clearly outperforms only for the first few epochs. This result is still important, as in many cases higher number of epochs are computationally unfeasable. For example, in [1] the authors show results with just 6 epochs. We perform the computations also for the big size dataset, but we could not establish any advantage for the QMC algorithms. Our belief is that this is due to the high number of classes in that case.

## 7    Conclusions and Directions for Future Work

The popularity of the randomized algorithms for training of Machine Learning models and in particular, Artificial Neural Networks, justifies the study of QMC approaches for improving the speed of the training and the accuracy of the final model. One particular approach, based on reordering of the samples from the dataset, is proposed in this paper and two algorithms that follow this approach are implemented and tested. The results show that the training accuracy is definitely improved, while the more important testing accuracy, while the trained models are tested against unseen data, depends on the size of the dataset. For smaller dataset the new algorithms definitely outperform the regular implementation of the Stochastic Gradient Descent with mini-batches, while for larger datasets the results depend on the number of training epochs. For smaller number of epochs the QMC algorithms show substantial improvement, and for larger number of epochs, the results are not definitive. Regardless of the fact that the positive results are obtained for relatively small datasets, the testing curves seem to converge to a different number, better for the QMC, i.e., the improvement from QMC is not only in terms of faster training, but also in obtaining a model with better performance. The algorithms are easy to integrate in a typical training workflow as the required changes affect only the dataset loading and can be thought of as a pre-processing step, i.e., the effects on computational time are negligible. We noticed also visible advantage of using full Owen scrambling for the Sobol sequences versus the less computationally intensive Matousek scrambling. This results seems to suggest the importance of the particular choice of low-discrepancy sequences for the performance of the method. The QMC algorithms reorder the samples in a way, that is related to their classes. When the number of classes increase, the advantage of QMC seems to decrease. However, we expect that if we could reorder the samples in a way that groups samples with close properties together, in whatever sense, the performance of the training will again be improved. This generic approach to application of QMC in the setting of stochastic gradient descent or other similar training methods is suitable for further research.

**Acknowledgments.** The work of the first author (E.A.) has been supported by Bulgarian Ministry of Education and Science (contract D01-205/23.11.2018) under the National Scientific Program Information and Communication Technologies for a Single Digital Market in Science, Education and Security (ICTinSES), approved by DCM # 577/17.08.2018.

The work of the second author (T.G.) has been carried out in the framework of the National Science Program "Environmental Protection and Reduction of Risks of Adverse Events and Natural Disasters", approved by the Resolution of the Council of Ministers № 577/17.08.2018 and supported by the Ministry of Education and Science (MES) of Bulgaria (Agreements: № DO1-322/18.12.2019 and № D01-363/17.12.2020).

The work of the fourth author (S.I.) has been supported by the National Center for High-performance and Distributed Computing (NCHDC), part of National Roadmap of RIs under grant No. D01-387/18.12.2020.

We acknowledge the provided access to the e-infrastructure of the Centre for Advanced Computing and Data Processing, with the financial support by the Grant No BG05M2OP001-1.001-0003, financed by the Science and Education for Smart Growth Operational Program (2014-2020) and co-financed by the European Union through the European structural and investment funds.

# References

1. Abed, M.H., Al-Asfoor, M., Hussain, Z.M.: Architectural heritage images classification using deep learning with CNN. In: Proceedings of 2nd International Workshop on Visual Pattern Extraction and Recognition for Cultural Heritage Understanding co-located with 16th Italian Research Conference on Digital Libraries (IRCDL 2020), Bari, Italy, 29 January 2020. CEUR Workshop Proceedings, vol. 2602, pp. 1–12. CEUR-WS.org (2020). http://ceur-ws.org/Vol-2602/paper1.pdf
2. Atanassov, E., Ivanovska, S., Karaivanova, A.: Optimization of the direction numbers of the Sobol sequences. In: Dimov, I., Fidanova, S. (eds.) HPC 2019. SCI, vol. 902, pp. 145–154. Springer, Cham (2021). https://doi.org/10.1007/978-3-030-55347-0_13
3. Software for generating the Sobol sequences by BRODA. https://www.broda.co.uk/software.html. Accessed 7 Feb 2021
4. Buduma, N., Locascio, N.: Fundamentals of Deep Learning: Designing Next-Generation Machine Intelligence Algorithms, 1st edn. O'Reilly Media Inc., Sebastopol (2017)
5. Deng, J., Dong, W., Socher, R., Li, L.J., Li, K., Fei-Fei, L.: Imagenet: a large-scale hierarchical image database. In: 2009 IEEE Conference on Computer Vision and Pattern Recognition, pp. 248–255. IEEE (2009)
6. Glasserman, P.: Monte Carlo Methods in Financial Engineering. Springer, New York (2004). https://doi.org/10.1007/978-0-387-21617-1
7. He, K., Zhang, X., Ren, S., Sun, J.: Deep residual learning for image recognition. In: 2016 IEEE Conference on Computer Vision and Pattern Recognition (CVPR), pp. 770–778 (2016). https://doi.org/10.1109/CVPR.2016.90
8. Keller, A., Keirsbilck, M.V.: Artificial neural networks generated by low discrepancy sequences. CoRR abs/2103.03543 (2021). https://arxiv.org/abs/2103.03543
9. Niederreiter, H.: Random Number Generation and quasi-Monte Carlo Methods. Society for Industrial and Applied Mathematics, Philadelphia (1992)

10. ImageNet ResNet Tensorflow2.0. https://github.com/Apm5/ImageNet_ResNet_Tensorflow2.0
11. ResNet V2. https://tfhub.dev/google/imagenet/resnet_v2_50/classification/4
12. Sobol, I.M., Asotsky, D., Kreinin, A., Kucherenko, S.: Construction and comparison of high-dimensional Sobol' generators. Wilmott **2011**, 64–79 (2011)
13. de Sousa, C.A.R.: An overview on weight initialization methods for feedforward neural networks. In: 2016 International Joint Conference on Neural Networks (IJCNN), pp. 52–59 (2016). https://doi.org/10.1109/IJCNN.2016.7727180

# A PGAS-Based Implementation for the Parallel Minimum Spanning Tree Algorithm

Vahag Bejanyan$^{(\boxtimes)}$ and Hrachya Astsatryan

Institute for Informatics and Automation Problems of the National Academy of Sciences of the Republic of Armenia, 1, Paruyr Sevak Street, 0014 Yerevan, Armenia
bejanyan.vahag@protonmail.com, hrach@sci.am

**Abstract.** The minimum spanning tree is a critical problem for many applications in network analysis, communication network design, and computer science. The parallel implementation of minimum spanning tree algorithms increases the simulation performance of large graph problems using high-performance computational resources. The minimum spanning tree algorithms generally use traditional parallel programming models for distributed and shared memory systems, like Massage Passing Interface or OpenMP. Furthermore, the partitioned global address space model offers new capabilities in the form of asynchronous computations on distributed shared memory, positively affecting the performance and scalability of the algorithms. The paper aims to present a new minimum spanning tree algorithm implemented in a partitioned global address space model. The experiments with diverse parameters have been conducted to study the efficiency of the asynchronous implementation of the algorithm.

**Keywords:** Minimum spanning tree · PGAS model · Parallel algorithms · Large-scale graphs · HPC

## 1 Introduction

Large scale graph analysis plays an emerging role in social network analysis, business analytics, brain network analysis, or physics [6,14]. Graph analysis consists of exploring properties connected with its vertices (V) and edges (E).

$$G = (V, E) \tag{1}$$

The number of vertices reaches billions in large-scale graphs. Consequently, the complexity of algorithms grows exponentially, making the usage of High-Performance Computational (HPC) resources inevitable [5]. Many common graph-theoretical problems are compute-intensive, such as the shortest path, minimum spanning tree (MST), or maximum flow algorithms simulations [9,16].

The complexity growth of computational problems purse to explore and find new solutions and approaches to addressing these problems from a new perspective, find more scalable and HPC solutions.

© Springer Nature Switzerland AG 2022
I. Lirkov and S. Margenov (Eds.): LSSC 2021, LNCS 13127, pp. 431–438, 2022.
https://doi.org/10.1007/978-3-030-97549-4_49

MST has an essential role in weighted undirected and connected graph analysis, often used in optimal communication network construction or cluster analysis. A spanning tree of the graph is a subgraph, a tree that connects all the vertices. Two vertices $V_i$ and $V_j$ are connected if they satisfy the following requirement:

$$0 < |i - j| <= 2. \tag{2}$$

Each vertex between any vertex pair $(V_i, V_j)$ is assigned a weight i + j. While constructing a communication network between nodes, MST can select such a combination of the edges to minimize communication cost or material needed to build such a network. Also, MST may divide nodes into clusters by keeping a set of connected components and merging them by minimum weight edge until a necessary number of clusters is found. This approach of clustering almost directly maps to several algorithms, such as the algorithm of Kruskal [8].

In the meantime, the large-scale graph processing algorithms are memory intensive demanding innovative memory management strategies to handle the amount of big data. The implementation of novel computation models is critical to study graph processing algorithms. The Partitioned Global Address Space (PGAS) introduces novelty in partitioned shared memory. Partition has an affinity to some specific computation node to exploit locality of reference and achieve better performance. The PGAS model is in a basis for UPCXX [2], a C++ framework using asynchronous remote procedure calls (RPC), distributed objects, and global pointers. The role of UPCXX to study the MST problem in a PGAS model is essential, as UPCXX features partition the graph nodes between computation nodes or RPCs, as distributed objects. UPCXX may establish communication between different computation nodes, such as candidate edge transmission between master and worker nodes.

The paper aims to present MST algorithm implemented within the PGAS model and perform scalability and performance evaluation. The algorithm has been implemented in the scope of the open-source PGAS based graph algorithms library [4]. The article is organized in the following way. A brief discussion of the related work is presented in Sect. 2. Section 3 offers the graph representation in the PGAS model and the implementation of the proposed algorithm. The experimental results are discussed in Sect. 4, while the conclusion is presented in Sect. 5.

## 2    Related Work

MST is a well-studied problem in graph theory for several decades. Many solvers and implementations have been proposed to address this problem, such as Prim [13], Kruskal [8], or Borvuka [11] algorithms.

Prim's algorithm is operated iteratively started with random initial vertex.

Then iteratively, it adds edge with the minimum weight into the MST until MST isn't full. Kruskal algorithm works similarly. The main difference between Prim's and Kruskal's algorithms is that Kruskal's algorithm may also operate on non-connected graphs. Therefore, instead of growing a single MST, it increases the forest of the MSTs. Boruvka's algorithm is another widely used greedy algo-rithm acting as a basis for different distributed algorithms. This algorithm starts with a forest of the cheapest edges connected to the nodes. Then on each itera-tion, it connects trees grown so far by connecting them with the cheapest edge common to both of them. After each iteration, the tree's number within each con-nected component to at most half of this former value is in such an approach. The performance of the above-mentioned modest algorithms depends on the input graph and the data structures. The parallel implementation of MST algorithms generally relies on shared-memory programming models, like OpenMP (Open Multi-Processing) [3, 12].

The memory limitation of shared-memory computers directly affects the sizes of the evaluated graphs. Therefore, distributed MST algorithms may help over-come the challenge using standardized parallel programming libraries based on asynchronous models [7, 10], like Massage Passing Interface (MPI).

As the performance of MST simulations depends on the memory size, the large distributed memory mutualization with a shared memory abstraction, such as RDMA (Remote Direct Memory Access) or PGAS, may impact the perfor-mance. The implementation of the MST algorithm in the PGAS memory model addresses the increasing performance limitations.

## 3   Methodology

The suggested methodology for parallel implementation of MST algorithm for large graphs using PGAS model consists of a graph analysis API layer on the top of cloud resources, partitioned global address space, and HPC framework layers (see Fig. 1).

The Cloud resources layer relies on the IaaS (Infrastructure as a Service) cloud service of the Armenian hybrid research computing platform [15].

The global memory address space focuses on the PGAS model for paral-lel computations by creating a shared memory space using the local memories of VMs of IaaS cloud infrastructure. As its parallel execution model, PGAS uses Single Program Multiple Data. A fixed number of program instances are launched at each API call by assigning a rank per program instance. Each RPC can be given an explicit rank to perform calls. All the communication through RPCs is done asynchronously. Each RPC returns a future to explicitly block on the calling side or get a return value. An example of such communication is a transfer of the candidate MST edge to the master node.

The distributed objects provided by UPCXX have been implemented for graph data structures to store graph vertices. It is assumed that $G = (V, E)$ is a connected graph with distinct edge weights and a unique MST. Besides, it is required unique $id$ assigned to each graph vertex. To store the graph data structure internally slightly modified adjacency list is used. Instead of storing pointers to the neighbors, each vertex stores list of unique $id$s of the neighbors. Such $id$ is then used to retrieve the global pointer to that neighbor from the vertex store. The global UPCXX points to a shared memory segment of the

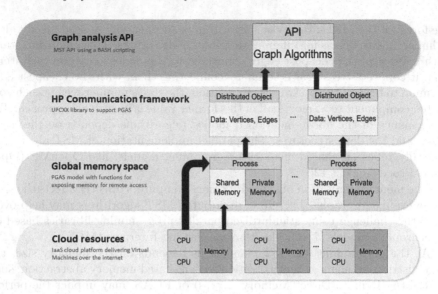

**Fig. 1.** Architecture overview diagram.

process. Vertices are distributed among all the computation nodes in a hash table to store vertex *id* as key and global pointer as value. Such distribution provides more significant locality and better performance for memory operations because computation with each vertex is done on the vertex's node to which the vertex has an affinity.

The master node decides on a globally consistent partial solution of the algorithm in a centralized way, while computational nodes keep local vertices added to the MST. The core of the current graph analyzes API is the following suggested MST algorithm.

In the initialization step, the *min*() function has an empty MST. The worker node finds the minimal weight edge using *min*() of its graph cut and adds it into the MST. The *mincan*() serves a similar purpose. The only difference is that *mincan*() finds an edge with a minimal weight such that it connects the new node to the MST. After the *mincan*() finish, the candidate MST edge is transferred to the master node using *reduce*() operation. Reduce operation applies binary operation provided to it to the stream of the candidate MST edges. The end returns a globally minimal weight edge to the master node, which adds an edge to the MST.

## 4   Experiments

The small, medium, and large instances of the Armenian IaaS cloud with the following configurations have been used for the experiments to customize several clusters: Intel Xeon 1.99 GHz processor, 4 GB RAM, Ubuntu 20.04.1 OS,

---

**Algorithm 1:** PGAS MST algorithm

---

**Result**: $MST$
$A_i = \emptyset$
$A_i = A_i \cup min(e_i)$
barrier()
**while** $MSTIsNotFull$ **do**
  $\quad candidate_i = mincan(e_i)$
  $\quad global_{min} = reduce(master, candidate_i, minop)$
  $\quad$ barrier()
  $\quad$ **if** $MasterNode$ **then**
    $\quad\quad |$ rpc(add $global_{min}$ into $MST$)
  $\quad$ **end**
**end**

---

**Table 1.** The graph configurations

| Parameter | Value |
|---|---|
| Density (%) | 1, 5, 10, 15, 70, 80, 90 |
| Vertex | 5000–15000 |
| Edge | 500000–45000000 |

UPCXX version 2020.10.0, and GCC 9.3.0 compiler with C++11/14/17 standard features.

Table 1 shows the densities, vertices, and edges studied in the experiments. The suggested graph generation algorithm ensures that generated graph is connected. Unique ids are assigned to the vertices of the graph. The distribution of edges in graphs is uniformly random, and all edge weights are unique. Figure 2 shows the behavior of the algorithm with the increasing number of densities.

Results show that in a single node setup running time of the algorithm tends to increase. At the same time, starting from the density equal to 15% running time of the distributed run of the algorithm tends to decrease. The experiments show high correlation coefficients both for single ($R^2 = 0.935$) and double ($R^2 = 0.815$) node simulations.

Figure 4 illustrates the behavior of the graph generation algorithm with increasing densities. As with the MST algorithm, experimentation was done with the single and double node setups. Results show that in the single node setup generation time grows very slowly with the increase of the densities in contrast to double node setup case, in which generation time grows very fast because of communication cost, which is bigger than the computation cost for the generation part (Fig. 3).

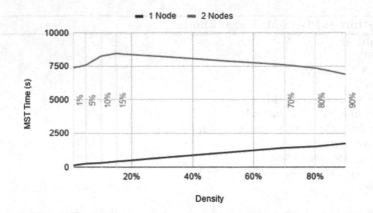

**Fig. 2.** CPU time (in seconds) for MST algorithm with increasing number of densities

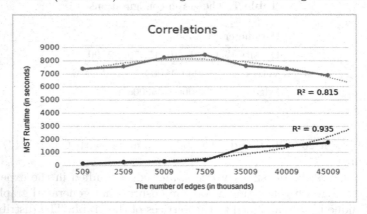

**Fig. 3.** Correlation between densities and run-time for experimental setups

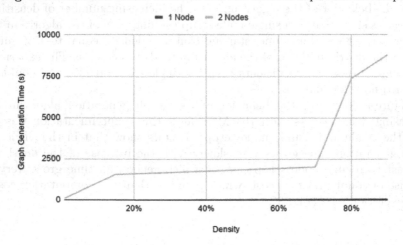

**Fig. 4.** CPU time (in seconds) for distributed graph generation with an increasing number of densities

# 5   Conclusion

The article presents a distributed algorithm for finding MST in a PGAS model. The suggested algorithm has been studied with a connected weighted graph with different vertices, edges, and densities. The experimental results show that in a single-core setup run-time of the algorithm has a strong tendency to increase, while simultaneously, a double-core setup run-time of the algorithm starts to decrease at some point. There are high correlation coefficients both for single ($R^2 = 0.93$) and double ($R^2 = 0.81$) node simulations.

It is planned to develop performance and energy-efficient algorithms for distributed large graphs in the PGAS model focusing on chunk-sizes and communications [1]. The communication cost in the PGAS model will be analyzed and optimized. The results will be used to design and implement new algorithms in the PGAS model for graph clustering, centrality, link analysis, and shortest path calculation.

**Acknowledgements.** The paper is supported by the European Union's Horizon 2020 research infrastructures programme under grant agreement No 857645, project NI4OS Europe (National Initiatives for Open Science in Europe).

# References

1. Astsatryan, H., Narsisian, W., Kocharyan, A., Da Costa, G., Hankel, A., Oleksiak, A.: Energy optimization methodology for e-infrastructure providers. Concurr. Comput. Pract. Exp. **29**(10), e4073 (2017)
2. Bachan, J., et al.: UPC++: a high-performance communication framework for asynchronous computation. In: IEEE International Parallel and Distributed Processing Symposium (IPDPS), pp. 963–973. IEEE (2019). https://doi.org/10.1109/IPDPS.2019.00104
3. Bader, D.A., Cong, G.: Fast shared-memory algorithms for computing the minimum spanning forest of sparse graphs. J. Parallel Distrib. Comput. **66**(11), 1366–1378 (2006). https://doi.org/10.1016/j.jpdc.2006.06.001
4. Bejanyan, V., Astsatryan, H.: MST PGAS algorithm (2021). https://github.com/lnikon/pgas-graph
5. Caíno-Lores, S., Carretero, J., Nicolae, B., Yildiz, O., Peterka, T.: Toward high-performance computing and big data analytics convergence: the case of spark-DIY. IEEE Access **7**, 156929–156955 (2019). https://doi.org/10.1109/ACCESS.2019.2949836
6. Du, Z., Yin, Z., Liu, W., Bader, D.: On accelerating iterative algorithms with CUDA: a case study on conditional random fields training algorithm for biological sequence alignment. In: IEEE International Conference on Bioinformatics and Biomedicine Workshops (BIBMW), pp. 543–548. IEEE (2010). https://doi.org/10.1109/BIBMW.2010.5703859
7. Gallager, R.G., Humblet, P.A., Spira, P.M.: A distributed algorithm for minimum-weight spanning trees. ACM Trans. Program. Lang. Syst. (TOPLAS) **5**(1), 66–77 (1983). https://doi.org/10.1145/357195.357200
8. Kruskal, J.B.: On the shortest spanning subtree of a graph and the traveling salesman problem. Proc. Am. Math. Soc. **7**(1), 48–50 (1956). https://doi.org/10.1090/S0002-9939-1956-0078686-7

9. Madduri, K., Bader, D.A., Berry, J.W., Crobak, J.R.: An experimental study of a parallel shortest path algorithm for solving large-scale graph instances. In: Proceedings of the Ninth Workshop on Algorithm Engineering and Experiments (ALENEX), pp. 23–35. SIAM (2007). https://doi.org/10.1137/1.9781611972870.3

10. Mazeev, A., Semenov, A., Simonov, A.: A distributed parallel algorithm for the minimum spanning tree problem. In: Sokolinsky, L., Zymbler, M. (eds.) PCT 2017. CCIS, vol. 753, pp. 101–113. Springer, Cham (2017). https://doi.org/10.1007/978-3-319-67035-5_8

11. Nešetřil, J., Milková, E., Nešetřilová, H.: Otakar Borůvka on minimum spanning tree problem translation of both the 1926 papers, comments, history. Discret. Math. **233**(1–3), 3–36 (2001)

12. Olson, C.F.: Parallel algorithms for hierarchical clustering. Parallel Comput. **21**(8), 1313–1325 (1995)

13. Prim, R.C.: Shortest connection networks and some generalizations. Bell Syst. Tech. J. **36**(6), 1389–1401 (1957). https://doi.org/10.1002/j.1538-7305.1957.tb01515.x

14. Riedy, E.J., Meyerhenke, H., Ediger, D., Bader, D.A.: Parallel community detection for massive graphs. In: Wyrzykowski, R., Dongarra, J., Karczewski, K., Waśniewski, J. (eds.) PPAM 2011. LNCS, vol. 7203, pp. 286–296. Springer, Heidelberg (2012). https://doi.org/10.1007/978-3-642-31464-3_29

15. Shoukourian, Y.H., Sahakyan, V.G., Astsatryan, H.V.: E-infrastructures in Armenia: virtual research environments. In: Ninth International Conference on Computer Science and Information Technologies Revised Selected Papers, pp. 1–7. IEEE (2013). https://doi.org/10.1109/CSITechnol.2013.6710360

16. Yang, C., Wang, Y., Owens, J.D.: Fast sparse matrix and sparse vector multiplication algorithm on the GPU. In: IEEE International Parallel and Distributed Processing Symposium Workshop, pp. 841–847. IEEE (2015). https://doi.org/10.1109/IPDPSW.2015.77

# Comparison of Different Methods for Multiple Imputation by Chain Equation

Denitsa Grigorova[1,2]($\boxtimes$)(iD), Demir Tonchev[1], and Dean Palejev[1,3](iD)

[1] Big Data for Smart Society (GATE) Institute, Sofia University,
125 Tsarigradsko Shosse, Bl. 2, 1113 Sofia, Bulgaria
{denitsa.grigorova,demir.tonchev,dean.palejev}@gate-ai.eu
[2] Faculty of Mathematics and Informatics, Sofia University,
5 James Bourchier Blvd., 1164 Sofia, Bulgaria
[3] Institute of Mathematics and Informatics, Bulgarian Academy of Sciences,
Acad. G. Bonchev Street, Bl. 8, 1113 Sofia, Bulgaria

**Abstract.** Missing data is a common problem when analysing real-world data from many different research fields such as biostatistics, sociology, economics etc. Three types of missing data are typically defined: missing completely at random (MCAR), missing at random (MAR), and missing not at random (MNAR). Ignoring observations with missingness could lead to serious bias and inefficiency, especially when the number of such cases is large compared to the sample size. One popular technique for solving the missing data issue is multiple imputation (MI).

There are two general approaches to MI. One is joint modelling which draws missing values simultaneously for all incomplete variables from a multivariate distribution. The other is the fully conditional specification (FCS, also known as MICE), which imputes variables one at a time from a series of univariate conditional distributions. For each incomplete variable FCS draws from a univariate density conditional on the other variables included in the imputation model.

In this work we define a computationally efficient numerical simulation framework for data generation and evaluation of different imputation methods. We consider different FCS imputation methods along with traditional ones under different scenarios for the parameters of the models - percentage of missingness, data dimensionality, different combination of categorical and numerical predictors and different correlation between the covariates. Our results are based on synthetic data generated on HPC cluster and show the optimal imputation methods in the different cases according to two scoring techniques.

**Keywords:** Fully conditional specification · Missing data · Multiple imputation

© Springer Nature Switzerland AG 2022
I. Lirkov and S. Margenov (Eds.): LSSC 2021, LNCS 13127, pp. 439–446, 2022.
https://doi.org/10.1007/978-3-030-97549-4_50

# 1   Introduction

Missing data is a common problem in real-world data analysis. Typically three types of missing mechanisms are defined, e.g. as discussed in [14]: missing completely at random (MCAR), missing at random (MAR), and missing not at random (MNAR). MCAR indicates that the probability of missingness does not depend on the unobserved and observed data. The MAR mechanism is defined when conditional on the observed data, the probability of data being missing does not depend on the unobserved data. MCAR is a special case of MAR. The MNAR case is when conditional on the observed data, the probability of data being missing still depends on the unobserved data. The simplest, but naïve solution is to ignore all observations with missing values for one or more predictor variables. This approach is known as complete case analysis. When there are not many missing cases and when the data are MCAR, the complete case analysis works well although the efficiency could be lower. Different approaches, among which is multiple imputation (MI), are developed for solving the issue in the case of MAR. MI is a method for filling in the missing values resulting in several complete data sets, then analysing the complete data sets and combining the results according to Rubin's rule as described in [8,13]. The main groups of variables that should be included in the imputation model are: variables in the main model(s) of interest, variables that are correlated with the imputed variable, variables that are associated with the missingness of the imputed variable and the dependent variable(s).

In this article we consider one of the methods for multiple imputation called fully conditional specification (FCS) as described in [12] which works also in the case of categorical variables as opposed to the other method for multiple imputation called joint modelling, described in [13] which assumes multivariate normal distribution for the variables to be imputed. The first method is also known as multiple imputation by chained equations (MICE). It is implemented in several software packages, for example the R package mice [5] and the Python library Autoimpute [7]. Practical tips and guidelines for implementing FCS MI are described in [9]. That article demonstrates the application of FCS MI in support of a large epidemiological study evaluating national blood utilization patterns in a sub-Saharan African country.

We compare several FCS MI methods including predictive mean matching and local residual draws. The book by van Buuren [4] gives a comprehensive review and description of different FCS MI methods implemented in the R package mice. As described in that book, to apply the predictive mean matching (pmm) procedure for each missing value we start by selecting a small set of donors with typically 3, 5 or 10 elements out of all complete cases that have predicted values closest to the one of the missing entry. The underlying assumption is that the distribution of the missing data is the same as the observed values of the donors. Then we randomly draw one of the donors and replace the missing value with the donor's value.

An example of another imputation model is local residual draws (lrd). It is analogous to the pmm procedure but lrd adds the donor's residual to the recipient's linear predictor. Article [11] gives a description of both imputation models and provides information about the available implementations. An article that compares the joint modelling multiple imputation approach and the fully conditional specification approach for clustered data is [10]. Paper [6] describes a comparison of wide range of MI methods available in standard software packages via simulation and on real longitudinal data.

In this article we present a comparison of single imputation and several multiple imputation FCS methods using adjusted $R^2$ and a score that measures the averaged and standardized squared distance of the estimates of the coefficients from the true values of the coefficients in a linear regression model calculated on simulated cross-sectional data.

In the Methods section we present a detailed description of the data generation process and the application of the imputation techniques. The Results section describes the optimal imputation methods according to adjusted $R^2$ and a metric-based score that we adopted for the comparison of the different methods and our comments on these results. We conclude with ideas of future work.

## 2  Methods

Our pipeline starts with data generation, followed by applications of different imputation methods and analysis of the results.

### 2.1  Data Generation

Our numerical simulations include three steps - synthetic data generation under different scenarios, application of several imputation methods, and evaluation of two scores using the imputed data in comparison with data without missing observations. There are many possible set-ups for the parameters included in the model and in this work we consider only several basic ones.

The generated data consist of predictor variables of two possible types (numerical and categorical) and one dependent numerical variable. Our script uses input information about the total number of the observations $(200, 400, 600, 800$ and $1000)$ and the percentage of missing data $(5, 10$ and $15\%)$. The code is implemented in R and is able to create groups of clustered and correlated within each cluster categorical or numerical variables although in this work we consider up to two groups. Our synthetic data include three cases according to the types of the predictors. In the first case the predictors consist of two groups of numerical variables, in the second - of two groups of categorical variables, and the third case consists of one group of each type. Our code can generate groups of different sizes, however for simplicity we consider cases where the number of the variables in each group is equal and take values of $2, 4, 6, 8,$ and $10$.

For generating numerical data we use multivariate normal distribution (zero mean, variance one, and random correlation matrix within each cluster) and we shift and scale it by predetermined means and standard deviations. When generating the categorical data we apply discretization of the multivariate normal distribution using the input parameters, i.e. the number of levels of each variable

(2 for variables from the first group and 3 for the second group in the case when a second group of categorical variables is present) and the thresholds for the discretization which also define the proportions of each of the levels.

For the generation of the dependent variable we use linear model with the predictors that are already generated and input information about the value of the standard deviation of the error terms in the linear model (in the simulations presented here it is fixed to 2). When we construct the linear model we explicitly specify if the predictor is categorical and if that is the case dummy variables are created in the model (for example if we have three-level categorical predictor we need two dummy variables).

We also create a mask file with zeros and ones with the same dimension as the predictors with values of one indicating the missing observations. In our set-up there are no missing values in the outcome variable. The generation of the mask is independent of the observed and the unobserved data and therefore our mechanism of missingness is completely at random. We also produce a configuration file that contains the names of the variables and their respective type (numerical, binomial or multinomial) that is used to correctly set the imputation procedures in the subsequent Python script.

## 2.2 Imputation and Comparisons

The Python package `Autoimpute` is used for the imputation. It has different methods (or strategies as referred to in the library API) for imputation. When applying MI methods for a given dataset one typically generates several different instances of imputed data and creates an aggregate estimate of the parameters of interest. To significantly reduce the computational resources that are needed for carrying on the simulations we generate 100 imputed single-instance datasets (with different seeds). From them we created 1000 aggregates of sizes 5 and 10 on which we perform the main analysis of interest. We should note that there is some weak dependence among the 1000 values that are calculated based on the procedure described, however it could be ignored for the purpose of our investigation.

The script loads the full data, applies the mask indicating the missing data and then creates 100 iterations for each set of imputation strategies. After applying each imputation strategy a linear regression using `Statsmodels` [16] is fit and several values based on the fitted regression are reported. The different imputation methods for the different parameter types are as follows:

- numerical: median, least squares, stochastic least squares, Bayesian least squares, pmm, lrd;
- binomial: mode, binary logistic regression, Bayesian binary logistic;
- multinomial: mode, multinomial logistic regression.

Here we give a very short description of the imputation methods. For the specific method the missing values are imputed with:

- least squares: prediction using the line of best fit given a set of predictors;
- stochastic least squares: predictions using the least squares regression line of best fit plus a random draw from the normal error distribution;
- Bayesian least squares: samples from the posterior predictive distribution of each missing data point;
- pmm: predictions using a combination of Bayesian approach to least squares and least squares itself. The method is the one described in the Introduction section;
- lrd: similar to pmm, also described in the Introduction section;
- binary logistic: predictions using logistic regression with two classes;
- Bayesian binary logistic: samples from the posterior predictive distribution of each missing point using the Bayesian approach to logistic regression;
- multinomial logistic regression: predictions from the regression model with the same name.

We should note that some of these methods appear in the `mice` package under different names.

Out of those, the following are actual MI strategies: stochastic least squares, Bayesian least squares, pmm, lrd, Bayesian binary logistic. The others fill in only one deterministic value (similarly to the mode or the median) for each missing observation (this is called single imputation), but we have included them for comparison purposes. For simplicity, in this work we always use the same method and not combinations of different methods within each variable type. In the case of a combination between numerical and binomial predictors we mix and match the different imputation types for the different types of variables. When using the class consisting of mode and median, we only match them with imputations from the same class.

Our Python code generates several output values: estimates of the coefficients of the fitted linear regression model; standard errors of the coefficient estimates; $p$-values for testing whether the coefficients are statistically significant; 95% confidence intervals for the coefficients; $R^2$; adjusted $R^2$; AIC (Akaike information criterion, [1]) and BIC (Bayesian information criterion, [15]). In this article we consider only two scores – adjusted $R^2$ and a metric-based score comparing the original regression coefficients and their estimates when utilizing imputed data. That score is defined as $m = k^{-1} \sum_{i=1}^{k} (\theta_i - \hat{\theta}_i)^2 / |\theta_i|$, where $\theta_i$ is the true value of the regression parameter (in our simulations it can take value in $\{-2, -1, 1, 2\}$), $\hat{\theta}_i$ is the estimate of the corresponding regression parameter and $k$ is the total number of the regression coefficients in the model. The metric-based score $m$ is averaged and standardized characteristic of how far the estimates are from the true values of the regression coefficients in the model and is similar to $\chi^2$ test statistic. It allows us to compare models with different number of predictors.

## 3   Results

Because of space constrains, the tables with our aggregated results are provided as Supplementary Material in .csv format at www.math.bas.bg/~palejev/MIS. At the same page we have provided descriptions of the different columns of the tables.

We divide the imputation techniques into two groups according to the speed of convergence of the imputation procedure - the fast group consists of the least squares, stochastic least squares, binary logistic, multinomial logistic methods and the slow group consists of the Bayesian least squares, lrd, pmm, Bayesian binary logistic methods.

We notice that in cases in which we have only categorical variables, the performance for the score $m$ (described in the previous Section) is very similar for binary logistic and Bayesian binary logistic, except for some cases when the percentage of missingness is smaller. In instances when the two methods perform similarly, for computational reasons we recommend using binary logistic. We observe that when we compare the adjusted $R^2$, the binary logistic and Bayesian binary logistic yield close results. In all cases utilizing the mode for imputation performs the worst overall and we do not recommend using it.

When we have only numerical variables, lrd and pmm perform generally the best in all cases regarding the score $m$. When for example the percentage of missingness is 10%, stochastic least squares also performs fine. For the adjusted $R^2$ almost all methods, except for the median, perform similarly.

In cases when we have both categorical and numerical variables, we observe several scenarios. When the number of predictors is small and the percentage of missingness is also small we observe that most methods perform similarly in terms of the score $m$ and therefore we recommend the faster methods. When the number of predictors increases and the number of observations also increases but the missingness remains small then pmm, lrd and stochastic least squares perform better. When the missingness is larger and the number of observations is moderate then pmm, lrd and stochastic least squares perform better when comparing the values of $m$. In the other cases when the missingness is larger the faster methods perform better. For adjusted $R^2$ almost all methods, except for the median and in many cases lrd, perform similarly. When the missingness is large, lrd performs the worst regarding both the score $m$ and adjusted $R^2$. In the case of mixed categorical and numerical variables we do not recommend using the mixed mode and median method.

As expected, the standard deviation of the values of $m$ and of the adjusted $R^2$ each differ by a factor of about $\sqrt{2}$ when comparing the simulations within each score for 5 and 10 imputed data sets.

When there are categories with two levels the binary logistic method might work better as in this case it could separate the two categories correctly whereas the Bayesian binary logistic method gives some, although small, probability to the incorrect level of the variable yielding somewhat worse results.

When we have a small number of categorical predictors we have discretisation of $m$, leading in some cases to a standard deviation of almost 0 that may not be detected empirically resulting in an empirical standard deviation of 0. We also observe that some methods (such as binary logistic and least squares) that are single imputation procedures nevertheless fill in different values for different seeds which leads to slightly different results in some iterations.

## 4 Discussion

The FCS MI approach is relatively easy to apply but a primary disadvantage is that it does not have the same theoretical justification as other imputation techniques. In particular, fitting a series of conditional distributions, as it is done using the series of regression models, may not be consistent with a proper joint distribution. Another disadvantage is that it may produce biased estimates under MNAR [3].

In our simulations we have considered an MCAR mechanism. Future research is needed under the MAR scenario. Simulation of such mechanism of missingness is implemented for example in the function ampute in R, [5].

Our simulations have demonstrated that in many situations the performance of various MI methods is very similar when comparing the adjusted $R^2$. Therefore adjusted $R^2$ is not a suitable score for evaluating the optimality of these methods. As an extension of the current research we are planning to use other scores or metrics besides the considered score $m$. These could include AIC, BIC, and mean square error. Another approach that we might deploy would be to use out-of-sample test sets.

**Acknowledgements.** The result presented in this paper is part of the GATE project. The project has received funding from the European Union's Horizon 2020 WIDESPREAD-2018-2020 TEAMING Phase 2 programme under Grant Agreement No. 857155 and Operational Programme Science and Education for Smart Growth under Grant Agreement No. BG05M2OP001-1.003-0002-C01.

The numerical simulations were performed on the Avitohol supercomputer at IICT-BAS described in [2]. The computational resources and infrastructure were provided by NCHDC – part of the Bulgarian National Roadmap of RIs, with the financial support by Grant No DO1 - 387/18.12.2020.

## References

1. Akaike, H.: A new look at the statistical model identification. IEEE Trans. Autom. Control **19**(6), 716–723 (1974)
2. Atanassov, E., Gurov, T., Ivanovska, S., Karaivanova, A.: Parallel Monte Carlo on Intel MIC architecture. Procedia Comput. Sci. **108**, 1803–1810 (2017). International Conference on Computational Science, ICCS 2017, 12–14 June 2017, Zurich, Switzerland
3. Azur, M., Stuart, E., Frangakis, C., Leaf, P.: Multiple imputation by chained equations: what is it and how does it work? Int. J. Methods Psychiatr. Res. **20**(1), 40–49 (2011)
4. van Buuren, S.: Flexible Imputation of Missing Data, 2nd edn. CRC Press, Boca Raton (2018)

5. van Buuren, S., Groothuis-Oudshoorn, K.: mice: multivariate imputation by chained equations in R. J. Stat. Softw. **45**(3), 1–67 (2011)
6. Huque, M.H., Carlin, J.B., Simpson, J.A., Lee, K.J.: A comparison of multiple imputation methods for missing data in longitudinal studies. BMC Med. Res. Methodol. **18**(1), 168 (2018)
7. Kearney, J., Barkat, S., Bose, A.: Python package for analysis and implementation of imputation methods (2019). https://pypi.org/project/autoimpute/
8. Little, R., Rubin, D.: Statistical Analysis with Missing Data. Wiley Series in Probability and Mathematical Statistics. Probability and Mathematical Statistics, Wiley (2002)
9. Liu, Y., De, A.: Multiple imputation by fully conditional specification for dealing with missing data in a large epidemiologic study. Int. J. Stat. Med. Res. **4**(3), 287–295 (2015)
10. Mistler, S.A., Enders, C.K.: A comparison of joint model and fully conditional specification imputation for multilevel missing data. J. Educ. Behav. Stat. **42**(4), 432–466 (2017)
11. Morris, T.P., White, I.R., Royston, P.: Tuning multiple imputation by predictive mean matching and local residual draws. BMC Med. Res. Methodol. **14**, 75 (2014)
12. Raghunathan, T.E., Lepkowski, J.M., Hoewyk, J.V., Solenberger, P.: A multivariate technique for multiply imputing missing values using a sequence of regression models. Surv. Pract. **27**(1), 85–95 (2001)
13. Rubin, D.B.: Multiple Imputation for Nonresponse in Surveys. Wiley, Hoboken (1987)
14. Rubin, D.B.: Inference and missing data. Biometrika **63**(3), 581–592 (1976)
15. Schwarz, G.: Estimating the dimension of a model. Ann. Stat. **6**(2), 461–464 (1978)
16. Seabold, S., Perktold, J.: Statsmodels: econometric and statistical modeling with python. In: 9th Python in Science Conference (2010)

# Monte Carlo Method for Estimating Eigenvalues Using Error Balancing

Silvi-Maria Gurova$^{(\boxtimes)}$ and Aneta Karaivanova

Institute of Information and Communication Technologies,
Bulgarian Academy of Sciences, Acad. G. Bonchev Street, bl. 25A, Sofia, Bulgaria
{smgurova,anet}@parallel.bas.bg

**Abstract.** Monte Carlo (MC) power iterations are successfully applied for estimating extremal eigenvalue especially those of large sparse matrices. They use truncated Markov chain simulations for estimating matrix-vector products. The iterative MC methods contain two type of errors-systematic (a truncation error) and stochastic (a probable error). The systematic error depends on the number of iterations and the stochastic error depends on the probabilistic nature of the MC method. In this paper we propose a new version of the MC power iterations using balancing of both errors to determine the optimal length of the chain. Numerical results for estimating the largest eigenvalue are also presented and discussed.

**Keywords:** Monte Carlo method · Markov chain · Eigenvalue

## 1 Introduction

Although some of the first publications in the field of Monte Carlo methods of the 50s and 60s of the 20th century refer to problems in linear algebra, they do not consider the problem of approximate computation of eigenvalues. However, this problem is very important as many applications in different fields can be reduced to finding the eigenvalues and eigenvectors of a matrix $A$. If the matrix is very large, we are rarely interested in all eigenvalues. Often the properties of the system described by the matrix are determined by the largest and the smallest eigenvalues. These two values also determine the condition number of the matrix.

A Monte Carlo method for finding the dominant eigenvalue was proposed by Sobol in 1973 [10] and for the smallest - by Mikhailov in 1987 [1]. The power method with Monte Carlo iterations was proposed by Dimov and Karaivanova in a series of publications during the period 1996–1999 [4,5]. The use of the resolvent matrix to find the smallest eigenvalue was proposed by Dimov and Karaivanova in 1998 [5]. The first publications in the literature presenting quasi-Monte Carlo methods for finding extreme eigenvalues are by Karaivanova and Mascagni in the period 2001–2003 [8,9]. In the subsequent publications of Atanassov, Karaivanova, Ivanovska, Chi, Mascagni and others [2,3], the use of

© Springer Nature Switzerland AG 2022
I. Lirkov and S. Margenov (Eds.): LSSC 2021, LNCS 13127, pp. 447–455, 2022.
https://doi.org/10.1007/978-3-030-97549-4_51

various low discrepancy sequences (generalized, randomized) in the Monte Carlo methods for extreme eigenvalues has been proposed and studied.

In this paper we reconsider the original method of MC power iterations and propose a variant with error balancing which permits us to achieve the desired accuracy with smaller number of computations. The paper is organized as follows: after Introduction, the section Background presents the deterministic power method, the MC power method and their convergence and complexity. Section 3 presents MC method for estimating eigenvalues using error balancing. The method implementation and some numerical results are presented in the Sect. 4. Finally, we give some conclusions.

## 2    Background and Related Research

### 2.1    Formulation

Let $A = \{a_{ij}\}_{i,j=1}^n \in R^{n \times n}$ be a given (non-singular) matrix. Consider the eigenvalue problem: find $\lambda(A)$ such that:

$$Au = \lambda(A)u, \tag{1}$$

We denote a matrix polynomial of degree $k$ by the equation:

$$p_k(A) = \sum_{i=0}^k c_i.A^i, \quad c_i \in R. \tag{2}$$

### 2.2    Computing the Extreme Eigenvalues

The well-known Power method [7,11] gives an estimate for the dominant eigenvalue $\lambda_1$. Let $A$ be an $n \times n$ matrix with real elements $a_{ij}$. Denote the matrix eigenvectors by $\{u_j\}_{j=1}^n$ and suppose that

$$|\lambda_1| > |\lambda_2| \geq \ldots \geq |\lambda_{n-1}| > |\lambda_n|.$$

Using the Power method for estimating the eigenvalue we get

$$\lambda_{max}^{(k+1)} = \lambda_1 \frac{a_1 u_{i1} + \sum_{j=2}^n \left(\frac{\lambda_j}{\lambda_1}\right)^{k+1} a_j u_{ij}}{a_1 u_{i1} + \sum_{j=2}^n \left(\frac{\lambda_j}{\lambda_1}\right)^k a_j u_{ij}}, \tag{3}$$

$$\lambda_{max}^{(k+1)} = \lambda_1 + O\left(\left|\frac{\lambda_2}{\lambda_1}\right|^k\right). \tag{4}$$

When $A$ is a symmetric matrix and has real eigenvalues and eigenvectors

$$\mu_{k+1} = \lambda_1 \frac{|a_1|^2 + \sum_{j=2}^n |a_j|^2 \left(\frac{\lambda_j}{\lambda_1}\right)^{2k+1}}{|a_1|^2 + \sum_{j=2}^n |a_j|^2 \left|\frac{\lambda_j}{\lambda_1}\right|^{2k}} \tag{5}$$

$$\mu_{k+1} = \lambda_1 + O\left(\left|\frac{\lambda_2}{\lambda_1}\right|^{2k}\right) \tag{6}$$

There are two deterministic numerical methods that can efficiently compute only the extremal eigenvalues - the *Power method* and *Lanczos-type methods*. (Note that, the Lanczos method is applicable only to *symmetric* eigenvalue problems [7]).

*Computational Complexity:* If $k$ iterations are required for convergence, the number of arithmetic operations is $O(kn^2)$ for the Power method and $O(n^3 + kn^2)$ for both the Inverse and Inverse shifted power method.

## 2.3   The Monte Carlo Method

Let $h = \{h_i\}_{i=1}^n$ and $f = \{f_i\}_{i=1}^n$ be two arbitrary vectors in $R^n$. We create a stochastic process using the matrix $A$ and those two vectors. We denote with $(h, f) = \sum_{i=1}^n h_i \cdot f_i$ the scalar product of both vectors. The problem we are interested in evaluating is the scalar products of the following type:

$$(h, p_m(A)f). \tag{7}$$

Considering a special case $p_m(A) = A^m$, the form (7) becomes

$$(h, A^m f). \tag{8}$$

Then the problem of approximate computing of the largest eigenvalue can be reformulated as:

$$\lambda_1 \approx \lim_{m \to \infty} \frac{(h, A^m f)}{(h, A^{m-1} f)}.$$

If we are interested to compute also the smallest eigenvalue of a matrix $A$, we work with its *resolvent* matrix $R_q = [I - qA]^{-1} \in R^{n \times n}$ [5].

If $|q\lambda| < 1$, $R_q$ may be expanded as a series via the binomial theorem:

$$p_\infty = p(A) = \sum_{i=0}^{\infty} q^i C_{i+m-1}^i A^i = [I - qA]^{-m} = R_q^m, \tag{9}$$

The eigenvalues of the matrices $R_q$ and $A$ are connected by the equality $\mu = \frac{1}{1-q\lambda}$ and the eigenvectors of the two matrices coincide[1].

Applying the Power method [5], leads to the following iterative processes:

$$\lambda^{(m)} = \frac{(h, A^m f)}{(h, A^{m-1} f)} \xrightarrow[m \to \infty]{} \lambda_{max}, \tag{10}$$

$$\mu^{(m)} = \frac{([I - qA]^{-m} f, h)}{([I - qA]^{-(m-1)} f, h)} \xrightarrow[m \to \infty]{} \mu_{max} = \frac{1}{1 - q\lambda}. \tag{11}$$

---

[1] If $q > 0$, the largest eigenvalue $\mu_{max}$ of the resolvent matrix corresponds to the largest eigenvalue, $\lambda_{max}$ of the matrix $A$ but if $q < 0$, then $\mu_{max}$, corresponds to the smallest eigenvalue, $\lambda_{min}$ of the matrix $A$.

To construct the MC method we define Markov chain

$$k_0 \to k_1 \to \cdots \to k_i \to \ldots, \qquad (1 \le k_i \le n)$$

with initial density vector $p = \{p_\alpha\}_{\alpha=1}^n$, where $Pr(k_i = \alpha) = p_\alpha$ and the transition density matrix, $P = \{p_{\alpha\beta}\}_{\alpha,\beta=1}^n$, where $Pr(k_j = \beta | k_{j-1} = \alpha) = p_{\alpha\beta}$.

Define the following random variables $W_j$ using the recursion formula:

$$W_0 = \frac{h_{k_0}}{p_{k_0}}, \quad W_j = W_{j-1} \frac{a_{k_{j-1}k_j}}{p_{k_{j-1}k_j}}, \quad j = 1, 2 \ldots . \tag{12}$$

This has the desired expected values:

$$E[W_i f_{k_i}] = (h, A^i f), \quad i = 1, 2, \ldots, \tag{13}$$

$$E[\sum_{i=0}^{\infty} q^i C_{i+m-1}^i W_i f(x_i)] = (h, [I - qA]^{-m} f), \quad m = 1, 2, \ldots, \tag{14}$$

and allows us to estimate the desired eigenvalues as:

$$\lambda_{max} \approx \frac{E[W_i f_{k_i}]}{E[W_{i-1} f_{k_{i-1}}]}. \tag{15}$$

and

$$\lambda \approx \frac{1}{q} \left(1 - \frac{1}{\mu^{(m)}}\right) = \frac{E[\sum_{i=1}^{\infty} q^{i-1} C_{i+m-2}^{i-1} W_i f(x_i)]}{E[\sum_{i=0}^{\infty} q^i C_{i+m-1}^i W_i f(x_i)]}. \tag{16}$$

We remark that in (15) the length of the Markov chain is equal to the number of iterations, $i$, in the Power method. However, in (16) the length of the Markov chain is equal to the number of terms in truncated series for the resolvent matrix. In this second case the parameter $m$ corresponds to the number of power iterations.

## 2.4   Convergence and Complexity

Consider a random variable $\theta^i = W_i f_{ki}$, that has a mathematical expectation formula (13). The Monte Carlo error obtained when computing a matrix-vector product is well-known to be:

$$|h^T A^i f - \frac{1}{N} \sum_{s=1}^{N} (\theta^i)_s| \approx c.Var(\theta^i)^{1/2} N^{-1/2},$$

where $c$ is a constant, the variance is $Var(\theta^i) = E[(\theta^i)^2] - (E[\theta^i])^2$ and

$$E[\theta^i] = E[\frac{h_{k_0}}{p_{k_0}} W_i f_{k_i}]$$

$$= \sum_{k_0=1}^{n} \frac{h_{k_0}}{p_{k_0}} p_{k_0} \sum_{k_1=1}^{n} \cdots \sum_{k_i=1}^{n} \frac{a_{k_0 k_1} \cdots a_{k_{m-1} k_m}}{p_{k_0 k_1} \cdots p_{k_{m-1} k_m}} p_{k_0 k_1} \cdots p_{k_{m-1} k_m}.$$

Using importance sampling, we define the initial and transition densities as follows: $p_\alpha = \frac{|h_\alpha|}{\sum_{\alpha=1}^n |h_\alpha|}$; $p_{\alpha\beta} = \frac{|a_{\alpha\beta}|}{\sum_{\beta=1}^n |a_{\alpha\beta}|}$, $\alpha = \overline{1,n}$.

In this case we get the following estimation for the variance:

$$Var[\theta^i] = E[(h_{k_0} W_m f_{k_m})^2] - (E[h_{k_0} W_m f_{k_m}])^2 \leq (E[(h_{k_0} W_m f_{k_m})^2]$$

$$\leq \sum_{i=1}^n |a_{k_0 i}| \cdot \sum_{i=1}^n |a_{k_1 i}| \ldots \sum_{i=1}^n |a_{k_{m-1} i}|,$$

for $f$ and $h$ are normalized.

The special case is when the elements of $A$ are positive and the rows' sums of $A$ are constant, i.e. $\sum_{j=1} a_{ij} = a$, $i = \overline{1,n}$ and if all the elements of the vector $f$ are constant, then $Var[\theta^i] = 0$ [6,9].

## 3 Monte Carlo Method for Estimating Eigenvalues Using Error Balancing

We want to estimate the value of the form $(h, A^m f)$, with a given probability $Pr < 1$, so that the error is smaller than a given positive $\varepsilon$

$$\left| (h, A^m f) - \frac{1}{N} \sum_{s=1}^N (\theta^m)_s \right| \leq \varepsilon \qquad (17)$$

The stochastic error $(R_{m,N})$ depends on two factors - the sample size $N$ and the variance of the random variable (r.v.). Once we have fixed the r.v. whose mathematical expectations coincides with the desired value, the error depends on the sample size. The systematic error $(R_{m,s})$ comes from the approximate calculation of the scalar product for a given $m$. Next, in the case of Monte Carlo power iterations, we have to find the smallest value of $m$ in order to decrease the computational complexity.

The error balancing will give us: $R_{m,N} = R_{m,s} = \frac{\varepsilon}{2}$.

When a mapping procedure is applied one may assume that a constant $\alpha$ exist, $0 < \alpha < 1$, such that $\alpha \geq |g_i^{(m)}| \times ||A||$ for any $i$ and $m$.

$$\frac{\varepsilon}{2} \leq \frac{(|g_i^{(m)}|.||A||)^{m+1}||f||}{1 - |g_i^{(m)}|.||A||} \leq \frac{\alpha^{m+1}||f||}{1-\alpha} \qquad (18)$$

and for $m$ should be chosen a smaller natural number, so that

$$\frac{|log\delta|}{|log\alpha|} - 1 \leq m, \quad \delta = \frac{\varepsilon(1-\alpha)}{2||f||}. \qquad (19)$$

The problem of the error estimation is complicated because from one side the matrix-vector scalar product is computed approximately and from the other - the value of the eigenvalue is estimated by division of two approximately computed quantities. The problem is even more complex when we use the resolvent matrix. One has to mention that the computational complexity of the algorithm for computing the largest eigenvalue also depends on the matrix condition number.

# 4  Method Implementation and Numerical Results

The numerical results are based on the MC method that estimates the largest eigenvalue by the Power method applied to a formula (10) using important sampling to define $p_\alpha$ and $p_{\alpha\beta}$

$$\lambda_{\max} = \lim_{m \to \infty} \frac{E\{sign(a_{\alpha_{m-1}\alpha_m})||a_{\alpha_{m-1}}||f_{\alpha_m}\}}{\{Ef_{\alpha_{m-1}}\}}. \tag{20}$$

The computational formula:

$$\lambda_{\max} \approx \frac{1}{\sum_{i=1}^{N} f_{\alpha_{m-1}}^{(i)}} \sum_{i=1}^{N} sign(a_{\alpha_{m-1}\alpha_m}^{(i)})||a_{\alpha_{m-1}}^{(i)}||f_{\alpha_m}^{(i)}, \tag{21}$$

where the upper subscript $(i)$ denotes the $i^{\text{th}}$ realisation of the Markov chain, so that $f_{\alpha_{m-1}}^{(i)}$ is the value of the corresponding element of vector $f$ after the $(m-1)^{\text{st}}$ jump in the $i^{\text{th}}$ Markov chain and $||a_{\alpha_{m-1}}^{(i)}||$ is the corresponding vector norm of the row which element is the last visited in the $i^{\text{th}}$ Markov chain after the $m^{\text{th}}$ jump and $N$ is the total number of Markov chain performed.

Various numerical tests have been performed in order to test the applicability of the proposed algorithm. We have used randomly generated matrices with different size. The Fig. 1 illustrates the MC theory that the algorithm convergence depends mostly on the number of chains. After 10000 walks (Markov chains) the estimations of dominant eigenvalue using different values of $m$ (number of jumps in the Markov chain) are very similar (Fig. 1 and Fig. 2). The proposed variant of the algorithm with error balancing permits us to use the smallest value of $m$ (which in this case is equal to 4) and thus to decrease the complexity. We have performed various tests. Table 1 shows the results of the proposed algorithm for finding the smallest $m$ for three randomly generated matrices with different size.

**Table 1.** MC power iterations: Estimated length of walks using error balancing for randomly generated sparse symmetric matrices with size 128, 1024, 2000.

| Matrix size | $n = 128$ | $n = 1024$ | $n = 2000$ |
|---|---|---|---|
| Length of Markov chain | 5 | 6 | 8 |
| Relative error for estimating $\lambda_{max}$ | 0.0424 | 0.0472 | 0.0779 |

---

**Algorithm 1:** Pseudo code for computing the Power MC method

---

1: INPUT: $A \in R^{n \times n}, h, f \in R^n, N$ {Markov chain realisation}, $n, g^{(m)}, \varepsilon$
2: COMPUTE: $p_\alpha = \frac{|h_\alpha|}{\sum_{\alpha=1}^n |h_\alpha|}; p_{\alpha\beta} = \frac{|a_{\alpha\beta}|}{\sum_{\beta=1}^n |a_{\alpha\beta}|}$
3: COMPUTE: $B \leftarrow \frac{A}{g^{(m)}}$
4: COMPUTE: $b_k \leftarrow \sum_{j=1}^n |b_{kj}|, \quad 1 \le k \le n$
5: COMPUTE: $a \leftarrow \max_{1 \le k \le n} \{b_k\}$
6: COMPUTE: $m \leftarrow$ using formula (19)
7: Input: $WN1 = 0, WN2 = 0$
8: **for** i=1, N **do**
9:     Call rand($\gamma$)
10:     Compute: $k_0 = max_{1 \le k \le n}\{k| \gamma > \sum_{j=1}^k p_j\}$
11:     $W = 1, W1 = 1, W2 = 1,$ jump=1
12:     **while** jump $\neq m$ **do**
13:         $k = k_0, l = 1, Sum = p_{k1}$
14:         Call rand($\gamma$)
15:         **while** $\gamma > Sum$ **do**
16:             $l = l + 1$
17:             $Sum = Sum + p_{kl}$
18:         **end while**
19:         $k_0 = l$
20:         jump = jump + 1
21:     **end while**
22:     $W2 = f(k_0)$
23:     $k_1 = k_0, l = 1, Sum = p_{k_1 1}$
24:     Call rand($\gamma$)
25:     **while** $\gamma > Sum$ **do**
26:         $l = l + 1$
27:         $Sum = Sum + p_{k_1 l}$
28:     **end while**
29:     $k_0 = l$
30:     $W \leftarrow sign\{b_{k_1 k_0}\} \times b_{k_1}$
31:     $W1 \leftarrow W \times f(k_0)$
32:     $WN1 \leftarrow WN1 + W1$
33:     $WN2 \leftarrow WN2 + W2$
34: **end for**
35: OUTPUT: $\lambda_{\max} = \frac{WN1}{WN2} \times g^{(m)}$

---

Algorithm 1 describes the power MC method for estimating the largest eigenvalue of non-singular symmetric matrix $A$ using formula (21). Here we denote with $W1$ one realization of the r.v. in the numerator of a formula (20) with $(m + 1)$ jumps in the Markov chain and with $W2$ one realization of the r.v. in the denominator of (20) with $m$ jumps, respectively. Quantities $WN1$ and $WN2$ accumulate all realizations of $W1$ and $W2$, respectively.

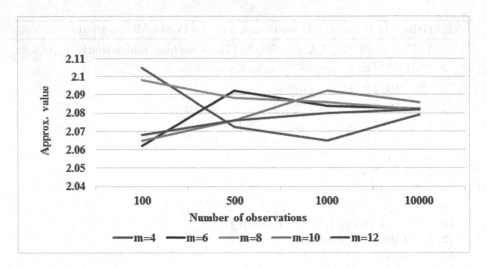

**Fig. 1.** Estimating the approximate extremal eigenvalue with different breaksteps of the Markov chain

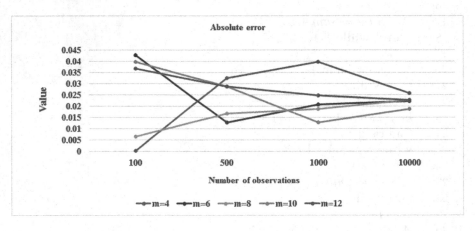

**Fig. 2.** The absolute error depending on the number of observations and the breaksteps of the Markov chain

## 5    Conclusions

In this paper we presented and studied a refined algorithm for implementing Monte Carlo power iterations in order to find approximately the extreme eigenvalues of a given matrix. Modifying the original algorithm using the procedure for balancing the statistical and systematic errors improves the algorithm complexity and saves computational time. This is very important for applications where the matrix size is growing in real time and the estimation of the extreme eigenvalues has to be performed many times. Our future work is to apply the proposed algorithm for geophysical applications.

**Acknowledgements.** This work has been accomplished with the partial support by the Grant No. BG05M2OP001-1.001-0003, financed by the Science and Education for Smart Growth Operational Program (2014-2020) and it has been partially supported by the National Geoinformation Center (part of National Roadmap of RIs) under grants No. DO1-404/18.12.2020 and DO1-282/17.12.2019.

# References

1. Lobo, P.D.C., Mikhailov, M., Ozisik, M.N.: On the complex eigenvalues of Luikov system of equations. Drying Technol. **5**(2), 273–286 (1987). https://doi.org/10.1080/07373938708916540

2. Alexandrov, V., Davila, D., Esquivel-Flores, O., Karaivanova, A., Gurov, T., Atanassov, E.: On Monte Carlo and Quasi-Monte Carlo for matrix computations. In: Lirkov, I., Margenov, S. (eds.) LSSC 2017. LNCS, vol. 10665, pp. 249–257. Springer, Cham (2018). https://doi.org/10.1007/978-3-319-73441-5_26

3. Alexandrov, V., Esquivel-Flores, O., Ivanovska, S., Karaivanova, A.: On the pre-conditioned Quasi-Monte Carlo algorithm for matrix computations. In: Lirkov, I., Margenov, S.D., Waśniewski, J. (eds.) LSSC 2015. LNCS, vol. 9374, pp. 163–171. Springer, Cham (2015). https://doi.org/10.1007/978-3-319-26520-9_17

4. Dimov, I.T., Karaivanova, A.N.: Iterative Monte Carlo algorithms for linear algebra problems. In: Vulkov, L., Waśniewski, J., Yalamov, P. (eds.) WNAA 1996. LNCS, vol. 1196, pp. 150–160. Springer, Heidelberg (1997). https://doi.org/10.1007/3-540-62598-4_89

5. Dimov, I., Karaivanova, A.: Parallel computations of eigenvalues based on a Monte Carlo approach. Monte Carlo Methods Appl. **4**(1), 33–52 (1998)

6. Dimov, I.T., Philippe, B., Karaivanova, A., Weihrauch, C.: Robustness and applicability of Markov chain Monte Carlo algorithms for eigenvalue problems. Appl. Math. Model. **32**(8), 1511–1529 (2008). https://doi.org/10.1016/j.apm.2007.04.012. ISSN 0307-904X

7. Golub, G.H., Van Loon, C.F.: Matrix Computations. The Johns Hopkins University Press, Baltimore (1996)

8. Mascagni, M., Karaivanova, A.: Matrix computations using quasirandom sequences. In: Vulkov, L., Yalamov, P., Waśniewski, J. (eds.) NAA 2000. LNCS, vol. 1988, pp. 552–559. Springer, Heidelberg (2001). https://doi.org/10.1007/3-540-45262-1_65

9. Mascagni, M., Karaivanova, A.: A parallel Quasi-Monte Carlo method for solving systems of linear equations. In: Sloot, P.M.A., Hoekstra, A.G., Tan, C.J.K., Dongarra, J.J. (eds.) ICCS 2002. LNCS, vol. 2330, pp. 598–608. Springer, Heidelberg (2002). https://doi.org/10.1007/3-540-46080-2_62

10. Sobol, I.M.: Monte Carlo Numerical Methods. Nauka, Moscow (1973).(in Russian)

11. Isaacson, E., Keller, H.B.: Analysis of Numerical Methods. Dover Publications, Mineola; Wiley, New York (1996). ISBN 0-486-68029-0

# Multi-lingual Emotion Classification Using Convolutional Neural Networks

Alexander Iliev[1,2]($\boxtimes$), Ameya Mote[2], and Arjun Manoharan[2]

[1] Institute of Mathematics and Informatics, Bulgarian Academy of Sciences,
Sofia, Bulgaria
ailiev@berkeley.edu
[2] SRH University Berlin, Charlottenburg, Berlin, Germany
{3104966,3105614}@stud.srh-campus-berlin.de

**Abstract.** Emotions play a central role in human interaction. Interestingly, different cultures have different ways to express the same emotion, and this motivates the need to study the way emotions are expressed in different parts of the world to understand these differences better. This paper aims to compare 4 emotions namely, anger, happiness, sadness, and neutral as expressed by speakers from 4 different languages (Canadian French, Italian, North American English and German - Berlin Deutsche) using modern digital signal processing methods and convolutional neural networks.

**Keywords:** Digital culture · Multimedia systems · Information retrieval · Audio analysis · Speech processing · Multi-culture comparison · Emotion recognition · Convolutional neural networks

## 1 Introduction

In a new social environment, emotions are one-way people use as hint to correctly behave in front of other observers. Negative emotions like anger or sadness by the viewers tells whether the person performing the task in doing it correctly or not. Positive emotions signal the opposite of the former [1]. William James, in 1890, classified emotions into 4 basic types. They are namely: Fear, Grief, Love, and Rage. Paul Ekman, later classified emotions into 6 basic types and this classification is recognized worldwide in the following set of emotions: happiness, anger, sadness, disgust, fear, and surprise.

Although there is wide consensus on the identity of these basic emotions, the way they are expressed in different parts of the world is unique to the people inhabiting those regions. In a globally connected world with constant interaction between individuals of vastly different cultural backgrounds, it becomes important to study the differences in the way emotions are expressed and document them such that, we as a collective human species can understand and work

Bulgarian Ministry of Education and Science.

I. Lirkov and S. Margenov (Eds.): LSSC 2021, LNCS 13127, pp. 456–463, 2022.
https://doi.org/10.1007/978-3-030-97549-4_52

through these variabilities. Proper understanding of emotional ques forms the bedrock of a healthy and successful human interaction and thus warrants the need to better understand subtle differences in the variabilities of expression of these emotions. Through this paper, we seek to understand the various nuances of emotional expression using speech data as expressed by native speakers of different languages. This paper is a follow up paper to the one published by the authors at DIPP: 2020 in the relevant area of research. In this work, we have added more datasets and hence hope to make this work more expansive, while improving on the techniques used in our original paper.

## 2    Dataset

Our dataset consists of speech data of various actors enacting different emotions in Canadian French [2], Italian [3], North American English [4], Berlin Deutsche [5] respectively. The datasets contain audio samples of both male and female speakers. The ages of the people were random, and no particular emphasis was given to this metric. Due to unavailability of a single source, the datasets were procured from different sources, but effort was taken to standardize it in terms of recording quality and audio channel (Table 1).

A large dataset is required to achieve good results with convolutional neural networks. As our datasets where not large enough, we used data augmentation techniques to enlarge it. Data augmentation refers to the technique of enlarging datasets by adding slight modifications to the original data and is a widely practised trick in the data science community. Since our data consists of audio files, we used the following techniques to enlarge our data [6]:

Pitch Tuning: The pitch of the audio was shifted to create a different sample but with the same audio. This augmented dataset had different pitch characteristics but expressed the same emotions as of the original audio samples.

Addition of White Noise: A random white noise was introduced into the audio sample. On a superficial level, humans ears distinguish these sounds as a background buzz.

**Table 1.** Database description.

| Dataset | Male | Female | Emotions | Audio quality | Total sentences |
|---|---|---|---|---|---|
| Canadian French | 6 | 6 | Anger, Disgust, Fear, Happiness, Sad, Neutral | 48 KHz, Mono | 443 |
| EMOVO Italian | 3 | 3 | Anger, Disgust, Fear, Happiness, Sad, Neutral | 48 KHz, Mono | 510 |
| North American | 12 | 12 | Anger, Disgust, Fear, Happiness, Sad, Neutral, Calm | 48 KHz, Mono | 1440 |
| Berlin Deutsche | 5 | 5 | Anger, Disgust, Fear, Happiness, Sad, Neutral | 16 KHz, Mono | 535 |

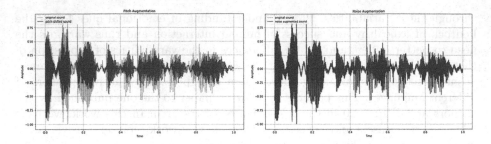

**Fig. 1.** Original vs pitch and noise shifted sample

**Table 2.** Data augmentation.

| Dataset | Number of samples before augmentation | Number of samples after augmentation |
|---|---|---|
| Canadian French | 299 | 897 |
| EMOVO Italian | 286 | 858 |
| North American | 299 | 897 |
| Berlin Deutsche | 301 | 603 |

From Fig. 1, we see that although the two signals are visibly different, but they still have the same overall characteristics. The difference introduced by the data augmentation techniques is enough to make the augmented data have different features in comparison to the their originals different without altering the characteristics we are interested it and thus we effectively enlarge our dataset (Table 2).

After enlarging the dataset, the samples needed to be transformed in such relevant information could be captured. For this purpose, we decided to use the Mel Frequency Cepstrum associated with each audio file as the feature vector representing that audio clip (Figs. 2 and 3).

Detailing it further, the raw file in 3a, German male happy is converted into the frequency scale using the Fast Fourier Transform. The image 3b shows the audio sample in the frequency spectrum. The audio is then again converted into the Mel frequency scale as shown by the Spectrogram in image 4a. The conversion gives us the Mel Frequency Cepstral Coefficients (MFCC). These features are

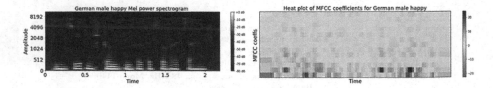

**Fig. 2.** Figure (2a-Berlin Deutsche male happy Mel power spectrogram), Figure (2b-Heat plot for MFCC coefficients for Berlin Deutsche male happy)

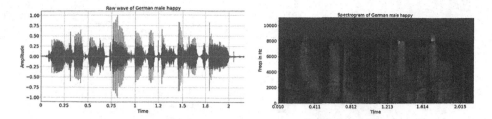

**Fig. 3.** Figure (3a-Raw wave of Berlin Deutsche male happy), Figure (3b-Spectogram of Berlin Deutsche male happy)

utilized for training the model. 4b gives us the variation of the audio utterances at different points of time, i.e., 2.5 s of 65 blocks.

## 3   Model

### 3.1   Mel Frequency Spectrum

For achieving a good classification technique, it always becomes utmost important to have similar data available. The datasets used herein have been gathered from various sources and after applying a data augmentation are now somewhat like each other. To achieve more similarity and to gather necessary features we use MFCC [7]. It is well documented how Human ears cannot perceive about 1000 Hz. For this purpose, Mel frequency can be used via 2 filters: below 1 kHz, they are spaced linearly, whereas above its logarithmic equations are used [8]. The equation for the frequency is given by,

Displayed equations are centered and set on a separate line.

$$F(m) = [1000/log_{10}(2)] * [log_{10}(1 + f/1000)], \tag{1}$$

F(m) stands for Mel-Frequency and f stands for Frequency in Hz.

$$MFCCm = \sum_{k=1}^{1} 3(log_{10}E_K)[m(k - 1/2)\pi/20], \tag{2}$$

Here m stands for the total number of Mel-Frequency cepstral coefficients and Ek stands the total number for the energy output of the Kth filter (Fig. 4).

### 3.2   Neural Network Description

To ensure that MFCC's capture the features associated with emotions, we trained a convolutional neural network and tested if our hypothesis, we needed a technique which could give us good results. For this, Convolutional Neural Network was used [9]. The 13 MFCC features chosen were deemed capable to compare the emotions across different cultures to each other. To achieve high accuracy, it is suggested to use many layers and that was incorporated.

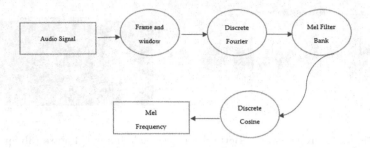

**Fig. 4.** MFCC

1. Conv1d: There are in total 8 convolutional layers in our network.
2. Activation: One activation layer was associated with each of the conv1d layers. Of the, thus 8 layers, 7 of them are ReLU activation layers and the activation layer associated with the final layer is 'Softmax' layer.
3. Batch Normalisation: As the training set is fed into the network as mini batches, a batch normalization layer was implemented to normalize the inputs of each mini batch.
4. Dropout: This technique is used to remove the problem of overfitting. A few features are randomly dropped so that the model does not completely learn everything. This helps the model in learning more important features and ignore other random features. Dropout layer was added after the 2nd and the 6th layer [10]
5. MaxPooling: Maxpooling is a regularization technique and also serves as a dimension reduction technique. It combines all the important features together and passes on the most relevant patterns and helps the model can learn faster.
6. Flatten: Since the model requires a single vector, flatten unravels the input matrix into one long feature vector.
7. Dense: The final layer of the network was a fully connected layer with a sigmoid activation layer [11].

### 3.3   Train and Test Samples

The datasets for the three languages French, Italian and North American English were split into training and test sets as discussed in [12] for the four emotions: happy, angry, sad, neutral. An extended including the additional Berlin Deutsche dataset is shown above (Table 3).

As compared to the previous paper [12], adding another language helps in understanding how robust the CNN model is. Earlier, after comparing all 3 languages, on an average, it was found out that 71% of all emotions are similar to each other. While the results being good, It really did not offer much in terms of complexity. As a result, in the present work, it was decided to use another language set and compare it with other languages to understand if a

**Table 3.** Berlin deutsche train and test samples.

| Emotion | Male | Female | Train | Test |
|---------|------|--------|-------|------|
| Happy | 57 | 75 | 46(Male), 60(Female) | 11(Male), 15(Female) |
| Anger | 102 | 111 | 81(Male), 89(Female) | 21(Male), 22(Female) |
| Sad | 51 | 51 | 41(Male), 41(Female) | 10(Male), 10(Female) |
| Neutral | 72 | 84 | 57(Male), 67(Female) | 15(Male), 17(Female) |

higher order complexity could be achieved with the same CNN model. Thus, The Berlin Deutsche set was employed.

## 4  Results

Prior to using data augmentation, we had achieved an average accuracy of 53% which was deemed unacceptable. Low accuracy is usually related to underfitting and this motivated our decision to enlarge our dataset using data augmentation techniques. A network trained on the enlarged dataset achieved a frequency of 71.10% on the Canadian French Dataset, 79.07% on the Italian dataset, 73.89% on the North American Dataset and finally 70% on the Berlin Deutsche Dataset and therefore we concluded that a neural network trained on MFCC values of

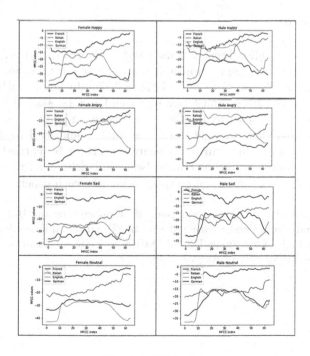

**Fig. 5.** MFCC plots

audio files was able to accurately predict. The MFCC parameters are thus an effective way to gauge the accuracy of the emotions when compared to each other.

The plots as shown in Fig. 5, depicts means of the individual MFCC coefficients obtained for all the cultures for each point. Start from the first graph, it can be deduced that all the cultures express the emotion 'Happy' in their own ways and no emotion is like each other across any gender. The same could be said about the emotion 'Angry' as no 2 emotions are like each other. However, there is a small hint of Berlin Deutsche and the North American datasets being like each regarding the 2 emotions mentioned. This could be seen via the MFCC plots wherein the 2 cultures intersect each other at 2 points. A welcome break is seen in the emotion 'Sad' wherein it can be said that for the gender 'Female', Italian and the North American emotions are in some ways similar as them.

MFCC values overlap each other at certain points. Berlin Deutsche dataset although remains totally different to the other 3 cultures combined. For the same emotion, the gender 'Male' expresses likeliness across each dataset as there is an overlap at many points. Here again, Berlin Deutsche expresses the emotion in a different way. The emotion 'Neutral' again shows the same trend as seen in 'Sadness' , but here for the gender 'Female' the 2 cultures showing similar traits are the Italian and the North American dataset. While for the gender 'Male', all cultures express the emotion 'Neutral' in the same way as to each other with the Berlin Deutsche following suit.

## 5   Discussion

Overall, For the Canadian French, Italian and the North American, emotions are expressed in a way that are recognizable across these cultures. So, an individual from a cultural background of any of these 3 should not have high difficulty in striking a conversation. Same however could not be said about the Berlin Deutsche Dataset. People from a German background would probably find it difficult to express their thoughts in a way that would lead to desired consequences. Having said that, further research needs to be carried across other cultures to see more significant comparisons and deduce as to whether there strikes a possibility of a group of cultures being like each other.

**Acknowledgements.** This work was partially supported by the Bulgarian Ministry of Education and Science under National Scientific Program "Information and Communication Technologies for a Single Digital Market in Science, Education and Security", approved by DCM No 577, 17 August 2018.

# References

1. Hareli, S., Kafestios, K., Hess, U.: A cross-cultural study on emotion expression and the learning of social norms. J. Front. Psychol. **6**, 1501 (2015)
2. Gournay, P., Lahaie, O., Lefebvre, R.: A Canadian French emotional speech dataset. In: 9th ACM Multimedia Systems Conference, Amsterdam, pp. 399–402 (2018)
3. Costantini, G., Iaderola, I., Paoloni, A., Todisco, M.: Emovo corpus: an Italian emotional speech database. In: International Conference on Language Resources and Evaluation, pp. 3501–3504. Association of Computing Machinery (2014)
4. Livingstone, S.R., Russo, F.: The Ryerson audio-visual database of emotional speech and song (RAVDESS): a dynamic, multimodal set of facial and vocal expressions in North American English. J. PLoS ONE **13**(5), e0196391 (2018)
5. Burkhardt, F., Paeschke, A., Rolfes, M., Sendlmeier, W.F., Weiss, B.: A database of German emotional speech. In: 9th European Conference on Speech Communication and Technology, pp. 1517–1520. Interspeech 2005 - Eurospeech (2005)
6. Rebai, I., BenAYED, Y., Mahdi, W., Lorre, J.-P.: Improving speech recognition using data augmentation and acoustic model fusion. J. Procedia Comput. Sci. **112**, 316–322 (2017)
7. Iliev, A.I., Stanchev, P.L.: Glottal attributes extracted from speech with application to emotion driven smart systems. In: Proceedings of the 10th International Joint Conference on Knowledge Discovery, Knowledge Engineering and Knowledge Management, Seville, Spain, pp. 297–302 (2018)
8. Iliev, A.I.: Emotion Recognition From Speech. Lambert Academic Publishing (2012)
9. Albawi, S., Mohammed, T.A., Al-Zawi, S.: Understanding of a convolutional neural network. In: 2017 International Conference on Engineering and Technology (ICET), Antalya, pp. 1–6 (2017)
10. Srivastava, N., Hinton, G., Krizhevsky, A., Sutskever, I., Salakhutdinov, R.: Dropout: a simple way to prevent neural networks from overfitting. J. Mach. Learn. Res. **15**(1), 1929–1958 (2014)
11. Hue, A.: Dense or Convolutional Neural Network. Medium
12. Iliev, A., Mote, A., Manoharan, A.: Cross-cultural emotion recognition and comparison using convolutional neural networks. In: Digital Presentation and Preservation of Cultural and Scientific Heritage Conference Proceedings, vol. 10. Institute of Mathematics and Informatics, Sofia (2020)

# On Parallel MLMC for Stationary Single Phase Flow Problem

Oleg Iliev[1,3]($\boxtimes$), N. Shegunov[1], P. Armyanov[2], A. Semerdzhiev[2], and I. Christov[2]

[1] Fraunhofer ITWM, Kaisersautern, Germany
{iliev,shegunov}@itwm.fraunhofer.de
[2] Sofia University, Sofia, Bulgaria
[3] Institute of Mathematics and Informatics, Bulgarian Academy of Sciences, Sofia, Bulgaria

**Abstract.** Many problems that incorporate uncertainty often requires solving a Stochastic Partial Differential Equation. Fast and efficient methods for solving such equations are of particular interest for computational fluid dynamics. Efficient methods for uncertainty quantification in porous media flow simulations are Multilevel Monte Carlo sampling based algorithms. They rely on sample drawing from a probability space. The error is quantified by the root mean square error. Although computationally they are significantly faster than the classical Monte Carlo, parallel implementation is necessity for realistic simulations. The problem of finding optimal processor distribution is considered NP-complete. In this paper, a stationary single-phase flow through a random porous medium is studied as a model problem. Although simple, it is well-established problem in the field, that shows well the computational challenges involving MLMC simulation. For this problem different dynamic scheduling strategies exploiting three-layer parallelism are examined. The considered schedulers consolidate the sample to sample time differences. In this way, more efficient use of computational resources is achieved.

**Keywords:** UQ · MLMC · Parallel

## 1 Mathematical Model

Due to the limited length of this presentation, we refer to the overview [1,4] and to our previous papers [6,9] for motivation and literature review, and skip the discussion on them here.

Consider steady state single phase flow in random porous media in a unit cube, with domain $D = (0,1)^d, d = 2,3$, and prescribed pressure drop from left boundary to the right boundary.

$$-\nabla \cdot [k(x,\omega)\nabla p(x,\omega)] = 0 \text{ for } x \in D = (0,1)^d, \omega \in \Omega.$$
$$p_{x=0} = 1 \quad p_{x=1} = 0 \quad \partial_n p = 0 \text{ on other boundaries.}$$

(1)

© Springer Nature Switzerland AG 2022
I. Lirkov and S. Margenov (Eds.): LSSC 2021, LNCS 13127, pp. 464–471, 2022.
https://doi.org/10.1007/978-3-030-97549-4_53

Both the coefficient $k(x, \omega)$ and the solution $p(x, \omega)$ are subject to uncertainty, characterized by the random vector $\omega$ from a properly defined random space $(\Omega, F, P)$. The coefficient $k(x, \omega)$ describes the permeability field within the domain and the solution $p(x, \omega)$ describes the steady pressure distribution under pressure drop. An object of interest for this model is the mean quantity of the total flux through the unit cube:

$$Q(x, \omega) := \int_{x=1} k(x, \omega) \partial_n p(x, \omega) dx. \qquad (2)$$

Uncertainty quantification (UQ) for Eq. (1) is extremely challenging task. Common way to overcome this limitation is to use simple designs for $k(x, \omega)$, that expresses as well as possible the data. One model, that has been studied extensively is the log-normal distribution for $k(x, \omega)$: $C(x, y) = \sigma^2 exp(-||x - y||_2/\lambda)$. Here $\sigma$ denotes the standard deviation and $\lambda$ is the correlation length. To quantify the uncertainty in the model a Multilevel Monte Carlo (MLMC) method is used. Let $\{M_l : l = 0 \dots L\} \in N$ be increasing sequence of numbers called levels, with corresponding quantities $\{Q_{M_l}\}_{l=0}^{L}$, for $l = 0 \dots L$. Defining $Y_l = Q_{M_l} - Q_{M_{l-1}}$ and setting $Y_0 = Q_{M_0}$, the following expansion for $E[Q_M]$ can be formulated $E[Q_M] = E[Q_{M_0}] + \sum_{l=1}^{L} E[Q_{M_l} - Q_{M_{l-1}}] = \sum_{l=0}^{L} E[Y_l]$. The terms in the above expression are approximated using standard MC independent estimators, with $N_l$ samples: $\widehat{Y_l} = N_l^{-1} \sum_i^{N_l} (Q_{M_l}^{(i)} - Q_{M_{l-1}}^{(i)})$. Then the multilevel Monte Carlo estimator is defined as: $\widehat{Q}_{M,N}^{ML} = \sum_{l=1}^{L} \widehat{Y_l}$. Analogously to MC method, the root mean square error is defined as: $e(\widehat{Q}_{M,N}^{ML})^2 = \sum_{l=0}^{L} N_l^{-1} V[Y_l] + (E[Q_M - E[Q]])^2$.

The definition of levels is critical for the performance of the algorithm. It is problem dependant and it is subject of intensive research. Different techniques have been considered in the past. In [3], the levels are defined as mixed MsFEM. In [6] the levels are defined as a spatial resolution of the discretization grid. In this work the multilevel Monte Carlo construction in [6] and [9] is followed and the levels are defined as a spatial resolution of the distribution grid. This means that for a given estimator $Y_l$ the equation is solved on a $2^z \times 2^z$ grid for the fine level and $2^{z-1} \times 2^{z-1}$ for the coarser grid. Estimated number of samples that needs to be performed by the estimator, is computed by minimizing the total computational time. This gives relation between the empirical variance and the number of samples that needs to be performed on a given level:

$$N_l = \lceil \lambda \sqrt{(v_l/t_l)} \rceil \text{ where } \lambda = \frac{1}{\epsilon^2} \sum_{l=0}^{L} \sqrt{(v_l/t_l)} \qquad (3)$$

The number of samples that needs to be drawn per a level in order to achieve desired accuracy is not known. In practice a few samples are drawn first, to obtain an initial estimate for the number of needed samples. At a next stage this estimate can be refined.

To obtain a representation of the permeability field on the coarser grids (levels) a simplified renormalization is used. This technique achieves good variance approximation on the coarser levels and leads to efficient MLMC algorithms [5–7,9]. Cell centered finite volume method is used for discretization of each of the deterministic PDEs.

## 2   Parallel Model

MLMC approach by design defines three distinct parallel layers [2]: (i) Parallelizing the solution for each sample (deterministic PDE); (ii) Parallelizing the solution of all samples on a given MLMC level; (iii) Parallelizing for all or several MLMC level estimators simultaneously. The most efficient and flexible strategies will be those, taking advantage of all of the three layers of parallelism.

## 3   Performance Parameters

The design of the scheduling strategy requires different parameters to be taken into account. The most prominent is the number of the different estimators, the samples per estimator for a current *estimate-solve* cycle, and the time to compute one sample for a given estimator. In the case of a sample solved by multiple processes, the parallel solver inefficiencies contribute to the overall time lost for communication and synchronization. The efficiency and the predictability of the underlying solver is crucial for the performance of the scheduler. Modern multigrid solvers scales very well under reasonable assumptions [2]. Assuming there is no parallel computation overhead and no inefficiency lost due to load imbalances, the theoretical minimum computational time is given by:

$$T^{min} = \frac{1}{P^{all}} \sum_{l=0}^{L} N_l E[t_l] \tag{4}$$

In the case of a single process per sample per level, the minimum time can be computed directly by recording the time to solve a sample. In the case of more than one processor per problem, the minimal time can only be estimated. Assume $\theta$ is a measure of how effective the underlying parallel sample solver algorithm is. Then the time to compute a sample in parallel can be expressed as: $C_l = \theta_l C_{min}^l$. Here $C_{min}^l$ is the time to compute a single sample on level $l$ on $P_{min}^l$ processors. Rewriting Eq. (4) by substituting $t_l$ with $C_l$, the equation becomes:

$$T^{min} = \frac{1}{P^{all}} \sum_{l=0}^{L} N_l E[C_l] = \sum_{l=0}^{L} N_l E[\theta_l C_{min}^l] = \sum_{l=0}^{L} N_l \theta_l E[C_{min}^l] \tag{5}$$

In the case when $\theta_l = 1$ for all levels the two formulations are identical. In the case when more than one process is assigned to a sample on level $l$, to compute the minimal time $\theta_l$ needs to be determined. This can be done in a pre-processing phase, by computing the $\theta_l$ function for a given scalability window $\{P^l_{min}, \ldots, P^l_{max}\}$, where $P_{max}$ is the maximum number of processors that achieve the desired threshold efficiency. And $P^l_{min}$ is the minimal number of processes that are able to solve the problem by fully utilizing the memory capacity. By setting $C^p_l$ to be the time to compute one sample on level $l$ by $p$, $\theta_l$ becomes: $\theta_l = C^p_l / C^{min}_l$. To define the parallel efficiency of a given level of $MLMC$, the relative inefficiency is computed by substituting the computational time $C^{comp}_l$ with the minimal level time $T^{min}_l$ and the resulting fraction is subtracted from 1:

$$Eff_l = 1 - (C^{comp}_l - T^{min}_l)/T^{min}_l \tag{6}$$

Expressing (6) in terms of $\theta$ and $C^p_l$ becomes:

$$Eff_l(\theta, p) = 1 - \frac{(C^{comp}_l - N_l\theta_l E[C^l_{min}])}{(N_l\theta_l E[C^l_{min}])} \tag{7}$$

Finally the MLMC efficiency is defined as a sum over the levels:

$$Eff(\theta) = 1 - \frac{(\sum_l C^{comp}_l - \sum_l N_l\theta_l E[C^l_{min}])}{\sum_l(N_l\theta_l E[C^l_{min}])} \tag{8}$$

where $l \in \{0, \ldots L\}$.

## 3.1   Dynamic Scheduler

This scheme uses greedy approach and adopts during the simulation. It is very flexible as it can be combined with different strategies for pure MC estimation Please refer to [8] for more details. The design uses all of the parallel layers to schedule a sample computation. A prerequisite for the strategy is the availability of solution times and variance statistics across the levels. This information can be obtained in two ways: as a precomputed step or as a compute step by computing first cycle with non optimal distribution of processors. To simplify the algorithm and without loss of generality consider three-level MLMC construction. Let $N_i, i = \{0, 1, 2\}$, be the number of required realizations per Monte Carlo estimator - $\widehat{Y_l}$, where $N_0$ is the number on the coarsest estimator $\widehat{Y_0}$. Let $p_i$ be the number of processes allocated per $\widehat{Y_i}$, $p^g_{l_i}$ the respective group size of processes working on a single realization, with $t_i$ be respective time constants, for solving a single problem once on a single process and finally $P^{total}$ be the total number of available processes. Then the total CPU compute time for the current *Estimate-Solve* cycle, can be estimated as:

$$T^{total}_{CPU} = N_0 t_0 + N_1 t_1 + N_2 t_2 \tag{9}$$

Then the optimal compute time per processor is:

$$T^p_{CPU} = \frac{T^{total}_{CPU}}{P^{total}} \tag{10}$$

By dividing the CPU time needed for a $\widehat{Y}_i$ with $T^p_{CPU}$, a continuous value for the number of processes for a given MC estimator is obtained.

$$p_i^{ideal} := \frac{N_i t_i}{T^p_{CPU}} \text{ for } i = \{0, 1, 2\} \tag{11}$$

Lets further assume that all of the available processors must be distributed on all of the estimators $Y_i$. Then by rounding down to integer, the $p_i^{ideal}$ value for processors distribution across the levels can be obtained

$$p_i := \left\lfloor p_i^{ideal} \right\rfloor, \text{ for } i = \{0, 1, 2\} \tag{12}$$

Depending on the scheme that will be used for parallel computing on the estimator $Y_i$, additional restrictions may be imposed for $p_i$. Till now the only considered case is the distribution of all of the available processors to work simultaneously on all of the estimators. This may not be the optimal strategy. It is rarely the case because of the strong imbalance of work between the estimators. To find a reasonable strategy of all possible combinations of unions of elements from the Estimator power set is considered. This ensures that cases when all of the processors are allocated on the coarsest level $\{Y_0\}$ and then levels $\{Y_1, Y_2\}$ are concurrently computed then just trying to balance the processors on all the estimators together $\{Y_0, Y_1, Y_2\}$. For the parallel strategy on the estimator, depending on the scenario, allows different schemes to be used. To incorporate the sample to sample compute time fluctuations, two approaches are considered:

**Interrupted Dynamic Strategy.** This approach is very similar to the processor interruptions. The algorithm starts as a standard dynamic strategy. A heuristic assumption is made that either there is no sample to sample computational differences, or if existing, they will balanced out. During the parallel computation, due to load imbalances and sample to sample fluctuations, a group of MPI process among all of the MPI processes completes computation, before the others. Upon completion this group sends a signal to a part of the other groups, or all of the groups, that are still computing, informing them that it is in an idle state. Upon receiving the signal, the computing groups break the computation (interrupt the current computation). When all those message exchanges completes and all of the groups are in an idle state and rescheduling is done. For this type of optimization two strategies are considered. The first type is the *local interruption*, done within an Estimator and the second type of interruption is the *global* one - computation on all MPI processes on all the levels stops and a rescheduling is performed.

In practice, the signals are modeled as a message exchange between the groups by a master-slave approach. A group sends a message to the designated processor that it is in an idle state. The master process takes responsibility to inform the other groups and synchronize the data between them. The messages carry a small amount of meta-information and the dominating part of the time needed for the process to complete will be the latency of the system.

**Job Queue Dynamic Strategy.** This procedure simulates the idea of job dispatching, or task-based parallelism in the multi-thread environment, adopted to MPI message system. First an optimal distribution of processors per Estimator is obtained by Eq. (11). For each Estimator, using the master-slave programming paradigm, one of the available MPI processes set to be master, and the others are set to be slaves. The master process acts as a dispatcher that assigns work to the slaves. Each of the slaves performs the given tasks and reports back to the master for more work. In this way each process is busy working regardless of the time for computing a sample. This is at the expense of a large number of small message exchanges.

## 4    Numerical Experiments

Tests were performed on the *Beehive cluster* in *Fraunhofer ITWM*. Each node consists of 48 cores, with 185 *GB* memory. The CPU model is *Intel(R) Xeon(R) Gold 6132 CPU @ 2.60 GHz.*

On Fig. 1, time to solution histograms are presented. The stochastic parameters considered are with increasing variance from left to right. Each test contains 1000 samples. The more uncertainty in the system, leads to more difficult overall computation. The minimum and maximum time to compute a sample slowly increase, but overall distance stays relatively unchanged- at around 2 s. In all three cases the data is very dispersed. The maximum time deviation value is at around 25% of the mean time.

On Fig. 2, an investigation of the efficiency is done for the different scheduler types. The parallel efficiency is measured under Eq. (8). For this tests, the finest grid of the MLMC algorithm is $2^{10} \times 2^{10}$ number of cells. This is approximately $10^6$ unknowns. The number of levels of MLMC is set to 4. On Fig. 2a) parallel efficiency for problem with parameters $\sigma = 2$, $\lambda = 0.3$, $\epsilon = 1e - 3$ is considered. On Fig. 2b) a more computationally intensive problem is plotted, with generating parameters $\sigma = 2.25$, $\lambda = 0.4$, $\epsilon = 1e - 3$. For both experiments each sample is computed by a single processor. Figure 2c) considers the same problem as

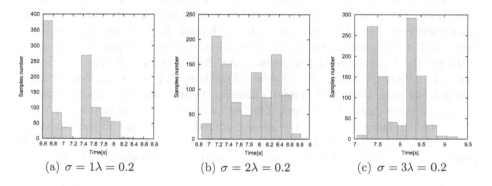

(a) $\sigma = 1 \lambda = 0.2$        (b) $\sigma = 2 \lambda = 0.2$        (c) $\sigma = 3 \lambda = 0.2$

**Fig. 1.** Time histogram of solution times.

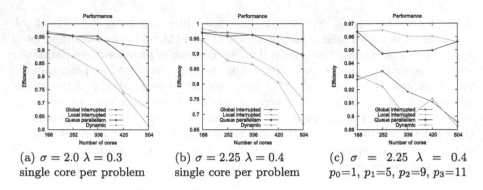

(a) $\sigma = 2.0$ $\lambda = 0.3$
single core per problem

(b) $\sigma = 2.25$ $\lambda = 0.4$
single core per problem

(c) $\sigma = 2.25$ $\lambda = 0.4$
$p_0=1$, $p_1=5$, $p_2=9$, $p_3=11$

**Fig. 2.** Efficiency of different schedulers for Laplace equation

Fig. 2b), but with different processor distribution. Each problem on a given level is solved by: $p_0 = 1, p_1 = 5, p_2 = 9, p_3 = 11$ processes respectively. For this processor distribution the efficiency function is $Eff(1, 3.6, 7.35, 9.00)$.

Figure 2a) and 2b) shows that local interrupted approach achieves best efficiency for a relatively small number of processors. These efficiency values, nonetheless, are not kept, when the number of processors increases. In the case of 504 cores, the effectiveness drops from 7 to 8% better than the pure dynamic approach. There are two main reasons for that: the increased number of processors leads to larger load imbalances between the different levels when the number of processors per level is estimated. Even if the algorithm finds a good processor distribution that leads to small waiting time per processor level groups, this time has much more impact on the effectiveness, compared to the same waiting time on a smaller number of processors. Simply, there are more idling processors. The other reason is a large number of messages exchanged at a single point in time. These messages must be sent through the network, that can lead to message flooding the communication canal. This problem becomes apparent when the queue method is compared to the local one. Although there are much more messages that are exchanged between the processors, all of them have small byte sizes and are spread in time. This leads to better overall efficiency for the queue scheduler. The best performing algorithm that is considered for the case of a single processor per problem is the global technique. It shows small performance degradation when the number of processors is increased. Figure 2c) shows that utilizing the third layer of parallelism and thus counting all layers of parallel execution. This leads to very efficient algorithms. All of the considered methods achieve more than 0.89 units of efficiency. In this case, the local interrupted technique outperforms the others. The increased effectiveness comes again from the fact that groups of processors are considered, rather than all the processors, by the dynamic scheduler. Part of the imbalances is offloaded to the underlying parallel algorithm used for the generalization of a single sample (Fig. 3).

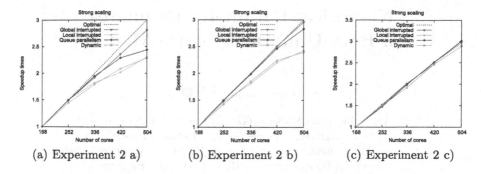

(a) Experiment 2 a)  (b) Experiment 2 b)  (c) Experiment 2 c)

**Fig. 3.** Strong scalability for experiments on Fig. 2

# References

1. Cliffe, K., Giles, M., Scheichl, R., Teckentrup, A.: Multilevel Monte Carlo methods and applications to elliptic PDEs with random coefficients. Comput. Vis. Sci. **14**, 3–15 (2010)
2. Drziga, D., Gmeiner, B., Rude, U., Scheichl, R., Wohlmuth, B.: Scheduling massively parallel multigrid for multilevel Monte Carlo methods. SIAM J. Sci. Comput. **39**(5), S873–S897 (2017)
3. Efendiev, Y., Iliev, O., Kronsbein, C.: Multilevel Monte Carlo methods using ensemble level mixed MsFEM for two-phase flow and transport simulations. Comput. Geosci. **17**(5), 833–850 (2013). https://doi.org/10.1007/s10596-013-9358-y
4. Giles, M.B.: Multilevel Monte Carlo methods. Acta Numer. **24**, 259–328 (2015). https://doi.org/10.1017/S096249291500001X
5. Graham, G., Kuo, F., Nuyens, D., Scheichl, R., Sloan, I.: Quasi-Monte Carlo methods for elliptic PDEs with random coefficients and applications. J. Comput. Phys. **230**, 3668–3694 (2011)
6. Iliev, O., Mohring, J., Shegunov, N.: Renormalization based MLMC method for scalar elliptic SPDE. In: Lirkov, I., Margenov, S. (eds.) LSSC 2017. LNCS, vol. 10665, pp. 295–303. Springer, Cham (2018). https://doi.org/10.1007/978-3-319-73441-5_31
7. Lunati, I., Bernard, D., Giudici, M., Parravicini, G., Ponzini, G.: A numerical comparison between two upscaling techniques: non-local inverse based scaling and simplified renormalization. Adv. Water Resour. **24**, 913–929 (2001)
8. Shegunov, N., Iliev, O.: On dynamic parallelization of multilevel Monte Carlo algorithm. Cybern. Inf. Technol. **20**, 116–125 (2020)
9. Zakharov, P., Iliev, O., Mohring, J., Shegunov, N.: Parallel multilevel Monte Carlo algorithms for elliptic PDEs with random coefficients. In: Lirkov, I., Margenov, S. (eds.) LSSC 2019. LNCS, vol. 11958, pp. 463–472. Springer, Cham (2020). https://doi.org/10.1007/978-3-030-41032-2_53

# Numerical Parameter Estimates of Beta-Uniform Mixture Models

Dean Palejev[1,2]([✉])(iD)

[1] Big Data for Smart Society (GATE) Institute, Sofia University,
125 Tsarigradsko Shosse, Bl. 2, 1113 Sofia, Bulgaria
[2] Insititute of Mathematics and Informatics, Bulgarian Academy of Sciences,
Acad. G. Bonchev Street, Bl. 8, 1113 Sofia, Bulgaria
`dean.palejev@gate-ai.eu`, `dean.palejev@math.bas.bg`

**Abstract.** When analyzing biomedical data, researchers often need to apply the same statistical test or procedure to many variables resulting in a multiple comparisons setup. A portion of the tests are statistically significant, their unadjusted $p$-values form a spike near the origin and as such they can be modeled by a suitable Beta distribution. The unadjusted $p$-values of the non-significant tests are drawn form a uniform distribution in the unit interval. Therefore the set of all unadjusted $p$-values can be represented by a beta-uniform mixture model. Finding the parameters of that model plays an important role in estimating the statistical power of the subsequent Benjamini-Hochberg correction for multiple comparisons. To empirically investigate the properties of some parameter estimation procedures we carried out a series of computationally intensive numerical simulations on a high-performance computing facility. As a result of these simulations, in this article we have identified the overall optimal method for estimating the mixture parameters. We also show an asymptotic property of one of the parameter estimates.

**Keywords:** Beta-uniform mixture · Benjamini-Hochberg procedure · Multiple comparisons

## 1 Introduction

Scientific studies that produce multidimensional data have become common in the fields of genetics, molecular biology, etc. In such setups researchers often need to apply the same statistical test or procedure to many candidate variables in order to extract a subset of features for which there is a significant difference between two groups. Common example is the differential expression analysis in which one needs to find the genes with overall different expression levels between two groups often representing the patients and the controls in a study. In this situation one needs to subsequently apply a multiple comparisons correction procedure to the test $p$-values. The most popular such procedure is the Benjamini-Hochberg method described in [3], which controls the false discovery rate at a predetermined level. The article introducing that method is among the most cited scientific papers as reported in [13].

© Springer Nature Switzerland AG 2022
I. Lirkov and S. Margenov (Eds.): LSSC 2021, LNCS 13127, pp. 472–479, 2022.
https://doi.org/10.1007/978-3-030-97549-4_54

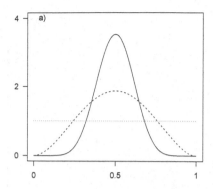 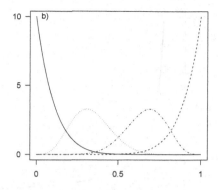

**Fig. 1.** Beta densities for different values of the parameters. Panel a) solid line: $B(10,10)$, dashed: $B(3,3)$, dotted: $B(1,1)$. Panel b) solid: $B(1,10)$, dashed: $B(10,1)$, dotted: $B(5,10)$, dot-and-dashed: $B(10,5)$.

An important question when planning biomedical research studies that result in a multiple comparison setup is the statistical power, for example as described in [10]. One approach to estimating the necessary sample size in order to achieve the desired statistical power is to model the distribution of the unadjusted for multiple comparisons $p$-values. The significant ones form a peak near the origin. Because of the definition of $p$-value under the null hypothesis, the non-significant ones are uniformly distributed in the unit interval. Therefore all unadjusted for multiple comparisons $p$-values are typically modeled as a mixture of a two distributions – one representing the significant $p$-values and a uniform distribution in the unit interval, representing the non-significant ones as described in [11].

The Beta distribution is a versatile way to describe random variables taking values in the unit interval. The typical representation of its probability density function is $x^{\alpha-1}(1-x)^{\beta-1}/B(\alpha,\beta)$ for $x$ in the unit interval, with both $\alpha, \beta > 0$ and the Beta function being $B(\alpha,\beta) = \Gamma(\alpha)\Gamma(\beta)/\Gamma(\alpha+\beta)$. Depending on its parameters the density of the Beta distribution can take different shapes and some examples are shown in Fig. 1.

Suitable parameters of the Beta distribution representing the unadjusted significant $p$-values are small values of $\alpha$ and values of $\beta$ that are close to 1. Some examples are shown in Fig. 2 where only a small interval near the origin is depicted. In these cases the density is only slightly affected by relatively small changes of the second parameter. This is also shown in Fig. 2 where the densities for B(0.1, 0.8) and B(0.1, 1) are very close.

General Betamix model based on the EM method [5] is proposed in [9]. Different parameterization of the Beta distribution is proposed in [7], it uses the mean $\mu = \alpha/(\alpha+\beta) \in (0,1)$ and also a precision parameter $\phi = \alpha+\beta > 0$. This choice of parameters is used in some of the software packages that we use in this work, however it might be more intuitive in cases of more balanced parameters where the mean is well within the unit interval rather than towards the origin as in the cases considered here.

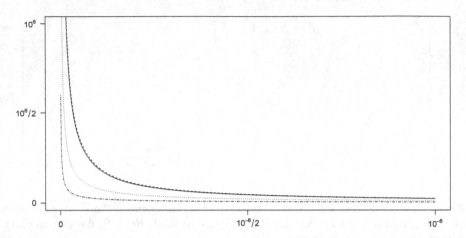

**Fig. 2.** Beta densities near the origin for different values of the parameters. Solid line: $B(0.1, 0.8)$, dashed: $B(0.1, 1)$, dotted: $B(0.2, 1)$, dot-and-dashed: $B(0.3, 1)$.

## 2    Methods

In this work we are interested in the quality of the parameter estimates of beta-uniform mixtures as derived by different algorithms. We consider a collection of beta-uniform mixture models $\gamma B(\alpha, \beta) + (1 - \gamma)U(0, 1)$, where $\gamma \in (0, 1)$ represents the proportion of significant tests. We draw 2400 times $n$ observations from each mixture. The values used in our simulations are as follows: $\alpha \in \{0.1, 0.15, 0.2, 0.25, 0.3\}$, $\beta \in \{0.8, 0.9, 1, 1.1, 1.2\}$, $\gamma \in \{0.05, 0.1, 0.2\}$ and $n \in \{500, 1000, 2000, 5000, 10000, 20000\}$.

We use three models to estimate the parameters of the beta-uniform mixture. The first, which we call general, represents the data as a mixture of two beta distributions, $\gamma B(\alpha, \beta) + (1 - \gamma)B(\alpha_2, \beta_2)$, where $\gamma \in (0, 1)$. The beta-uniform mixture from which we have generated the data is a particular case of the general model with parameters $\alpha_2 = \beta_2 = 1$, because $B(1, 1)$ is the uniform distribution in the unit interval.

The second model is the original one from which the data is generated, $\gamma B(\alpha, \beta) + (1 - \gamma)U(0, 1)$, where $\gamma \in (0, 1)$. We refer to it as the exact model. The third model is a restriction of the original model when $\beta = 1$ and it is $\gamma B(\alpha, 1) + (1 - \gamma)U(0, 1)$, where $\gamma \in (0, 1)$. Naturally we call this model restricted.

We estimate the parameters of each of the datasets using the three models described above. The parameter estimations using either the general and the exact model are done using the R package betareg, originally described in [4]. The fitting under these two models is done using its function betamix described in [8]. These packages implement mixture fitting procedures using maximum likelihood based on the EM algorithm. They use the Beta parametrization based on mean and precision that is described above and for consistency we subsequently transform the parameters to the common definition of Beta density. The parameter estimations for the restricted model are done using the fitBumModel function from the R package `BioNet` [2,6]. That function implements the model defined in [11].

When applying the general or the exact model, the estimation process does not always converge or sometimes it converges to local extreme points that are far away from the model parameters. In these cases we discard the missing or infeasible values. In practice in these situations we could restart the estimation with different starting points, however this would not affect our results within the simulation framework. There is a different number of datasets that are discarded for the different parameter values. Because of that for all parameter combinations we only keep 1000 randomly selected datasets for which the process converges to a feasible value.

## 3   Results

First we are interested in the performance of the three methods when estimating the parameter $\alpha$ of the Beta distribution that models the unadjusted significant $p$-values. For a given value of $\alpha$ we calculate the normalized differences between that value and the average estimates of that parameter based on the 1000 repetitions. Each histogram in Fig. 3 represents the normalized differences for one of the three methods that we consider. Within each method, for a fixed set of parameters $\alpha, \beta, \gamma$ and $n$, we set $\hat{\alpha}$ to be the empirical mean of the 1000 estimates of $\alpha$ under the particular method for the given set of parameters. We then calculate the normalized difference between $\alpha$ and its estimates as $(\hat{\alpha} - \alpha)/\alpha$. Aggregating over the parameters $\alpha, \beta, \gamma$ and $n$ within each method gives us the normalized values depicted in the respective histograms for these methods. Both the general and the exact model on average overestimate the parameter. The means of the three sets of normalized distances for the general, exact and restricted method are respectively $0.061, 0.038$, and $0.004$. Therefore the restricted method outperforms the other two methods when estimating the first parameter of the Beta distribution that is part of the beta-uniform mixture.

The restricted model uses fixed value of $\beta = 1$. For the other two methods, we can similarly consider the normalized distances between the average estimates of $\beta$ and the respective true values. The results are shown on Fig. 4 and we observe that both the general and the exact method perform similarly.

Next, we are interested in the estimates of the mixing coefficient $\gamma$. Similarly to the normalization described above, we calculate the normalized distances between the average estimates of $\gamma$ and the respective true values. We also note that these distances are the same as those for the sizes of the smaller clusters or of the larger clusters respectively. The results are shown in Fig. 5. Clearly both the general and the exact model underestimate the proportion of significant $p$-values and label on average approximately half of them as truly significant.

Finally we are interested in the asymptotic convergence of the estimates of $\alpha$ under the restricted model when $n$ increases and the other parameters are fixed. We observe that there is a logarithmic decay of the standard deviation of the estimates of $\alpha$. Figure 6 shows these results for three different sets of

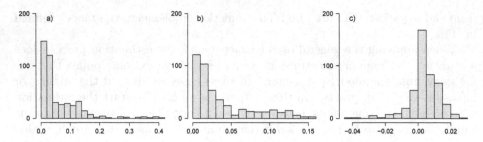

**Fig. 3.** Histograms of the normalized differences between the average estimates of $\alpha$ and the respective true values. Panel a) shows the results for the general model, panel b) exact model, panel c) restricted model.

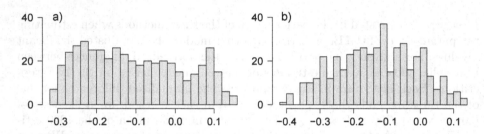

**Fig. 4.** Histograms of the normalized differences between the average estimates of $\beta$ and the respective true values. Panel a) shows the results for the general model, panel b) exact model.

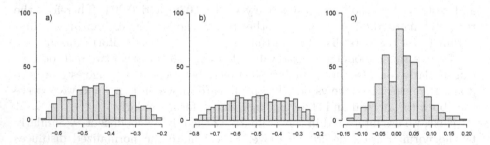

**Fig. 5.** Histograms of the differences between the average estimates of $\alpha$ and the respective true values. Panel a) shows the results for the general model, panel b) exact model, panel c) restricted model.

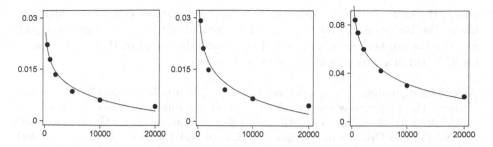

**Fig. 6.** Asymptotic convergence of the average estimates of $\alpha$ under the restricted model. Panel a) $\alpha = 0.1, \beta = 0.8, \gamma = 0.05$, panel b) $\alpha = 0.2, \beta = 1, \gamma = 0.2$, panel c) $\alpha = 0.3, \beta = 0.8, \gamma = 0.05$.

parameters. The logarithmic fit is created using the R function nls. The $x$ axis in this Figure shows the values of $n$ and the $y$ axis shows the standard deviations of the estimates of $\alpha$ under the restricted model. This observation allows us to efficiently predict the standard deviation of the estimates by calculating it based on simulated data only for few relatively small values of $n$ and then extrapolating for larger values.

Due to space constrains, a table with the aggregated information based on our simulations is given as Supplementary Material at www.math.bas.bg/~palejev/ BUM.

## 4    Conclusions

We have demonstrated that when $\alpha$ is small and $\beta$ is close to 1, the restricted model $\gamma B(\alpha, 1) + (1 - \gamma)U(0, 1)$ implemented in the BioNet package in R outperforms the general model $\gamma B(\alpha, \beta) + (1 - \gamma)B(\alpha_2, \beta_2)$ and the exact model $\gamma B(\alpha, \beta) + (1 - \gamma)U(0, 1)$ when estimating the parameters $\alpha$ and $\gamma$ of the beta-uniform model. Both the general and the exact model overestimate the parameter $\alpha$ and significantly underestimate the parameter $\gamma$. Due to its specifics, the restricted model does not provide an estimate for $\beta$ as it is fixed to 1. However as shown in Fig. 2 when that parameter is close to 1 and $\alpha$ is small, changes in $\beta$ do not change the density of the Beta distribution by a lot. Therefore we recommend using the restricted model to obtain reliable estimates of the main parameters of the model. Of course, that conclusion holds only in our setup when the Beta distribution models the unadjusted $p$-values of the significant tests in multiple comparisons setup. In other situations, for example when there are two Beta distributions neither of which has a shape of a peak near one of the ends of the unit interval, nor is close to the uniform distribution in the unit interval, the general model would work better.

This work may be extended in several ways. We can estimate the bias of the estimates of $\alpha$ under both the general and the exact model and perhaps adjust for that bias. Article [12] underlines a problem with possible values at either end of the unit interval, resulting in an infinite log-likelihood function. Although under the beta-uniform mixture model the probability of such events is 0, real-world data can take such values and the model described in that article might be included in a subsequent extension of this work.

**Acknowledgement.** The results presented in this article are part of the GATE project. The project has received funding from the European Union Horizon 2020 WIDESPREAD-2018-2020 TEAMING Phase 2 programme under Grant Agreement No. 857155 and Operational Programme Science and Education for Smart Growth under Grant Agreement No. BG05M2OP001-1.003-0002-C01. The numerical simulations were performed on the Avitohol supercomputer at IICT-BAS described in [1]. The computational resources and infrastructure were provided by NCHDC – part of the Bulgarian National Roadmap of RIs, with the financial support by Grant No. DO1 - 387/18.12.2020.

# References

1. Atanassov, E., Gurov, T., Ivanovska, S., Karaivanova, A.: Parallel Monte Carlo on Intel MIC architecture. Procedia Comput. Sci. **108**, 1803–1810 (2017). https://doi.org/10.1016/j.procs.2017.05.149
2. Beisser, D., Klau, G.W., Dandekar, T., Mueller, T., Dittrich, M.: BioNet: an R-package for the functional analysis of biological networks. Bioinformatics **26**(8), 1129–1130 (2010). https://doi.org/10.1093/bioinformatics/btq089
3. Benjamini, Y., Hochberg, Y.: Controlling the false discovery rate: a practical and powerful approach to multiple testing. J. R. Stat. Soc. Ser. B (Methodol.) **57**(1), 289–300 (1995). https://doi.org/10.2307/2346101
4. Cribari-Neto, F., Zeileis, A.: Beta regression in R. J. Stat. Softw. **34**(2), 1–24 (2010). https://doi.org/10.18637/jss.v034.i02
5. Dempster, A.P., Laird, N.M., Rubin, D.B.: Maximum likelihood from incomplete data via the EM algorithm. J. R. Stat. Soc. Ser B (Methodol.) **39**(1), 1–38 (1977). https://doi.org/10.1111/j.2517-6161.1977.tb01600.x
6. Dittrich, M.T., Klau, G.W., Rosenwald, A., Dandekar, T., Mueller, T.: Identifying functional modules in protein-protein interaction networks: an integrated exact approach. Bioinformatics **24**(13), i223–i231 (2008). https://doi.org/10.1093/bioinformatics/btn161
7. Ferrari, S., Cribari-Neto, F.: Beta regression for modelling rates and proportions. J. Appl. Stat. **31**(7), 799–815 (2004). https://doi.org/10.1080/0266476042000214501
8. Grün, B., Kosmidis, I., Zeileis, A.: Extended beta regression in R: shaken, stirred, mixed, and partitioned. J. Stat. Softw. **48**(11), 1–25 (2012). https://doi.org/10.18637/jss.v048.i11
9. Ji, Y., Wu, C., Liu, P., Wang, J., Coombes, K.R.: Applications of beta-mixture models in bioinformatics. Bioinformatics **21**(9), 2118–2122 (2005). https://doi.org/10.1093/bioinformatics/bti318
10. Palejev, D., Savov, M.: Estimating the statistical power of the Benjamini-Hochberg procedure. In: Dimov, I., Fidanova, S. (eds.) HPC 2019. SCI, vol. 902, pp. 298–308. Springer, Cham (2021). https://doi.org/10.1007/978-3-030-55347-0_26
11. Pounds, S., Morris, S.W.: Estimating the occurrence of false positives and false negatives in microarray studies by approximating and partitioning the empirical distribution of p-values. Bioinformatics **19**(10), 1236–1242 (2003). https://doi.org/10.1093/bioinformatics/btg148

12. Schröder, C., Rahmann, S.: A hybrid parameter estimation algorithm for beta mixtures and applications to methylation state classification. Algorithms Mol. Biol. **12**, 21 (2017). https://doi.org/10.1186/s13015-017-0112-1
13. Van Noorden, R., Maher, B., Nuzzo, R.: The top 100 papers. Nature **514**(7524), 550–553 (2014). https://doi.org/10.1038/514550a

# Large-Scale Computer Simulation of the Performance of the Generalized Nets Model of the LPF-algorithm

Tasho D. Tashev[1](✉), Alexander K. Alexandrov[1], Dimitar D. Arnaudov[2], and Radostina P. Tasheva[2]

[1] Institute of Information and Communication Technologies, Bulgarian Academy of Sciences, Acad. Georgi Bonchev Street, Block 2, 1113 Sofia, Bulgaria
{ttashev,akalexandrov}@iit.bas.bg
[2] Technical University-Sofia, 8 Kliment Ohridski Blvd., 1000 Sofia, Bulgaria
{dda,rpt}@tu-sofia.bg

**Abstract.** Large-scale simulation of the throughput (TP) of an existing LPF-algorithm (Longest Port First) for crossbar switch are presented. The throughput for Generalized Nets (GN) model of algorithm is studied for uniform independent and identically distributed Bernoulli traffic. The presented simulations are executed on the AVITOHOL supercomputer located at the IICT, Bulgaria. The modeling of the TP utilizes LPF for a switch with $N \in [2, 60]$ commutation field size. Problems arise due to the time complicatedness of the implementation of the LPF-algorithm ($O(N^{4.7})$). It is necessary to reduce the time complexity without introducing distortions into the results of simulation. One variant of the LPF with simplified random selection of a starting cell of weighting-matrix are discussed.

**Keywords:** Large-scale simulation · Modeling · Generalized nets · Switch node

## 1 Introduction

Crossbar switches use alternative paths for data transfer between their ingress and egress. If the information packets are sent without delay or losses the case can be considered as ideal. The control unit of the switch is responsible for the computation of a non-conflict schedule. (Fig. 1 - Switching Logic) [1,2].

The calculation of this schedule seems to be an NP-complete problem [3]. The existing approaches to solve this task may be realized by using various formal instruments [4,5]. With the increasing volume of global information traffic a need of new solutions (algorithms) has arisen. Their effectiveness is unknown and requires a careful examination. Developers face similar problems when it is necessary to delocalize evolutionary algorithms [6], algorithms for evaluating railway transport [7], as well as for the rapidly evolving modern networks [8,9].

© Springer Nature Switzerland AG 2022
I. Lirkov and S. Margenov (Eds.): LSSC 2021, LNCS 13127, pp. 480–486, 2022.
https://doi.org/10.1007/978-3-030-97549-4_55

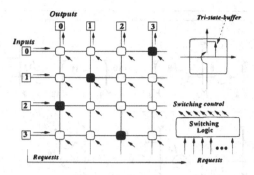

**Fig. 1.** Control unit in crossbar switch [1]

In order to calculate a non-blocking schedule we have constructed an algorithm Minimum of Maxima (called MiMa) [10]. The Generalized Nets (GN) apparatus [11] aids specification of this algorithm. A comparison between the throughputs (TP) of the MiMa-algorithm and the LPF-algorithm (Longest Port First) [12] leads us to an assessment of their effectiveness. Different algorithms for TP calculation are proposed as to successfully design the switch. This TP quantity depends on the load traffic for a certain algorithm. After we have chosen the traffic model, the TR depends only on the input intensity $\rho$ [4].

When calculating the upper boundary of the TP, the first step is to build a detailed GN algorithmic LPF model. Then the data of high-performance calculations - TP simulations - are submitted as input data for the numerical procedure [13]. We use the AVITOHOL supercomputer as a tool for TP modeling. The simulated algorithm LPF [12] is intended for a switch field size $n \times n$ from $2 \times 2$ to $150 \times 150$ and allows up to 1 billion simulations. Using the family of templates [13] allows implementing chosen load traffic [4] and makes it possible to embody an intensity $\rho = 100\%$ for each input line.

## 2    LPF-algorithm for Non-blocking Schedule

To represent requests to transfer packets through the $n \times n$ input/output switch lines, a $R$ matrix is constructed (called the request matrix). Thus, the flow of packets, which is unidirectional, transmitted from ingress to egress lines, is specified using the $R$. If there is more than one conflict in any $R$ row (column), the number of requests will be more than 1 [4].

Using GN we synthesize a model of algorithm LPF. In result, the first matrix-solution $Q1$ will consist of elements with maximum conflict weight. The following matrices $Q2, \ldots, Qr$ are calculated with the same constraints. The ending matrix will comprise of only non-blocking requests in $R$. The TP depends on the choices of maximum conflicting elements. If two and more elements have equal and maximum weight a random choice is made between them [12].

A form of a graphical presentation of the GN model is shown below (Fig. 2). As a first step the token enters the first position l1. The initial token represents

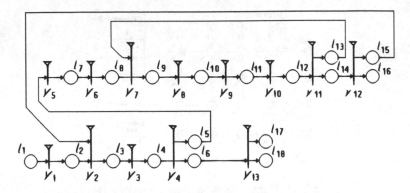

**Fig. 2.** GN-model graphical representation (LPF)

the transfer requirements (all packets have the same size). The token has initial characteristic "$ch_0 = (pr^1ch_0 \; pr^2ch_0) = \; < n, R >$" and $n$ is the number of input lines. The end of the LPF execution is evidenced by the arrival of the token at the l18 location. In Place l16, at the end, matrices for commutation have accumulated - $Q1, Q2, \ldots, Qr$.

With the help of the Generalized Nets [11,15] it is possible to effectively describe the implemented information processes during the LPF algorithm operation. The description of the phases requires at least two cycles.

The formal specification of the LPF-algorithm is given in [14]. Each of the transitions has the same priority and the same applies to tokens. The GN-analysis shows that the transition Y8 corresponds to the most complex operation in time. This transition is responsible for randomly selecting only one of several requests with the same weighting factor. Additional information (k, r,...) is collected in the tokens, which is used to calculate the average execution time, the average TR, etc. ..

## 3    Computer Simulations

Following certain rules [15], we convert the GN model into a computer program. For coding we use the Vfort free access package [16]. The compilation is performed using the AVITOHOL supercomputer owned by the Bulgarian Academy of Sciences. The binary code is executed locally in AVITOHOL under the Red Hat Linux [17]. The resource used is 16 CPUs and the operating time is limited to 144 h.

A family of patterns [13] simulated uniform traffic with intensity $\rho = 100\%$ type Bernoulli i.i.d. 1 000 000 simulations for pattern Uni-1,2,4,8 are executed, for each dimension $n$ in the range from $2 \times 2$ to $60 \times 60$. In Fig. 3 on the left the resulting average throughput (TP) is shown, on the right—the differences for numerical procedure [13].

Next, we calculate ratio $\delta_1$ between differences res1, res2 of TP for adjacent patterns - the result is depicted in Fig. 4. The estimation of $\delta_1$ vary around

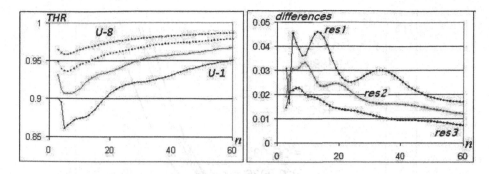

**Fig. 3.** Throughput for Uni-1,2,4,8 and differences 1,2,3

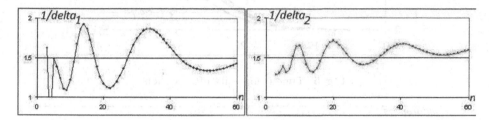

**Fig. 4.** Ratio $1/\delta_1$, $1/\delta_2$ between differences and time complexity

$1/(1,5)$. The expected delta value is $1/(1,41) = 1/(m^{1/2}) = 2^{-1/2}$ for a sufficiently accurate calculation of the upper bound for TP. Obviously it will be necessary to obtain a $\delta_6$ or $\delta_7$, i.e. we will need simulations for Uni-256. Will the available AVITOHOL resource be sufficient to perform the necessary modeling? It depends on the LPF-time complexity.

Figure 5 presents the time complexity - $O(n^{4,7})$. This is quite a high complexity. To get $\delta_6$ you will have to sacrifice accuracy - there will be about 1000 simulations available. Are there other ways to maintain accuracy?

## 4   Disscussion

The size $n$ varies from 2 to 60 for each dimension $(n \times n)$ of pattern $Uni - 16$. The resulting throughput is averaged by data for 1 million runs. Then we calculated the differences $res - 4$ and $res - 3$ and ratio $\delta_3$. The obtained results are shown in Fig. 6. The values of $\delta_1$ vary around $1/(1,5)$. What can we say about the tendency of the values of $\delta_2$ and $\delta_3$?

In our TP simulations for the PIM algorithm in the same case (i.i.d. B uni, $m = 2$) the trend $\delta_j$ tends towards $1/1.41$. We need to ask ourselves the question:

**Question:** Does the dependence $\delta_{j+1} = \phi(\delta_j)$ tend here to the same constant $m^{-1/2} = 2^{-1/2}$? Do these fluctuations in delta values occur due to internal connections in the LPF algorithm?

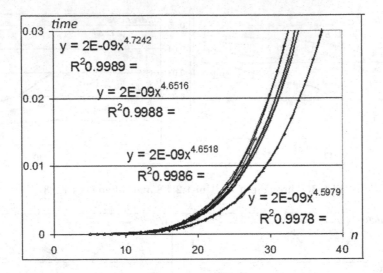

**Fig. 5.** Time complexity of execution

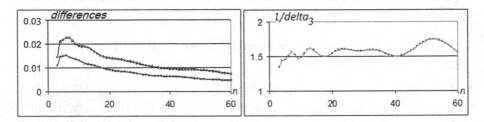

**Fig. 6.** Differences res3, res4 and ratio $1/\delta_3$ between them

In order to answer, simulations for TP are required at values of index $i$ of the template over 100. This will require an increase in simulation time. We can count on a doubling of the allowed time limit. But the selection of a request with a maximum weighting factor in the GN-model is done in a very costly way. This is done to match the mathematical model in the [10]. This entails high computational complexity. Should we parallel this difficult operation? Or, instead of $n^2$ sorting out arrangement, can we get the same with $n$? To reduce the time complexity instead of 4.7 to 3.7? Its for further study.

## 5    Conclusion

Herein, a Generalized Nets model of the algorithm LPF for calculation of a schedule for crossbar communication node is used. The large-scale computer simulation of the throughput of node is fulfilled. The load traffic is independent and identically distributed (Bernoulli uniform).

Using a numerical procedure, parameters are calculated that describe the tendency of the throughput to approach its bound when the algorithm LPF is applied. Difficulties of simulation implementation occur due to its high complexity - ($O(n^{4.7})$) or higher. Reducing the time complexity of the algorithm is under discussion.

**Acknowledgment.** We acknowledge the provided access to thee-infrastructure of the NCHDC - part of the Bulgarian National Roadmap on RIs, which is supported financially by the Grant No. D01-387/18.12.2020.

# References

1. Cheung, S.Y.: CrossBar. Department of Computer Science, Emory University, Atlanta, Georgia, USA. http://www.mathcs.emory.edu/-cheung/Courses/355/Syllabus/90-parallel/CrossBar.html. Accessed 10 Apr 2021
2. Meng, J., Gebara, N., Ng, H.-C., Costa, P., Luk, W.: Investigating the feasibility of FPGA-based network switches. In: 2019 IEEE 30th International Conference on Application-Specific Systems Architectures and Processors (ASAP), vol. 2160-052X, IEEE Publications, pp. 218–226 (2019)
3. Chen, T., Mavor, J., Denyer, P., Renshaw, D.: Traffic routing algorithm for serial superchip system customization. IEE Proc. **137**(1), 65–73 (1990)
4. Rojas-Cessa, R.: Interconnections for Computer Communications and Packet Networks. CRC Press, Taylor & Francis Group, Boca Raton (2017)
5. Kolchakov, K., Monov, V.: An approach to weights distribution of requests in two algorithms for non-conflict scheduling. In: Problems of Engineering Cybernetics and Robotics, vol. 70, pp. 21–26, Prof. Marin Drinov Academic Publishing House, Sofia (2018)
6. Balabanov, T., Zankinski, I., Barova, M.: Distributed evolutionary computing migration strategy by incident node participation. In: Lirkov, I., Margenov, S.D., Waśniewski, J. (eds.) LSSC 2015. LNCS, vol. 9374, pp. 203–209. Springer, Cham (2015). https://doi.org/10.1007/978-3-319-26520-9_21
7. Hensel, S., Marinov, M.A.: Comparison of time warping algorithms for rail vehicle velocity estimation in low-speed scenarios. Metrol. Measure. Syst. J. **24**(1), 161–173 (2017)
8. Nedyalkov, I.: An original and simple method for studying the performance of a VoIP network. In: 2020 XI National Conference with International Participation (ELECTRONICA), pp. 1–4 (2020)
9. Otsetova-Dudin, E., Markov, K.: Mobility factor in new generations wireless networks. In: 2020 IEEE 10th International Conference on Intelligent Systems (IS), pp. 601–605 (2020)
10. Tashev, T., Marinov, M., Monov, V., Tasheva, R.: Modeling of the MiMa-algorithm for crossbar switch by means of Generalized Nets. In: 2016 IEEE 8th International Conference on Proceedings of Intelligent Systems (IS), Sofia, Bulgaria, pp. 593–598. IEEE Publications (2016)
11. Atanassov, K.: Generalized nets as a tool for the modelling of data mining processes. In: Sgurev, V., Yager, R.R., Kacprzyk, J., Jotsov, V. (eds.) Innovative Issues in Intelligent Systems. SCI, vol. 623, pp. 161–215. Springer, Cham (2016). https://doi.org/10.1007/978-3-319-27267-2_6

12. Mekkittikul, A., McKeown, N.: A practical algorithm to achieve 100% throughput in input-queued switches. Proc. IEEE INFOCOM **1998**, 792–799 (1998)
13. Tashev, T., Monov, V.: A numerical study of the upper bound of the throughput of a crossbar switch utilizing MiMa-algorithm. In: Dimov, I., Fidanova, S., Lirkov, I. (eds.) NMA 2014. LNCS, vol. 8962, pp. 295–303. Springer, Cham (2015). https://doi.org/10.1007/978-3-319-15585-2_33
14. Tashev, T., Marinov, M., Tasheva, R., Alexandrov, A.: Generalized nets model of the LPF-algorithm of the crossbar switch node for determining LPF-execution time complexity. In: AIP Conference Proceedings, vol. 2333(1), pp. 090039-1–090039-8 (2021). https://doi.org/10.1063/5.0042856
15. Tashev, T.: Computering simulation of schedule algorithm for high performance packet switch node modelled by the apparatus of generalized nets. In: Proceedings of the 11th International Conference on Computer Systems and Technologies CompSysTech 2010, pp. 240–245. Association for Computing Machinery, NY, United States (2010)
16. Vabishchevich, P.: VFort. http://www.nomoz.org/site/629615/vfort.html. Accessed 15 Apr 2021
17. HPC and Data Centre at IICT-BAS. http://www.hpc.acad.bg/system-1/. Accessed 2021

# Contributed Papers

# A New Error Estimate for a Primal-Dual Crank-Nicolson Mixed Finite Element Using Lowest Degree Raviart-Thomas Spaces for Parabolic Equations

Fayssal Benkhaldoun[1] and Abdallah Bradji[1,2]([⊠])

[1] LAGA, University of Paris 13, Paris, France
fayssal@math.univ-paris13.fr
[2] LMA Laboratory, University of Annaba, Annaba, Algeria
abdallah.bradji@gmail.com, abdallah.bradji@etu.univ-amu.fr
https://www.math.univ-paris13.fr/~fayssal/,
https://www.i2m.univ-amu.fr/perso/abdallah.bradji/

**Abstract.** We consider the heat equation as a model for parabolic equations. We establish a fully discrete scheme based on the use of Primal-Dual Lowest Order Raviart-Thomas Mixed method combined with the Crank-Nicolson method. We prove a new convergence result with convergence rate towards the "velocity" $P(t) = -\nabla u(t)$ in the norm of $L^2(H_{\mathrm{div}})$, under assumption that the solution is smooth. The order is proved to be two in time and one in space. This result is obtained thanks to a new well developed discrete a priori estimate. The convergence result obtained in this work improves the existing one for PDMFEM (Primal-Dual Mixed Finite Element Method) for Parabolic equations which states the convergence towards the velocity in only the norm of $L^\infty\left(\left(L^2\right)^d\right)$, see [6, Theorem 2.1, p. 54].

This work is an extension of [1] which dealt with new error estimates of a MFE scheme of order one in time. It is also motivated by the work [9] in which a full discrete Crank Nicolson scheme based on another MFE approach, different from the one we use here, is established in the two dimensional space.

**Keywords:** Parabolic equations · Lowest Order Raviart-Thomas Mixed Finite Elements · Crank-Nicolson method · New error estimate

**MSC2010:** 65M60 · 65M12

## 1 Problem to Be Solved and Motivation

Let us consider the following heat problem:

$$u_t(\boldsymbol{x}, t) - \Delta u(\boldsymbol{x}, t) = f(\boldsymbol{x}, t), \qquad (\boldsymbol{x}, t) \in \Omega \times (0, T), \tag{1}$$

Supported by MCS team (LAGA Laboratory) of the "Université Sorbonne- Paris Nord".

I. Lirkov and S. Margenov (Eds.): LSSC 2021, LNCS 13127, pp. 489–497, 2022.
https://doi.org/10.1007/978-3-030-97549-4_56

where $\Omega$ is an open bounded polyhedral subset in $\mathbb{R}^d$, with $d \in \mathbb{N}^* = \mathbb{N} \setminus \{0\}$, $T > 0$ , and $f$ is a given function.
An initial condition is given by:

$$u(\boldsymbol{x}, 0) = u^0(\boldsymbol{x}), \qquad \boldsymbol{x} \in \Omega, \tag{2}$$

and, for the sake of simplicity, we consider the homogeneous Dirichlet boundary conditions

$$u(\boldsymbol{x}, t) = 0, \qquad (\boldsymbol{x}, t) \in \partial\Omega \times (0, T), \tag{3}$$

where, we denote by $\partial\Omega = \overline{\Omega} \setminus \Omega$ the boundary of $\Omega$.

Heat Eq. (1) is used in several areas of applications. It is also the prototypical parabolic partial differential equation which is in turn involved in many different models like Navier–Stokes and reaction–diffusion systems. There are several numerical methods devoted to parabolic equations, e.g. Finite Volumes, Finite Elements, and Mixed Finite Element, see for instance [3, 5, 7, 8] and references therein. In this note, we consider PDMFEM (Primal- Dual Mixed Finite Element Method) for (1)–(3) using Lowest Order Raviart-Thomas Elements as discretization in space and Crank-Nicolson method as discretization in time. We establish a fully discrete scheme and we provide new error estimate towards the "velocity" $P(t) = -\nabla u(t)$ in the norm of $L^2(H_{\mathrm{div}})$. This error estimate improves and extends the one proved in [6, (2.4), Theorem 2.1, p. 54] which states the error estimate for MFEM for parabolic equations towards the velocity in only the norm of $L^\infty \left( (L^2)^d \right)$.

## 2    Space and Time Discretizations

Let $\mathcal{T}_h$ be a family of triangulations of $\overline{\Omega}$. For the sake of simplicity, we consider a reference polyhedron $\hat{K}$ which is the unit $d$-simplex. We define the lowest Raviart-Thomas Mixed Finite Elements (see [7, Subsect. 7.2.2, pp. 235–236]):

$$V_h^{\mathrm{div}} = \{v \in H_{\mathrm{div}}(\Omega) : \quad v|_K \in \mathbb{D}_0, \quad \forall K \in \mathcal{T}_h\} \tag{4}$$

and

$$W_h = \{p \in L^2(\Omega) : \quad p|_K \in \mathbb{P}_0, \quad \forall K \in \mathcal{T}_h\}, \tag{5}$$

where $\mathbb{P}_0$ is the space of constant functions and

$$\mathbb{D}_0 = (\mathbb{P}_0)^d \oplus \boldsymbol{x}\mathbb{P}_0.$$

The time discretization is given by a constant time step $k = \dfrac{T}{N+1}$, where $N \in \mathbb{N}^*$. Throughout this paper, we use the notation $[\![L, M]\!]$, with $M, N \in \mathbb{N}$ and $L \leq M$, to denote the set of all the natural numbers $n \in \mathbb{N}$ such that $n \in [L, M]$, that is $[\![L, M]\!] = [L, M] \cap \mathbb{N}$. We shall denote $t_n = nk$, for $n \in [\![0, N+1]\!]$. We

introduce the operator $\partial^1$ of the discrete temporal derivative $\partial^1 v^n = \dfrac{v^n - v^{n-1}}{k}$.
We need to use the discrete second time derivative

$$\partial^2 v^{n+1} = \partial^1(\partial^1 v^{n+1}) = \frac{\partial^1 v^{n+1} - \partial^1 v^n}{k} = \frac{v^{n+1} - 2v^n + v^{n-1}}{k^2}.$$

We shall use the notation $\partial^0 v^n = v^n$. We shall denote by $v^{n-\frac{1}{2}}$ the following arithmetic mean value between the two time levels $n-1$ and $n$:

$$v^{n-\frac{1}{2}} = \frac{v^n + v^{n-1}}{2}. \tag{6}$$

Throughout this paper, the letter $C$ stands for a positive constant independent of the parameters of discretizations.

## 3 Principles of a Mixed Finite Element Scheme for (1)–(3)

The approach of Mixed Finite Element method we consider in this paper is based on a weak formulation involving the two spaces $H_{\mathrm{div}}(\Omega)$ and $L^2(\Omega)$ and its principles can be sketched as follows for the Poisson's equation: for a given $F \in L^2(\Omega)$

$$-\Delta U(\boldsymbol{x}) = F(\boldsymbol{x}), \qquad \boldsymbol{x} \in \Omega, \tag{7}$$

with homogeneous boundary conditions

$$U(\boldsymbol{x}) = 0, \qquad \boldsymbol{x} \in \partial\Omega. \tag{8}$$

In some applications, we are interested to get approximation for

$$P = -\nabla U. \tag{9}$$

The Primal Dual Mixed Formulation of problem (7)–(8) (see for instance [3, p. 415]) is: Find $(P, U) \in H_{\mathrm{div}}(\Omega) \times L^2(\Omega)$ such that

$$(\varphi, \mathrm{div}P)_{L^2(\Omega)} = (\varphi, F)_{L^2(\Omega)}, \qquad \forall \varphi \in L^2(\Omega) \tag{10}$$

and

$$(\psi, P)_{L^2(\Omega)} = (\mathrm{div}\psi, U)_{L^2(\Omega)}, \qquad \forall \psi \in H_{\mathrm{div}}(\Omega). \tag{11}$$

This approach followed for elliptic equations can also be applied to obtain the following mixed formulation for (1)–(3) (see for instance [6, p. 53]), for each $t \in (0, T)$, find $(p(t), u(t)) \in H_{\mathrm{div}}(\Omega) \times L^2(\Omega)$ such that

$$(u(t), \varphi)_{L^2(\Omega)} + (\varphi, \mathrm{div}p(t))_{L^2(\Omega)} = (\varphi, f(t))_{L^2(\Omega)}, \qquad \forall \varphi \in L^2(\Omega), \tag{12}$$

$$(\psi, p(t))_{L^2(\Omega)} = (\mathrm{div}\psi, u(t))_{L^2(\Omega)}, \qquad \forall \psi \in H_{\mathrm{div}}(\Omega), \tag{13}$$

and

$$u(0) = u^0. \tag{14}$$

We now set the formulation of an implicit mixed finite volume scheme for problem (1)–(3). Such formulation is based on the use of the primal-dual mixed formulation (12)–(14). The unknowns of this scheme are the set of the couples

$$\left\{ (p_h^n, u_h^n) \in V_h^{\mathrm{div}} \times W_h; n \in [\![0, N+1]\!] \right\}.$$

These unknowns are expected to approximate the set of the unknowns

$$\left\{ (-\nabla u(t_n), u(t_n)); \quad n \in [\![0, N+1]\!] \right\}.$$

The scheme can be formulated as:

– For any $n \in [\![0, N]\!]$ and for all $\varphi \in W_h$ :

$$\left( \partial^1 u_h^{n+1}, \varphi \right)_{L^2(\Omega)} + \left( \nabla \cdot p_h^{n+\frac{1}{2}}, \varphi \right)_{L^2(\Omega)} = \left( \frac{f(t_{n+1}) + f(t_n)}{2}, \varphi \right)_{L^2(\Omega)}, \tag{15}$$

– For any $n \in [\![0, N+1]\!]$:

$$(p_h^n, \psi)_{L^2(\Omega)^d} = (u_h^n, \nabla \cdot \psi)_{L^2(\Omega)}, \qquad \forall \psi \in V_h^{\mathrm{div}}, \tag{16}$$

where

$$\left( \nabla \cdot p_h^0, \varphi \right)_{L^2(\Omega)} = \left( -\Delta u^0, \varphi \right)_{L^2(\Omega)}, \qquad \forall \varphi \in W_h. \tag{17}$$

It is useful to note that by taking $n = 0$ in Eq. (16), we get

$$\left( p_h^0, \psi \right)_{L^2(\Omega)^d} = \left( u_h^0, \nabla \cdot \psi \right)_{L^2(\Omega)}, \qquad \forall \psi \in V_h^{\mathrm{div}}. \tag{18}$$

Using the techniques of the proof of [6, Theorem 2.1, p. 54], the following $L^\infty \left( \left( L^2 \right)^d \right)$–error estimate can be proved:

$$\max_{n=0}^{N+1} \| p_h^n + \nabla u(t_n) \|_{L^2(\Omega)^d} \le C(k^2 + h). \tag{19}$$

Our aim is to improve the error estimate (19) in the sense that the norm $L^\infty \left( \left( L^2 \right)^d \right)$ is replaced by $L^2(H_{\mathrm{div}})$.

## 4   Convergence Analysis of Scheme (15)–(17)

**Theorem 1 (New error estimate for scheme (15)–(17)).** *Let $\Omega$ be a polyhedral open bounded subset of $\mathbb{R}^d$, where $d \in \mathbb{N} \setminus \{0\}$. Assume that the solution of (1)–(3) satisfies $u \in \mathcal{C}^3([0, T]; \mathcal{C}^3(\overline{\Omega}))$. Let $k = \dfrac{T}{N+1}$, where $N \in \mathbb{N} \setminus \{0\}$. We shall denote $t_n = nk$, for $n \in [\![0, N+1]\!]$. Let $\mathcal{T}_h$ be a family of triangulations of $\overline{\Omega}$ and $V_h^{\mathrm{div}}$ and $W_h$ be the lowest order Raviart-Thomas spaces given* *respectively by* (4) *and* (5). *Then, there exists a unique solution* $((p_h^n, u_h^n))_{n=0}^{N+1} \in$ $(V_h^{\mathrm{div}} \times W_h)^{N+2}$ *for scheme* (15)–(17) *and the following* $L^2(H_{\mathrm{div}})$–*error estimate holds:*

$$\left( \sum_{n=0}^{N+1} k \| \mathrm{div} p_h^n + \Delta u(t_n) \|_{L^2(\Omega)}^2 \right)^{\frac{1}{2}} \leq C(k^2 + h). \tag{20}$$

To prove Theorem 1, we need to use a new discrete *a priori estimate* which is stated in the following lemma:

**Lemma 1 (New a priori estimate)** *Under the same hypotheses of Theorem 1, we assume that there exists* $\left( (\eta_D^n)_{n=0}^{N+1}, (\overline{\eta}_D^n)_{n=0}^{N+1} \right) \in \left( V_h^{\mathrm{div}} \right)^{N+2} \times W_h^{N+2}$ *such that* $\overline{\eta}_h^0 = 0$ *and, for given* $\left( \mathcal{S}^{n+1} \right)_{n=0}^{N} \in L^2(\Omega)^{N+1}$

– *For any* $n \in [\![0, N]\!]$, *for all* $\varphi \in W_h$:

$$\left( \partial^1 \overline{\eta}_h^{n+1}, \varphi \right)_{L^2(\Omega)} + \left( \mathrm{div} \eta_h^{n+\frac{1}{2}}, \varphi \right)_{L^2(\Omega)} = \left( \mathcal{S}^{n+1}, \varphi \right)_{L^2(\Omega)}, \tag{21}$$

– *For any* $n \in [\![0, N+1]\!]$:

$$(\eta_h^n, \psi)_{L^2(\Omega)^d} = (\overline{\eta}_h^n, \mathrm{div} \psi)_{L^2(\Omega)}, \qquad \forall \psi \in V_h^{\mathrm{div}}. \tag{22}$$

*Then, the following* $L^2(H_{\mathrm{div}})$–*a priori estimate holds:*

$$\left( \sum_{n=0}^{N+1} k \| \mathrm{div} \eta_h^n \|_{L^2(\Omega)}^2 \right)^{\frac{1}{2}} \leq C\mathcal{S}, \tag{23}$$

*where* $\mathcal{S} = \max\limits_{n=0}^{N} \| \mathcal{S}^{n+1} \|_{L^2(\Omega)}$.

*Proof.* Taking $\varphi = \partial^1 \overline{\eta}_h^{n+1}$ in (21) implies that

$$\| \partial^1 \overline{\eta}_h^{n+1} \|_{L^2(\Omega)}^2 + \left( \mathrm{div} \eta_h^{n+\frac{1}{2}}, \partial^1 \overline{\eta}_h^{n+1} \right)_{L^2(\Omega)} = \left( \mathcal{S}^{n+1}, \partial^1 \overline{\eta}_h^{n+1} \right)_{L^2(\Omega)}. \tag{24}$$

Writing (22) in the level $n+1$ and acting $\partial^1$ on the both sides of the result yield

$$\left( \partial^1 \eta_h^{n+1}, \psi \right)_{L^2(\Omega)^d} = \left( \partial^1 \overline{\eta}_h^{n+1}, \mathrm{div} \psi \right)_{L^2(\Omega)}, \qquad \forall \psi \in V_h^{\mathrm{div}}. \tag{25}$$

Taking now $\psi = \eta_h^{n+\frac{1}{2}}$ in (25) yields

$$\left( \partial^1 \eta_h^{n+1}, \eta_h^{n+\frac{1}{2}} \right)_{L^2(\Omega)^d} = \left( \partial^1 \overline{\eta}_h^{n+1}, \mathrm{div} \eta_h^{n+\frac{1}{2}} \right)_{L^2(\Omega)}. \tag{26}$$

From (24) and (26), we deduce that

$$\| \partial^1 \overline{\eta}_h^{n+1} \|_{L^2(\Omega)}^2 + \left( \partial^1 \eta_h^{n+1}, \eta_h^{n+\frac{1}{2}} \right)_{L^2(\Omega)^d} = \left( \mathcal{S}^{n+1}, \partial^1 \overline{\eta}_h^{n+1} \right)_{L^2(\Omega)}. \tag{27}$$

Multiplying both sides of (27) by $2k$, using the equality $(a+b)(a-b) = a^2 - b^2$, and summing the result over $n \in [\![0, J]\!]$, for all $J \in [\![0, N]\!]$, lead to

$$2 \sum_{n=0}^{J} k \|\partial^1 \overline{\eta}_h^{n+1}\|_{L^2(\Omega)}^2 + \|\eta_h^{J+1}\|_{L^2(\Omega)}^2 - \|\eta_h^0\|_{L^2(\Omega)}^2 = 2 \sum_{n=0}^{J} k \left( \mathcal{S}^{n+1}, \partial^1 \overline{\eta}_h^{n+1} \right)_{L^2(\Omega)}.$$

(28)

From the hypothesis $\overline{\eta}_h^0 = 0$ and Eq. (22), we deduce that $\eta_h^0 = 0$. Gathering this with (28) and taking $J = N$ imply that

$$\sum_{n=0}^{N+1} k \|\partial^1 \overline{\eta}_h^n\|_{L^2(\Omega)}^2 \leq \sum_{n=1}^{N+1} k \left( \mathcal{S}^n, \partial^1 \overline{\eta}_h^n \right)_{L^2(\Omega)}.$$

Gathering this with the Cauchy Schwarz inequality and inequality implies that

$$\left( \sum_{n=0}^{N+1} k \|\partial^1 \overline{\eta}_h^n\|_{L^2(\Omega)}^2 \right)^{\frac{1}{2}} \leq C\mathcal{S}.$$

(29)

On the other hand, using the Cauchy Schwarz inequality, Eq. (21) implies that, for all $\varphi \in W_h$

$$\left( \text{div} \eta_h^{n+1}, \varphi \right)_{L^2(\Omega)} \leq \left( \|\mathcal{S}^{n+1}\|_{L^2(\Omega)} + \|\partial^1 \overline{\eta}_h^{n+1}\|_{L^2(\Omega)} \right) \|\varphi\|_{L^2(\Omega)}.$$

This gives, for all $\varphi \in W_h \setminus \{0\}$

$$\frac{\left( \text{div} \eta_h^{n+1}, \varphi \right)_{L^2(\Omega)}}{\|\varphi\|_{L^2(\Omega)}} \leq \|\mathcal{S}^{n+1}\|_{L^2(\Omega)} + \|\partial^1 \overline{\eta}_h^{n+1}\|_{L^2(\Omega)}.$$

(30)

Taking $\varphi = \text{div} \eta_h^{n+1}$ in (30) (this is feasible since $\text{div} V_h^{\text{div}} \subset W_h$)

$$\|\text{div} \eta_h^{n+1}\|_{L^2(\Omega)} \leq \|\mathcal{S}^{n+1}\|_{L^2(\Omega)} + \|\partial^1 \overline{\eta}_h^{n+1}\|_{L^2(\Omega)}.$$

(31)

This gives, thanks to inequality $(a+b)^2 \leq 2(a^2 + b^2)$ and to the definition of $\mathcal{S}$

$$\|\text{div} \eta_h^{n+1}\|_{L^2(\Omega)}^2 \leq C \left( (\mathcal{S})^2 + \|\partial^1 \overline{\eta}_h^{n+1}\|_{L^2(\Omega)}^2 \right).$$

(32)

Multiplying both sides of (32) by $k$, summing the result over $n \in [\![0, N]\!]$, and using estimate (29) yield

$$\left( \sum_{n=0}^{N} k \|\text{div} \eta_h^{n+1}\|_{L^2(\Omega)}^2 \right)^{\frac{1}{2}} \leq C\mathcal{S}.$$

Which is the desired estimate (23). This completes the proof of Lemma 1.     □

## Proof Sketch for Theorem 1

**1. Existence and uniqueness for the scheme** (15)–(17). The existence and uniqueness can be justified successively on $n$ and using the fact that these schemes lead to systems with $\mathcal{N} \times \mathcal{N}$ square matrices, with $\mathcal{N} = \dim V_h^{\mathrm{div}} + \dim W_h$.

**2. Proof of estimate** (20). Let us consider the following auxiliary problem: Find $(\bar{p}_h^n, \bar{u}_h^n) \in V_h^{\mathrm{div}} \times W_h$ such that

$$(\mathrm{div}\,\bar{p}_h^n, \varphi)_{L^2(\Omega)} = (-\Delta u(t_n), \varphi)_{L^2(\Omega)}, \qquad \forall \varphi \in W_h \tag{33}$$

and

$$(\bar{p}_h^n, \psi)_{L^2(\Omega)} = (\bar{u}_h^n, \mathrm{div}\,\psi)_{L^2(\Omega)}, \qquad \forall \psi \in V_h^{\mathrm{div}}. \tag{34}$$

The following error estimate are proved in [7, (7.2.26), p. 237], for all $n \in [\![0, N+1]\!]$

$$\|\nabla u(t_n) + \bar{p}_h^n\|_{H_{\mathrm{div}}(\Omega)} + \|u(t_n) - \bar{u}_h^n\|_{L^2(\Omega)} \le Ch. \tag{35}$$

Writing (33) in the level $n+1$ and taking the mean value of the result with (33) yield

$$\left(\mathrm{div}\,\bar{p}_h^{n+\frac{1}{2}}, \varphi\right)_{L^2(\Omega)} = \left(-\frac{\Delta u(t_{n+1}) + \Delta u(t_n)}{2}, \varphi\right)_{L^2(\Omega)}, \qquad \forall \varphi \in W_h \tag{36}$$

Subtracting (36) and (34) from respectively (15) and (16) and using (1) yield, for all $\varphi \in W_h$

$$\left(\partial^1 u_h^{n+1}, \varphi\right)_{L^2(\Omega)} + \left(\mathrm{div}\,\eta_h^{n+\frac{1}{2}}, \varphi\right)_{L^2(\Omega)} = \left(\frac{u_t(t_{n+1}) + u_t(t_n)}{2}, \varphi\right)_{L^2(\Omega)} \tag{37}$$

and

$$(\eta_h^n, \psi)_{L^2(\Omega)^d} = (\bar{\eta}_h^n, \mathrm{div}\,\psi)_{L^2(\Omega)}, \qquad \forall \psi \in V_h^{\mathrm{div}}, \tag{38}$$

where $\eta_h^n = p_h^n - \bar{p}_h^n \in V_h^{\mathrm{div}}$ and $\bar{\eta}_h^n = u_h^n - \bar{u}_h^n \in W_h$.
Subtracting $\left(\partial^1 \bar{u}_h^{n+1}, \varphi\right)_{L^2(\Omega)}$ from both sides of (37) and using (1) yields

$$\left(\partial^1 \bar{\eta}_h^{n+1}, \varphi\right)_{L^2(\Omega)} + \left(\mathrm{div}\,\eta_h^{n+\frac{1}{2}}, \varphi\right)_{L^2(\Omega)} = (\mathcal{S}^{n+1}, \varphi)_{L^2(\Omega)}, \qquad \forall \varphi \in W_h, \tag{39}$$

where

$$\mathcal{S}^{n+1} = \frac{u_t(t_{n+1}) + u_t(t_n)}{2} - \partial^1 \bar{u}_h^{n+1}. \tag{40}$$

In addition to this, taking $n = 0$ in (33)–(34), using the fact that $u(0) = u^0$ (subject of (2)), comparing the result with (17)–(18), and using the uniqueness proved in the above item **1.** of this proof imply that $p_h^0 = \bar{p}_h^0$ and $u_h^0 = \bar{u}_h^0$. This implies that

$$\eta_h^0 = 0 \qquad \text{and} \qquad \bar{\eta}_h^0 = 0. \tag{41}$$

From (39) and (41) we deduce that $\left((\eta_{\mathcal{D}}^n)_{n=0}^{N+1}, (\overline{\eta}_{\mathcal{D}}^n)_{n=0}^{N+1}\right)$ satisfies the hypotheses of Lemma 1. Applying now (23) of Lemma 1 imply

$$\left(\sum_{n=0}^{N+1} k\|\mathrm{div}\eta_h^n\|_{L^2(\Omega)}^2\right)^{\frac{1}{2}} \leq C\mathcal{S}. \tag{42}$$

To get estimates on $\mathcal{S}$, we apply the discrete operators $\partial^1$ on the both sides of (33)–(34) to get, for all $n \in [\![1, N+1]\!]$

$$\left(\mathrm{div}\partial^1\overline{p}_h^n, \varphi\right)_{L^2(\Omega)} = \left(-\Delta\partial^1 u(t_n), \varphi\right)_{L^2(\Omega)}, \qquad \forall\varphi \in W_h \tag{43}$$

and

$$\left(\partial^1\overline{p}_h^n, \psi\right)_{L^2(\Omega)} = \left(\partial^1\overline{u}_h^n, \mathrm{div}\psi\right)_{L^2(\Omega)}, \qquad \forall\psi \in V_h^{\mathrm{div}}. \tag{44}$$

This implies that $(\partial^1\overline{p}_h^n, \partial^1\overline{u}_h^n) \in V_h^{\mathrm{div}} \times W_h$ satisfies the same scheme (33)–(34) but with right hand side $-\Delta\partial^1 u(t_n)$ instead of $-\Delta u(t_n)$. This allows to apply error estimates (35) to get

$$\|\partial^1 u(t_n) - \partial^1\overline{u}_h^n\|_{L^2(\Omega)} \leq Ch. \tag{45}$$

The following estimate can be justified using Taylor expansion, for $u \in \mathcal{C}^3([0, T])$

$$\left\|\partial^1 u(t_n) - \frac{u_t(t_{n+1}) + u_t(t_n)}{2}\right\|_{\mathcal{C}(\overline{\Omega})} \leq Ck^2. \tag{46}$$

Gathering now the triangle inequality, (40), and (46) implies that

$$\mathcal{S} = \max_{n=0}^{N} \|\mathcal{S}^{n+1}\|_{L^2(\Omega)} \leq C\left(k^2 + h\right). \tag{47}$$

This with (42) yield

$$\left(\sum_{n=0}^{N+1} k\|\mathrm{div}\eta_h^n\|_{L^2(\Omega)}^2\right)^{\frac{1}{2}} \leq C\left(k^2 + h\right). \tag{48}$$

Using the triangle inequality, the fact that $\Delta u(t_n) + \mathrm{div}p_h^n = \Delta u(t_n) + \mathrm{div}\overline{p}_h^n + \mathrm{div}\eta_h^n$ and estimates (35) and (48) yield the desired estimate (20).

This completes the proof of Theorem 1.                                    □

## 5    Conclusion and Perspectives

We established a numerical scheme using Lowest Order Raviart-Thomas Mixed Finite Element methods as discretization in space and Crank-Nicolson finite difference method as discretization in time. We proved a new convergence result towards $-\nabla u(t)$ in the norm of $L^2(H_{\mathrm{div}})$, under assumption that the solution is smooth. The order in time is two and is one in space. One of the main perspectives in the near future is to extend this result to a large class of mixed finite elements.

# References

1. Benkhaldoun, F., Bradji, A.: Two new error estimates of a fully discrete primal-dual mixed finite element scheme for parabolic equations in any space dimension (2020, Submitted)
2. Braess, D.: Finite Elements. Theory, Fast Solvers and Applications in Solid Mechanics. Translated from German by Larry L. Schumaker, 3rd edn. Cambridge University Press, Cambridge (2007)
3. Ciarlet, P.G.: The Finite Element Method for Elliptic Problems. Classics in Applied Mathematics, vol. 40. Society for Industrial and Applied Mathematics (SIAM), Philadelphia (2002)
4. Chatzipandtelidis, P., Lazarov, R.D., Thomée, V.: Error estimates for a finite volume element method for parabolic equations on convex polygonal domain. Numer. Methods Partial Differ. Equ. **20**, 650–674 (2004)
5. Eymard, R., Gallouët, T., Herbin, R.: Finite Volume Methods. In: Ciarlet, P.G., Lions, J.L. (eds.) Handbook of Numerical Analysis, North-Holland, Amsterdam, vol. 7, pp. 723–1020 (2000)
6. Johnson, C., Thomee, V.: Error estimates for some mixed finite element methods for parabolic type problems. RAIRO Anal. Numér. **15**(1), 41–78 (1981)
7. Bartels, S.: Numerical Approximation of Partial Differential Equations. TAM, vol. 64. Springer, Cham (2016). https://doi.org/10.1007/978-3-319-32354-1
8. Thomée, V.: Time discretization by the discontinuous Galerkin method. In: Galerkin Finite Element Methods for Parabolic Problems. LNM, vol. 1054, pp. 126–148. Springer, Heidelberg (1984). https://doi.org/10.1007/BFb0071799
9. Weng, Z., Feng, X., Huang, P.: A new mixed finite element method based on the Crank-Nicolson scheme for the parabolic problems. Appl. Math. Model. **36**(10), 5068–5079 (2012)

# A Finite Volume Scheme for a Wave Equation with Several Time Independent Delays

Fayssal Benkhaldoun[1] , Abdallah Bradji[1,2](✉) , and Tarek Ghoudi[1]

[1] LAGA, University of Paris 13, Paris, France
{fayssal,ghoudi}@math.univ-paris13.fr
[2] LMA Laboratory, University of Annaba, Annaba, Algeria
abdallah.bradji@gmail.com, abdallah.bradji@etu.univ-amu.fr
https://www.math.univ-paris13.fr/~fayssal/
https://www.i2m.univ-amu.fr/perso/abdallah.bradji/

**Abstract.** We establish a new finite volume scheme for a second order hyperbolic equation with several time independent delays in any space dimension. This model is considered in [7] where some exponential stability estimates are proved and in [8] which dealt with the oscillation of the solutions. The delays are included in both the exact solution and its time derivative. The scheme uses, as space discretization, SUSHI (Scheme using Stabilization and Hybrid Interfaces) developed in [5]. We first prove the existence and uniqueness of the discrete solution. We subsequently, develop a new discrete *a priori estimate*. Thanks to this *a priori estimate*, we prove error estimates in several discrete seminorms.

This work is an extension and improvement of our recent work [2] which dealt with the finite volume approximation of the wave equation but with only one delay which is included in the time derivative of the exact solution.

**Keywords:** Several time independent delays · SUSHI · Apriori estimates

**MSC2010:** 65M08 · 65M12 · 65M15 · 35L20

## 1 Problem to be Solved and Motivation

We consider the following second order hyperbolic equation with several time delays (see [7, Page 1563] and [8,9]):

$$u_{tt}(x,t) - \Delta u(x,t) + \alpha_0 u_t(x,t) + \alpha_1 u_t(x,t-\tau_1) + \alpha_2 u_t(x,t-\tau_2)$$
$$+ \beta_0 u(x,t) + \beta_1 u(x,t-\tau_3) + \beta_2 u(x,t-\tau_4) = f(x,t), \qquad (x,t) \in \Omega \times (0,T), \quad (1)$$

Supported by MCS team (LAGA Laboratory) of the "Université Sorbonne- Paris Nord".

I. Lirkov and S. Margenov (Eds.): LSSC 2021, LNCS 13127, pp. 498–506, 2022.
https://doi.org/10.1007/978-3-030-97549-4_57

where $\Omega$ is an open polygonal bounded subset in $\mathbb{R}^d$, $f$ is a given function defined on $\Omega \times (0, T)$, and $T > 0$, $\alpha_0, \alpha_1, \alpha_2, \beta_0, \beta_1, \beta_2 \geq 0$, $\tau_1, \tau_2, \tau_3, \tau_4 > 0$ (the time delays) are given.

Initial conditions are defined by

$$u(\boldsymbol{x}, 0) = u^0(\boldsymbol{x}) \quad \text{and} \quad u_t(\boldsymbol{x}, t) = u^1(\boldsymbol{x}, t), \qquad \boldsymbol{x} \in \Omega, \quad -\tau \leq t \leq 0, \qquad (2)$$

where $u^0$ and $u^1$ are two given functions defined respectively on $\Omega$ and $\Omega \times (-\tau, 0)$ with $\tau = \max\{\tau_1, \tau_2, \tau_3, \tau_4\}$.

Homogeneous Dirichlet boundary conditions are given by

$$u(\boldsymbol{x}, t) = 0, \qquad (\boldsymbol{x}, t) \in \partial\Omega \times (0, T). \qquad (3)$$

Delay differential equations appear in many applications, see [1,6,9]. However, there is huge literature on the numerical methods for ordinary and partial differential equations but there is a lack of these numerical methods for equations with Delay Partial Differential Equations, cf. [1, Pages 9–19]. In this contribution, we consider a finite volume method based on the uses of SUSHI [5] for the model (1)–(3) in any space dimension. Such model is considered for instance in [7, (1.12)–(1.16), Page 1563] where some exponential stability estimates are proved. Equation (1) generalizes then the wave equation treated in our previous works [3,4] and differs in several terms: presence of $u_t(t)$ and $u(t)$, and their delays. This work is also an extension of the previous work [2] which dealt with the wave equation but with only one delay which is included in the time derivative of the exact solution. Such differences between the present work and previous ones needed to provide a new suitable formulation for a finite volume scheme and to adapt carefully the techniques in order to prove a discrete stability as well as an optimal convergence order. The first aim in this work is to establish a new implicit finite volume scheme for (1)–(3) in which several delays are involved in the exact solution and its time derivative. The second aim is to justify the well-posedness of this scheme and prove its convergence in several norms.

## 2    Space and Time Discretizations and Some Preliminaries

**Definition 1 (Space discretization, cf. [5]).** *Let $\Omega$ be a polyhedral open bounded subset of $\mathbb{R}^d$, where $d \in \mathbb{N} \setminus \{0\}$, and $\partial\Omega = \overline{\Omega} \setminus \Omega$ its boundary. A discretization of $\Omega$, denoted by $\mathcal{D}$, is defined as the triplet $\mathcal{D} = (\mathcal{M}, \mathcal{E}, \mathcal{P})$, where:*

1. *$\mathcal{M}$ is a finite family of non empty connected open disjoint subsets of $\Omega$ (the "control volumes") such that $\overline{\Omega} = \cup_{K \in \mathcal{M}} \overline{K}$. For any $K \in \mathcal{M}$, let $\partial K = \overline{K} \setminus K$ be the boundary of $K$; let $\mathrm{m}(K) > 0$ denote the measure of $K$ and $h_K$ denote the diameter of $K$.*
2. *$\mathcal{E}$ is a finite family of disjoint subsets of $\overline{\Omega}$ (the "edges" of the mesh), such that, for all $\sigma \in \mathcal{E}$, $\sigma$ is a non empty open subset of a hyperplane of $\mathbb{R}^d$, whose $(d-1)$-dimensional measure is strictly positive. We also assume that, for all $K \in \mathcal{M}$, there exists a subset $\mathcal{E}_K$ of $\mathcal{E}$ such that $\partial K = \cup_{\sigma \in \mathcal{E}_K} \overline{\sigma}$. For*

any $\sigma \in \mathcal{E}$, we denote by $\mathcal{M}_\sigma = \{K, \sigma \in \mathcal{E}_K\}$. We then assume that, for any $\sigma \in \mathcal{E}$, either $\mathcal{M}_\sigma$ has exactly one element and then $\sigma \subset \partial\Omega$ (the set of these interfaces, called boundary interfaces, denoted by $\mathcal{E}_{\text{ext}}$) or $\mathcal{M}_\sigma$ has exactly two elements (the set of these interfaces, called interior interfaces, denoted by $\mathcal{E}_{\text{int}}$). For all $\sigma \in \mathcal{E}$, we denote by $\boldsymbol{x}_\sigma$ the barycentre of $\sigma$. For all $K \in \mathcal{M}$ and $\sigma \in \mathcal{E}$, we denote by $\boldsymbol{n}_{K,\sigma}$ the unit vector normal to $\sigma$ outward to $K$.

3. $\mathcal{P}$ is a family of points of $\Omega$ indexed by $\mathcal{M}$, denoted by $\mathcal{P} = (\boldsymbol{x}_K)_{K \in \mathcal{M}}$, such that for all $K \in \mathcal{M}$, $\boldsymbol{x}_K \in K$ and $K$ is assumed to be $\boldsymbol{x}_K$–star-shaped, which means that for all $\boldsymbol{x} \in K$, the property $[\boldsymbol{x}_K, \boldsymbol{x}] \subset K$ holds. Denoting by $d_{K,\sigma}$ the Euclidean distance between $\boldsymbol{x}_K$ and the hyperplane including $\sigma$, one assumes that $d_{K,\sigma} > 0$. We then denote by $\mathcal{D}_{K,\sigma}$ the cone with vertex $\boldsymbol{x}_K$ and basis $\sigma$.

The time discretization is performed with a constrained time step-size $k$ such that $\dfrac{\tau}{k} \in \mathbb{N}$. We set then $k = \dfrac{\tau}{M}$, where $M \in \mathbb{N} \setminus \{0\}$. Denote by $N$ the integer part of $\dfrac{T}{k}$, i.e. $N = \left[\dfrac{T}{k}\right]$. We shall denote $t_n = nk$, for $n \in [\![-M, N]\!]$. As particular cases $t_{-M} = -\tau$, $t_0 = 0$, and $t_N \leq T$. One of the advantages of this time discretization is that the point $t = 0$ is a mesh point which is suitable since we have Eq. (1) defined for $t \in (0, T)$ and the second initial condition in (2) is defined for $t \in (-\tau, 0)$. We denote by $\partial^1$ and $\partial^2$ the discrete first and second time derivatives given by $\partial^1 v^{j+1} = \dfrac{v^{j+1} - v^j}{k}$ and $\partial^2 v^{j+1} = \partial^1(\partial^1 v^{j+1})$.

Throughout this paper, the letter $C$ stands for a positive constant independent of the parameters of discretizations.

We define the discrete space $\mathcal{X}_{\mathcal{D},0}$ as the set of all $v = \left((v_K)_{K\in\mathcal{M}}, (v_\sigma)_{\sigma\in\mathcal{E}}\right)$, where $v_K, v_\sigma \in \mathbb{R}$ and $v_\sigma = 0$ for all $\sigma \in \mathcal{E}_{\text{ext}}$. Let $H_\mathcal{M}(\Omega) \subset L^2(\Omega)$ be the space of functions which are constant on each control volume $K$ of the mesh $\mathcal{M}$. For all $v \in \mathcal{X}_\mathcal{D}$, we denote by $\Pi_\mathcal{M} v \in H_\mathcal{M}(\Omega)$ the function defined by $\Pi_\mathcal{M} v(\boldsymbol{x}) = v_K$, for a.e. $\boldsymbol{x} \in K$, for all $K \in \mathcal{M}$. To analyze the convergence, we consider the size of the discretization $\mathcal{D}$ defined by $h_\mathcal{D} = \sup\{\operatorname{diam}(K), K \in \mathcal{M}\}$ and the regularity of the mesh $\theta_\mathcal{D} = \max\left(\displaystyle\max_{\sigma\in\mathcal{E}_{\text{int}}, K,L\in\mathcal{M}} \frac{d_{K,\sigma}}{d_{L,\sigma}}, \displaystyle\max_{K\in\mathcal{M},\sigma\in\mathcal{E}_K} \frac{h_K}{d_{K,\sigma}}\right)$.

The scheme we shall present is based on the discrete gradient of [5]. For $u \in \mathcal{X}_\mathcal{D}$, we define, for all $K \in \mathcal{M}$

$$\nabla_\mathcal{D} u(\boldsymbol{x}) = \nabla_K u + \left(\frac{\sqrt{d}}{d_{K,\sigma}}\left(u_\sigma - u_K - \nabla_K u \cdot (\boldsymbol{x}_\sigma - \boldsymbol{x}_K)\right)\right)\boldsymbol{n}_{K,\sigma}, \quad (4)$$

$$\text{a.e. } \boldsymbol{x} \in \mathcal{D}_{K,\sigma},$$

where $\nabla_K u = \dfrac{1}{m(K)} \displaystyle\sum_{\sigma\in\mathcal{E}_K} m(\sigma)(u_\sigma - u_K)\boldsymbol{n}_{K,\sigma}$. We define the bilinear form $\langle \cdot, \cdot \rangle_F$ defined on $\mathcal{X}_{\mathcal{D},0} \times \mathcal{X}_{\mathcal{D},0}$ by $\langle u, v \rangle_F = \displaystyle\int_\Omega \nabla_\mathcal{D} u(\boldsymbol{x}) \cdot \nabla_\mathcal{D} v(\boldsymbol{x}) d\boldsymbol{x}$.

## 3  Formulation of a New Finite Volume Scheme for the Problem (1)–(3)

We now set a new implicit finite volume scheme for (1)–(3). The unknowns of this scheme are $\{u_{\mathcal{D}}^n; n \in [\![-M, N]\!]\}$ which are expected to approximate the unknowns $\{u(t_n); n \in [\![-M, N]\!]\}$.

1. **Approximation of initial conditions** (2). The discretization of initial conditions (2) can be performed as: Find $u_{\mathcal{D}}^n$ for $n \in [\![-M, 0]\!]$ such that for all $v \in \mathcal{X}_{\mathcal{D},0}$, for all $n \in [\![-M + 1, 0]\!]$

$$\langle u_{\mathcal{D}}^0, v \rangle_F = -\left(\Delta u^0, \Pi_{\mathcal{M}} v\right)_{L^2(\Omega)} \quad \text{and} \tag{5}$$

$$\langle \partial^1 u_{\mathcal{D}}^n, v \rangle_F = -\left(\Delta u^1(t_n), \Pi_{\mathcal{M}} v\right)_{L^2(\Omega)}. \tag{6}$$

2. **Approximation of** (1) **and** (3). For any $n \in [\![0, N-1]\!]$, find $u_{\mathcal{D}}^{n+1} \in \mathcal{X}_{\mathcal{D},0}$ such that

$$\left(\partial^2 \Pi_{\mathcal{M}} u_{\mathcal{D}}^{n+1}, \Pi_{\mathcal{M}} v\right)_{L^2(\Omega)} + \langle u_{\mathcal{D}}^{n+1}, v \rangle_F + \alpha_0 \left(\partial^1 \Pi_{\mathcal{M}} u_{\mathcal{D}}^{n+1}, \Pi_{\mathcal{M}} v\right)_{L^2(\Omega)}$$

$$+ \alpha_1 \left(\partial^1 \Pi_{\mathcal{M}} u_{\mathcal{D}}^{n+1-M_1}, \Pi_{\mathcal{M}} v\right)_{L^2(\Omega)} + \alpha_2 \left(\partial^1 \Pi_{\mathcal{M}} u_{\mathcal{D}}^{n+1-M_2}, \Pi_{\mathcal{M}} v\right)_{L^2(\Omega)}$$

$$+ \beta_0 \left(\Pi_{\mathcal{M}} u_{\mathcal{D}}^{n+1}, \Pi_{\mathcal{M}} v\right)_{L^2(\Omega)} + \beta_1 \left(\Pi_{\mathcal{M}} u_{\mathcal{D}}^{n+1-M_3}, \Pi_{\mathcal{M}} v\right)_{L^2(\Omega)}$$

$$+ \beta_2 \left(\Pi_{\mathcal{M}} u_{\mathcal{D}}^{n+1-M_4}, \Pi_{\mathcal{M}} v\right)_{L^2(\Omega)} = (f(t_{n+1}), \Pi_{\mathcal{M}} v)_{L^2(\Omega)}, \forall v \in \mathcal{X}_{\mathcal{D},0}, \tag{7}$$

where, for $i \in \{1, 2, 3, 4\}$, $M_i = \left[\dfrac{\tau_i}{k}\right]$. We assume that

$$k < \min\{\tau_1, \tau_2, \tau_3, \tau_4\}. \tag{8}$$

This implies that $M_i \geq 1$, for all $i \in \{1, 2, 3, 4\}$.

## 4  Convergence Analysis of Scheme (5)–(7)

In addition to the new formulation (5)–(7) of a scheme approximating the second order hyperbolic problem with several delays (1)–(3), we present also its existence, uniqueness, and error estimates when the exact solution is smooth.

**Theorem 1.** *(New error estimates for (5)–(7)) Let $\Omega$ be a polyhedral open bounded subset of $\mathbb{R}^d$, where $d \in \mathbb{N} \setminus \{0\}$. Assume that the solution of (1)–(3) satisfies $u \in \mathcal{C}^3([-\tau, T]; \mathcal{C}^2(\overline{\Omega}))$, where $\tau$ is the maximum of the delays $\{\tau_1, \tau_2, \tau_3, \tau_4\}$ appearing in the wave Eq. (1) with delays. Let $k = \dfrac{\tau}{M}$ satisfying (8), where $M \in \mathbb{N} \setminus \{0\}$. Denote by $N$ the integer part of $\dfrac{T}{k}$. We shall denote $t_n = nk$, for $n \in [\![-M, N]\!]$. Let $\mathcal{D} = (\mathcal{M}, \mathcal{E}, \mathcal{P})$ be a discretization in the sense of Definition 1. Assume that $\theta_{\mathcal{D}}$ satisfies $\theta \geq \theta_{\mathcal{D}}$. Let $\nabla_{\mathcal{D}}$ be the discrete gradient given by (4). Then, there exists a unique solution $(u_{\mathcal{D}}^n)_{n=-M}^N \in \mathcal{X}_{\mathcal{D},0}^{M+N+1}$ for scheme (5)–(7) and the following error estimates hold:*

$-$ $L^\infty(L^2)$ and $L^\infty(H_0^1)$ error estimates.

$$\max_{n=0}^{n=N}\|u(t_n) - \Pi_{\mathcal{M}}u_{\mathcal{D}}^n\|_{L^2(\Omega)} + \max_{n=0}^{n=N}\|\nabla u(t_n) - \nabla_{\mathcal{D}}u_{\mathcal{D}}^n\|_{L^2(\Omega)} \le C(k + h_{\mathcal{D}}). \quad (9)$$

$-$ $W^{1,\infty}(L^2)$-estimate.

$$\max_{n=-M+1}^{n=N}\|u_t(t_n) - \Pi_{\mathcal{M}}\partial^1 u_{\mathcal{D}}^n\|_{L^2(\Omega)} \le C(k + h_{\mathcal{D}}). \quad (10)$$

To prove Theorem 1, we need to use the following new discrete *a priori* estimate:

**Lemma 1.** *(New discrete a priori estimate) Under the same hypotheses of Theorem 1, assume that there exists* $(\eta_{\mathcal{D}}^n)_{n=-M}^N \in (\mathcal{X}_{\mathcal{D},0})^{M+N+1}$ *such that for all* $n \in [\![0, N-1]\!]$ *and for all* $v \in \mathcal{X}_{\mathcal{D},0}$

$$\left(\partial^2 \Pi_{\mathcal{M}}\eta_{\mathcal{D}}^{n+1}, \Pi_{\mathcal{M}}v\right)_{L^2(\Omega)} + \langle\eta_{\mathcal{D}}^{n+1}, v\rangle_F + \alpha_0\left(\partial^1 \Pi_{\mathcal{M}}\eta_{\mathcal{D}}^{n+1}, \Pi_{\mathcal{M}}v\right)_{L^2(\Omega)}$$

$$+ \alpha_1\left(\partial^1 \Pi_{\mathcal{M}}\eta_{\mathcal{D}}^{n+1-M_1}, \Pi_{\mathcal{M}}v\right)_{L^2(\Omega)} + \alpha_2\left(\partial^1 \Pi_{\mathcal{M}}\eta_{\mathcal{D}}^{n+1-M_2}, \Pi_{\mathcal{M}}v\right)_{L^2(\Omega)}$$

$$+ \beta_0\left(\Pi_{\mathcal{M}}\eta_{\mathcal{D}}^{n+1}, \Pi_{\mathcal{M}}v\right)_{L^2(\Omega)} + \beta_1\left(\Pi_{\mathcal{M}}\eta_{\mathcal{D}}^{n+1-M_3}, \Pi_{\mathcal{M}}v\right)_{L^2(\Omega)}$$

$$+ \beta_2\left(\Pi_{\mathcal{M}}\eta_{\mathcal{D}}^{n+1-M_4}, \Pi_{\mathcal{M}}v\right)_{L^2(\Omega)}$$

$$= \left(\mathcal{S}^{n+1}, \Pi_{\mathcal{M}}v\right)_{L^2(\Omega)}, \quad (11)$$

*where* $\mathcal{S}^{n+1} \in L^2(\Omega)$, *for all* $n \in [\![0, N-1]\!]$. *Then, the following estimate holds, for all* $J \in [\![1, N]\!]$

$$\mathbb{E}_{\mathcal{D}}^J \le C\left((\mathcal{S})^2 + \max_{n=-M}^0 \|\Pi_{\mathcal{M}}\eta_{\mathcal{D}}^n\|_{L^2(\Omega)}^2\right.$$

$$\left. + \max_{n=1-M}^0 \|\partial^1 \Pi_{\mathcal{M}}\eta_{\mathcal{D}}^n\|_{L^2(\Omega)}^2 + \|\nabla_{\mathcal{D}}\eta_{\mathcal{D}}^0\|_{L^2(\Omega)}^2\right), \quad (12)$$

*where*

$$\mathbb{E}_{\mathcal{D}}^n = \|\partial^1 \Pi_{\mathcal{M}}\eta_{\mathcal{D}}^n\|_{L^2(\Omega)}^2 + \|\nabla_{\mathcal{D}}\eta_{\mathcal{D}}^n\|_{L^2(\Omega)}^2 \text{ and } \mathcal{S} = \max_{n=0}^{N-1}\|\mathcal{S}^{n+1}\|_{L^2(\Omega)}.$$

**Proof Sketch for Lemma 1**

Let us sketch the main ideas of the proof of Lemma 1. Taking $v = \partial^1\eta_{\mathcal{D}}^{n+1}$ in (11), using rules similar to $x(x - y) = (x - y)^2/2 + x^2/2 - y^2/2$, using the fact that $\alpha_0 \ge 0$, and multiplying the result by $2k$ to get, for all $n \in [\![0, N-1]\!]$

$$\overline{\mathbb{E}}_{\mathcal{D}}^{n+1} - \overline{\mathbb{E}}_{\mathcal{D}}^n \le (\mathbb{T}_1^n + \Pi_{\mathcal{M}}(\mathbb{T}_2^n + \mathbb{T}_3^n + \mathbb{T}_4^n + \mathbb{T}_5^n), \Pi_{\mathcal{M}}v)_{L^2(\Omega)}, \quad (13)$$

where, with $v = \partial^1\eta_{\mathcal{D}}^{n+1}$

$$\mathbb{T}_1^n = 2k\mathcal{S}^{n+1}, \quad \mathbb{T}_2^n = -2k\alpha_1\partial^1\eta_{\mathcal{D}}^{n+1-M_1}, \quad \mathbb{T}_3^n = -2k\alpha_2\partial^1\eta_{\mathcal{D}}^{n+1-M_2}$$

$$\mathbb{T}_4^n = -2k\beta_1\eta_{\mathcal{D}}^{n+1-M_3}, \quad \mathbb{T}_5^n = -2k\beta_2\eta_{\mathcal{D}}^{n+1-M_4}$$

and

$$\overline{\mathbb{E}}_{\mathcal{D}}^n = \|\partial^1 \Pi_{\mathcal{M}} \eta_{\mathcal{D}}^n\|_{L^2(\Omega)}^2 + \|\nabla_{\mathcal{D}} \eta_{\mathcal{D}}^n\|_{L^2(\Omega)}^2 + \beta_0 \|\Pi_{\mathcal{M}} \eta_{\mathcal{D}}^n\|_{L^2(\Omega)}^2.$$

Summing (13) over $n \in [\![0, J-1]\!]$, where $J \in [\![1, N]\!]$, and using the Cauchy Schwarz inequality yield

$$\overline{\mathbb{E}}_{\mathcal{D}}^J \leq \overline{\mathbb{E}}_{\mathcal{D}}^0 + \sum_{n=0}^{J-1} \left(\mathbb{T}_1^n + \Pi_{\mathcal{M}}\left(\mathbb{T}_2^n + \mathbb{T}_3^n + \mathbb{T}_4^n + \mathbb{T}_5^n\right), \Pi_{\mathcal{M}} v\right)_{L^2(\Omega)}. \qquad (14)$$

Using the Cauchy Schwarz and Young inequality $ab \leq \epsilon a^2 + b^2/\epsilon$, and the discrete Poincaré inequality [5, Lemma 5.4], we get the following estimate

$$\overline{\mathbb{E}}_{\mathcal{D}}^J \leq C \left(\overline{\mathbb{E}}_{\mathcal{D}}^0 + (\mathcal{S})^2 + \max_{n=1-M}^0 \|\partial^1 \Pi_{\mathcal{M}} \eta_{\mathcal{D}}^n\|_{L^2(\Omega)}^2 + \max_{n=1-M}^0 \|\Pi_{\mathcal{M}} \eta_{\mathcal{D}}^n\|_{L^2(\Omega)}^2 \right)$$

$$+ \frac{k}{2\tau} \sum_{n=1}^J \|\partial^1 \Pi_{\mathcal{M}} \eta_{\mathcal{D}}^n\|_{L^2(\Omega)}^2 + C \sum_{n=1}^{J-1} k \|\nabla_{\mathcal{D}} \eta_{\mathcal{D}}^n\|_{L^2(\Omega)}^2. \qquad (15)$$

Gathering this estimate with the fact that $k/(2\tau) = 1/(2M) \leq 1/2$ and definition of $\overline{\mathbb{E}}_{\mathcal{D}}^n$ yield

$$\overline{\mathbb{E}}_{\mathcal{D}}^J \leq C \sum_{n=1}^{J-1} k \overline{\mathbb{E}}_{\mathcal{D}}^n + C \left((\mathcal{S})^2 \max_{n=1-M}^0 \|\partial^1 \Pi_{\mathcal{M}} \eta_{\mathcal{D}}^n\|_{L^2(\Omega)}^2 \right.$$

$$\left. + \max_{n=1-M}^0 \|\Pi_{\mathcal{M}} \eta_{\mathcal{D}}^n\|_{L^2(\Omega)}^2 + \|\nabla_{\mathcal{D}} \eta_{\mathcal{D}}^0\|_{L^2(\Omega)}^2 \right). \qquad (16)$$

Applying a discrete version for the Gronwall lemma (see [3, Lemma 4.7, Page 1303]) to this inequality and using the fact that $\beta_0 \geq 0$ yield the desired estimate (12). $\square$

**Proof sketch for Theorem 1**

**1. Existence and uniqueness for scheme** (5)–(7). The existence and uniqueness can be justified successively on $n$ and using the fact that (5) and (7) lead to systems with square matrices.

**2. Proof of estimates** (9)–(10). To prove (9)–(10), we compare (5)–(7) with the following auxiliary scheme: For any $n \in [\![-M, N]\!]$, find $\bar{u}_{\mathcal{D}}^n \in \mathcal{X}_{\mathcal{D},0}$ such that

$$\langle \bar{u}_{\mathcal{D}}^n, v \rangle_F = (-\Delta u(t_n), \Pi_{\mathcal{M}} v)_{L^2(\Omega)}, \quad \forall v \in \mathcal{X}_{\mathcal{D},0}. \qquad (17)$$

**2.1. Comparison between the solutions of** (17) **and** (1)–(3). The following convergence results hold, see [2–5]:

– Discrete $L^\infty(L^2)$ and $L^\infty(H^1)$–error estimates. For all $n \in [\![-M, N]\!]$

$$\|u(t_n) - \Pi_{\mathcal{M}} \bar{u}_{\mathcal{D}}^n\|_{L^2(\Omega)} + \|\nabla u(t_n) - \nabla_{\mathcal{D}} \bar{u}_{\mathcal{D}}^n\|_{(L^2(\Omega))^d} \leq C h_{\mathcal{D}}. \qquad (18)$$

– $\mathcal{W}^{j,\infty}(L^2)$–error estimate, for $j \in \{1, 2\}$.

$$\max_n = -M + 2^N \|u_{tt}(t_n) - \partial^2 \Pi_{\mathcal{M}} \bar{u}_{\mathcal{D}}^n\|_{L^2(\Omega)}$$

$$+ \max_{n=-M+1}^N \|u_t(t_n) - \partial^1 \Pi_{\mathcal{M}} \bar{u}_{\mathcal{D}}^n\|_{L^2(\Omega)} \leq C(h_{\mathcal{D}} + k). \qquad (19)$$

**2.2. Comparison between the schemes** (5)–(7) **and** (17)**.** Let us define the error $\eta_{\mathcal{D}}^n = u_{\mathcal{D}}^n - \bar{u}_{\mathcal{D}}^n$. Comparing (17) with the first scheme in (5) and using the fact that $u(0) = u^0$ (see (2)) imply that $\eta_{\mathcal{D}}^0 = 0$. Writing scheme (17) in the level $n + 1$ and subtracting the result from (7) to get, for all $n \in [\![0, N - 1]\!]$ and for all $v \in \mathcal{X}_{\mathcal{D},0}$

$$\left(\partial^2 \Pi_{\mathcal{M}} u_{\mathcal{D}}^{n+1}, \Pi_{\mathcal{M}} v\right)_{L^2(\Omega)} + \langle \eta_{\mathcal{D}}^{n+1}, v\rangle_F + \alpha_0 \left(\partial^1 \Pi_{\mathcal{M}} u_{\mathcal{D}}^{n+1}, \Pi_{\mathcal{M}} v\right)_{L^2(\Omega)}$$

$$+ \alpha_1 \left(\partial^1 \Pi_{\mathcal{M}} u_{\mathcal{D}}^{n+1-M_1}, \Pi_{\mathcal{M}} v\right)_{L^2(\Omega)} + \alpha_2 \left(\partial^1 \Pi_{\mathcal{M}} u_{\mathcal{D}}^{n+1-M_2}, \Pi_{\mathcal{M}} v\right)_{L^2(\Omega)}$$

$$+ \beta_0 \left(\Pi_{\mathcal{M}} u_{\mathcal{D}}^{n+1}, \Pi_{\mathcal{M}} v\right)_{L^2(\Omega)} + \beta_1 \left(\Pi_{\mathcal{M}} u_{\mathcal{D}}^{n+1-M_3}, \Pi_{\mathcal{M}} v\right)_{L^2(\Omega)}$$

$$+ \beta_2 \left(\Pi_{\mathcal{M}} u_{\mathcal{D}}^{n+1-M_4}, \Pi_{\mathcal{M}} v\right)_{L^2(\Omega)} = (f(t_{n+1}) + \Delta u(t_{n+1}), \Pi_{\mathcal{M}} v)_{L^2(\Omega)}.$$

Let us subtract the following term from the both sides of the previous equation

$$(\Pi_{\mathcal{M}} \mathbb{T}_6, \Pi_{\mathcal{M}} v)_{L^2(\Omega)},$$

with

$$\mathbb{T}_6 = \partial^2 \bar{u}_{\mathcal{D}}^{n+1} + \alpha_0 \partial^1 \bar{u}_{\mathcal{D}}^{n+1} + \alpha_1 \partial^1 \bar{u}_{\mathcal{D}}^{n+1-M_1} + \alpha_2 \partial^1 \bar{u}_{\mathcal{D}}^{n+1-M_2}$$

$$+ \beta_0 \bar{u}_{\mathcal{D}}^{n+1} + \beta_1 \bar{u}_{\mathcal{D}}^{n+1-M_3} + \beta_2 \bar{u}_{\mathcal{D}}^{n+1-M_4}.$$

We subsequently replace $f(t_{n+1}) + \Delta u(t_{n+1})$ by its value which stems from (1), we get, for all $v \in \mathcal{X}_{\mathcal{D},0}$

$$\left(\partial^2 \Pi_{\mathcal{M}} \eta_{\mathcal{D}}^{n+1}, \Pi_{\mathcal{M}} v\right)_{L^2(\Omega)} + \langle \eta_{\mathcal{D}}^{n+1}, v\rangle_F + \alpha_0 \left(\partial^1 \Pi_{\mathcal{M}} \eta_{\mathcal{D}}^{n+1}, \Pi_{\mathcal{M}} v\right)_{L^2(\Omega)}$$

$$+ \alpha_1 \left(\partial^1 \Pi_{\mathcal{M}} \eta_{\mathcal{D}}^{n+1-M_1}, \Pi_{\mathcal{M}} v\right)_{L^2(\Omega)} + \alpha_2 \left(\partial^1 \Pi_{\mathcal{M}} \eta_{\mathcal{D}}^{n+1-M_2}, \Pi_{\mathcal{M}} v\right)_{L^2(\Omega)}$$

$$+ \beta_0 \left(\Pi_{\mathcal{M}} \eta_{\mathcal{D}}^{n+1}, \Pi_{\mathcal{M}} v\right)_{L^2(\Omega)} + \beta_1 \left(\Pi_{\mathcal{M}} \eta_{\mathcal{D}}^{n+1-M_3}, \Pi_{\mathcal{M}} v\right)_{L^2(\Omega)}$$

$$+ \beta_2 \left(\Pi_{\mathcal{M}} \eta_{\mathcal{D}}^{n+1-M_4}, \Pi_{\mathcal{M}} v\right)_{L^2(\Omega)} = \left(\mathcal{S}^{n+1}, \Pi_{\mathcal{M}} v\right)_{L^2(\Omega)}, \tag{20}$$

with

$$\mathcal{S}^n = u_{tt}(t_n) - \partial^2 \Pi_{\mathcal{M}} \bar{u}_{\mathcal{D}}^n + \alpha_0 (u_t(t_n) - \partial^1 \bar{u}_{\mathcal{D}}^{n+1}) + \alpha_1 (u_t(t_n - \tau_1) - \partial^1 \Pi_{\mathcal{M}} \bar{u}_{\mathcal{D}}^{n-M_1})$$

$$+ \alpha_2 (u_t(t_n - \tau_2) - \partial^1 \bar{u}_{\mathcal{D}}^{n+1-M_2}) + \beta_0 (u(t_n) - \Pi_{\mathcal{M}} \bar{u}_{\mathcal{D}}^n)$$

$$+ \beta_1 (u(t_n - \tau_3) - \Pi_{\mathcal{M}} \bar{u}_{\mathcal{D}}^{n-M_3}) + \beta_2 (u(t_n - \tau_4) - \Pi_{\mathcal{M}} \bar{u}_{\mathcal{D}}^{n-M_4}). \tag{21}$$

Since $(\eta_{\mathcal{D}}^n)_{n=-M}^N \in (\mathcal{X}_{\mathcal{D},0})^{N+M+1}$ is satisfying (20), hence it satisfies the hypothesis (11) of Lemma 1. Applying now the discrete *a priori estimate* (12) of Lemma 1 and using the property $\eta_{\mathcal{D}}^0 = 0$ and error estimates (18)–(19) together with the triangle inequality and some convenient Taylor expansions to get, for all $J \in [\![1, N]\!]$

$$\mathbb{E}_{\mathcal{D}}^J \le C \left( (h_{\mathcal{D}} + k)^2 + \max_{n=1-M}^0 \|\partial^1 \Pi_{\mathcal{M}} \eta_{\mathcal{D}}^n\|_{L^2(\Omega)}^2 + \max_{n=-M}^0 \|\Pi_{\mathcal{M}} \eta_{\mathcal{D}}^n\|_{L^2(\Omega)}^2 \right). \tag{22}$$

Let us estimate the terms $\|\partial^1 \Pi_{\mathcal{M}} \eta_{\mathcal{D}}^n\|_{L^2(\Omega)}$ involved in rhs (the right hand side) of (22). We have, for all $n \in [\![1 - M, 0]\!]$, $\partial^1 \Pi_{\mathcal{M}} \eta_{\mathcal{D}}^n = \partial^1 \Pi_{\mathcal{M}} u_{\mathcal{D}}^n - u_t(t_n) + u_t(t_n) - \partial^1 \Pi_{\mathcal{M}} \bar{u}_{\mathcal{D}}^n$. This with the triangle inequality and estimate (19) yield

$$\|\partial^1 \Pi_{\mathcal{M}} \eta_{\mathcal{D}}^n\|_{L^2(\Omega)} \leq \|\partial^1 \Pi_{\mathcal{M}} u_{\mathcal{D}}^n - u_t(t_n)\|_{L^2(\Omega)} + C(h_{\mathcal{D}} + k). \tag{23}$$

Since, for all $n \in [\![1 - M, 0]\!]$, $\partial^1 u_{\mathcal{D}}^n$ satisfies (see the second scheme in (5)) the same scheme (17) but with $u^1(t_n)$ in the rhs instead of $u(t_n)$, we are able then to apply estimates (18) on the second scheme in (5) to get, for all $n \in [\![1 - M, 0]\!]$,

$$\|u^1(t_n) - \Pi_{\mathcal{M}} \partial^1 u_{\mathcal{D}}^n\|_{L^2(\Omega)} + \|\nabla u^1(t_n) - \nabla_{\mathcal{D}} \partial^1 u_{\mathcal{D}}^n\|_{(L^2(\Omega))^d} \leq C h_{\mathcal{D}}.$$

This with the fact that $u^1(t_n) = u_t(t_n)$, for all $n \in [\![1 - M, 0]\!]$ (see (2)), imply that

$$\|u_t(t_n) - \Pi_{\mathcal{M}} \partial^1 u_{\mathcal{D}}^n\|_{L^2(\Omega)} + \|\nabla u_t(t_n) - \nabla_{\mathcal{D}} \partial^1 u_{\mathcal{D}}^n\|_{(L^2(\Omega))^d} \leq C h_{\mathcal{D}}. \tag{24}$$

Gathering this and (23) implies that $\max\limits_{n=1-M}^{n=0} \|\partial^1 \Pi_{\mathcal{M}} \eta_{\mathcal{D}}^n\|_{L^2(\Omega)} \leq C(h_{\mathcal{D}} + k)$. On the other hand, this last estimate yields also an estimate for $\max\limits_{n=-M}^{0} \|\Pi_{\mathcal{M}} \eta_{\mathcal{D}}^n\|_{L^2(\Omega)}$ which appears on the rhs of (22), since $\eta_{\mathcal{D}}^0 = 0$, we have $\eta_{\mathcal{D}}^n = \sum\limits_{j=n+1}^{0} -k \partial^1 \eta_{\mathcal{D}}^j$, we then use the triangle inequality together with the proved estimate $\max\limits_{n=1-M}^{n=0} \|\partial^1 \Pi_{\mathcal{M}} \eta_{\mathcal{D}}^n\|_{L^2(\Omega)} \leq C(h_{\mathcal{D}} + k)$. These estimates on $\max\limits_{n=1-M}^{n=0} \|\partial^1 \Pi_{\mathcal{M}} \eta_{\mathcal{D}}^n\|_{L^2(\Omega)}$ and $\max\limits_{n=-M}^{0} \|\Pi_{\mathcal{M}} \eta_{\mathcal{D}}^n\|_{L^2(\Omega)}$ together with (22) yield

$$\|\partial^1 \Pi_{\mathcal{M}} \eta_{\mathcal{D}}^J\|_{L^2(\Omega)}^2 + \|\nabla_{\mathcal{D}} \eta_{\mathcal{D}}^J\|_{L^2(\Omega)}^2 \leq C(h_{\mathcal{D}} + k)^2, \quad \forall J \in [\![1, N]\!]. \tag{25}$$

Using now (18), (19), and the discrete Poincaré inequality [5, Lemma 5.4], and (25) yield the desired estimates (9)–(10) when $n \in [\![1, N]\!]$. The case of (10) when $n \in [\![-M + 1, 0]\!]$ is a result of (24). The case when $n = 0$ in (9) can be deduced from the property $\eta_{\mathcal{D}}^0 = 0$ and (18). $\qquad\square$

## 5   Conclusion and Perspectives

We developed a new finite volume scheme for a second hyperbolic equation with several time independent delays. These delays are involved in both the exact solution and its time derivative. The model we considered is studied for instance in [7,8]. A full convergence analysis is provided for the finite volume scheme. We plan to extend the present results to second order hyperbolic equations with several time dependent delays and to the delayed coupled system [8].

# References

1. Bellen, A., Zennaro, M.: Numerical Methods for Delay Differential Equations. Numerical Mathematics and Scientific Computation. Oxford University Press (2003)
2. Benkhaldoun, F., Bradji, A.: Note on the convergence of a finite volume scheme for a second order hyperbolic equation with a time delay in any space dimension. In: Klöfkorn, R., Keilegavlen, E., Radu, F.A., Fuhrmann, J. (eds.) FVCA 2020. SPMS, vol. 323, pp. 315–324. Springer, Cham (2020). https://doi.org/10.1007/978-3-030-43651-3_28
3. Bradji, A.: Convergence analysis of some high-order time accurate schemes for a finite volume method for second order hyperbolic equations on general nonconforming multidimensional spatial meshes. Numer. Meth. Partial Differ. Equ. **29**(4), 1278–1321 (2013)
4. Bradji, A.: A theoretical analysis of a new finite volume scheme for second order hyperbolic equations on general nonconforming multidimensional spatial meshes. Numer. Meth. Partial Differ. Equ. **29**(1), 1–39 (2013)
5. Eymard, R., Gallouët, T., Herbin, R.: Discretization of heterogeneous and anisotropic diffusion problems on general nonconforming meshes. IMA J. Numer. Anal. **30**(4), 1009–1043 (2010)
6. Kuang, Y.: Delay Differential Equations: with Applications in Population Dynamics. Mathematics in Science and Engineering, vol. 191. Academic Press, Boston (1993)
7. Nicaise, S., Pignotti, C.: Stability and instability results of the wave equation with a delay term in the boundary or internal feedbacks. SIAM J. Control Optim. **45**(5), 1561–1585 (2006)
8. Parhi, N., Kirane, M.: Oscillatory behaviour of solutions of coupled hyperbolic differential equations. Analysis **14**(1), 43–56 (1994)
9. Raposo, C., Nguyen, H., Ribeiro, J.-O., Barros, V.: Well-posedness and exponential stability for a wave equation with nonlocal time-delay condition. Electron. J. Differ. Equ. **2017**(279), 1–11 (2017)
10. Zhang, Q., Zhang, C.: A new linearized compact multisplitting scheme for the nonlinear convection-reaction-diffusion equations with delay. Commun. Nonlinear Sci. Numer. Simul. **18**(12), 3278–3288 (2013)

# Recovering the Time-Dependent Volatility in Jump-Diffusion Models from Nonlocal Price Observations

Slavi G. Georgiev$^{(\boxtimes)}$ and Lubin G. Vulkov

Department of Applied Mathematics and Statistics, FNSE, University of Ruse,
Ruse, Bulgaria
{sggeorgiev,lvalkov}@uni-ruse.bg

**Abstract.** This paper is devoted to a recovery of time-dependent volatility under jump-diffusion processes assumption. The problem is formulated as an inverse problem: given nonlocal observations of European option prices, find a time-dependent volatility function such that the theoretical option prices match the observed ones in an optimal way with respect to a prescribed cost functional. We propose a variational adjoint equation approach to derive the gradients of the functionals. A finite difference formulation of the 1D inverse problem is discussed.

**Keywords:** Jump-diffusion model · Implied volatility ·
Time-dependent inverse problem · Adjoint equation optimization ·
Nonlocal observation

## 1 Introduction

Although globally used in option pricing, the Black–Scholes model is not able to reflect the evolution of stock returns in the real world. Since the Black–Scholes era, numerous deterministic and stochastic volatility models have been proposed to price and hedge options. A jump-diffusion model, for example, which allows jumps in the underlying asset price, is more accurate. What is more, the empirical studies of real market data suggest that volatility is not constant, but depends on the remaining time to maturity. Under this jump-diffusion framework, we employ the pricing partial differential equation with the nonlocal integral term.

The jump-diffusion model, introduced by Merton in [8], is the most widely spread one. In this model the process of the asset price is a sum of a jump process with lognormally distributed jump magnitude and a normal Brownian motion. A semi-closed analytical formula is helpful not only by its own, but it can serve as a benchmark to check the consistency of various numerical methods, especially concerning multidimensional problems [1]. What is more, reasonable volatility smiles could be obtained by accurately calibrating jump-diffusion models, i. e. determining the parameters of the Wiener and Poisson processes, or solving *inverse problems*.

© Springer Nature Switzerland AG 2022
I. Lirkov and S. Margenov (Eds.): LSSC 2021, LNCS 13127, pp. 507–514, 2022.
https://doi.org/10.1007/978-3-030-97549-4_58

The models proposed in [1] suggest a local volatility function, incorporated with jump processes. Kou [6] proposed a model with double exponential distribution of the log-jump size. Currently, the Merton's and Kou's jump-diffusion models are the most used ones, and the development of robust numerical methods seems to be of high interest, see e. g. [10].

Volatility, which measures the amount of randomness, is widely used in risk and portfolio management. Furthermore, it is the only parameter that is not directly observable on the financial markets. In general, it can be estimated from the quoted option prices. The volatility value, implied by an observed option price, is called the *implied volatility*. The interest of reliable *recovery* of the volatility from observed market data has not faded away over time for both traders and scientists [4].

The problems of selection and calibration are central to the financial theory and practice. They are treated as ill-posed inverse problems, often by a suitable regularization type. For example, the authors of [9] consider the volatility as an unknown parameter in a pricing model under jump-diffusion and solve an inverse problem using Tikhonov regularization. The nature of ill-posedness of the inverse problems in case of time-dependent volatility in option pricing is commented in [5].

The paper is organized as follows. In Sect. 2 we introduce the 1D and 2D Merton's and Kou's jump-diffusion models and formulate the inverse problems for them. We derive the adjoint equations to the 1D models and use them to calculate the gradient of the cost functional in Sect. 3 as well as apply a similar procedure for the 2D problems. Finite difference adjoint equations for the 1D case are discussed in Sect. 4. The concluding remarks are presented in the last section.

## 2    The Pricing Models and the Inverse Problems

### 2.1    The Direct Problems

Let the price of an underlying asset be denoted by $S$ and make the assumption that its movement pursues the jump-diffusion dynamics described by the stochastic differential equation as follows:

$$dS = (\nu(t) - \lambda\kappa)Sdt + \sigma(t)SdW + (\eta - 1)Sdq, \tag{1}$$

where $dW$ is an increment of the standard Gauss-Wiener process and $dq$ is an increment of the independent Poisson process with a deterministic jump intensity $\lambda$. Further in (1) $\sigma(t)$ is the volatility, $\nu(t)$ is the drift rate; and $\eta - 1$ is an impulse function producing a jump from $S$ to $S\eta$ and $\kappa := \int_0^\infty (\eta - 1)f(\eta)d\eta = \mathbb{E}[\eta - 1]$, where $f(\eta)$ is the probability density function of the jump amplitude $\eta$, respecting $\int_0^\infty f(\eta)d\eta = 1$. $\mathbb{E}[\cdot]$ indicates the expectation operator.

Let the value of a European contingent claim with strike price $K$ on the underlying asset $S$ and time $t$ be denoted by $V(S, t)$. Also let $\tau := T - t$ be the remaining time to maturity and $x := \ln S$, $y := \ln \eta$. Then

$\frac{\partial v}{\partial \tau} = \mathscr{L}v + \lambda Q_L(x; v)$, where $\mathscr{L}$ and $Q_L$ are the differential and integral operators

$$\mathscr{L}v(x,\tau) := \frac{\sigma^2(\tau)}{2}\left(\frac{\partial^2 v}{\partial x^2} - \frac{\partial v}{\partial x}\right) + (r(\tau) - \lambda\kappa)\frac{\partial v}{\partial x} - (r(\tau) + \lambda)v; \quad Q_L = \int_{-L}^{L} v(x+y)\phi(y)\mathrm{d}y \quad (2)$$

with the initial and boundary conditions

$$v(x,0) = v_0(x) = V_0(e^x); \qquad v(-L,\tau) = 0, \quad v(L,\tau) = e^L - Ke^{-\int_0^\tau r(u)\mathrm{d}u} \quad (3)$$

and the respective Merton's and Kou's density functions are

$$\phi(y) = \frac{1}{\sqrt{2\pi}\breve{\sigma}}\exp\left(\frac{-(y-\breve{\mu})^2}{2\breve{\sigma}^2}\right) \text{ and } \phi(y) = p\breve{\eta}_1 e^{-\breve{\eta}_1 y}\mathbb{1}_{\{y\geqslant 0\}} + q\breve{\eta}_2 e^{\breve{\eta}_2 y}\mathbb{1}_{\{y<0\}},$$

where $\breve{\mu}$, $\breve{\sigma}$ are the parameters of the Merton's normal distribution and $\breve{\eta}_{1,2}$ define the Kou's double-exponential distribution.

Similarly to the 1D case, after applying the logarithmic substitutions $x_1 = \ln S_1$ and $x_2 = \ln S_2$, we solve the two-asset time-dependent volatility jump-diffusion equation

$$\frac{\partial v}{\partial \tau} = \frac{1}{2}\sigma^2(\tau)\underbrace{\left(\sigma_1^2\frac{\partial^2 v}{\partial x_1^2} + 2\rho(\tau)\sigma_1\sigma_2\frac{\partial^2 v}{\partial x_1\partial x_2} + \sigma_2^2\frac{\partial^2 v}{\partial x_2^2} - \frac{\sigma_1^2}{2}\frac{\partial v}{\partial x_1} - \frac{\sigma_2^2}{2}\frac{\partial v}{\partial x_2}\right)}_{M(v)}$$

$$+ (r(\tau) - \lambda\kappa_1)\frac{\partial v}{\partial x_1} + (r(\tau) - \lambda\kappa_2)\frac{\partial v}{\partial x_2} - (r(\tau) + \lambda)v$$

$$+ \lambda\int_{-L_1}^{L_1}\int_{-L_2}^{L_2} v(x_1+y_1, x_2+y_2)\phi(y_1, y_2)\mathrm{d}y_1\mathrm{d}y_2 \quad (4)$$

for $(x_1, x_2, \tau) \in (-\infty, +\infty) \times (-\infty, +\infty) \times (0, T]$, where $v(x_1, x_2, \tau) = V(e^{x_1}, e^{x_2}, \tau)$, $y_1 = \ln\eta_1, y_2 = \ln\eta_2$, $\phi(y_1, y_2) = f(e^{y_1}, e^{y_2})e^{y_1+y_2}$, where $f(\eta_1, \eta_2)$ is the Merton's lognormal or Kou's double-exponential bivariate density [2].

We use the payoff functions and boundary conditions described in [4].

## 2.2   The Inverse Problems

**Problem 1 (P1).** Suppose that all coefficients in (2) except $\sigma(\tau)$ are available and $v(x, \tau)$ is the option premium that satisfies (2)–(3). Then

$$\frac{1}{4L}\int_{-L}^{L} v(x, \tau)\mathrm{d}x = \varphi(\tau), \quad (0 \leqslant \tau \leqslant T), \quad (5)$$

where $\varphi(\tau)$ is a given function of time for all $\tau$.

*Recover the functions $v(x, \tau)$ and $\sigma(\tau)$ if provided with (5).*

The notation $\frac{1}{2L}\int_{-L}^{L} v(x, \tau)\mathrm{d}x$ represents the *average option premium* corresponding to options with the same strike price and different asset log-prices $x \in [-L, L]$.

Following the theory and applications of the inverse problems, see e. g. [7], we solve **P1** considering the cost (error) functional

$$\min_{\sigma} \left\{ J(\sigma) = \frac{1}{4LT} \int_0^T \int_{-L}^L \left( v(x, \tau; \sigma) - \varphi(\tau) \right)^2 dx d\tau \right\}. \tag{6}$$

**Problem 2 (P2)** (*Rectangle observation*). Suppose that all the coefficients in equation (4) except $\sigma(\tau)$ are available and $v(x_1, x_2, \tau)$ is the option premium that satisfies (4). For $\Omega = [-2L_1, 2L_1] \times [-2L_2, 2L_2]$ assume that

$$\frac{1}{2S(\Omega)} \iint_\Omega v(x_1, x_2, \tau) dx_1 dx_2 = \varphi(\tau), \quad (0 \leqslant \tau \leqslant T) \tag{7}$$

is observed, and $\varphi(\tau)$ is a given function of time for all $\tau$.

Recover the functions $v(x, \tau)$ and $\sigma(\tau)$ if provided with (7).

In the 2D case we minimize the cost functional

$$\min_{\sigma} \left\{ J(\sigma) = \frac{1}{2S(\Omega)T} \int_0^T \iint_\Omega \left( v(x_1, x_2, \tau; \sigma) - \varphi(\tau) \right)^2 dx_1 dx_2 d\tau \right\}. \tag{8}$$

## 3 Inverse Problems Solution

### 3.1 1D Sensitivity Problem

In sensitivity analysis it is assumed that when $\sigma(\tau)$ undergoes an increment $\varepsilon \delta \sigma(\tau)$, $\varepsilon > 0$ is small, the price changes by an amount $\varepsilon \delta v(x, \tau)$. We replace $\sigma(\tau)$ by $\sigma(\tau) + \varepsilon \delta \sigma(\tau)$ and $v(x, \tau)$ by $v(x, \tau; \sigma + \varepsilon \delta \sigma)$ in the direct problem (2) and differentiate with respect to $\varepsilon$ and we set $\varepsilon = 0$ in the result. Define the sensitivity function $u(x, \tau; \sigma)$ with respect to $\sigma(\tau)$

$$u(x, \tau; \sigma) = \frac{\partial v(x, \tau; \sigma)}{\partial \sigma} \delta \sigma$$

and the initial-boundary value sensitivity problem is given by

$$\frac{\partial u}{\partial \tau} = \frac{1}{2} \sigma^2(\tau) \left( \frac{\partial^2 u}{\partial x^2} - \frac{\partial u}{\partial x} \right) + (r(\tau) - \lambda k) \frac{\partial u}{\partial x} - (r(\tau) + \lambda) u + \lambda Q_L(x; u)$$

$$+ \sigma(\tau) \delta \sigma \left( \frac{\partial^2 v}{\partial x^2} - \frac{\partial v}{\partial x} \right); \quad u(x, 0) = 0; \quad u(-L, \tau) = 0, \quad u(L, \tau) = 0. \tag{9}$$

**Theorem 1.** *The gradient of the cost functional* (6) *is given by*

$$J'(\sigma) = \frac{1}{2LT} \int_0^T \int_{-L}^L \sigma(\tau) \left( \frac{\partial^2 v}{\partial x^2} - \frac{\partial v}{\partial x} \right) q(x, \tau) dx d\tau,$$

*where $q(x, \tau)$ is the solution of the adjoint equation*

$$\frac{\partial q}{\partial \tau} + \frac{1}{2}\sigma^2(\tau)\frac{\partial^2 q}{\partial x^2} - \left(r(\tau) - \frac{1}{2}\sigma^2(\tau) - \kappa\lambda\right)\frac{\partial q}{\partial x} - (r(\tau) + \lambda)q$$

$$+ \lambda \int_{-L}^{L} q(x - y, \tau)f(y)dy - (v(x, \tau; \sigma) - \varphi(\tau)) = 0 \qquad (10)$$

*with final and boundary conditions*

$$q(x, T) = 0, \quad q(-L, \tau) = q(L, \tau) = 0. \qquad (11)$$

*Remark 1.* The explicit expression for the gradient $J'(\sigma)$ of the error functional (6) could be exploited in such an approach that the inverse problem **(P1)** to be solved through minimization of the functional.

*Outline of the Proof.* Let us take the Gateaux derivative of the cost functional (6) with respect to the function $\sigma$

$$\frac{d}{d\varepsilon}J(\sigma + \varepsilon\delta\sigma)|_{\varepsilon=0} = J'(\sigma)\delta\sigma = \frac{1}{LT}\int_{0}^{T}\int_{-L}^{L}\sigma(\tau)(v(x, \tau; \sigma) - \varphi(\tau))\frac{\partial v}{\partial \sigma}\delta\sigma dxd\tau. \qquad (12)$$

Let $q(x, \tau)$ be a test function with final condition $q(x, T) = 0$. Then, in view of the zero initial condition (9) for $u$ and the zero final condition (11) for $q$, we have

$$\int_{-L}^{L}\int_{0}^{T}\frac{d}{d\tau}(u(x, \tau)q(x, \tau))d\tau dx = \int_{-L}^{L}\int_{0}^{T}\left(\frac{\partial u}{\partial \tau}(x, \tau)q(x, \tau) + u(x, \tau)\frac{\partial q}{\partial \tau}(x, \tau)\right)dxd\tau = 0. \qquad (13)$$

Next, we place the expression for $\frac{\partial u}{\partial \tau}(x, \tau)$ from (9) into (13) and integrate by parts to obtain:

$$0 = \int_{0}^{T}\int_{-L}^{L}\left[\frac{1}{2}\sigma^2(\tau)\left(\frac{\partial^2 u}{\partial x^2} - \frac{\partial u}{\partial x}\right) + (r(\tau) - \lambda\kappa)\frac{\partial u}{\partial x} - (r(\tau) + \lambda)u + \lambda Q_L(x; u)\right]qdxd\tau$$

$$+ \int_{0}^{T}\int_{-L}^{L}\left[\sigma(\tau)\delta\sigma\left(\frac{\partial^2 v}{\partial x^2} - \frac{\partial v}{\partial x}\right)\right]qdxd\tau + \int_{-L}^{L}\int_{0}^{T}u\frac{\partial q}{\partial \tau}d\tau dx.$$

Now, gathering in all the above separate parts, we obtain

$$0 = \int_{-L}^{L}\int_{0}^{T}u\left(-\frac{1}{2}\sigma^2(\tau)\frac{\partial^2 q}{\partial x^2} + \left(r(\tau) - \frac{1}{2}\sigma^2(\tau) - \kappa\lambda\right)\frac{\partial q}{\partial x} + (r(\tau) + \lambda)q + \frac{\partial q}{\partial \tau}\right.$$

$$\left. - \lambda \int_{-L}^{L}q(x - y, \tau)f(y)dy\right)d\tau dx + \int_{-L}^{L}\int_{0}^{T}(-\sigma(\tau)\delta\sigma\left(\frac{\partial^2 v}{\partial x^2} - \frac{\partial v}{\partial x}\right)q)d\tau dx.$$

Adding and subtracting $(v(x, \tau; \sigma) - \varphi(\tau))u$ inside the integral and setting the first part to zero, we arrive at the adjoint Eq. (10). The remaining part of the same integral is also zero:

$$\int_{-L}^{L}\int_{0}^{T}\left(-\sigma(\tau)\delta\sigma\left(\frac{\partial^2 v}{\partial x^2} - \frac{\partial v}{\partial x}\right)q + (v(x, \tau; \sigma) - \varphi(\tau))\right)ud\tau dx = 0.$$

Then, using (12), we obtain the expression for the gradient $J'(\sigma)$.     □

## 3.2    2D Sensitivity Problem

Now we follow the sensitivity analysis of 1D case to obtain for the sensitivity

$$\frac{\partial u}{\partial \tau} = \frac{1}{2}\sigma^2(\tau)M(u) + (r(\tau) - \lambda\kappa_1)\frac{\partial u}{\partial x_1} + (r(\tau) - \lambda\kappa_2)\frac{\partial u}{\partial x_2}$$
$$- (r(\tau) + \lambda)u + \lambda Q_\Omega(x_1, x_2; u) + \sigma(\tau)\delta\sigma M(v) \qquad (14)$$

with initial and boundary conditions

$$u(x_1, x_2, 0) = 0, \quad u(\pm L_1, x_2, \tau) = 0, \quad u(x_1, \pm L_2, \tau) = 0. \qquad (15)$$

**Theorem 2.** *The gradient of the cost functional (8) is given by*

$$J'(\sigma) = \frac{1}{S(\Omega)T} \int_0^T \iint_\Omega \sigma(\tau)M(v)q(x_1, x_2, \tau)\mathrm{d}x_1\mathrm{d}x_2\mathrm{d}\tau, \qquad (16)$$

*where $q(x_1, x_2, \tau)$ is the solution to the adjoint equation*

$$\frac{\partial q}{\partial \tau} + \frac{1}{2}\sigma^2(\tau)M(q) + (r(\tau) - \lambda\kappa_1)\frac{\partial q}{\partial x_1} + (r(\tau) - \lambda\kappa_2)\frac{\partial q}{\partial x_2} - (r(\tau) + \lambda)q$$
$$+ \lambda \iint_\Omega q(x_1 - y_1, x_2 - y_2)f(y_1, y_2)\mathrm{d}y_1\mathrm{d}y_2 - \big(v(x_1, x_2, \tau; \sigma) - \varphi(\tau)\big) = 0$$

*with final and boundary conditions*

$$q(x_1, x_2, T) = 0; \quad q(\pm L_1, x_2, \tau) = q(x_1, \pm L_2, \tau) = 0. \qquad (17)$$

*Outline of the Proof.* We follow the line of the proof of Theorem 1. First we calculate the Gateaux derivative of the cost functional (8) with respect to the function $\sigma(\tau)$

$$\frac{\mathrm{d}}{\mathrm{d}\varepsilon}J(\sigma + \varepsilon\delta\tau)|_{\varepsilon=0} = J'(\sigma)\delta\sigma = \frac{1}{S(\Omega)T}\int_0^T\iint_\Omega \sigma(\tau)\big(v(x_1, x_2, \tau; \sigma) - \varphi(\tau)\big)u\mathrm{d}x_1\mathrm{d}x_2\mathrm{d}\tau. \qquad (18)$$

Let $q(x_1, x_2, \tau)$ be a test function with zero final condition $q(x_1, x_2, T) = 0$. Then, in view of the zero initial condition (15) for $u$ and the zero final condition (17) for $q$, we have

$$0 = \iint_\Omega\left(\int_0^T \frac{\partial u}{\partial \tau}(x_1, x_2, \tau)q(x_1, x_2, \tau) + u(x_1, x_2, \tau)\frac{\partial q}{\partial \tau}(x_1, x_2, \tau)\mathrm{d}\tau\right)\mathrm{d}x_1\mathrm{d}x_2. \qquad (19)$$

Next, we place the expression for $\frac{\partial u}{\partial \tau}(x_1, x_2, \tau)$ from (14) into (19), integrate by parts with respect to the space variables, and add and subtract $\big(v(x_1, x_2, \tau; \sigma) - \varphi(\tau)\big)u$ to obtain:

$$\iint_\Omega\int_0^T \Big(-\sigma(\tau)\delta\sigma M(v)q + \big(v(x_1, x_2, \tau; \sigma) - \varphi(\tau)\big)\Big)u\mathrm{d}\tau\mathrm{d}x_1\mathrm{d}x_2 = 0.$$

In analogous way to the 1D case, after taking (18) into account, we arrive at the formula (16) for the gradient $J'(\sigma)$. $\qquad\square$

## 4    Finite Difference Adjoint Equations Formulation

In this section we discuss finite difference approximations of the cost functional gradient via adjoint equations. We follow [3]. First the spatial grid is defined. We compute (2), (3) over the interval $I \equiv [-L, L]$. Since the integral term $Q_L(x, \tau)$ introduces a shift in the computational domain, we need a way to calculate the values of $v(x, \tau)$ outside $I$. This is why we introduce the generalized domain $G = [-2L, 2L]$: $G \supset I$ and we compute $v(x, \tau)$ for $x \in G \setminus I$ employing the asymptotic behaviour of the option price [3]. Then the number of the total spatial nodes is $N = 4n + 1$. The equidistant grid over $G$ is

$$x_1 = -2L, \ldots, x_{n+1} = -L, \ldots, x_{2n+1} = 0, \ldots, x_{3n+1} = L, \ldots, x_{4n+1} = 2L$$

with constant step size $h = x_{i+1} - x_i$ for $i = 1, \ldots, 4n$. There are $2n + 1$ nodes $x_i$ in the interval $I$: from $i = n + 1$ to $i = 3n + 1$.

For sake of keeping things simple, we make use of a uniform temporal mesh

$$\overline{\omega}_\tau = \omega_\tau \cup \{T\} = \{\tau^j = j\Delta\tau, \ j = 0, \ldots, J, \ \Delta\tau J = T\}.$$

Using a left difference for the first order derivative $\partial/\partial x$, then the difference scheme for (2) with initial and boundary conditions (3) could be written for $w := v$ in the canonical form as

$$w_{n+1}^{j+1} = 0, \quad w_{3n+1}^{j+1} = e^L - Ke^{-\int_0^{\tau^{j+1}} r(s)ds},$$

$$-A_i w_{i-1}^{j+1} + C_i w_i^{j+1} - B_i w_{i+1}^{j+1} = F_i \text{ for } i = n + 2, \ldots, 3n,$$

$$A_i = \frac{(\sigma^{j+1})^2}{2h^2} + \frac{(\sigma^{j+1})^2 - (r^{j+1} - \lambda\kappa)}{h}, \quad F_i = \frac{w_i^j}{\Delta\tau^j} + \lambda \sum_{k=-n}^{n} c_k w_{i+k}^j \phi_k,$$

$$B_i = \frac{(\sigma^{j+1})^2}{2h^2}, \quad C_i = \frac{1}{\Delta\tau^j} + \frac{(\sigma^{j+1})^2}{h^2} + \frac{(\sigma^{j+1})^2 - (r^{j+1} - \lambda\kappa)}{h} + (r^{j+1} + \lambda),$$

where $\sigma^{j+1} = \sigma(\tau^{j+1})$, $\phi_k = \phi(x_{2n+1+k})$ and $c_k$ are the weights, calculated according to the trapezoidal or Simpson's rule, for instance.

The same scheme applies to the computation of $w := q$, but $q_{3n+1}^{j+1} = 0$ and the adjoint equation is backward. This simply means that the iterations over the time layers have to be conducted from the final layer $j = J$, corresponding to the terminal condition (11), going down to the first layer $j = 0$.

The approximation of the gradient $J'(\sigma)$ from Theorem 1 in terms of the adjoint equation for $q(x, \tau)$ could be written as

$$J'(\sigma^j) \cong \sum_{j=1}^{J} \sum_{i=n+1}^{3n+1} \sigma^j \left( \frac{v_{i-1}^j - 2v_i^j + v_{i+1}^j}{h^2} - \frac{v_i^j - v_{i-1}^j}{h} \right) q_i^j \Delta x_i \Delta\tau^j,$$

where $\Delta x_i$ and $\Delta\tau^j$ are the respective weights.

The second way to compute the gradient $J'(\sigma)$ is through the *sensitivity* equation method.

# 5    Conclusion

In this paper we considered the problem of recovering the time-dependent volatility assuming a jump-diffusion framework. The inverse problems are approached by employing an adjoint state optimization algorithm. In order to compute the implied volatility, an adjoint PDE is solved and a cost functional is minimized which make the developed algorithms easy to implement and computationally efficient. Moreover, they are proved to be correct and robust. Another advantage of our approach is that it is independent of the numerical methods used. It is also applicable when dealing with multidimensional problems.

The obtained analytical-algorithmic results in the current work are a starting point for further computational realization of the proposed approach.

Here we considered options of European type, but it is possible to extend and improve the algorithms for computing the implied volatility of more complex options within jump-diffusion models.

**Acknowledgements.** This research is supported by the Bulgarian National Science Fund under Project DN 12/4 "Advanced analytical and numerical methods for nonlinear differential equations with applications in finance and environmental pollution" from 2017.

# References

1. Andersen, L., Andreasen, J.: Jump-diffusion processes: volatility smile fitting and numerical methods for option pricing. Rev. Deriv. Res. **4**, 231–262 (2000)
2. Clift, S.S., Forsyth, P.A.: Numerical solution of two asset jump diffusion models for option valuation. Appl. Num. Math. **58**, 743–782 (2008)
3. Georgiev, S.G., Vulkov, L.G.: Computation of the unknown volatility from integral option price observations in jump-diffusion models. Math. Comp. Sim. **188**, 591–608 (2021)
4. Georgiev, S.G., Vulkov, L.G.: Computation of time-dependent implied volatility from point observations for European options under jump-diffusion models. In: AIP Conference Proceedings, vol. 2172, p. 070006 (2019)
5. Hofmann, B., Hein, T.: Well-posed and ill-posed situations in option pricing problems when the volatility is purely time-dependent. Proc. Appl. Math. Mech. **3**, 450–451 (2003)
6. Kou, S.G.: A jump-diffusion model option pricing. Manag. Sci. **48**(8), 1086–1101 (2002)
7. Marchuk, G.I., Agoshkov, V.I., Shutyaev, V.P.: Adjoint Equations and Perturbation Algorithms in Nonlinear Problems. CRC Press, Boca Raton (1996)
8. Merton, R.C.: Option pricing when underlying stock returns are discontinuous. J. Fin. Econ. **3**, 125–144 (1976)
9. Neisy, A., Salmany, K.: An inverse finance problem for estimation of the volatility. Comp. Math. Math. Phys. **53**(1), 63–77 (2013)
10. Pironneau, O.: Pricing Futures by Deterministic Methods. Laboratoire Jacques Louis Lions, Université Pierre & Marie Curie, Paris (2012)

# On the Solution of Contact Problems with Tresca Friction by the Semismooth* Newton Method

Helmut Gfrerer[1], Jiří V. Outrata[2], and Jan Valdman[2,3(✉)]

[1] Institute of Computational Mathematics, Johannes Kepler University Linz, Linz, Austria
helmut.gfrerer@jku.at
[2] Czech Academy of Sciences, Institute of Information Theory and Automation, Prague, Czech Republic
outrata@utia.cas.cz
[3] Department of Mathematics, Faculty of Science, University of South Bohemia, České Budějovice, Czech Republic
jvaldman@prf.jcu.cz

**Abstract.** An equilibrium of a linear elastic body subject to loading and satisfying the friction and contact conditions can be described by a variational inequality of the second kind and the respective discrete model attains the form of a generalized equation. To its numerical solution we apply the semismooth* Newton method by Gfrerer and Outrata (2019) in which, in contrast to most available Newton-type methods for inclusions, one approximates not only the single-valued but also the multi-valued part. This is performed on the basis of limiting (Morduchovich) coderivative. In our case of the Tresca friction, the multi-valued part amounts to the subdifferential of a convex function generated by the friction and contact conditions. The full 3D discrete problem is then reduced to the contact boundary. Implementation details of the semismooth* Newton method are provided and numerical tests demonstrate its superlinear convergence and mesh independence.

**Keywords:** Contact problems · Tresca friction · Semismooth* Newton method · Finite elements · Matlab implementation

## 1 Introduction

In [3] the authors developed a new, so-called semismooth* Newton-type method for the numerical solution of an inclusion

$$0 \in F(x), \tag{1}$$

J. Valdman—The work of the 2nd and the corresponding author was supported by the Czech Science Foundation (GACR), through the grant GF19-29646L.

© Springer Nature Switzerland AG 2022
I. Lirkov and S. Margenov (Eds.): LSSC 2021, LNCS 13127, pp. 515–523, 2022.
https://doi.org/10.1007/978-3-030-97549-4_59

where $F : \mathbb{R}^n \rightrightarrows \mathbb{R}^n$ is a closed-graph multifunction. In contrast to existing Newton-type method $F$ is approximated on the basis of the limiting (Mordukhovich) normal cone to the graph of $F$, computed at the respective point. Under appropriate assumptions, this method exhibits local superlinear convergence and, so far, it has been successfully implemented to the solution of a class of variational inequalities (VIs) of the first and second kind, cf. [3] and [4]. This contribution is devoted to the application of the semismooth* method to the discrete 3D contact problem with Tresca friction which is modelled as a VI of the second kind. Therefore the implementation can be conducted along the lines of [4]. The paper has the following structure: In Sect. 2 we describe briefly the main conceptual iterative scheme of the method. Section 3 deals with the considered discrete contact problem and Sect. 4 concerns the suggested implementation and one computational benchmark.

We employ the following notation. For a cone $K$, $K^0$ stands for its (negative) polar and for a multifunction $F : \mathbb{R}^n \rightrightarrows \mathbb{R}^n$, $\mathrm{dom}\, F$ and $\mathrm{gph}\, F$ denote its domain and its graph, respectively. The symbol "$\xrightarrow{A}$" means the convergence within the set $A$, $\|B\|_F$ denotes the Frobenius norm of a matrix $B$ and $\mathcal{B}_\delta(x)$ signifies the $\delta-$ ball around $x$.

## 2   The Semismooth* Newton Method

For the reader's convenience we recall fist the definition of the tangent cone and the limiting (Mordukhovich) normal cone.

**Definition 1.** Let $A \subset \mathbb{R}^n$ be closed and $\bar{x} \in A$. Then,

(i) the cone $T_A(\bar{x}) := \{u \in \mathbb{R}^n | \exists t_k \searrow 0, u_k \to u \text{ such that } \bar{x} + t_k u_k \in A \,\forall k\}$ is called the (Bouligand) tangent cone to $A$ at $\bar{x}$;

(ii) The cone $N_A(\bar{x}) := \{x^* \in \mathbb{R}^n | \exists x_k \xrightarrow{A} \bar{x}, x_k^* \to x^* \text{ such that } x_k^* \in (T_A(x_k))^0 \,\forall k\}$ is called the limiting (Mordukhovich) normal cone to $A$ at $\bar{x}$.

The latter cone will be extensively used in the sequel. Let us assign to a pair $(\tilde{x}, \tilde{y}) \in \mathrm{gph}\, F$ two $[n \times n]$ matrices $A, B$ such that their $i$th rows, say $u_i^*, v_i^*$, fulfill the condition

$$(u_i^*, -v_i^*) \in N_{\mathrm{gph}\, F}(\tilde{x}, \tilde{y}), \qquad i = 1, 2, \ldots, n. \tag{2}$$

Moreover, let $\mathcal{A}F(\tilde{x}, \tilde{y})$ be the set of matrices $A, B$ satisfying (2) and

$$\mathcal{A}_{reg}F(\tilde{x}, \tilde{y}) := \{(A, B) \in \mathcal{A}F(\tilde{x}, \tilde{y}) | A \text{ is non-singular}\}.$$

The general conceptual iterative scheme of the semismooth* Newton method is stated in Algorithm 1 below.

---

**Algorithm 1.** Semismooth* Newton-type method for generalized equations

---

1: Choose a starting point $^0x$, set the iteration counter $k := 0$.
2: If $0 \in F(^kx)$, stop the algorithm.
3: Approximation step: compute $(\hat{x}, \hat{y}) \in \mathrm{gph}\, F$ close to $(^kx, 0)$ such that $\mathcal{A}_{\mathrm{reg}} F(\hat{x}, \hat{y}) \neq \emptyset$.
4: Newton step: select $(A, B) \in \mathcal{A}_{\mathrm{reg}} F(\hat{x}, \hat{y})$ and compute the new iterate

$$^{k+1}x = \hat{x} - A^{-1}B\hat{y}.$$

5: Set $k := k + 1$ and go to 2.

---

Let $\bar{x}$ be a (local) solution of (1). Since $^kx$ need not to belong to dom $F$ or $0$ need not to be close to $F(^kx)$ even if $^kx$ is close to $\bar{x}$; one performs in step 3 an approximate projection of $(^kx, 0)$ onto $\mathrm{gph}\, F$. Therefore the step 3 is called the *approximation step*. The *Newton step* 4 is related to the following fundamental property, according to which the method has been named.

**Definition 2** [3]. *Let $(\tilde{x}, \tilde{y}) \in \mathrm{gph}\, F$. We say that $F$ is semismooth\* at $(\tilde{x}, \tilde{y})$ provided that for every $\epsilon > 0$ there is some $\delta > 0$ such that the inequality*

$$|\langle x^*, x - \tilde{x} \rangle + \langle y^*, y - \tilde{y} \rangle| \leq \epsilon \|(x, y) - (\tilde{x}, \tilde{y})\| \, \|(x^*, y^*)\| \tag{3}$$

*is valid for all $(x, y) \in \mathcal{B}_\delta(\tilde{x}, \tilde{y})$ and for all $(x^*, y^*) \in N_{\mathrm{gph}\, F}(x, y)$.*

If we assume that $F$ is semismooth\* at $(\bar{x}, 0)$, then it follows from (3) that for every $\epsilon > 0$ there is some $\delta > 0$ such that for every $(x, y) \in \mathrm{gph}\, F \cap \mathcal{B}_\delta(\bar{x}, 0)$ and every pair $(A, B) \in \mathcal{A}_{\mathrm{reg}} F(x, y)$ one has

$$\|(x - A^{-1}By) - \bar{x}\| \leq \epsilon \|A^{-1}\| \, \|(A \vdots B)\|_F \, \|(x, y) - (x, 0)\|,$$

cf. [3, Proposition 4.3]. This is the background for the Newton step in Algorithm 1.

Finally, concerning the convergence, assume that $F$ is semismooth\* at $(\bar{x}, 0)$ and there are positive reals $L, \kappa$ such that for every $x \notin F^{-1}(0)$ sufficiently close to $\bar{x}$ the set of quadruples $(\hat{x}, \hat{y}, A, B)$, satisfying the conditions

$$\|(\hat{x} - \bar{x}, \hat{y})\| \leq L\|x - \bar{x}\|, \tag{4}$$

$$(A, B) \in \mathcal{A}_{\mathrm{reg}} F(\hat{x}, \hat{y}), \tag{5}$$

$$\|A^{-1}\| \, \|(A \vdots B)\|_F \leq \kappa \tag{6}$$

is nonempty. Then it follows from [3, Theorem 4.4], that Algorithm 1 either terminates at $\bar{x}$ after a finite number of steps or converges superlinearly to $\bar{x}$ whenever $^0x$ is sufficiently close to $\bar{x}$.

The application of the semismooth\* Newton methods to a concrete problem of type (1) requires thus the construction of an approximation step and the Newton step which fulfill conditions (4)–(6).

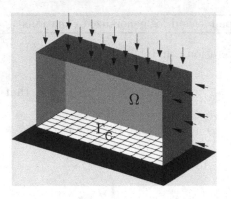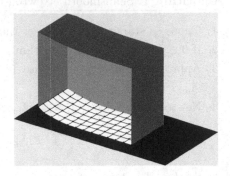

**Fig. 1.** The left picture depicts an undeformed elastic prism occupying domain $\Omega$ with the left (blue) face attached (Dirichlet condition) and some surface tractions applied to the right and top faces (depicted in green). They press the contact face $\Gamma_C$ against the (red) rigid plane foundation. Example of the resulting deformed body is depicted in the right picture. Front faces are not visualized. (Color figure online)

## 3  The Used Model

The fundamental results concerning unilateral contact problems with Coulomb friction have been established in [8]. The infinite-dimensional model of the contact problem with Tresca friction in form of a variational inequality of the second kind can be found, e.g., in [5,10]. These friction models can be used, for instance, in the numerical simulation of technological processes in metal forming [6]. Other related friction-type contact problems are described, e.g., in [9].

We assume that an elastic prism occupying domain $\Omega$ is pressed against a rigid plane foundation (cf. Fig. 1). The full three-dimensional domain $\Omega$ is discretized by a mesh of brick elements and consists of $n$ nodes (vertices). The finite element method using trilinear basis functions is then applied to approximate a displacement field vector $u \in \mathbb{R}^{3n}$ in each mesh node. Entries of $u$ are ordered in such a way that $u = (u^1, u^2, \ldots, u^n)$ and the $j$th node is associated with the pair $u^j = (u_\tau^j, u_\nu^j) \in \mathbb{R}^2 \times \mathbb{R}$ of its *tangential* and *normal displacements*, respectively.

A sparse stiffness matrix $K \in \mathbb{R}^{3n \times 3n}$ and the loading (column) vector $l \in \mathbb{R}^{3n}$ are first assembled and then both condensed to incorporate zero displacements in Dirichlet nodes corresponding to the (blue) Dirichlet boundary. Secondly, all nodes not lying in the (bottom) contact face $\Gamma_C$ are eliminated by the *Schur complement* technique and the *Cholesky factorization* resulting in a dense matrix $\tilde{A} \in \mathbb{R}^{3p \times 3p}$ and a vector $\tilde{b} \in \mathbb{R}^{3p}$, where $p \ll n$ is the number of $\Gamma_C$ nodes excluding Dirichlet boundary nodes.

At last, all local $3 \times 3$ blocks of $\tilde{A}$ and all $3 \times 1$ blocks of $\tilde{b}$ are expanded to $4 \times 4$ blocks and $4 \times 1$ blocks, respectively, in order to incorporate the non-penetrability condition

$$0 \in u_\nu^i + N_{\mathbb{R}_+}(\lambda^i),$$

where $\lambda^i \in \mathbb{R}_+$ is the *Lagrange multiplier* associated with non-penetrability constraint. Here and in the following, we assume that $i = 1, \ldots, p$. In this way, we obtain a dense regular matrix $A \in \mathbb{R}^{4p \times 4p}$ and a (column) vector $b \in \mathbb{R}^{4p}$.

Finally, let us simplify the notation via

$$x_{12}^i = (x_1^i, x_2^i) = u_\tau^i \in \mathbb{R}^2, \qquad x_3^i = u_\nu^i \in \mathbb{R}, \qquad x_4^i = \lambda^i \in \mathbb{R}$$

to define a vector of unknowns $x = (x^1, x^2, \ldots, x^p) \in \mathbb{R}^{4p}$.

Following the development in [1], our model attains the form of generalized equation (GE)

$$0 \in f(x) + \widetilde{Q}(x), \tag{7}$$

where the single-valued function $f : \mathbb{R}^{4p} \to \mathbb{R}^{4p}$ is given by $f(x) = Ax - b$ and the multifunction $\widetilde{Q} : \mathbb{R}^{4p} \rightrightarrows \mathbb{R}^{4p}$ by

$$\widetilde{Q}(x) = \bigtimes_{i=1}^p Q^i(x^i) \quad \text{with } Q^i(x^i) = \left\{ \begin{bmatrix} -\phi\, \partial\|x_{12}^i\| \\ 0 \\ N_{\mathbb{R}_+}(x_4^i) \end{bmatrix} \right\}$$

with $\phi \geq 0$ being the *friction coefficient*. GEs of the type (7) have been studied in [4] and so all theoretical results derived there are applicable. For our approach it is also important that the Jacobian $\nabla f(\bar{x})$ is positive definite.

## 4    Implementation of the Semismooth* Method

In order to facilitate the approximation step we will solve, instead of GE (7), the enhanced system

$$0 \in \mathcal{F}(x, d) = \begin{bmatrix} f(x) + \widetilde{Q}(d) \\ x - d \end{bmatrix} \tag{8}$$

in variables $(x, d) \in \mathbb{R}^{4p} \times \mathbb{R}^{4p}$. Clearly, $\bar{x}$ is a solution of (7) if and only if $(\bar{x}, \bar{x})$ is a solution of (8).

In the approximation step we suggest to solve for all $i$ consecutively the next three low-dimensional strictly convex optimization problems:

$$\text{(i)} \quad \underset{v \in \mathbb{R}^2}{\text{minimize}}\ \frac{1}{2}\langle v, v\rangle + \langle f_{12}^i({}^k x), v\rangle + \phi\|{}^k x_{12}^i + v\|,$$

$$\text{(ii)} \quad \underset{v \in \mathbb{R}}{\text{minimize}}\ \frac{1}{2}\langle v, v\rangle + f_3^i({}^k x) \cdot v,$$

$$\text{(iii)} \quad \underset{{}^k x_4^i + v \geq 0}{\text{minimize}}\ \frac{1}{2}\langle v, v\rangle + {}^k x_3^i \cdot v,$$

obtaining thus their unique solutions $\hat{v}_{12}^i, \hat{v}_3^i, \hat{v}_4^i$, respectively. These solutions can be ordered in vectors $\hat{v}^i = (\hat{v}_{12}^i, \hat{v}_3^i, \hat{v}_4^i) \in \mathbb{R}^4$ and all together in a vector

$$\hat{v} = (\hat{v}^1, \hat{v}^2, \ldots, \hat{v}^p) \in \mathbb{R}^{4p}.$$

Thereafter we compute the outcome of the approximation step via

$$\hat{x} = {}^k x, \qquad \hat{d} = {}^k x + \hat{v}, \qquad \hat{y} = (-\hat{v}, -\hat{v}).$$

Clearly $(\hat{x}, \hat{d}, \hat{y}) \in \text{gph}\,\mathcal{F}$ and, using the theory [4, Section 4], it is possible to show that condition (4) is fulfilled.

In the Newton step we put

$$A = I, \qquad B = \begin{bmatrix} I & 0 \\ 0 & G \end{bmatrix} D^{-1}, \qquad \text{where } D = \begin{bmatrix} \nabla f(\hat{x}) & -H \\ I & G \end{bmatrix}. \tag{9}$$

In (9), $I$ is an identity matrix and block diagonal matrices $G, H$ attain the form

$$G = \text{diag}(G^1, G^2, \ldots, G^p), \qquad H = \text{diag}(H^1, H^2, \ldots, H^p),$$

where the diagonal blocks $G^i, H^i \in \mathbb{R}^{4\times4}$ have the structure

$$G^i = \begin{bmatrix} G_1^i & & \\ & 1 & \\ & & G_2^i \end{bmatrix}, \qquad H^i = \begin{bmatrix} H_1^i & & \\ & 0 & \\ & & H_2^i \end{bmatrix}$$

and submatrices $G_1^i, H_1^i \in \mathbb{R}^{2\times2}$ and scalar entries $G_2^i, H_2^i \in \mathbb{R}$ are computed in dependence on values $\hat{d}_{12}^i \in \mathbb{R}^2$ and $\hat{d}_4^i \in \mathbb{R}$ as follows:

- If $\hat{d}_{12}^i = 0$ (sticking), we put $G_1^i = 0, H_1^i = I$, otherwise we put

$$G_1^i = I, \qquad H_1^i = \frac{\phi}{\|\hat{d}_{12}^i\|^3} \begin{bmatrix} (\hat{d}_2^i)^2 & -\hat{d}_1^i \hat{d}_2^i \\ -\hat{d}_1^i \hat{d}_2^i & (\hat{d}_1^i)^2 \end{bmatrix}.$$

- If $\hat{d}_4^i = 0$ (no contact or weak contact), we put $G_2^i = 0, H_2^i = 1$, otherwise we put $G_2^i = 1, H_2^i = 0$.

This choice ensures that matrices $(I, B)$ with $B$ given by (9) fulfill conditions (5), (6) with $F$ replaced by $\mathcal{F}$.

**Stopping Rule.** It is possible to show (even for more general Coulomb friction model [1]) that there is a Lipschitz constant $c_L > 0$ such that, $\|(\hat{x}, \hat{d}) - (\bar{x}, \bar{x})\| \leq c_L \|\hat{y}\|$, whenever the output of the approximation step lies in a sufficiently small neighborhood of $(\bar{u}, \bar{u}, 0)$. It follows that, with a sufficiently small positive $\varepsilon$, the condition

$$\|\hat{v}\| \leq \varepsilon, \tag{10}$$

tested after the approximation step, may serve as a simple yet efficient stopping rule.

**Computational Benchmark.** We assume that the domain $\Omega = (0, 2) \times (0, 1) \times (0.1, 1)$ is described by elastic parameters $E = 2.1 \cdot 10^9$ (Young's modulus), $\nu = 0.277$ (Poisson's ratio) and subject to surface tractions

$$f = (-5 \cdot 10^8, 0, 0) \quad \text{on the right-side face,} \tag{11}$$
$$f = (0, 0, -1 \cdot 10^8) \quad \text{on the top face.} \tag{12}$$

and the friction coefficient $\phi = 1$. The domain is uniformly divided into $e_x \cdot e_y \cdot e_z$ hexahedra (bricks), where $e_x = \lceil 4 \cdot 2^{\ell/2} \rceil, e_y = \lceil 2 \cdot 2^{\ell/2} \rceil, e_z = \lceil 2 \cdot 2^{\ell/2} \rceil$ are numbers of hexahedra along with coordinate axis, $\ell$ denotes the mesh level of refinement and $\lceil \cdot \rceil$ the ceiling function. Consequently the number of $\Omega$ nodes $n$ and the number of $\Gamma_C$ nodes $p$ read

$$n(\ell) = (e_x + 1) \cdot (e_y + 1) \cdot (e_z + 1), \qquad p(\ell) = e_x \cdot (e_y + 1),$$

respectively. Table 1 reports on the performance of the whole method for various meshes assuming zero initial approximation $^0x = 0$ and the stopping criterion $\epsilon = 10^{-6}$. We can clearly see that the number of iterations of the semismooth*

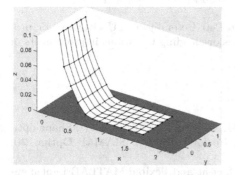 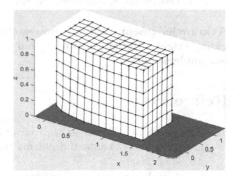

**Fig. 2.** The left picture depicts the deformed contact boundary and the right figure shows the corresponding deformed elastic prism, both pictures together with the (red) rigid plane foundation. (Color figure online)

**Table 1.** Performance of the MATLAB solver.

| Level ($\ell$) | Nodes ($n$) | Assembly of K (s) | Cholesky & Schur (s) | nodes ($p$) | Semismooth* time (s) | Solver items |
|---|---|---|---|---|---|---|
| 2 | 225 | 0.031 | 0.003 | 40 | 0.017 | 6 |
| 3 | 637 | 0.094 | 0.021 | 84 | 0.047 | 6 |
| 4 | 1377 | 0.141 | 0.092 | 144 | 0.101 | 6 |
| 5 | 4056 | 0.516 | 0.701 | 299 | 0.507 | 7 |
| 6 | 9537 | 1.297 | 3.968 | 544 | 1.928 | 7 |
| 7 | 27072 | 3.156 | 32.110 | 1104 | 9.734 | 7 |
| 8 | 70785 | 18.672 | 1242.211 | 2112 | 48.275 | 8 |

Newton method (displayed in the last column) only slightly increase with the mesh size. This behaviour shows that the method is mesh-independent.

Figure 2 visualizes a deformed contact boundary together with a deformation of the full domain $\Omega$ obtained by post-processing. Displacements of non-contact boundary nodes are then obtained from a linear system of equations with the matrix $K$ and the vector $l$. All pictures and running times were produced by our MATLAB code available for downloading and testing at

https://www.mathworks.com/matlabcentral/fileexchange/91005.

It is based on original codes of [1] and its performance is further enhanced by a vectorized assembly of $K$ using [2].

*Remark 1.* To a similar 3D contact problem with Tresca friction, a special variant of the classical semismooth Newton method has been applied in [7].

**Concluding Remarks and Further Perspectives.** The choice (9) of matrices $A, B$ in the Newton step of the method is not unique and may be used to simplify the linear system in the Newton step. The convergence may be further accelerated by an appropriate scaling in the approximation step.

**Acknowledgment.** Authors are grateful to Petr Beremlijski (TU Ostrava) for providing original Matlab codes of [1] and discussions leading to various improvements of our implementation.

# References

1. Beremlijski, P., Haslinger, J., Kočvara, M., Kučera, R., Outrata, J.V.: Shape optimization in 3D contact problems with Coulomb friction. SIAM J. Optim. **20**, 416–444 (2009)
2. Čermák, M., Sysala, S., Valdman, J.: Efficient and flexible MATLAB implementation of 2D and 3D elastoplastic problems. Appl. Math. Comput. **355**, 595–614 (2019)
3. Gfrerer, H., Outrata, J.V.: On a semismooth* Newton method for solving generalized equations. SIAM J. Optim. **31**(1), 489–517 (2021)
4. Gfrerer, H., Outrata, J.V., Valdman, J.: On the application of the semismooth* Newton method to variational inequalities of the second kind. arXiv:2007.11420 (2020)
5. Haslinger, J., Hlaváček, I., Nečas, J.: Numerical methods for unilateral problems in solid mechanics. In: Ciarlet, P.G., Lions, J.L. (eds.) Handbook of Numerical Analysis, pp. 313–485. Elsevier Science (1996)
6. Kikuchi, N., Oden, T.: Contact Problems in Elasticity: A Study of Variational Inequalities and Finite Element Methods. Society for Industrial and Applied Mathematics (1995)
7. Kučera, R., Motyčková, K., Markopoulos, A., Haslinger, J.: On the inexact symmetrized globally convergent semi-smooth Newton method for 3D contact problems with Tresca friction: the R-linear convergence rate. Optim. Meth. Softw. **35**(1), 65–86 (2020)

8. Nečas, J., Jarušek, J., Haslinger, J.: On the solution of the variational inequality to the Signorini problem with small friction. Boll. Unione Mat. Ital. V. Ser. B **17**, 796–811 (1980)
9. Neittaanmäki, P., Repin, S., Valdman, J.: Estimates of deviations from exact solutions of elasticity problems with nonlinear boundary conditions. Russ. J. Numer. Anal. Math. Model. **28**(6), 597–630 (2013)
10. Outrata, J.V., Kočvara, M., Zowe, J.: Nonsmooth Approach to Optimization Problems with Equilibrium Constraints. Kluwer Academic Publishers, Dordrecht (1998)

# Fitted Finite Volume Method for Unsaturated Flow Parabolic Problems with Space Degeneration

Miglena N. Koleva[✉] and Lubin G. Vulkov

University of Ruse, 8 Studentska Street, 7017 Ruse, Bulgaria
{mkoleva,lvalkov}@uni-ruse.bg

**Abstract.** In the present work, we discuss a question of correct boundary conditions and adequate approximation of parabolic problems with space degeneration in porous media. To the Richards equation, as a typical problem, we apply a time discretization, linearize the obtained nonlinear problem and introduce correct boundary conditions. Then, we develop fitted finite volume method to get the space discretization of the model problem. A graded space mesh is also deduced. We illustrate experimentally that the proposed method is efficient in the case of degenerate permeability.

## 1 Introduction

Over the past decade unsaturated flow has become one of the most important and active topics of the research. This is because the prediction of moisture flow under transient conditions is important in engineering practice when considering such practical problems as the design of shallow foundation, pavements and the stability of unsaturated soil slopes [11,13,16,17,20]. The basic equations for saturated flow are

$$S_s \frac{\partial h}{\partial t} = -\frac{\partial u}{\partial y} - \frac{\partial v}{\partial z}, \quad u = -K_s \frac{\partial h}{\partial y}, \quad v = -K_s \left( \frac{\partial h}{\partial z} + 1 \right),$$

where $S_s$ is the specific storage coefficient, $h$ is the pressure head, $t$ is the time, $y$ and $z$ are the spatial coordinates ($z$ is the vertical one), $u$ and $v$ are Darcy's velocities in the $y$ and $z$ directions, $K_s$ is the saturate hydraulic conductivity.

In this paper we concentrate on the important particular case of 1D Richards' equation [14] of water content on time in porous medium:

$$\frac{\partial u}{\partial t} = L(x,t)[u], \quad L(x,t)[u] = \frac{\partial}{\partial x} \left( k(x)a(u) \left( \frac{\partial u}{\partial x} + g \right) \right), \quad (1)$$

where $u = u(t,x)$ stands for the water content, $x \in \Omega = (0,L)$, $L > 0$, $t \in (0,T_f)$, $K(x,u) = k(x)a(u)$ is the hydraulic conductivity function (LT$^{-1}$), where the

© Springer Nature Switzerland AG 2022
I. Lirkov and S. Margenov (Eds.): LSSC 2021, LNCS 13127, pp. 524–532, 2022.
https://doi.org/10.1007/978-3-030-97549-4_60

permeability $k(x)$ can degenerate, i.e. $k(x) \geq 0$ and $k(0) = 0$, in the fluid melt, $g \geq 0$ is the gravity constant and $a(u)$ satisfies the ellipticity condition

$$(1 + |u|^s)^{-1} \leq a(u) \leq 1, \quad u \in \mathbb{R}, \quad s > 0. \tag{2}$$

We consider the following initial and boundary conditions for (1)

$$u(x,0) = u^0(x), \quad x \in \Omega, \quad u(0,t) = 0, \quad u(L,t) = u_r(t), \quad t \in [0, T_f]. \tag{3}$$

The Eq. (1) is a non-linear and boundary *degenerate* (at $x = 0$) *parabolic* equation, governed by non-linear physical relationships, described by soil-water characteristics curves [10, 13]. Since the lack of closed-form solution, numerical methods are required for accurate unsaturated flow simulation capability.

Numerous numerical methods for solving classical Richards' equation in its different forms are available in the literature. For example, Picard's and Newton's linearizations based schemes are proposed in [4, 5, 15, 20], discretization strategies as finite elements, finite volume, finite difference or hybrid combinations, see [6, 8] and reference therein. In our previous work [12] quasilinearization finite difference scheme for the model problem is constructed.

The aim of the paper is to present and examine efficient numerical approach for solving Richards' equation with a degenerate absolute permeability. To this end, we develop fitted finite volume method (FFVM) for the linearized problem.

FFVM is introduced in [18, 19] for resolving the degeneration in Black-Scholes equation. The idea is to apply finite volume method and local approximation of the solution, determined by two-points boundary value problems.

The remaining part of the paper is organized as follows. In Sect. 2 we develop the numerical method, based on the linearization of the semidiscrete (in time) problem and FFVM for the space approximation. We argue correctness of the boundary conditions. In the next section we discuss the implementation of the discrete scheme and present numerical results. In the last section we give concluding remarks.

## 2   Numerical Method

The approximation of the model problem (1)–(3) involves two main steps: linearization and discretization. At the first step we apply quasilinearization approach and discuss the correctness of the boundary conditions. Then we unfold FFVM for the linearized problem.

### 2.1   Linearization

In the time interval $J = [0, T_f]$, we introduce an uniform mesh $\omega_\tau = \{t_n = n\tau, \ n = 0, 1, \ldots, N, \ \tau = T_f/N\}$ and denote by $u^n(x)$ the function $u(t_n, x)$ from $\Omega \times J$. The weighted ($0 \leq \sigma \leq 1$) time discretization applied to the Eq. (1) reads

$$\frac{u^{n+1}(x) - u^n(x)}{\tau} = \sigma L(x, t_{n+1})[u^{n+1}] + (1 - \sigma)L(x, t_n)[u^n]. \tag{4}$$

Let us denote $v^{n+1} = \partial u^{n+1}/\partial x$, $a'(u) = da/du$ and suppose that $u_0^{n+1}$ and $v_0^{n+1}$ are the initial approximations of the corresponding exact values of $u^{n+1}$ and $v^{n+1}$ at the $(n+1)$-th time layers. Consider $u^{n+1}$ and $v^{n+1}$ as independent variables and applying quasilinearization technique [2], we obtain

$$L(x, t_{n+1})[u^{n+1}] = \frac{\partial}{\partial x}\Big(L_1(x, t_{n+1})[u^{n+1}]\Big) + \frac{\partial}{\partial x}L_0(x, t_{n+1}), \tag{5}$$

$$L_1(x, t_{n+1})[u^{n+1}] = k(x)\Big(a(u_0^{n+1})v^{n+1} + a'(u_0^{n+1})(v_0^{n+1} + g)u^{n+1}\Big),$$

$$L_0(x, t_{n+1}) = k(x)\Big(a(u_0^{n+1})g - a'(u_0^{n+1})(v_0^{n+1} + g)u_0^{n+1}\Big).$$

Substituting (5) in (4) we get

$$\frac{u^{n+1} - u^n}{\tau} - \sigma \frac{\partial\Big(L_1(x, t_{n+1})[u^{n+1}]\Big)}{\partial x} = (1-\sigma)L(x, t_n)[u^n] + \sigma \frac{\partial L_0(x, t_{n+1})}{\partial x}. \tag{6}$$

## 2.2   Correct Boundary Conditions

Applying the substitution $\chi(x) = \exp\left(\int_0^x \frac{a'(u_0^{n+1})(v_0^{n+1} + g)}{a(u_0^{n+1})}d\varsigma\right)$ in (6), we derive

$$-\sigma\frac{\partial}{\partial x}\left(\chi(x)k(x)a(u_0^{n+1})\frac{\partial \widehat{u}}{\partial x}\right) + \chi(x)\left(\frac{1}{\tau} - \sigma\frac{\partial\big(k(x)a'(u_0^{n+1})(v_0^{n+1} + g)\big)}{\partial x}\right)\widehat{u}$$
$$= \chi(x)\left(\frac{u^n}{\tau} + (1-\sigma)L(x, t_n)[u^n] + \sigma\frac{\partial L_0(x, t_{n+1})}{\partial x}\right), \quad \text{where } \widehat{u} := u^{n+1}(x). \tag{7}$$

Following the rate of degeneracy of $k(x)$ at $x = 0$, we will discuss appropriate left boundary condition for (7). Since the function $a(u)$ is a strongly positive, we consider the equation prototype of (7):

$$-\frac{d}{dx}\left(k(x)\frac{\partial \widehat{w}(x)}{\partial x}\right) + \widehat{w}(x) = w(x), \tag{8}$$

where $k(x) \in C[0, L] \cap C^1(0, L]$, $k(x) > 0$ $\forall x \in (0, L]$, $k(0) = 0$ and $\widehat{w}(L) = w_r$.

In many papers, the degeneracy is measured by the real parameter $\lambda_k := \sup\limits_{0 < x \leq L} \frac{x|k'(x)|}{k(x)}$. A simple consequence from the properties of $k(x)$ is the inequality $k(x) \geq k(L)x^{\lambda_k}$ $\forall$ $x \in [0, L]$. Further, we distinguish two basic cases: (i) $\lambda_k \in (0, 1)$ - *weak degeneracy*; (ii) $\lambda_k \in [1, 2]$ - *strong degeneracy* .

It is proved in [3] that for strong degeneracy, left Dirichlet boundary conditions cannot be described for (8), respectively (7). Actually, for $\lambda_k \geq 1$, there is no $H_{\mathrm{loc}}^2(0, L)$ solution of (7). In this work we consider weak degeneracy.

Following the proofs of Theorems 1.1, 1.2 in [3] one can establish the next results for (8) (respectively (7)) with Dirichlet right boundary condition.

**Theorem 1.** *Given $k(x)$ with $\lambda_k \in (0,1)$ and $w \in L^2(0,L)$, for each time level $t_n$, there exists unique function $\widehat{w} \in H^2_{loc}(0,L]$, satisfying (8), $\widehat{w}(L) = w_r$ with the following properties:* **(i)** $\lim\limits_{x\to 0^+} \widehat{w}(x) = 0$; **(ii)** $\widehat{w} \in C^{0,1-\lambda_k}[0,L]$ *with* $\|\widehat{w}\|_{C^{0,1-\lambda_k}} \leq C\|w\|_{L^2}$; **(iii)** $k(x)\widehat{w}' \in H^1(0,L)$ *with* $\|k(x)\widehat{w}'\|_{H^1} \leq C\|w\|_{L^2}$; **(iv)** $x^{\lambda_k-1}\widehat{w} \in H^1(0,L)$ *with* $\|x^{\lambda_k-1}\widehat{w}\|_{H^1} \leq C\|w\|_{L^2}$; **(v)** $k(x)\widehat{w} \in H^2(0,L)$ *with* $\|k(x)\widehat{w}\|_{H^2} \leq C\|w\|_{L^2}$, *where the constant $C$ depends only on $\lambda_k$.*

## 2.3 Fitted Finite Volume Method

In this section we discretizy (6) in space, implementing FFVM [18,19].

Consider the nonuniform mesh $\overline{\omega}_h$ in $\Omega$:

$$\overline{\omega}_h = \{x_i = x_{i-1} + h_i, \; i = 1,\ldots,M, \; x_0 = 0, \; x_M = L\}, \quad |h| = \max_{1\leq i\leq M} h_i$$

and denote by $U_i^n$ and $(U_0)_i^n$ the numerical solution of $u$ and $u_0$ at point $(t_n, x_i)$.

Beside the primal mesh $\overline{\omega}_h$, we define a dual mesh $\mathring{\omega}_h$:

$$x_{-1/2} = x_0 < x_{1/2} < x_1 < x_{3/2} < x_2 < \cdots < x_{M-1} < x_{M-1/2} < x_M = x_{M+1/2},$$

where $x_{i-1/2}$ is the midpoint of the interval $(x_{i-1}, x_i)$ for each $i = 1, 2, \ldots, M$.

Integrating (6) over the volumes $(x_{i-1/2}, x_{i+1/2})$, for $i = 1, 2, \ldots, M$, we get

$$\frac{1}{\tau} \int_{x_{i-1/2}}^{x_{i+1/2}} u^{n+1} dx - \sigma L_1(x, t_{n+1})[u^{n+1}]\Big|_{x_{i-1/2}}^{x_{i+1/2}} = \frac{1}{\tau} \int_{x_{i-1/2}}^{x_{i+1/2}} u^n dx$$

$$+ \sigma L_0(x, t_{n+1})\Big|_{x_{i-1/2}}^{x_{i+1/2}} + (1-\sigma)\Big[k(x)a(u^n)(v^n + g)\Big]_{x_{i-1/2}}^{x_{i+1/2}}. \tag{9}$$

Let $\alpha^{n+1}(x) = k(x)a(u_0^{n+1})$, $\beta^{n+1}(x) = k(x)a'(u_0^{n+1})(v_0^{n+1} + g)$ and denote the numerical flux by $\rho(u^{n+1}) = \alpha^{n+1}(x)\partial u^{n+1}/\partial x + \beta^{n+1}(x)u^{n+1}$.

We approximate $\rho^{n+1} = \rho(u^{n+1})$ in the interval $I_i = (x_i, x_{i+1})$, $i = 0, 1, \ldots, M-1$ by a constant $\rho_i^{n+1}$, which satisfies the following boundary value problem:

$$[\rho_i^{n+1}(z^{n+1}(x))]' := \Big(\alpha_{i+1/2}^{n+1}(z^{n+1})' + \beta_{i+1/2}^{n+1}z^{n+1}\Big)' = 0, \quad x \in I_i,$$

$$z^{n+1}(x_i) = u_i^{n+1}, \quad z^{n+1}(x_{i+1}) = u_{i+1}^{n+1}, \tag{10}$$

where $\alpha_{i+1/2}^{n+1} = \alpha^{n+1}(x_{i+1/2})$, $\beta_{i+1/2}^{n+1} = \beta^{n+1}(x_{i+1/2})$.

Solving the problem (10) exactly, we obtain

$$z^{n+1} = \frac{u_{i+1}^{n+1} - u_i^{n+1}}{e^{-\mu_i^{n+1}x_{i+1}} - e^{-\mu_i^{n+1}x_i}}e^{-\mu_i^{n+1}x} + \frac{e^{-\mu_i^{n+1}h_{i+1}}u_i^{n+1} - u_{i+1}^{n+1}}{e^{-\mu_i^{n+1}h_{i+1}} - 1},$$

$$\rho_i^{n+1} = \beta_{i+1/2}^{n+1}\frac{e^{-\mu_i^{n+1}h_{i+1}}u_i^{n+1} - u_{i+1}^{n+1}}{e^{-\mu_i^{n+1}h_{i+1}} - 1}, \quad \mu_i^{n+1} = \frac{\beta_{i+1/2}^{n+1}}{\alpha_{i+1/2}^{n+1}}. \tag{11}$$

Let $l_i = x_{i+1/2} - x_{i-1/2}$. We approximate the integrals in (9) by the mid point rule, replace the quantity $v_0^{n+1}$ by the corresponding second order central finite differences, use (11) and for $i = 1, 2, \ldots, M - 1$, we get

$$\frac{1}{\tau} U_i^{n+1} - \frac{\sigma}{l_i} \left( \mathcal{P}_{i+1}^{n+1} - \mathcal{P}_i^{n+1} \right) = \mathcal{F}_i, \quad i = 1, 2, \ldots, M - 1, \quad (12)$$

$$\mathcal{F}_i = \frac{1}{\tau} U_i^n + \frac{1-\sigma}{l_i} \left( \mathcal{K}_{i+1/2}^n - \mathcal{K}_{i-1/2}^n \right) - \frac{\sigma}{l_i} \left( \mathcal{A}_{i+1/2}^0 - \mathcal{A}_{i-1/2}^0 \right),$$

$$\mathcal{P}_i^{n+1} = \beta_{i-1/2}^{n+1} \frac{e^{-\mu_{i-1}^{n+1} h_i} U_{i-1}^{n+1} - U_i^{n+1}}{e^{-\mu_{i-1}^{n+1} h_i} - 1},$$

$$\mathcal{K}_{i+\frac{1}{2}}^n = k_{i+\frac{1}{2}} a_{i+\frac{1}{2}}^n \left( \frac{U_{i+1}^n - U_i^n}{h_{i+1}} + g \right), \quad (\mathcal{A}_0')_i^{n+1} = a'((U_0)_i^{n+1}),$$

$$\mathcal{A}_{i+\frac{1}{2}}^0 = k_{i+\frac{1}{2}} \left( -g a_{0_{i+\frac{1}{2}}}^{n+1} + (\mathcal{A}_0')_{i+\frac{1}{2}}^{n+1} \left( \frac{(U_0)_{i+1}^{n+1} - (U_0)_i^{n+1}}{h_{i+1}} + g \right) (U_0)_{i+\frac{1}{2}}^{n+1} \right).$$

where $f_{i+\frac{1}{2}}^n = 0.5(f(U_{i+1}^n) + f(U_i^n))$, $k_{i+\frac{1}{2}} = k(x_{i+\frac{1}{2}})$, $v_{i+\frac{1}{2}}^n = 0.5(v_{i+1}^n + v_i^n)$. The discretization (12) is completed with initial and boundary conditions

$$\begin{aligned} U_i^0 &= u^0(x_i), \quad i = 1, 2, \ldots, M - 1, \\ U_0^{n+1} &= 0, \quad U_M^{n+1} = u_r(t_{n+1}), \quad n = 0, 1, \ldots, N - 1, \end{aligned} \quad (13)$$

If $\mu_i \to 0$, we have

$$\lim_{\mu_i \to 0} \frac{\beta_{i+1/2}^{n+1}}{e^{-\mu_i^{n+1} h_{i+1}} - 1} = \alpha_{i+1/2}^{n+1} \lim_{\mu_i \to 0} \frac{\mu_i^{n+1}}{e^{-\mu_i^{n+1} h_{i+1}} - 1} = -\frac{\alpha_{i+1/2}^{n+1}}{h_{i+1}}. \quad (14)$$

**Lemma 1.** *The coefficient matrix of the FFVS (12)–(14) is an M-matrix.*

*Proof (outline).* The proof is based on the observation that $\mathrm{sign}(\beta_{i-1/2}^n) = \mathrm{sign}(\mu_i^n)$ and $\beta_{i-1/2}^n (e^{-\mu_{i-1}^n h_i} - 1)^{-1} < 0$, $i = 1, 2, \ldots, M$, $n = 1, 2, \ldots, N$.

## 3  Implementation and Numerical Results

In this section we discuss the realization and efficiency of the proposed numerical scheme. We set the following model parameters $k(x) = x^\alpha$, $0 < \alpha < 1$, $g = 10$.

*Iteration Algorithm.* At each time level, we initiate iteration procedure, replacing $U_0^{n+1}$ by the solution at old iteration $U_0^{n+1,(m)}$, $m = 0, 1, \ldots$, $U_0^{n+1,(0)} = U^n$. The iteration process continue up to reaching the desired precision *tol* (*tol* $= 1.e - 6$ for the present numerical tests.)

*Graded Mesh.* Let $\epsilon^{n+1} = u^{n+1} - \overline{u}^{n+1}$, where $\overline{u}^{n+1}$ is the solution of the interpolation problem $([\rho_i^{n+1}(\overline{u}^{n+1}(x))]' = 0$, $\overline{u}^{n+1}(x_i) = u_i^{n+1}$, $\overline{u}^{n+1}(x_{i+1}) = u_{i+1}^{n+1}$, $x \in I_i$. Following the same line of considerations as in [19], we deduce

$$\epsilon^{n+1}(x_{i+1/2}) = \frac{g(x_{i+1/2})}{\alpha_{i+1/2}^{n+1}} \left[ O(\delta_i^2) + O(h_i\delta_i) \right], \quad \delta_i = x_{i+1/2} - x_i,$$

where $[\rho_i^{n+1}(\epsilon^{n+1}(x))]' = g(x)$, $\epsilon^{n+1}(x_{i+1}) = \epsilon^{n+1}(x_i) = 0$, $x \in I_i$.

Let us consider grid nodes close to the generation. From (2) and $\delta_i^2/x_{i+1/2} = O(\delta_i)$, we get

$$\epsilon^{n+1}(x_{i+1/2}) = \frac{O(\delta_i) + O(h_i)}{x_{i+1/2}^{\alpha-1}}. \tag{15}$$

We seek a graded mesh

$$x_i = L \left( \frac{i}{M} \right)^\gamma, \quad \gamma > 1, \tag{16}$$

such that $\epsilon^{n+1}(x_{i+1/2}) = O(M^{-2})$. Substituting (16) in (15), taking into account that $\alpha - 1 < 0$ and applying the mean value theorem, we get $\gamma = 2/(1 - \alpha)$. Note that in the case $1 \le \alpha < 2$, we will derive $\gamma = 2/(2 - \alpha)$.

*Numerical Examples.* We consider Gardner one-parameter exponential model [9] $a(u) = e^{\alpha_g u}$, with pore size distribution parameter $\alpha_g$ and van Genutchen-Mualem model with pore size distribution indices $n = 2$, $m = 1/n$ [10,14]:

$$a(u) = \begin{cases} (1 + u^2)^{-5/4}(\sqrt{1 + u^2} + u)^{-2}, & u > 0, \\ 1, & u \le 0. \end{cases}$$

For the numerical tests we consider the problem (1)–(3), $L = 1$ with exact solution, obtained by adding residual term in the right-hand side of (1) and setting initial and right boundary conditions according to the exact solution. Taking into account the results in [1,3] we chose the exact solution $u(x,t) = 2e^{-t/2}x^\beta(1 - \sin \pi x)$, $0 < \beta < 1$. The error $E_M = \max_{0 \le i \le M} |U_i^{n+1} - u(x_i, t^{n+1})|$ and convergence rate $CR = \log_2[E_M/E_{2M}]$ are computed at final time $T_f = 1$, fixed ratio $\tau/|h| = 1$ and $\sigma = 0.5$. For all runs we use graded mesh (16). We observe that when the spatial derivative of the solution goes faster to the infinity at $x = 0$, the order of convergence stabilizes for fine mesh. To speed up this process, we set $\gamma = 2.5/(1 - \alpha)$ in (16), instead of $\gamma = 2/(1 - \alpha)$.

*Example 1 (Gardner model).* Let $\alpha_g = 0.01$ [7]. In Table 1 we compare the results, obtained by FFVM (12)–(14) and Crank-Nicolson discretization of (1)–(3), realized by Picard iteration process (PIS) for different values of $\alpha$ and $\beta$. We may conclude that the accuracy of both schemes is $O(\tau^2 + M^{-2})$, but for solutions, which spatial derivative goes faster to the infinity at $x = 0$, the error produced by FFVM is much smaller than the error of PIS.

**Table 1.** Errors and order of convergence, Gardner model

| $M$ | FFVM | | | | PIS | | | |
|---|---|---|---|---|---|---|---|---|
| | $\alpha = 0.9$, $\beta = 0.3$ | | $\alpha = 0.9$, $\beta = 0.1$ | | $\alpha = 0.9$, $\beta = 0.3$ | | $\alpha = 0.9$, $\beta = 0.1$ | |
| | $E_M$ | $CR$ | $E_M$ | $CR$ | $E_M$ | $CR$ | $E_M$ | $CR$ |
| 160 | 9.312e−3 | | 2.053e−3 | | 9.297e−3 | | 7.387e−3 | |
| 320 | 2.201e−3 | 2.081 | 5.219e−4 | 1.976 | 2.197e−3 | 2.081 | 2.212e−3 | 1.740 |
| 640 | 5.426e−4 | 2.020 | 1.324e−4 | 1.979 | 5.416e−4 | 2.020 | 4.939e−4 | 2.163 |
| 1280 | 1.360e−4 | 1.996 | 3.292e−5 | 2.008 | 1.357e−4 | 1.996 | 1.645e−4 | 1.586 |
| 2560 | 3.375e−5 | 2.010 | 8.254e−6 | 1.996 | 3.369e−5 | 2.010 | 4.252e−5 | 1.952 |
| 5120 | 8.450e−6 | 1.998 | 2.063e−6 | 2.002 | 8.434e−6 | 1.998 | 7.785e−6 | 2.450 |

*Example 2 (Genutchen-Mualem model).* We illustrate that although the function $a(u)$ is not smooth for van Genutchen-Mualem model, the proposed method is still efficient. In Table 2 we give the computational results with FFVM and PIS. We observe that in contrast to FFVM, PIS is efficient only for very fine meshes and produce much more iterations in order to reach the desired precision.

On Fig. 1 we depict the effect of the grading mesh on the accuracy of the numerical solution.

**Table 2.** Errors and convergence rates, $\alpha = 0.9$, $\beta = 0.1$, Genutchen-Mualem model

| $M$ | FFVM | | | PIS | | |
|---|---|---|---|---|---|---|
| | $E_M$ | $CR$ | $Iter$ | $E_M$ | $CR$ | $Iter$ |
| 160 | 1.704e−1 | | 58.010 | 1.869 | | 879.428 |
| 320 | 5.656e−2 | 1.592 | 11.780 | 7.714e−1 | 1.277 | 1368.307 |
| 640 | 1.541e−2 | 1.876 | 5.115 | 4.764e−1 | 0.695 | 1317.846 |
| 1280 | 3.974e−3 | 1.955 | 3.038 | 1.995e−3 | 7.900 | 44.692 |
| 2560 | 9.802e−4 | 2.019 | 3.000 | 4.889e−4 | 2.027 | 15.243 |
| 5120 | 2.464e−4 | 1.992 | 2.127 | 1.228e−4 | 1.993 | 7.810 |

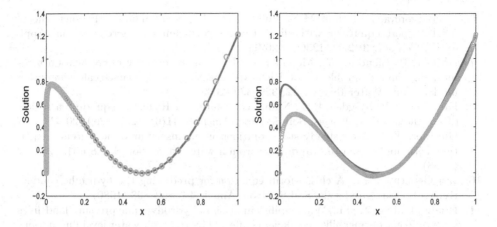

**Fig. 1.** Exact (*solid line*) and numerical (*circles*) solution by FFVM for Genutchen-Mualem model on graded (*left*) and uniform (*right*) mesh, $M = 400$

## 4   Conclusions

Based on the analytical results in [3], in this paper we introduce correct boundary conditions for a class of space-degenerate nonlinear parabolic equations of unsaturated flows. Then, we construct efficient numerical method to overcome the degeneracy and nonlinearity of the model equations. Numerical experiments show that the proposed algorithm provides accurate and reliable results.

**Acknowledgements.** This work is supported by the Bulgarian National Science Fund under the Project DN 12/4 "Advanced analytical and numerical methods for nonlinear differential equations with applications in finance and environmental pollution", 2017 and Bilateral Project KP/Russia 06/12 "Numerical methods and algorithms in the theory and applications of classical hydrodynamics and multiphase fluids in porous media", 2020.

## References

1. Arbogast, T., Taicher, A.L.: A linear degenerate elliptic equation arising from two-phase mixtures. SIAM J. Numer. Anal. **54**(5), 3105–3122 (2016)
2. Bellman, R., Kalaba, R.: Quasilinearization and Nonlinear Boundary-Value Problems. Elsevier Publishing Company, New York (1965)
3. Castro, H., Wang, H.: A singular Sturm-Liouville equation under homogeneous boundary conditions. J. Funct. Anal. **261**, 1542–1590 (2011)
4. Casulli, V., Zanolli, P.: A nested Newton-type algorithm for finite volume methods solving Richards' equation in mixed form. SIAM J. Sci. Comput. **32**, 2255–2273 (2010)
5. Celia, M., Boulout, F., Zarba, R.L.: A general mass-conservative numerical solution for the unsaturated flow equation. Water Resour. Res. **26**(7), 1483–1496 (1990)

6. Chernogorova, T., Koleva, M.N., Vulkov, L.G.: Exponential finite difference scheme for transport equations with discontinuous coefficients in porous media. Appl. Math. Comput. **392**(1), 125691 (2021)
7. Dostert, P., Efendiev, Y., Mohanty, B.: Efficient uncertainty quantification techniques in inverse problems for Richards' equation using coarse-scale simulation models. Adv. Water Resour. **32**, 329–339 (2009)
8. Farthing, M.W., Ogden, F.L.: Numerical solution of Richards' equation: a review of advances and challenges. Soil Sci. Soc. Amer. J. **81**(6), 1257–1269 (2017)
9. Gardner, W.R.: Some steady-state solutions of the unsaturated moistureflow equation with application to evaporation from a water table. Soil Sci. **85**(4), 228–232 (1958)
10. van Genuchten, M.: A closed-form equation for predicting the hydraulic conductivity of unsaturated soils. Soil Sci. Soc. Am. J. **44**, 892–898 (1980)
11. Huang, F., Luo, X., Liu, W.: Stability analysis of hydrodynamic pressure landslides with different permeability coefficients affected by reservoir water level fluctuations and rainstorms. Water **9**(7), 450 (2017)
12. Koleva, M.N., Vulkov, L.G.: Weighted time-semidiscretization Quasilinearization method for solving Rihards' equation. In: Lirkov, I., Margenov, S. (eds.) LSSC 2019. LNCS, vol. 11958, pp. 123–130. Springer, Cham (2020). https://doi.org/10.1007/978-3-030-41032-2_13
13. Ku, C.Y., Liu, C.Y., Xiao, J.E., Yeih, W.: Transient modeling of flow in unsaturated soils using a novel collocation meshless method. Water **9**(12), 954 (2017)
14. Misiats, O., Lipnikov, K.: Second-order accurate finite volume scheme for Richards' equation. J. Comput. Phys. **239**, 125–137 (2013)
15. Mitra, K., Pop, I.S.: A modified L-scheme to solve nonlinear diffusion problems. Comput. Math. Appl. **77**(6), 1722–1738 (2019)
16. Richards, L.A.: Capillary conduction of liquids through porous mediums. Physics **1**(5), 318–333 (1931)
17. Sinai, G., Dirksen, C.: Experimental evidence of lateral flow in unsaturated homogeneous isotropic sloping soil due to rainfall. Water Resour. Res. **42**, W12402 (2006)
18. Wang, S.: A novel fitted finite volume method for the Black-Scholes equation governing option pricing. IMA J. Numer. Anal. **24**(4), 699–720 (2004)
19. Wang, S., Shang, S., Fang, Z.: A superconvergence fitted finite volume method for Black-Sholes equation governing European and American options. Numer. Meth. Part. Differ. Equat. **31**(4), 1190–1208 (2014)
20. Zadeh, K.S.: A mass-conservative switching algorithm for modeling fluid flow in variably saturated porous media. J. Comput. Phys. **230**, 664–679 (2011)

# Minimization of p-Laplacian via the Finite Element Method in MATLAB

Ctirad Matonoha[1], Alexej Moskovka[2(✉)], and Jan Valdman[3,4]

[1] The Czech Academy of Sciences, Institute of Computer Science,
Pod Vodárenskou věží 2, 18207 Prague 8, Czechia
`matonoha@cs.cas.cz`
[2] Faculty of Applied Sciences, Department of Mathematics, University of West
Bohemia, Technická 8, 30614 Pilsen, Czechia
`alexmos@kma.zcu.cz`
[3] The Czech Academy of Sciences, Institute of Information Theory and Automation,
Pod Vodárenskou věží 4, 18208 Prague 8, Czechia
`jan.valdman@utia.cas.cz`
[4] Department of Computer Science, Faculty of Science, University of South Bohemia,
Branišovská 1760, 37005 České Budějovice, Czechia

**Abstract.** Minimization of energy functionals is based on a discretization by the finite element method and optimization by the trust-region method. A key tool to an efficient implementation is a local evaluation of the approximated gradients together with sparsity of the resulting Hessian matrix. Vectorization concepts are explained for the p-Laplace problem in one and two space-dimensions.

**Keywords:** Finite elements · Energy functional · Trust-region methods · p-Laplace equation · MATLAB code vectorization

## 1 Introduction

We are interested in a (weak) solution of the p-Laplace equation [5,8]:

$$
\begin{aligned}
\Delta_p u &= f && \text{in } \Omega, \\
u &= g && \text{on } \partial\Omega,
\end{aligned}
\tag{1}
$$

where the p-Laplace operator is defined as $\Delta_p u = \nabla \cdot \left( |\nabla u|^{p-2} \nabla u \right)$ for some power $p > 1$. The domain $\Omega \in \mathbb{R}^d$ is assumed to have a Lipschitz boundary $\partial\Omega$, $f \in L^2(\Omega)$ and $g \in W^{1-1/p,p}(\partial\Omega)$, where $L$ and $W$ denote standard Lebesque and Sobolev spaces. It is known that (1) represents an Euler-Lagrange equation corresponding to a minimization problem

$$
J(u) = \min_{v \in V} J(v), \qquad J(v) := \frac{1}{p} \int_\Omega |\nabla v|^p \, dx - \int_\Omega f \, v \, dx,
\tag{2}
$$

C. Matonoha was supported by the long-term strategic development financing of the Institute of Computer Science (RVO:67985807). A. Moskovka announces the support of the Czech Science Foundation (GACR) through the grant 18-03834S and J. Valdman the support by the Czech-Austrian Mobility MSMT Grant: 8J21AT001.

I. Lirkov and S. Margenov (Eds.): LSSC 2021, LNCS 13127, pp. 533–540, 2022.
https://doi.org/10.1007/978-3-030-97549-4_61

where $V = W_g^{1,p}(\Omega) = \{v \in W^{1,p}, v = g \text{ on } \partial\Omega\}$ includes Dirichlet boundary conditions on $\partial\Omega$. The minimizer $u \in V$ of (2) is known to be unique for $p > 1$.

Due to the high complexity of the p-Laplace operator (with the exception of the case $p = 2$ which corresponds to the classical Laplace operator), the analytical handling of (1) is difficult. The finite element method [2,3] can be applied as an approximation of (2) and results in a minimization problem

$$J(u_h) = \min_{v \in V_h} J(v), \qquad J(v) := \frac{1}{p} \int_\Omega |\nabla v|^p \, dx - \int_\Omega f \, v \, dx \qquad (3)$$

formulated over the finite-dimensional subspace $V_h$ of $V$. We consider for simplicity the case $V_h = P^1(\mathcal{T})$ only, where $P^1(\mathcal{T})$ is the space of nodal basis functions defined on a triangulation $\mathcal{T}$ of the domain $\Omega$ using the simplest possible elements (intervals for $d = 1$, triangles for $d = 2$, tetrahedra for $d = 3$). The subspace $V_h$ is spanned by a set of $n_b$ basis functions $\varphi_i(x) \in V_h, i = 1, \ldots, n_b$ and a trial function $v \in V_h$ is expressed by a linear combination

$$v(x) = \sum_{i=1}^{n_b} v_i \, \varphi_i(x), \qquad x \in \Omega,$$

where $\bar{v} = (v_1, \ldots, v_{n_b}) \in \mathbb{R}^{n_b}$ is a vector of coefficients. The minimizer $u_h \in V_h$ of (3) is represented by a vector of coefficients $\bar{u} = (u_1, \ldots, u_{n_b}) \in \mathbb{R}^{n_b}$ and some coefficients of $\bar{u}, \bar{v}$ related to Dirichlet boundary conditions are prescribed.

In this paper, the first-order optimization methods are combined with FEM implementations [1,7] in order to solve (3) efficiently. These are the quasi-Newton (QN) and the trust-region (TR) methods [4] that are available in the MATLAB Optimization Toolbox. The QN methods only require the knowledge of $J(v)$ and is therefore easily applicable. The TR methods additionally require the numerical gradient vector

$$\nabla J(\bar{v}) \in \mathbb{R}^{n_b}, \quad \bar{v} \in \mathbb{R}^{n_b}$$

and also allow to specify a sparsity pattern of the Hessian matrix $\nabla^2 J(\bar{v}) \in \mathbb{R}^{n_b \times n_b}, \bar{v} \in \mathbb{R}^{n_b}$, i.e., only positions (indices) of nonzero entries. The sparsity pattern is directly given by a finite element discretization.

We compare four different options:

- option 1 : the TR method with the gradient evaluated directly via its explicit form and the specified Hessian sparsity pattern.
- option 2 : the TR method with the gradient evaluated approximately via central differences and the specified Hessian sparsity pattern.
- option 3 : the TR method with the gradient evaluated approximately via central differences and no Hessian sparsity pattern.
- option 4 : the QN method.

Clearly, option 1 is only applicable if the exact form of gradient is known while option 2 with the approximate gradient is only bounded to finite elements discretization and is always feasible. Similarly to option 2, option 3 also operates with the approximate form of gradient, however the Hessian matrix is not specified. Option 3 serves as an intermediate step between options 2 and 4. Option 4 is based on the Broyden-Fletcher-Goldfarb-Shanno (BFGS) formula.

## 2  One-Dimensional Problem

The p-Laplace equation (1) can be simplified as

$$\left(|u_x|^{p-2}u_x\right)_x = f \qquad \text{in } \Omega = (a, b) \tag{4}$$

and the energy as

$$J(v) := \frac{1}{p}\int_a^b |v_x|^p\, dx - \int_\Omega f\, v\, dx. \tag{5}$$

Assume for simplicity an equidistant distribution of $n+2$ discretization points ordered in a vector $(x_0, \ldots, x_{n+1}) \in \mathbb{R}^{n+2}$, where $x_i := a + i\,h$ for $i = 0, 1, \ldots, n+1$ and $h := (b-a)/(n+1)$ denotes an uniform length of all sub-intervals. It means that $x_0 = a, x_{n+1} = b$ are boundary nodes.

There are $n + 2 = n_b$ well-known hat basis functions $\varphi_0(x), \ldots, \varphi_{n+1}(x)$ satisfying the property $\varphi_i(x_j) = \delta_{ij}, i, j = 0, \ldots, n + 1$, where $\delta$ denotes the Kronecker symbol.

Then, $v \in V_h$ is a piecewise linear and globally continuous function on $(a, b)$ represented by a vector of coefficients $\bar{v} = (v_0, \ldots, v_{n+1}) \in \mathbb{R}^{n+2}$. The minimizer $u_h \in V_h$ is similarly represented by a vector $\bar{u} = (u_0, \ldots, u_{n+1}) \in \mathbb{R}^{n+2}$. Dirichlet boundary conditions formulated at both interval ends imply $v_0 = u_0 = g(a), v_{n+1} = u_{n+1} = g(b)$, where boundary values $g(a), g(b)$ are prescribed.

It is convenient to form a mass matrix $M \in \mathbb{R}^{(n+2)\times(n+2)}$ with entries

$$M_{i,j} = \int_a^b \varphi_{i-1}(x)\,\varphi_{j-1}(x)\, dx = h \cdot \begin{cases} 1/3, & i-j \in \{1, n+2\} \\ 2/3, & i = j \in \{2, \ldots, n+1\} \\ 1/6, & |i-j| = 1 \\ 0, & \text{otherwise} \end{cases}. \tag{6}$$

If we assume that $f \in V_h$ is represented by a (column) vector $\bar{f} = (f_0, \ldots, f_{n+1}) \in \mathbb{R}^{n+2}$, then the linear energy term reads exactly

$$\int_a^b f v\, dx = \bar{f}^T M \bar{v} = \bar{b}^T \bar{v} = \sum_{i=0}^{n+1} b_i v_i,$$

where $\bar{b} = (b_0, \ldots, b_{n+1}) = \bar{f}^T M \in \mathbb{R}^{n+2}$.

The gradient energy term is based on the derivative $v_x$ which is a piecewise constant function and reads

$$v_x|_{(x_{i-1}, x_i)} = (v_i - v_{i-1})/h, \qquad i = 1, \ldots, n+1.$$

Now, it is easy to derive the following minimization problem:

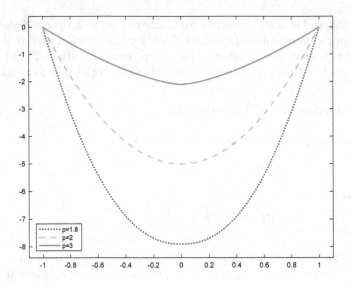

**Fig. 1.** Solutions for $p \in \{1.8, 2, 3\}, \Omega = (-1, 1), f = -10$ and Dirichlet boundary conditions $u(-1) = u(1) = 0$.

*Problem 1 (p-Laplacian in 1D with Dirichlet conditions at both ends).*
    Find $u = (u_1, \ldots, u_n) \in \mathbb{R}^n$ satisfying

$$J(u) = \min_{v \in \mathbb{R}^n} J(v), \qquad J(v) = \frac{1}{p \, h^{p-1}} \sum_{i=1}^{n+1} |v_i - v_{i-1}|^p - \sum_{i=0}^{n+1} b_i v_i, \qquad (7)$$

where values $v_0 := g(a), v_{n+1} := g(b)$ are prescribed.

Note that the full solution vector reads $\bar{u} = (g(a), u, g(b)) \in \mathbb{R}^{n+2}$, where $u \in \mathbb{R}^n$ solves Problem 1 above.

Figure 1 illustrates discrete minimizers $\bar{u}$ for $(a, b) = (-1, 1)$, $f = -10$ and $p \in \{1.8, 2, 3\}$ assuming zero Dirichlet conditions $u(a) = u(b) = 0$. Recall that the exact solution $u$ is known in this simple example.

Table 1 depicts performance of all four options for the case $p = 3$ only, in which the exact energy reads $J(u) = -\frac{16}{3}\sqrt{10} \approx -16.8655$. The first column of every option shows evaluation time, while the second column provides the total number of linear systems to be solved (iterations), including rejected steps. Clearly, performance of options 1 and 2 dominates over options 3 and 4.

## 3    Two-Dimensional Problem

The Eq. (1) in 2D has the form

$$\nabla \cdot \left( \left[ \left( \frac{\partial u}{\partial x} \right)^2 + \left( \frac{\partial u}{\partial y} \right)^2 \right]^{\frac{p-2}{2}} \nabla u \right) = f \qquad \text{in } \Omega \qquad (8)$$

**Table 1.** MATLAB performance in 1D for $p = 3$. Times are given in seconds.

| n | Option 1: Time | Iters | Option 2: Time | Iters | Option 3: Time | Iters | Option 4: Time | Iters |
|---|---|---|---|---|---|---|---|---|
| 1e1 | 0.01 | 8 | 0.01 | 6 | 0.02 | 6 | 0.02 | 17 |
| 1e2 | 0.03 | 12 | 0.05 | 11 | 0.49 | 11 | 0.29 | 94 |
| 1e3 | 0.47 | 37 | 0.50 | 15 | 96.22 | 14 | 70.51 | 922 |

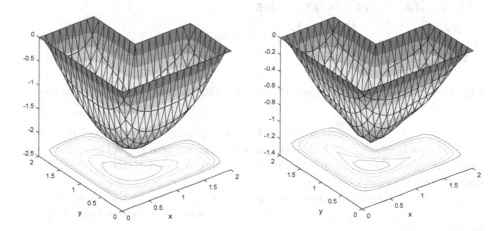

**Fig. 2.** Numerical solutions with contour lines for $p = 1.8$ (left) and $p = 3$ (right) and a L-shape domain $\Omega$, $f = -10$ and zero Dirichlet boundary conditions on $\partial\Omega$.

and the corresponding energy reads

$$J(v) := \frac{1}{p} \iint_{\Omega} \left( |v_x|^p + |v_y|^p \right) \mathrm{d}x\mathrm{d}y - \iint_{\Omega} f\,v\,\mathrm{d}x\mathrm{d}y. \tag{9}$$

Assume a domain $\Omega \in \mathbb{R}^2$ with a polygonal boundary $\partial\Omega$ is discretized by a regular triangulation of triangles [3]. The sets $\mathcal{T}$ and $\mathcal{N}$ denote the sets of all triangles and their nodes (vertices) and $|\mathcal{T}|$ and $|\mathcal{N}|$ their sizes, respectively. Let $\mathcal{N}_{dof} \subset \mathcal{N}$ be the set of all internal nodes and $\mathcal{N} \setminus \mathcal{N}_{dof}$ denotes the set of boundary nodes.

A trial function $v \in V_h = P_1(\mathcal{T})$ is a globally continuous and linear scalar function on each triangle $T \in \mathcal{T}$ represented by a vector of coefficients $\bar{v} = (v_1, \ldots, v_{|\mathcal{N}|}) \in \mathbb{R}^{|\mathcal{N}|}$. Similarly the minimizer $u_h \in V_h$ is represented by a vector of coefficients $\bar{u} = (u_1, \ldots, u_{|\mathcal{N}|}) \in \mathbb{R}^{|\mathcal{N}|}$. Dirichlet boundary conditions imply

$$v_i = u_i = g(N_i), \qquad \text{where } N_i \in \mathcal{N} \setminus \mathcal{N}_{dof}, \tag{10}$$

and the function $g : \partial\Omega \to \mathbb{R}$ prescribes Dirichlet boundary values.

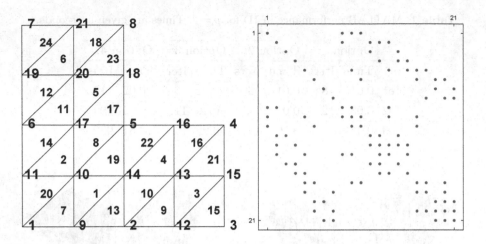

**Fig. 3.** A triangular mesh (left) and the corresponding Hessian sparsity pattern (right).

*Example 1.* A triangulation $\mathcal{T}$ of the L-shape domain $\Omega$ is given in Fig. 3 (left) in which $|\mathcal{T}| = 24, |\mathcal{N}| = 21$. The Hessian sparsity pattern (right) can be directly extracted from the triangulation: it has a nonzero value at the position $i, j$, if nodes $i$ and $j$ share a common edge.

The set of internal nodes that appear in the minimization process reads $\mathcal{N}_{dof} = \{N_{10}, N_{13}, N_{14}, N_{17}, N_{20}\}$, while the remaining nodes belong to the boundary $\partial\Omega$.

For an arbitrary node $N_k$, $k \in \{1, 2, \ldots, |\mathcal{N}|\}$ we define a global basis function $\varphi_k$ which is linear on every triangle and holds $\varphi_k(N_l) = \delta_{kl}, l \in \{1, 2, \ldots, |\mathcal{N}|\}$. Note that with these properties all global basis functions are uniquely defined.

Similarly to 1D, assume $f \in V_h$ is represented by a (column) vector $\bar{f} \in \mathbb{R}^{|\mathcal{N}|}$, and introduce a (symmetric) mass matrix $M \in \mathbb{R}^{|\mathcal{N}| \times |\mathcal{N}|}$ with entries $M_{i,j} = \iint_\Omega \varphi_i \varphi_j \, \mathrm{d}x\mathrm{d}y$. Then it holds $\iint_\Omega fv \, \mathrm{d}x\mathrm{d}y = \sum_{i=1}^{|\mathcal{N}|} b_i v_i$, where $b = \bar{f}^T M \in \mathbb{R}^{|\mathcal{N}|}$.

Next, for an arbitrary element $T_i \in \mathcal{T}$, $i \in \{1, 2, \ldots, |\mathcal{T}|\}$, denote $\varphi^{i,1}, \varphi^{i,2}, \varphi^{i,3}$ all three local basis functions on the $i$-th element and let $\varphi_x^{i,j}, \varphi_y^{i,j}, j \in \{1, 2, 3\}$ be the partial derivatives with respect to 'x' and 'y' of the $j$-th local basis function on the $i$-th element, respectively. In order to formulate the counterpart of (7) in two dimensions, we define gradient vectors $v_{x,el}, v_{y,el} \in \mathbb{R}^{|\mathcal{T}|}$ with entries

$$v_{x,el}^i = \sum_{j=1}^{3} \varphi_x^{i,j} v^{i,j}, \qquad v_{y,el}^i = \sum_{j=1}^{3} \varphi_y^{i,j} v^{i,j},$$

where $v^{i,j}$ is the value of $v$ in the $j$-th node of the $i$-th element.

**Table 2.** MATLAB performance in 2D for $p = 3$. Times are given in seconds.

| | Option 1: | | Option 2: | | Option 3: | | Option 4: | |
|---|---|---|---|---|---|---|---|---|
| $|\mathcal{N}_{dof}|$ | Time | Iters | Time | Iters | Time | Iters | Time | Iters |
| 33 | 0.04 | 8 | 0.05 | 8 | 0.15 | 8 | 0.06 | 19 |
| 161 | 0.20 | 10 | 0.29 | 9 | 3.19 | 9 | 0.56 | 31 |
| 705 | 0.75 | 9 | 1.17 | 9 | 70.59 | 9 | 12.89 | 64 |
| 2945 | 3.30 | 10 | 5.02 | 9 | – | – | 388.26 | 133 |
| 12033 | 16.87 | 12 | 24.07 | 10 | – | – | – | – |
| 48641 | 75.32 | 12 | 107.38 | 10 | – | – | – | – |

With these substitutions we derive the 2D counterpart of Problem 1:

**Problem 2 (p-Laplacian in 2D).** Find a minimizer $u \in \mathbb{R}^{|\mathcal{N}|}$ satisfying

$$J(u) = \min_{v \in \mathbb{R}^{|\mathcal{N}|}} J(v), \quad J(v) = \frac{1}{p} \sum_{i=1}^{|T|} |T_i| \left( |v_{x,el}^i|^p + |v_{y,el}^i|^p \right) - \sum_{i=1}^{|\mathcal{N}|} b_i v_i \quad (11)$$

with prescribed values $v_i = g(N_i)$ for $N_i \in \mathcal{N} \setminus \mathcal{N}_{dof}$.

Figure 2 illustrates numerical solutions for the L-shape domain from Fig. 3, for $f = -10$ and $p \in \{1.8, 3\}$. Table 2 depicts performance of all options for $p = 3$. Similarly to 1D case (cf. Table 1), performance of options 1 and 2 clearly dominates over options 3 and 4. Symbol '-' denotes calculation which ran out of time or out of memory. The exact solution $u$ is not known in this example but numerical approximations provide the upper bound $J(u) \approx -8.1625$.

### 3.1   Remarks on 2D Implementation

As an example of our MATLAB implementation, we introduce below the following block describing the evaluation of formula (11):

```
1  function e=energy(v)
2      v_elems=v(elems2nodes);
3      v_x_elems=sum(dphi_x.*v_elems,2);
4      v_y_elems=sum(dphi_y.*v_elems,2);
5      intgrds=(1/p)*sum(abs([v_x_elems v_y_elems]).^p,2);
6      e=sum(areas.*intgrds) - b'*v;
7  end
```

The whole code is based on several matrices and vectors that contain the topology of the triangulation and gradients of basis functions. Note that these objects are assembled effectively by using vectorization techniques from [1,7] once and do not change during the minimization process. These are (with dimensions):

`elems2nodes` $|\mathcal{T}| \times 3$ - for a given element returns three corresponding nodes
`areas` $|\mathcal{T}| \times 1$ - vector of areas of all elements, `areas(i)` $= |T_i|$
`dphi_x` $|\mathcal{T}| \times 3$ - partial derivatives of all three basis functions with respect to $x$ on every element
`dphi_y` $|\mathcal{T}| \times 3$ - partial derivatives of all three basis functions with respect to $y$ on every element

The remaining objects are recomputed in every new evaluation of the energy:

`v_elems` $|\mathcal{T}| \times 3$ - where `v_elems(i,j)` represents $v^{i,j}$ above
`v_x_elems` $|\mathcal{T}| \times 1$ - where `v_x_elems(i)` represents $v^i_{x,el}$ above
`v_y_elems` $|\mathcal{T}| \times 1$ - where `v_y_elems(i)` represents $v^i_{y,el}$ above

The evaluation of the energy above is vital to option 4. For other options, exact and approximate gradients of the discrete energy (11) are needed, but not explained in detail here. Additionally, for options 1 and 2, the Hessian pattern is needed and is directly extracted from the object `elems2nodes` introduced above.

**Implementation and Outlooks.** Our MATLAB implementation is available at

https://www.mathworks.com/matlabcentral/fileexchange/87944

for download and testing. The code is designed in a modular way that different scalar problems involving the first gradient energy terms can be easily added. Additional implementation details on evaluation of exact and approximate gradients will be explained in the forthcoming paper.

We are particularly interested in further vectorization of current codes resulting in faster performance and also in extension to vector problems such as nonlinear elasticity. Another goal is to exploit line search methods from [6].

# References

1. Anjam, I., Valdman, J.: Fast MATLAB assembly of FEM matrices in 2D and 3D: edge elements. Appl. Math. Comput. **267**, 252–263 (2015)
2. Barrett, J.W., Liu, J.G.: Finite element approximation of the p-Laplacian. Math. Comput. **61**(204), 523–537 (1993)
3. Ciarlet, P.G.: The Finite Element Method for Elliptic Problems. SIAM, Philadelphia (2002)
4. Conn, A.R., Gould, N.I.M., Toint, P.L.: Trust-Region Methods. SIAM, Philadelphia (2000)
5. Drábek P., Milota J.: Methods of Nonlinear Analysis: Applications to Differential Equations, 2nd edn. Birkhauser (2013)
6. Lukšan, L., et al.: UFO 2017. Interactive System for Universal Functional Optimization. Technical Report V-1252. Prague, ICS AS CR (2017). http://www.cs.cas.cz/luksan/ufo.html
7. Rahman, T., Valdman, J.: Fast MATLAB assembly of FEM matrices in 2D and 3D: nodal elements. Appl. Math. Comput. **219**, 7151–7158 (2013)
8. Lindqvist, P.: Notes of the p-Laplace Equation (second edition), report 161: of the Department of Mathematics and Statistics. University of Jyvaäskylä, Finland (2017)

# Quality Optimization of Seismic-Derived Surface Meshes of Geological Bodies

P. Popov[1]($\boxtimes$), V. Iliev[2], and G. Fitnev[2]

[1] Institute of Information and Communication Technologies,
Bulgarian Academy of Sciences, Sofia, Bulgaria
ppopov@ppresearch.bg
[2] PPResearch Ltd., Sofia, Bulgaria
{viliev,gfitnev}@ppresearch.bg

**Abstract.** Availability of 3D datasets of geological formations present a number of opportunities for various numerical simulations provided quality meshes can be extracted for the features of interest. We present a particular technique designed to generate an initial levelset-based triangulation of geological formations such as salt volumes, turbidites, faults and certain types of shallow horizons. We then work directly with the underlying voxel data to improve the mesh quality so that the resulting triangulation is suitable for numerical simulations involving PDEs, while approximating well enough the underlying (implicit) levelset. We apply our algorithm on typical Gulf of Mexico formations including turbidite reservoirs and multiple salt domes. We demonstrate that the resulting meshes are of high quality and can be directly used in coupled poroelastic reservoir simulations.

## 1 Introduction

The standard approach in reservoir simulation and seismic image processing is to work with regular or quasi-regular corner point grids [1]. Over the past few years the use of unstructured meshes has been gaining ground in the reservoir simulation community [4,5,7]. Reservoir simulators such as *Intersect* that work with general unstructured grids are also becoming more popular. This presents a unique problem of generating simulation models of geological structures using general unstructured meshes.

Surface representation of geobodies such a salts, reservoirs, faults, etc. is usually done semi-manually [7]. Geological data tend to be defined on very high resolution grids, typically of very poor aspect ratio. Even after significant image processing fine-scale features are very rich and widespread. Reservoirs in particular tend to have very thin layers, with sharp edges, holes and small-scale tubular interconnections between layers. The ultimate goal of this work is to run flow simulations which critically depend on mesh connectivity, however are not sensitive to small-scale features, e.g. edges as long as they do not influence the topology of the mesh. For that reason, the main goal of mesh extraction

© Springer Nature Switzerland AG 2022
I. Lirkov and S. Margenov (Eds.): LSSC 2021, LNCS 13127, pp. 541–551, 2022.
https://doi.org/10.1007/978-3-030-97549-4_62

for flow/geomechanical simulations is to preserve and maintain the iso-surface topology, eliminate topologically insignificant small features, including edges, and critically, produce meshes with the least amount of elements, preferably no small angles, in-plane or dihedral, and, naturally, without self intersections.

The marching cube algorithm [6], originally designed for medical imaging, is uniquely suited to automatically reconstruct an polygonal mesh from geological data. This can then be followed by standard mesh optimization techniques [2] to derive a FEM quality mesh. The issue of surface consistency [3,8] however is critical for FEM applications and needs to be properly addressed. For example [3] resolve the issue by reconstructing the contour first, with sub-sampling where needed. The authors additionally develop a sharp-feature preserving scheme. The generated meshes are very large, a situation typically addressed by applying the marching cubes on adaptive grids [9,10] which results in crack that need to be patched. Note that crack-patching is not a feasible approach in geological formations. The fine-scale features are typically very rich and volume layers can happen to be very thin with the data defined on very poor aspect ratio cells. This will result in introducing a large number of very-bad aspect ratio elements which will likely have intersection issues with other parts of the extracted surface.

In this work we propose to use seismic cubes directly in extracting the surfaces of relevant geobodies. Seismic velocity cubes can be used to extract high contrast objects such as salt bodies, density cubes allow extraction of shallow horizons and shale/clay markers can be used to extract reservoir surfaces. The data invariably comes on regular rectangular grids. We use trilinear interpolation of the data on each cell and then use a levelset in conjunction with a marching cube method [3,6] in order to extract an initial polygonal surface. We use criteria similar to that of [8] to resolve face inconsistencies. Similarly to [3] we build the iso-curves on faces first, followed by triangular reconstruction in the interior of a cell. The triangulation in the interior of the cell is designed to limit the number of elements while still producing consistent surfaces. This is driven by the fact that geological data comes in rather high resolution with very rich local features that are not required for FEM computations. So, rather than subdivide cells to extract finer features [3] our main focus is producing topologically consistent initial meshes. We have also designed our marching cube algorithm to handle the case of entire edges belonging to the levelset, which in the context of a trilinear interpolation happens when two vertices from the same edge are exactly on the surface. This is a situation that frequently occurs when the initial data is an integer volume marker. Rather than artificially shift the data to avoid such situations it is important to handle them directly.

The second step of our algorithm is a radical optimization of the initial mesh with the goal of producing an FEM quality mesh which retains the original topology, however eliminates fine-scale topologically insignificant features. This surface is then converted to a triangular surface and optimized using standard techniques [2] to produce FEM quality meshes. The main feature of this work was designing criteria which would preserve the topology of the mesh, prevent self-intersections and eliminate small features, while producing FEM quality triangles.

We present and example of a surface mesh extracted from a clay marker of a complex turbidite reservoir with multiple thin layers with complex interconnections. The algorithm has so far been applied to single-volume data, however generalizations to multiple volumes, e.g. as present in lithology markers, are straightforward [11]. An online demo of this algorithm can be found at: http://websegy.ppresearch.com.

## 2  Initial Surface Extraction

In this section we discuss the initial task of extracting a triangular mesh from a levelset function defined in a regular (sugarcube) grid. Assume the data is given at the vertices of a regular Cartesian grid. The simpler approach is to split each individual rectangular cuboid (in 3D) into tetrahedrons in a conforming way across the entire grid. It can be done in a number of non-unique ways by splitting each cuboid into five or six tetrahedrons. The non-uniqueness leads to non-unique mesh representations of the underlying dataset, see Fig. 1. As can be seen this leads to meshes with different topology compared to the data. Another reason to avoid working with tetrahedral support is the typically very large size of the initial tessellation. It needs to be maintained in memory in order to support the mesh optimization step. Binary/ternary trees are required to support geometric search and projection operations which can easily become prohibitive in terms of memory and/or speed.

We instead choose to create a triangular representation of the levelset defined by trilinear interpolation of the data using a marching cubes algorithm [3, 6]. Let $Q_k$, $k = 1 \ldots N_x N_y N_z$ be a single rectangular cuboid from a grid $N_x \times N_y \times N_z$. Let also $\phi_i$, $i = 1, \ldots, 8$ be the standard trilinear basis on $Q$ and let $F_i$ be

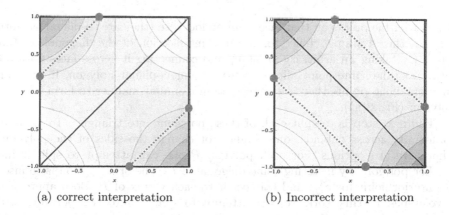

(a) correct interpretation                    (b) Incorrect interpretation

**Fig. 1.** Mesh dependency of a levelset when different triangular partitions of the underlying regular quads are used (2D example, plotted are contours of $f(x,y) = xy$), one leading to correct (a) and the other to incorrect (b) interpretation of then underlying bilinear function.

the values of the data specified at vertices of $Q$. The trilinear interpolation has the form:

$$f(x) = \sum_{i=1}^{8} F_i \phi_i(x), \tag{1}$$

and define the levelset surface for a level $f(x) = l$ as:

$$S_{Q_k}^l = \{\forall x | f(x) = l\}. \tag{2}$$

It is clear that the union $S^l = \bigcup_{k=1}^{N_x N_y N_z} S_{Q_k}^l$ is a conforming, uniquely defined surface. The trilinear interpolation provides a rich representation of sugarcube data, as seen from Figs. 2. Generating an initial triangulation, however, is not trivial.

The initial mesh is generated in two steps. First, we create (non-planar) polygons which we then split into triangles. For each $Q_k$ we first find all *levelset vertices* $\{V_i\}$, where $V_i = e \cap S_{Q_k}^l$, $e$ being an edge of $Q_k$. As the trilinear map (1) is linear on each edge this is a fast and trivial operation. Next, we connect those vertices into one or more polygons which respect the topology of $S_{Q_k}^l$. We take an arbitrary levelset vertex and connect it to another one that shares the same face. Assuming all $F_i \neq l$, there are either two or four levelset vertices on a face (Fig. 2). When there are only two it is trivial to connect them. Having four levelset vertices on a face implies that the signs of $F_i - l$ on that face are positive/negative in a checkerboard pattern. In that case one has to determine which two pairs are connected. We need a point $p$ that is between the two branches of the hyperbola in Fig. 1. Then, the points $V_i$ and $V_j$ are connected by a segment on that faces iff:

$$\left( f\left( \frac{V_i + V_j}{2} \right) - l \right) (f(p) - l) < 0 \tag{3}$$

If we order the points on that face counterclockwise (they are co-planar) to form a quad, then a simple choice for $p$ is the intersection of the diagonals of that quad. By taking an arbitrary point $V_1$ and connecting it across faces until one returns to the same point allows to from a non-coplanar polygon. If there are any remaining points they are connected in a similar manner to form a second polygon (Fig. 2(a)).

The next step is to split each of these polygons into triangles. To maintain conformity across cuboids, one should not change the sides of the polygons. There are two options. For each polygon $P$ one can attempt to split it into smaller polygons by inserting some diagonal of $P$. The other option is to insert an interior point for $S_{Q_k}^l$ and connect it to each vertex of $P$. Both approaches have issues. In this work we first attempt to split $P$ into two polygons with smaller number of vertices by finding a diagonal whose midpoint is close enough to $S_{Q_k}^l$, usually some fraction of the characteristic length-scale of $Q_k$. We do not insert diagonals if they create very small triangle angles (less then 1 deg). This is done until we get a triangular decomposition. If, however, inserting a diagonal is not desirable we select an interior point. The simplest way is to project the

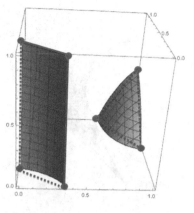
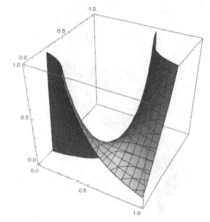

(a) 7 levelset vertices, two surfaces     (b) 7 levelset vertices, single surface

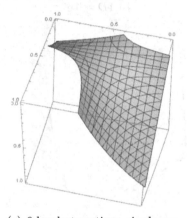
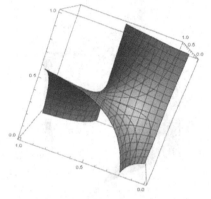

(c) 6 levelset vertices, single sur-     (d) 7 levelset vertices, single surface
face

**Fig. 2.** Level sets of trilinear interpolation functions on a unit cube. All feature one or more faces with four levelset vertices. The first example (a) is what one usually imagines, however often multiple levelset vertices are connected by a single, highly complex surface (b)–(d).

centroid of $P$ onto the surface (2) (see next section for the projection operation). However in certain cases this projection can easily fall outside of $Q_k$. Then we take the centroid of $Q_k$ and connect it alternatively with the vertices of $Q_k$. We intersect each such segment with $S_{Q_k}^l$ and choose the intersection point which minimizes the distance to the centroid of $P$. Note that inserting an interior point generates the most amount of individual triangles from a polygon and this should be avoided to keep the size of initial triangulations manageable.

The preceding discussion assumes all $F_i \neq l$. If a single $F_i = l$ then this is an isolated point (a vertex of $Q_k$) which need not be included to the triangulation. If,

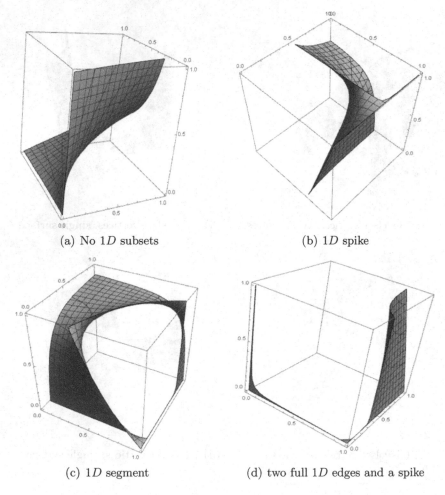

(a) No $1D$ subsets

(b) $1D$ spike

(c) $1D$ segment

(d) two full $1D$ edges and a spike

**Fig. 3.** Examples of one or more cuboid edges entirely contained in $S_{Q_k}^l$.

however, an edge $e$ of $Q_k$ is contained in $S_{Q_k}^l$, then several special cases arise. Let $f_1$ and $f_2$ be the two faces containing $e$. The cases when $f_1 \subset S_{Q_k}^l$ or $f_2 \subset S_{Q_k}^l$ are simple to handle, so assume otherwise. Then, if neither of the two edges opposing $e$ in $f_1$ and $f_2$ respectively has an interior levelset vertex $V_i$ we still have a simple case to handle (Fig. 3(a)). However, if one or both opposing edges contain an interior levelset vertex, then $S$ turns partially one-dimensional on $e$, see Fig. 3(b) and (c). Both situation can be combined, for example in Fig. 3(d) where two of the edges are purely one dimensional (no opposing levelset vertex) and the third is not. These cases need to be detected and the $1D$ portion of $S_{Q_k}^l$ removed.

# 3    Search and Projection Operations

The resulting initial triangulation $T^I$ formed by the union of triangulations on each $Q_k$ should be optimized. The process must ensure all vertices of $T^I$ belong strictly to $S^l$ and that $T^I$ represents $S^l$ accurately enough. Using $S^l$ has several advantages when compared to working with using $T^I$ as a definition of the surface (which is inevitable if one splits the regular grid into tetrahedrons). To optimize a mesh one needs to asses distance to $S^l$ and compute normals at points that belong to $S^l$. The second is a trivial operation. The first requires that one is able to project a point $p$ onto $S^l$. To do that, we need to be able to project $p$ on an individual piece $S^l_{Q_k}$. This is done by Lagrange multipliers. Given a cell $Q_k$ and a point $p$, minimize:

$$\mathcal{L}(x, \lambda) = \|x - p\|^2 - \lambda f(x). \tag{4}$$

One has to be careful to perform the above operation in the local coordinates of $Q_k$. If the projection $x \notin Q_k$ then one has to use two multipliers to project $p$ on each face of $Q_k$ and see which one minimizes the distance to $p$. If a projection on particular face $f$ falls outside of that face then test the distance between $p$ and each levelset vertex $V_i$ on the edges of $f$. In practice a projection (4) almost always is inside $Q_k$ and the remaining computational work to set faces and edges is not needed often.

Computing the projection onto $S^l$ is now performed as follows:

1. Determine the cuboid $Q_k$ to which $p$ belongs.
2. Traverse the neighbors of $Q_k$ in ascending distance from $Q_k$ and project on each one of them and maintain the projection which minimizes the distance to $p$
3. Stop the traverse whenever the triangle inequality shows that no further minimization is theoretically possible.

If for example $p$ is close to the center of $Q_k$ and the cell is not of bad aspect ratio, then one rarely has to test any neighbors. If $p$ is close a face, but not a vertex of $Q_k$ one typically has to test the neighboring cell. Being close to an edge usually involves 4 tests, and being closet to a vertex 8. This picture however, can be degraded considerably if the grid is of bad aspect ratio.

The above procedure is not memory intensive. Testing if a point belongs to $S^l$ is trivial (requires the location of $Q_k$ which is fast), computing normals is straightforward and projecting points involves a few nonlinear minimizations. In contrast, if one was working with $T^I$ as surface definition, rather than (the implicit) $S^l$, then determining if a point belongs to the tessellation and projecting onto the tessellation requires complex and memory intensive $K - N$ trees. It would also require to keep $T^I$ in memory during the entire optimization step which is also memory intensive.

# 4    Mesh Optimization

We follow a standard strategy of combining collapses, swaps and smooths to optimize the initial mesh. In this we attempt to improve:

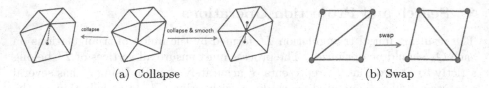

(a) Collapse                          (b) Swap

**Fig. 4.** The collapse (a) and swap (b) operations.

1. triangle minimal edge length,
2. minimal triangle angle,
3. dihedral angle between mesh elements,
4. distance to $S^l$, measured from an edges's midpoint to its projection onto $S^l$,
5. deviation between mesh normals and surface normals.

Each criteria has an associated target quality function $b_h(e)$, $b_a(\tau)$, $b_d(e)$, $b_\rho(e)$, $b_n(v, e)$ respectively. We choose each $b_i$, to be 1 if it the target criteria is precisely satisfied, less then 1 it the criteria improved or greater than 1 if it degraded.

Each operation affects several triangles and for each individual $b_i$ we record the worst, best and average values before and after. A single criteria improved if:

1. its new value is less then 1, or
2. the previous values is bad (>1) but the maximum decreased, or
3. if the previous value is bad (>1) but the new average is good (<1).

In general we want all criteria to improve. For collapses and swaps we use special rules that temporarily reduce quality.

*Edge Collapse.* We attempt to collapse an edge at either of its endpoints, at a distance of 0.2, 0.5 (midpoint) and 0.8 along the length of the edge. Other than the endpoints the remaining options combine a collapse with a smooth (Fig. 4(a)). All affected edges are checked for self-intersections. If a collapse is possible and mesh quality improves the operation is performed.

*Edge Swap.* We check if an edge swap will introduce self-intersections. If not, then the quality is assessed and if it improves the operation is performed (Fig. 4(b)).

*Point Smooth.* Given a mesh vertex, we compute the centroid of the patch of elements containing it and then project it onto $S^l$. Then check if moving the vertex to this projection will introduce self intersections. If yes, the projected point is relaxed towards the original point and the check is performed again until no self intersections are induced. The new location is accepted if the mesh quality improved.

The three local operations are used in an global optimization scheme.

## 5    Numerical Examples

We present a numerical example involving reconstructing a turbidite reservoir in the Gulf of Mexico. The original data for the reservoir is defined on a highly

**Fig. 5.** A slice of the marker cube used to extract the surface.

anisotropic grid [864 × 912 × 4601, each cell having dimensions 50 × 50 × 2.5 m. That data represents a shale/clay continuous marker in the range [0, 1] (Fig. 5). The data was subjected to extensive smoothing and data processing which included Gaussian blurs, removal of small tubular structures and small edges. This was done to improve the connectivity of the volumes (details can be found here http://ppresearch.com/seam/preupper.php). The initial extraction has 9.95 M triangles. The mesh shown in Fig. 6 is the final result of optimizations with a mesh size of 400 m, an anisotropic distance to surface criteria of 100 × 100 × 5 m, a preference to eliminate small dihedral angles over small triangle angles and no restrictions on the deviations of the mesh normals.

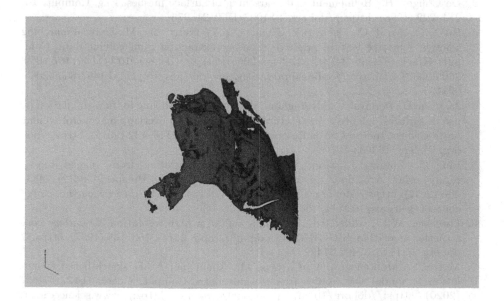

**Fig. 6.** FEM optimized surface mesh extraction for GoM turbidite reservoir

| Summary | | | Quality | | | |
|---|---|---|---|---|---|---|
| | | | | Min | Max | Avarage |
| Elements | 349980 | | Minimal Angle (deg) | 1.174 | 59.96 | 37.31 |
| Points | 169089 | | Dihedral Angle (deg) | 1.0157 | 180 | 149.1 |
| Disjoint Domains | 72 | | Edge (meters) | 1.026 | 411.3 | 81.45 |

**Fig. 7.** Quality of surface mesh extraction

A summary of the mesh quality is shown in Figs. 7. The mesh was used successfully in a series of FEM based geomechanical simulations.

## 6   Conclusions

We have demonstrated a viable strategy to generate optimized surface meshes from seismic data suitable for FEM computations. Further investigations can be undertaken to parallelize the algorithm. Much work is still needed to figure the optimal mesh improvement criteria, especially with respect to mesh curvature and dihedral angles.

## References

1. Chen, Z., Huan, G., Ma, Y.: Computational Methods for Multiphase Flows in Porous Media. SIAM, Philadelphia (2006)
2. De Cougny, H.: Refinement and coarsening of surface meshes. Eng. Comput. **14**, 214–222 (1998). https://doi.org/10.1007/BF01215975
3. Ho, C.C., Wu, F.C., Chen, B.Y., Chuang, Y.Y., Ouhyoung, M.: Cubical marching squares: adaptive feature preserving surface extraction from volume data. Comput. Graph. Forum **24**(3), 537–545 (2005). https://doi.org/10.1111/j.1467-8659.2005.00879.x. https://onlinelibrary.wiley.com/doi/abs/10.1111/j.1467-8659.2005.00879.x
4. Jackson, D., Percival, J., Mostaghimi, R.: Reservoir modeling for flow simulation by use of surfaces, adaptive unstructured meshes, and an overlapping-control-volume finite-element method. SPE Reservoir Eval. Eng. **18**, 115–132 (2015). https://doi.org/10.2118/163633-PA
5. Liu, Q., Zhang, J., Gao, H.: Reverse-time migration from rugged topography using irregular, unstructured mesh. Geophys. Prospect. **65**(2), 453–466 (2017). https://doi.org/10.1111/1365-2478.12415. https://www.earthdoc.org/content/journals/10.1111/1365-2478.12415
6. Lorensen, W.E., Cline, H.E.: Marching cubes: a high resolution 3D surface construction algorithm. SIGGRAPH Comput. Graph. **21**(4), 163–169 (1987). https://doi.org/10.1145/37402.37422
7. Motta, S., Montenegro, A., Gattass, M., Roehl, D.: A 3D sketch-based formulation to model salt bodies from seismic data. Comput. Geosci. **142**, 104457 (2020). https://doi.org/10.1016/j.cageo.2020.104457. https://www.sciencedirect.com/science/article/pii/S0098300419306983

8. Nielson, G.M., Hamann, B.: The asymptotic decider: resolving the ambiguity in marching cubes. In: Proceedings of the 2nd Conference on Visualization 1991, VIS 1991, pp. 83–91. IEEE Computer Society Press, Washington, DC (1991)
9. Shu, R., Zhou, C., Kankanhalli, M.S.: Adaptive marching cubes. Vis. Comput. **11**, 202–217 (1995)
10. Wilhelms, J., Van Gelder, A.: Octrees for faster isosurface generation. ACM Trans. Graph. **11**(3), 201–227 (1992). https://doi.org/10.1145/130881.130882
11. Wu, Z., Sullivan Jr., J.M.: Multiple material marching cubes algorithm. Int. J. Numer. Methods Eng. **58**(2), 189–207 (2003). https://doi.org/10.1002/nme.775. https://onlinelibrary.wiley.com/doi/abs/10.1002/nme.775

# Author Index

Alexandrov, Alexander K.  480
Alexandrov, Boian S.  351, 359
Alihodžić, Adis  201, 209
Andrä, H.  387
Apostolov, Stoyan  31, 257
Armyanov, P.  464
Arnaudov, Dimitar D.  480
Astsatryan, Hrachya  431
Atanassov, E.  421
Axelsson, Owe  91

Bădică, Amelia  217
Bădică, Costin  217
Bayleyegn, Teshome  101
Beck, Andrea  410
Bejanyan, Vahag  431
Benkhaldoun, Fayssal  489, 498
Bradji, Abdallah  489, 498
Buchfink, Patrick  402
Buligiu, Ion  217

Capuani, Rossana  297
Chahin, Malek  201
Chervenkov, Hristo  172
Christov, I.  464
Ciora, Liviu Ion  217
Corella, Alberto Domínguez  306
Cowen, Lenore J.  3
Čunjalo, Fikret  201, 209

Dimitrov, Yuri  31
Dimov, Ivan  156, 180, 188, 257
Djidjev, Hristo N.  351
Dobrinkova, Nina  226
Dreesen, Philippe  333
Drenchev, Ludmil  132

Falcone, Maurizio  322
Faragó, István  101, 188
Fidanova, Stefka  234
Fitnev, G.  541
Fokina, Daria  369

Gadzhev, Georgi  109, 124
Ganev, Kostadin  109
Gavrilenko, Pavel  378
Georgiev, D.  421
Georgiev, Ivan  148
Georgiev, Krassimir  148, 188
Georgiev, Slavi G.  40, 507
Georgieva, Irina  49
Georgieva, Ivelina  109
Georgieva, Rayna  156, 180, 257
Gfrerer, Helmut  515
Ghoudi, Tarek  498
Gkotsis, Ilias  226
Gochakov, Alexander  164
Grigorova, Denitsa  439
Gurov, T.  421
Gurova, Silvi-Maria  447
Gushchin, Valentin A.  117

Haasdonk, Bernard  378, 402, 410
Hahn, Georg  351
Harizanov, Stanislav  57
Havasi, Ágnes  101, 188
Hofreither, Clemens  49
Hu, Xiaozhe  3

Iliev, Alexander  456
Iliev, Oleg  369, 378, 464
Iliev, V.  541
Ilieva, Nevena  132
Ishteva, Mariya  333
Ivanov, Vladimir  124
Ivanovska, S.  421

Karaivanova, Aneta  447
Katsaros, Evangelos  226
Katzarov, Ivaylo  132
Keil, Tim  16
Köbler, J.  387
Koleva, Miglena N.  524
Kondakov, Vasilii G.  117
Konopleva, Viktoriia  164
Korotov, Sergey  140
Kosturski, Nikola  57

Křížek, Michal  140
Kurz, Marius  410

Langer, Ulrich  314
Lin, Junyuan  3
Liolios, Konstantinos  148
Lirkov, Ivan  57
Logofătu, Doina  217
Lukáš, Dalibor  91

Magino, N.  387
Manoharan, Arjun  456
Manzini, Gianmarco  342
Margenov, Svetozar  57, 71
Marigonda, Antonio  297
Matonoha, Ctirad  533
Meiser, Clemens  393
Mogentale, Marta  297
Moskovka, Alexej  533
Mote, Ameya  456
Mucherino, Antonio  242
Müller, R.  387

Neytcheva, Maya  91

O'Malley, Daniel  351
Ohlberger, Mario  16, 378
Oseledets, Ivan  369
Ostromsky, Tzvetan  156, 180
Outrata, Jiří V.  515

Pacifico, Agnese  322
Palejev, Dean  439, 472
Pelofske, Elijah  351
Penenko, Alexey  164
Penenko, Vladimir  164
Penev, Kalin  250
Pesare, Andrea  322
Popov, P.  541
Poryazov, Stoyan  156, 180, 257
Prodanov, Dimiter  65
Pyanova, Elza  164

Roeva, Olympia  234
Röhrle, Oliver  402

Santin, Gabriele  410
Schafelner, Andreas  314
Schindler, Felix  378
Schneider, M.  387
Schuster, Thomas  393
Semerdzhiev, A.  464
Shegunov, N.  464
Shen, Yue  3
Shuva, Shahnewaz  402
Simidchiev, Alexandar  275
Skau, Erik  342, 359
Skodras, Georgios  148
Slavchev, Dimitar  71
Spiridonov, Valery  172
Stanev, Valentin  359

Takeuchi, Ichiro  359
Tashev, Tasho D.  480
Tasheva, Radostina P.  480
Todorov, Venelin  31, 156, 180, 257
Toktaliev, Pavel  378
Tonchev, Demir  439
Tranev, Stoyan  266
Traneva, Velichka  266
Truong, Duc P.  342
Tsvetova, Elena  164

Vabishchevich, Petr N.  81, 287
Valdman, Jan  515, 533
Vangara, Raviteja  342
Veliov, Vladimir M.  306
Vulkov, Lubin G.  40, 507, 524
Vutov, Yavor  57

Wald, Anne  393
Welschinger, F.  387
Wenzel, Tizian  378, 410
Wu, Kaiyi  3

Youssef, Maha  378

Zhivkov, Petar  275
Zlatev, Zahari  156, 180, 188
Zvorničanin, Enes  209